Essentials of Chemical Biology

Essentials of Chemical Biology

Edited by
Oliver Stone

Willford Press
www.willfordpress.com

Published by Willford Press,
118-35 Queens Blvd., Suite 400,
Forest Hills, NY 11375, USA

ISBN: 978-1-68285-405-1

Cataloging-in-Publication Data

Essentials of chemical biology / edited by Oliver Stone.
p. cm.
Includes bibliographical references and index.
ISBN 978-1-68285-405-1
1. Biochemistry. I. Stone, Oliver.
QD415 .E87 2018
572--dc23

For information on all Willford Press publications
visit our website at www.willfordpress.com

Willford Press

Printed in China.

Contents

Preface

Chemical biology tries to answer questions related to the field of biology, by directly examining living systems at the chemical level. As a subject it studies the applications of chemical tools in relation to the study of biological systems. Some of the fields studied under the umbrella of chemical biology are proteomics, glycobiology, enzymology, etc. This book contains some path-breaking studies in the field of chemical biology. Coherent flow of topics, student-friendly language and extensive use of examples make this book an invaluable source of knowledge. It aims to serve as a resource guide for students and experts alike and contribute to the growth of the discipline.

The researches compiled throughout the book are authentic and of high quality, combining several disciplines and from very diverse regions from around the world. Drawing on the contributions of many researchers from diverse countries, the book's objective is to provide the readers with the latest achievements in the area of research. This book will surely be a source of knowledge to all interested and researching the field.

In the end, I would like to express my deep sense of gratitude to all the authors for meeting the set deadlines in completing and submitting their research chapters. I would also like to thank the publisher for the support offered to us throughout the course of the book. Finally, I extend my sincere thanks to my family for being a constant source of inspiration and encouragement.

Editor

1

AAA+ chaperones and acyldepsipeptides activate the ClpP protease via conformational control

Malte Gersch[1,*,†], Kirsten Famulla[2,*], Maria Dahmen[1], Christoph Göbl[1,3], Imran Malik[2], Klaus Richter[1], Vadim S. Korotkov[1], Peter Sass[2], Helga Rübsamen-Schaeff[4], Tobias Madl[1,3,5], Heike Brötz-Oesterhelt[2,†,§] & Stephan A. Sieber[1,§]

The Clp protease complex degrades a multitude of substrates, which are engaged by a AAA+ chaperone such as ClpX and subsequently digested by the dynamic, barrel-shaped ClpP protease. Acyldepsipeptides (ADEPs) are natural product-derived antibiotics that activate ClpP for chaperone-independent protein digestion. Here we show that both protein and small-molecule activators of ClpP allosterically control the ClpP barrel conformation. We dissect the catalytic mechanism with chemical probes and show that ADEP in addition to opening the axial pore directly stimulates ClpP activity through cooperative binding. ClpP activation thus reaches beyond active site accessibility and also involves conformational control of the catalytic residues. Moreover, we demonstrate that substoichiometric amounts of ADEP potently prevent binding of ClpX to ClpP and, at the same time, partially inhibit ClpP through conformational perturbance. Collectively, our results establish the hydrophobic binding pocket as a major conformational regulatory site with implications for both ClpXP proteolysis and ADEP-based anti-bacterial activity.

[1] Department of Chemistry and Center for Integrated Protein Science Munich (CIPSM), Technische Universität München, Lichtenbergstraße 4, 85748 Garching, Germany. [2] Institute for Pharmaceutical Biology and Biotechnology, University of Düsseldorf, Universitätsstraße 1, 40225 Düsseldorf, Germany. [3] Institute of Structural Biology, Helmholtz Zentrum München, 85764 Neuherberg, Germany. [4] AiCuris GmbH & Co. KG, Friedrich-Ebert-Straße 475, 42117 Wuppertal, Germany. [5] Institute of Molecular Biology & Biochemistry, Center of Molecular Medicine, Medical University of Graz, 8010 Graz, Austria. * These authors contributed equally to this work. § These authors jointly supervised this work. † Present address: Medical Research Council Laboratory of Molecular Biology, Francis Crick Avenue, Cambridge Biomedical Campus, Cambridge CB2 0QH, UK (M.G.) Interfaculty Institute for Microbiology and Infection Medicine, Department for Microbial Bioactive Compounds, University of Tuebingen, Auf der Morgenstelle 28, 72076 Tuebingen, Germany (H.B-O). Correspondence and requests for materials should be addressed to H.B.-O. (email: heike.broetz-oesterhelt@uni-tuebingen.de) or to S.A.S. (email: Stephan.sieber@mytum.de).

The proper balance of protein synthesis and protein degradation on a cellular level is of pivotal importance in all kingdoms of life and its perturbation is currently intensively investigated for therapeutic application in the context of cancer, inflammation and infectious diseases[1–3]. Recently, two opposite strategies based on the chemical stimulation or inhibition of protein degradation have been shown to either kill bacteria or silence their pathogenesis[4]. β-lactones are covalent inhibitors of the bacterial serine protease ClpP[5,6]. Inhibition of ClpP enzymatic activity halts the degradation of proteins involved in virulence regulation, which results in attenuated pathogenesis[7,8]. In contrast, ADEPs are a class of natural product-derived cyclic acyldepsipeptides that act as potent antibiotics through the dysregulation and activation of ClpP[9,10]. In an unperturbed cell, proteolysis occurs through binding of a AAA+ chaperone such as ClpX to the apical sides of the barrel-shaped, proteolytic component ClpP whereupon a functional ClpXP protease complex with a continuous substrate channel is formed[11–15]. In this complex, ClpX engages substrate proteins prone to degradation, unfolds them and threads them into the ClpP proteolytic chamber where they are subsequently degraded[16–18]. Crystal structures of the ClpP barrel in different conformations have been solved, including an active extended and an inactive compressed conformation, however, how the associated structural dynamics are regulated is currently not known[6,13,19,20]. ClpP alone is able to cleave only small peptides, with binding of ClpX being strictly required for the degradation of proteins[21]. ADEPs eliminate this regulation by binding at the junction of the ClpP subunits in a hydrophobic pocket that is also used by ClpX[22,23]. Binding of ADEPs causes a conformational change in the N-terminal region of ClpP whereupon the axial pore of the protease is enlarged allowing proteins to access the ClpP active sites within the degradation chamber (Fig. 1a)[24,25]. Proteomic experiments showed that this uncontrolled proteolysis leads to the depletion of several essential proteins such as the bacterial cytoskeleton protein FtsZ, which in turn causes impaired cell division and ultimately cell death[26,27]. Since this unique antibiotic mechanism exploits a cellular machinery and activates it for destruction, ADEPs were recently shown to be active also against bacterial persister cells and established biofilms of pathogenic *Staphylococcus aureus*[27].

Here we show that the conformation of the ClpP barrel is controlled through the hydrophobic pocket on the protease surface. ClpP is arrested in the active, extended conformation on engagement with ADEP or chaperone, which leads to a stimulation of catalytic activity. Our results reveal an additional layer of regulation in the ClpXP system: the link between ClpP barrel conformation and occupation of the allosteric sites. This demonstrates that the allosteric binding pocket of ClpP functions as a conformational switch that not only controls the accessibility of the active sites but also their activity.

Results

Characterization of the SaClpP–ADEP interaction. We set out to investigate the activation of ClpP from *S. aureus* (SaClpP) by ADEPs and to analyze the corresponding effects on protease function and conformation. We started by screening a small compilation of ADEP derivatives[28], which we found to be equally potent in inducing SaClpP proteolysis (Supplementary Fig. 1a–c). All biochemical and structural assays were performed with ADEP7 (Fig. 1b) unless noted otherwise, which showed a slightly higher affinity than the widely used reference compound ADEP4 towards both C-terminally Strep-tagged and native SaClpP protein (Supplementary Fig. 1d–f). Wild-type SaClpP was purified as a tetradecamer with 14 identical ADEP binding sites. A detailed concentration-dependent analysis of ADEP7-induced fluorescein isothiocyanate (FITC)–casein degradation by SaClpP exhibited a sigmoidal behaviour and was thus adequately described by the Hill equation (Fig. 1c)[25,29]. Data analysis yielded an affinity constant of 3.1 μM and a Hill parameter of 2.0, indicative of positive cooperativity and consistent with previous studies on *Escherichia coli* ClpP[25,29]. Isothermal titration calorimetry (ITC) experiments in which ADEP7 was added to a solution of SaClpP showed nonstandard behaviour[30] (Fig. 1d) with the reaction becoming more exothermic during the initial increase of the ADEP7:SaClpP ratio. This can be explained by a lower binding affinity of the initial ADEP molecules and thus confirms positive cooperativity. We next reverted the conditions and titrated SaClpP into a solution of ADEP7 thereby starting with a fully saturated SaClpP (Fig. 1e). Here data analysis yielded an ADEP7:SaClpP molar ratio of 1.0 and a K_d of 2.1 μM, which likely describes the binding of an ADEP7 molecule to a SaClpP tetradecamer that already has several ADEP molecules bound (see Supplementary Fig. 2 for replicate and control experiments and Supplementary Table 1 for a compilation of parameters obtained from ITC experiments).

We further determined the degree of stabilization in a thermal shift assay (Fig. 1f). Addition of ADEP7 led to a drastic increase in the protein melting temperature and thus SaClpP folding stability with an half-maximal effective concentration of 2.7 μM. Collectively, these data confirm an affinity of ADEP7 to SaClpP in the 2–3 μM range, slightly higher than the values of 0.82 and 0.37 μM determined for *Bacillus subtilis* ClpP (BsClpP, Supplementary Fig. 3a) and *E. coli* ClpP (EcClpP)[25], respectively. We therefore compared the binding sites based on available structure and sequence data and identified three residues that were different in SaClpP (Supplementary Fig. 3b,c). We generated SaClpP mutant proteins where these three residues were replaced by the corresponding amino acids of BsClpP/EcClpP and found that exchanging histidine 83 for phenylalanine was sufficient for increasing the affinity of SaClpP for ADEP7 to a level comparable to BsClpP (Supplementary Fig. 3d–f). Notably, the mutation also caused partial heptamer formation as evidenced by size exclusion chromatography. Importantly, this oligomerization defect was completely abrogated by ADEP7 binding, indicating that ADEPs exhibit conformational control over ClpP. The basis for this conformational control as well as for the observed cooperativity was unknown and we therefore decided to carry out a series of in-depth structural and functional studies.

ADEP accelerates ClpP catalysis. At first, we investigated how binding of ADEP to SaClpP influences protease catalysis. In a simplified view, the cleavage of a peptide substrate can be described as a two-step process in which first the catalytic serine is acylated by a nucleophilic attack at the scissile amide bond, and second the acyl-ester intermediate is hydrolyzed by a water molecule to regenerate the free enzyme (Fig. 2a). To dissect the mechanism of catalysis, we made use of customized β-lactones that were previously described and characterized as SaClpP inhibitors[5,31]. These compounds can be viewed as electrophilic, stripped-down substrate mimetics in which the R^1 substituent resembles the S1 amino acid side chain and which otherwise lack an amide backbone or prime site substituents[32].

The nucleophilic attack of the active site serine at the β-lactone carbonyl leads to ring opening and acylation. We quantified this step by measuring catalytic efficiencies (that is, observed rate constant per inhibitor concentration values, $k_{obs}/[I]$) for five different β-lactones both in the presence and absence of ADEP7 (Fig. 2b,c, Supplementary Fig. 4). The values spanned a range of 3 orders of magnitude and correlated well with the length of the R^1

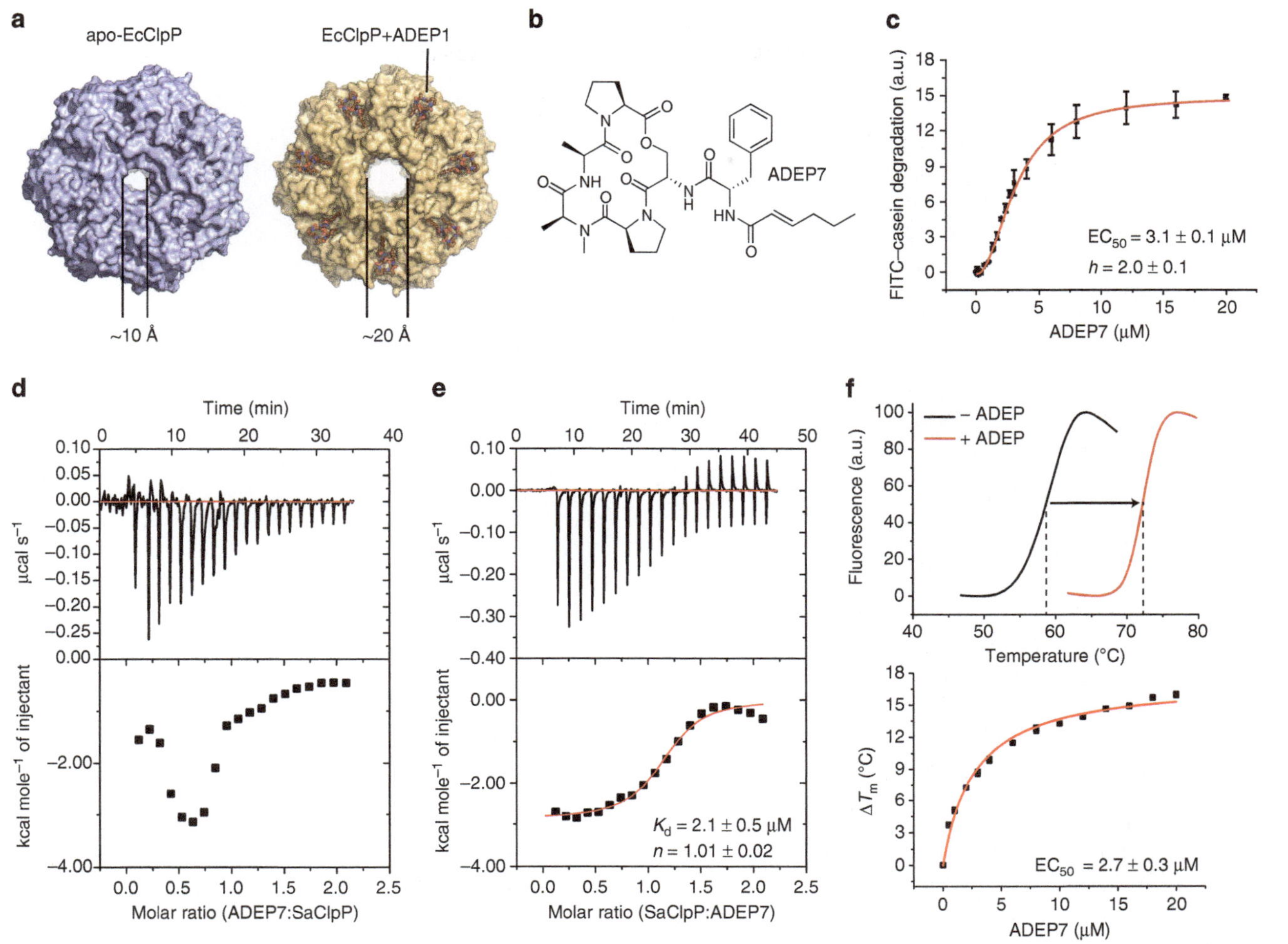

Figure 1 | ADEP cooperatively binds and stabilizes SaClpP. (**a**) Binding of ADEP to the hydrophobic pocket on the apical surface of ClpP causes a conformational change that opens the axial substrate entry pore of ClpP (PDB IDs: 1TYF, 3MT6). (**b**) Structure of ADEP7. (**c**) ADEP7-induced degradation of FITC-casein by SaClpP (0.5 μM) indicates positive cooperativity (Hill coefficient $h = 2.0 \pm 0.1$; $EC_{50} = 3.1 \pm 0.1$ μM). (**d**) ITC titration of ADEP7 (500 μM) into a solution of SaClpP (50 μM) shows positive cooperative binding, which prevents standard data analysis. (**e**) ITC titration of SaClpP (617 μM) into a solution of ADEP7 (62 μM). Parameters obtained from this experiment are: stoichiometry factor $n = 1.01 \pm 0.02$; $K_d = 2.1 \pm 0.5$ μM; $\Delta H = -3369 \pm 100$ cal mol^{-1}; $\Delta S = 13.8$ cal mol^{-1} K^{-1}. The stoichiometry factor refers to the ratio of monomeric SaClpP and ADEP7. A value of 1 is equivalent to 14 ADEP molecules binding per $SaClpP_{14}$. (**f**) Thermal shift assays reveal strong stabilization of SaClpP folding by ADEP7 in a concentration-dependent manner (melting temperature of free SaClpP: 58.3 °C; ADEP7 concentration in upper panel: 12 μM; $EC_{50} = 2.7 \pm 0.3$ μM). Plotted data in **c** and **f** are mean ± s.d. ($N = 3$). Parameters determined from curve fits in **c**,**e** and **f** are given with fitting errors.

substituent as established previously[31]. In addition, we confirmed the site of modification by peptide mass spectrometry to be the catalytic serine 98 (Fig. 2d). Comparing the catalytic efficiencies in the presence of a saturating amount of ADEP7 to the respective value of free SaClpP yielded a general increase of a factor of 2, irrespective of the size of the β-lactone (Fig. 2c). Previous structural studies have revealed that ADEP enlarges the axial pore of ClpP from a diameter of roughly 10 to 20 Å, and pore opening is currently viewed as the sole reason for ADEP activity[24,25]. Our data suggest an additional stimulation of catalytic turnover since the increase in catalytic efficiency was found to be independent of the size of the β-lactone inhibitor and was also obtained with the smallest compound VK292, which has a maximum diameter of ~7 Å.

To quantitatively assess the second reaction step, the hydrolysis of the acyl-enzyme intermediate, we incubated SaClpP with high concentrations of lactones D3 and U1, diluted the reaction mixture and followed the hydrolysis of the acyl-ester intermediate by intact-protein mass spectrometry both in the presence and absence of ADEP7 (Fig. 2e). ADEP again dramatically increased the reaction velocity roughly by twofold by decreasing the half-life time ($T_{1/2}$) of the hydrolysis of D3 from 11.1 to 4.8 h. The behaviour of lactone U1 is different in that it reacts more slowly with the enzyme, but once acylation has occurred it exhibits particularly slow hydrolysis ($T_{1/2} > 40$ h). While the reason for this β-lactone-inherent difference is unknown, ADEP7 released this impediment and caused even faster hydrolysis with a $T_{1/2}$ of 2.38 h. Since the hydrolysis reaction occurs within the catalytic chamber, the state of the N-terminal residues (that is, the diameter of the axial pore) is not expected to affect the hydrolysis velocity (for coordinated conformational changes, see the Discussion section). Hence these results unequivocally demonstrate allosteric stimulation of ClpP catalytic activity by ADEP, independent of axial pore enlargement.

We next examined if this stimulation of activity is accompanied by functional changes of the active site. To this end, we analyzed the turnover of a fluorogenic substrate peptide (Suc-LY-AMC) and observed that saturating ADEP only affected the catalytic rate constant $k_{cat,app}$, while the affinity constant $K_{M,app}$ remained unchanged (Supplementary Fig. 5a). Moreover, we

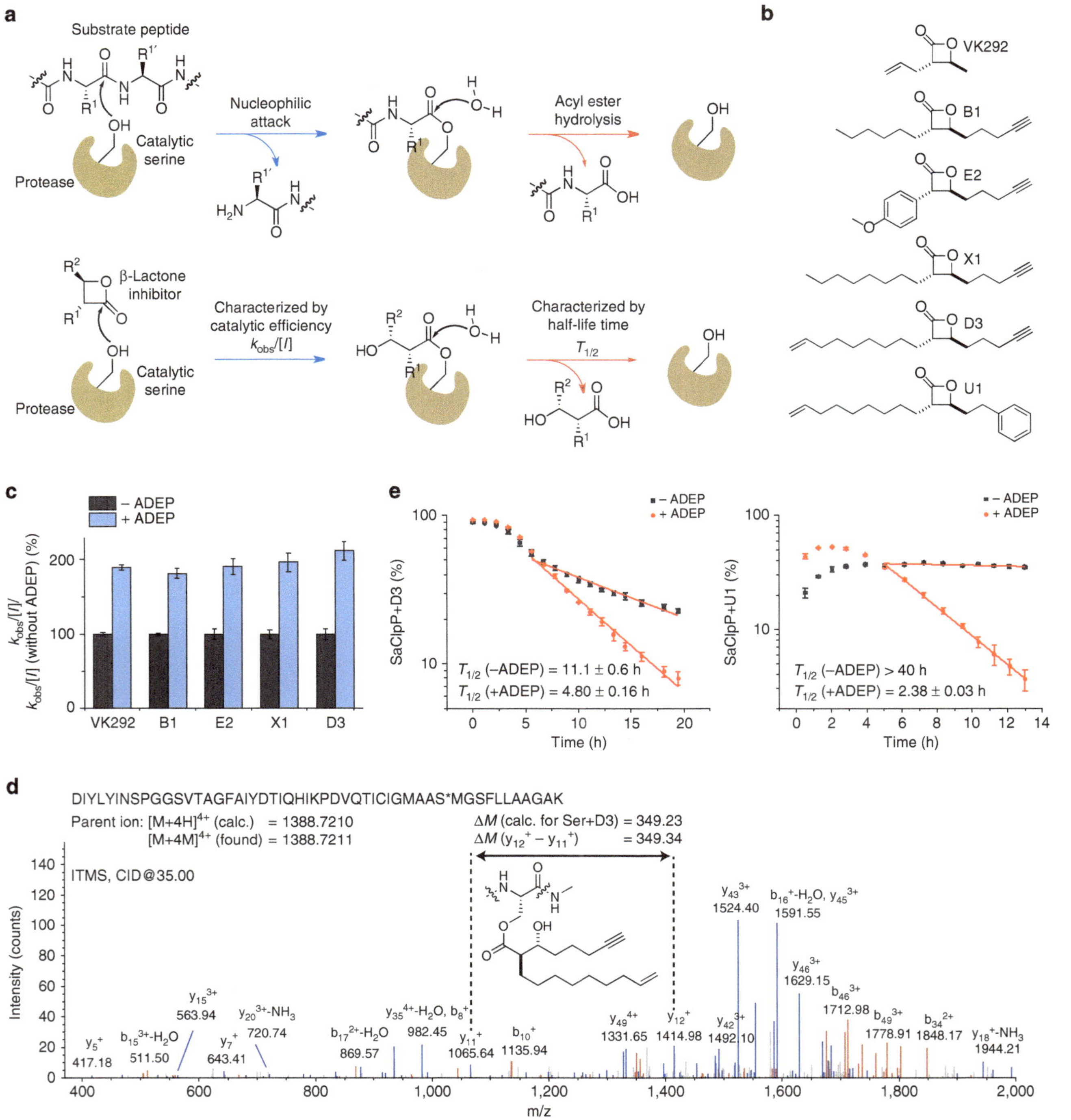

Figure 2 | ADEP accelerates ClpP catalysis. (**a**) Simplified schematic representations of the reactions of the SaClpP active site serine with a substrate peptide (upper panel) and with a β-lactone (lower panel). (**b**) Structures of β-lactones used in this study. (**c**) Catalytic efficiencies for the reactions of SaClpP (0.5 μM) with five β-lactones in the presence or absence of saturating ADEP7 (12 μM). For each lactone, values were normalized to the respective catalytic efficiency in the absence of ADEP7 for easy comparison. See Supplementary Fig. 4 for raw data, error bars denote fitting errors. (**d**) CID fragmentation spectrum of the peptide indicated bearing a covalent modification of β-lactone D3 at the catalytic serine 98 of SaClpP (y ions, blue; b ions, red). For a complete list of identified fragments, refer Supplementary Table 2. (**e**) ADEP7 accelerated the acyl-ester hydrolysis step. SaClpP (10 μM) was incubated for 5 min with saturating concentrations of lactone D3 and U1 (50 μM), then diluted fivefold and treated with ADEP7 (12 μM) or DMSO, followed by incubation at 37 °C. The degree of covalent modification of SaClpP was quantified via intact-protein mass spectrometry after several time points (three independent experiments, mean ± s.d.). CID, Collision-Induced Dissociation.

screened a variety of β-lactones with diverse substituents for their potential to acylate SaClpP in the presence and absence of ADEP7; however, ADEP did not apparently change the preference of SaClpP for specific β-lactone probes (Supplementary Fig. 5b,c). In addition, we synthesized all four stereoisomers of lactone U1 and looked for changes in binding through intact-protein mass spectrometry. Only the (*S,S*)-configured lactone reacted with SaClpP with both *cis*-configured lactones showing marginal modification of the protease (Supplementary Fig. 5d,e). These results indicate that the

ADEP-induced increase of catalytic activity is not due to a change in geometry of the ClpP substrate pocket. Consistent with this, previous co-crystal structures of both BsClpP and EcClpP revealed no difference in the orientation of the active site catalytic triad on ADEP1 binding and ascribed the effect of ADEP binding to a change in the nearby N-terminal region[24,25]. These observations excluded that ADEP binding induces a re-orientation of active site residues and raises the question of the molecular mechanism of enhanced catalysis.

ADEP allosterically controls ClpP barrel conformation. An explanation for the unusual ADEP-induced activation in line with all previous experiments would be a stimulation of ClpP through conformational restriction of the complex into a more active form. The ClpP protease is known to be highly dynamic[20] and structural information on SaClpP is available for an active extended[19], an inactive compact[33] and an inactive compressed[6] conformation[34]. These conformations differ in the height of the barrel ($\sim$10 nm for the extended and $\sim$9 nm for the compact and compressed conformations), in the alignment of the catalytic triad residues and in the orientation of the E-helix in the handle region (Fig. 3a and Supplementary Fig. 6a). Protein NMR studies with EcClpP[35] have revealed the presence of at least two states in solution, which is supported by molecular dynamics simulations[36] and normal mode analysis[37] suggesting that ClpP samples different conformations. We sought to elucidate the structural basis of ADEP allostery by studying a panel of mutant proteins, which adopt different conformational states. We introduced mutations at all residues of the catalytic triad—S98, H123 and D172 (H123A and H123N proteins showed an increased tendency to aggregation and were therefore excluded from further studies). We also mutated the oligomerization sensor residues R171 and D170 (as well as nearby T169), which engage in inter-ring salt bridging depending on the conformational state and thereby facilitate a functional coupling of tetradecamer formation to the catalytic triad[19,38,39]. Analytical size exclusion chromatography showed a clustering of these mutants into three distinct groups (Fig. 3b). While the R171A and T169A proteins were clearly heptameric as reported previously[19,31], the D172N and D170A proteins migrated only slightly slower than wild-type, tetradecameric SaClpP. We confirmed the integrity of all protein samples by a Coomassie gel as well as by intact-protein mass spectrometry to exclude abnormal autocatalytic processing (Supplementary Table 3). We also confirmed the assigned oligomerization states by size exclusion chromatography–multi-angle light scattering (SEC–MALS) analysis (Supplementary Fig. 6b). In addition, we validated these data by analytical ultracentrifugation, which also indicated a different sedimentation of the D172N mutant compared with wild-type SaClpP (Fig. 3c). With the molecular masses of the subunits as well as the oligomerization unchanged, the smaller appearance in the size exclusion experiment of D172N proteins could result from a more compact/compressed tetradecameric conformation.

To correlate ClpP conformations with both peptidase and proteolytic activity, we functionally characterized the mutant enzymes in the presence of ADEP (Fig. 3d,e). Both mutants of the active site aspartate (D172A and D172N) containing a weakened charge-relay system of the catalytic triad were catalytically inactive alone. Unexpectedly, they gained proteolytic activity in the presence of ADEP, with undistinguishable activity of D172N and wild type in a FITC–casein degradation assay. Moreover, ADEP stimulated hydrolysis of the fluorogenic peptide Suc-LY-AMC in the D172N but not in the D172A mutant.

We speculated that subtle changes near the oligomerization sensor residues of ClpP such as D172N might trigger the adoption of a compacted conformation as the lowest energy state. To investigate the mutation- and ADEP-induced changes in more detail on a structural level, we applied small angle X-ray scattering (SAXS; Fig. 3f,g). A comparison of the inter-atomic pair distance distribution functions (PDDF, $P(R)$) of wild-type and D172N SaClpP clearly showed a more compacted conformation of the D172N protein. The data also demonstrate that the oligomeric assemblies in wild-type and D172N SaClpP have the same mass (identical integral of the $P(R)$ in Fig. 3f, see Table 1), confirming different conformations rather than oligomerization states. Addition of either ADEP4 or ADEP7 led to an increase in the size of both the wild-type and the D172N protein as revealed by a shift in the $P(R)$ function, and the conformations became virtually indistinguishable in the ADEP-bound state. The ADEP-induced shift of the $P(R)$ function as well as of the radius of gyration, R_g, in the D172N sample is fully in agreement with the shift predicted by analyzing the structures of the compressed and the extended conformation (Fig. 3h; $\Delta R_g^{\text{(compressed->extended), calculated}} = 1.8$ Å; $\Delta R_g^{\text{(SaClpP-D172N->+ADEP4), observed}} = 1.7$ Å). Moreover, low-resolution models inferred from the data support the conclusion of an ADEP-mediated shift from a compressed to an extended conformation in the D172N protein (Fig. 3i). Notably, also the wild-type protein either bound to ADEP4 or ADEP7 showed a small increase in size. The inter-ring bridge mutants T169A and R171A were clearly heptameric in SAXS analysis and failed to assemble to functional tetradecameric complexes on addition of ADEP, which is in line with their lack of catalytic activity (Supplementary Fig. 6c–f). SAXS data were recorded three times with independently purified protein samples including tag-free and C-terminally Strep-tagged protein, and the described changes were consistently visible.

It has been speculated that the compressed conformation precludes ADEP binding due to a closed binding pocket[33]. Consistently, we found a rough threefold lower affinity of SaClpP-D172N for ADEP7 by ITC while the affinity towards SaClpP-S98A was less affected (Supplementary Fig. 2e,f). In addition, we validated the SAXS results with dynamic light scattering (DLS), which significantly showed a small ADEP-induced increase in the wild-type protein radius and a larger ADEP-induced increase in the D172N sample, with both ADEP-bound proteins being equal in size (Fig. 3j). Since it is difficult to correlate static crystal structures to the data inferred from solution-based methods, the possibility remains that the compressed structure we are observing in solution may not entirely represent that described by X-ray crystallography. However, when taken together, these data clearly demonstrate how binding of ADEP exhibits conformational control on the ClpP barrel. Occupation of the regulatory allosteric site reverts the catalytically incompetent, compressed conformation of the D172N mutant and locks the wild-type protein in an extended conformation, explaining the ADEP-induced increase in catalytic activity.

Substoichiometric occupation leads to partial inhibition. When we analyzed the concentration-dependent response of SaClpP peptidase activity to ADEP7, we unexpectedly found biphasic behaviour. While saturating amounts of ADEP stimulated SaClpP peptidase activity as described above, small ADEP concentrations corresponding to three to four ADEP molecules per $SaClpP_{14}$ showed partial inhibition (Fig. 4a). This partial inhibition was also observed with ADEP4, as well as with tag-free SaClpP protein (Supplementary Fig. 7a–c). Moreover, a similar phenomenon was observed with the D172N mutant whose low residual peptidase activity was further diminished by adding a small amount of ADEP (Fig. 4b). Consistent with the wild-type SaClpP data, no such effect was observed during proteolysis (Fig. 4c). We analyzed the respective SaClpP:ADEP ratios via

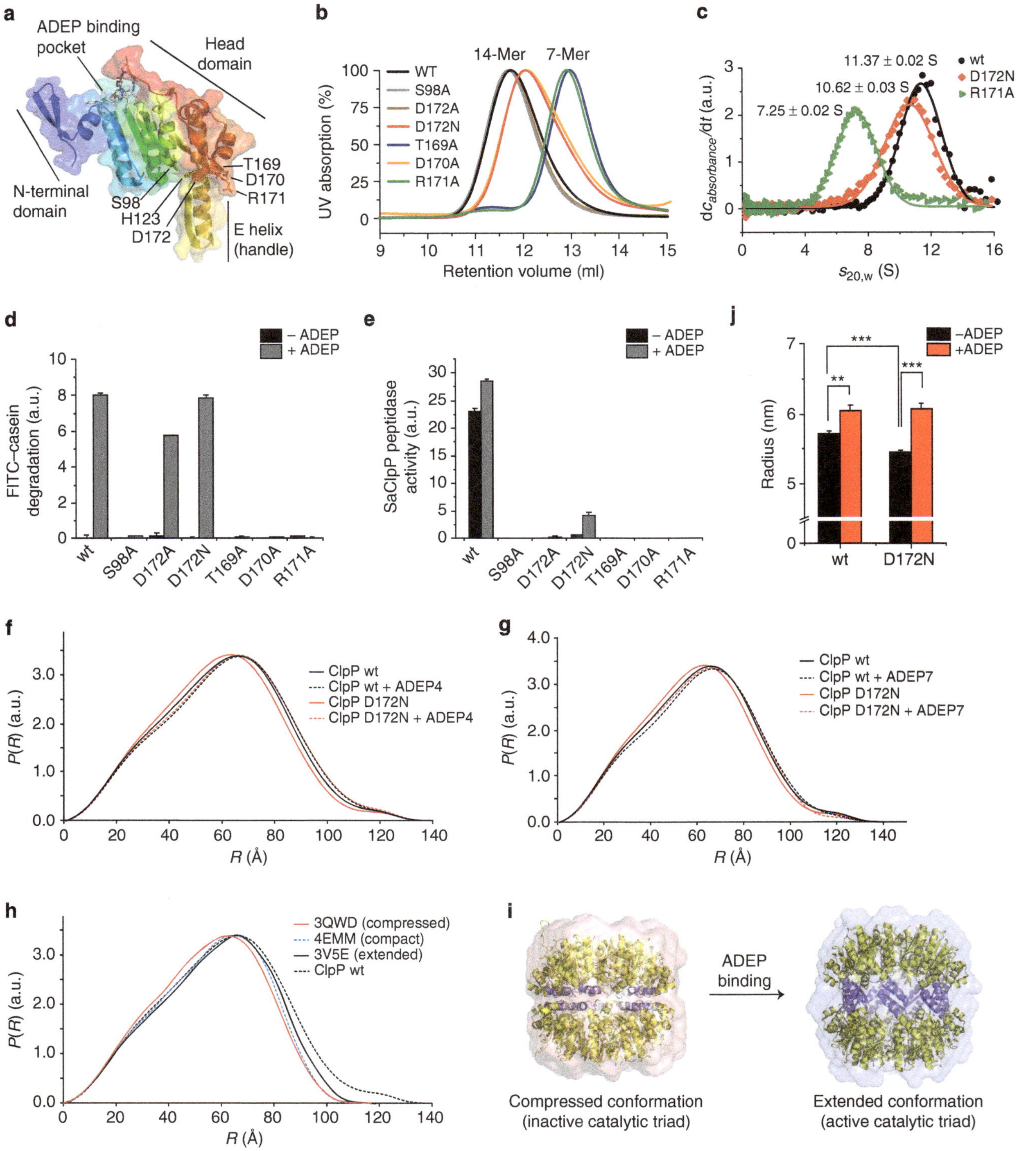

Figure 3 | Binding of ADEP to ClpP induces a conformational switch to the active, extended ClpP barrel conformation. **(a)** Cartoon representation of a ClpP monomer showing the position of the ADEP binding site, the catalytic triad (S98, H123, D172) and the oligomerization sensor R171 (PDB ID: 3MT6). **(b)** Analytical size exclusion chromatograms of wild-type and mutant SaClpP. **(c)** Analytical ultracentrifugation results for selected mutants. Sedimentation coefficients were obtained as the centres of Gauss functions whose fitting errors are given. **(d)** FITC-casein degradation activity of wild-type and mutant SaClpP (1 μM) in the presence or absence of ADEP7 (12 μM). Mean ± s.d. of initial slopes are given. **(e)** Peptidase activity of wild-type and mutant SaClpP (0.5 μM) in the presence or absence of ADEP7 (12 μM). Mean ± s.d. of initial slopes are given. **(f)** Inter-atomic pair distance distribution functions (PDDF) of tag-free SaClpP proteins treated with either DMSO or ADEP4 (1:1 mixture, 380 μM, 0.6% DMSO). **(g)** PDDF of tag-free SaClpP with/without ADEP7 (1:1 mixture, 380 μM, 0.6% DMSO). **(h)** Comparison of measured (ClpP wt) and predicted (3QWD, 4EMM, 3V5E) inter-atomic pair distance distribution functions. **(i)** *Ab initio* low-resolution models of SAXS scattering curves from SaClpP-D172N superimposed with SaClpP in the compressed conformation (left), as well as from SaClpP-D172N + ADEP4 superimposed with SaClpP in the extended conformation (right). **(j)** Dynamic light scattering results of wild type and D172N SaClpP treated with either DMSO or ADEP7 (1:1 mixture, 310 μM, 3% DMSO). Please note the axis break. Mean ± s.e.m.; $^{**}P<0.01$, $^{***}P<0.001$ (independent two-sample *t*-tests, $N=10$, with each of these 10 results established from 10 technical replicates, equal variances not assumed). UV, ultraviolet.

analytical ultracentrifugation and found that a substoichiometric amount of ADEP slightly decreased the sedimentation coefficient of wild-type ClpP, whereas this effect was reverted at higher ADEP concentrations (Fig. 4d). Similarly, the sedimentation coefficient of D172N first decreased with small amounts of ADEP and increased to a higher value than the free protein on further addition of ADEP (Fig. 4e). We hypothesize that binding of the initial ADEP molecules to a tetradecameric ClpP induces structural perturbations owing to a conformationally heterogeneous assembly, which would also account for the observed positive cooperativity.

Table 1 | SAXS data and analysis of SaClpP proteins.

Sample	Stoichiometry	R_g (Å)	Molecular mass (kDa)*
SaClpP wt	—	46.4 ± 0.1	292
SaClpP wt + ADEP4	1:1	46.7 ± 0.1	305
SaClpP wt + ADEP7	1:1	46.3 ± 0.1	296
SaClpP-D172N	—	44.7 ± 0.1	303
SaClpP-D172N + ADEP4	1:1	46.4 ± 0.1	300
SaClpP-D172N + ADEP7	1:1	45.6 ± 0.1	296
SaClpP wt (Strep)	—	49.0 ± 0.1	320
SaClpP wt (Strep) + ADEP7	1:1	49.2 ± 0.1	330
SaClpP-D172N (Strep)	—	47.4 ± 0.1	324
SaClpP-D172N (Strep) + ADEP7	1:1	48.8 ± 0.1	330
SaClpP T169A	—	40.7 ± 0.1	167
SaClpP T169A + ADEP7	1:1	41.8 ± 0.1	160
SaClpP R171A	—	39.5 ± 0.1	170
SaClpP R171A + ADEP7	1:1	40.2 ± 0.1	180

Samples were treated either with ADEP or with the respective amount of DMSO.
*The molecular mass was determined from the scattering intensity at zero angle ($I(0)$) using BSA as reference. Expected molecular masses of native SaClpP and Strep-tagged SaClpP are 301 kDa and 316 kDa, respectively.
ADEP, acyldepsipeptide; BSA, bovine serum albumin; DMSO, dimethylsulphoxide; Rg, radius of gyration; SaClpP, ClpP from *Staphylococcus aureus*; SAXS, small angle X-ray scattering; wt, wild type.

Despite a sequence homology of ~60% (Supplementary Fig. 3b), ClpP proteins from different species are remarkably different in their oligomerization behaviour. We studied BsClpP, which under our purification conditions using affinity and size exclusion chromatography in glycerol containing buffer yielded a tetradecameric and a monomeric fraction (Supplementary Fig. 7d–i)[10]. In analogy to SaClpP, the intrinsically active tetradecameric BsClpP showed partial inhibition of peptidase activity at low ADEP concentrations while the monomeric BsClpP was inactive in the absence of ADEP and was activated with an even lower ADEP concentration than the tetradecameric BsClpP (Supplementary Fig. 7e–g). Similar results were obtained with *Listeria monocytogenes* ClpP2 (LmClpP2) tetradecamer and heptamer (Supplementary Fig. 7h,i)[40]. These data provide evidence that the hydrophobic ADEP binding pocket acts as a major conformational regulatory site across organisms, which allows for differential regulation of protease activity depending on the degree of occupation per tetradecamer.

SaClpX and ADEP share a ClpP activation mechanism. The ADEP-bound conformation of ClpP has been suggested to serve as a model for the Clp–ATPase-bound state of ClpP since ADEP

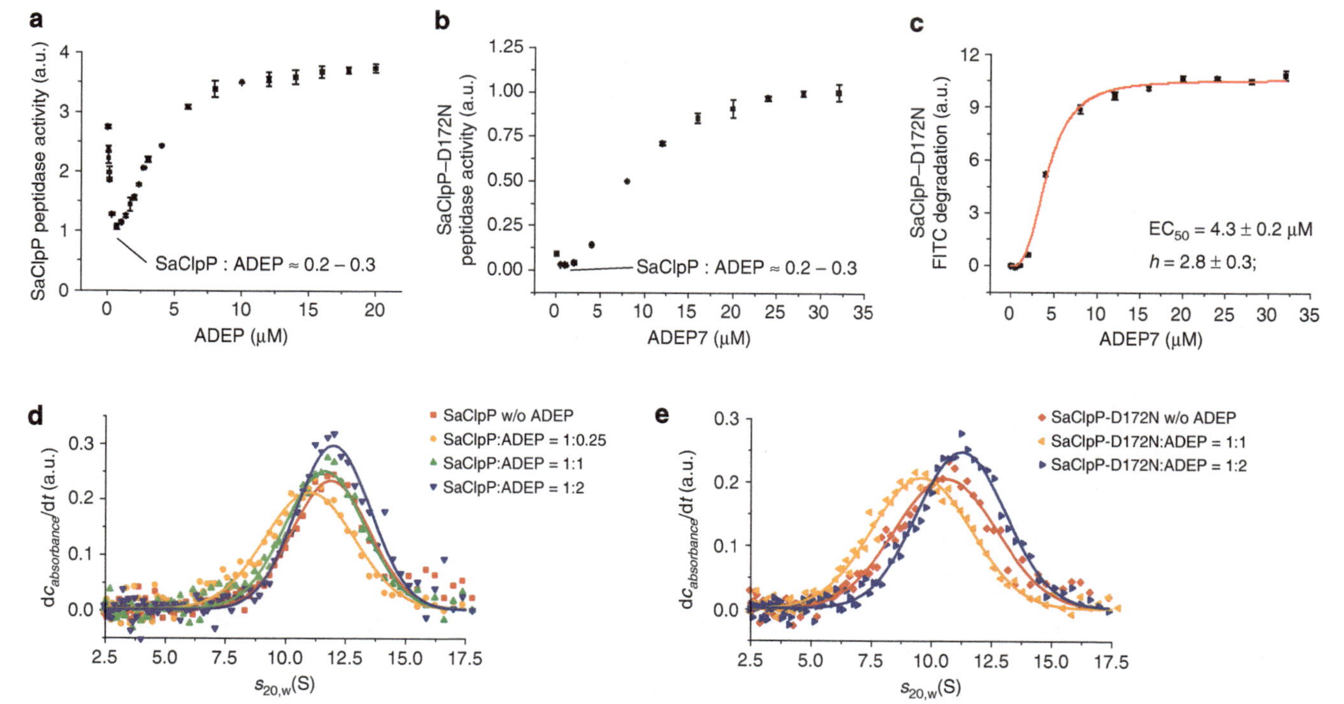

Figure 4 | Substoichiometric amounts of ADEP cause partial inhibition of ClpP. (**a**) Concentration-dependent partial inhibition or stimulation of wild-type SaClpP peptidase activity by ADEP7. The ratio of ADEP bound to ClpP was estimated from the concentrations at the minimum and a K_d of 2.1 μM. A ratio of 0.2-0.3 corresponds to 2.8-4.2 ADEP molecules per $SaClpP_{14}$. (**b**) Concentration-dependent inhibition or stimulation of SaClpP-D172N peptidase activity by ADEP7. In line with the compressed conformation of SaClpP-D172N, the initial peptidase activity is lower than that of the wild-type protein. The ratio of ADEP bound to ClpP was estimated from a K_d of 6.1 μM. (**c**) FITC-casein degradation by SaClpP-D172N indicates positive cooperativity ($h = 2.8 \pm 0.3$; $EC_{50} = 4.3 \pm 0.2$ μM). (**d**) Analytical ultracentrifugation results of SaClpP (7.5 μM) treated with DMSO (0.6% (v/v)) or ADEP7 (2, 7.5 and 15 μM). (**e**) Analytical ultracentrifugation results of SaClpP-D172N (7.5 μM) treated with DMSO (0.6% (v/v)) or ADEP7 (7.5 and 15 μM). Plotted data in **a**–**c** are mean ± s.d. ($N = 3$). Parameters determined from curve fits in **c** are given with fitting errors.

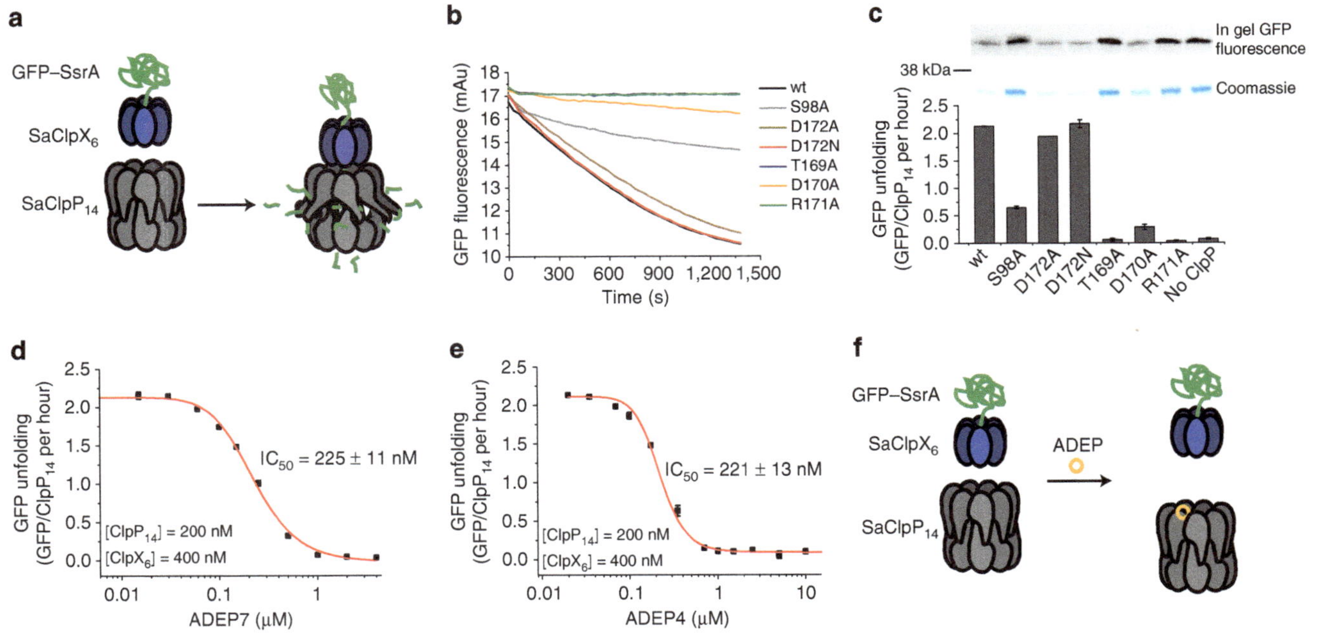

Figure 5 | ClpX activates ClpP in a similar manner and is competed by ADEP. (**a**) Schematic of the SaClpX-mediated unfolding of SsrA-tagged GFP and its subsequent degradation by SaClpP. (Note, the precise path of product peptide release is unclear and under debate.) (**b**) SaClpXP–GFP assay. The initial slopes of the decrease in GFP fluorescence of the fluorescence-time courses were quantified and are displayed in the bar graph. A Coomassie-stained gel of these samples showing GFP after 3 h at 30 °C indicates that wild-type, D172A, D172N, and D170A SaClpP are capable of protein degradation, while S98A, T169A and R171A SaClpP are not. This is consistent with in-gel GFP fluorescence measurements. See Supplementary Fig. 7n,o for uncropped gel images. (**c,d**) ADEP7 and ADEP4 disrupt SaClpXP-mediated GFP unfolding with an IC_{50} value comparable to the concentration of $SaClpP_{14}$. (**e**) Schematic showing how the binding of ADEP to $SaClpP_{14}$ prevents the binding of $ClpX_6$ and thus GFP unfolding. Mean ± s.d. are given in all panels ($N = 3$). IC_{50} values are given with fitting errors.

and the IGF loop of ClpX are assumed to share the same binding site. We therefore reasoned that ClpX might be able to exert conformational control on ClpP in a similar way. We thus tested all protein mutants in a SaClpXP assay in which we monitored the fluorescence of a green fluorescent protein (GFP) variant to which a C-terminal degradation tag (SsrA tag) was appended (Fig. 5a,b)[17]. To differentiate between GFP unfolding and degradation, the assay mixture was analyzed by Coomassie gel after 3 h. The D172N mutant showed GFP degradation activity similar to the wild-type enzyme, which is in full agreement with the results of ADEP-induced casein degradation, indicating that SaClpX is also able to revert the compressed conformation of the D172N protein. The intermediate assay signal of the S98A mutant protein was assigned as unfolding activity without proteolysis[41], whereas the heptameric oligomerization of the T169A and R171A mutants consistent with the SAXS data precluded activation by SaClpX and thus GFP unfolding. In a competition assay, ADEP7 and ADEP4 inhibited GFP unfolding by SaClpXP and thus binding of SaClpX to ClpP with IC_{50} values of 225 and 221 nM (Fig. 5c). This value is in the range of the assay concentration of tetradecameric SaClpP (200 nM), consistent with a shared binding site of ADEP and ClpX[10]. The IC_{50}, furthermore, suggests that the binding of a single ADEP molecule to only one SaClpP apical surface is sufficient to reduce SaClpXP proteolysis by half (Fig. 5e).

Discussion

Multicomponent protease complexes require the precise coordination of several processes including substrate engagement, unfolding, degradation and product release. To this end, the different parts of the complexes need to be functionally coupled in an accurate and versatile manner. Moreover, information must be coded in structural form and transmitted through allosteric binding events. Since uncontrolled proteolysis poses a threat to cellular viability, reliable and multi-layered regulation is essential for these machineries.

The ClpXP protease is a prime example of such a complex since the dynamics of its components have been studied and a framework of reference structures exist[42,43], though a structure of the entire complex is lacking[34,44–46]. Biochemical experiments have shown that ClpX interacts with ClpP primarily through docking of its IGF loop into a hydrophobic pocket on the surface of ClpP that is also used by ADEPs to dysregulate ClpP[22–24]. The N-termini of ClpP have been shown to gate the entry into the degradation chamber by adopting different conformations, where the 'down' conformation as a hydrophobic plug closes and the 'up' conformation opens the entry portal[47,48]. Although ADEPs have already been shown to regulate these gated pores, further activating influences of the antibiotic had not been considered. Here we show that ADEPs additionally stimulate ClpP catalysis through cooperative binding. In previously published studies on ADEP-ClpP co-crystal structures this allosteric activation had been missed, since it is not based on changes of active site geometry but on conformational dynamics. We find that ADEPs are able to revert the catalytically incompetent compressed conformation of the D172N ClpP mutant through switching its conformation to the extended state, which exhibits an aligned catalytic triad (contrary to the compressed conformation, where the triad is not aligned). Similarly, ClpX activates ClpP-D172N for proteolysis, implicating that a chaperone is also able to induce the conformational switch of the ClpP-D172N barrel conformation necessary for catalysis. Importantly, we observe that also wild-type ClpP adopts an extended state by addition of ADEP. Our biochemical and structural data thus converge at the view that occupation of the hydrophobic pocket arrests the dynamic ClpP barrel in its catalytically competent conformation, which contributes to sustained peptidase as well as proteolytic activity.

Conformational arrest[49,50] is a strong regulation principle in ClpP as it even compensates for the lack of a catalytic D172 residue and facilitates proteolytic processing with a Ser–His catalytic dyad. This interpretation is corroborated by a previous study on functional coupling in the ClpXP complex, where occupation of the ClpP active site by a covalent inhibitor increased the affinity of the peptidase for the ATPase[22]. Our results now establish the bidirectionality of this cross-talk and implicate ClpP conformations and catalytic activity in this regulation. Studies on the human ClpP[51,52] showed that hClpP is heptameric in solution and that its catalytically competent tetradecameric state can be triggered by addition of ClpX from *E. coli*[53]. These data suggested a conformational coupling of the hydrophobic pocket with the handle domain. Our data structurally expand this finding, showing that this conformational control is also present within tetradecameric ClpP. Further indications came from hydrogen-exchange mass spectrometry experiments where ADEPs were used as ClpX mimetics and ADEP binding induced rigidification of ClpP in the equatorial handle region[48]. Moreover, mutations in the N-terminal region were used to propose that substrate access might coordinate with protease active site reactivity[54]. Most recently, ADEP was shown to contribute to ClpP activation in the heterooligomeric ClpP1P2 system from *Mycobacterium tuberculosis*, but only in the presence of an additional activating peptide[55].

ClpX interacts with ClpP via two distinct loop structures. ADEP is an IGF-loop model that demonstrates the sophisticated mechanism by which the ATPase exerts conformational control on the peptidase. The hydrophobic pocket at the surface of ClpP serves as the major regulatory 'button' that the IGF loop presses to stabilize the extended, catalytically active conformation of ClpP and simultaneously to put the N-terminal loops of ClpP in the upward position[56], thereby opening the gated pore. Shielding the open ClpP pore with its own pore-2 loops, ClpX establishes a secluded channel for substrate passage into the catalytic cavity. Coupling the two conformational movements to the same allosteric site and regulatory event ensures that catalytic triads are only active, when the pore is securely covered.

The ClpXP complex is not homogeneous during operation. In addition to the symmetry mismatch of hexameric ClpX and the axially heptameric ClpP, ClpX was shown to exhibit conformational heterogeneity based on the nucleotide state of the ATPase domain where coordinated hydrolysis by two to four subunits is coupled to substrate translocation[57,58]. While we provide evidence that full occupation of the allosteric site on ClpP leads to activation, we also show that a smaller degree of occupation causes partial inhibition through conformational perturbation. It is currently unclear how the symmetry-mismatched interaction of ClpX and ClpP is mediated and how many hydrophobic pockets on one apical side of a ClpP heptamer are occupied by one ClpX hexamer. We find it intriguing that in a SaClpXP assay with concentrations corresponding to two ClpX hexamers bound to the two apical sides of ClpP (that is, a $ClpX_6$:$ClpP_{14}$:$ClpX_6$ complex as observed with electron microscopy)[53,59,60], the addition of one ADEP molecule per ClpP tetradecamer was sufficient to reduce the substrate unfolding activity by half (Fig. 5c,d). This suggests that occupation of already one of the seven hydrophobic pockets on a ClpP ring is incompatible with ClpX binding, either through cooperativity-induced conformational heterogeneity or through direct collision with ClpX. Moreover, we are fascinated by the unexpected finding that an allosteric activator molecule is also capable of reducing the activity of the catalytic complex when binding at sub-stoichiometric conditions. This allows for a putative model in which ClpP activity is differentially regulated dependent on the respective ClpX binding state. While the relevance of this phenomenon in ClpXP proteolysis is unclear, we find it important to note that low concentrations of ADEP in a medicinal setting may lead to partial ClpP inhibition and thus bacteria with reduced pathogenicity[7].

Collectively, our results shed light on the way ClpP reciprocates activation by ClpX and ADEP. We provide mechanistic insights into how ADEPs and chaperones exhibit conformational control on ClpP and how this interaction goes beyond active site accessibility, but also influences protease catalysis. We provide evidence that the hydrophobic pocket acts as a major regulatory site of ClpP and that the binding of ADEP or ClpX to this site allows for differential regulation of protease activity and conformation depending on the degree of occupation per tetradecamer. Further experiments should investigate if this allosteric layer of regulation is present also in other multi-component proteases such as the proteasome.

Methods

Cloning and protein purification. C-terminally Strep-tagged ClpP from *S. aureus* NCTC 8325 (SaClpP)[19] and ClpP from *L. monocytogenes* (LmClpP2)[40] were purified through affinity and size exclusion chromatography from expression in *E. coli* BL21(DE3). Tag-free wild-type SaClpP was purified through anion exchange, hydrophobic interaction and size exclusion chromatography[39]. Expression constructs of mutant proteins were created using the QuikChange methodology using primers listed in the Supplementary Table 4 (refs 19,39). Expression was carried out in BL21(DE3) or SG1146a cells and purification proceeded as described for the wild-type proteins. Strain SG1146a was a gift from S. Gottesman (NIH, Bethesda, USA). Monomeric concentrations of ClpP are given unless otherwise noted.

BsClpP was expressed and purified as follows. *E. coli* SG1146a was transformed with pClpP11 (ref. 61). Cultures (1 l) were grown at 37 °C until an OD_{600} of 0.4–0.6 was reached. Then expression of the C-terminal His-tagged fusion protein was induced with 1 mM isopropyl-β-D-thiogalactoside and cultures were grown at 28 °C over night. Cells were lysed in buffer A (50 mM sodium phosphate, 300 mM NaCl, 20% (v/v) glycerol, 20 mM imidazole; pH 8.0) using a Precellys homogenisor (PeqLab). Supernatants were mixed with Ni-NTA matrix (Qiagen) at 4 °C over night and subjected to plastic columns (Thermo Scientific). ClpP was eluted with buffer B (50 mM sodium phosphate, 300 mM NaCl, 20% (v/v) glycerol, 250 mM imidazole; pH 8.0). The eluted proteins were concentrated with Amicon Ultra Centrifugal Filters with 10 kDa cutoff (Merck-Millipore) and subjected to gel filtration on an ÄKTA purifier system with a Superdex 200 16/60 column (GE Healthcare). Tetradecamers and monomers of ClpP were eluted with buffer C (50 mM sodium phosphate, 300 mM NaCl, 20% (v/v) glycerol; pH 8.0). Our reported purifications of BsClpP in the absence of glycerol and in the presence of dithiothreitol (DTT)[61] yielded predominately monomeric ClpP[62]. In the presence of glycerol, both tetradecameric and monomer species were obtained (Supplementary Fig. 7d). Monomeric BsClpP could be converted to tetradecameric BsClpP through addition of ADEP (Supplementary Fig. 7h,i).

Enhanced GFP tagged for ClpXP degradation, enhanced GFP–ssrA (eGFP–ssrA), originates from eGFP (protein ID C5MKY7) and was cloned as a fusion protein with an ssrA tag (AANDENYALAA) at the C terminus. An expression clone was assembled using the Gateway cloning strategy (Invitrogen) with pDonr201 as donation vector and pDest007 (ref. 63) as expression vector. The expression construct was transformed into *E. coli* KY2266 cells. Expression was carried out in 4 l LB media after induction with anhydrotetracycline (0.2 mg/l) at an OD_{600} of 0.5 for 4 h at 37 °C. The pellet was washed with PBS and resuspended in ice-cold lysis/wash buffer (100 mM Tris, 150 mM NaCl, 1 mM EDTA, pH 8.0). The cells were lysed using a Constant Cell Disruption system. The lysate was cleared via centrifugation (38,000 g, 45 min, 4 °C). Protein purification was achieved on an Äkta Purifier 10 system (GE Healthcare). Affinity chromatography was carried out with a StrepTrap HP 5 ml column. eGFP–ssrA containing elution fractions were pooled, concentrated and purified with a HiLoad 16/60 Superdex 200 pg gel filtration column in GF buffer (100 mM NaCl, 20 mM Tris, 10% (v/v) glycerol, pH 7.0).

Tag-free ClpX from *S. aureus* NCTC 8325 (SaClpX) was used in all experiments and obtained as follows. The gene encoding ClpX (geneID 3919696) was amplified using primers listed in Supplementary Table 4. An expression construct was assembled using the Gateway cloning strategy with pDonr207 as donation vector and pET300 as expression vector. The expression construct encoding N-terminally His_6-tagged, full-length SaClpX with an N-terminal TEV site was transformed into chemically competent BL21(DE3) cells. Expression was carried out in 1 l LB media after induction with isopropyl-β-D-thiogalactoside (0.5 mM) at an OD_{600} of 0.6 for

15 h at 25 °C. The cell pellet was washed with PBS and resuspended in ice-cold lysis buffer (50 mM Hepes pH 7.5, 300 mM KCl, 15% (v/v) glycerol, 1 mM DTT). The cells were lysed with a Constant Cell Disruption system and subsequent sonification (3 × 30 s, 80%, Bandelin sonoplus). $MgCl_2$ (5 mM) was added and the suspension was cleared through centrifugation (36,000g, 30 min, 4 °C). Affinity purification was carried out on an ÄKTA Purifier 10 chromatography system with a Ni-NTA Superflow cartridge (5 ml). The column was equilibrated in lysis buffer (+ 5 mM $MgCl_2$), the lysate was loaded and it was washed with 10 column volumes lysis buffer (+ 5 mM $MgCl_2$, + 40 mM imidazole). Elution was carried out with a steep gradient in elution buffer (lysis buffer + 5 mM $MgCl_2$ + 500 mM imidazole). EDTA (2 mM) and TEV protease (500 µl of 1.7 mg ml^{-1} stock) were added to the pooled elution fractions and it was incubated at 4 °C over night. Cleavage was monitored by intact-protein mass spectrometry and SDS-PAGE. If necessary, a second addition of TEV protease was carried out followed by incubation at 4 °C over night. The sample was concentrated with a 10 kDa Amicon ultra centrifugal filter to reduce the imidazole content, diluted in breaking buffer (10 ml final volume, final imidazole: 40 mM). $MgCl_2$ (2 mM) was added and the solution was passed through a pre-equilibrated Ni-NTA column. The flow-through was collected, three times buffer-exchanged with a 10 kDa Amicon ultracentrifugation device in lysis buffer (+ 2 mM $MgCl_2$) and frozen in aliquots. A gel filtration step was omitted since analytical runs showed no signs of aggregated protein.

Peptidase activity assay. ADEP (1 µl of 100 × stock in dimethylsulphoxide (DMSO)) was placed into black flat-bottom 96-well plates. Solutions of SaClpP (90 µl, final concentration: 0.5–1 µM) in assay buffer E (100 mM Hepes pH 7.0, 100 mM NaCl) were added and incubated at 32 °C for 15 min. Then Suc-LY-AMC solution (10 ×, final concentration: 200 µM; also preheated to 32 °C) in assay buffer B was added to the wells (10 µl each, assay volume: 100 µl). The reaction was monitored by following the increase of fluorescence (excitation: 380 nm, emission: 440 nm), which was recorded with one measurement per minute using an infinite M200Pro plate reader (Tecan) at 32 °C. Slopes were calculated with Microsoft Excel from the fluorescence over time plots via linear regression, whereby a 5-min lag period at the beginning of the measurement was excluded. All data were recorded in triplicate and in at least two independent experiments. S.d. was calculated from triplicate measurements. Normalization was achieved through measurement of the fluorescence of a set of completely cleaved Suc-LY-AMC samples with different concentrations (0.2–6 µM). LmClpP2 assay buffer (100 mM Hepes pH 7.0, 100 mM KCl) was used for LmClpP2 activity measurements with addition of glycerol (15% v/v) where indicated. BsClpP assay buffer (50 mM Tris-HCl, 25 mM $MgCl_2$, 100 mM KCl, 2 mM DTT; pH 8.0) was used for BsClpP activity measurements (1 µM BsClpP; 100 µl reaction volume; 0.5 µl DMSO/ADEP 200 ×; excitation/emission: 380/460 nm).

Catalytic efficiency measurements. Serial β-lactone inhibitor dilutions (1 µl of 100 × stock in DMSO) were added to wells and Suc-LY-AMC (50 µl, 2 ×, final concentration: 200 µl) in assay buffer A was added. ADEP (12 µM from 10 mM stock) or the respective amount of DMSO was added to a SaClpP solution (2 ×, final concentration: 1 µM) in assay buffer A and everything was brought to 32 °C. Subsequently, 50 µl of the protein solution was added to the wells and fluorescence was recorded as described above with a 20-s interval. Kinetic constants were obtained assuming pseudo-first-order kinetics by fitting the curves to

$$F(t) = F_0 + A(1 - e^{-k_{obs}t})$$

wherein $F(t)$ denotes the fluorescence-time course, F_0 denotes the initial fluorescence, A denotes the slope of the uninhibited control samples and k_{obs} denotes the rate constant. Catalytic efficiencies were determined as slopes from $k_{obs}/[I]$ plots[31].

FITC-casein assay. ADEP (1 µl of 100 × stock in DMSO) was added to wells of black flat-bottomed 96-well plates. Protein dilution (80 µl, 1.2 ×, final concentration: 1 µM) was added and incubated at 37 °C for 15 min. Casein mix (20 µl, 5 ×, final concentration of mixed casein: 0.24 mg ml^{-1}, final concentration of FITC-casein: 0.048 mg ml^{-1}) was added and the reaction was followed in an infinite M200Pro plate reader with excitation at 494 nm and detection at 521 nm at 37 °C. Initial slopes (usually 0–300 s) were used for data analysis. All data were referenced against the slope of casein solution without addition of protease. Fitting was performed with OriginPro (Microcal) using the Michaelis–Menten equation, wherein instead of a K_m the half-maximal effective concentration EC_{50} was used.

Isothermal titration calorimetry. All ITC experiments were performed on a MicroCal iTC200 system (GE Healthcare) in 20 mM Hepes pH 7.0, 100 mM NaCl with a maximum of 2% (v/v) DMSO at 25 °C and with constant stirring at 1,000 r.p.m. Prior to experiments, the protein was gel filtrated into 20 mM Hepes pH 7.0, 100 mM NaCl, concentrated if necessary using 10 kDa Nanosep centrifugational devices (Pall) and the ligand was dissolved in the exact same buffer from a 50 mM stock in DMSO. DMSO concentrations of syringe and cell samples were matched if necessary by addition of pure DMSO. The experiment was started after equilibration for 300 s with a first injection of 0.4 µl that was discarded during the analysis. A typical experiment consisted of 20 subsequent injections with a 2 µl injection volume into a cell filled with 200 µl sample. Each injection was made over a period of 4 s with a 2–3-min interval between subsequent injections. Power was recorded at 'high' gain setting, with a reference power of 10 µcal s^{-1} and a 5-s filter period. Data analysis including baseline correction and evaluation was carried out with OriginPro 8.5ITC. Due to cooperativity and thus the nonstandard form of some ITC curves, no fits based on simplified models were obtained for ligand into protein experiments. In the case of protein into ligand experiments, fits were carried out taking into account all injections until the maximum heat occurred and the two subsequent injections. We assume this to reflect the binding of ADEP to a ClpP protein with already some ADEP molecules bound. The binding characteristics of the initial ADEP molecules binding to free ClpP are likely to be different.

Thermal shift assay. ADEP (0.5 µl of 100 × stocks in DMSO) was placed into wells of white 96-well PCR plates. Sypro Orange (1 ×) was added to a SaClpP (1 µM) solution in buffer D and 50 µl were added to each well. The plate was sealed and fluorescence was recorded in a CFX96 Real-Time System (Bio-Rad) while heating from 20 to 80 °C in 0.3 °C steps. Data analysis was performed with CFX Manager software and OriginPro. Melting temperatures were referenced against the melting temperature of DMSO-treated SaClpP.

Intact-protein mass spectrometry. Samples were diluted down to 1–2 µM protein concentration in buffer D. Aliquots (4–2 µl) were desalted with a Massprep online desalting cartridge (Waters) according to the manufacturer's procedure on a Dionex UltiMate 3000 HPLC system and subsequently measured on a Thermo LTQ FT Ultra mass spectrometer with electron spray ionization. Promass Deconvolution software (Thermo Scientific) was used for data analysis and deconvolution (input range: 500–2,000 *m/z*; output range: 20,000–30,000 Da; peak width = 3; merge width = 0.3; smooth width = 7; number of smooths = 2). Thermo Scientific Xtract software was used for verification. For the time-course experiments, SaClpP (10 µM) was incubated for 5 min at room temperature with saturating concentrations of lactone D3 and U1 (50 µM). It was then diluted fivefold and treated with ADEP7 (12 µM) or DMSO, followed by incubation at 37 °C. The degree of covalent modification of SaClpP was quantified via intact-protein mass spectrometry after several time points (three independent experiments). Mass intensities were normalized against the sum of intensities of the free protein and the modified protein. $T_{1/2}$ was calculated from linear fits of the logarithmic plots.

SEC-MALS. Analytical size exclusion chromatography was carried out on an ÄKTA Purifier 10 chromatography system with a calibrated Superdex 200 10/300 GL column in buffer D (20 mM Hepes pH 7.0, 100 mM NaCl) at 4 °C with a flow of 0.5 ml min^{-1}. Samples (300 µl of a 1 mg ml^{-1} SaClpP solution or 1 ml of a 250 µg ml^{-1} solution) were loaded into a 500 µl or 1 ml sample loop and elution was monitored by ultraviolet absorption at 280 nm. Ultraviolet traces were referenced against the salt peak in the conductivity trace and normalized to the highest signal for easy comparison. SEC-MALS analysis was carried out on an Agilent 1200 Series chromatography system equipped with a Superdex 200 10/300 GL column and coupled to a DAWN Heleos II MALS detector as well as a Optilab rEX refractive index detector (Wyatt Technology). Samples (100 µl of 2 mg ml^{-1} protein solutions) were loaded from an autosampler and analyzed in buffer D (20 mM Hepes pH 7.0, 100 mM NaCl; used for SaClpP proteins and LmClpP2) with a flow of 0.5 ml min^{-1} or in buffer F (50 mM Tris pH 8.0, 200 mM NaCl, 20% glycerol; used for LmClpP2 + glycerol and BsClpP proteins) with a flow of 0.4 ml min^{-1}. Masses and errors were derived from analysis in Astra 6.1 (Wyatt Technology) and calibration with BSA.

Analytical ultracentrifugation. Sedimentation velocity experiments were performed in a ProteomeLab XL-A (Beckman Coulter, USA) instrument equipped with a ultraviolet/visible-detection unit. Protein was detected at 280 nm. Runs were performed at 34,000 r.p.m. (93,220g) at 20 °C. Proteins were spun in a Ti50 rotor equipped with seven sample cells and one reference counterbalance cell. Samples of mutant and wild-type SaClpP were analyzed in buffer D with different concentrations (7.5–80 µM). Samples of mutant and wild-type SaClpP with ADEP were analyzed in buffer D + 0.6% (v/v) DMSO with a protein concentration of 7.5 µM. Scanning intervals were set to 0.003 cm. All data were verified by at least two independent experiments. Data analysis was performed using the 'Time Course' function of the SedView programme. In addition, all the sedimentation experiments were evaluated with the UltraScan II software.

Small angle X-ray scattering. SAXS data for solutions of SaClpP wild-type and mutant samples free and bound to ADEP were recorded on an in-house SAXS instrument (SAXSess mc2, Anton Paar, Graz, Austria) equipped with a Kratky camera, a sealed X-ray tube source and a two-dimensional Princeton Instruments PI · SCX:4300 (Roper Scientific) CCD detector. The scattering patterns were measured with a 90-min exposure time (540 frames, each 10 s) for several solute concentrations in the range 1.8–8.2 mg ml^{-1}. Data for each sample was recorded at three different concentrations, that is, undiluted, at half concentration and one-fourth concentration. As no changes in the shape of the SAXS curves were observed for all samples recorded at different dilutions, the highest concentration with the best signal-to-noise ratio was selected to prepare the figures and for further analysis. Tag-free or C-terminal Strep-tagged proteins were purified by size exclusion chromatography prior to measurement. Samples were taken from the central peak fractions and treated with either ADEP4/7 (50 mM stocks in DMSO, final concentration range: 84–380 μM corresponding to a 1:1 ratio of ADEP:SaClpP) or the respective amount of DMSO (0.6% (v/v) max). Radiation damage was excluded based on a comparison of individual frames of the 90-min exposures, where no changes were detected. A range of momentum transfer of $0.012 < s < 0.63$ Å^{-1} was covered ($s = 4\pi\sin(\theta)/\lambda$, where 2θ is the scattering angle and $\lambda = 1.5$ Å is the X-ray wavelength). All SAXS data were analyzed with the package ATSAS (version 2.5). The data were processed with the SAXSQuant software (version 3.9), and desmeared using the programmes GNOM[64] and GIFT[65]. The forward scattering, $I(0)$, the radius of gyration, R_g, the maximum dimension, D_{max}, and the inter-atomic distance distribution functions, ($P(R)$), were computed with the programme GNOM. The masses of the solutes were evaluated by comparison of the forward scattering intensity with that of a human serum albumin reference solution (molecular mass 69 kDa). To generate *ab initio* shape models, a total number of 20 models were calculated using the programme DAMMIF[66] and aligned and averaged using the programme DAMAVER[67]. C72 and C7 symmetry was defined for tetradecameric and heptameric SaClpP, respectively. The *ab initio* shape models were aligned with crystal structures of SaClpP (PDB IDs 3V5E and 3QWD) using the programme SUPCOMB[68].

Dynamic light scattering. Protein samples in buffer D (322 μM) with ADEP (322 μM) or DMSO (3%) were loaded into disposable plastic cuvettes. DLS was measured on a Wyatt DynaPro Nanostar laser photometer (662.3 nm) at 25 °C. For each sample, 10 measurements were performed, which consisted of 10 acquisitions each with an acquisition time of 5 s. Data were fit with Dynamics version 7 software to a multimodal spheric model. *P* values were calculated in OriginPro using an independent two-sample *t*-test ($N = 10$) with equal variances not assumed.

SaClpXP assay. Degradation assays were performed in PZ buffer (25 mM Hepes, 200 mM KCl, 5 mM $MgCl_2$, 1 mM DTT, 10% (v/v) glycerol, pH 7.6) with 60 μl reaction volume at 30 °C. GFP fluorescence was monitored in white, flat-bottom well plates (Greiner) by exciting at 465 nm and measuring emission at 535 nm using an Infinite F200 Pro (Tecan). Degradation reactions contained 0.4 μM $SaClpX_6$, 0.2 μM $ClpP_{14}$, 0.36 μM GFP–ssrA and an ATP regeneration system (4 mM ATP, 16 mM creatine phosphate, 20 U ml^{-1} creatine phosphokinase). 0.6 μl (1% of reaction volume) ADEP was added in different concentrations in DMSO. All reaction partners except the substrate were pre-incubated for 10 min at 30 °C. GFP–ssrA was added to start the reaction. All data were recorded in triplicate measurements. GFP unfolding activity was derived as initial slopes in fluorescence time courses. Reactions were quenched after 3 h by addition of SDS sample buffer, and samples were analyzed both by denaturing and weakly denaturing SDS–PAGE for Coomassie and in-gel GFP fluorescence analysis, respectively.

Binding site identification. SaClpP (10 μM) was treated with β-lactone D3 (100 μM) in buffer D and incubated for 1 h at 25 °C. It was checked by protein mass spectrometry for complete modification and the sample was then buffer-exchanged to 50 mM ammonium bicarbonate (pH 7.3). Following a trypsin digest, peptides were analyzed on a LTQ Orbitrap XL mass spectrometer (Thermo Scientific)[39]. Data analysis was performed with the SEQUEST algorithm and the *S. aureus* NCTC 8325 proteome via Bioworks software allowing for covalent modification with D3 (monoisotopic mass: 262.1933 amu) on serine, threonine, cysteine, histidine and lysine residues.

***In vitro* activity-based protein profiling.** SaClpP (1 μM) was incubated with alkyne-tagged β-lactones (50 μM) in buffer D for 1 h at room temperature in a total volume of 100 μl, and the entire procedure of click-chemistry-mediated attachment of a fluorophore, separation via SDS-PAGE and fluorescence scanning have been described previously[39]. See the Supplementary Methods and Supplementary Figs 8–12 for the synthesis of the four stereoisomers of β-lactone U1.

References

1. Drag, M. & Salvesen, G. S. Emerging principles in protease-based drug discovery. *Nat. Rev. Drug Discov.* **9,** 690–701 (2010).
2. Huber, E. M. & Groll, M. Inhibitors for the immuno- and constitutive proteasome: current and future trends in drug development. *Angew. Chem. Int. Ed.* **51,** 8708–8720 (2012).
3. Turk, B. Targeting proteases: successes, failures and future prospects. *Nat. Rev. Drug Discov.* **5,** 785–799 (2006).
4. Brötz-Oesterhelt, H. & Sass, P. Bacterial caseinolytic proteases as novel targets for antibacterial treatment. *Int. J. Med. Microbiol.* **304,** 23–30 (2014).
5. Böttcher, T. & Sieber, S. A. ß-Lactones as specific inhibitors of ClpP attenuate the production of extracellular virulence factors of Staphylococcus aureus. *J. Am. Chem. Soc.* **130,** 14400–14401 (2008).
6. Geiger, S. R., Böttcher, T., Sieber, S. A. & Cramer, P. A conformational switch underlies ClpP protease function. *Angew. Chem. Int. Ed.* **50,** 5749–5752 (2011).
7. Frees, D., Qazi, S. N., Hill, P. J. & Ingmer, H. Alternative roles of ClpX and ClpP in *Staphylococcus aureus* stress tolerance and virulence. *Mol. Microbiol.* **48,** 1565–1578 (2003).
8. Böttcher, T. & Sieber, S. A. Structurally refined β-lactones as potent inhibitors of devastating bacterial virulence factors. *Chembiochem* **10,** 663–666 (2009).
9. Brötz-Oesterhelt, H. *et al.* Dysregulation of bacterial proteolytic machinery by a new class of antibiotics. *Nat. Med.* **11,** 1082–1087 (2005).
10. Kirstein, J. *et al.* The antibiotic ADEP reprogrammes ClpP, switching it from a regulated to an uncontrolled protease. *EMBO Mol. Med.* **1,** 37–49 (2009).
11. Katayama-Fujimura, Y., Gottesman, S. & Maurizi, M. R. A multiple-component, ATP-dependent protease from *Escherichia coli. J. Biol. Chem.* **262,** 4477–4485 (1987).
12. Maurizi, M. R. *et al.* Sequence and structure of Clp P, the proteolytic component of the ATP-dependent Clp protease of *Escherichia coli. J. Biol. Chem.* **265,** 12536–12545 (1990).
13. Wang, J., Hartling, J. A. & Flanagan, J. M. The structure of ClpP at 2.3 Å resolution suggests a model for ATP-dependent proteolysis. *Cell* **91,** 447–456 (1997).
14. Sauer, R. T. *et al.* Sculpting the proteome with AAA(+) proteases and disassembly machines. *Cell* **119,** 9–18 (2004).
15. Sauer, R. T. & Baker, T. A. AAA+ proteases: ATP-fueled machines of protein destruction. *Annu. Rev. Biochem.* **80,** 587–612 (2011).
16. Gottesman, S., Roche, E., Zhou, Y. & Sauer, R. T. The ClpXP and ClpAP proteases degrade proteins with carboxy-terminal peptide tails added by the SsrA-tagging system. *Genes Dev.* **12,** 1338–1347 (1998).
17. Weber-Ban, E. U., Reid, B. G., Miranker, A. D. & Horwich, A. L. Global unfolding of a substrate protein by the Hsp100 chaperone ClpA. *Nature* **401,** 90–93 (1999).
18. Martin, A., Baker, T. A. & Sauer, R. T. Protein unfolding by a AAA+ protease is dependent on ATP-hydrolysis rates and substrate energy landscapes. *Nat. Struct. Mol. Biol.* **15,** 139–145 (2008).
19. Gersch, M., List, A., Groll, M. & Sieber, S. A. Insights into the structural network responsible for oligomerization and activity Of the bacterial virulence regulator caseinolytic protease P (ClpP). *J. Biol. Chem.* **287,** 9484–9494 (2012).
20. Liu, K., Ologbenla, A. & Houry, W. A. Dynamics of the ClpP serine protease: a model for self-compartmentalized proteases. *Crit. Rev. Biochem. Mol. Biol.* **49,** 400–412 (2014).
21. Maurizi, M. R., Thompson, M. W., Singh, S. K. & Kim, S. H. Endopeptidase Clp: ATP-dependent Clp protease from *Escherichia coli. Methods Enzymol.* **244,** 314–331 (1994).
22. Joshi, S. A., Hersch, G. L., Baker, T. A. & Sauer, R. T. Communication between ClpX and ClpP during substrate processing and degradation. *Nat. Struct. Mol. Biol.* **11,** 404–411 (2004).
23. Kim, Y. I. *et al.* Molecular determinants of complex formation between Clp/Hsp100 ATPases and the ClpP peptidase. *Nat. Struct. Mol. Biol.* **8,** 230–233 (2001).
24. Lee, B.-G. *et al.* Structures of ClpP in complex with acyldepsipeptide antibiotics reveal its activation mechanism. *Nat. Struct. Mol. Biol.* **17,** 471–478 (2010).
25. Li, D. H. S. *et al.* Acyldepsipeptide antibiotics induce the formation of a structured axial channel in ClpP: a model for the ClpX/ClpA-bound state of ClpP. *Chem. Biol.* **17,** 959–969 (2010).
26. Sass, P. *et al.* Antibiotic acyldepsipeptides activate ClpP peptidase to degrade the cell division protein FtsZ. *Proc. Natl Acad. Sci. USA* **108,** 17474–17479 (2011).
27. Conlon, B. P. *et al.* Activated ClpP kills persisters and eradicates a chronic biofilm infection. *Nature* **503,** 365–370 (2013).
28. Hinzen, B. *et al.* Medicinal chemistry optimization of acyldepsipeptides of the enopeptin class antibiotics. *ChemMedChem* **1,** 689–693 (2006).
29. Leung, E. *et al.* Activators of cylindrical proteases as antimicrobials: identification and development of small molecule activators of ClpP protease. *Chem. Biol.* **18,** 1167–1178 (2011).
30. Freiburger, L. A., Auclair, K. & Mittermaier, A. K. Elucidating protein binding mechanisms by variable-c ITC. *Chembiochem* **10,** 2871–2873 (2009).
31. Gersch, M. *et al.* The mechanism of caseinolytic protease (ClpP) inhibition. *Angew. Chem. Int. Ed.* **52,** 3009–3014 (2013).
32. Gersch, M., Kreuzer, J. & Sieber, S. A. Electrophilic natural products and their biological targets. *Nat. Prod. Rep.* **29,** 659–682 (2012).
33. Zhang, J. *et al.* Structural switching of *Staphylococcus aureus* Clp protease—a key to understanding protease dynamics. *J. Biol. Chem.* **286,** 37590–37601 (2011).
34. Alexopoulos, J. A., Guarne, A. & Ortega, J. ClpP: a structurally dynamic protease regulated by AAA+ proteins. *J. Struct. Biol.* **179,** 202–210 (2012).
35. Sprangers, R., Gribun, A., Hwang, P., Houry, W. & Kay, L. E. Quantitative NMR spectroscopy of supramolecular complexes: dynamic side pores in ClpP are important for product release. *Proc. Natl Acad. Sci. USA* **102,** 16678–16683 (2005).

36. Ye, F. *et al.* Helix unfolding/refolding characterizes the functional dynamics of *Staphylococcus aureus* Clp protease. *J. Biol. Chem.* **288,** 17643–17653 (2013).
37. Kimber, M. S. *et al.* Structural and theoretical studies indicate that the cylindrical protease ClpP samples extended and compact conformations. *Structure* **18,** 798–808 (2010).
38. Gribun, A. *et al.* The ClpP double ring tetradecameric protease exhibits plastic ring-ring interactions, and the N termini of its subunits form flexible loops that are essential for ClpXP and ClpAP complex formation. *J. Biol. Chem.* **280,** 16185–16196 (2005).
39. Gersch, M., Kolb, R., Alte, F., Groll, M. & Sieber, S. A. Disruption of oligomerization and dehydroalanine formation as mechanisms for ClpP protease inhibition. *J. Am. Chem. Soc.* **136,** 1360–1366 (2014).
40. Zeiler, E. *et al.* Vibralactone as a tool to study the activity and structure of the ClpP1P2 complex from *listeria monocytogenes*. *Angew. Chem. Int. Ed.* **50,** 11001–11004 (2011).
41. Kim, Y. I., Burton, R. E., Burton, B. M., Sauer, R. T. & Baker, T. A. Dynamics of substrate denaturation and translocation by the ClpXP degradation machine. *Mol. Cell* **5,** 639–648 (2000).
42. El Bakkouri, M. *et al.* Structural insights into the inactive subunit of the apicoplast-localized caseinolytic protease complex of *Plasmodium falciparum*. *J. Biol. Chem.* **288,** 1022–1031 (2013).
43. Lee, B. G., Kim, M. K. & Song, H. K. Structural insights into the conformational diversity of ClpP from *Bacillus subtilis*. *Mol. Cell* **32,** 589–596 (2011).
44. Stinson, B. M. *et al.* Nucleotide binding and conformational switching in the hexameric ring of a AAA+ machine. *Cell* **153,** 628–639 (2013).
45. Martin, A., Baker, T. A. & Sauer, R. T. Distinct static and dynamic interactions control ATPase-peptidase communication in a AAA+ protease. *Mol. Cell* **27,** 41–52 (2007).
46. Glynn, S. E., Martin, A., Nager, A. R., Baker, T. A. & Sauer, R. T. Structures of asymmetric ClpX hexamers reveal nucleotide-dependent motions in a AAA+ protein-unfolding machine. *Cell* **139,** 744–756 (2009).
47. Bewley, M. C., Graziano, V., Griffin, K. & Flanagan, J. M. The asymmetry in the mature amino-terminus of ClpP facilitates a local symmetry match in ClpAP and ClpXP complexes. *J. Struct. Biol.* **153,** 113–128 (2006).
48. Sowole, M. A., Alexopoulos, J. A., Cheng, Y. Q., Ortega, J. & Konermann, L. Activation of ClpP protease by ADEP antibiotics: insights from hydrogen exchange mass spectrometry. *J. Mol. Biol.* **425,** 4508–4519 (2013).
49. Boehr, D. D., Nussinov, R. & Wright, P. E. The role of dynamic conformational ensembles in biomolecular recognition. *Nat. Chem. Biol.* **5,** 789–796 (2009).
50. Zorn, J. A. & Wells, J. A. Turning enzymes ON with small molecules. *Nat. Chem. Biol.* **6,** 179–188 (2010).
51. Kang, S. G., Maurizi, M. R., Thompson, M., Mueser, T. & Ahvazi, B. Crystallography and mutagenesis point to an essential role for the N-terminus of human mitochondrial ClpP. *J. Struct. Biol.* **148,** 338–352 (2004).
52. Kang, S. G. *et al.* Functional proteolytic complexes of the human mitochondrial ATP-dependent protease, hClpXP. *J. Biol. Chem.* **277,** 21095–21102 (2002).
53. Kang, S. G., Dimitrova, M. N., Ortega, J., Ginsburg, A. & Maurizi, M. R. Human mitochondrial ClpP is a stable heptamer that assembles into a tetradecamer in the presence of ClpX. *J. Biol. Chem.* **280,** 35424–35432 (2005).
54. Jennings, L. D., Bohon, J., Chance, M. R. & Licht, S. The ClpP N-terminus coordinates substrate access with protease active site reactivity. *Biochemistry* **47,** 11031–11040 (2008).
55. Schmitz, K. R., Carney, D. W., Sello, J. K. & Sauer, R. T. Crystal structure of Mycobacterium tuberculosis ClpP1P2 suggests a model for peptidase activation by AAA+ partner binding and substrate delivery. *Proc. Natl Acad. Sci. USA* **111,** E4587–E4595 (2014).
56. Effantin, G., Maurizi, M. R. & Steven, A. C. Binding of the ClpA unfoldase opens the axial gate of ClpP peptidase. *J. Biol. Chem.* **285,** 14834–14840 (2010).
57. Maya, S. *et al.* The ClpXP protease unfolds substrates using a constant rate of pulling but different gears. *Cell* **155,** 636–646 (2013).
58. Maillard, R. A. *et al.* ClpX(P) generates mechanical force to unfold and translocate its protein substrates. *Cell* **145,** 459–469 (2011).
59. Grimaud, R., Kessel, M., Beuron, F., Steven, A. C. & Maurizi, M. R. Enzymatic and structural similarities between the *Escherichia coli* ATP-dependent proteases, ClpXP and ClpAP. *J. Biol. Chem.* **273,** 12476–12481 (1998).
60. Ortega, J., Lee, H. S., Maurizi, M. R. & Steven, A. C. Alternating translocation of protein substrates from both ends of ClpXP protease. *EMBO J.* **21,** 4938–4949 (2002).
61. Turgay, K., Hahn, J., Burghoorn, J. & Dubnau, D. Competence in *Bacillus subtilis* is controlled by regulated proteolysis of a transcription factor. *EMBO J.* **17,** 6730–6738 (1998).
62. Kirstein, J. *et al.* Adaptor protein controlled oligomerization activates the AAA+ protein ClpC. *EMBO J.* **25,** 1481–1491 (2006).
63. Ober, M., Muller, H., Pieck, C., Gierlich, J. & Carell, T. Base pairing and replicative processing of the formamidopyrimidine-dG DNA lesion. *J. Am. Chem. Soc.* **127,** 18143–18149 (2005).
64. Svergun, D. I. Determination of the regularization parameter in indirect-transform methods using perceptual criteria. *J. Appl. Crystallogr.* **25,** 495–503 (1992).
65. Bergmann, A., Fritz, G. & Glatter, O. Solving the generalized indirect Fourier transformation (GIFT) by Boltzmann simplex simulated annealing (BSSA). *J. Appl. Crystallogr.* **33,** 1212–1216 (2000).
66. Franke, D. & Svergun, D. I. DAMMIF, a program for rapid ab-initio shape determination in small-angle scattering. *J. Appl. Crystallogr.* **42,** 342–346 (2009).
67. Volkov, V. V. & Svergun, D. I. Uniqueness of ab initio shape determination in small-angle scattering. *J. Appl. Crystallogr.* **36,** 860–864 (2003).
68. Kozin, M. B. & Svergun, D. I. Automated matching of high- and low-resolution structural models. *J. Appl. Crystallogr.* **34,** 33–41 (2001).

Acknowledgements

We are grateful to Deutsche Forschungsgemeinschaft [(Emmy Noether program MA 5703/1-1, to T.M.), FOR 854 (to H.B.-O. and P.S.), SFB 1035 (to S.A.S)], CIPSM (to S.A.S.) and the European Research Council (ERC starting grant to S.A.S.) for funding the research. This work was further supported by the Bavarian Ministry of Sciences, Research and the Arts (Bavarian Molecular Biosystems Research Network to T.M.). We acknowledge the Leibniz Supercomputing Centre (LRZ, www.lrz.de) for providing computing time on the Linux Cluster. We thank Mona Wolff and Katja Bäuml for technical assistance; Hamed Kooshapur and Michael Sattler for access to the ITC calorimeter; Michael Groll for access to the DLS laser photometer; as well as Berthold Hinzen, Holger Paulsen and Siegfried Raddatz for ADEP synthesis.

Author contributions

M.G., K.F., M.D., I.M., P.S. produced the proteins and performed the biochemical experiments. C.G. and T.M. performed SAXS measurements and modelling; M.G. and V.S.K. synthesized β-lactones; K.R. performed ultracentrifugation experiments; H.R.-S. provided ADEPs. S.A.S., H.B.-O. and M.G. designed the experiments and wrote the manuscript. All authors analyzed the data, discussed the results and provided input for the preparation of the manuscript.

Additional information

Competing financial interests: H.R.-S. reports holding stock options in AiCuris. All other authors declare no competing financial interests.

2

Conformational heterogeneity of the calmodulin binding interface

Diwakar Shukla[1,2,3,*], Ariana Peck[4,*] & Vijay S. Pande[1,2]

Calmodulin (CaM) is a ubiquitous Ca^{2+} sensor and a crucial signalling hub in many pathways aberrantly activated in disease. However, the mechanistic basis of its ability to bind diverse signalling molecules including G-protein-coupled receptors, ion channels and kinases remains poorly understood. Here we harness the high resolution of molecular dynamics simulations and the analytical power of Markov state models to dissect the molecular underpinnings of CaM binding diversity. Our computational model indicates that in the absence of Ca^{2+}, sub-states in the folded ensemble of CaM's C-terminal domain present chemically and sterically distinct topologies that may facilitate conformational selection. Furthermore, we find that local unfolding is off-pathway for the exchange process relevant for peptide binding, in contrast to prior hypotheses that unfolding might account for binding diversity. Finally, our model predicts a novel binding interface that is well-populated in the Ca^{2+}-bound regime and, thus, a candidate for pharmacological intervention.

[1] Department of Chemistry, Stanford University, Stanford, California 94305, USA. [2] SIMBIOS NIH Center for Biomedical Computation, Stanford University, Stanford, California 94305, USA. [3] Department of Chemical and Biomolecular Engineering, University of Illinois at Urbana-Champaign, Urbana, Illinois 61801, USA. [4] Department of Biochemistry, Stanford University School of Medicine, Stanford, California 94305, USA. * These authors contributed equally to this work. Correspondence and requests for materials should be addressed to V.S.P. (email: pande@stanford.edu).

Calcium signalling plays a pivotal role in diverse physiological processes, including lymphocyte activation, neuronal firing, muscle contraction and apoptosis[1]. Under resting conditions, cells maintain a 20,000-fold gradient in Ca^{2+} concentration across the plasma membrane that is rapidly dissipated and restored after stimulation[2]. As a consequence, hundreds of Ca^{2+}-binding proteins have evolved whose affinity for Ca^{2+} collectively spans the nM to mM range[2,3]. Broadly, these can be classified into proteins that act as Ca^{2+} sinks to buffer the intracellular environment and adaptor proteins that act as Ca^{2+} sensors, undergoing large conformational changes upon binding to carry out distinct effector functions in the absence and presence of Ca^{2+} (ref. 4).

Calmodulin (CaM) is the prototypical eukaryotic Ca^{2+} sensor and regulates a diverse set of proteins, including G-protein-coupled receptors, small conductance potassium (SK) channels and CaM kinases[4,5]. Structurally, CaM is comprised of two lobes connected by a short linker[6]. Each lobe, in turn, contains a pair of EF hand motifs that bind two Ca^{2+} ions cooperatively[7]. Extensive NMR and small-angle X-ray scattering studies have shown that CaM is a dynamic protein, with its inter-domain linker exhibiting considerable flexibility and the C-lobe undergoing conformational exchange in the μs regime[8,9]. Comparison of high-resolution structures of apo (Ca^{2+}-unbound) and holo (Ca^{2+}-bound) CaM clarified the functional role of these dynamics by indicating that Ca^{2+} binding induces large-scale structural rearrangements that expose each lobe's hydrophobic interface to CaM-binding partners[6,10] (Fig. 1). These binding partners commonly feature a pair of bulky hydrophobic residues that serve as anchors to the hydrophobic interface presented by each lobe of Ca^{2+}-bound CaM, which can accommodate variable spacing between these anchor residues due to the flexibility of the inter-lobe linker[11].

However, this prevailing paradigm of CaM function has been challenged by the discovery of proteins that preferentially bind to apo CaM and tune its affinity for Ca^{2+} (refs 5,12). Furthermore, CaM-binding partners share no sequence homology[4], and attempts to categorize them based on structural features are riddled with exceptions[11,13]. Thus, a holistic framework that accounts for CaM's binding diversity is lacking. Prior computational work has focused on CaM equilibrium dynamics, with particular emphasis on dissecting the temperature- and Ca^{2+}-dependence of the transition rates among the unfolded, apo- and holo-states and their relative populations[14,15]. While these studies have helped resolve conflicting interpretations of NMR and single-molecule force spectroscopy studies by distinguishing between states that are difficult to resolve experimentally[14,16,17], they have been too coarse-grained to address the mechanism of these transitions at atomistic detail. Furthermore, only the question of Ca^{2+}-binding has been addressed, while the structural basis underlying CaM's ability to interact specifically with a diverse set of binding partners has not been broached computationally.

Molecular dynamics (MD) simulations have been harnessed to study the dynamics of complex biological systems at atomic resolution and predict the mechanisms that underlie transitions among biologically important states[18–20]. Two challenges faced by MD simulations are the need for a framework to deal with the high dimensionality of the resulting data set and the requirement of highly specialized hardware to obtain atomistic dynamics of proteins with timescales comparable to the conformational process under investigation[21]. Markov state models (MSMs) are a powerful analytical tool that overcomes both obstacles by merging massively parallel short simulations into a unified, more accessible model that captures rare events whose timescales exceed those of the individual simulations[21,22].

In this study, we constructed MSMs from MD simulations of the C-lobe of CaM (C-CaM) in the Ca^{2+}-bound and unbound regimes to explore the contributions of a single domain of CaM to binding diversity. In particular, we performed MD for a total aggregate simulation time of 700 μs with an all-atom description of protein and an explicit representation of water. This study is in contrast to previous computational investigations of CaM conformational dynamics, which relied either on coarse-grained models[14] or short (ns) simulations[23,24]. Our MSM-based approach resolved the experimentally observed apo to holo and unfolding transitions as orthogonal processes, while comparison with available high-resolution structures indicated that the folded ensemble of C-CaM populates multiple, topologically distinct binding interfaces that may account for binding specificity and diversity by enabling conformational selection. In addition, our models predict novel binding interfaces that could be stabilized by structurally matched targets, thus disrupting downstream signalling by sequestering CaM from the

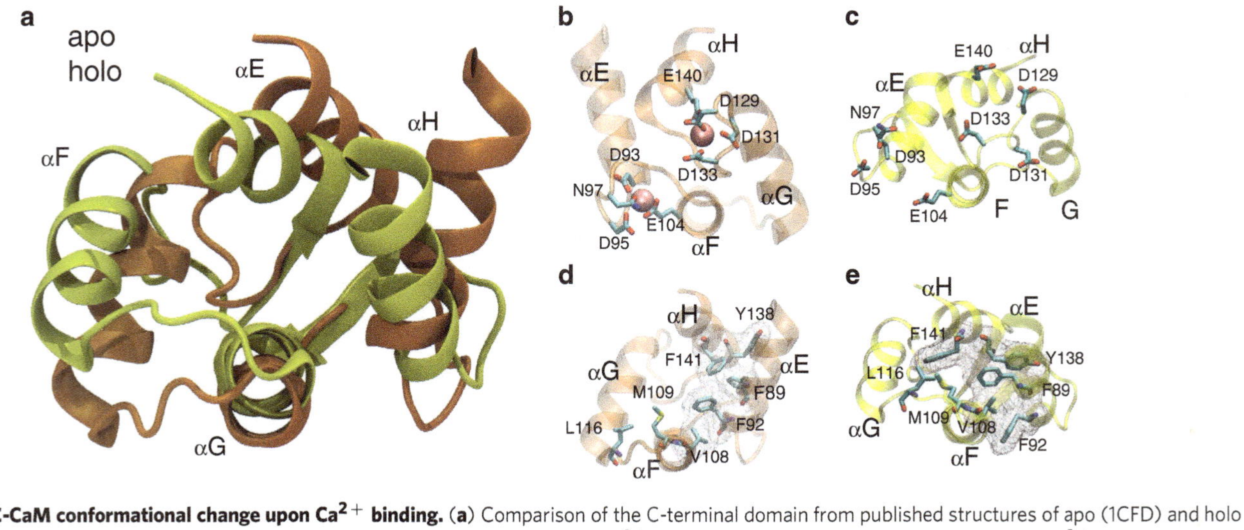

Figure 1 | C-CaM conformational change upon Ca^{2+} binding. (**a**) Comparison of the C-terminal domain from published structures of apo (1CFD) and holo (1CLL) CaM indicates the large-scale structural rearrangements that occur when Ca^{2+} binds. (**b**,**c**) In particular, repositioning of the Ca^{2+}-ligating residues from solvent-exposed rotamers renders each EF hand competent to bind Ca^{2+}. (**d**,**e**) Additionally, binding of Ca^{2+} induces repacking in the hydrophobic core and binding interface, enabling compact clustering of the aromatic residues outlined by the mesh.

active pool of signalling molecules. Given that CaM is a pivotal hub in signalling pathways that are frequently deregulated in disease[25,26], a sophisticated framework for understanding target recognition by CaM could help make it a tractable candidate for pharmacological activation or inhibition.

Results

Distinct exchange processes characterize apo- and holo-C-CaM. To investigate the contributions of C-CaM to CaM's binding diversity, we ran MD simulations of this domain in the absence and presence of Ca^{2+}. Inspection of the corresponding high-resolution structures of apo- and holo-CaM from which these simulations were started demonstrates the gross structural rearrangements that occur upon Ca^{2+} binding (Fig. 1a)[27,28]. In addition to repositioning of the residues that ligate the Ca^{2+} ions in each of the Ca^{2+}-binding sites (Fig. 1b,c), Ca^{2+} binding is accompanied by reorientation of the EF hand α-helices to achieve a more compact clustering of the four aromatic residues in the C-CaM hydrophobic core (Fig. 1d,e). However, these are static depictions that offer atomic-level structural information but minimal insight into the dynamics that govern each state or the mechanism underlying their exchange. MD simulations allowed us to bridge this gap, preserving the atomic-level detail afforded by high-resolution structures while providing extensive kinetic information that can be unified and interpreted through the framework of MSMs.

Constructing MSMs that accurately recapitulate MD data requires a valid decomposition of phase space. We partitioned the conformational landscape explored by apo- and holo-C-CaM using time-structure-based independent component analysis (tICA), a distance metric generated by projecting high-dimensional data onto degrees of freedom (in this case, distances in the contact map) that decorrelate slowly[29]. MSMs were then built from the clustered data at a Markovian lag-time (Supplementary Fig. 1) to analyse the dominant pathways of conformational exchange. Building MSMs from the tIC-partitioned phase space allowed us to analyse the rates and pathways of conformational change that govern apo- and holo-C-CaM dynamics. The first eigenvector of the transition probability matrix defining each MSM provides an estimate of the equilibrium populations. In the absence of bound Ca^{2+}, conformations similar to the solution structure of apo-C-CaM dominate, with the holo-like state accounting for a minor but significant population ($P\sim0.14$), similar to the minor 5–10% population observed in NMR studies[9,27,30]. These values are within 1 kcal of each other and thus difficult to distinguish experimentally or computationally (Supplementary Fig. 2a). By contrast, most of the population in the Ca^{2+}-bound regime is concentrated in a state that resembles the crystal structure of holo-C-CaM ($P\sim0.93$) (Supplementary Fig. 2b). To further validate these computational models, we estimated NMR order parameters from each MSM ensemble and found good agreement with experimental values (Supplementary Fig. 3 and Supplementary Tables 1–2)[30–32]. These order parameters measure local flexibility, so the high correlation suggests that the MSM ensembles accurately reproduce solution-state dynamics. We additionally verified that the conformational landscape explored by apo-C-CaM was invariant to force field choice by re-running our simulations in a different force field. The population distributions along order parameters characteristic of the apo- to holo-like transition and local unfolding were similar between apo-C-CaM simulations run in distinct force fields, and equally distant to the population distribution of the holo system (Supplementary Fig. 4 and Supplementary Table 3).

Sampling from this equilibrium description recovers the dynamic processes and the timescales at which they occur. In the absence of Ca^{2+}, exchange with the holo-like state occurs with a k_{ex} of $\sim26\,\mu s$, within the range of 18–150 μs that has been experimentally determined for the isolated C-terminal domain (Fig. 2a)[30,33–37]. Transitions to the holo-like state coincide with aromatic stacking and formation of the second Ca^{2+}-binding site, as assessed by inter-residue distances between Phe92 and Phe141 and between the Ca^{2+} ligands Asp129 and Asp133, respectively. On the other hand, the first Ca^{2+}-binding site remains poorly formed even upon transition to the holo-like state, which likely accounts for its lower affinity for Ca^{2+} (ref. 30) and structurally distinguishes the holo-like and holo states. Exchange along these order parameters correlates well with the second eigenvector (Supplementary Fig. 5a), indicating that the apo- to holo-like transition is the slowest dynamic process in the absence of Ca^{2+}.

By contrast, motions in the loop connecting the EF hands account for the slowest dynamic process of the holo system, with a relaxation timescale of $\sim10\,\mu s$ (Fig. 2b). Specifically, twisting of this region reorients Leu116 away from solvent, allowing it to pack against Met124 and Met109, while adjacent Thr117 flips away from the hydrophobic cluster in this region (Supplementary Figs 5b, 6). These dynamics are not characteristic of the apo- to holo-like transition. Similarly, the conformational change processes that distinguish the apo MSM do not typify the excursions from the dominant population in the holo MSM, indicating that that the dynamics in the Ca^{2+}-bound and unbound regimes are distinct (Fig. 2).

Computationally discovered metrics reveal complex dynamics. The ability of our MSMs to recover experimentally determined populations and rates indicates that tICA is a robust method for discretizing the phase space explored by apo- and holo-C-CaM. The dominant tICs—that is, those with the largest eigenvalues in the time-lag covariance matrix—represent the slowest decorrelating orthogonal linear combinations of distances in the contact map[29]. Thus, deducing the molecular correlates of the tICs provides insight into the most important degrees of freedom in the conformational landscape of each system. Projecting the conformations visited by apo- and holo-C-CaM onto their respective tICs confirmed that within each system, the top tICs capture distinct dynamic processes (Fig. 3a,b). In the apo system, the first and second tICs correspond to exchange with a holo-like form and a partially unfolded state, respectively (Supplementary Fig. 7a), consistent with prior studies that identified these as the dominant dynamics in the unliganded regime[14,17,30]. Projecting the conformational space explored by holo-C-CaM onto these same tICs resulted in a density map that overlaid well with the holo-like state explored by apo-C-CaM, but there was no indication that holo-C-CaM populates an apo-like state or undergoes the cracking experienced by the Ca^{2+}-unbound state (Fig. 3c).

Instead, the dominant tICs that characterize holo-C-CaM correspond to different rearrangements of the hydrophobic core (Supplementary Fig. 7b). However, these tICs are specific to the holo system, and the first tIC in particular fails to capture the dynamics of the holo-like state of apo-C-CaM (Fig. 3d), again underscoring that dynamics of C-CaM in the Ca^{2+}-bound and unbound regimes are distinct. Visualizing the dominant tIC of each system as a contact map confirms this: the region that most contributes to the top tIC is localized to the G helix and inter-EF hand loop region for holo-C-CaM but more distributed throughout the protein for apo-C-CaM (Fig. 3e,f and Supplementary Fig. 8). The complexity of these contact maps

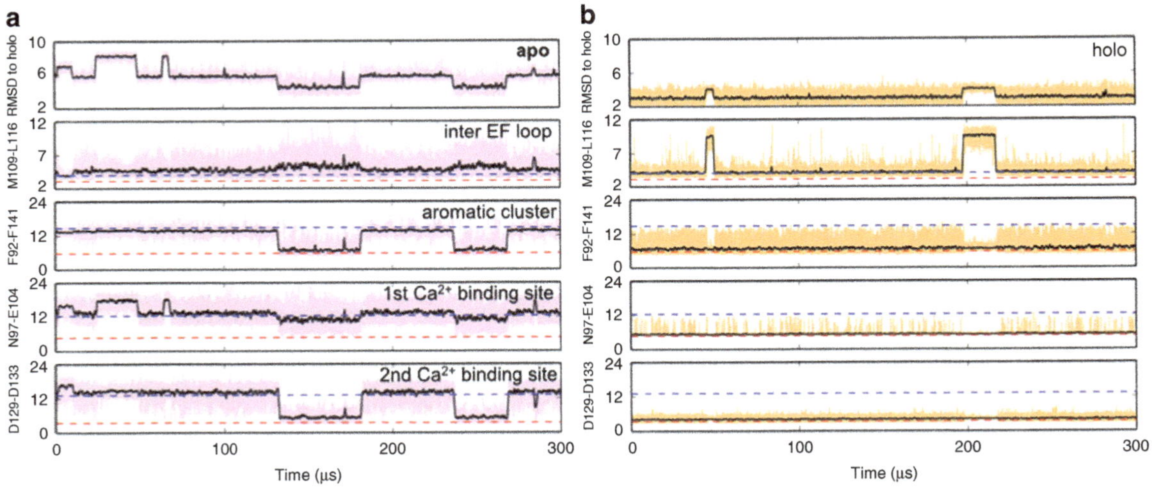

Figure 2 | Kinetics of conformational change. Sampling from the MSM captures conformational changes and the timescales at which they occur. (**a**) In apo-C-CaM, transition to a holo-like state coincides with aromatic stacking and formation of the second, but not the first Ca^{2+}-binding site. (**b**) Holo-C-CaM experiences fluctuations in the inter-EF hand loop region but does not show sustained fluctuations in the aromatic cluster or either Ca^{2+}-binding site. All distances are shown in Å.

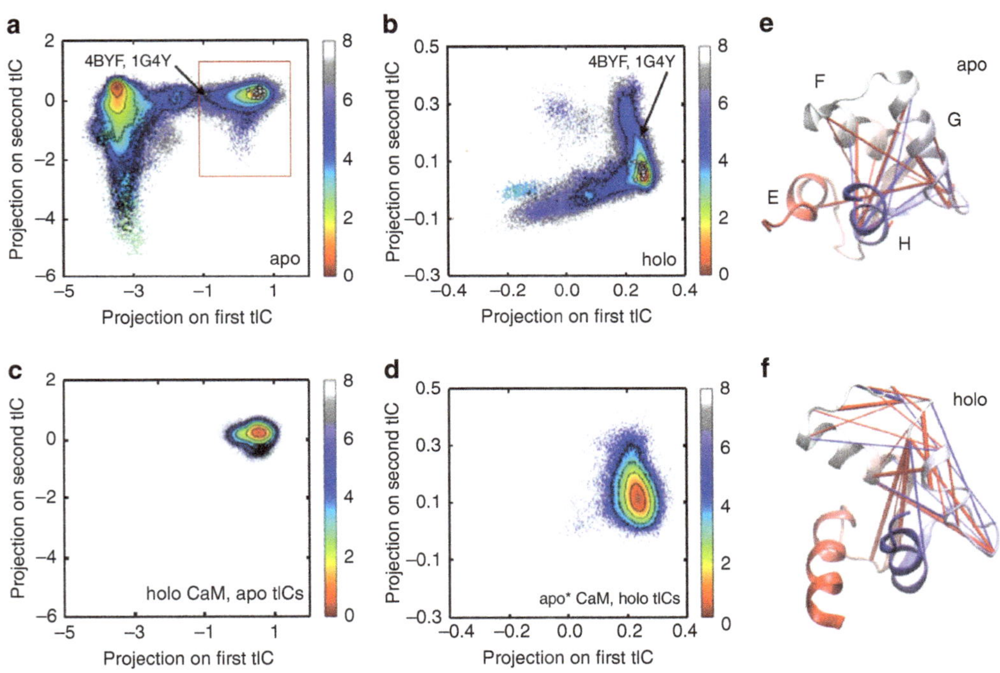

Figure 3 | The top tICs capture distinct dynamic processes. The energy landscapes of (**a**) apo- and (**b**) holo-C-CaM were generated by projecting the system's conformations, weighted by their MSM probabilities, onto their respective tICs. Comparison was made to available experimental structures by projecting the C-lobe of these published structures onto the relevant tICs (white dots). (**c**) Projection of the holo conformations on the apo tICs and (**d**) the holo-like basin in the apo system (boxed in red in **a**) on the holo tICs underscores their different dynamics. Free energy values are reported in kcal mol^{-1}. The 25 inter-residue distances whose dynamics most contribute to the top (**e**) apo and (**f**) holo tIC are visualized on cartoon representations of the published structures of apo- and holo-C-CaM. Red and blue lines denote positive and negative tIC values, respectively.

underscores the need for order parameters that can capture such intricate and non-uniform dynamics. Indeed, the commonly used order parameter of root-mean-square distance (RMSD) provides much poorer separation of the apo- and holo-like basins in the Ca^{2+}-unbound regime, and fails to resolve dynamic exchange in the Ca^{2+}-bound regime entirely (Supplementary Fig. 9).

Hydrophobic repack dominates the apo to holo-like transition. Applying transition pathway theory to MSMs predicts the ensemble of pathways between states and estimates their associated fluxes. In the apo MSM, a structural rearrangement characteristic of the transition to the holo-like state is compact stacking of four aromatic residues in the hydrophobic core, which is impeded in the apo state by packing of Val108 between Phe89 and Phe92 (Fig. 1d,e). Over 60% of the flux passes through an intermediate, in which packing of Val108 against Phe89 (but not Phe92) is disrupted, but only in the holo-like state does the helix containing Val108 shift sufficiently to accommodate the compact aromatic cluster (Fig. 4a). Repacking of the core orients Phe89

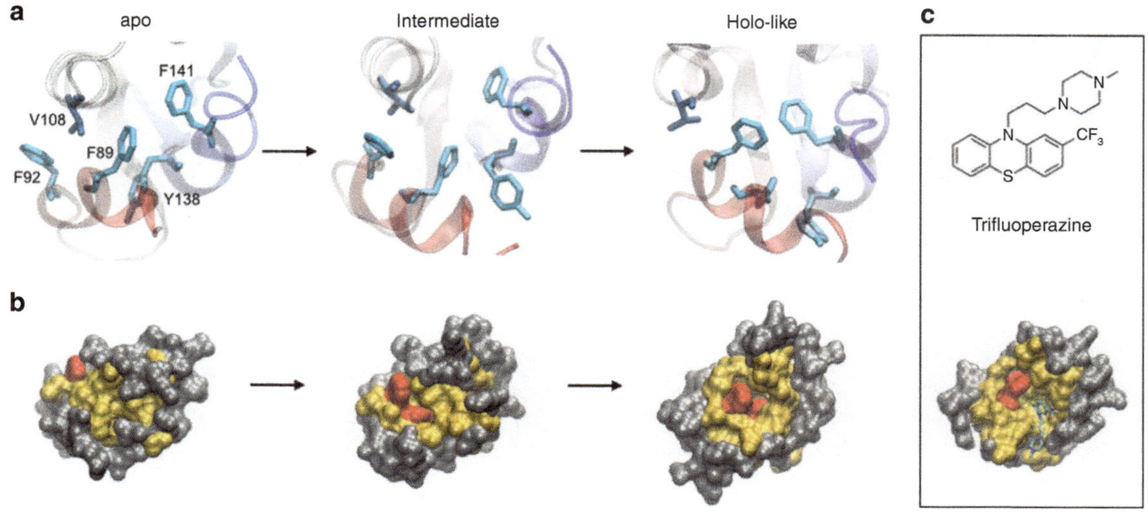

Figure 4 | Hydrophobic repack determines the topology of the binding interface. (**a**) Packing of Val108 between Phe92 and Phe89 is disrupted during the apo- to holo-like transition, which enables the formation of a compact aromatic cluster in the C-CaM core. (**b**) Surface representations of the apo, intermediate, and holo-like states show that repack exposes distinct surface topologies. Phe, all other hydrophobic, and polar residues coloured red, yellow and grey, respectively. (**c**) The Phe-lined cavity formed in the holo-like state is commonly accessed by bulky aromatic rings of binding partners, as shown in the crystal structure of holo CaM in complex with trifluoperazine. Only the C-terminal domain of this structure (1CTR) is shown.

and Phe141 in the bottom of a hydrophobic cavity, an arrangement that mimics the surface topology of Ca^{2+}-bound C-CaM and is late to form along the principle transition paths (Fig. 4b). Interestingly, a common structural motif in CaM's diverse binding partners is the presence of a bulky aromatic feature that accesses this hydrophobic pocket and stacks against the Phe residues that line one side of the cavity (Fig. 4c)[38,39]. Thus, rearrangements in the hydrophobic core facilitate the presentation of a specific surface topology that contributes to binding specificity.

Another intriguing question about the mechanism underlying this transition is whether unfolding plays a role. An unfolding process on the μs timescale has been observed in T-jump spectroscopic studies[17], and although exchange with the holo-like state rather than the unfolded state is predicted to dominate at low temperatures, the unfolded population is still estimated to be 1% in this regime[14]. Our MSM of apo-C-CaM populates a landscape that is qualitatively similar to the one explored by the coarse-grained Gö model of Chen *et al.*[14] (Supplementary Fig. 10). However, the higher resolution of our simulations reveal that unfolding is not global, but rather a localized phenomenon that primarily involves unwinding of helix G or deformation of the second Ca^{2+}-binding site (Fig. 5a). Less than 3% of the flux between the apo- and holo-like states visits a partially unfolded intermediate (using lower cutoffs of 11 and 15 Å for the Met124-Ala128 and Ala128-Gly134 inter-Cα distances, respectively, to categorize states as partially unfolded). This indicates that unfolding is largely irrelevant for the apo- to holo-like transition, despite prior speculation that intrinsic disorder within C-CaM might be a mechanism underlying its diverse binding repertoire[14]. Furthermore, we find no evidence of partial unfolding in the restricted dynamics of holo-C-CaM (Fig. 5b), consistent with the observation of increased stability in the Ca^{2+}-bound regime[40]. Because force field choice has been shown to impact secondary structure stability, we compared simulations run in different force fields to ensure that the observed local unfolding was invariant to this choice. As has been previously observed[41], median helical content is slightly higher in the CHARMM36 simulations than the Amber99sb-ildn trajectories (Supplementary Fig. 4c). However, these distributions are broad and overlap between the different force fields. Furthermore, the conclusions drawn about local unfolding are made based on specific inter-residue distances, which show similar distributions between force fields, rather than helical content (Supplementary Fig. 4a,b).

Apo- and holo-C-CaM populate topologically distinct ensembles. Given the unimportance of unfolding to the ability of C-CaM to bind structurally dissimilar substrates, we sought to identify a mechanism that could explain this binding diversity by comparing published structures of CaM-target complexes to our MSMs. Mapping 25 of these high-resolution structures to tIC space revealed that C-CaM explores the configurations stabilized by its peptide and small molecule-binding partners even in their absence (Fig. 6). Because our simulations started only from the published apo and holo structures, recovering these other conformations adopted in the presence of CaM-binding partners was not guaranteed. Instead, this pre-existing heterogeneity suggests that conformational selection may facilitate CaM's binding diversity. As expected, most of the structures mapped to the holo-like and central basins in apo- and holo-tIC space, respectively (Fig. 3a,b). The two outliers were the structures of CaM in complex with myosin and an SK channel splice variant (4BYF and 1G4Y, respectively); in both crystal structures, the C-lobes were found to be apo, consistent with their physiological role[42,43].

Curiously, in apo-tIC space these outliers mapped to an intermediary position between the apo- and holo-like basins rather than fully in the former basin despite the absence of Ca^{2+} (Fig. 6a). As described above, transition between these basins involves extensive structural reorganization, including formation of the second Ca^{2+}-binding site and hydrophobic repack. Because the first tIC encompasses both rearrangements, we examined the extent to which these features are coupled in the dynamics of apo C-CaM. Mutual information calculations provide a measure of long-range coupling between residues by computing the correlation between the motions of their backbone and side-chain torsion angles[44]. Applying this analysis to apo-C-CaM indicated extensive allosteric communication between the second Ca^{2+}-binding site

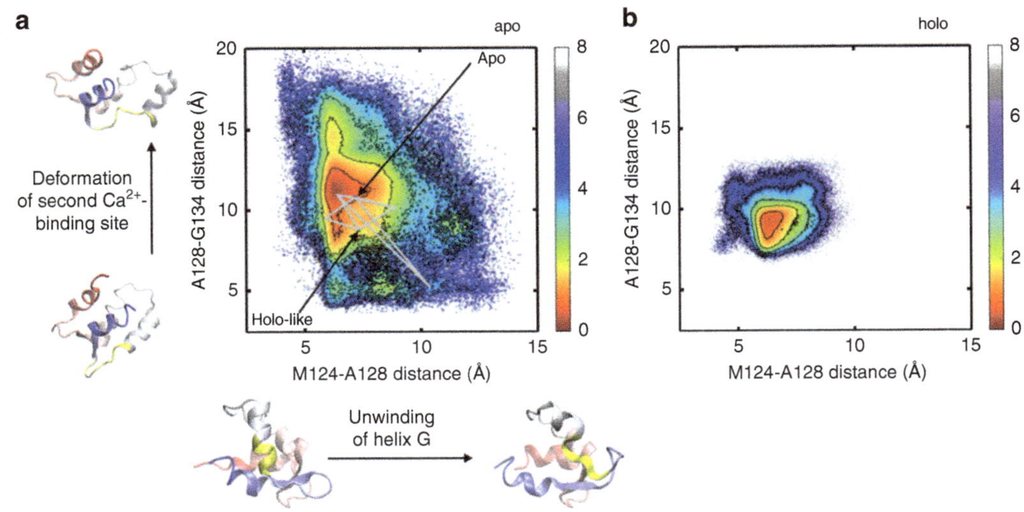

Figure 5 | Local unfolding occurs in apo but not holo-C-CaM. The energy landscapes of (**a**) apo and (**b**) holo-C-CaM as a function of the Met124-Ala128 and Ala128-Gly134 inter-Cα distances, which correlate with distortions in helix G and the second Ca^{2+}-binding site, respectively. The regions that span these residues are highlighted in yellow in the cartoon representations below and to the side of the plot in (**a**). Conformations were weighted by their MSM probabilities, and free energy values are reported in kcal mol^{-1}. In **a**, the five paths with the greatest flux between the apo- and holo-like states are plotted as grey lines.

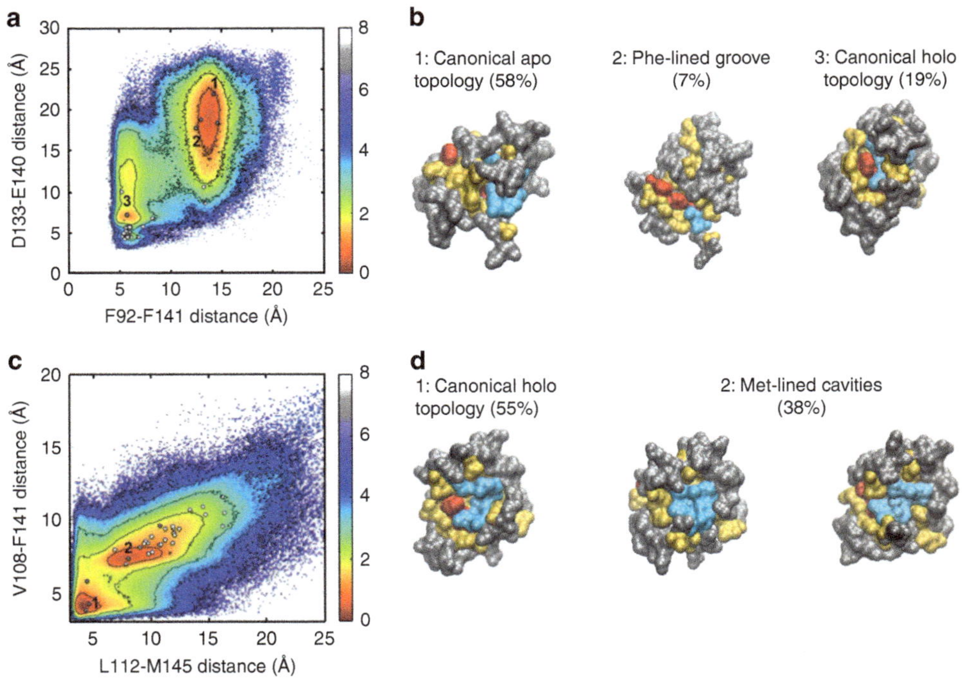

Figure 6 | C-CaM presents an ensemble of hydrophobic interfaces in the absence of binding partners. (**a**) Energy landscape of apo-C-CaM as a function of the Phe92-Phe141 and Asp133-Glu140 inter-residue distances, which correspond to aromatic clustering and formation of the second Ca^{2+}-binding site, respectively. White and grey dots represent these values for published structures and topological cluster centres, respectively. (**b**) Surface representations of the cluster centres indicated by number in **a**, with Phe, Met, all other hydrophobic, and polar residues coloured red, cyan, yellow and grey, respectively. (**c**,**d**) As in (**a**,**b**), except for holo-C-CaM, with inter-residue distances between hydrophobic residues at the binding interface (specifically, Leu112-Met145 and Val108-Phe141) used as order parameters. Free energy values are reported in kcal mol^{-1}.

and the hydrophobic residues that constitute the binding interface (Supplementary Fig. 11). In particular, the aromatic residues Phe89 and Tyr138 (Supplementary Fig. 12) show a strong coupling with the charged residues (Asp129, Asp131 and Asp133) that ligate the Ca^{2+} ion. By contrast, the first Ca^{2+}-binding site, which has lower affinity for Ca^{2+} (ref. 30), does not show a strong dynamical coupling with any region of the protein (Supplementary Fig. 10).

The majority of flux between the apo- and holo-like states passes through a reaction tube, in which these features change in a concerted fashion, while configurations in which the Ca^{2+}-binding sites are fully formed but the aromatic cluster is not (and vice versa) are energetically destabilized by several kT (Supplementary Fig. 13). Consistent with this, splice variants of an SK channel that are known to differentially tune the affinity of the C-lobe for Ca^{2+} stabilize distinct hydrophobic surface topologies. While the splice variant that enhances Ca^{2+} binding interacts with the canonical Phe-lined cavity, the SK2-a splice variant that abolishes Ca^{2+} binding stabilizes a less compact Phe-lined groove that is also observed in the structure of apo CaM in complex with myosin (Supplementary Fig. 14)[5,42,43]. An exception to the concerted rearrangements in the hydrophobic network and Ca^{2+}-binding sites was observed in the NMR structure of CaM in complex with phosphorylated nitric oxide synthase, which features the canonical holo-pocket topology but has poorly formed Ca^{2+}-binding sites (Supplementary Fig. 14). However, in this case it has been suggested that proximity of the phosphate to the second Ca^{2+}-binding site may prevent closure of the charged residues that ligate Ca^{2+}, thus decoupling the allosteric network by an unusual mechanism[12].

Despite the sequence dissimilarity between the regions of myosin and SK2-a that bind C-CaM, the Phe-lined grooves they stabilize are strikingly similar (Supplementary Fig. 14)[42,43]. To predict which other surface topologies are well-represented and thus amenable to conformational selection, we clustered configurations from equilibrium sampling of the apo and holo MSMs based on RMSD of the 10 residues that constitute the hydrophobic network in C-CaM. In the case of the apo system, cluster centres that resembled the canonical apo and holo topologies were recovered, with similar populations ($P\sim0.58$ and 0.19, respectively,) to the apo- and holo-like states in the original MSM. Additionally, a Phe-lined groove similar to the one stabilized by myosin and SK2-a was observed at lower frequency ($P\sim0.07$) (Fig. 6a,b and Supplementary Fig. 14).

In the case of the holo system, the canonical Phe-lined cavity is most commonly observed ($P\sim0.56$), while a significant alternative topology features a Met-lined pocket ($P\sim0.38$) (Fig. 6c,d). Although this latter topology is not observed in the repertoire of currently published CaM structures, we predict that it could be readily stabilized by conformational selection based on its distinctive topology and high population in the Ca^{2+}-bound regime. This prediction is bolstered by the experimental observation of a binding interface that, in terms of the relative positioning and exposure of the Phe and Met clusters, is intermediary between the canonical Phe-lined cavity and our predicted Met-lined pocket (Supplementary Fig. 15). In this case, an anti-microtubular agent stabilizes an unusual topology featuring an exposed aromatic cluster adjacent to a Met-bordered depression[45], supporting the potential for other small molecules to sequester alternative topologies presented by holo-C-CaM. Interestingly, the Phe-lined cavity and Met-lined pocket we observe in our simulations are not only sterically but also chemically distinct. As noted above, the Phe cavity provides the opportunity for aromatic stacking and thus commonly accommodates the indole or benzyl rings of CaM-binding partners[38,39]. Similarly, the greater polarizability of Met compared with more canonical hydrophobic residues is expected to endow the Met pocket with distinct binding opportunities.

Discussion

Despite the wealth of structural information of CaM in complex with a variety of its binding partners, a unified framework that accounts for its ability to interact specifically with a diverse repertoire of proteins and small molecules has been lacking. In this study, we use MSMs to join massively parallel, short MD simulations of the isolated C-lobe in the Ca^{2+}-bound and unbound regimes to arrive at a more holistic understanding of the dynamics that give rise to CaM's functional plasticity.

Constructing MSMs requires partitioning phase space with distance metrics that distinguish important degrees of freedom in a protein's dynamic landscape. The complexity of the structural features captured by tICA suggests that even for a single domain of a small protein, simple order parameters are insufficient to describe the intricate dynamics of conformational change. Linear combinations of several simple order parameters have already been used to study conformational change; however, the order parameters in these studies were chosen based on already available structural information[18,19,46,47]. In this study, we show that computationally discovered order parameters can readily distinguish between the conformational space explored by apo- and holo-C-CaM, despite being agnostic to available structural information. Furthermore, the tICA method, which identifies slowly decorrelating degrees of freedom in the contact map, provides a superior separation of dynamic processes that are too subtle or complex for more global metrics like RMSD.

Using this unbiased approach to identify the dominant dynamics of C-CaM, we investigated the origins of CaM's ability to bind structurally diverse protein targets. Recently, intrinsic disorder has gained considerable attention as a mechanism for generating binding diversity: initial contacts formed between an intrinsically disordered protein and different substrates seed the formation of distinct secondary structures[48]. However, we find that the unfolding process of apo-C-CaM previously observed[14,17] is irrelevant to the apo- to holo-like transition and thus off-pathway for target binding. In contrast to the induced fit mechanism of intrinsic disorder, our simulations indicate that conformational sub-states within the folded ensemble of apo-C-CaM present distinct surface topologies that may facilitate binding of diverse targets. Although a direct comparison of target binding to different sub-states of CaM would be required to determine whether induced fit or conformational selection dominates, our simulations indicate that the pre-requisite heterogeneity for the former mechanism exists. Furthermore, the sub-states' relative populations predicted by our models can be used to estimate the fold-increase in binding rate required for one mechanism to outcompete the other. Prior studies investigating the process of peptide binding have primarily focused on dynamics of the holo state[31,49,50], but our models indicate that the apo ensemble also presents binding interfaces stabilized in crystal structures of CaM-peptide complexes. Furthermore, allosteric communication between the hydrophobic network and the Ca^{2+}-binding sites provides an opportunity for binding of a target protein to tune the affinity of CaM for Ca^{2+} and vice versa[5,12,51].

Hydrophobic interactions are commonly viewed as relatively non-specific compared with salt bridges and hydrogen bond networks, but the hydrophobic binding interfaces presented by the apo- and holo/holo-like structures are sterically and chemically distinct and thus can provide specificity. Given the observed conformational selection, we used our MSMs to identify alternative binding interfaces that are well-populated by apo- and holo-C-CaM that we predict could be readily stabilized by a structurally matched substrate. As expected, we recovered the Phe-lined cavity and groove topologies that have been observed in crystal structures of CaM-target complexes at high frequency[5,52,53]. In addition, we identified a well-populated topology featuring a distinctive Met-lined cavity in the holo MSM that, to our knowledge, is not represented in the current repertoire of published CaM structures.

A necessary consideration in the face of the observed conformational heterogeneity is the choice of force field. The Amber-99sb-ildn force field used in this study has been shown to have complex effects on stability, tending to destabilize helical structures but predicting overly compact non-native conformations on folding pathways because of poor solvation[41,54,55]. However, these studies also suggest that such systemic errors primarily affect simulations of small peptide systems, while good agreement is found between experiment and the ensembles of larger model systems[41,56]. Given that our models likewise reproduce experimental structural information, predict substantial heterogeneity in non-helical regions shown previously to be highly flexible, and are similar between different force fields, we do not expect force field choice to affect the conclusions drawn in this study.

Given the distinct kinetic and thermal properties of the two lobes of CaM[9,40], it will be of interest to assess the extent to which conformational heterogeneity in N-CaM contributes to full CaM's binding diversity. NMR experiments indicate that C-CaM is more dynamic than N-CaM[9], so it is possible that the ensemble of binding interfaces of the former contributes most to binding diversity, while binding of the latter primarily enhances the affinity of this interaction. Another possibility is that N-CaM features a distinct ensemble of binding interfaces that could, in combination with the ensemble populated by C-CaM, be used as an opportunity to effect functional bifurcation, a phenomenon that has already been reported in CaM's physiological function[57,58]. CaM plays a pivotal role in signal transduction pathways that are frequently perturbed in disease states[25,26]; therefore, a nuanced understanding of its binding modes could facilitate the rational targeting of this signalling hub for pharmacological intervention.

Methods

Simulation details for apo CaM. The starting structures were taken from the crystal structure of apo CaM (PDB: 1CFD (ref. 27)). The all-atom structures were then solvated in a cubic box with box length of ~65 Å with TIP3P (ref. 59) water molecules such that water extended at least 10 Å away from the surface of the protein; 24 Na^+ ions and 12 Cl^- ions were added to the system to neutralize the charge. Covalent bonds involving hydrogen atoms were constrained with the LINear Constraint Solver (LINCS)[60] and particle mesh Ewald[61] was used to treat long-range electrostatic interactions. The structures obtained after an initial equilibration for 1 ns at constant temperature and pressure and with constraints on the heavy atom positions were used as the starting conformation for the distributed MD simulations. Production MD simulations were carried out at constant temperature and pressure of 298 K and 1 atm respectively, with a time step of 2 fs. The Amber99sb-ildn (ref. 62) force field was used for protein and ions. Distributed MD simulations were performed using GROMACS[63] on the Folding@home[64] computing platform. For the first set of simulations, 5,000 simulations were started from the two structures (2,500 each) for a total aggregate simulation time of 200 µs. Adaptive sampling algorithms[65] based on an MSM built from the initial round of sampling were used to improve the sampling the conformational landscape of apo CaM. Specifically, the initial structures (500 states with lowest population) for adaptive sampling rounds were chosen from the 2,000 state MSM built using RMSD metric and the simulation data from the first round. In all, we performed a total of 12,184 simulations with a total duration of 455 µs. Trajectory snapshots were recorded every 100 ps. The distribution of total number of simulations versus the simulation length for apo CaM is shown in Supplementary Fig. 16a.

Simulation details for holo-CaM. The starting structure for the holo-CaM was taken from PDB: 1CLL. The force field parameters for the Ca^{2+} ions were taken from the Amber99sb-ildn (ref. 62) force field. The all-atom structures were then solvated in a cubic box with box length of ~60 Å with TIP3P water molecules, such that water extended at least 10 Å away from the surface of the protein; 12 Na^+ ions and 4 Cl^- ions were added to the system to neutralize the charge. A total of 9,995 distributed MD simulations were performed using GROMACS on the Folding@home platform for a total duration of 256 µs. Simulation protocols employed for holo-CaM were similar to the protocols described above for the apo CaM. The distribution of total number of simulations versus the simulation length for holo-CaM is shown in Supplementary Fig. 16b. Preservation of Ca^{2+}-binding site geometry was confirmed by calculating the RMSD of the Ca^{2+}-ligating residues between the holo-crystal structure and each conformation weighted by its MSM probability (Supplementary Fig. 17).

MSM construction. The theoretical framework underlying MSMs has been detailed extensively elsewhere[22] so will briefly be summarized herein. MSMs are an application of discrete-space master equations and describe the kinetic network that underlies a particular partitioning of phase space. The phase space explored by biomolecular systems can be discretized by dimensionality reduction along specific order parameters. In this study, we discretized phase space using tICA[29], which is a variant of principal component analysis that defines a kinetically motivated, projection-based distance metric. The tICA method computes the time-lag correlation matrix, whose eigenvectors represent linear combinations of the most slowly decorrelating degrees of freedom in a system. To apply the tICA method to C-CaM, we chose the degrees of freedom to correspond to the minimum distances in the contact map between the heavy atoms of all residue pairs separated by at least three residues. Representing each conformation as a vector of its pairwise residue distances and projecting these vector representations onto tIC components (eigenvectors of the time-lag correlation matrix) mapped the system to a reduced dimensionality space. The distance between conformations $|A\rangle$ and $|B\rangle$ in this reduced tIC space was then calculated by

$$d(|A\rangle,|B\rangle)=\left\|\mathbf{P}^T|\mathbf{A}\rangle-\mathbf{P}^T|\mathbf{B}\rangle\right\|_2 \quad (1)$$

where $P^T|X$ denotes the projection of the vector representation of conformation X onto a $d \times N$ matrix P, whose columns are the N slowest tIC components and row-length is determined by the size of the contact map, and $\|\cdot\|_2$ is the N-dimensional Euclidean norm. For apo- and holo-C-CaM, the time-lag correlation matrices were calculated with a delta time of 40 ns, and the reduced phase space was computed by projection onto the slowest 20 tICs and clustered into 100 states using the k-centres algorithm.

A kinetic network describing the rates of transitions in this partitioned space was then constructed by computing the transition probability matrix at variable lag-times. The transition probability matrix, $T(\tau)$, contains the probability of transitioning from state i to state j in the time interval τ, obtained by counting the number of transitions, n_{ij}, observed between time t and $t+\tau$ and then normalizing to the sum of all transitions from state i. Detailed balance is enforced by symmetrizing the transition probability matrix by the maximum likelihood estimate. The first eigenvector of the transition probability matrix corresponds to the equilibrium distribution; subsequent eigenvectors correspond to dynamic processes in the system. Their eigenvalues, μ, are related to the implied timescales, $1/k$, of these dynamic processes:

$$k = -\tau/ln(\mu) \quad (2)$$

Examining the slowest eigenvalues as a function of lag-time for convergence behaviour identifies a Markovian (memoryless) lag-time. MSMs were constructed at these lag-times (20 and 15 ns for apo- and holo-C-CaM, respectively). Transition Path Theory, which has been outlined elsewhere[66], was used to determine the pathways and their associated fluxes between specific states. Conformational landscapes were generated by summing over the microstates of each model using the following equation:

$$W(x,y) = k_B T \ln\left[\sum_{i=1}^{N} \pi_i h_i(x,y)\right] \quad (3)$$

where N is the number of microstates, π_i is the equilibrium probability of state i and $h_i(x,y)$ is the normalized histogram of the order parameters x and y restricted to microstate i. MSMs were built using the MSMbuilder[67] software.

Generation and analysis of MSM trajectories. Representative trajectories that recapitulate the MSM can be generated by applying a kinetic Monte Carlo sampling scheme to the transition probability matrix[19]. Given an MSM trajectory in state i at time t, the subsequent state visited at $t+\tau$ (a period of one lag-time) was determined by sampling from the multinomial distribution given by T_i. For each frame of the resulting trajectory, a random conformation belonging to the relevant state was selected and from this a time-series of the observables was calculated. For both apo- and holo-C-CaM, the trajectories were started from the most populated state at equilibrium.

Comparison with the following structures from the Protein Data Bank was made in Figs 3,6: 1CKK, 1CM1, 1G4Y, 1IQ5, 1IWQ, 1K90, 1NIW, 1NID, 1XA5, 1YRT, 2BBM, 2F3Y, 2HQW, 2KDU, 2M55, 2MG5, 2MGU, 2W73, 2YGG, 3BXL, 3G43, 3SJQ, 4BYF and 4EHQ. In each case, only residues E82-A147 from a single chain of CaM were analysed, and missing C-terminal residues were modelled in using the Crystallographic Object-Oriented Toolkit (Coot, v 0.7.1) (ref. 68) or Modeller, v 9.14 (ref. 69) where necessary.

Macrostate models of the most populated binding interfaces were built by using hierarchical clustering[70] to partition each MSM trajectory into a five state model based on RMSD of the following residues that comprise the hydrophobic binding interface: Phe89, Phe92, Ile100, Leu105, Val108, Met109, Leu112, Leu116, Met124, Ala128, Val136, Phe141, Met144 and Met145.

References

1. Weinstein, H. & Mehler, E. L. Ca^{2+}-binding and structural dynamics in the functions of calmodulin. *Annu. Rev. Physiol.* **56,** 213–236 (1994).
2. Clapham, D. E. Calcium signaling. *Cell* **131,** 1047–1058 (2007).
3. Berridge, M. J., Lipp, P. & Bootman, M. D. The versatility and universality of calcium signalling. *Nat. Rev. Mol. Cell Biol.* **1,** 11–21 (2000).

4. Chin, D. & Means, A. R. Calmodulin: a prototypical calcium sensor. *Trends Cell Biol.* **10,** 322–328 (2000).
5. Zhang, M. *et al.* Structural basis for calmodulin as a dynamic calcium sensor. *Structure* **20,** 911–923 (2012).
6. Babu, Y. S. *et al.* Three-dimensional structure of calmodulin. *Nature* **315,** 37–40 (1984).
7. Linse, S., Helmersson, A. & Forsén, S. Calcium binding to calmodulin and its globular domains. *J. Biol. Chem.* **266,** 8050–8054 (1991).
8. Heidorn, D. B. & Trewhella, J. Comparison of the crystal and solution structures of calmodulin and troponin C. *Biochemistry* **27,** 909–915 (1988).
9. Tjandra, N., Kuboniwa, H., Ren, H. & Bax, A. Rotational dynamics of calcium-free calmodulin studied by 15N-NMR relaxation measurements. *Eur. J. Biochem.* **230,** 1014–1024 (1995).
10. Finn, B. E. *et al.* Calcium-induced structural changes and domain autonomy in calmodulin. *Nat. Struct. Mol. Biol.* **2,** 777–783 (1995).
11. Hoeflich, K. P. & Ikura, M. Calmodulin in action: diversity in target recognition and activation mechanisms. *Cell* **108,** 739–742 (2002).
12. Piazza, M., Taiakina, V., Guillemette, S. R., Guillemette, J. G. & Dieckmann, T. Solution structure of calmodulin bound to the target peptide of endothelial nitric oxide synthase phosphorylated at Thr495. *Biochemistry* **53,** 1241–1249 (2014).
13. Yamauchi, E., Nakatsu, T., Matsubara, M., Kato, H. & Taniguchi, H. Crystal structure of a MARCKS peptide containing the calmodulin-binding domain in complex with Ca^{2+}-calmodulin. *Nat. Struct. Mol. Biol.* **10,** 226–231 (2003).
14. Chen, Y.-G. & Hummer, G. Slow conformational dynamics and unfolding of the calmodulin C-terminal domain. *J. Am. Chem. Soc.* **129,** 2414–2415 (2007).
15. Tripathi, S. & Portman, J. J. Inherent flexibility determines the transition mechanisms of the EF-hands of calmodulin. *Proc. Natl. Acad. Sci. USA* **106,** 2104–2109 (2009).
16. Stigler, J. & Rief, M. Calcium-dependent folding of single calmodulin molecules. *Proc. Natl. Acad. Sci. USA* **109,** 17814–17819 (2012).
17. Rabl, C.-R., Martin, S. R., Neumann, E. & Bayley, P. M. Temperature jump kinetic study of the stability of apo-calmodulin. *Biophys. Chem.* **101,** 553–564 (2002).
18. Kohlhoff, K. *et al.* Cloud-based simulations on Google Exacycle reveal ligand modulation of GPCR activation pathways. *Nat. Chem.* **6,** 15–21 (2014).
19. Shukla, D., Meng, Y., Roux, B. & Pande, V. S. Activation pathway of Src kinase reveals intermediate states as targets for drug design. *Nat. Commun.* **5,** 3397 (2014).
20. Lawrenz, M., Shukla, D. & Pande, V. S. Cloud computing approaches for prediction of ligand binding poses and pathways. *Sci. Rep.* **5,** 7918 (2015).
21. Lane, T. J., Shukla, D., Beauchamp, K. A. & Pande, V. S. To milliseconds and beyond: challenges in the simulation of protein folding. *Curr. Opin. Struct. Biol.* **23,** 58–65 (2013).
22. Shukla, D., Hernndez, C. X., Weber, J. K. & Pande, V. S. Markov state models provide insights into dynamic modulation of protein function. *Acc. Chem. Res.* **48,** 414–422 (2015).
23. Komeiji, Y., Ueno, Y. & Uebayasi, M. Molecular dynamics simulations revealed Ca^{2+}-dependent conformational change of calmodulin. *FEBS Lett.* **521,** 133–139 (2002).
24. Shepherd, C. M. & Vogel, H. J. A molecular dynamics study of Ca^{2+}-calmodulin: Evidence of interdomain coupling and structural collapse on the nanosecond timescale. *Biophys. J.* **87,** 780–791 (2004).
25. Hait, W. N. & Lazo, J. Calmodulin: a potential target for cancer chemotherapeutic agents. *J. Clin. Oncol.* **4,** 994–1012 (1986).
26. O'Day, D. H. & Myre, M. A. Calmodulin-binding domains in Alzheimer's disease proteins: extending the calcium hypothesis. *Biochem. Biophys. Res. Commun.* **320,** 1051–1054 (2004).
27. Kuboniwa, H. *et al.* Solution structure of calcium-free calmodulin. *Nat. Struct. Biol* **2,** 768–776 (1995).
28. Chattopadhyaya, R., Meador, W. E., Means, A. R. & Quiocho, F. A. Calmodulin structure refined at 1.7 Å resolution. *J. Mol. Biol.* **228,** 1177–1192 (1992).
29. Schwantes, C. R. & Pande, V. S. Improvements in Markov state model construction reveal many non-native interactions in the folding of NTL9. *J. Chem. Theory Comput.* **9,** 2000–2009 (2013).
30. Malmendal, A., Evenäs, J., Forsén, S. & Akke, M. Structural dynamics in the C-terminal domain of calmodulin at low calcium levels. *J. Mol. Biol.* **293,** 883–899 (1999).
31. Marlow, M. S., Dogan, J., Frederick, K. K., Valentine, K. G. & Wand, A. J. The role of conformational entropy in molecular recognition by calmodulin. *Nat. Chem. Biol.* **6,** 352–358 (2010).
32. Bowman, G. R. & Geissler, P. L. Extensive conformational heterogeneity within protein cores. *J. Phys. Chem. B* **118,** 6417–6423 (2014).
33. Evenäs, J., Forsn, S., Malmendal, A. & Akke, M. Backbone dynamics and energetics of a calmodulin domain mutant exchanging between closed and open conformations. *J. Mol. Biol.* **289,** 603–617 (1999).
34. Evenäs, J., Malmendal, A. & Akke, M. Dynamics of the transition between open and closed conformations in a calmodulin C-terminal domain mutant. *Structure.* **9,** 185–195 (2001).
35. Weininger, U. *et al.* Protein conformational exchange measured by 1H R1pÂrelaxation dispersion of methyl groups. *J. Biomol. NMR* **57,** 47–55 (2013).
36. Lundström, P. & Akke, M. Quantitative analysis of conformational exchange contributions to 1Hâ'15N multiple-quantum relaxation using field-dependent measurements. Time scale and structural characterization of exchange in a calmodulin C-terminal domain mutant. *J. Am. Chem. Soc.* **126,** 928–935 (2004).
37. Lundström, P. & Akke, M. Microsecond protein dynamics measured by 13C rotating-frame spin relaxation. *Chembiochem.* **6,** 1685–1692 (2005).
38. Cook, W. J., Walter, L. J. & Walter, M. R. Drug binding by calmodulin: crystal structure of a calmodulin-trifluoperazine complex. *Biochemistry* **33,** 15259–15265 (1994).
39. Fallon, J. L. *et al.* Crystal structure of dimeric cardiac L-type calcium channel regulatory domains bridged by Ca^{2+} calmodulins. *Proc. Natl. Acad. Sci. USA* **106,** 5135–5140 (2009).
40. Masino, L., Martin, S. R. & Bayley, P. M. Ligand binding and thermodynamic stability of a multidomain protein, calmodulin. *Protein Sci.* **9,** 1519–1529 (2000).
41. Lindorff-Larsen, K. *et al.* Systematic validation of protein force fields against experimental data. *PLoS ONE* **7,** e32131 (2012).
42. Münnich, S., Taft, M. H. & Manstein, D. J. Crystal structure of human myosin 1c -- the motor in GLUT4 exocytosis: Implications for Ca^{2+} regulation and 14-3-3 binding. *J. Mol. Biol.* **426,** 2070–2081 (2014).
43. Schumacher, M. A., Rivard, A. F., Bächinger, H. P. & Adelman, J. P. Structure of the gating domain of a Ca^{2+}-activated K^{+} channel complexed with Ca^{2+}/calmodulin. *Nature* **410,** 1120–1124 (2001).
44. McClendon, C., Friedland, G., Mobley, D., Amirkhani, H. & Jacobson, M. Quantifying correlations between allosteric sites in thermodynamic ensembles. *J. Chem. Theory Comput.* **5,** 2486–2502 (2009).
45. Horváth, I. *et al.* The structure of the complex of calmodulin with KAR-2: a novel mode of binding explains the unique pharmacology of the drug. *J. Biol. Chem.* **280,** 8266–8274 (2005).
46. Vanatta, D. K., Shukla, D., Lawrenz, M. & Pande, V. S. A network of molecular switches controls the activation of the two-component response regulator ntrc. *Nat. Commun.* **6,** 7283 (2015).
47. Shukla, D., Lawrenz, M. & Pande, V. S. Elucidating ligand-modulated conformational landscape of gpcrs using cloud-computing approaches. *Methods Enzymol.* **557,** 551–572 (2015).
48. Dyson, H. J. & Wright, P. E. Intrinsically unstructured proteins and their functions. *Nat. Rev. Mol. Cell Biol.* **6,** 197–208 (2005).
49. Fiorin, G., Pastore, A., Carloni, P. & Parrinello, M. Using metadynamics to understand the mechanism of calmodulin/target recognition at atomic detail. *Biophys. J.* **91,** 2768–2777 (2006).
50. Hoang, J. & Prosser, R. S. Conformational selection and functional dynamics of calmodulin: a (19)F nuclear magnetic resonance study. *Biochemistry* **53,** 5727–5736 (2014).
51. Peersen, O. B., Madsen, T. S. & Falke, J. J. Intermolecular tuning of calmodulin by target peptides and proteins: differential effects on Ca^{2+} binding and implications for kinase activation. *Protein Sci.* **6,** 794–807 (1997).
52. Aoyagi, M., Arvai, A. S., Tainer, J. A. & Getzoff, E. D. Structural basis for endothelial nitric oxide synthase binding to calmodulin. *EMBO J.* **22,** 766–775 (2003).
53. Liu, Y. *et al.* Crystal structure of calmodulin binding domain of Orai1 in complex with Ca^{2+}-calmodulin displays a unique binding mode. *J. Biol. Chem.* **287,** 43030–43041 (2012).
54. Best, R. B. & Hummer, G. Optimized molecular dynamics force fields applied to the helix-coil transition of polypeptides. *J. Phys. Chem. B* **113,** 9004–9015 (2009).
55. Best, R. B., Zheng, W. & Mittal, J. Balanced protein-water interactions improve properties of disordered proteins and non-specific protein association. *J. Chem. Theory Comput.* **10,** 5113–5124 (2014).
56. Beauchamp, K. A., Lin, Y. S., Das, R. & Pande, V. S. Are protein force fields getting better? a systematic benchmark on 524 diverse NMR measurements. *J. Chem. Theory Comput.* **8,** 1409–1414 (2012).
57. DeMaria, C. D., Soong, T. W., Alseikhan, B. A., Alvania, R. S. & Yue, D. T. Calmodulin bifurcates the local Ca^{2+} signal that modulates P/Q-type Ca^{2+} channels. *Nature* **411,** 484–489 (2001).
58. Rodríguez-Castañeda, F. *et al.* Modular architecture of Munc13/calmodulin complexes: dual regulation by Ca^{2+} and possible function in short-term synaptic plasticity. *EMBO J.* **29,** 680–691 (2010).
59. Jorgensen, W., Chandrasekhar, J., Madura, J., Impey, R. & Klein, M. Comparison of simple potential functions for simulating liquid water. *J. Chem. Phys.* **79,** 926 (1983).

60. Hess, B. *et al.* LINCS: a linear constraint solver for molecular simulations. *J. Comput. Chem.* **18,** 1463–1472 (1997).
61. Darden, T., York, D. & Pedersen, L. Particle mesh Ewald: An Nlog(N) method for Ewald sums in large systems. *J. Chem. Phys.* **98,** 10089 (1993).
62. Lindorff-Larsen, K. *et al.* Improved side-chain torsion potentials for the Amber ff99SB protein force field. *Proteins* **78,** 1950–1958 (2010).
63. Hess, B., Kutzner, C., van der Spoel, D. & Lindahl, E. Gromacs 4: algorithms for highly efficient, load-balanced, and scalable molecular simulation. *J. Chem. Theory Comput.* **4,** 435–447 (2008).
64. Shirts, M. & Pande, V. Screen savers of the world unite! *Science* **290,** 1903–1904 (2000).
65. Bowman, G., Ensign, D. & Pande, V. Enhanced modeling via network theory: Adaptive sampling of Markov state models. *J. Chem. Theory Comput.* **6,** 787–794 (2010).
66. Metzner, P., Schütte, C. & Vanden-Eijnden, E. Transition path theory for markov jump processes. *Multiscale Model Simul.* **7,** 1192–1219 (2009).
67. Beauchamp, K. A. *et al.* MSMBuilder2: modeling conformational dynamics on the picosecond to millisecond scale. *J. Chem. Theory Comput.* **7,** 3412–3419 (2011).
68. Emsley, P., Lohkamp, B., Scott, W. & Cowtan, K. Features and development of Coot. *Acta Crystallogr. D Biol. Crystallogr.* **66,** 486–501 (2010).
69. Webb, B. & Sali, A. . *Protein Structure Prediction* 1–15 (Springer, 2014).
70. Yao, Y. *et al.* Hierarchical Nyström methods for constructing Markov state models for conformational dynamics. *J. Chem. Phys.* **138,** 174106 (2013).

Acknowledgements

D.S. and V.S.P. acknowledge support from the Simbios, NIH Center for Biomedical Computation (NIH U54 Roadmap GM072970). V.S.P. acknowledges NIH (R01GM62828). D.S. is supported by the Biomedical Data Science Initiative Fellowship from Stanford University School of Medicine. A.P. is supported by the National Science Foundation Graduate Fellowship. We thank the users of the Folding@Home distributed computing project and the Blue Waters sustained-petascale computing project for providing computing time for this study.

Author contributions

D.S. and A.P. contributed equally to this project. V.S.P. and D.S. conceived the project. D.S. performed simulations. D.S. and A.P. analysed the data. A.P. and D.S. made figures and wrote the manuscript with inputs from V.S.P.

Additional information

Competing financial interests: The authors declare no competing financial interests.

3

Triangular prism-shaped β-peptoid helices as unique biomimetic scaffolds

Jonas S. Laursen[1,†], Pernille Harris[1], Peter Fristrup[1] & Christian A. Olsen[1,†]

β-Peptoids are peptidomimetics based on *N*-alkylated β-aminopropionic acid residues (or *N*-alkyl-β-alanines). This type of peptide mimic has previously been incorporated in biologically active ligands and has been hypothesized to be able to exhibit foldamer properties. Here we show, for the first time, that β-peptoids can be tuned to fold into stable helical structures. We provide high-resolution X-ray crystal structures of homomeric β-peptoid hexamers, which reveal right-handed helical conformations with exactly three residues per turn and a helical pitch of 9.6–9.8 Å between turns. The presence of folded conformations in solution is supported by circular dichroism spectroscopy showing length- and solvent dependency, and molecular dynamics simulations provide further support for a stabilized helical secondary structure in organic solvent. We thus outline a framework for future design of novel biomimetics that display functional groups with high accuracy in three dimensions, which has potential for development of new functional materials.

[1] Department of Chemistry, Technical University of Denmark, Kemitorvet 207, DK-2800 Kongens Lyngby, Denmark. † Present address: Department of Drug Design and Pharmacology, Center for Biopharmaceuticals, University of Copenhagen, Universitetsparken 2, DK-2100 Copenhagen, Denmark (J.S.L.; C.A.O.). Correspondence and requests for materials should be addressed to C.A.O. (email: cao@sund.ku.dk).

The ability to mimic or complement native protein and peptide folds using non-natural oligomers endowed with resistance towards proteolysis holds promise for the design of chemotypes with valuable applications in medical and materials sciences. Oligomeric architectures able to adopt well-defined folding patterns have been coined foldamers[1] and several peptidomimetic designs have demonstrated this type of behaviour[2–5], including β-peptides[6–8] and peptoids (*N*-alkylglycines)[9,10]. The β-peptoids, first described as dimers and trimers by Hamper *et al.*[11], combine the structural features of the two latter mentioned foldamers, and have been suggested by theoretical methods to be able to adopt helical conformations[12]. Bulky *N*-alkyl side chains with chirality in the α-position have been shown to induce helical secondary structures in α-peptoids with alternating handedness depending on stereochemical configuration[13–15], and *cis*-amide-inducing α-chiral side chains have also been shown to induce helicity in aromatic oligoamides[16]. However, circular dichroism (CD)-spectroscopical evaluation of oligomers containing β-peptoid units with α-chiral side chains have proven inconclusive thus far[17,18], and high-resolution structures have been lacking, except for head-to-tail cyclized species[19]. Nevertheless, β-peptoid units have been incorporated in oligomers showing non-haemolytic anti-microbial[20–24], antiplasmodial[25] and cell-penetrating[26] activities, as well as excellent stability towards proteolytic enzymes[22].

Inspired by studies performed using α-peptoid model systems[27,28], the tunability of the *cis–trans*-amide bond equilibria of the tertiary amide bonds in β-peptoids have also been investigated[29,30], and building on these insights, here we design β-peptoid oligomers containing *N*-(*S*)-1-(1-naphthyl)ethyl side chains for structural evaluation, and report the first high-resolution X-ray crystal structures of helical β-peptoid oligomers, which should enable informed, structure-based design of foldameric materials of this type in the future.

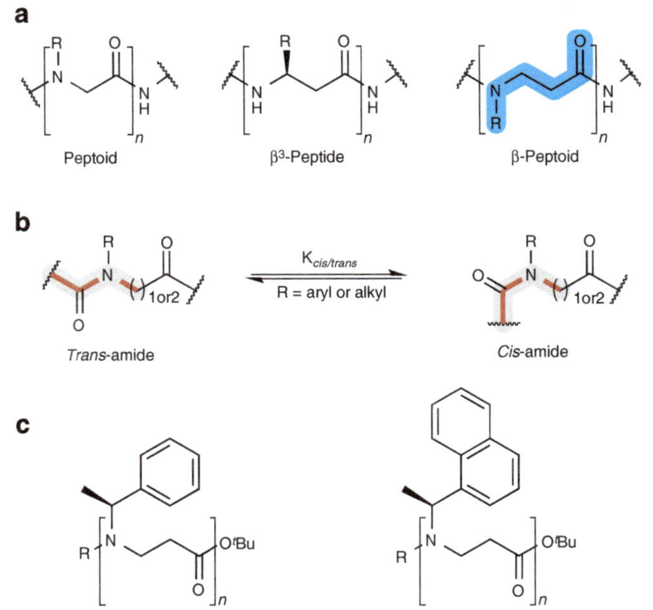

n = 2: 1a (R = Ac), 1b (R = $COCF_3$)
n = 3: 2a (R = Ac), 2b (R = $COCF_3$)
n = 4: 3a (R = Ac), 3b (R = $COCF_3$)
n = 5: 4a (R = Ac), 4b (R = $COCF_3$)
n = 6: 5a (R = Ac), 5b (R = $COCF_3$)

n = 2: 6a (R = Ac), 6b (R = $COCF_3$)
n = 3: 7a (R = Ac), 7b (R = $COCF_3$)
n = 4: 8a (R = Ac), 8b (R = $COCF_3$)
n = 5: 9a (R = Ac), 9b (R = $COCF_3$)
n = 6: 10a (R = Ac), 10b (R = $COCF_3$), 10c (R = H)

Figure 1 | Chemical structures. (**a**) Generic structures of the backbone architectures of peptoids (that is, α-peptoids), β^3-peptides (β^2- and disubstituted β-peptides are not shown) and β-peptoids. (**b**) Depiction of the equilibrium of *transoid* and *cisoid* amide conformations in peptoid residues. (**c**) β-Peptoid oligomers prepared for structural and spectroscopic evaluation in this study.

Results

Compound design. Since high-resolution structures of α-peptoids have been successfully obtained for oligomers of just 4–5 residues in length, we envisioned that a hexamer might be sufficient to acquire stabilized folding. Two different α-chiral side chains were chosen (Fig. 1c): the *N*-(*S*)-1-phenylethyl (**1–5**), which has been studied in both α- and β-peptoids, and the *N*-(*S*)-1-(1-naphthyl)ethyl (**6–10**), which was applied in an α-peptoid tetramer structure solved by X-ray crystallography[31]. The *N*-(*S*)-1-phenylethyl side chain, although shown to successfully induce helical conformations in α-peptoids[13–15,32], does not induce significant control of the amide bond *cis–trans* equilibrium. It is therefore not entirely surprising that β-peptoid oligomers of this type have not previously been unambiguously shown to establish robust secondary structures. In recent monomer-based model systems, however, we found that the *trans*-amide conformation was preferred upon trifluoro-acetylation of *N*-(*S*)-1-phenylethyl monomers[30]. The effect of this substitution was therefore included in the present investigation (**1b–5b**) along with the acetylated control compound series **1a–5a**. The second series contained the *N*-(*S*)-1-(1-naphthyl)ethyl side chain (**6a–10d**), which was first introduced in β-peptoids[33], and later found to strongly induce *cis*-amide conformations in model systems[28,30] as well as in helical[31] and ribbon-shaped[34] α-peptoids.

All compounds were prepared by subunit synthesis in solution, essentially as previously described[35] (Fig. 2). For the considerably sterically congested naphthyl-containing compounds, in particular, we found the necessary reaction times to increase with the growing length of the oligomer, and found that fragment-based coupling may be necessary for efficient preparation of higher oligomers[20] (further synthetic considerations *vide infra*).

***Cis–trans*-amide rotamers**. ^{1}H nuclear magnetic resonance (NMR) spectra of the oligomers exhibited significant signal overlapping due to the identical side chains throughout each series as would be expected. We therefore focused on the shifts corresponding to the side chain methyne protons to determine overall *cis–trans*-amide bond ratios (Supplementary Figs 1–5; Table 1). NMR data showed no significant trend on the overall *cis–trans* ratios for the compounds in the *N*-(*S*)-1-phenylethyl series (**1–5**). In the *N*-(*S*)-1-(1-naphthyl)ethyl series (**6–10**), however, increases in $K_{cis-trans}$ values could be observed upon elongation of the oligomers, as particularly evident for the constants determined for the hexamers in benzene (Table 1). Unfortunately, the two highest oligomers of this series were not sufficiently soluble in acetonitrile to obtain NMR spectra of sufficient quality, but especially for the trifluoroacetylated series, there appears to be a slight increase in $K_{cis-trans}$ from dimer to tetramer. Taken together, these values may indicate propensity for the oligomers to adopt length- as well as solvent-dependent secondary structures, although high-resolution measurements are required to determine whether this is the case.

X-ray crystallography. Gratifyingly, we were able to crystallize compounds **10a**, **10b** and **10c** by slow evaporation of MeOH–$CHCl_3$ or benzene solutions of the compounds. Diffraction-quality crystals were obtained for **10a** and **10c** from MeOH–$CHCl_3$, and their structures were solved by X-ray single-crystal

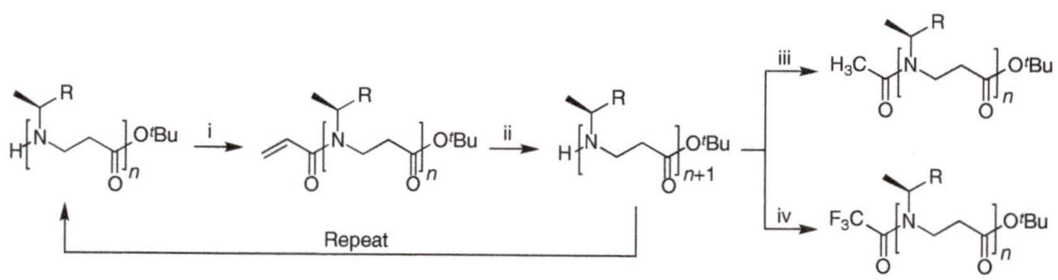

Figure 2 | Oligomerization and N-terminal functionalization of β-peptoids. Reagents and conditions: (i) acryloyl chloride, NEt_3, THF, 0 °C; (ii) (*S*)-1-phenylethylamine or (*S*)-1-(1-naphthyl)ethylamine, MeOH, 50 °C; (iii) AcCl, NEt_3, 0 °C; (iv) $(CF_3CO)_2O$, NEt_3, CH_2Cl_2, 0→20 °C.

Table 1 | Overall *cis-trans*-amide bond ratios[a].

***N*-(*S*)-1-phenylethyl series**										
Compound	**1a**	**1b**	**2a**	**2b**	**3a**	**3b**	**4a**	**4b**	**5a**	**5b**
CD_3CN	0.8	0.3	0.8	0.4	0.8	0.5	0.7	0.5	0.7	0.5
C_6D_6	0.6	0.3	0.6	0.3	0.6	0.3	0.5	0.3	0.4	0.3
***N*-(*S*)-1-(1-Naphthyl)ethyl series**										
Compound	**6a**	**6b**	**7a**	**7b**	**8a**	**8b**	**9a**	**9b**	**10a**	**10b**
CD_3CN	5.2	5.4	5.2	6.0	5.9	7.3	—[b]	—[b]	—[b]	—[b]
C_6D_6	4.8	4.0	4.4	4.0	5.6	4.3	6.4	4.9	11.0	10.0

[a]$K_{cis\text{-}trans}$ values determined by ^1H NMR spectroscopy. [b]Limited solubility of this compound did not allow for NMR analysis in acetonitrile-d_6.

structure determination to 1.05 and 1.00 Å resolution, respectively (see Supplementary Figs 6–8 for additional representations).

The N-terminally non-acylated compound **10c** was solved first and revealed a helical conformation of precisely three residues per turn and a pitch of ~9.6 Ångström, with only the N-terminal naphthyl side chain twisted away from the helical axis (Fig. 3a; Supplementary Data 1). The remaining naphthyl groups were highly organized along each of the three faces of the helix to give an equilateral triangle when viewed down the helical axis (Fig. 3b,c). The main helix thus adopts a triangular prism shape, not taking the N-terminal side chain into consideration. We envisioned, based on our monomer studies, that acetylation of the N terminus would also force this amide into a *cisoid* configuration, and hoped that this would direct the final side chain into alignment along the third face of the molecule. The X-ray crystal structure of compound **10a** indeed confirmed this (Fig. 3d; Supplementary Data 2), as this compound crystallized in the *R*3 space group with only one -CH_2CON(1-(1-naphthyl)ethyl)CH_2- fragment in the asymmetric unit. Thus, the crystal symmetry generates infinitely long, parallel chains, and this renders the *t*-butyl group invisible, due to the 1:6 ratio of *t*-butyl to 1-(1-naphthyl)ethyl.

This results in crystal packing, where the helical segments are arranged in a head-to-tail manner to give regularly elongated triangular threads (Fig. 3g) with a helical pitch of 9.8 Å (Fig. 3d). To reveal the complete molecule, including the N-terminal-acetamido and C-terminal *t*-butyl ester groups, data sets were collected on several crystals, but attempts to solve the structure in *P*1 space group were unsuccessful. This packing issue was not observed for compound **10c** due to the positioning of the N-terminal side chain, and therefore enabled visualization of the full structure including the C-terminal *t*-butyl (Fig. 3f).

The torsion angles in the two helices are very similar (Table 2), and the extended structures compare reasonably well with one of the helical conformations previously suggested by density functional theory calculations, albeit with a ~20-degree difference in the ω and θ angles (Supplementary Table 1)[12]. To the best of our knowledge, however, the present work provides the first experimental demonstration of the existence of this helical type. Most previously determined helical conformations of a β-peptidic nature have been shown to contain stabilizing hydrogen bond networks[7], and investigation of oligomers of homologated proline residues indicated *trans*-amide-containing conformations[36]. Our structures are thus unique to the field of foldamers. The highly ordered, tight packing of the side chains along the three faces of the helical axis (Fig. 3e) combined with their strong *cis*-amide-inducing properties[30], suggests that this type of side chain provides a particularly strong stabilization of this novel secondary-structure motif.

Furthermore, intramolecular distances indicate CH–π interactions between the backbone methylene group and side chain naphthyl groups, which may also contribute to stabilization and protection of the helical backbone from solvation (Supplementary Fig. 9; Supplementary Table 2). This is in stark contrast to the X-ray crystal structure reported for an α-peptoid tetramer containing the same side chains, where the naphthyl groups pointed away from the helical axis[31]. In agreement with that structure, however, as well as the findings from our previous monomer investigations, we did not see evidence for stabilizing $n \rightarrow \pi^*_{Ar}$ interactions in our crystal structure. On the other hand, our *cis*-amide conformations were in agreement with the recently described $n \rightarrow \sigma^*$ orbital overlap between carbonyl and side chain methyne atoms observed for peptoids[37].

CD and fluorescence spectroscopy. We then evaluated the compounds by CD spectroscopy. The spectra obtained for the acetylated *N*-(*S*)-1-phenylethyl series (Fig. 4a,b) were similar to those previously reported for β-peptoids containing this side chain, whereas the *N*-(*S*)-1-(1-naphthyl)ethyl series (**6a–10a**) revealed a minimum at 224–228 nm and a maximum at 215–218 nm (Fig. 4c). At first glance, we were surprised that these Cotton effects decreased with increasing length of the oligomer. However, both the minimum and the maximum were present in the CD spectrum of an acetyled monomer and not of a non-acetylated monomer (Supplementary Fig. 10), which indicates that these signals are not related to secondary-structure formation but rather a signature of the particular naphthyl-containing amide motif itself. Starting at the tetramer length (**8a**), on the

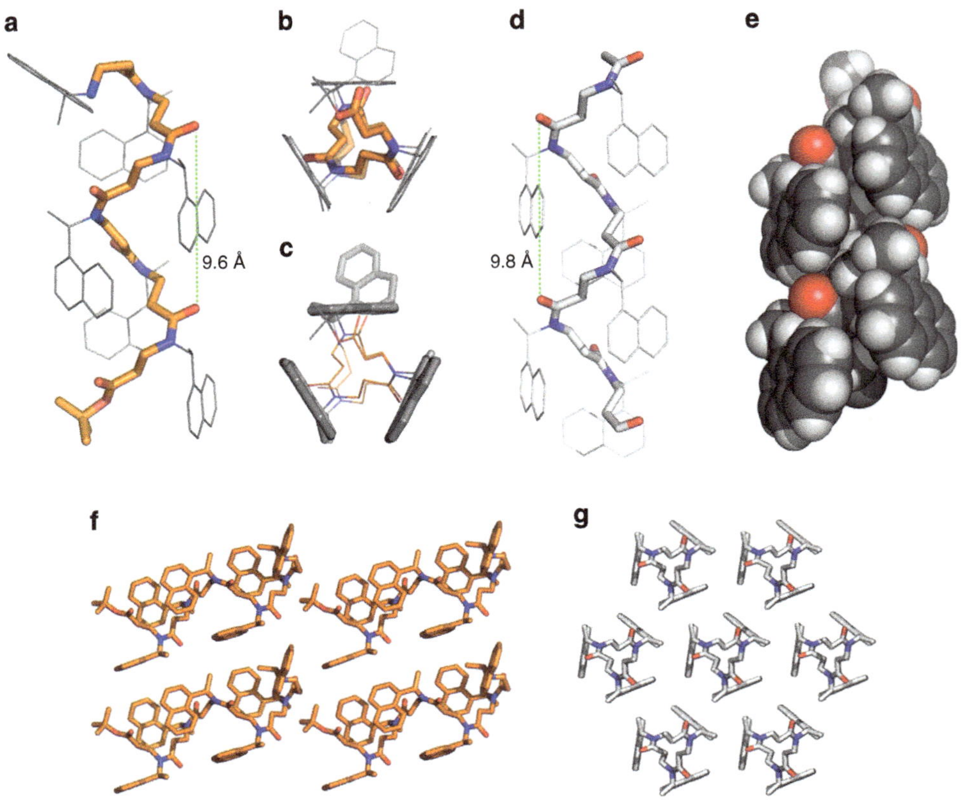

Figure 3 | X-ray crystal structures. (**a**) Side view of compound **10c**. (**b**,**c**) End views of compound **10c** from the C terminus. The backbone is shown in orange and hydrogen atoms are omitted for clarity. (**d**) Compound **10a** showing the backbone helix as sticks and side chains as lines. (**e**) Spacefilling representation of compound **10a** showing the packing of the naphthyl groups along the phases of the triangular prism-shaped conformation. (**f**) Crystal packing of compound **10c** viewed perpendicular to the helical axis. (**g**) Crystal packing of compound **10a** viewed down the helical axis to demonstrate the highly ordered arrangement. Hydrogen atoms are omitted for clarity except for the spacefilling representation in **e**. Representations of X-ray crystal structures were prepared using the PyMOL Molecular Graphics System, Schrödinger, LLC.

Table 2 | Torsion angles in the helical structures.

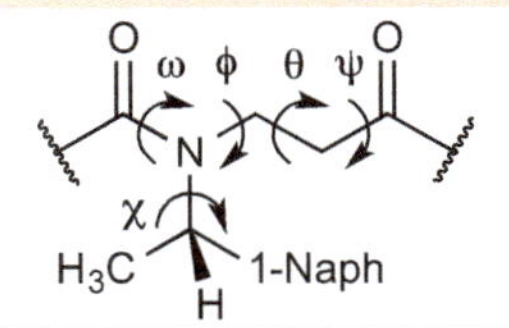

Angle	ϕ	θ	ψ	ω	χ_1*	χ_2†
10a‡	96.3°	172.5°	−175.3°	−13.8°	53.6°	−80.4°
10c‡	97.4°	166.0°	−173.9°	−13.4°	56.3°	−73.8°

*Measured by the naphthyl substituent.
†Measured by the methyl substituent.
‡Measured at residue 5 in the structure to give representative values for a residue within the helix. See Supplementary Table 1 in the Supplementary Information for dihedral angles of additional residues.

other hand, a positive signal appeared near 232 nm. This signal intensified with growing length of the oligomers, suggesting that this peak may be indicative of length-dependent secondary-structure formation.

We also obtained CD spectra of the trifluoroacetylated oligomers, which exhibited the same overall spectral shapes (Fig. 4d). Notably, six residues were necessary to obtain a positive signal at 232 nm in this series, suggesting that trifluoroacetylation has a negative effect on secondary-structure formation in acetonitrile. In our monomer study, we observed stabilization of the *cis*-amide conformation upon trifluoroacetylation compared with acetylation[30], but this apparently does not translate into a stabilizing effect on the oligomer secondary structure. This may be explained by a steric clash between the bulkier trifluoroacetyl group and the first backbone methylene group at the N terminus.

To assess whether the positive signal at 232 nm is indicative of folding, we also collected CD spectra of both **10a** and **10b** at temperatures in the range 20–75 °C (Supplementary Fig. 11). The gradual decrease in signal intensity at 232 nm upon heating was indeed in accordance with temperature-mediated denaturation. Importantly, the remaining spectral shape was not affected significantly, which is in accordance with our hypothesis that this is not related to secondary structure. Furthermore, spectra recorded upon cooling were identical to those obtained before heating, indicating refolding of the structures (Supplementary Fig. 12). In addition, we investigated the effect of solvent polarity by obtaining CD spectra of compounds **9a**, **10a** and **10b** in the presence of varying amounts of MeOH (Supplementary Fig. 13). These spectra also showed a decrease in the signal at 232 nm, which is in agreement with denaturation of our structure at gradually increasing MeOH concentrations. Finally, we obtained CD spectra at varying concentrations of **10a** (Supplementary Fig. 14), which showed a slight increase of the signal at 232 nm upon dilution. While this was somewhat surprising to us, it suggests that aggregation does not play a role in the formation of secondary structure.

Aggregation of naphthyl-containing side chains in peptoids was recently shown to give rise to CD signals that correlated with fluorescent excimer bands[38]; thus, to rule out a similar origin of the maxima at 232 nm, we subjected oligomer **10a** and dimer **6a** to fluorescence measurements. In acetonitrile, the hexameric

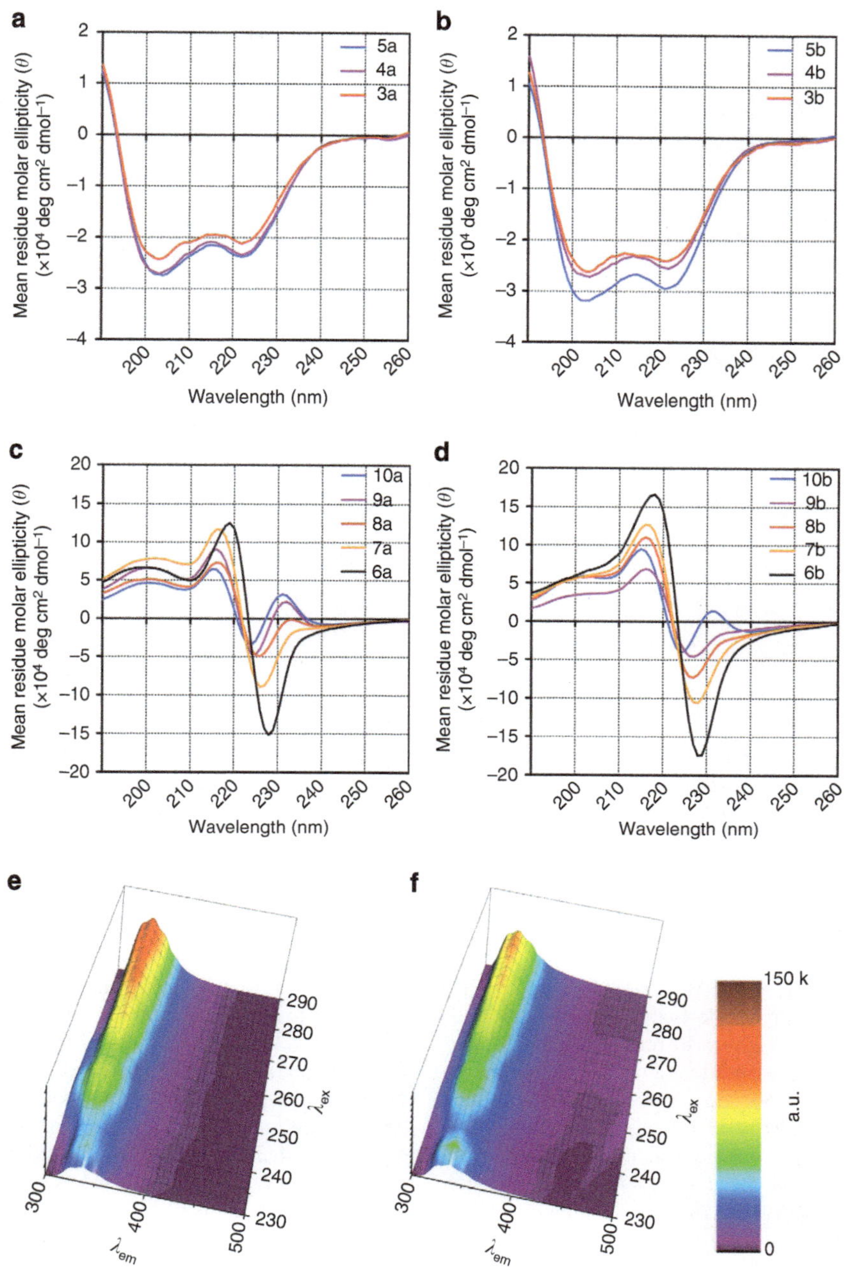

Figure 4 | CD and fluorescence spectra. (**a**) CD spectra of compounds **3a–5a**; (**b**) CD spectra of compounds **3b–5b**; (**c**) CD spectra of compounds **6a–10a**; and (**d**) CD spectra of compounds **6b–10b** measured at room temperature in acetonitrile (60 μM). (**e**) Fluorescence emission spectra recorded for compound **10a** in acetonitrile or (**f**) in MeOH-acetonitrile (3:2). Spectra were obtained at room temperature using 60-μM solutions, and the intensities are shown in arbitrary units (a.u.).

oligomer displayed a single maximum at ~336 nm (Fig. 4e) and no change in spectral shape was observed upon addition of methanol, which might promote aggregation due to poorer solubility (Fig. 4f). An excimer band would be expected to appear at a longer wavelength in the emission dimension if aggregates formed, as previously reported for peptoids containing the same type of side chains (392 nm)[38]. This was not the case for either compound **10a** or the dimer **6a** (Supplementary Fig. 15), suggesting that the CD maxima observed at 232 nm for our longer oligomers are not a result of excimer effects caused by naphthyl group aggregation.

Structure–folding relationships. Acylation of the N-terminus appeared to have a pronounced effect on folding propensity as indicated by the two X-ray crystal structures (**10a** versus **10c**), as well as the CD spectra (for example, **9a** versus **9b**). These observations in addition to an interest in being able to introduce additional functionalities in future investigations of the applicability of these oligomers as scaffolds for multivalent display prompted us to prepare modified analogues of **10a**. Owing to the challenges associated with synthesis of oligomers by subunit assembly in solution above the hexamer length, we explored two additional strategies to prepare analogues. First, we attempted to transfer the subunit approach to solid support using a ChemMatrix resin with a Rink linker to give oligomers containing C-terminal amides (Fig. 5a). Due to increasing solubility difficulties encountered during purification for acetylated oligomers upon growing length (that is, **10d**), we prepared oligomers with free N termini by

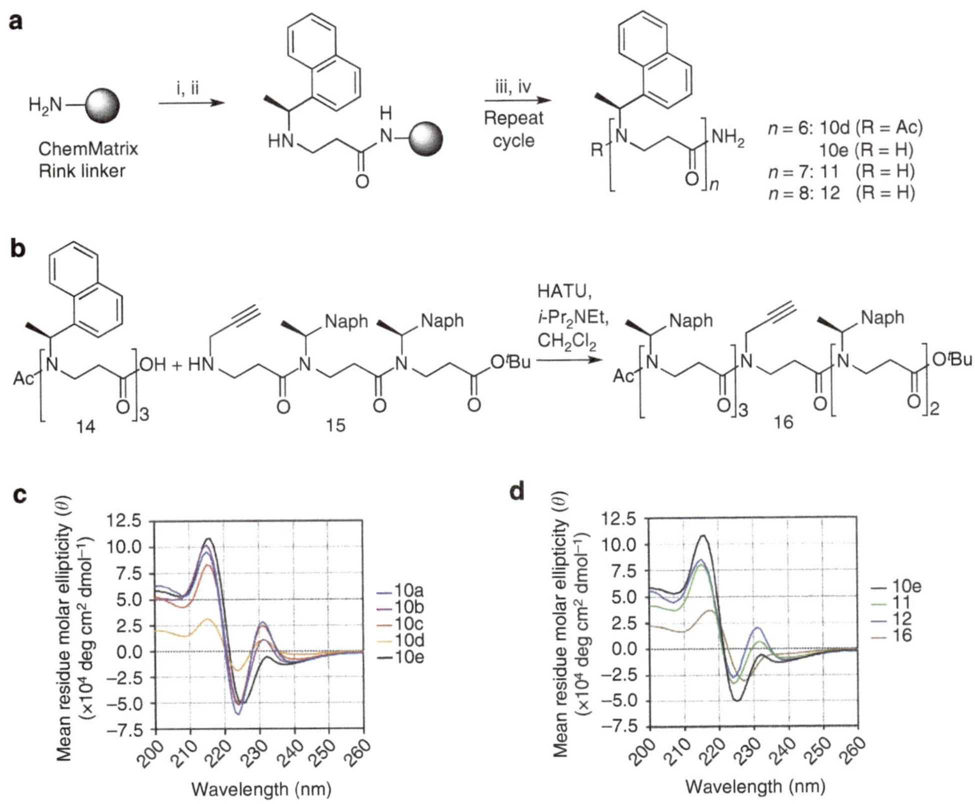

Figure 5 | Synthesis and CD spectra of additional analogues. (**a**) Solid-phase synthesis of homo-oligomers. Reagents and conditions: (i) acryloyl chloride (6 equiv), i-Pr_2NEt (12 equiv), DMF, 2 h, RT; (ii) (*S*)-1-(1-naphthyl)ethylamine (6 equiv), MeOH, 16 h, 50 °C; (iii) only for compound **10d**: AcCl (6 equiv), *i*-Pr2NEt (12 equiv), DMF, 2 h, RT; (iv) trifluoroacetic acid CH_2Cl_2 (1:1). (**b**) Fragment coupling in solution to give hetero-oligomeric hexamer **16** (49% isolated yield). Reagents and conditions: HATU (1.1 equiv), i-Pr_2NEt (2 equiv), CH_2Cl_2, 16 h, RT. (**c**,**d**) CD spectra measured in MeCN (60 μM).

solid-phase synthesis (**10e**, **11** and **12**). This strategy afforded sufficient material for CD spectroscopic evaluation, but the crude purities observed by analytical high-performance liquid chromatography (HPLC) were not appropriate for synthesis of hetero-oligomers (Supplementary Fig. 16). Thus, it was decided to explore a fragment-based coupling strategy to introduce an alkyne-containing side chain, which could be easily functionalized by Cu(I)-catalyzed Huisgen azide-alkyne cycloaddition[39,40] after oligomerization. Indeed, compound **16** was obtained in good yield by solution-phase coupling of two trimer building blocks (Fig. 5b), showing that β-peptoid oligomers may be obtained efficiently by fragment coupling in solution as was also previously demonstrated on solid phase[20].

This additional collection of compounds was evaluated by CD spectroscopy to investigate structural influence on the hypothesized structure indicating Cotton effect at ~232 nm (Fig. 5c). As evident from the series of different hexamers **10a–e**, the combination of N-terminal acetylation and a C-terminal *tert*-butyl ester of compound **10a** gave rise to the most robust signal at 232 nm. Compound **10c**, containing a *tert*-butyl group at the C-terminus and a free N-terminus, displayed an intense Cotton effect at 232 nm similar to compound **10a** and more intense than compound **10d**, which suggests that the bulky C-terminal esterification is more important for helix formation and/or stabilization than N-terminal acetylation (Fig. 5c).

Compound **10e**, which has a C-terminal amide and free N-terminus, exhibited the lowest intensity at 232 nm in this series. However, this signal was gradually rescued upon elongation to heptamer (**11**) and octamer (**12**) length, indicating cooperative length-dependent secondary-structure formation (Fig. 5d).

Finally, we were interested in investigating whether side chains could be mutated to other residues than *N*-(*S*)-1-(1-naphthyl) ethyl with retention of structural integrity, which would allow for further functionalization of these novel scaffolds. We therefore prepared compound **16** as described above. The CD spectrum of this compound, however, indicated a complete collapse of the secondary structure (Fig. 5d) by exhibiting a shape similar to trimer **7a**. While this supports our previously mentioned hypothesis that tight packing of the naphthyl side chains along the helix serves to stabilize the structure, it also shows that more elaborate synthetic efforts are required to achieve funtionalized oligomers with retention of the helical structure. We therefore envision that functional groups should be introduced at the methyl position of the side chains instead.

CD spectroscopy of peptide analogues, however, is not sufficient to derive specific secondary structures[41], and although we find it highly likely that our oligomers adopt helical conformations in solution based on our collective results including X-ray, we stress that our current NMR and CD data do not unambiguously verify the presence of such secondary structures. Therefore, we also performed molecular dynamics simulations to address the folding propensity of hexamer **10a** in solution.

Molecular dynamics simulations. To lend further theoretical support to the presence and stability of helical structures in solution, we carried out a series of molecular dynamics simulations on hexamer **10a** using a graphics processing unit (GPU)-accelerated version of Desmond (v 3.9). The hexamer was built in a helical conformation in maestro v. 9.8 (Schrödinger, LLC, New

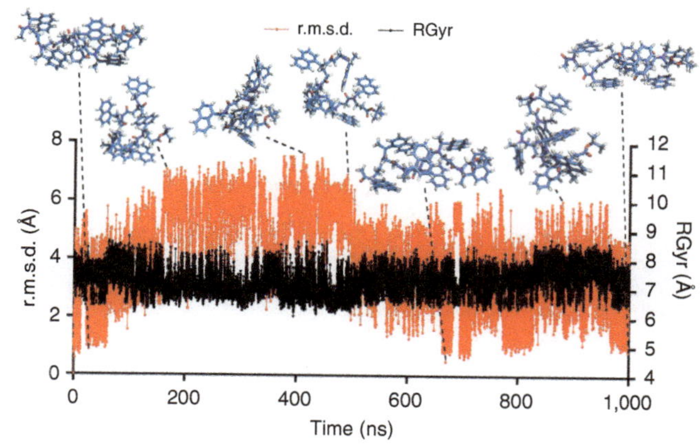

Figure 6 | Molecular dynamics simulation of compound 10a in acetonitrile. The axes show root mean square deviation (r.m.s.d.) and gyrational radius (RGyr) during a 1-μs molecular dynamics (MD) simulation at 300 K. Representative conformations are shown to illustrate the folded and unfolded states of the helical structure throughout the simulation. The simulation is uploaded as Supplementary Movie 1.

York, NY, 2014) and treated as previously described[42] (see the Methods section for details).

To investigate the influence of temperature, we carried out a series of 250-ns simulations at 200, 250, 300, 350 and 400 K, respectively (Supplementary Figs 17–21). These simulations revealed significant temperature dependence with the 'low-temperature' regimes (200 and 250 K) showing only few conformations, while the structural ensemble displayed high heterogeneity in the high-temperature regimes (350 and 400 K). In the simulation carried out at 200 K, the hexamer was dominated by a meta-stable folded conformation (root mean squared deviation (r.m.s.d.) ~4 Å), which was populated for more than 100 ns (Supplementary Fig. 17). When the temperature was raised to 250 K, the structure alternated between the folded and unfolded conformation significantly faster with the meta-stable conformations having lifetimes of 10–30 ns (Supplementary Fig. 18).

The simulation at 300 K exhibited intermediary fluctuations in r.m.s.d. and gyrational radius (RGyr), with conformations similar to the starting point (r.m.s.d. ~ 2 Å) appearing throughout the simulation (Supplementary Fig. 19). To investigate conformational space at this temperature in more detail, we also performed a longer (1 μs) simulation where low r.m.s.d. and high Rgyr, which is in agreement with helical conformation, were observed within the first 60-ns interval (Fig. 6). Then bent conformations with r.m.s.d.s of up to >6 Å and slightly decreased RGyr (<7 versus 7.5 Å for the helical structure) were observed. A pattern of unfolding and refolding occurred throughout the simulation, with the conformations gradually diverging from the initial X-ray structure until approximately the 500-ns time point. Interestingly, significantly lower r.m.s.d. values reappeared in the later part of the simulation, and overlaying one of the final structures with the X-ray crystal structure of compound **10a** revealed an r.m.s.d. for all backbone atoms of just 0.837 Å (Fig. 7).

Discussion

We report the first examples of high-resolution structures of linear oligomeric β-peptoids, which are also the longest structures of any peptoid oligomer determined by X-ray crystallography to date. These novel helical structures show definitively that the β-peptoids qualify as a valid addition to the already rich ensemble of foldamer designs. Our crystal structures revealed highly regular equilateral triangular prism-shaped conformations in the solid state, which was achieved by synthesis of homomeric hexamers containing the highly *cis*-amide-inducing side chain, (*S*)-1-(1-naphthyl)ethylamine. Furthermore, ^{1}H NMR and CD spectroscopic data supported the existence of length-, solvent- and temperature-dependent secondary structures in solution in organic solvents. Although the determination of high-resolution structures of oligomeric β-peptoids in solution will be an interesting future challenge, we here performed molecular dynamics simulations in acetonitrile, which provided further evidence for the formation of helical conformations in solution. Taken together, we demonstrate control of β-peptoid folding for the first time, which now opens the possibility of taking advantage of β-peptoids in design of novel structurally well-defined biomimetic materials. Furthermore, a preliminary structure–folding relationship investigation suggested that a bulky ester substituent at the C-terminus is important for stable folding and, importantly, showed that mutation of an (*S*)-1-(1-naphthyl) ethylamine side chain to a readily functionalizable acetylene-containing side chain was incompatible with helicity, as indicated by CD spectroscopy. We are therefore currently investigating various strategies to introduce alternative functionalities through substitution at the methyl substituent position of our side chains to ensure structural integrity of these novel scaffolds. We envision that functionalized helical β-peptoids have potential applicability in biomedical research. Well-defined positioning of amino acid side chain functionalities in three dimensions is important for disruption of protein–protein interactions, which account for a number of notoriously difficult-to-inhibit drug targets. The structures provided herein may serve as useful abiotic scaffolds with proteolytic stability in such endeavours. Furthermore, we envision that helical β-peptoids may serve as scaffolds for highly accurate multivalent display of drug molecules in three dimensions to achieve increased binding affinities against targets with multiple binding sites.

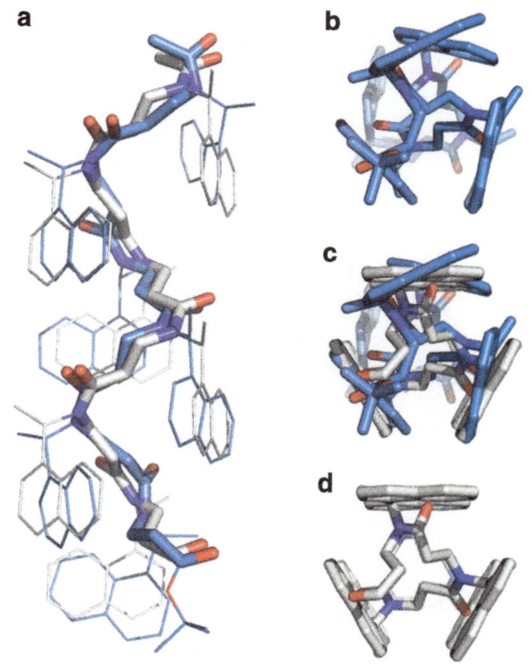

Figure 7 | Overlays of conformations of compound 10a. (**a**) Side view of the X-ray crystal structure shown in light grey and the conformation after 1 μs MD simulation shown in blue. Only the backbones are shown as sticks to highlight the similarity. (**b-d**) End views shown as stick representations. The r.m.s.d. was calculated for all backbone atoms in structural overlays prepared using the 'pair-fit' command in MacPyMOL, Schrödinger Inc.

Methods

General. All chemicals and solvents were analytical grade and used without further purification. Vacuum liquid chromatography (VLC) was performed on silica gel 60 (particle size 0.015 – 0.040 mm). Ultra high-performance liquid chromatography (UPLC)-mass spectrometry (MS) analyses were performed on a Phenomenex Kinetex column (1.7 μm, 50 × 2.10 mm) using a Waters Acquity ultra-HPLC system. A gradient with eluent I (0.1% HCOOH in water) and eluent II (0.1% HCOOH in acetonitrile) rising linearly from 0 to 95% of II during $t = 0.00$–5.20 min was applied at a flow rate of 1 ml min^{-1}. Preparative HPLC purification was performed on a C18 Phenomenex Luna column (5 μm, 100 Å, 250 × 20 mm) using an Agilent 1260 LC system equipped with a diode array ultraviolet detector and an evaporative light-scattering detector (ELSD). A gradient with eluent III (water – MeCN – TFA, 95:5:0.1) and eluent IV (0.1% TFA in acetonitrile) rising linearly from 45 to 95% of IV during $t = 5 - 35$ min, and isocratically at 95% from $t = 35$–55 min was applied at a flow rate of 20 ml min^{-1}. Analytical HPLC was performed on a C18 phenomenex Luna column (3 μm, 100 Å, 150 × 4.60 mm) using an Agilent 1100 series system equipped with a diode array ultraviolet detector. A gradient using eluent III and eluent IV rising linearly from 0 to 80% of IV during $t = 2 - 10$ min, then from 80 to 95% during $t = 10$–27 min and finally isocratically at 95% IV from $t = 27$–33 min was applied at a flow rate of 1 ml min^{-1}. High-resolution LC-DAD-MS was performed on an Agilent 1,100 system equipped with a photodiode array detector (DAD) and coupled to a LCT orthogonal time-of-flight mass spectrometer (Waters-Micromass, Manchester, UK) with Z-spray electrospray ionization. Two-dimensional NMR spectra were recorded on a Bruker Ascend 400 at 400 MHz for ^{1}H and 100 MHz for ^{13}C. All spectra were recorded at 298 K. Heteronuclear single-quantum coherence spectra were recorded with a relaxation delay of 1.5 s before each scan, a spectral width of 4,800 × 16,600, collecting 4 FIDs and 1,000 × 256 data points. All spectra were recorded at 298 K. Chemical shifts are reported in p.p.m. relative to deuterated solvent peaks as internal standards (δH, C_6D_6 7.16 p.p.m.; δC, C_6D_6 128.06 p.p.m., δH, CD_3CN 1.94 p.p.m.; δC, CD_3CN 1.32 p.p.m.).

General procedure for oligomerization of β-peptoids. The β-peptoids were oligomerized using a procedure closely resembling that reported by Taillefumier *et al.*[35]. Briefly, the β-peptoid (1.0 equiv) and NEt_3 (1.2 equiv) were dissolved in THF (0.05 M) and cooled to 0 °C. Then acryloyl chloride (1.4 equiv) was added and the reaction was stirred for 1 h. The reaction mixture was filtered and the solids were washed with EtOAc. The combined filtrates were concentrated *in vacuo* yielding crude acrylamide. The crude acrylamide was dissolved in MeOH (0.1 M), the desired primary amine (2.0 eqiuv) was added and the reaction was stirred for 18 h at 50 °C. In case of the penta- and hexamer of compounds containing *N*-(*S*)-1-(1-naphthyl)ethyl side chains, additional amine (2.0 equiv) and stirring for 72 h were needed for the reaction to go to completion. Furthermore, the acrylamide of the *N*-(*S*)-1-(1-naphthyl)ethyl-containing pentamer was poorly soluble in MeOH, therefore MeOH–CH_2Cl_2 (1:3, 0.015 M) was used for the final aza-Michael step. Purification was performed using vacuum liquid silica gel chromatography applying a gradient of MeOH in CH_2Cl_2 from 0→5%.

General procedure for N-terminal functionalization of β-peptoid oligomers. Acetylation of β-peptoid oligomers was achieved by dissolving the β-peptoid oligomer (1.0 eqiuv) in CH_2Cl_2 (0.02 M) at 0 °C and then adding Et_3N (1.4 equiv) and acetyl chloride (1.2 equiv) under stirring for 1 h. Trifluoroacetylation of β-peptoid oligomers required addition of Et_3N (4.0 equiv) and trifluoroacetic anhydride (5.0 equiv) to the β-peptoid solution, which was stirred for 1 h at 0 °C, followed by ambient temperature for 18 h. For both acetylated and trifluoroacetylated compounds, purification could be achieved by evaporation of the solvent *in vacuo* followed by preparative reversed-phase HPLC. Vacuum liquid silica gel chromatographic (VLC) purification of compounds **6a–10b** applying a gradient of 0.2% MeOH in CH_2Cl_2 from 0→5% was unsuccessful. However, crude **10a** and **10b** could be redisolved in *N,N*-dimethylformamide (DMF), and precipitated upon addition of water. The solid product contained only minor impurities, which could be removed using VLC with the above-mentioned gradient. All compounds were purified to >95% homogeneity, as judged by UPLC-MS and analytical HPLC (230 nm). HPLC traces for all oligomers are shown in Supplementary Figs 22 and 23. ^{1}H NMR spectra are included as Supplementary Figs 24–44 and high-resolution mass spectrometry (HRMS) data are reported in Supplementary Table 3.

Subunit synthesis of oligomers 10d, 10e, 11 and 12. A solution of acryloyl chloride (6 equiv) and *i*-Pr_2NEt (12 equiv) in DMF (5 ml) were added to a Rink-amide-functionalized Chem-Matrix resin (1 equiv). After shaking for 3 h, the resin was drained and subsequently washed with DMF (3 × 3 ml), MeOH (3 × 3 ml), and CH_2Cl_2 (3 × 3 ml). Then a solution of (*S*)-1-(1-naphthyl)ethylamine (6 equiv) in MeOH (5 ml) was added to the resin that was shaken at 50 °C for 18 h, and then washed as described above. After nine coupling cycles, the crude product was cleaved from the resin using 50% TFA in CH_2Cl_2 for 2 × 30 min. Due to incomplete couplings it was possible to collect oligomers of 6–8 residues using preparative HPLC (Supplementary Fig. 16). All compounds were purified to >95% homogeneity as judged by UPLC-MS and analytical HPLC (230 nm). ^{1}H NMR spectra are included as Supplementary Figs 45–48 and HRMS data are reported in Supplementary Table 4.

Compound 14. The methylester-protected β-peptoid oligomer could be synthesized in the same manner as the *tert*-butylesters, starting from methylacrylate. The trimeric oligomer (514 mg, 0.68 mmol, 1 equiv) was dissolved in MeOH (7 ml), and 1 M aq. LiOH (2.7 ml, 2.7 mmol, 4 equiv) was added; the resulting mixture was stirred for 18 h at room temperature. The reaction mixture was poured into CH_2Cl_2 (100 ml) and washed with 1 M aq. HCl (100 ml), and the aqueous phase was then back extracted with CH_2Cl_2 (3 × 100 ml). The combined organic phases were dried ($MgSO_4$), filtered and concentrated *in vacuo*. The compound was used without further purification. Yield: 450 mg (95%); UPLC-MS, $t_R = 2.48$ min, *m/z* [M + H] + calcd for $C_{47}H_{50}N_3O_5^+$, 736.4; found 736.9.

Compound 15. The propargyl-containing β-peptoid oligomer could be synthesized by the general procedure described above, using propargyl amine in the third Michael addition. Purification was performed using vacuum liquid silica gel chromatography (2 × 6 cm, CH_2Cl_2–MeOH, 0.2% gradient from 0→5%) to give the product as a white solid. Yield: 819 mg (19% overall), UPLC-MS: $t_R = 2.00$ min, *m/z* 634.4 ([M + H]$^+$, $C_{20}H_{34}N_3O_4^+$, calcd 634.8).

Compound 16. Compound **14** (219 mg, 0.32 mmol, 1 equiv), HATU (134 mg, 0.35 mmol, 1.1 equiv) and *i*-Pr_2NEt (111 μl, 0.64 mmol, 2 equiv) were dissolved in CH_2Cl_2 (25 ml) and preincubated for 10 min at room temperature. Then, compound **15** (200 mg, 0.32 mmol, 1 equiv) was added and the resulting reaction mixture was allowed to stir for 2 h. The solvent was evaporated *in vacuo* and the crude compound was purified to >95% homogeneity using preparative HPLC. Yield 220 mg (49%). UPLC-MS: $t_R = 3.37$ min, *m/z* [M + H]$^+$ calcd for $C_{83}H_{88}N_7O_7^+$, 1,351.7; found 1,351.6. HPLC: $t_R = 19.25$ min (>95%). HRMS: *m/z* [M + Na]$^+$ calcd for $C_{87}H_{93}N_6O_8Na^+$, 1,373.7025; found, 1,373.7019 (ΔM = 0.2 p.p.m.). See Supplementary Fig. 49 for a copy of the ^{1}H NMR spectrum.

Crystallization. Selected peptoid hexamers (~30 mg) were dissolved in a minimum amount of $CHCl_3$, and MeOH was added until precipitation could be observed; sufficient $CHCl_3$ was then added to give clear solutions. The solutions were left for slow evaporation at room temperature, and crystals were generally formed within a week. The crystals could be removed from the solvent and washed using MeOH. The hexamers (30 mg) also crystalized upon slow evaporation from benzene; however, these specimens were of insufficient quality for X-ray diffraction.

X-ray crystallography. Several crystals of each compound were tested before suitable crystals were found. Measurements were performed at room temperature as cryo cooling turned out to destroy the crystals. Data collection was performed on an Agilent Supernova Diffractometer using CuKα radiation. For crystal **10a**, data to 1.00 Å resolution were included and for crystal **10c**, data to 1.05 Å resolution were included. Data were processed and scaled using the *CrysAlisPro* software (Agilent Technologies). Details of the data collection may be found in Supplementary Data 1 and 2.

Compound **10c** crystallized in space group *P*1 with one molecule in the asymmetric unit. The structure was solved using SHELXS and refined using SHELXL[43]. Due to the rather low resolution, it turned out, however, that geometric restraints had to be included in the refinement. All naphthyl groups were imposed with FLAT and DELU restraints. The N-terminal naphthyl group had rather large thermal parameters and was restrained further using RIGU and ISOR restraints. Furthermore, the *tert*-butyl group was restrained using DELU, and all six residues were restrained to be similar using a SAME command. In the rather large voids between the helices we observed disordered electron density. The three largest peaks were modelled as oxygen with occupancy of 0.5. Compound **10a**, on the other hand, crystallized in space group *R*3 with one peptoid residue in the asymmetric unit. The structure was solved using SHELXS and refined in SHELXL[43]. Also here, geometric restraints had to be included in the refinement and the naphthyl group was imposed with FLAT and DELU restraints. The peptoid oligomer was revealed when crystallographic symmetry was imposed. However, the crystal structure generates infinitely long chains and the *tert*-butyl group was not visible, although for every sixth naphthyl group a *tert*-butyl group must be present. Hydrogen atoms were included on ideal positions using riding coordinates for both crystals. The data quality did not allow for determination of the absolute configurations, which were imposed in accordance with the stereochemistry of the building blocks used in the synthesis.

CD spectroscopy. Spectra were acquired with an Olis DSM 10 CD spectrophotometer (Olis Inc., Bogart, GA, USA), equipped with a Quantum Systems temperature control module, or on a Jasco 810 spectrometer equipped with a Julabo AWC 100 cooling unit. Spectra were obtained using 1-mm quartz cuvettes, and compound solutions in MeCN were prepared using dry weight of the lyophilized material followed by dilution to give the desired concentrations and solvent combination. The data are averages of 3–5 successive accumulations. Spectra were recorded in millidegree units, corrected for solvent contributions and normalized to mean residue ellipticity $[\theta] = 100\psi/lcn$, where ψ is the signal in millidegrees, l is the path length (0.1 cm), c is the concentration in mM and n is the number of peptoid amide bonds.

Fluorescence scanning. Samples were prepared based on dry weight of lyophilized compounds **6a** and **10a**, respectively. Fluorescence was measured at ambient temperature in a black non-binding NUNC 96-well plate using a Perkin-Elmer (Enspire) multimode plate reader by varying the excitation wavelength (230–290 nm) and measuring emmision (300–500 nm) in steps of 3 nm in both dimensions. The data were plotted using the OriginPro 9.1 software (OriginLab Corporation, Northampton, MA, USA).

Molecular modelling. The structure of **10a**, based on the X-ray crystal structure, was solvated in a 15-$Å^3$ custom-built acetonitrile box, and the resulting system was minimized with the default presimulation protocol in Desmond (Desmond Molecular Dynamics System, version 3.4, D. E. Shaw Research, New York, NY, 2013; *Maestro-Desmond Interoperability Tools, version 3.4*, Schrödinger, 2013) consisting of (i) minimization with solute restraints, (ii) minimization without restraints, (iii) Berendsen NVT simulation at $T = 10$ K with small time steps and restraints on heavy solute atoms, (iv) Berendsen NPT simulation at $T = 10$ K with restraints on heavy solute atoms, (v) Berendsen NPT simulation with restraints on heavy solute atoms and (vi) unrestrained Berendsen NPT simulation[44]. Then, NPT simulations with periodic boundary conditions were carried out for 250 ns with the Nose–Hoover chain thermostat[45,46] set to 300 K and a relaxation time of 1 ps. The Martyna–Tobias–Klein barostat[47] was used to regulate pressure to 1 bar with isotropic coupling and a relaxation time of 2 ps. Equations of motion were integrated using the RESPA integrator[48] with bonded, near and far, time steps of 2, 2 and 6 femtoseconds, respectively. Non-bonded interactions were subjected to a cutoff of 9 Å. Long-range electrostatics were treated with the smooth-particle mesh Ewald method[49] with a tolerance of 10^{-9}. The OPLS-2005 force field[50,51] was used to describe both **10a** and solvent. All atomic coordinates and corresponding molecular mechanics energies were saved at 25-ps intervals for the entire 250-ns simulation. The structure obtained after the presimulation protocol for the reference simulation at 300 K was also used as an initial structure for additional 250-ns simulations carried out at 200, 250, 350 and 400 K to ease comparison between the different temperatures. Finally, a 1-μs simulation was carried out at 300 K as well. The molecular dynamics trajectories were analysed using visual molecular dynamics[52].

References

1. Gellman, S. H. Foldamers: A manifesto. *Acc. Chem. Res.* **31,** 173–180 (1998).
2. Goodman, C. M., Choi, S., Shandler, S. & DeGrado, W. F. Foldamers as versatile frameworks for the design and evolution of function. *Nat. Chem. Biol.* **3,** 252–262 (2007).
3. Horne, W. S. & Gellman, S. H. Foldamers with heterogeneous backbones. *Acc. Chem. Res.* **41,** 1399–1408 (2008).
4. Olsen, C. A. Peptoid-peptide hybrid backbone architectures. *ChemBioChem.* **11,** 152–160 (2010).
5. Horne, W. S. Peptide and peptoid foldamers in medicinal chemistry. *Exp. Opin. Drug Discov.* **6,** 1247–1262 (2011).
6. Cheng, R. P., Gellman, S. H. & DeGrado, W. F. beta-Peptides: from structure to function. *Chem. Rev.* **101,** 3219–3232 (2001).
7. Seebach, D., Beck, A. K. & Bierbaum, D. J. The world of beta- and gamma-peptides comprised of homologated proteinogenic amino acids and other components. *Chem. Biodivers.* **1,** 1111–1239 (2004).
8. Seebach, D. & Gardiner, J. Beta-peptidic peptidomimetics. *Acc. Chem. Res.* **41,** 1366–1375 (2008).
9. Simon, R. J. *et al.* Peptoids: a modular approach to drug discovery. *Proc. Natl Acad. Sci. USA* **89,** 9367–9371 (1992).
10. Butterfoss, G. L., Renfrew, P. D., Kuhlman, B., Kirshenbaum, K. & Bonneau, R. A preliminary survey of the peptoid folding landscape. *J. Am. Chem. Soc.* **131,** 16798–16807 (2009).
11. Hamper, B. C., Kolodziej, S. A., Scates, A. M., Smith, R. G. & Cortez, E. Solid phase synthesis of beta-peptoids: N-substituted beta-aminopropionic acid oligomers. *J. Org. Chem.* **63,** 708–718 (1998).
12. Baldauf, C., Guenther, R. & Hofmann, H.-J. Helices in peptoids of alpha- and beta-peptides. *Phys. Biol.* **3,** S1–S9 (2006).
13. Kirshenbaum, K. *et al.* Sequence-specific polypeptoids: a diverse family of heteropolymers with stable secondary structure. *Proc. Natl Acad. Sci. USA* **95,** 4303–4308 (1998).
14. Wu, C. W., Sanborn, T. J., Huang, K., Zuckermann, R. N. & Barron, A. E. Peptoid oligomers with alpha-chiral, aromatic side chains: sequence requirements for the formation of stable peptoid helices. *J. Am. Chem. Soc.* **123,** 6778–6784 (2001).
15. Wu, C. W., Sanborn, T. J., Zuckermann, R. N. & Barron, A. E. Peptoid oligomers with alpha-chiral, aromatic side chains: effects of chain length on secondary structure. *J. Am. Chem. Soc.* **123,** 2958–2963 (2001).
16. Tanatani, A. *et al.* Helical structures of N-alkylated poly(p-benzamide)s. *J. Am. Chem. Soc.* **127,** 8553–8561 (2005).
17. Norgren, A. S., Zhang, S. & Arvidsson, P. I. Synthesis and circular dichroism spectroscopic investigations of oligomeric beta-peptoids with alpha-chiral side chains. *Org. Lett.* **8,** 4533–4536 (2006).
18. Olsen, C. A., Lambert, M., Witt, M., Franzyk, H. & Jaroszewski, J. W. Solid-phase peptide synthesis and circular dichroism study of chiral beta-peptoid homooligomers. *Amino Acids* **34,** 465–471 (2008).
19. Roy, O., Faure, S., Thery, V., Didierjean, C. & Taillefumier, C. Cyclic beta-peptoids. *Org. Lett.* **10,** 921–924 (2008).
20. Shuey, S. W., Delaney, W. J., Shah, M. C. & Scialdone, M. A. Antimicrobial beta-peptoids by a block synthesis approach. *Bioorg. Med. Chem. Lett.* **16,** 1245–1248 (2006).
21. Olsen, C. A. *et al.* Alpha-peptide/beta-peptoid chimeras. *Org. Lett.* **9,** 1549–1552 (2007).
22. Olsen, C. A. *et al.* Antimicrobial, hemolytic, and cytotoxic activities of beta-peptoid-peptide hybrid oligomers: improved properties compared to natural AMPs. *ChemBioChem.* **11,** 1356–1360 (2010).
23. Jahnsen, R. D., Frimodt-Møller, N. & Franzyk, H. Antimicrobial activity of peptidomimetics against multidrug-resistant Escherichia coli: a comparative study of different backbones. *J. Med. Chem.* **55,** 7253–7261 (2012).
24. Andreev, K. *et al.* Guanidino groups greatly enhance the action of antimicrobial peptidomimetics against bacterial cytoplasmic membranes. *Biochim. Biophys. Acta Biomembr.* **1838,** 2492–2502 (2014).
25. Vedel, L. *et al.* Antiplasmodial and prehemolytic activities of alpha-peptide-beta-peptoid chimeras. *ChemBioChem.* **8,** 1781–1784 (2007).
26. Foged, C. *et al.* Cellular uptake and membrane-destabilizing properties of alpha -peptide/beta -peptoid chimeras: lessons for the design of new cell-penetrating peptides. *Biochim. Biophys. Acta Biomembr.* **1778,** 2487–2495 (2008).
27. Gorske, B. C., Bastian, B. L., Geske, G. D. & Blackwell, H. E. Local and tunable n->pi* interactions regulate amide isomerism in the peptoid backbone. *J. Am. Chem. Soc.* **129,** 8928–8929 (2007).
28. Gorske, B. C., Stringer, J. R., Bastian, B. L., Fowler, S. A. & Blackwell, H. E. New strategies for the design of folded peptoids revealed by a survey of noncovalent interactions in model systems. *J. Am. Chem. Soc.* **131,** 16555–16567 (2009).
29. Caumes, C., Roy, O., Faure, S. & Taillefumier, C. The click triazolium peptoid side chain: a strong cis-amide inducer enabling chemical diversity. *J. Am. Chem. Soc.* **134,** 9553–9556 (2012).
30. Laursen, J. S., Engel-Andreasen, J., Fristrup, P., Harris, P. & Olsen, C. A. Cis-trans amide bond rotamers in beta-peptoids and peptoids: evaluation of stereoelectronic effects in backbone and side chains. *J. Am. Chem. Soc.* **135,** 2835–2844 (2013).
31. Stringer, J. R., Crapster, J. A., Guzei, I. A. & Blackwell, H. E. Extraordinarily robust polyproline type I peptoid helices generated via the incorporation of alpha-chiral aromatic N-1-naphthylethyl side chains. *J. Am. Chem. Soc.* **133,** 15559–15567 (2011).
32. Armand, P. *et al.* NMR determination of the major solution conformation of a peptoid pentamer with chiral side chains. *Proc. Natl Acad. Sci. USA* **95,** 4309–4314 (1998).
33. Bonke, G. *et al.* Dimeric building blocks for solid-phase synthesis of alpha-peptide-beta-peptoid chimeras. *Synthesis* **15,** 2381–2390 (2008).
34. Crapster, J. A., Guzei, I. A. & Blackwell, H. E. A peptoid ribbon secondary structure. *Angew. Chem. Int. Ed.* **52,** 5079–5084 (2013).
35. Hjelmgaard, T. *et al.* Convenient solution-phase synthesis and conformational studies of novel linear and cyclic alpha,beta-alternating peptoids. *Org. Lett.* **11,** 4100–4103 (2009).
36. Abele, S., Vogtli, K. & Seebach, D. Oligomers of beta(2)- and of beta(3)-homoproline: What are the secondary structures of beta-peptides lacking H-bonds? *Helv. Chim. Acta* **82,** 1539–1558 (1999).
37. Gorske, B. C., Nelson, R. C., Bowden, Z. S., Kufe, T. A. & Childs, A. M. "Bridged" n->pi* interactions can stabilize peptoid helices. *J. Org. Chem.* **78,** 11172–11183 (2013).
38. Fuller, A. A., Yurash, B. A., Schaumann, E. N. & Seidl, F. J. Self-association of water-soluble peptoids comprising (S)-N-1-(naphthylethyl)glycine residues. *Org. Lett.* **15,** 5118–5121 (2013).
39. Tornøe, C. W., Christensen, C. & Meldal, M. Peptidotriazoles on solid phase: [1,2,3]-triazoles by regiospecific copper(I)-catalyzed 1,3-dipolar cycloadditions of terminal alkynes to azides. *J. Org. Chem.* **67,** 3057–3064 (2002).
40. Rostovtsev, V. V., Green, L. G., Fokin, V. V. & Sharpless, K. B. A stepwise huisgen cycloaddition process: copper(I)-catalyzed regioselective "ligation" of azides and terminal alkynes. *Angew. Chem. Int. Ed.* **41,** 2596–2599 (2002).

41. Glattli, A., Daura, X., Seebach, D. & van Gunsteren, W. F. Can one derive the conformational preference of a beta-peptide from its CD spectrum? *J. Am. Chem. Soc.* **124,** 12972–12978 (2002).
42. Wixe, T., Christensen, N. J., Lidin, S., Fristrup, P. & Warnmark, K. A short designed semi-aromatic organic nanotube--synthesis, chiroptical characterization, and host properties. *Org. Biomol. Chem.* **12,** 8930–8941 (2014).
43. Sheldrick, G. M. A short history of SHELX. *Acta Crystallogr. A* **64,** 112–122 (2008).
44. Berendsen, H. J. C., Postma, J. P. M., Vangunsteren, W. F., Dinola, A. & Haak, J. R. Molecular-dynamics with coupling to an external bath. *J. Chem. Phys.* **81,** 3684–3690 (1984).
45. Nose, S. A unified formulation of the constant temperature molecular-dynamics methods. *J. Chem. Phys.* **81,** 511–519 (1984).
46. Hoover, W. G. Canonical dynamics: equilibrium phase-space distributions. *Phys. Rev. A* **31,** 1695–1697 (1985).
47. Martyna, G. J., Tobias, D. J. & Klein, M. L. Constant-pressure molecular-dynamics algorithms. *J. Chem. Phys.* **101,** 4177–4189 (1994).
48. Tuckerman, M., Berne, B. J. & Martyna, G. J. Reversible multiple time scale molecular-dynamics. *J. Chem. Phys.* **97,** 1990–2001 (1992).
49. Essmann, U. *et al.* A smooth particle mesh Ewald method. *J. Chem. Phys.* **103,** 8577–8593 (1995).
50. Tiradorives, J. & Jorgensen, W. L. Molecular-dynamics of proteins with the Opls potential functions - simulation of the 3rd domain of silver pheasant ovomucoid in water. *J. Am. Chem. Soc.* **112,** 2773–2781 (1990).
51. Banks, J. L. *et al.* Integrated modeling program, applied chemical theory (IMPACT). *J. Comput. Chem.* **26,** 1752–1780 (2005).
52. Humphrey, W., Dalke, A. & Schulten, K. VMD: visual molecular dynamics. *J. Mol. Graph.* **14,** 33–38 (1996).

Acknowledgements

We thank Ms. Anne Hector and Dr Charlotte H. Gotfredsen for assistance with NMR spectroscopy, Ms. Tina Gustafsson for technical assistance with UPLC-MS and Ms. Lise Lotte Berring for assistance with X-ray data collection. We thank Dr S.W. Pedersen and Professor B.B. Kragelund (University of Copenhagen) for assistance with CD spectroscopy and Dr A.S. Madsen for help with the fluorescence scanning. This work was financially supported by DTU (PhD fellowship to J.S.L.), the Danish Independent Research Council|Natural Sciences (Steno Grant no. 10-080907, C.A.O.; Sapere Aude Grant no. 11-105487, P.F.), the Carlsberg Foundation (P.H.; Grant no. 2013_01_0333 to C.A.O.; and Grant no. 2013_01_0489 to P.F.) and the University of Copenhagen (Center for Biopharmaceuticals grant). C.A.O. is a Lundbeck Foundation Fellow.

Author contributions

J.S.L. and C.A.O. designed the research, analyzed the data and wrote the manuscript with input from all authors. J.S.L. performed syntheses, NMR spectroscopy, CD spectroscopy, compound crystallization and fluorescence measurements. P.H. solved the X-ray crystal structures. P.F. performed the molecular dynamics calculations.

Additional information

Accession codes: Coordinates and structure factors for the X-ray crystal structures of compounds **10a** (1054162) and compound **10c** (1054163), respectively, are deposited in the Cambridge Crystallographic Data Centre under the accession numbers mentioned above.

Competing financial interests: The authors declare no competing financial interests.

4

FR171456 is a specific inhibitor of mammalian NSDHL and yeast Erg26p

Stephen B. Helliwell[1], Shantanu Karkare[1], Marc Bergdoll[2], Alain Rahier[2], Juliet R. Leighton-Davis[1], Celine Fioretto[1], Thomas Aust[1], Ireos Filipuzzi[1], Mathias Frederiksen[1], John Gounarides[3], Dominic Hoepfner[1], Andreas Hofmann[1], Pierre-Eloi Imbert[1], Rolf Jeker[1], Richard Knochenmuss[1], Philipp Krastel[1], Anais Margerit[1], Klaus Memmert[1], Charlotte V. Miault[1], N. Rao Movva[1], Alban Muller[1], Hans-Ulrich Naegeli[1], Lukas Oberer[1], Vivian Prindle[4], Ralph Riedl[1], Sven Schuierer[1], Jessica A. Sexton[3], Jianshi Tao[4], Trixie Wagner[1], Hong Yin[3], Juan Zhang[1], Silvio Roggo[1], Stefan Reinker[1] & Christian N. Parker[1]

FR171456 is a natural product with cholesterol-lowering properties in animal models, but its molecular target is unknown, which hinders further drug development. Here we show that FR171456 specifically targets the sterol-4-alpha-carboxylate-3-dehydrogenase (*Saccharomyces cerevisiae*—Erg26p, *Homo sapiens*—NSDHL (NAD(P) dependent steroid dehydrogenase-like)), an essential enzyme in the ergosterol/cholesterol biosynthesis pathway. FR171456 significantly alters the levels of cholesterol pathway intermediates in human and yeast cells. Genome-wide yeast haploinsufficiency profiling experiments highlight the *erg26/ERG26* strain, and multiple mutations in *ERG26* confer resistance to FR171456 in growth and enzyme assays. Some of these *ERG26* mutations likely alter Erg26 binding to FR171456, based on a model of Erg26. Finally, we show that FR171456 inhibits an artificial Hepatitis C viral replicon, and has broad antifungal activity, suggesting potential additional utility as an anti-infective. The discovery of the target and binding site of FR171456 within the target will aid further development of this compound.

[1] Novartis Institutes for BioMedical Research, Novartis Campus, Basel, CH-4056, Switzerland. [2] Institut de Biologie Moléculaire des Plantes, CNRS, Unité Propre de Recherche 2357, Strasbourg cedex 67083, France. [3] Novartis Institutes for BioMedical Research, 250 Massachusetts Avenue, Cambridge, Massachusetts, 02139, USA. [4] GNF, San Diego, California 92121, USA. Correspondence and requests for materials should be addressed to S.H. (email: stephen.helliwell@novartis.com).

The cholesterol biosynthesis pathway, also known as the mevalonate pathway, is an essential cellular pathway that generates cholesterol—a major component of the plasma membrane. It represents a complex but highly regulated pathway, with the enzyme 3-hydroxy-3-methylglutaryl-coenzyme A reductase (HMGCR) being the most tightly regulated component controlling entry into the cholesterol pathway[1]. Cholesterol plays an important role in maintaining the plasma membrane integrity and dysregulation of this pathway has been shown to be a major cause of cardiovascular disease, thus contributing to mortality and morbidity worldwide[2].

To counter cholesterol pathway imbalances in human disease, many molecules have been developed that target sterol biosynthesis enzymes. Statins target HMGCR, bisphosphonates target farnesyl diphosphate synthase, zaragozic acid and quinuclidines (3-(biphenyl-4-yl)-3-hydroxyquinuclidine) target squalene synthase. Further down the pathway, allylamines target squalene epoxidase, azoles target lanosterol 14α-demethylase, morpholines target sterol C_8–C_7 isomerase/sterol reductase and azasterol targets sterol 24-C-methyltransferase[3–7]. Statins have been extremely successful in treating hypercholesterolemia but a significant clinical population experience side effects that prevent continuous or further use[8].

Most other sterol pathway inhibitors have proven to be unsuitable for widespread clinical application due to detrimental physiological side effects. A number of these agents target fungal-specific stages of the pathway and have found application as anti-fungals. However, their poor anti-fungal spectrum, and the development of resistance to these anti-fungal treatments limits their usefulness. There is thus a clinical need for inhibitors of other components of the cholesterol pathway.

Sterol-4-α-carboxylate 3-dehydrogenase, decarboxylating (NSDHL; often referred to as 3β-hydroxysteroid dehydrogenase/C4 decarboxylase [3βHSD/D]), is conserved amongst eukaryotes and lies in the cholesterol pathway. NSDHL is distal to lanosterol synthase, and catalyses NAD^+-dependent oxidative decarboxylation of 4α-carboxysterol intermediates involved in the C-4 demethylation process of sterol precursors to produce the corresponding 3-keto, C-4-decarboxylated products[9,10]. The NSDHL equivalent in *S. cerevisiae*—Erg26p, encoded by the essential gene *ERG26*, also acts in the lanosterol–distal section of the yeast ergosterol synthesis pathway[11]. 3βHSD/D from *Arabidopsis thaliana* has been characterized and its enzymology has been studied[12], and essential catalytic and binding residues have been identified[12,13]. According to a homology model and biochemical studies of the *A. thaliana* enzyme[13], Tyr159 and Lys163 are oriented near the 3β-hydroxyl group of the substrate and directly involved in the dehydrogenation process, while Arg326 forms an essential salt bridge with the 4α-carboxyl group of the substrate. The Asp39 residue is thought to contact the hydroxyl groups of the adenosine-ribose ring of NAD^+. These crucial residues are highly conserved across plant, fungal and mammalian enzymes.

This report describes the identification of the previously described natural product FR171456 (refs 14,15) using bioactivity guided fractionation directed by inhibition of a Hepatitis C viral (HCV) replicon assay[16]. We used a range of complementary methods in three organisms to demonstrate that FR171456 targets the Erg26p/NSDHL enzyme of the sterol biosynthesis pathway. This is the first compound known to inhibit this enzyme specifically, and therefore represents a useful tool for chemical biologists. Since FR171456 targets a previously untargeted node in the sterol pathway these results may spur the development of a novel class of compounds with utility in hypercholesterolemia or fungal infection[17]. This discovery is entirely consistent with the compound's effects on cholesterol in rats and rabbits[14,15].

Results

Metabolomics suggests that FR171456 inhibits NSDHL. Bioactivity guided fractionation identified FR171456 with an IC_{50} of 6.3 nM in a Huh-7 cell-based assay that measures HCV replicon activity[16]. Although potent in the replicon assay the compound did not affect the proliferation of the replicon-carrying Huh-7 cells at concentrations up to 4 μM under these assay conditions (Fig. 1a). The compound did not affect the proliferation of two other mammalian cell lines, HepG2 and K562, except at very high concentration (80 and 36 μM, respectively). In a screen to profile compound activity against 503 cancer cell lines only five cell lines were sensitive to FR171456 at an A_{max} < 50% and IC_{50} < 5 μM (Supplementary Fig. 1)[18].

FR171456 was previously identified as a compound that blocks cholesterol synthesis at or after the squalene synthesis step of the pathway[14]. Various stages in the lifecycle of HCV require a normally functioning cholesterol biosynthesis pathway[19], suggesting that FR171456's effect on cholesterol synthesis is the reason why it scored in the HCV replicon assay. In an attempt to understand which sterol (and other) metabolites are altered by FR171456 a metabolite-profiling experiment was conducted using cells from the original replicon assay exposed to vehicle or four concentrations of FR171456 (Fig. 1c,f, Methods, supplementary Fig. 2). Peaks in the mass spectra with statistically significant differences in intensities between the high (80 nM) and low (0 nM) concentration samples were selected by *P* values < 0.01 (Student's *t*-test, Bonferroni corrected) and ranked by fold changes. Metabolite identities were assigned to those peaks whose *m*/*z* ratio corresponds to the mass of known metabolites (Supplementary Table 1).

Analysis of the most significantly increased metabolite peaks in this set revealed one at m/z 465.3339 Da (Fig. 1c, Supplementary Table 1). This corresponds to the mass of the known substrate of NSDHL, 4-β-methylzymosterol-4-α-carboxylate[11,20]. The dose-dependent increase in intensities of this peak is consistent with inhibition of the NSDHL enzyme. A second peak at 467.3496 could represent 3-β-hydroxy 4β-methyl 5α-cholest-7-ene 4-α -carboxylate, that differs from 4-β-methylzymosterol-4-α-carboxylate by two hydrogens (Supplementary Table 1). This may represent an alternative reduced form of the substrate for NSDHL that would not normally be produced by these cells, but accumulates on blockage of the biosynthetic flux by FR171456 treatment. Reviewing this dataset for other known sterols revealed that calcidiol (*m*/*z* 401.3418), a cholesterol metabolite situated downstream of NSDHL, is decreased in a FR171456 dose-dependent manner, consistent with a reduction of cholesterol synthesis (Fig. 1f). No other sterol masses could be unequivocally assigned. Consistent with this data are the interpretation that FR171456 inhibits sterol synthesis between NSDHL and Calcidol synthase in Huh-7 cells.

Interestingly, three additional metabolites were identified as increasing in a dose-dependent manner. The peak at *m*/*z* 463.3183 may correspond to Pfaffic acid, but the other two masses, *m*/*z* 489.3314 and *m*/*z* 1198.8982 cannot be assigned. The accumulation of these peaks at doses that correspond well to the activity of the compound in the HCV replicon assay suggests that these metabolites are possibly linked to the cholesterol synthesis pathway, as this pathway has been implicated in replication of the HCV replicon[19].

FR171456 is antifungal and inhibits ergosterol synthesis. FR171456 demonstrated broad antifungal activity in a growth culture turbidity assay with standard minimum inhibitory concentration[21,22], in the range of 9–144 μM with *Aspergillus terreus* as the most susceptible species (Supplementary Table 2). To assess whether FR171456 had effects on fungal ergosterol synthesis similar to those in mammalian cells (ergosterol is the fungal/plant equivalent of cholesterol), ergosterol pathway intermediates were measured following labelling of *C. albicans* with ^{13}C-glucose, ^{13}C-acetate and treatment with increasing doses of FR171456 (0–200 μM). The well-characterized Erg11p (Lanosterol 14-α demethylase) inhibitor fluconazole was used as a control (Supplementary Fig. 3). Consistent with its known activity, fluconazole caused an accumulation in ^{13}C-labelled lanosterol and a decrease in both ^{13}C-labelled zymosterol and ergosterol concentrations. FR171456 caused a dose-dependent reduction in zymosterol and ergosterol production, and an increase in lanosterol (Fig. 1b,d,e). At the highest tested concentration of FR171456, the amount of ^{13}C-labelled ergosterol was increased compared with the vehicle control for reasons that are not clear. These data are consistent with the results from the mammalian cell metabolomics and previous work[14] and suggests that FR171456 inhibits a target downstream of lanosterol which may be NSDHL/Erg26p.

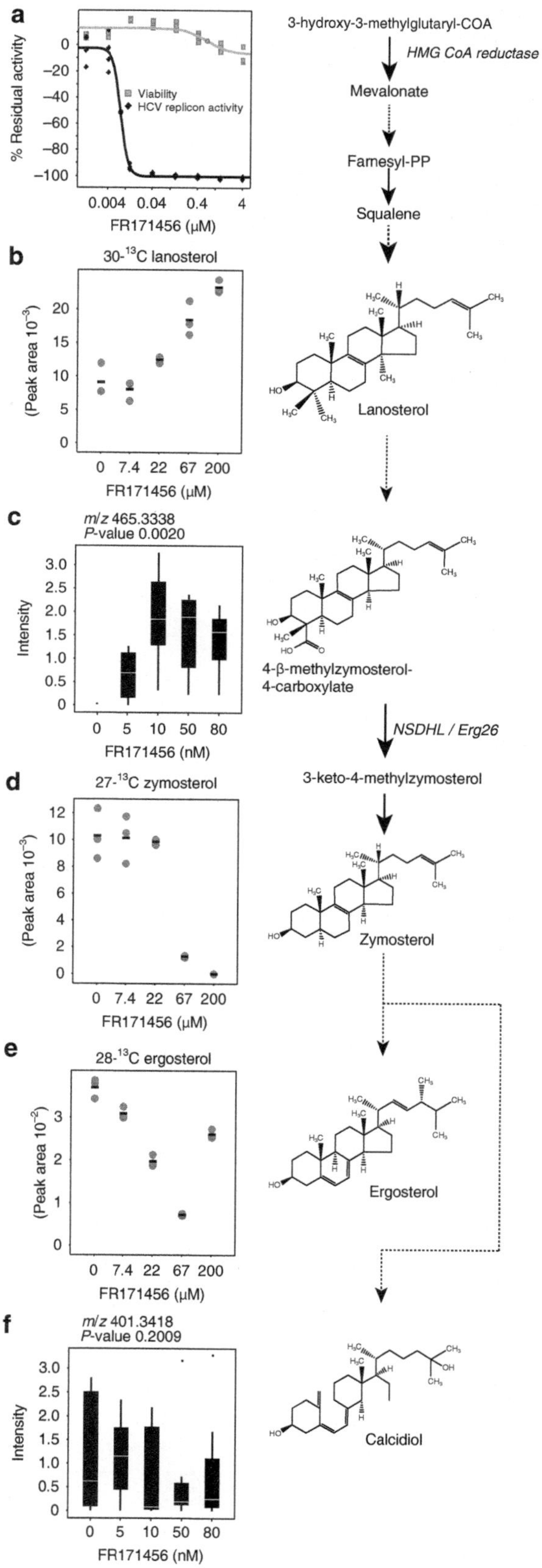

To further evaluate the potential use of FR171456 as an anti-fungal, we assessed FR171456 activity on *R. oryzae* and *F. solani* (clinical pathogens) in combination with two clinical antifungal agents, Posaconazole and Amphotericin B. No synergy was detected with posaconazole on either microbe, nor was it detected with Amphotericin B on *R. oryzae* (Loewe synergy score <1 in all three cases, data not shown). However, a small but significant effect was observed when treating *R. oryzae* with a combination of Amphotericin B and FR171456 (Supplementary Fig. 4), with a Loewe synergy score of 2.7 (where anything >1 is synergistic[23]). Interestingly, at several doses of FR171456 that alone are relatively ineffective (6.67 and 20 μM), the amount of Amphotericin B required for maximal inhibition is halved.

Genomic profiling identifies Erg26p as the FR171456 target. FR171456 inhibited growth of *S. cerevisiae* (BY4743 strain) with an IC_{50} of 14 μM, but neither FR171456 derivative (Compound 1, Compound 2, see Methods) demonstrated any significant growth inhibition at concentrations up to 200 μM (Fig. 2a,b). Haploinsufficiency (HIP)/homozygous (HOP) profiling[24] was performed using FR171456 at a concentration that would cause a 30% inhibition (IC_{30}) of growth in a standard overnight culture (Fig. 2c). In the HIP profile, the ERG26 heterozygous strain (*erg26Δ/ERG26*), shows the most significant hypersensitivity to FR171456. In addition, the SET6 heterozygote is also significantly hypersensitive. The SET6 strain often scores when profiling compounds that inhibit ergosterol synthesis in *S. cerevisiae*[25]. Erg26p functions in a complex with Erg25p and Erg27p (ref. 10), but strains heterozygous for *ERG25* and *ERG27* were indistinguishable from wild type in the HIP profile. This may suggest that Erg26p is the limiting factor in the enzyme complex, and also suggests high specificity of FR171456 for Erg26p. In addition, HIP and HOP was repeated using different concentrations of FR171456, revealing altered sensitivity of the ERG26 strain in a dose-dependent manner (Fig. 2d).

Figure 1 | FR171456 inhibits HCV replication and alters the levels of multiple sterol pathway intermediates. (**a**) The effect of increasing doses of FR171456 on Huh-7 human cell proliferation (grey symbols) and the HCV replicon (black symbols) were measured and plotted as a dose-response curve. (**b**–**f**) The sterol pathway is shown highlighting selected products/substrates of pathway enzymes that could be detected as significantly altered in either Huh-7 cells treated for 48 h with increasing doses of FR171456 (**c**,**f**; black candlestick - bars and sticks represent 25/75th percentile and 1.58 times the interquartile range, respectively, from 15 separate measurements), or *C. albicans* cells treated with increasing doses of FR171456 for 14 h (**b**,**d** and **e**; grey circles and black bar represent the values and mean respectively from three biological replicates.

Two homozygous diploid deletion strains demonstrated significant hypersensitivity to FR171456, YOR1 (*yor1Δ/yor1Δ*) and YPR090w (*ypr090wΔ/ypr090wΔ*). *YOR1* encodes a known multidrug resistance pump for which FR171456 is likely a substrate. *YPL090w* encodes an ORF of unknown function whose deletion confers sensitivity to miconazole, an ergosterol synthesis inhibitor, for reasons that are as yet unclear[26]. HMG-CoA reductase catalyses the conversion of HMG-CoA to mevalonate, the rate-limiting step in sterol biosynthesis. Strains heterozygous for either one of the two genes encoding the yeast HMGCR (*HMG1* and *HMG2* [27]), and therefore carrying three out of four copies of HMGCR show weak but significant resistance to FR171456 (Fig. 2c). Furthermore, the HMG1 HOP strain carrying two from four copies of the HMGCR is also resistant (Fig. 2c). In these situations the effects of FR171456 are presumably partially remediated by a reduction in flux through the pathway, perhaps because an increase in sterol intermediate(s) between HMGCR and Erg26 is toxic.

Assessing the relative fitness of the ERG26 HIP strain across HIP profiles produced using 1,800 structurally unique compounds[24] demonstrates that FR171456 is the only compound that has a significant effect on this strain (Fig. 2e). These results indicate that Erg26p is the major target of FR171456 in *S. cerevisiae*.

FR171456-resistant *S. cerevisiae* Erg26p mutants. The isolation of yeast mutants that confer resistance has been used to identify the target of novel compounds and their binding pockets[28,29]. In a semi-biased approach to identify critical sites on Erg26p that potentially interact with FR171456, we randomly mutagenized a haploid yeast strain lacking two major multidrug-resistance pumps and selected FR171456-resistant colonies on solid, rich, media containing FR171456 at a concentration that prevented wild-type colony formation (50 μM). For 42 resistant clones, we sequenced only the *ERG26* coding region—from these, 35 had single nucleotide alterations within *ERG26*. These mutations cause single-amino-acid changes in any one of only eight amino acids within the coding region of *ERG26* gene (Supplementary Fig. 5, Supplementary Table 3). The other 7 resistant colonies were wild type for *ERG26* and were not pursued further.

To confirm that the observed *ERG26* mutations are dominant and sufficient to confer FR171456 resistance, but retain the essential function of Erg26p, we assessed the growth of *erg26Δ* haploid yeast expressing wild-type *ERG26* or one of seven mutant alleles inserted at another locus in rich liquid media. All these mutations caused an increase in the IC_{50} compared with the isogenic control (Fig. 3a, Supplementary Fig. 6). It was not possible to generate a haploid strain carrying a single copy of *ERG26* encoding the Gly90Ser mutation after multiple attempts. Thus seven of the eight mutations within the coding region of *ERG26* are both dominant for FR171456 resistance and retain sufficient enzymatic function to support life. This additional genetic data set confirms that Erg26p is the target of FR171456.

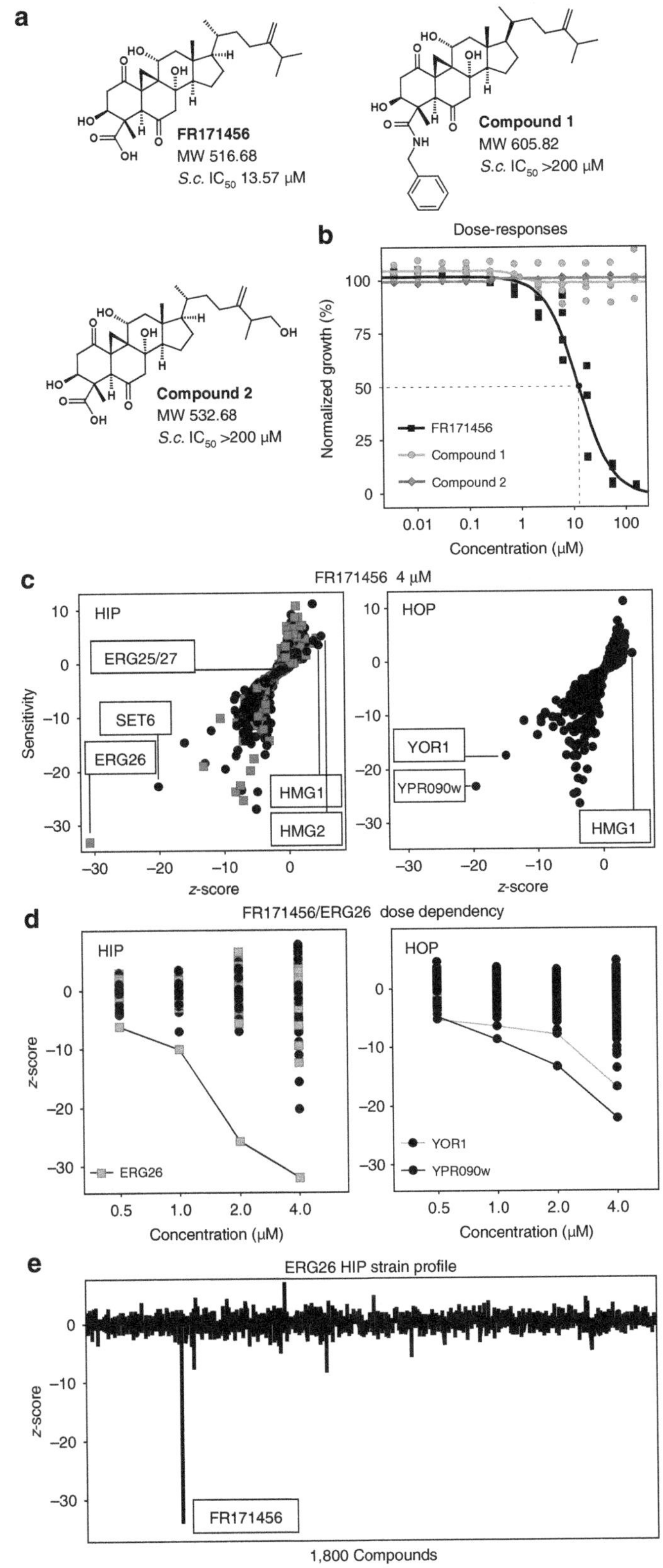

Figure 2 | FR171456 targets Erg26p in *S. cerevisiae*. (**a**) Structures of FR171456, and two related triterpenes (compound 1, compound 2). (**b**) Duplicate 12-point Dose response curves of wild-type diploid yeast with three compounds in **a**. Strain BY4743 was grown for 14 h at 30 °C in YPD media with 2% DMSO containing a dilution series of FR171456: black; Compound-1: light grey circles, and Compound-2: light grey squares. *Y* axis: OD_{600} absorbance indicating yeast growth, *x* axis: compound concentrations (log10 μM). (**c**) Haploinsufficiency and homozygous profiling (HIP resp. HOP) for FR171456 at 4 μM. Each mutant strain in the pool is represented by a circle (non-essential genes) or a square (essential genes). The, *y* axis represents strain sensitivity and *x* axis represents *z*-score (specificity). In (**d**), the z-scores of the HIP pool strains (*y* axis) plotted against increasing doses of FR171456 (*x* axis; μM). (**e**) Relative compound sensitivity profile of the *erg26Δ/ERG26* strain across HIP experiments performed with 1,800 diverse small molecule yeast proliferation inhibitors.

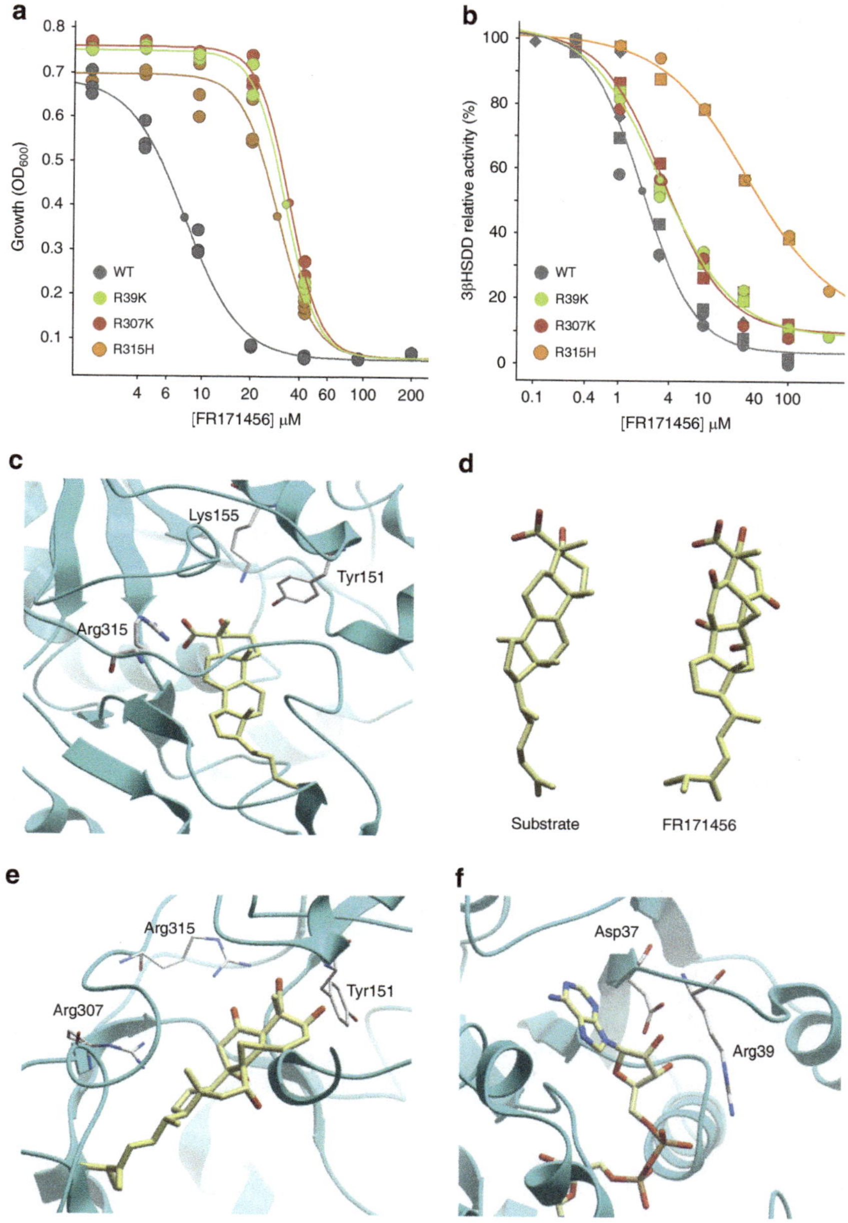

Figure 3 | FR171456 inhibits Erg26p *in vitro* (**a**) Dose-response curve for yeast haploid strains with single copy of ERG26 carrying single point mutations conferring resistance to FR171456 (see also Supplementary Fig. 6). *Y* axis: OD_{600} absorbance indicating growth, *x* axis: compound concentrations, dose response performed in triplicate. (**b**) FR171456 inhibits *S. cereviasiae* Erg26p enzyme activity as demonstrated by microsome-based biochemical assay, dose response performed in triplicate (wt) or duplicate (mutants). *Y* axis: relative reaction velocity compared with *A. thaliana* 3BHSDD, *x* axis: FR171456 concentration. (**c**) Homology modeling of *S. cerevisiae* Erg26p with close up of active site with substrate and key interacting amino acids Tyr151, Lys155 and Arg315. (**d**) 3D structure of Erg26p substrate and FR171456 (see also Supplementary Fig. 9). (**e**) similar to (**c**) but with FR171456 docked, with Tyr151, Arg307 and Arg315 highlighted. Arg307Lys and Arg315His are changes giving resistance to FR171456 (**f**) Homology modelling of *S. cerevisiae* Erg26p with close up of NAD binding site with Arg37 and Arg39 highlighted. Arg39Lys mutants are FR171456 resistant.

FR171456 inhibits Erg26p enzymatic activity. Metabolic and genetic evidence suggests that Erg26p is the target of FR171456, presumably by directly inhibiting its enzymatic function. To test this hypothesis an *in vitro* microsomal Erg26p activity assay[30] was used. Erg26p activity was directly assessed in the presence of increasing concentrations of FR171456 (10 nM–0.1 mM) in the presence of 150 μM carboxysterol substrate (Fig. 1c) and the cofactor NAD^+ (0.5 mM) followed by GC-MS detection of the product 4α-methyl-cholest-8,24-dien-3-one. Figure. 3b shows that FR171456 strongly inhibits Erg26p activity *in vitro*. These results allowed the concentration of FR171456 required to reduce the reaction velocity to half (I_{50}) to be determined: $I_{50} = 2.2 \pm 0.5$ μM.

K_m is an apparent dissociation constant for an enzyme–substrate complex and K*i* is the dissociation constant for an enzyme-inhibitor complex. The inhibition constant (K_i apparent) [31] value for Erg26p/FR171456 was 1.35 ± 0.3 μM after fitting to a simple Michaelis-Menten mechanism inhibition equation. The K_m value of the carboxysterol substrate for the wild-type yeast Erg26p has been previously measured to be 550 μM (ref. 32). The I_{50}/K_m ratio is used to compare the relative affinity towards the enzyme[31], so FR171456 is a strong *in vitro* inhibitor of Erg26p with an approximate I_{50}/K_m ratio of 4×10^{-3} for the carboxysterol substrate. Thus, this is the first reported *in vitro* enzyme activity that demonstrates that FR171456 has a high affinity for Erg26p and is a strong inhibitor of Erg26p.

Erg26p inhibition requires the 4α-carboxyl group of FR171456. Previous work with *A. thaliana* 3βHSD/D (NSDHL) indicates that the formation of an Arg326-substrate carboxylate salt bridge is required for productive binding of the substrate in the plant enzyme active site[13]. This is in accordance with the strict substrate specificity of the 3bHSD/D that requires a free C-4 carboxyl group[12]. The Erg26p equivalent of *A. thaliana* Arg326 is Arg315. Substitution of the 4α-carboxylic group of FR171456 by a carbobenzylamido group (Compound-1) decreased inhibition by > 300 fold (I_{50} > 0.5 mM) confirming the requirement for the free C-4-carboxylic group for Erg26p inhibition by FR171456. This data suggests that this specific interaction is conserved between the plant and yeast enzymes.

FR171456 docking into homology model of *S. cerevisiae* Erg26p. In the absence of a crystal structure for *S. cerevisiae* Erg26p, or its human and plant homologues, a homology model of the enzyme was built based on the crystal structure of *Pseudomonas aeruginosa* UDP-N-acetylglucosamine 4-epimerase complexed with UDP-N-acetylgalactosamine (PDB ID: 1SB8) (Supplementary Fig. 8a). The homology model also shows conserved structural domains with three other templates in Protein Data Bank (Supplementary Fig. 8b–d), but the 1SB8-derived model scored the best overall using multiple criteria (see Methods). To improve the accuracy of the model we determined the nuclear magnetic resonance (NMR) structure of the natural substrate following its purification from yeast inhibited with FR171456 (Methods, Fig. 3d, Supplementary Fig. 7).

The docked substrate and NAD^+ cofactor interactions with the *S. cerevisiae* Erg26p active site shown by the model were consistent with those previously determined in the case of the plant *A. thaliana* 3βHSD/D (ref. 13; Fig 3c,f). These include the conserved catalytic Tyr151xxxLys155 motif proximal to the 3β-hydroxyl group of the carboxysterol, the Arg315 residue proximal to the 4α-carboxylic group of the substrate, and the Asp37 residue bridging the two hydroxyl groups of the adenine ribose of NAD^+ in accordance with the NAD^+ versus $NADP^+$ cofactor specificity. Note that Erg26p Arg315 is equivalent to Arg326 in *A. thaliana*, and forms a similar interaction with the carboxylic group of the substrate.

The X-ray and NMR structures of FR171456 were solved, indicating significant similarity between the inhibitor and the substrate (Fig. 3d, Supplementary Fig. 9, Table 1). Taking this into account, and the significance of the inhibitor 4α-carboxylic group, FR171456 was docked in the carboxysterol binding pocket of *S. cerevisiae* Erg26p (Fig. 3e, Supplementary Fig. 10). The model suggests 3β-hydroxyl and 4α-carboxyl interactions with Tyr151 and Arg315, polar interactions between the 6-oxo group of FR171456 and Cys312, the 1-oxo group and Ser86, and hydrophobic interactions between Phe214 and the steroid nucleus of FR171456. Thus this model supports the hypothesis that FR171456 binds to Erg26p in the substrate-binding pocket.

Modelling FR171456 resistance mutations on Erg26p. The mutations that conferred resistance against FR171456, but remain functional as measured by yeast proliferation, were mapped onto the *S. cerevisiae* homology model. According to this model FR171456-resistant mutations are located as follows: Arg315His preserves the capacity of this catalytic residue to make a salt bridge with the substrate carboxyl group, preserving the catalytic

Table 1 | Crystal data and structure refinement for FR171456.

Empirical formula	$C_{31}H_{48}O_8$
Formula weight	548.69
Temperature	100(2) K
Wavelength	1.54178 Å
Crystal system	Monoclinic
Space group	P21
Unit cell dimensions	a = 7.447(2) Å b = 10.193(2) Å c = 19.229(3) Å
Volume	1,439.1(5) $Å^3$
Z	2
Density (calculated)	1.266 g cm^{-3}
Absorption coefficient	0.730 mm^{-1}
F(000)	596
Crystal size	0.30 × 0.18 × 0.09 mm^3
Theta range for data collection	2.33-68.20°
Index ranges	$-8 \leq h \leq 8$, $-11 \leq k \leq 10$, $-23 \leq l \leq 23$
Reflections collected	44,812
Independent reflections	5,008 [R(int) = 0.0347]
Completeness to theta = 68.20°	99.3%
Absorption correction	Semi-empirical from equivalents
Max. and min. transmission	0.9372 and 0.8108
Refinement method	Full-matrix least-squares on F^2
Data/restraints/parameters	5,008/13/399
Goodness-of-fit on F^2	1.048
Final R indices (I > 2sigma(I))	R1 = 0.0276, wR2 = 0.0704
R indices (all data)	R1 = 0.0279, wR2 = 0.0708
Absolute structure parameter	0.06(10)
Largest difference peak and hole	0.168 and −0.177 e.$Å^{-3}$

max., maximum; min., minimum.

activity of *S. cerevisiae* Erg26p, but presumably reducing the binding of FR171456. In the case of the human enzyme NSDHL, Arg315 is substituted by His, demonstrating that this functionally conservative mutation still allows productive binding of the carboxysterol substrate. A second conservative mutation also located in the active-site pocket, Arg307Lys, is also likely to retain a viable catalytic activity, but could perturb the FR171456 interactions with the active site (Fig. 3e).

Two proline mutations, Pro144Leu and Pro147Leu, are not located in the active site, although Pro147 is proximal to the catalytic Tyr151 residue. Pro147 is conserved in *P. aeruginosa* UDP-*N*-acetylglucosamine 4-epimerase complexed with UDP-*N*-acetylgalactosamine (1SB8). This could play a role in the global architecture of the active site, and possibly in differential accessibility of the substrate and inhibitor to their binding domain. The Arg39Lys mutation is located in the active-site pocket (Fig. 3f). This functionally conservative mutation is proximal to the 2-adenosine ribose hydroxyl groups and diphosphate group of NAD^+, and as such likely affects NAD^+ binding and indirectly the interactions of its nicotinamide amido group with FR171456, possibly reducing its affinity.

Thus many of the FR171456 resistance mutations were mapped close to the NAD^+ or the substrate-binding site. The presence of three active conservative mutations Arg39Lys, Arg307Lys, and Arg315His in proximity to NAD^+ and substrate-binding site are consistent with FR171456 binding the substrate-binding pocket thus inhibiting the enzymatic activity. To address this more directly, we tested microsome extracts from the *erg26Δ* haploid yeast strains expressing the Erg26p wild type, Arg39Lys, Arg307His and Arg315His mutants in increasing doses of FR171456 to assess Erg26p mutant activity (Fig. 3b). Whilst Arg39Lys and Arg307His shift the I_{50} 2-fold, Arg315His confers a

20-fold shift in I_{50}. This highlights the importance of the interaction between Arg315 and the carboxyl group of the substrate/inhibitor as one of the most critical in the Erg26p active site (Fig. 3c).

Discussion

One of the central challenges facing chemical biologists is to 'identify a small-molecule inhibitor for each individual function of all human proteins'[33]. In this study, FR171456 was reidentified due to its inhibitory activity in a phenotypic assay for HCV replicon, purified and its structure solved. Results from metabolomics, biochemistry, yeast genetics, and protein/inhibitor/substrate structure modelling, are all consistent with FR171456 specifically inhibiting yeast Erg26, and the mammalian homologue NSDHL, an enzyme distal to squalene synthase in the sterol biosynthesis pathway. This is the first identification of an inhibitor for Erg26p/NSDHL, and contributes to the collection of compounds targeting specific points/ nodes in the ergosterol/cholesterol biosynthesis pathway. The availability of a small molecule inhibitor of this node for the first time allows the inhibition of a node of sterol synthesis for which null mutations are unviable in yeast or mammalian systems. Such a tool compound will be useful to dissect the sterol pathway post-squalene in detail and suggests that NSDHL could be a drugable node[34].

In addition to representing a specific and potent molecular probe for NSDHL function, this molecule demonstrates several lead-like properties, most significantly showing *in vivo* efficacy on the target pathway. Hatori *et al* treated rats at up to 10 mg kg^{-1} in short term experiments and showed a significant reduction in hepatic sterol synthesis even at doses as low as 0.1 mg kg^{-1}, demonstrating a potent effect on the cholesterol pathway. Most importantly, these authors treated rabbits for 3 weeks, demonstrating a significant reduction in serum cholesterol already at the 0.001 mg kg^{-1} dose, with no adverse effects reported for this experiment even at doses up to 1 mg kg^{-1} (ref. 14). In addition, pharmacological profiling against a panel of 24 receptors/enzymes identified only one target with significant binding, the delta-opioid receptor, where an IC_{50} of 100 nM was observed in a radioligand competition assay. Thus FR171456 has reasonable chemical properties, efficacy and specificity *in vitro*, and efficacy *in vivo* with no adverse events reported in two rodent models, all key properties of a lead molecule.

It has been demonstrated previously that host lipid metabolism is important for the lifecycle of HCV for several reasons: (1) the low-density lipid (LDL) receptor is important for HCV entry; (2) the mevalonate pathway is required for HCV RNA replication; and (3) very low-density lipid (VLDL) is essential for co-secretion with virus particles[35]. The isolation of FR171456 in the HCV replicon assay provides further evidence that viral replication is modulated by host sterol and lipid metabolism[19,36–39]. Thus, FR17145 represents a potential starting point for novel anti HCV drugs, and NSDHL in turn a novel target.

The compound also exhibited reasonable antifungal activity against a wide spectrum of fungal species such as *S. cerevisiae, C. albicans, A. fumigatus* with minimum inhibitory concentration[21,22] in a range from 4 to 64 μg ml^{-1} (9–144 μM). Although these values do not make FR171456 a particularly potent small molecule, mammalian cell proliferation is unaffected by FR171456 in almost 500 cell lines tested *in vitro*[18]. In addition, in the original report describing FR171456, rabbits were treated with efficacious doses for 14 days with no reported side effects[14]. Taken together these data raise the hope that there may be a reasonable therapeutic window *in vivo* for compounds that inhibit this node in the sterol synthesis pathway.

Mutations in the NSDHL gene cause the human syndrome CHILD (congenital hemidysplasia with ichthyosiform nevus and limb defects), a dominant X-linked, male lethal disorder[40]. In recessive heterozygote males, the syndrome causes unilateral ichthyosiform skin lesions and limb reduction defects—known as CK syndrome[41]. For these pathologies, it has been suggested that not only cholesterol deficiency, but also accumulation of methyl sterols and carboxysterols causes disease[42]. Using FR171456 to cause the accumulation of upstream intermediates in relevant model systems should help to clarify this hypothesis. Indeed, the Huh7 cell metabolomics performed here demonstrated the accumulation of known and novel molecules in a dose-dependent manner, and the elucidation of the novel molecule structures may suggest mechanisms whereby NSDHL inhibition causes toxic intermediate accumulation.

Methods

Purification of FR171456. A strain of *Monodictys* Sp. (from the Novartis strain collection) was cultured in medium MP3-01.00 (20 g l^{-1} defatted soya, 20 g l^{-1} D-Mannitol, 1 ml active solution Nr 1901, which contained 4 g l^{-1} $ZnSO_4.7H_2O$), 0.1 g l^{-1} H_3BO_3, 5 g l^{-1} $FeSO_4.7H_2O$, 0.005 g l^{-1} KI),1 g l^{-1} $CoCl_2$, 0.2 g l^{-1} $CuSO_4$. $5H_2O$, 2 g l^{-1} $MnCl_2.4H_2O$, 1 g l^{-1} $(NH_4)_6(Mo_7O_2)_4.4H_2O$. The culture was started with a 2.5% inoculum of a preculture which had been grown for four days from a cryovial of the strain. A 2-l culture was then cultured for 6 days at 24 °C at 200 r.p.m. The broth was filtrated through Celite. The filter cake containing the mycelium was extracted with ethyl acetate resulting in 0.5 g of extract. The complete ethyl acetate extract was separated on a reversed phase column (50 × 200 mm column packed with Merck Lichrospher RP18, 12-μm particle size). The elution gradient was from 30% acetonitrile in water rising to 100% within 60 min, the mobile phase contained 0.01% trifluoroacetic acid (TFA). Combined fractions resulted in 9.9 mg of purified FR171476.

The structure of FR171476 was determined by infrared spectroscopy, mass spectroscopy (MS) and NMR spectroscopy. The structure has been previously reported[14]. FR171456 can be described chemically as (1R,2aR,3R,5aR,5bS,7aR,8S,9S,11aS,12aR)-1,5b,9-trihydroxy-2a,8-dimethyl-3-((R)-6-methyl-5-methyleneheptan-2-yl)-7,11-dioxohexadecahydrocyclopenta[a]cyclopropa[e]-phenanthrene-8-carboxylic acid.

A sample was analysed by 1H NMR (600 MHz, dimethyl sulfoxide (DMSO)-d6) δ p.p.m. 0.68 (s, 3 H) 0.89 (d, *J* = 6.59 Hz, 3 H) 0.98–1.00 (m, 6 H) 1.00 (s, 3 H) 1.07–1.16 (m, 2 H) 1.17–1.35 (m, 4 H) 1.37 (d, *J* = 5.85 Hz, 1 H) 1.44–1.55 (m, 1 H) 1.64 (q, *J* = 9.03 Hz, 1 H) 1.78–1.91 (m, 3 H) 1.93–1.99 (m, 1 H) 2.02–2.11 (m, 2 H) 2.14–2.25 (m, 2 H) 2.32 (dd, *J* = 16.83, 8.78 Hz, 1 H) 2.51 (dd, *J* = 12.44, 7.32 Hz, 1 H) 2.83 (dd, *J* = 16.83, 7.32 Hz, 1 H) 4.15 (*t*, *J* = 8.05 Hz, 1 H) 4.21 (s, 1 H) 4.38 (br. s., 1 H) 4.51 (d, *J* = 6.59 Hz, 1 H) 4.64 (s, 1 H) 4.70 (s, 1 H) 4.97 (br. s., 1 H) 5.23 (br. s., 1 H).

The absolute configuration of FR171456 has been identified by high resolution X-ray structure determination. The structure of the compound is shown in Fig. 3d and Supplementary Fig. 9. X-ray data for FR171456 is available in Table 1. This is the first report of the absolute 3D structure of this compound.

Synthesis of compound 1. Compound 1[(1R,2aR,3R,5aR,5bS,7aR,8S,9S,11aS,12aR)-N-benzyl-1,5b,9-trihydroxy-2a,8-dimethyl-3-((R)-6-methyl-5-methyleneheptan-2-yl)-7,11-dioxohexadecahydrocyclopenta[a]cyclopropa[e]phenanthrene-8-carboxamide] was prepared as follows. To a solution of FR171456 (30 mg, 0.058 mmol), benzylamine (4.45 μl, 0.04 mmol), 1-hydroxy-7-azabenzotriazole (HOAt; 120 μl, 0.06 mmol) in 2.5 ml DMF, EDC.HCl (12.6 mg, 0.064 mmol) was added at room temperature. The reaction mixture was stirred for 2.5 h at room temperature. Since only partial conversion was observed, additional benzylamine (4.45 μl, 0.04 mmol), HOAt (120 μl, 0.06 mmol) and EDC.HCl (12.6 mg, 0.064 mmol) were added to the reaction. After 16 h at room temperature the reaction mixture was diluted in 20 ml of ethyl acetate and 5 ml of water. The organic phase was extracted with saturated $NaHCO_3$ (2 × 5 ml) and brine (2 × 5 ml), dried over Na_2SO_4, and subsequently evaporated under reduced pressure. The crude product was separated by RP-HPLC on Luna C18 stationary phase and a gradient of the eluents (water/0.1% formic acid and acetonitrile/0.1% formic acid). Fractions containing compound-1 were combined, dried under reduced pressure and subsequently lyophilized to yield 9.4 mg (26.7%) of compound-1. Agilent 1100 LCMS analysis, MS + *m/z* = 606.31. NMR data were recorded at 26 °C on a Bruker AV-I-600 or AV-III-600 spectrometer, using a 1.7 mm TXI cryoprobe. ^1H NMR (600 MHz, DMSO-d_6): 1H NMR (600 MHz, DMSO-d6) δ p.p.m. 0.69 (s, 3 H) 0.90 (d, *J* = 5.85 Hz, 3 H) 1.00 (dd, *J* = 6.50 Hz, 6 H) 1.13 (d, *J* = 5.85 Hz, 2 H) 1.15 (s, 3 H) 1.17–1.36 (m, 4 H) 1.39 (d, *J* = 5.85 Hz, 1 H) 1.45–1.57 (m, 1 H) 1.64 (q, *J* = 9.51 Hz, 1 H) 1.84 (d, *J* = 14.60 Hz, 1 H) 1.86–1.92 (m, 2 H) 1.97 (dd, *J* = 14.64, 6.59 Hz, 1 H) 2.02 (d, *J* = 19.03 Hz, 1 H) 2.05–2.11 (m, 1 H) 2.16 (d, *J* = 18.30 Hz, 1 H) 2.18–2.26 (m, 1 H) 2.35 (dd, *J* = 16.47, 8.42 Hz, 1 H) 2.45–2.55 (m, 1 H) 2.82 (dd, *J* = 16.83, 6.59 Hz, 1 H) 4.11 (t, *J* = 7.68 Hz, 1 H) 4.28 (dd, *J* = 15.37, 5.85 Hz, 1 H) 4.41 (s, 2 H) 4.44 (dd, *J* = 15.40, 5.90 Hz, 1 H)

4.47 (d, $J=6.59$ Hz, 1 H) 4.66 (s, 1 H) 4.72 (s, 1 H) 5.01 (br. s., 1 H) 5.20 (br. s., 1 H) 7.18–7.22 (m, 1 H) 7.28 (t, $J=7.68$ Hz, 2 H) 7.40 (d, $J=7.32$ Hz, 2 H) 7.87 (t, $J=5.85$ Hz, 1 H)

Synthesis of compound 2. Compound 2[(1R,2aR,3R,5aR,5bS,7aR,8S,9S,11a-S,12aR)-1,5b,9-trihydroxy-3-((2R)-7-hydroxy-6-methyl-5-methyleneheptan-2-yl)-2a,8-dimethyl-7,11-dioxohexadecahydrocyclopenta[a]cyclopropa[e]phenanthrene-8-carboxylic acid] was prepared as follows. The biotransformation of FR171456 with the strain Nocardioides simplex ATCC 6946 was performed in 0.5-l flasks each containing 95 ml liquid medium of 4 g/l D-(+)-glucose, 10 g l^{-1} malt extract, 4 g l^{-1} yeast extract and distilled water (pH 7,5). Each flask of the main culture was inoculated with 5 ml of the preculture and incubated for 2 days at 28 °C, 220 r.p.m. Solutions of FR1714556 (2.2 mg, mmol) in 4.4 ml methanol and 10 ml of 1 M KH_2PO_4 were added to each of the 11 flasks. After incubation for three days at 28 °C/180 r.p.m. the fermentations were stopped by adding 100 ml isopropanol to each culture. The solvents were evaporated under reduced pressure and the remaining broth was lyophilized. The dry powder was suspended in water and extracted with ethyl acetate. The organic phase was dried over Na_2SO_4 and evaporated under reduced pressure, yielding 405 mg of crude product. It contained the desired product based on HPLC analysis and no starting material was detected. The crude product was separated by preparative reverse phase chromatography using an Ascentis phenyl stationary phase and a gradient of the eluents (water/0.1% formic acid and acetonitrile/1% water/0.1% formic acid). Fractions containing compound-2 were combined, dried under reduced pressure and subsequently lyophilized. The sample was separated again using the same method yielding 5 mg of compound-2. Agilent 1100 LCMS analysis, MS+ m/z = 479.3, fragmentation of parent ion observed. 1H NMR (600 MHz, DMSO-d6) δ p.p.m. 0.66 (s, 3 H) 0.87 (d, $J=6.24$ Hz, 3 H) 0.96 (d, $J=6.60$ Hz, 3 H) 0.98 (s, 3 H) 1.04–1.12 (m, 1 H) 1.15 (d, $J=5.87$ Hz, 1 H) 1.16–1.27 (m, 3 H) 1.28–1.33 (m, 1 H) 1.35 (d, $J=5.14$ Hz, 1 H) 1.44–1.54 (m, 1 H) 1.57–1.65 (m, 1 H) 1.76–1.88 (m, 3 H) 1.94 (dd, $J=14.67$, 6.60 Hz, 1 H) 2.00–2.08 (m, 2 H) 2.13 (dd, $J=12.84$, 6.97 Hz, 1 H) 2.18 (dd, $J=19.07$, 2.57 Hz, 1 H) 2.30 (dd, $J=17.24$, 8.80 Hz, 1 H) 2.44–2.53 (m, 1 H) 2.80 (dd, $J=17.24$, 7.34 Hz, 1 H) 3.18 (ddd, $J=10.27$, 7.70, 5.50 Hz, 1 H) 3.41 (dt, $J=10.36$, 5.27 Hz, 1 H) 4.12 (t, $J=7.89$ Hz, 1 H) 4.19 (s, 1 H) 4.36 (d, $J=2.57$ Hz, 1 H) 4.43 (t, $J=5.32$ Hz, 1 H) 4.48 (dd, $J=6.24$, 2.93 Hz, 1 H) 4.67 (s, 1 H) 4.69 (s, 1 H) 4.97 (d, $J=2.93$ Hz, 1 H) 5.19 (br. s, 1 H) 11.89 (br. s, 1 H). NMR data were recorded at 26 °C on a Bruker AV-I-600 or AV-III-600 spectrometer, using a 1.7 mm TXI cryoprobe.

Isolation of compound 3 by Erg26p substrate purification. In the fermentation experiment, a single yeast colony was grown in 10 ml YP with galactose (2%) & Raffinose (1%) for 5 h (OD 1.0) at 30 °C. An inoculum (70 µl) from the same culture was used to inoculate 400 ml YP-gal (2%) + Raffinose (1%). This culture was then grown overnight at 30 °C. The OD at this point was 0.7. This culture was used to seed 20-l fermenter maintained at 30 °C. When OD in the fermenter was 0.7, then 1466 nM (IC_{80}) of FR171456 was added to the fermenter. The cell pellets (3.8 g) were harvested after 7 h. Erg26p substrate (1.95 mg) was obtained by lipid extraction with ethylacetate and methanol.

The extraction protocol included suspension of collected yeast cells in 250 ml methanol: water mixture (9:1). The extract was defatted by extraction with 250 ml cyclohexane yielding 2.8 g defatted extract. The defatted extract was then subjected to chromatograph with a Sephadex LH20 (5 × 100 cm) column using methanol as the solvent. The fraction (205 mg) containing the desired compound was then further purified using reversed phase HPLC with a 30 × 250 mm Sunfire RP18 column (5 µm). The column was developed using the following solvents (A: 0.03% formic acid, B: acetonitrile containing 0.03% formic acid). The column was first washed with a solution of 70% B for 5 min followed by a gradient from 70% B to 100% B in 25 min yielding in 1.1 mg desired substrate.

1H-NMR was used to determine the structure of the substrate. The parameters for the instrument were: 660.13 MHz frequency, PROBHD: 1.7 MM CPTCI1, DMSO solvent, PULPPROG ZG30, TD 65536. The NMR data for the compound:^{1}H NMR (600 MHz, DMSO-d_6) δ p.p.m. 0.55 (s, 3 H) 0.91 (d, $J=6.60$ Hz, 3 H) 0.92 (s, 3 H) 0.95 (s, 3 H) 0.97–1.03 (m, 1 H) 1.09–1.16 (m, 1 H) 1.21–1.25 (m, 3 H) 1.27–1.31 (m, 3 H) 1.31–1.39 (m, 2 H) 1.43–1.54 (m, 4 H) 1.55 (s, 3 H) 1.63 (s, 3 H) 1.63–1.67 (m, 1 H) 1.73 (dd, $J=12.62$, 0.91 Hz, 1 H) 1.78–1.87 (m, 3 H) 1.88–2.06 (m, 5 H) 3.70 (dd, ν10.98, 3.66 Hz, 1 H) 5.06 (t, $J=6.59$ Hz, 1 H).

HCV replicon inhibitor screening. The inhibitor screening was performed in 1,536-well plate with 3,000 cells per well, using Huh-7 cells containing the HCV replicon subgenomic genotype 1b, maintained in DMEM, supplemented with (1 ×) L-glutamine (L-Glue), 1 × non-essential amino acids, 10% heat-inactivated fetal bovine serum (FBS), and 1 mg ml^{-1} G418, and incubated with compounds for 48 h at 37 °C, 5% CO_2. Compounds or microbial extracts were diluted in 90% DMSO and then diluted in culture media so that the addition of 10 nl of dilutant to the cells resulted in a final DMSO concentration of 0.18%. Following compound incubation, Resazurin solution (10 ×: 1 µg $µl^{-1}$ Resazurin in DMSO) diluted to 1 × was then added to a volume of 0.5 µl to each well and incubated for further 3 h at 37 °C, 5% CO_2 to monitor cell viability. Subsequently 15 µl, diluted 1:2 from stock of Steady- Glo (Promega) was added and the luminescence from the HCV replicon was monitored using the Envision (Envision US Lum, using 0.1 s per well measurement time).

Yeast strains and plasmids. BY4741 a *pdr5::kanMX snq2::kanMX* was constructed using standard techniques using strains obtained from Invitrogen (Cat no: 95400.BY4741); the heterozygous and homozygous deletion collections were also obtained from Invitrogen. pRS416 is a CEN URA3 shuttle vector. All cloning was performed using standard molecular biology techniques[43].

For ^{13}C- labelling studies, the wild type *C. albicans* strain SC5314 was obtained from ATCC (Manassas, VA) and was cultured at 30 °C in YPD broth (1% yeast extract, 2% peptone, 2% dextrose) or Synthetic Medium (3% glycerol, 0.1% glucose, 65 mM NH_4Cl, 1.7 mM NaCl, 7 mM KH_2PO_4, 1 mM $CaCl_2$, 0.5 mM $MgSO_4$).

Mutagenesis. For EMS mutagenesis experiments, 1×10^8 cells were plated on one standard YPD agar plate containing 50 µM FR171456 in 2% DMSO v/v following a standard EMS mutagenesis protocol. After 2 days incubation at 30 °C 48 representative colonies were picked and retested in an overnight liquid assay for growth in YPD with or without 50 µM FR171456 in 2% DMSO v/v. Genomic DNA was prepared from 42 resistant clones (+ one wild type) and for each the entire *ERG26* gene was sequenced following PCR amplification with high-fidelity polymerase. These mutations were then cloned with their native promoter into BYintURA and then transformed into BY4743 *erg26::kanMX/ERG26* diploid yeast strain and processed to create haploid strains carrying the BYIntURA3 vector with wild type or mutant ERG26 in the *erg26::KanMX* background. The FR171456 dose – response assay for the mutants was performed in a 96-well plate format as described [24].

Mass spectrometric analysis for global metabolomics profiling. Samples of cell pellets containing 8×10^6 cells each were lysed and extracted with 300 µl cold methanol (− 80 °C) in ultrasonic bath for 10 min and subsequently agitated for 5 min before centrifugation at 10,000 *g* for 5 min at 4 °C. The supernatants were removed and the remaining pellets were again extracted twice. The supernatants from three extractions were then combined and dried using a SpeedVac (Savant, Fisher Scientific Inc., USA) for 2 h without heating. The residues were resuspended in 200 µl of 80% MeOH and centrifuged to remove any non-dissolvable particles. Finally, 10 µl of the supernatant was then diluted by factor 100 in MeOH/H_2O (1:1) for mass spectrometric analysis.

The diluted sample (100 µl) was infused directly into a LTQ-Orbitrap mass spectrometer (Thermo Fisher Scientific, USA) using a HTC PAL autosampler (CTC Analytics, Zwingen, Switzerland) at a flow rate of 18 µl min^{-1}. A make-up flow, consisting of 70% ethanol and 0.1% formic acid, was delivered through a T-piece at 15 µl min^{-1}. Electrospray ionization and selected monitoring scan were applied. The total mass range *m/z* 100–2,000 was segmented into 48 scan events with a width of mass window *m/z* 50 each. Mass resolution at *m/z* 400 was 100,000.

Each sample was measured in triplicate in positive ionization mode in random order. For quality and instrument stability control, the samples were interspersed with test mix infusion and pooled samples. Blank spectra were also recorded and the observed mass peaks of solvent and impurities were then suppressed in subsequent analysis.

The measured raw spectra were imported with in-house software written in the Igor package (Wavemetrics Inc.) for alignment, peak detection, and integral normalization. The table of exact peak masses and intensities (Supplementary Data, file Supplementary Dataset 1) were exported into the statistical software R, version 2.15.1, for subsequent univariate statistics and pathway analysis. For peak identification, the KEGG database was searched for matches of the observed peaks with endogenous metabolite entries. Metabolite masses with adducts of H^+, Na^+, or K^+ were considered a hit and possible identity of a peak if the difference between its exact monoisotopic mass and the observed mass plus adducts was below 3 p.p.m.

Focused metabolomics in *C. albicans*. Approximately 1.25×10^8 *C. albicans* cells from a saturated YPD culture were inoculated into 100 ml Synthetic Medium (SM) and grown at 30 °C to an OD of 1.0. Cells were washed once in modified SM (25 mM KPO_4 pH 7.0, with glucose omitted) and suspended to 5 OD_{600} units per ml in modified SM.

Labelling reactions were performed in 12-ml glass tubes with PTFE-lined caps (Corning) and containing 1 ml cell suspension ($\sim 6 \times 10^7$ cells), 10 µl DMSO or 10 µl compound dissolved in DMSO, 12 µl 1M D-Glucose (U-^{13}C6 glucose), and 10 µl 1 M (1,2-^{13}C2) sodium acetate (Cambridge Isotope Laboratories, Inc). Before LC-MS analysis, dried lipid extracts were dissolved in 500 µl methanol, vortexed and sonicated for 10 min.

Squalene, ergosterol, lanosterol and zymosterol were quantified using the tandem mass spectrometer API4000 triple quadrupole (AB Sciex) with atmospheric pressure chemical ionization (APCI), interfaced to an Agilent LC.

Labelling reaction and lipid quantification. Cells ($\sim 6 \times 10^7$) and compounds were preincubated 15 min at 30 °C (without shaking) before the addition of labelled substrates. Labelling was performed at 30 °C with shaking for 4 h. Reactions were stopped by adding 1 ml 30% KOH, 1 ml 100% ethanol, and 0.2 ml 2% pyrogallol. Samples were mixed and then heated at 80 °C for 30 min, then cooled to room temperature. Non-saponifiable lipids were extracted two times in petroleum ether, washed with water, dried under nitrogen gas and stored at − 80 °C.

For compound calibration, 2 μM solutions in 0.1% formic acid were directly infused into the mass spectrometer. Ergosterol, lanosterol and zymosterol undergo dehydration and their molecular species are detected as $[M+H\text{-}H_2O]^+$ while squalene is detected as $[M+H]^+$. Supplementary Table 4 lists the multiple reaction monitoring parameters of the sterols and squalene, as well as the ^{13}C incorporated metabolites. The mass spectrometer source parameters are the following: CUR = 40, GS1 = 60, NC = 4.5. TEM = 450 CAD = 4. A mobile phase of 0.2% formic acid in methanol and isocratic run at 0.4 ml min^{-1} was employed to separate the metabolites on a Waters XBridge C18 3.5 μm 2.1 × 50 mm column. The run time was 4.5 min.

Metabolic Profiling for substrate accumulation in *S. cerevisiae*. Lipid extracts from yeast cells were prepared as described as above. Ergosterol and 4α-methyl-zymosterol-4-carboxylic acid were analysed on an UltiMate 3000 RS LC (Thermo Fisher Scientific, Waltham, MA, USA) coupled to a 4000 QTRAP (AB Sciex, Foster City, CA, USA), which was equipped with a ESI-Turbo Spray ion source (AB Sciex).

Chromatography was carried out on a Waters Acquity UPLC BEH C18, 1.7 μm, 1 × 50 mm column (Waters, Milford, MA, USA) with a flow rate of 200 μl min^{-1} and a gradient from 30% A (H_2O with 0.1% (v/v) formic acid) and 70% B (acetonitrile with 0.1% (v/v) formic acid) to 2% A and 98% B in 3.5 min. Solvent composition remained for 4 min at 2% A and 98% B before the column was reconditioned to 30% A and 70% B. The total runtime for one analysis was 12 min. To reduce ion source contaminations, a switch valve was used to direct the flow to waste for the first 3.5 min and the last 5.5 min of the LC method.

Ion source parameters and compound-dependent instrument parameters were optimized by infusing pure standard solutions of ergosterol and 4α-methylzymosterol-4-carboxylic acid, respectively. Ergosterol was detected in positive ion mode by selected reaction monitoring (SRM) using the mass transitions 379.3 $[M\text{-}H_2O+H]^+$ → 69.2 and 379.3 → 83.1 and 4α-methylzymosterol-4-carboxylic acid was detected using the mass transitions 425.3 $[M\text{-}H_2O+H]^+$ → 95.1, 425.3 → 109.1 and 425.3 → 159.1. All mass transitions were monitored for 65 msec per SRM scan with a unit/unit resolution for Q1/Q3. Analyst 1.5.1 (AB Sciex) was used for data analysis and peaks were smoothed with the Savitzky-Golay algorithm with a smoothing half width of 3 points.

HIP-HOP assay. The HIP assay was performed in 24-well plates (Greiner 662102), with 1.6 ml per well YPD. Experimental compounds were tested at $n = 2$ within the same plate at or close to their IC_{30} concentration. Each plate contained two no drug controls, one positive control (Benomyl), 10 experimental compounds in duplicates and one contamination control that received no cells. A standard experiment was 4 plates per 40 experimental compounds processed robotically without human intervention. YPD per compound filled wells were inoculated with ~ 250 yeast cells per strain (100 μl of a 1.5-OD_{600} per ml culture) from an overnight log phase preculture to start the experiment (Supplementary Fig. 2). The plates were pipetted with a standard 96 pipettor head by providing tip boxes preconfigured with a special tip pattern. Plates were incubated for 16 h in a robotic shaking incubator at 30 °C/550 r.p.m. allowing for ~ 5 doublings. ~ 250 yeast cells per strain (120 μl of a 1.2-OD_{600} per ml culture) were subsequently transferred into a preconfigured 24-well plate that was stored in a robotic plate reservoir at 4 °C until 30 min prior to its use where it was prewarmed to 30 °C. Once inoculated the new plate was incubated at 30 °C/550 r.p.m. to allow the next five yeast generations (generation 6–10) and the plate containing the first five doubling cultures was stored at 4 °C. This procedure was repeated two more times until the final plate containing the yeast with ~ 20 generations were stored at 4 °C.

The HOP assay was performed similar to the HIP experiment but the duration was reduced to approximately five doublings and no dilutions were necessary. Before the experiment, aliquots of the HOP pool were thawed and recovered for 3 h in YPD. The robotic system inoculated the wells prefilled with YPD and compound at the onset of the experiment with ~ 320 yeast cells/strain (110 μl of a 1.50-OD_{600} per ml culture). Plates were incubated for 16 h in a robotic shaking incubator at 30 °C/550 r.p.m. allowing for approximately five doublings and where then stored at 4 °C.

By applying a fixed time per dilution scheme to the HIP–HOP assay and diluting entire plates at fixed times we have simplified the procedure and managed to get optimal throughput. Careful IC_{30} determination and compound potency normalization resulted in very homogenous growth behaviours of the cultures in the 24 wells per plate. After each HIP–HOP experiment, growth curves recorded during the experiment were analysed.

gDNA extraction, TAG amplification and hybridization. An aliquot of 5 OD_{600} units of yeast per well from the HIP and HOP experiments were arrayed in 96-well plates, spun and the supernatant discarded. gDNA extraction was performed using the ChargeSwitch kit (Invitrogen #18000) in a partially automated process. One hundred and fifty microlitres per well of Zymolyase buffer (2 U Zymolyase, 50 ng RNase A, in 20 mM DTT and 20 mM Tris pH 7.5) were added and the cells were incubated at 37 °C/700 r.p.m. for 45 min. Three hundred microlitres per well lysis buffer (L18, Invitrogen) was added and the plate incubated at 56 °C/ 700 r.p.m. for 30 min. Two hundred microlitres per well ice-cold precipitation buffer (N2, Invitrogen) was added and the precipitate pelleted by centrifugation at 4 °C. The supernatant was saved to a new deep well plate (AB-0932, Abgene) suitable to lock onto the Invitrogen MagnaRack magnet and 40 μl per well of predispersed ChargeSwitch magnetic beads were added. The plate was incubated at room temperature for 5 min followed by incubation on the MagnaRack for 5 min to pellet the beads. All supernatant was carefully aspirated. The deep well plate was removed from the MagnaRack and 500 μl per well wash buffer (W12, Invitrogen) were added and mixed to disperse the beads. The washing was repeated three times and the wash buffer completely removed. Finally, 70 μl of elution buffer (E6, Invitrogen) were added. The beads were resuspended by mixing and the plate incubated for 10 min at room temperature. The beads were pelleted for one last time by incubation on the magnet for 5 min and the supernatant (containing the gDNA) was transferred to a new plate. The TAG PCR amplification and GenFlex Tag16K v2 hybridization protocol was used as described[24].

Processing of TAG16K v2 data. The raw probe intensity values of the CEL are summarized and normalized to tag intensities. Tags with low intensity values in control samples are removed by computing an intensity value threshold based on the comparison of the correlation between the logarithmic intensity ratios for uptags and downtags across different intensity ranges. The tag intensities are then averaged to obtain a strain intensity value. To measure the relative abundance of each strain with respect to the averages of the control samples we compute *MAD logarithmic* (*MADL*) scores for each compound/concentration combination. If we denote the logarithm of the ratio of the average intensity of the compound samples over the average intensity of the control samples as r_L, then the MADL s_L score is given as $(r_L\text{–med}(r_L))/\text{MAD}(r_L)$ where the median and MAD are computed over all strains in one sample. MADL scores can be viewed as robustly computed experiment-wise z-scores. We also compute the t-test P value, P, between the two replicates for a compound and the four to eight control replicates as a measure of the variability of the compound and control sample intensities across the experiment. The final (adjusted) score a_L is decreased for highly variable strains and computed as:

$$a_L = \min(0.05/p, 1) \cdot s_L \tag{1}$$

Then, we compute gene-wise z-scores (across all experiments) which are based on a robust parametric estimation of gene variability allowing for up to 15% outliers. To do this we consider the adjusted MADL scores $(a_1,..., a_n)$ of a strain over n experiments. Usually, the z-score transformation of score a_i is defined as $z_i = (a_i - \bar{a})/\sigma$ where $\bar{a}$ is the mean and σ the s.d. of the values a_i. However, due to the special nature of the HIP–HOP data it is advantageous to introduce a number of changes to the computation of the adjusted MADL z-score transformation. Since an (adjusted) MADL score of zero indicates a relative growth rate of a strain in the compound treated sample that is equal to the relative growth rate of the strain in the untreated control, we set $\bar{a} = 0$, for all strains. What remains is to estimate σ as a measure of the variability of the strain which is described in the following.

In order to avoid biasing the computation of the s.d. of a strain profile we allow at most five entries of the same compound/concentration combination in the data set. The samples of the sixth or greater occurance are discarded. We normalize the positive and negative scores separately; that is, we treat the positive and the negative scores as the halves of two separate distributions (artificially creating two perfectly symmetric distributions in this way). In the following we only consider the negative scores. Let the profile of the negative scores of a gene be $(s_1,..., s_n)$ where we assume that the scores are sorted in ascending order. We assume that the scores follow a normal distribution $N(\mu,\sigma)$. Note that since the distribution is perfectly symmetric around 0, $\mu = 0$; moreover, for each i, the expectation of the score s_i is the $i/(2n+1)$-quantile q_i of the standard normal distribution $N(0,1)$ times σ. In other words, s_i/q_i is an estimator $\hat{\sigma}_i$ for σ, for each i. These estimators are computed only for the indices i in the interval $(0.15n, 0.85n)$, that is, we consider only the middle 70% of the data to estimate σ. We assume that the scores s_i are drawn for a normal distribution only if the s.d. of the $\hat{\sigma}_i$ is at most 0.05 times the mean $\bar{\sigma}$ of $\hat{\sigma}_i$; in this case, the estimators $\hat{\sigma}_i$ are considered to be consistent and $\bar{\sigma}$ a good estimate of σ. If the s.d. is larger, we abandon the parametric approach and just set $\bar{\sigma}$ to be the s.d. of sample $(s_1,..., s_n, -s_1,..., -s_n)$. The estimate $\bar{\sigma}$ of σ obtained from the single-array data is used to compute the normalized z-score transformation $(a_1/\bar{\sigma}, ..., a_n/\bar{\sigma})$ of the adjusted MADL scores $(a_1,..., a_n)$.

Standard assay for yeast 3β-hydroxysteroid-dehydrogenase/C4-decarboxylase. Wild-type yeast (FL200) microsomes were prepared as follows. Yeast cells were disrupted by glass bead homogenization (0.45-mm diameter) in 100 mM phosphate buffer (pH 7.5) containing 1.5 mM reduced glutathione and 30 mM nicotinamide for 8 min at 0 °C[44]. The cell extract was obtained by centrifugation at 10,000 × g for 20 min. Microsomes were then isolated by centrifuging the cell extract supernatant at 100,000 × g for 90 min. The resulting microsomal pellet was resuspended in 100 mM phosphate buffer (pH 7.5) containing 3 mM reduced glutathione and 20% glycerol (v/v) using an Elvehjem–Potter homogenizer. Yeast microsomes (1.2 mg of protein) were incubated in the presence of 4α-carboxy-4β-methyl-cholest-8,24-dien-3β-ol (50–300 mM) and NAD^+ (0.5 mM) for 35–45 min at 30 °C. After addition of coprostanone (2–6 μg) as internal standard, the reactions were extracted as described previously[12]. The extract was analysed by TLC on silica gel and eluted twice with CH_2Cl_2. The fraction migrating between a standard of 4-desmethyl-sterone (R_F = 0.50) and 4,4-dimethyl-sterone (R_F = 0.70) was eluted and analysed using a GC-MS spectrometer (Agilent 5973 N).

In the case of inhibition assays, microsomes were incubated for 35–45 min at 30 °C in the presence of 4α-carboxy-4β-methyl-cholest-8,24-dien-3β-ol (50–300 mM) and a range of inhibitor concentration (0.01–100 μM) from which a dose–response curve was obtained allowing the corresponding I_{50} values to be determined.

GC-MS analysis for 3β-hydroxysteroid-dehydrogenase/C4-decarboxylase assay. The GC-MS spectrometer (Agilent 5973 N) was equipped with an 'on column' injector and a capillary column (30 × 0.25 mm) coated with DB5 (H_2 flow rate of 2 ml min^{-1}). The temperature program included a 30 °C min^{-1} increase from 60 °C to 240 °C, and followed by a 2 °C/min increase from 240 °C to 280 °C. The areas of the GC peaks of coprostanone (t_R = 1.000) and of the produced 4α-methyl-cholest-8,24-dien-3-one (t_R = 1.139), corrected from endogenous component of the same t_R (1.139) determined in the blank (boiled microsomes), allowed the rate of transformation of 4α-carboxy-4β-methyl-cholest-8,24-dien-3β-ol to be measured. Under these conditions, the estimated limit of detection of the 3βHSD/D activity was 0.1 nmol h^{-1} mg^{-1}.

Molecular modelling. A homology model was produced by using *P. aeruginosa UDP-N-*acetylglucosamine 4-epimerase complexed with UDP-N-acetylgalactosamine and NAD+ (1SB8) as a template, due to its highest primary sequence similarity (24%) compared with all the other structures in PDB. The sequence alignment was performed by ClustalX[45] and the homology model was constructed by MODELLER[46]. The model was then checked by QMEAN[47] (http://swissmodel.expasy.org/qmean/cgi/index.cgi).The structural comparison of the homology model with the other structures in the PDB was performed by DaliLite[48] (http://www.ebi.ac.uk/Tools/structure/dalilite/). All the macromolecular figures were generated in PyMol (version 1.3; Schrödinger, LLC).

The QMEAN score and the Z-score for the model are 0.475 and −3.46 respectively. QMEAN score is global score for the whole model and reflects the predicted model reliability ranging from 0 to 127. The QMEAN Z-score estimates the absolute quality of a homology model by relating it to reference structures solved by X-ray crystallography. These reference structures are a non-redundant subset of the PDB sharing less than 30% pairwise sequence identity among each other and are solved at a resolution below than 2 Å. It is an estimate of the 'degree of nativeness' of the structural features observed in a model. The scores are not ideal because of the absence of any other closely related protein structures to compare with the structure used to build the model with *P. aeruginosa* UDP-N-acetylglucosamine 4-epimerase.

The homology model shows conserved structural domains with other templates in Protein Data Bank such as UDP-galactose 4-epimerase mutant (PDB ID: 3AW9), Thermus thermophilus HB8 UDP-glucose 4-epimerase complex with NAD (PDB ID: 2P5U), DesIV from Streptomyces venezuelae with NAD and TYD bound (PDB ID: 1R66) (Supplementary Fig.3). The structural comparision of the homology model with the above-mentioned templates using DaliLit program generated Z-score. The Z-scores for these are 34.9 with 2.2A° RMSD, 36.9 with 1.7A° RMSD and 37.1 with 1.8A° RMSD respectively.

References

1. Brown, M. S. & Goldstein, J. L. Atherosclerosis. Scavenging for receptors. *Nature* **343,** 508–509 (1990).
2. Kingwell, B. A., Chapman, M. J., Kontush, A. & Miller, N. E. HDL-targeted therapies: progress, failures and future. *Nat. Rev. Drug Discov.* **13,** 445–464 (2014).
3. Mercer, E. I. Inhibitors of sterol biosynthesis and their applications. *Prog. Lipid Res.* **32,** 357–416 (1993).
4. Onyewu, C., Blankenship, J. R., Del Poeta, M. & Heitman, J. Ergosterol biosynthesis inhibitors become fungicidal when combined with calcineurin inhibitors against *Candida albicans, Candida glabrata,* and *Candida krusei. Antimicrob. Agents Chemother.* **47,** 956–964 (2003).
5. BarrettBee, K. & Dixon, G. Ergosterol biosynthesis inhibition: A target for antifungal agents. *Acta Biochim. Pol.* **42,** 465–479 (1995).
6. de Souza, W. & Rodrigues, J. C. F. Sterol Biosynthesis pathway as target for anti-trypanosomatid drugs. *Interdiscipl. Perspec. Infect. Dis.* **2009,** 642502 (2009).
7. Roemer, T. & Boone, C. Systems-level antimicrobial drug and drug synergy discovery. *Nat. Chem. Biol.* **9,** 222–231 (2013).
8. Lim, S., Oh, P. C., Sakuma, I. & Koh, K. K. How to balance cardiorenometabolic benefits and risks of statins. *Atherosclerosis* **235,** 644–648 (2014).
9. Baudry, K. *et al.* The effect of the erg26-1 mutation on the regulation of lipid metabolism in *Saccharomyces cerevisiae. J. Biol. Chem.* **276,** 12702–12711 (2001).
10. Mo, C., Valachovic, M., Randall, S. K., Nickels, J. T. & Bard, M. Protein–protein interactions among C-4 demethylation enzymes involved in yeast sterol biosynthesis. *Proc. Natl Acad. Sci. USA* **99,** 9739–9744 (2002).
11. Gachotte, D., Barbuch, R., Gaylor, J., Nickel, E. & Bard, M. Characterization of the *Saccharomyces cerevisiae* ERG26 gene encoding the C-3 sterol dehydrogenase (C-4 decarboxylase) involved in sterol biosynthesis. *Proc. Natl Acad. Sci. USA* **95,** 13794–13799 (1998).
12. Rahier, A., Darnet, S., Bouvier, F., Camara, B. & Bard, M. Molecular and enzymatic characterizations of novel bifunctional 3β-hydroxysteroid dehydrogenases/c-4 decarboxylases from *Arabidopsis thaliana. J. Biol. Chem.* **281,** 27264–27277 (2006).
13. Rahier, A., Bergdoll, M., Génot, G., Bouvier, F. & Camara, B. Homology modeling and site-directed mutagenesis reveal catalytic key amino acids of 3β-hydroxysteroid-dehydrogenase/C4-decarboxylase from *Arabidopsis. Plant Physiol.* **149,** 1872–1886 (2009).
14. Hatori, H. *et al.* FR171456, a novel cholesterol synthesis inhibitor produced by *Sporormiella minima* no. 15604 - II. Biological activities. *J. Antibiot.* **57,** 260–263 (2004).
15. Hatori, H. *et al.* FR171456, a novel cholesterol synthesis inhibitor produced by Sporormiella minima No. 15604 - I. Taxonomy, fermentation, isolation, physico-chemical properties. *J. Antibiot.* **57,** 253–259 (2004).
16. Bartenschlager, R. Hepatitis C virus replicons: potential role for drug development. *Nat. Rev. Drug Discov.* **1,** 911–916 (2002).
17. Vanden Bossche, H., Marichal, P. & Odds, F. C. Molecular mechanisms of drug resistance in fungi. *Trends Microbiol.* **2,** 393–400 (1994).
18. Barretina, J. *et al.* The Cancer Cell Line Encyclopedia enables predictive modelling of anticancer drug sensitivity. *Nature* **483,** 603–607 (2012).
19. Scheel, T. K. H. & Rice, C. M. Understanding the hepatitis C virus life cycle paves the way for highly effective therapies. *Nat. Med.* **19,** 837–849 (2013).
20. Baudry, K. *et al.* The effect of the erg26-1 mutation on the regulation of lipid metabolism in Saccharomyces cerevisiae. *J. Biol. Chem.* **276,** 12702–12711 (2001).
21. CLSI. *Reference Method For Broth Dilution Antifungal Susceptibility Testing Of Filamentous Fungi; Approved Standard* 2nd edn. (CLSIdocument M38-2A, 2008).
22. CLSI. *Reference Method For Broth Dilution Antifungal Susceptibility Testing of Yeasts; Approved Standard* 3rd edn. (CLSIdocument M27-A3, 2008).
23. Lehar, J. *et al.* Synergistic drug combinations tend to improve therapeutically relevant selectivity. *Nat. Biotechnol.* **27,** 659–666 (2009).
24. Hoepfner, D. *et al.* High-resolution chemical dissection of a model eukaryote reveals targets, pathways and gene functions. *Microbiol. Res.* **169,** 107–120 (2014).
25. Hoepfner, D. *et al.* An integrated approach for the identification and target validation of anti-fungal compounds active against Erg11p. *Antimicrob. Agents Ch.* **56,** 4233–4240 (2012).
26. Thevissen, K. *et al.* Miconazole induces changes in actin cytoskeleton prior to reactive oxygen species induction in yeast. *J. Biol. Chem.* **282,** 21592–21597 (2007).
27. Basson, M. E., Moore, R. L., Orear, J. & Rine, J. Identifying mutations in duplicated functions in *Saccharomyces cerevisiae*—recessive mutations in HMG-CoA reductase genes. *Genetics* **117,** 645–655 (1987).
28. Ho, C. H. *et al.* A molecular barcoded yeast ORF library enables mode-of-action analysis of bioactive compounds. *Nat. Biotechnol.* **27,** 369–377 (2009).
29. Heitman, J., Movva, N. R. & Hall, M. N. Targets for cell cycle arrest by the immunosuppressant rapamycin in yeast. *Science* **253,** 905–909 (1991).
30. Rahier, A., Darnet, S., Bouvier, F., Camara, B. & Bard, M. Molecular and enzymatic characterizations of novel bifunctional 3beta-hydroxysteroid dehydrogenases/C-4 decarboxylases from Arabidopsis thaliana. *J. Biol. Chem.* **281,** 27264–27277 (2006).
31. Chou, A. C. & Wilson, J. E. Rat brain hexokinase: glucose and glucose-6-phosphate binding sites and C-terminal amino acid of the purified enzyme. *Arch. Biochem. Biophys.* **165,** 628–633 (1974).
32. Rondet, S. Biosynthese des sterols: 4-Carboxy-sterol decarboxylases de plante superieure et de levure; identification, characterisation et purification. *PhD Thesis* (University of Strasbourg, 1998).
33. Schreiber, S. L. Small molecules: the missing link in the central dogma. *Nat. Chem. Biol.* **1,** 64–66 (2005).
34. Bautz, D. J., Broman, K. W. & Threadgill, D. W. Identification of a Novel Polymorphism in X-Linked Sterol-4-alpha-carboxylate 3-dehydrogenase (Nsdhl) Associated with Reduced HDL Cholesterol Levels in I/LnJ Mice. *G3(Bethesda)* **3,** 1819–1825 (2013).
35. Ye, J. Reliance of host cholesterol metabolic pathways for the life cycle of hepatitis C Virus. *PLoS Pathog.* **3,** e108 (2007).
36. Lyn, R. K. *et al.* Direct imaging of the disruption of hepatitis C virus replication complexes by inhibitors of lipid metabolism. *Virology* **394,** 130–142 (2009).
37. Rothwell, C. *et al.* Cholesterol biosynthesis modulation regulates dengue viral replication. *Virology* **389,** 8–19 (2009).
38. Syed, G. H., Amako, Y. & Siddiqui, A. Hepatitis C virus hijacks host lipid metabolism. *Trends Endocrin. Met.* **21,** 33–40 (2010).

39. Syed, G. H. & Siddiqui, A. Effects of Hypolipidemic Agent Nordihydroguaiaretic Acid on Lipid Droplets and Hepatitis C Virus. *Hepatology* **54,** 1936–1946 (2011).
40. Caldas, H. & Herman, G. E. NSDHL, an enzyme involved in cholesterol biosynthesis, traffics through the Golgi and accumulates on ER membranes and on the surface of lipid droplets. *Hum. Mol. Genet.* **12,** 2981–2991 (2003).
41. Herman, G. E. X-linked dominant disorders of cholesterol biosynthesis in man and mouse. *Biochim. Biophys. Acta* **1529,** 357–373 (2000).
42. McLarren, K. W. *et al.* Hypomorphic temperature-sensitive alleles of NSDHL Cause CK Syndrome. *Am. J. Hum. Genet.* **87,** 905–914 (2010).
43. Sambrook, J. & Green, M. R. *Molecular Cloning: A Laboratory Manual* (Cold Spring Harbor Laboratory Press, 2012).
44. Darnet, S. & Rahier, A. Enzymological properties of sterol-C4-methyl-oxidase of yeast sterol biosynthesis. *Biochim. Biophys. Acta* **1633,** 106–117 (2003).
45. Thompson, J. D., Gibson, T. J., Plewniak, F., Jeanmougin, F. & Higgins, D. G. The CLUSTAL_X windows interface: flexible strategies for multiple sequence alignment aided by quality analysis tools. *Nucleic Acids Res.* **25,** 4876–4882 (1997).
46. Sali, A. & Blundell, T. L. Comparative protein modeling by satisfaction of spatial restraints. *J. Mol. Biol.* **234,** 779–815 (1993).
47. Benkert, P., Kunzli, M. & Schwede, T. QMEAN server for protein model quality estimation. *Nucleic Acids Res.* **37,** 510–514 (2009).
48. Holm, L., Kaariainen, S., Rosenstrom, P. & Schenkel, A. Searching protein structure databases with DaliLite v.3. *Bioinformatics* **24,** 2780–2781 (2008).

Author contributions

S.B.H., S.K., A.R., S.Ro. designed and performed experiments, analysed the data and wrote the paper, D.H., A.H., T.W., S.Re. designed and performed experiments, analysed the data and prepared figures, I.F., P.-E.I., C.V.M., J.R.L-D, C.F., J.G., H.U.N., R.R., J.A.S., H.Y., V.P., J.T. designed and performed experiments, analysed the data, R.J., J.Z., T.A., P.K., A.M., K.M., L.O. designed and performed the experiments, M.F., M.B. performed modelling, analysed the data and prepared figures, S.S., R.K. designed the experiments and analysed the data, C.N.P. designed the experiments, analysed the data and wrote the paper N.R.M. created the concept and provided support for this work.

Additional information

Accession codes. Crystallographic data (excluding structure factors) have been deposited with the Cambridge Crystallographic Data Centre as supplementary publication number CCDC 976241. These data can be obtained free of charge from the Cambridge Crystallographic Data Centre via www.ccdc.cam.ac.uk/data_request/cif

Competing financial interests: The authors declare no competing financial interests.

5

Transformable liquid-metal nanomedicine

Yue Lu[1,2], Quanyin Hu[1,2], Yiliang Lin[3], Dennis B. Pacardo[1,2], Chao Wang[1,2], Wujin Sun[1,2], Frances S. Ligler[1], Michael D. Dickey[3] & Zhen Gu[1,2,4]

To date, numerous inorganic nanocarriers have been explored for drug delivery systems (DDSs). However, the clinical application of inorganic formulations has often been hindered by their toxicity and failure to biodegrade. We describe here a transformable liquid-metal nanomedicine, based on a core-shell nanosphere composed of a liquid-phase eutectic gallium-indium core and a thiolated polymeric shell. This formulation can be simply produced through a sonication-mediated method with bioconjugation flexibility. The resulting nanoparticles loaded with doxorubicin (Dox) have an average diameter of 107 nm and demonstrate the capability to fuse and subsequently degrade under a mildly acidic condition, which facilitates release of Dox in acidic endosomes after cellular internalization. Equipped with hyaluronic acid, a tumour-targeting ligand, this formulation displays enhanced chemotherapeutic inhibition towards the xenograft tumour-bearing mice. This liquid metal-based DDS with fusible and degradable behaviour under physiological conditions provides a new strategy for engineering theranostic agents with low toxicity.

[1] Joint Department of Biomedical Engineering, University of North Carolina at Chapel Hill and North Carolina State University, Raleigh, North Carolina 27695, USA. [2] Division of Molecular Pharmaceutics and Center for Nanotechnology in Drug Delivery, Eshelman School of Pharmacy, University of North Carolina at Chapel Hill, Chapel Hill, North Carolina 27599, USA. [3] Department of Chemical and Biomolecular Engineering, North Carolina State University, Raleigh, North Carolina 27695, USA. [4] Department of Medicine, University of North Carolina at Chapel Hill, Chapel Hill, North Carolina 27599, USA. Correspondence and requests for materials should be addressed to Z.G. (email: zgu@email.unc.edu).

The engineering flexibility provided by inorganic nanoparticles with tailorable shape, size, surface ligands and physical properties has enabled on-demand design of novel drug delivery systems, contrast agents and integrated systems for disease diagnosis and treatment[1–6]. The last decade has witnessed numerous efforts in developing inorganic nanoparticles capable of effectively targeting different diseases[7–11]. However, these formulations often fail to be useable due to systemic toxicity. For instance, targeted cancer therapy requires nanoparticles with relatively large sizes to minimize clearance[12] and enhance tumour retention[13,14], yet such inorganic particles often remain in the body for a long time because of their lack of biodegradability. To date, few studies have demonstrated how to engineer the physicochemical properties of inorganic nanoparticles to satisfy both target delivery and efficient elimination[15,16]. This design bottleneck has long existed and is impeding the clinical translation of therapy and diagnostics based on inorganic carriers.

We report here a novel liquid metal-based nanoscale formulation for drug delivery to achieve enhanced anticancer therapy. Taking advantage of the unique characteristics of a eutectic alloy of gallium indium (EGaIn, 75% Ga and 25% In by weight), this novel nanomedicine has a variety of merits in terms of simple fabrication, facile surface bioconjugation and the capability of fusion and degradation in a mildly acidic environment. EGaIn is a low-viscosity liquid at room temperature. Unlike Hg, it has low toxicity and has thus attracted considerable attention for applications of microfluidic systems, soft robotics and stretchable electronics[17–19]. To obtain EGaIn-based drug nanocarriers, an emulsion-like ligand-mediated procedure is simply applied through ultrasonication at room temperature. During sonication[20], the thiolated ligands readily assemble onto the surface of the EGaIn, competing with the oxidation process and facilitating control of the particle size[17,20–22]. The two ligands used here, thiolated (2-hydroxypropyl)-β-cyclodextrin (designated MUA-CD) and thiolated hyaluronic acid (designated *m*-HA), not only serve as capping agents during the formation of nanoscaled liquid-metal spheres but also play roles of drug-loading matrix and active targeting moiety, respectively. The final formulation (designated LM-NP/*L*) comprises three primary functional constituents: a CD-based drug-loading motif, the targeting ligand HA and an EGaIn core. The seven-membered sugar ring of the CD provides faithful loading sites for doxorubicin (Dox)[23], a model broad-spectrum chemotherapeutic drug. The HA moiety supports active tumour targeting towards the receptors including the CD44 receptor, which is overexpressed on the cell surface of a broad variety of tumours, including human cervical cancer and breast cancer[24]. After intravenous injection, LM-NP/*L* is expected to accumulate at tumour sites as a result of passive[25] and active targeting effects. Upon endocytosis, the LM-NPs/*L* are expected to fuse with each other in the mildly acidic endosome microenvironment[26], leading to the dissociation of Dox-containing ligands and therefore promote drug release. The aggregates of fused LM-NPs/*L* are expected to degrade due to the synergistic effect of oxidative corrosion. Furthermore, the main degradation product, Ga (III), can serve as anticancer agent itself to reverse drug resistance in drug-resistant cancer cells[27,28]. Taken together, this liquid metal-based nanomedicine is transformable: (1) it can be simply generated from a bulk material; (2) fusion happens to promote drug release in an acidic cellular environment; and (3) it can be biodegraded to reverse drug resistance and avoid potential systemic toxicity.

Results

Preparation of transformable liquid-metal nanomedicine. To prepare the liquid metal nanocarriers, EGaIn was added to a centrifuge tube filled with an ethanolic solution of thiolate CD and thiolate HA (volume ratio of EGaIn: ethanol = 1:150, Fig. 1a,b,e). After sonication, the largest particles precipitated within seconds, and the nanoscale colloids was removed from the tube and resuspended in water (Fig. 2a). Next, Dox-loaded LM-NP/*L* (LM-NP/Dox-*L*) was obtained by overnight incubation with triethanolamine-treated Dox. The mixture was washed to remove the excess Dox. The resulting LM-NP/Dox-*L* appeared as core–shell structured nanospheres with a diameter of ~107 nm, which was consistent with the results determined by dynamic light scattering tests (Fig. 2b). The Dox-loading capacity could be tuned by adjusting the ratio between MUA-CD and *m*-HA. In a typical formulation used in the following investigations, the Dox-loading capacity was determined as 24% (Dox weight ratio of the total weight of nanoparticles). Interestingly, the addition of *m*-HA could significantly decrease the average particle size of product as well as enhance the systemic stability, which could be attributed to the negative charge of the anionic *m*-HA (Supplementary Table 1).

To study the acid-triggered conformational transformation and subsequently promoted drug release, the release profile of Dox from LM-NP/Dox-*L* (Fig. 2c) under neutral (1 × PBS buffer, pH 7.4) and acidic (1 × PBS buffer, pH 5.0) conditions were investigated through the dialysis method[29]. A burst release of loaded Dox was clearly observed within the first 30 min in the acidic buffer, significantly different from the release curve recorded in the neutral buffer. Similarly, paclitaxel (PTX), which shows insignificant solubility change in the pH range of 2.5–8 (ref. 30), was loaded to LM-NP/*L* as a model drug. As expected, the release profile of LM-NP/*L* loaded with PTX also presented a high release rate within the first 30 min in the acidic buffer (Supplementary Fig. 4). The acidic buffer caused 72% of the loaded PTX to release within 4 h, which was significantly higher than the 28% in the neutral environment. Dox-loaded gold nanoparticles (GNPs/Dox) with an average diameter of 100 nm, where Dox was included into the MUA-CD ligands with the same Dox-loading capacity as LM-NP/Dox-*L*, were also applied as a control. Compared with GNP/Dox, LM-NP/Dox-*L* displayed a more potent acid-promoted Dox release at pH 5.0 (Supplementary Fig. 5). This phenomenon could be attributed to fusion of LM-NP/*L* triggered by the disruption of the oxidized shell of LM-NP/*L*, and the subsequent dissociation of surface ligands. Next, the conformational change of LM-NP/Dox-*L* with up to 4 h acid treatment (pH 5.0) was visualized using transmission electron microscopy (TEM) imaging (Fig. 2d). The initial fusion behaviour was captured within the first 5 min after acid treatment, and the aggregates formed by fusion increased in size over time. These results were in good agreement with the significant increase in size distribution upon acid treatment as measured by dynamic laser scattering (Fig. 2e). We hypothesized that the acid attacked the oxidized layer[20] on the surface of the LM-NP/*L* to trigger fusion of LM-NP/*L* (Supplementary Figs 8,9; Fig. 2f), which further led to the fusion of the liquid inner cores[31]. Furthermore, the degradation of LM-NP/Dox-*L* at pH 5.0 was clearly observed via the optical microscope (Fig. 2g) and TEM imaging (Fig. 2d) over time; while negligible difference was observed in the case of neutral buffer. The representative TEM image of LM-NPs/Dox-*L* immersed in acidic (pH 5.0) PBS buffer for 72 h showed evidence of degradation into the hollow polymeric shells dotted with shrunk metallic cores in the inner wall evidenced the occurrence of degradation. To further verify the degradation of LM-NPs, the change of Ga (III) ion concentration in PBS buffer over time was monitored by the inductively coupled plasma optical emission spectrometry. The Ga (III) concentration in the acidic buffer displayed a notable increase over time; while the degradation of LM-NPs in neutral solution was moderate (Fig. 2f). The degradation process of LM-NPs was then tracked and quantified by monitoring the change of light transmittance over time (Fig. 2h). The light transmittance of acidic LM-NP solution increased from 25 to 66%, which was in good agreement of the proposed degradation of LM-NPs in acidic environment. Of note, the main degradation product, Ga (III) ion, has been reported to reverse drug resistance in drug-resistant cancer cells[27,28], which could potentially generate synergistic effects and prevent drug resistance.

Furthermore, the nature of the EGaIn metal core and the verified acid-triggered fusion might also open the possibility of utilizing these transformable nanospheres for imaging contrast enhancement. After acid treatment (12 h in 1 × PBS buffer, pH 5.0), the homogeneous contrast enhancement caused by LM-NPs was disrupted, replaced by heterogeneous contrast within the imaged area with higher contrast at the locations of fused particles and lower contrast at spots in absence of LM-NPs (Fig. 2i). The resulting changes in contrast were quantified by studying the grey value profiles of representative areas in X-ray images (Fig. 2i). The grey values of the imaging area were approximately constant before acid treatment, indicating homogeneous imaging contrast. The disruptions of the homogeneous imaging contrast after acid treatment were supported by the appearance of peaks in the grey value plot. Similar fusion of LM-NPs accumulated at tumour site would potentially lead to enhanced contrast of tumour tissue during X-ray imaging.

Intracellular acid-triggered fusion and promoted Dox release. The intracellular delivery of LM-NP/Dox-*L* into HeLa cells was further explored using the confocal laser scanning microscopy (Fig. 3a). The fluorescence of Dox was clearly observed in HeLa cells after 1 h of incubation with LM-NP/Dox-*L*, indicating the cellular internalization of the nanospheres. When the incubation time was prolonged to 4 h, Dox was highly localized within the nuclei of HeLa cells, as indicated by the magenta fluorescence (Fig. 3a). The cellular uptake of LM-NP/Dox-*L* displayed an energy-dependent manner, and LM-NP/Dox-*L* mainly entered HeLa cells via macropinocytosis (Supplementary Fig. 10). Next, the endocytosis process of LM-NP/Dox-*L* was also visualized by TEM imaging (Figs 1c,d and 3b). HeLa cells were

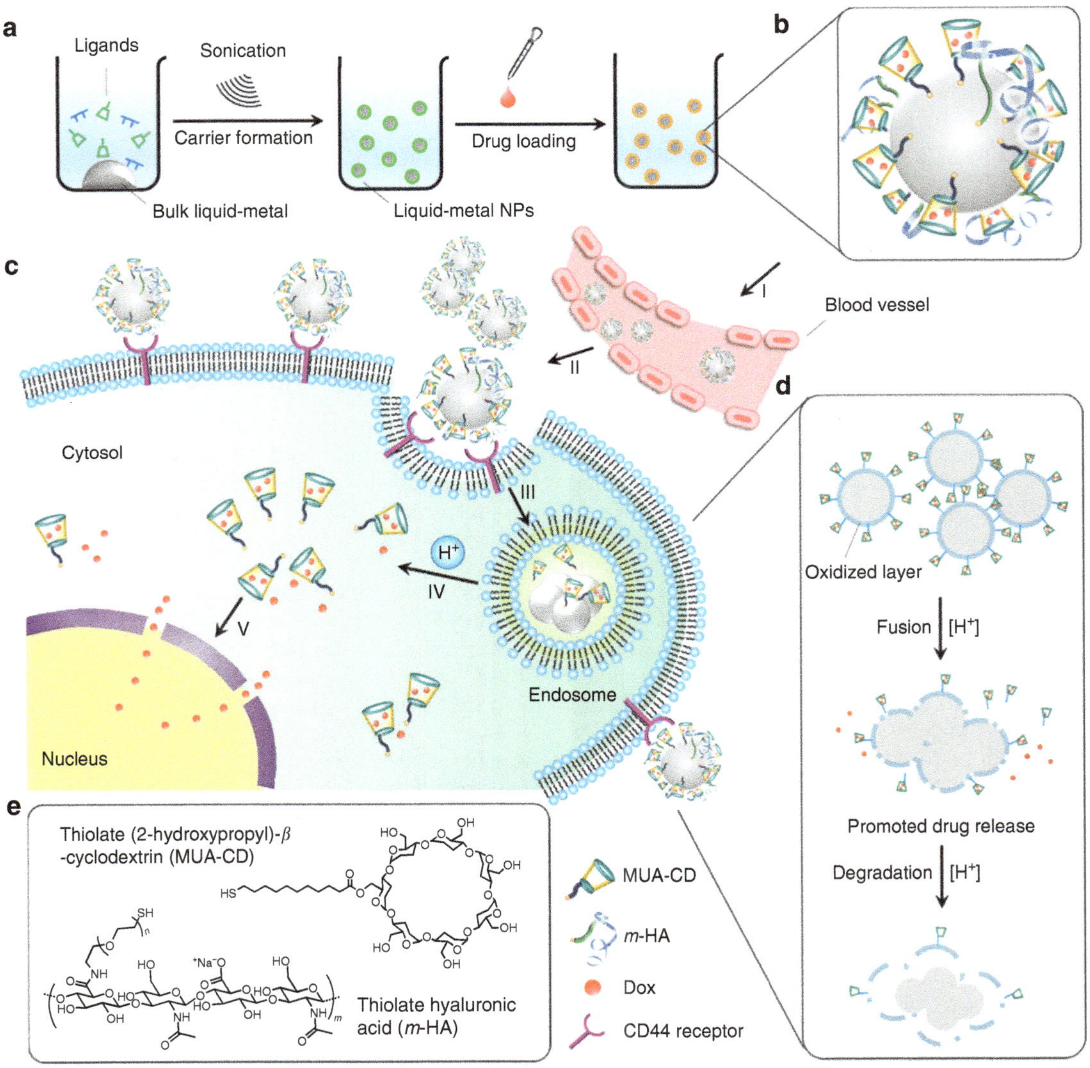

Figure 1 | Schematic design of the transformable liquid-metal delivery system. (**a**) Preparation route of LM-NP/Dox-*L*. (**b**) The main components of LM-NP/Dox-*L*: thiolated CD with Dox, HA-based targeting motif and an EGaIn core. (**c**) pH-responsive delivery of Dox by LM-NP/Dox-*L* to the nuclei for the targeted cancer therapy. (I) accumulation of LM-NP/Dox-*L* at the tumour site through passive and active targeting; (II) specific binding to the overexpressed receptors on the tumour cells; (III) receptor-mediated endocytosis; (IV) acid-triggered fusion of LM-NP/Dox-*L* and endosomal/lysosomal escape of Dox-containing ligands; (V) accumulation of Dox in the nucleus. (**d**) Acid-triggered fusion and degradation process of LM-NP/Dox-*L*. (**e**) Chemical structures of MUA-CD and *m*-HA.

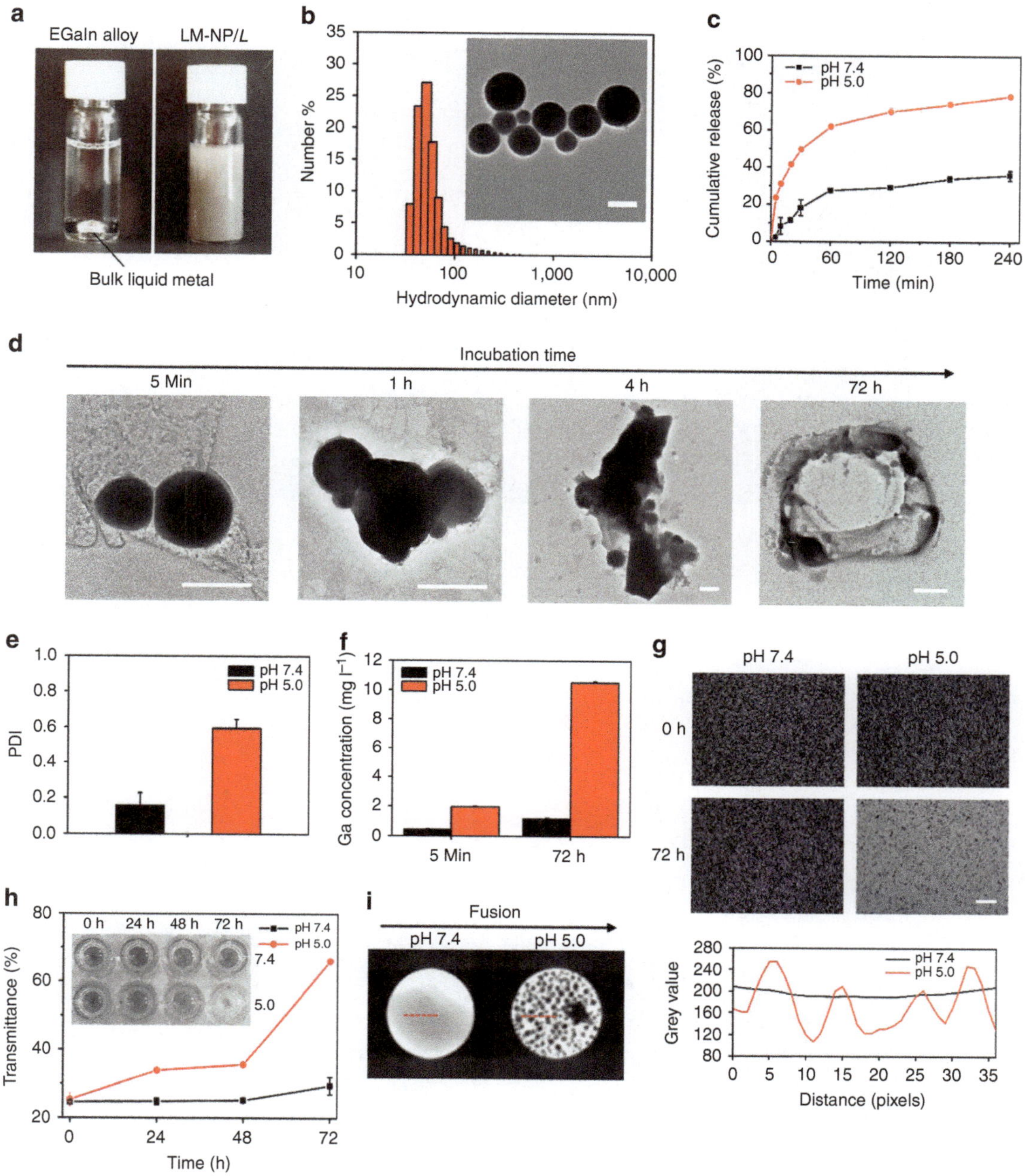

Figure 2 | Characterization of liquid-metal nanospheres. (**a**) EGaIn was ultrasonically dispersed in ligand containing ethanol mixture (shown schematically in Fig. 1a). (**b**) The hydrodynamic size of LM-NP/Dox-*L* measured by dynamic light scattering. Inset: TEM image of LM-NP/Dox-*L*. Scale bar, 100 nm. (**c**) The release profiles of LM-NP/Dox-*L* at different pH levels. (**d**) Representative TEM images of LM-NP/Dox-*L* after different time immersed in acidic (pH 5.0) PBS buffer. Scale bars, 100 nm (for 5 min); 100 nm (for 1 h); 100 nm (for 4 h); 400 nm (for 72 h). (**e**) Polydispersity index (PDI) of LM-NP/Dox-*L* immersed in neutral (pH 7.4) and acidic (pH 5.0) PBS buffer. (**f**) Changes of metal ion concentration under neutral and acidic environments. (**g**) Optical images of LM-NPs before and after acid treatment. Scale bar, 40 μm. (**h**) Light transmittance change over time. (**i**) X-ray images (left) of LM-NPs immersed in neutral (pH 7.4) and acidic (pH 5.0) PBS buffer (particles were imaged after 12 h treatment). The images were taken in a 96-well plate. Red lines indicate the areas quantitatively analysed. Grey value analysis (right) of representative area in X-ray images of LM-NPs. Error bars indicated s.d. ($n = 3$).

treated with LM-NP/Dox-*L* for different time durations. After 1 h incubation, fusion of LM-NPs was clearly observed in the acidic endosomes, agreeing with the *in vitro* results discussed above. After 4 h incubation, more fusion events of LM-NPs were spotted, together with few single LM-NPs dispersed in the cytosol. Furthermore, we quantified and visualized the fusion process intracellularly by lysing the cells incubated with LM-NPs/Dox-*L* at different incubation time points. The intracellular Ga/In ion concentrations were then quantified by inductively coupled plasma mass spectrometry. Importantly, the structures formed via fusion were also further validated using TEM and analysed with the titan scanning/TEM at an atomic scale. Similar to the LM-NPs/Dox-*L* treated with acid (Fig. 2d), the initial fusion behaviour was captured immediately upon cellular uptake, followed by the formation of larger aggregates and the eventual degradation, which was in good agreement with the significant increase of intracellular Ga (III) ion concentration over incubation time (Supplementary Fig. 1). The element mapping results revealed the fusion of the polymeric shells during the conformational changes of LM-NP/*L* (Fig. 3c). The scattering of

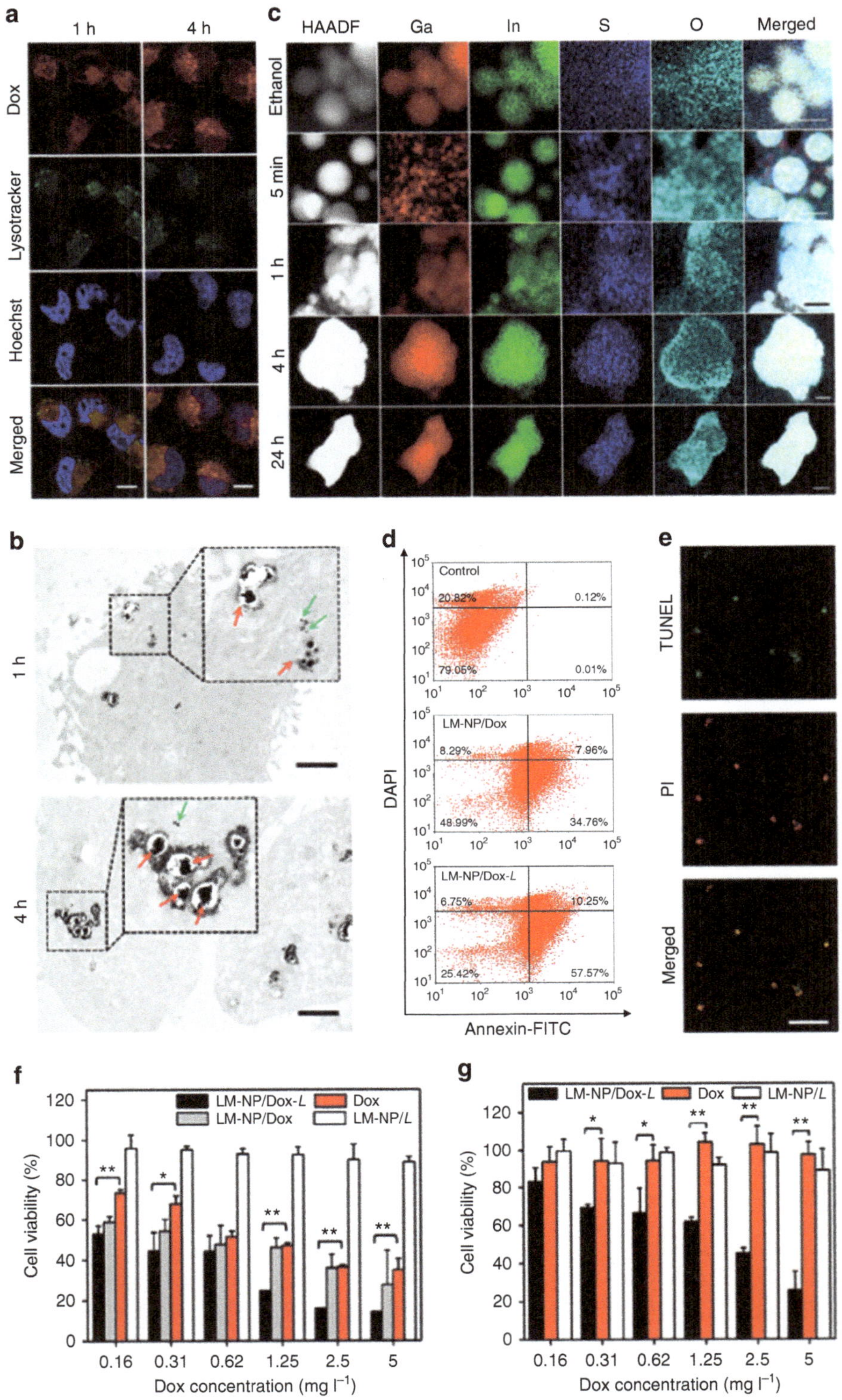

Figure 3 | Intracellular interaction and drug release. (**a**) Intracellular delivery of LM-NP/Dox-*L* towards HeLa cells at different time points observed by confocal laser scanning microscopy. The cells were incubated with LM-NP/Dox-*L* at 37 °C for 1 and 4 h, respectively. The late endosomes and lysosomes were stained with LysoTracker Green, and the nuclei were stained with Hoechst 33342. Scale bar, 10 μm. (**b**) Representative TEM images of HeLa cells incubated with LM-NP/Dox-*L* for 1 and 4 h. Red arrows show fused nanospheres; green arrows show dispersion of single nanosphere in the cytosol. Scale bar, 2 μm. (**c**) Element mapping results of intracellular LM-NPs/Dox-*L* collected from HeLa cells after different incubation times. Scale bars, 100 nm (for 5 min); 100 nm (for 1 h); 100 nm (for 4 h); 200 nm (for 24 h). (**d**) Flow cytometric analysis of HeLa cell apoptosis induced by LM-NP/Dox-*L* for 12 h using the Annexin V-FITC/4,6-diamidino-2-phenylindole (DAPI) staining. (**e**) HeLa cell apoptosis induced by LM-NP/Dox-*L* for 20 h using the APO-BrdU TUNEL assay. (**f**) *In vitro* cytotoxicity of LM-NP/Dox and LM-NP/Dox-*L* on HeLa cells for 24 h. Error bars indicate s.d. ($n = 4$). Note: the concentration of LM-NP/*L* is equal to the nanocarrier concentration of Dox-loaded formulations in each corresponding group; the error bars in **f** (group LM-NP/Dox-*L* at concentrations 1.25, 2.5 and 5 mg l^{-1}) are small. (**g**) *In vitro* cytotoxicity of LM-NP/Dox-*L* on Dox-resistant HeLa cells for 24 h. Error bars indicate s.d. ($n = 4$). Note: the concentration of LM-NP/*L* is equal to the nanocarrier concentration of Dox-loaded formulations in each corresponding group. *$P < 0.05$, **$P < 0.01$ compared with the Dox solution group (two-tailed Student's *t*-test).

Ga signal after 5-min incubation indicated the fast disruption of the oxidized shell. After 1 h of incubation, the increase of S and O signals in the background suggested the dissociation of thiolated surface ligands. The time-of-flight secondary ion mass spectrometry was further utilized to analyse the surface of LM-NP/*L* and intracellular LM-NP/*L* collected by lysing the cells internalized with particles (Supplementary Fig. 2). The coexistence of thiols and oxides was well evidenced by the overlap of the signals of GaO^- and GaS^- in LM-NP/*L*. The increase of HS^- signal in the background after 5 min incubation visualized the dissociation of thiolated surface ligands, indicating the initial stage of fusion behaviour. The remarkable increase in the background HS^- signal and the significant decreases in the signals of GaO^- and GaS^- clearly indicated that the intracellular fusion behaviour of LM-NP/*L* was a result of the synergistic action of the dissociation of thiolated surface ligands and the disruption of oxides.

The Annexin V-FITC/4,6-diamidino-2-phenylindole apoptosis detection and the terminal deoxynucleotidyl transferase dUTP nick end labelling (TUNEL) assay were performed to compare the apoptosis-inducing capabilities of LM-NP/Dox-*L* and free Dox solution. Annexin V-FITC labels the phosphatidylserine translocated to the extracellular membrane upon the initiation of apoptosis[32], while 4,6-diamidino-2-phenylindole has significantly higher staining efficiency with cells where the plasma membrane integrity has been compromised[33]. This combination allows the differentiation among the early apoptotic cells, the late apoptotic cells and the viable cells, which can be quantitatively determined by the flow cytometry. The total apoptotic ratio of HeLa cells incubated with LM-NP/Dox-*L* was 68%, which was significantly higher than 43% of HeLa cells incubated with LM-NP/Dox (Fig. 3d). In addition, the cells treated with LM-NP/Dox-*L* showed extensive apoptotic DNA fragmentation stained by the Alexa Fluor 488 as green fluorescence (Fig. 3e), substantiating

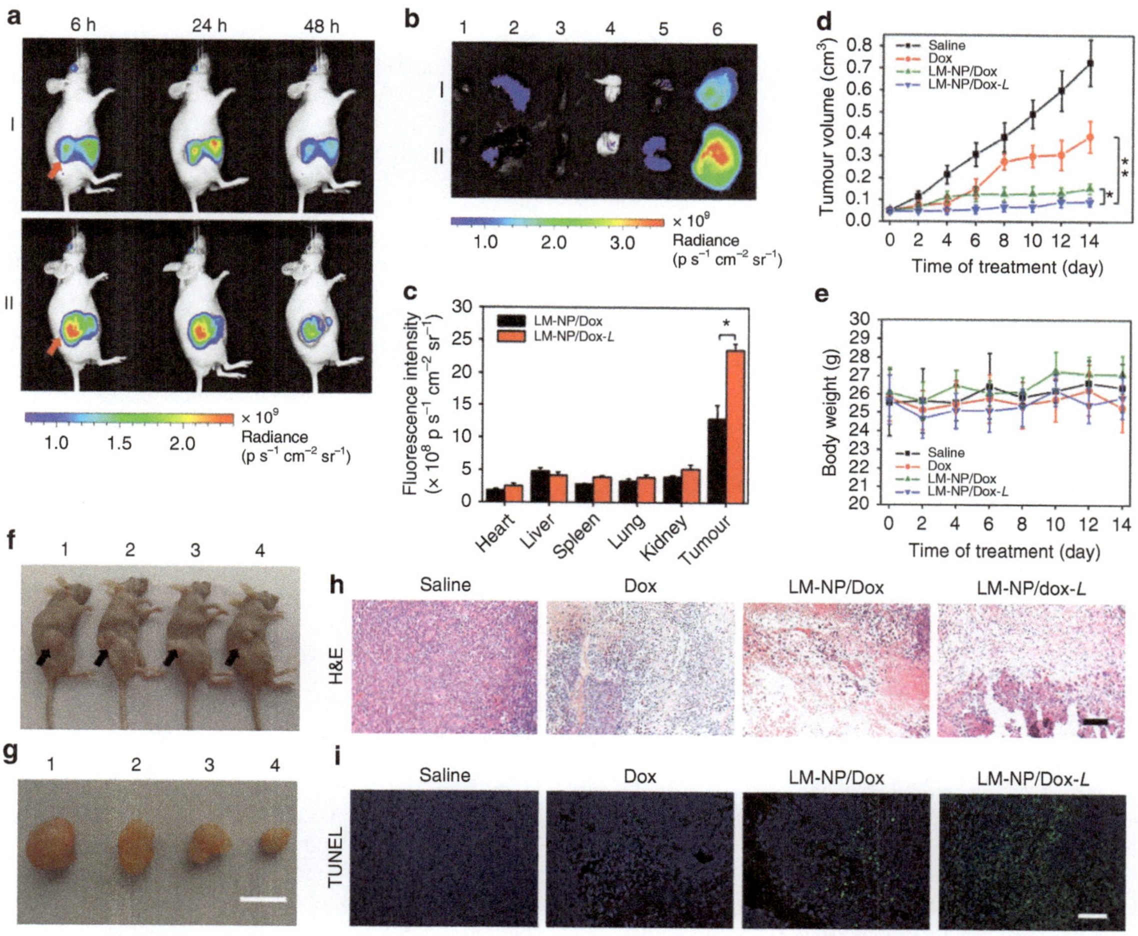

Figure 4 | Tumour targetability and antitumour activity. **(a)** *In vivo* fluorescence imaging of the HeLa tumour-bearing nude mice at 6, 24 and 48 h after intravenous injection of Cy5.5-(LM-NP/Dox) (I) and Cy5.5-(LM-NP/Dox-*L*) (II) at Cy5.5 dose of 30 nmol kg^{-1}. Arrows indicate the sites of tumours. **(b)** *Ex vivo* fluorescence imaging of the tumour and normal tissues collected from the HeLa tumour-bearing nude mice being killed at 48 h post injection. The numeric label for each organ is as follows: 1, heart; 2, liver; 3, spleen; 4, lung; 5, kidney; 6, tumour. **(c)** Region-of-interest analysis of fluorescent signals from the tumours and normal tissues. Error bars indicated s.d. ($n = 3$). $^*P < 0.05$ (two-tailed Student's *t*-test). **(d)** The HeLa tumour growth curves after intravenous injection of different formulations of Dox at a dose of 2 mg kg^{-1}. Error bars indicate s.d. ($n = 5$). $^*P < 0.05$, $^{**}P < 0.01$ (two-tailed Student's *t*-test). **(e)** The body weight variation of HeLa tumour-bearing mice during treatment. Error bars indicate s.d. ($n = 5$). **(f)** Representative images of the HeLa xenograft tumours of the mice after treatment with the studied Dox formulations at day 14. The numeric label for each mouse is as follows: 1, saline; 2, Dox; 3, LM-NP/Dox; 4, LM-NP/Dox-*L*. Arrows indicate the sites of tumours. **(g)** Representative images of HeLa xenograft tumours collected from the mice after treatment with different formulations at day 14. The numeric label for each tumour is as follows: 1, saline; 2, Dox; 3, LM-NP/Dox; 4, LM-NP/Dox-*L*. Scale bar, 1 cm. **(h)** Histological observation of the tumour tissues after treatment. The tumour sections were stained with haematoxylin and eosin (H&E). Scale bar, 100 μm. **(i)** Detection of apoptosis in the tumour tissues after treatment. The tumour sections were stained with fluorescein-dUTP (green) for apoptosis and Hoechst for the nuclei (blue). Scale bar, 50 μm.

that LM-NP/Dox-*L* is an effective intracellular delivery vehicle for enhanced apoptosis-inducing activity.

Next, the cytotoxicity of Dox-loaded nanospheres towards HeLa cells was evaluated with 3-(4, 5-dimethylthiazol-2-yl)-2, 5-diphenyltetrazolium bromide assay (Fig. 3f). Cell viability was dependent on both Dox concentration and incubation time, as expected. The half-maximal inhibitory concentration (IC_{50}) of LM-NP/Dox and LM-NP/Dox-*L* towards HeLa cells for 24 h treatment were 0.81 and 0.23 $mg\,l^{-1}$, respectively. LM-NP/Dox-*L* displayed higher cytotoxicity than free Dox solution ($IC_{50} = 1.33\,mg\,l^{-1}$), while blank LM-NPs/*L* without Dox showed insignificant cytotoxicity within the range of tested concentrations. At low Dox concentration (0.16 and 0.31 $mg\,l^{-1}$), LM-NP/Dox also had higher cytotoxicity than free Dox solution. It was suggested that the acid-promoted release of Dox achieved by liquid metal nanospheres provided higher cytotoxic activity towards cancer cells. It should be noted that LM-NP/Dox-*L* showed significantly enhanced cytotoxicity (3.5-fold for 24 h treatment) compared with LM-NP/Dox towards HeLa cells that overexpress CD44 receptor[34], which can be attributed to the HA-targeting moiety.

We further investigated *in vitro* cytotoxicity of LM-NP/Dox-*L* towards Dox-resistant HeLa cells. The results showed that LM-NP/Dox-*L* displayed significant cytotoxicity towards Dox-resistant HeLa cells ($IC_{50} = 2.46\,mg\,l^{-1}$), while free Dox solution did not show any obvious cytotoxicity even at the concentration of 5 $mg\,l^{-1}$ (Fig. 3g). This result is encouraging and offers guideline for future applications of LM-NPs in the combination treatment of cancer drug resistance.

Tumour targeting and antitumour efficacy of LM-NP/Dox-*L*. To evaluate the tumour-targeting capability of LM-NP/Dox-*L* (Fig. 1c,d), Cy5.5-labelled LM-NP/Dox-*L* (Cy5.5-(LM-NP/Dox-*L*)) was administrated intravenously into the HeLa tumour-bearing mice. Within a short time period, Cy5.5-(LM-NP/Dox-*L*) presented a stronger fluorescence signal in the tumour region compared with Cy5.5-labelled LM-NP/Dox (Cy5.5-(LM-NP/Dox)) without the HA-targeting moiety (Fig. 4a). As time increased, elevated fluorescence signals of Cy5.5-(LM-NP/Dox-*L*) were observed at the tumour site as compared with the normal tissues within 48 h post injection, indicating a notable tumour-targeting effect of the nanospheres. After 48 h post injection, the mice were killed, and the tumours as well as normal tissues were collected for *ex vivo* imaging study. The intensity of fluorescence signal of Cy5.5-(LM-NP/Dox-*L*) at tumour site was significantly higher than that of Cy5.5-(LM-NP/Dox) (Fig. 4b). On the basis of the quantitative region-of-interest analysis, the fluorescence intensity of Cy5.5-(LM-NP/Dox-*L*) at the tumour site was 1.82-fold that of Cy5.5-(LM-NP/Dox) (Fig. 4c). In addition, the fluorescence signal at the tumour site was three times higher than that in the liver or kidney (Fig. 4b,c). The potent tumour-targeting capability of LM-NP/Dox-*L* can be attributed to the combination of an enhanced permeability and retention effect and active targeting mechanisms.

The time-dependent biodistribution of LM-NPs/*L* was further quantified by inductively coupled plasma mass spectrometry. HeLa tumour-bearing nude mice were killed at 6, 24 and 48 h after intravenous injection of LM-NPs/Dox-*L* for tissue collection. The results further substantiated the potent tumour-targeting capability of LM-NPs/Dox-*L* (Supplementary Fig. 11). The pharmacokinetics of LM-NP/Dox-*L* administered intravenously into mice was evaluated by quantitatively monitoring the Dox concentration in blood plasma (Supplementary Fig. 12). The elimination half-life ($t_{1/2}$), the mean residence time and the area under the curve depicting the plasma drug concentration versus time were significantly higher than those of the free Dox solution, suggesting the capability of LM-NP/Dox-*L* to maintain a high drug concentration during a prolonged systemic circulation. The antitumour efficacy of LM-NP/Dox-*L* was further assessed in the HeLa tumour-bearing xenograft mice. Different Dox formulations displayed significant tumour inhibition effects compared with saline as a negative control after successive intravenous administration into the HeLa tumour-bearing mice (Fig. 4d, f, g). Both LM-NP/Dox and LM-NP/Dox-*L* displayed remarkably higher inhibition efficacy towards HeLa tumour growth than the free Dox solution, which primarily resulted from the tumour-targeting capability of these nanospheres. More importantly, a noticeable difference in tumour volume inhibition between LM-NP/Dox-*L* and LM-NP/Dox was observed (Fig. 4d,f,g), further demonstrating the enhanced active targeting capability of LM-NP/Dox-*L*. No significant change of mice body weights was observed during the treatment of LM-NP/Dox-*L* (Fig. 4e). The histological images using haematoxylin and eosin staining showed that a massive cancer cell remission occurred in the tumour tissue (Fig. 4h) after LM-NP/Dox-*L* administration, with no obvious pathological abnormalities in normal organs, including cardiomyopathy, the major toxic effect of Dox cancer treatment (Supplementary Fig. 13)[35]. Moreover, the images obtained using the *in situ* TUNEL assay showed the highest level of cell apoptosis in the tumour collected from the mice treated with LM-NP/Dox-*L* (Fig. 4i), indicating that the supreme tumour inhibition activity was attributable in part to the increased apoptosis induced by LM-NP/Dox-*L*. The anticancer behaviour of LM-NPs/*L* was also explored in tumour-bearing mice, where no significant anticancer effects were observed after intravenous injection of blank LM-NPs/*L* (Supplementary Fig. 14). Collectively, these results verified that HA-conjugated LM-NP/Dox-*L* efficiently accumulated at the tumour site, indicating effective receptor-mediated intracellular transport, and thereby achieved the optimal antitumour efficancy *in vivo*.

Toxicology evaluation of LM-NP/Dox-*L*. The toxicology of LM-NP/*L* to female Balb/c mice was investigated over 3 months (Fig. 5). The female Balb/c mice injected with LM-NP/*L* (45 $mg\,kg^{-1}$) were killed at 3, 7, 20, 40 and 90 days post injection for blood collection. The levels of important liver function markers, including the alanine aminotransferase, aspartate aminotransferase, alkaline phosphatase and the albumin concentration were all within reference ranges[36,37], indicating that no obvious hepatic toxicity was induced by LM-NP/*L* treatment (Fig. 5a,b). As an indicator of the kidney functions, the urea levels in the blood of treated mice were also within the normal range (Fig. 5c). For the haematological assessment, we selected white blood cells, red blood cells, haemoglobin, mean corpuscular volume, mean corpuscular haemoglobin, mean corpuscular haemoglobin concentration, platelet count and haematocrit (Fig. 5d–k). Although a vibration of platelets concentration was observed on day 3, the value itself still fell into a reference normal range[36]. In addition, a steady recovery in platelets level was observed over time. The platelets level fully recovered on day 20, with no significant difference compared with that of the age control group. All of the above indices in the LM-NP/*L*-treated groups were normal compared with the control groups. We also performed necropsy of the sacrificed mice, and no noticeable organ damage was observed (Fig. 5l). Tissues collected from the heart, brain and muscle were also investigated, and no obvious tissue injury was observed (Supplementary Fig. 15). Taken together, we concluded that LM-NP/*L* displayed no obvious toxicity at the treatment dose, which was highly encouraging for the application of LM-NP/*L* as nanomedicine.

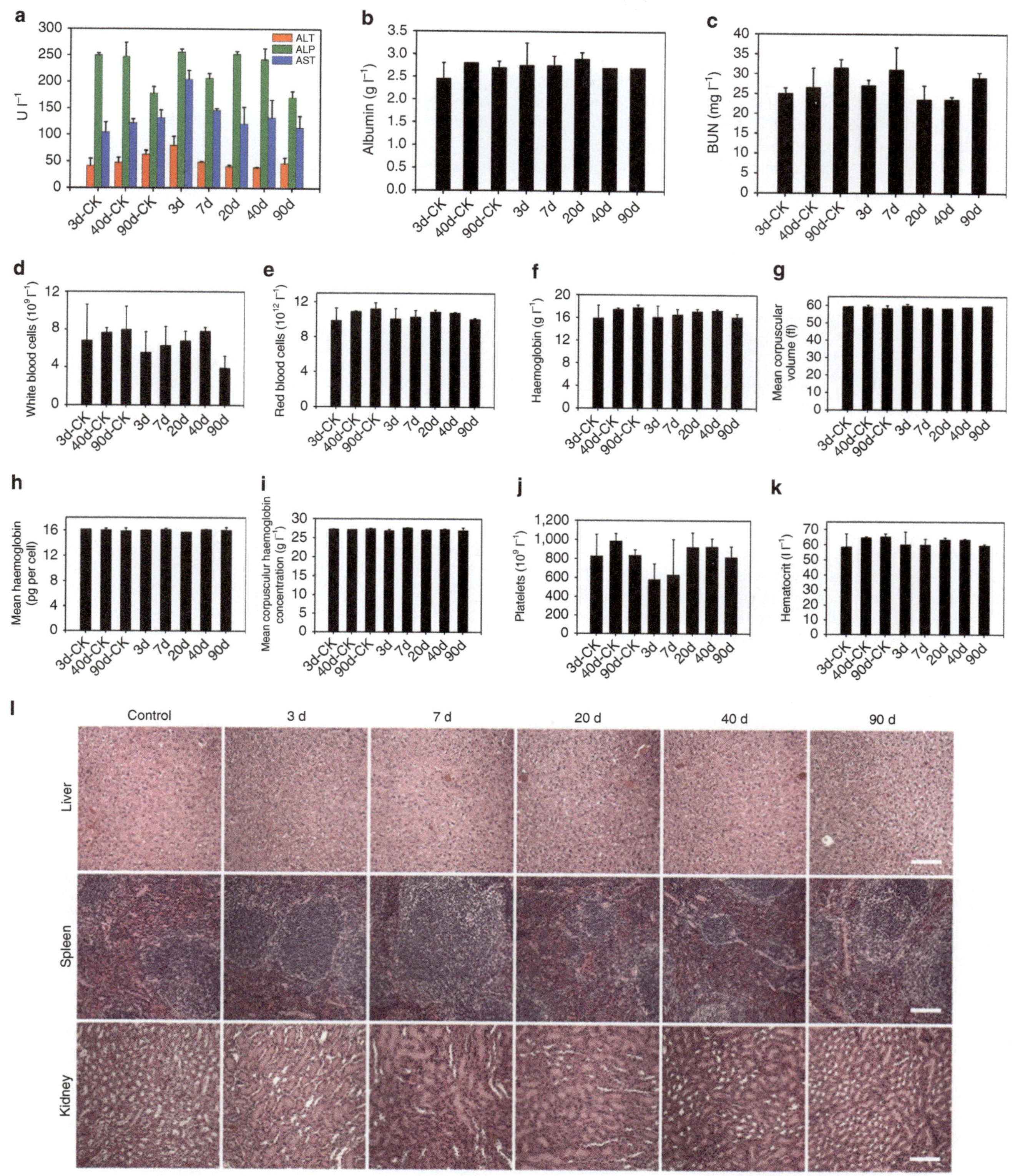

Figure 5 | Toxicology evaluation. Blood biochemistry and haematology data of female Balb/c mice treated with LM-NPs/*L* at the dose of 45 mg kg^{-1} (total nanocarrier dose used in antitumuor efficacy study) at 3, 7, 20, 40 and 90 d: (**a**) alanine aminotransferase (ALT), alkaline phosphatase (ALP) and aspartate aminotransferase (AST) levels in the blood at different time points after LM-NPs/*L* treatment. (**b,c**) Time-course changes of albumin concentration and blood urea nitrogen (BUN). (**d-k**) Time-course changes of white blood cells (**d**), red blood cells (**e**), platelets (**f**), haemoglobin (**g**), mean corpuscular volume (**h**), mean corpuscular haemoglobin (**i**), mean corpuscular haemoglobin concentration (**j**) and haematocrit (**k**) from control mice (CK) and LM-NPs/*L*-treated mice. Error bars indicate s.d. ($n = 3$). (**l**) Histology evaluation of the major organs (liver, spleen and kidney) collected from the control untreated mice and LM-NPs/*L*-injected mice at different time points post injection. Scale bars, 100 μm. d, days.

The single-dose, acute toxicity of LM-NP/*L* and LM-NP/Dox-*L* to female Balb/c mice was investigated by identifying the maximum tolerated dose (MTD). The estimated MTD of LM-NP/*L* was determined to be 700 mg kg^{-1}, suggesting low acute toxicity. The estimated MTD of LM-NP/Dox-*L* was found to be 55 mg kg^{-1} (Dox dose), 1.75-fold higher than that of free Dox solution. In addition, no obvious increase in IgE production was observed in LM-NP/*L*-injected mice, revealing no sign of allergic reactions (Supplementary Fig. 16).

Moreover, we have also monitored the time-dependent excretion to study the *in vivo* metabolism of LM-NPs/*L* (Supplementary Fig. 17). The results suggested that the clearance of LM-NPs/*L* was possibly through both fecal and renal excretions. The excreted concentrations of both gallium and indium were steadily decreased over time.

Discussion

We have developed a new liquid metal-based drug delivery platform for anticancer therapy. The formulation can be easily formed and tailored via ligand-mediated self-assembly using facile ultrasonication. The resulting liquid-metal nanospheres are able to fuse for promoting drug release and eventually degrade under a mild acidic environment. After fusing, these nanospheres also display a contrast enhancing capability when imaged by X-ray, suggesting potential as a theranostic reagent. Systematic investigation of toxicology of LM-NP/*L* revealed no obvious toxicity at the treatment dose, favouring its biomedical applications. Although there were no obvious signs of toxicity in these studies, we will further evaluate the systemic toxicity of the developed formulations, as well as other promising candidates made from different liquid metal, such as pure gallium. Moreover, considering other physical properties of liquid metal, including capability of dissolving iron or other metals, mechanical flexibility towards tissues and electrical conductivity, many new materials and scaffolds integrated with liquid metal for drug delivery and tissue engineering can be envisioned.

Methods

Materials. All chemicals were purchased from Sigma-Aldrich (St Louis, MO, USA) unless otherwise stated. Thiol–polyethylene glycol–amine (HS-PEG-NH_2) with the PEG molecular weight of 1,000 was purchased from Sigma-Aldrich and Nanocs Inc. (New York, NY, USA). Dox hydrochloride was purchased from BIOTANG Inc. (Lexington, MA, USA).

Preparation and characterization. For a typical LM-NP/*L* preparation, a small amount of EGaIn (80 μl) was added to a centrifuge tube (50 ml), which was filled to a total volume of 12 ml with ethanol containing 4 mg thiolate CD and 0.5 mg thiolate HA. After sonication, the largest particles precipitated within seconds, and the slurry was removed from the vial. The remained solution was further purified by mild centrifugation (1,000 r.p.m.) to remove relatively large particles. The resulting nanospheres were collected via centrifugation and resuspended in PBS buffer. To prepare LM-NP, EGaIn (80 μl) was added to a centrifuge tube (50 ml), which was filled to a total volume of 12 ml with ethanol containing 4 mg thiolate CD. After sonication, the largest particles precipitated within seconds, and the slurry was removed from the vial. The remained solution was further purified by mild centrifugation (1,000 r.p.m.) to remove relatively large particles. The resulting nanospheres were collected via centrifugation and resuspended in PBS buffer. Cy5.5-labelled nanospheres were prepared by adding Cy5.5-conjugated HS-PEG-NH_2 (molecular weight: 1,000) into the ethanol containing mixture. For loading of Dox, 1 ml of LM-NP/L was incubated with 100 μl of triethanolamine-treated Dox in dimethyl sulfoxide (5 mg ml^{-1}) with stirring at room temperature overnight. The resulting Dox-loaded LM-NP/*L* was extensively washed with PBS via centrifugation to remove the excess Dox. LM-NP/Dox, LM-NP/Rho and LM-NP/Rho-*L* were prepared following the same procedure. For TEM characterization, LM-NP/Dox was cast onto a TEM copper grid (300 mesh; Ted Pella). After drying in air, the sample was observed by TEM (JEM-2000FX, Hitachi) operating at 200 kV.

***In vitro* acid-triggered Dox and PTX release.** LM-NP/Dox and LM-NP/Dox-*L* (0.5 ml) were added into a dialysis tube (10K MWCO; Slide-A-Lyzer, Thermo Scientific) and dialysed against 14 ml of the PBS buffer solution at pH 5.0 or 7.4, and gently shaken at 37 °C in a shaker (New Brunswick Scientific) at 100 r.p.m. At predetermined time intervals, the total buffer solution was replaced with 14 ml of fresh buffer solution with the same pH. The fluorescence intensity of Dox released was measured at 596 nm with an excitation wavelength of 480 nm using a microplate reader (Infinite M200 PRO, Tecan). PTX release study was conducted following the same procedure. The concentration of PTX was determined by high-performance liquid chromatography (Hewlett Packard 1100).

Cell culture. HeLa cells were obtained from Tissue Culture Facility of UNC Lineberger Comprehensive Cancer Center and cultured in DMEM with 10% (v:v) fetal bovine serum, 100 U ml^{-1} penicillin and 100 μg ml^{-1} streptomycin in an incubator (Thermo Scientific) at 37 °C under an atmosphere of 5% CO_2 and 90% relative humidity. The cells were sub-cultivated approximately every 3 days at 80% confluence using 0.25% (w:v) trypsin at a split ratio of 1:5.

Intracellular trafficking. HeLa cells (1×10^5 cells per well) were seeded in a confocal microscopy dish (MatTek). After culture for 24 h, the cells were incubated with LM-NP/Dox-*L* (2 μM Dox concentration) at 37 °C for 1 h and 4 h, respectively, and then washed twice using PBS at 4 °C. Subsequently, the cells were stained with LyosTracker Green (50 nM) (Life Technologies) at 37 °C for 30 min and Hoechst 33342 (1 μg ml^{-1}; Life Technologies) at 37 °C for 10 min. Finally, the cells were washed using PBS twice at 4 °C and immediately observed using confocal laser scanning microscopy (LSM 710, Zeiss).

***In vitro* cytotoxicity.** HeLa cells (1×10^4 cells per well) were seeded in the 96-well plates. After culture for 24 h, the cells were exposed to the Dox solution and Dox-loaded LM-NPs with different concentrations of Dox for 24 h, followed by adding 20 μl of the 3-(4, 5-dimethylthiazol-2-yl)-2, 5-diphenyltetrazolium bromide solution (5 mg ml^{-1}). After 4 h of incubation, the medium was removed, and the cells were mixed with 150 μl of dimethyl sulphoxide. The absorbance was measured at a test wavelength of 570 nm and a reference wavelength of 630 nm by a microplate reader (Infinite M200 PRO, Tecan).

Apoptosis assay. Apoptosis of HeLa cells was detected using the APO-BrdU TUNEL Assay Kit (Life Technologies) and Annexin V-FITC Apoptosis Detection Kit (BD Biosciences). The cells (1×10^5 cells per well) were seeded in the six-well plates. After culture for 48 h, the cells were incubated with Dox-loaded LM-NPs for 12 h (Annexin V-FITC) or 20 h (TUNEL). The subsequent procedures were performed in accordance with the manufacturer's protocol. For Annexin V-FITC apoptosis detection, the cells were analysed by flow cytometry (BD FACSCalibur), while for the TUNEL assay the cells were observed by fluorescence microscope (IX71, Olympus).

Animals and tumour xenograft models. All animals were treated in accordance with the Guide for Care and Use of Laboratory Animals, approved by the Institutional Animal Care and Use Committee (IACUC) of North Carolina State University. To set-up the tumour xenograft model, the female nude mice (6 weeks, J:NU, The Jackson Laboratory) were subcutaneously inoculated in the back with 1×10^7 HeLa cells. The tumour size was monitored by a vernier caliper and the tumour volume (V) was calculated as $V = L \times W^2/2$, where L and W were the length and width of the tumour, respectively.

***In vivo* imaging study.** The animal study protocol was approved by the Institutional Animal Care and Use Committee at North Carolina State University and University of North Carolina at Chapel Hill. When the tumours reached to 200–400 mm^3, the mice were intravenously injected by Cy5.5-(LM-NP/Dox) and Cy5.5-(LM-NP/Dox-*L*) at Cy5.5 dose of 30 nmol kg^{-1}. Images were taken on the IVIS Lumina imaging system (Caliper, USA) at 6, 24 and 48 h post injection[29]. After the 48-h scanning, the mice were killed. The tumours as well as major organs were collected, weighed and subjected for *ex vivo* imaging. Region-of-interests were circled around the organs, and the fluorescence intensities were analysed by Living Image Software.

***In vivo* antitumour efficacy.** The tumour-bearing mice were weighed and randomly divided into different groups when the tumour volume reached to 50 mm^3. From day 0, the mice were intravenously injected with Dox solution (2 mg kg^{-1}), LM-NP/Dox (2 mg kg^{-1}), LM-NP/Dox-*L* (2 mg kg^{-1}) and saline as a negative control every other day for 12 days, and meanwhile the tumour size was measured. At day 14, the mice were killed, and the tumour as well as the heart were collected, weighed, washed with saline thrice and fixed in the 10% neutral-buffered formalin. For the haematoxylin and eosin staining, the formalin-fixed tumours and hearts were embedded in paraffin blocks and visualized by optical microscope (DM5500B, Leica). For the TUNEL apoptosis staining, the fixed tumour sections were stained by the *In Situ* Cell Death Detection Kit (Roche Applied Science) according to the manufacturer's protocol. Hoechst 33342 was used for nuclear counterstaining. The stained tumour slides were observed by fluorescence microscope (IX71, Olympus).

References

1. Liong, M. *et al.* Multifunctional inorganic nanoparticles for imaging, targeting, and drug delivery. *ACS Nano* **2,** 889–896 (2008).
2. Huang, H.-C., Barua, S., Sharma, G., Dey, S. K. & Rege, K. Inorganic nanoparticles for cancer imaging and therapy. *J. Control. Release* **155,** 344–357 (2011).
3. Langille, M. R., Personick, M. L., Zhang, J. & Mirkin, C. A. Defining rules for the shape evolution of gold nanoparticles. *J. Am. Chem. Soc.* **134,** 14542–14554 (2012).
4. Sun, T. *et al.* Engineered nanoparticles for drug delivery in cancer therapy. *Angew. Chem. Int. Ed. Engl.* **53,** 12320–12364 (2014).
5. Tian, B. *et al.* Three-dimensional, flexible nanoscale field-effect transistors as localized bioprobes. *Science* **329,** 830–834 (2010).
6. Mitragotri, S. *et al.* Accelerating the translation of nanomaterials in biomedicine. *ACS Nano* **9,** 6644–6654 (2015).
7. Peer, D. *et al.* Nanocarriers as an emerging platform for cancer therapy. *Nat. Nanotechnol.* **2,** 751–760 (2007).
8. Chow, E. K. *et al.* Nanodiamond therapeutic delivery agents mediate enhanced chemoresistant tumor treatment. *Sci. Transl. Med.* **3,** 73ra21 (2011).
9. Timko, B. P. *et al.* Near-infrared–actuated devices for remotely controlled drug delivery. *Proc. Natl Acad. Sci. USA* **111,** 1349–1354 (2014).
10. Xu, C., Wang, B. & Sun, S. Dumbbell-like Au-Fe(3)O(4) nanoparticles for target-specific platin delivery. *J. Am. Chem. Soc.* **131,** 4216–4217 (2009).
11. Kim, H. J. *et al.* Precise engineering of sirna delivery vehicles to tumors using polyion complexes and gold nanoparticles. *ACS Nano* **8,** 8979–8991 (2014).
12. Soo Choi, H. *et al.* Renal clearance of quantum dots. *Nat. Biotechnol.* **25,** 1165–1170 (2007).
13. Tang, L. *et al.* Investigating the optimal size of anticancer nanomedicine. *Proc. Natl Acad. Sci. USA* **111,** 15344–15349 (2014).
14. Mitragotri, S., Burke, P. A. & Langer, R. Overcoming the challenges in administering biopharmaceuticals: formulation and delivery strategies. *Nat. Rev. Drug Discov.* **13,** 655–672 (2014).
15. Choi, H. S. *et al.* Design considerations for tumour-targeted nanoparticles. *Nat. Nanotechnol.* **5,** 42–47 (2010).
16. Chou, L. Y. T., Zagorovsky, K. & Chan, W. C. W. DNA assembly of nanoparticle superstructures for controlled biological delivery and elimination. *Nat. Nanotechnol.* **9,** 148–155 (2014).
17. Boley, J. W., White, E. L. & Kramer, R. K. Nanoparticles: mechanically sintered gallium–indium nanoparticles. *Adv. Mater.* **27,** 2270–2270 (2015).
18. Khan, M. R., Eaker, C. B., Bowden, E. F. & Dickey, M. D. Giant and switchable surface activity of liquid metal via surface oxidation. *Proc. Natl Acad. Sci. USA* **111,** 14047–14051 (2014).
19. Dickey, M. D. Emerging applications of liquid metals featuring surface oxides. *ACS Appl. Mater. Interfaces* **6,** 18369–18379 (2014).
20. Hohman, J. N. *et al.* Directing substrate morphology via self-assembly: ligand-mediated scission of gallium–indium microspheres to the nanoscale. *Nano Lett.* **11,** 5104–5110 (2011).
21. Love, J. C., Estroff, L. A., Kriebel, J. K., Nuzzo, R. G. & Whitesides, G. M. Self-assembled monolayers of thiolates on metals as a form of nanotechnology. *Chem. Rev.* **105,** 1103–1169 (2005).
22. Ladd, C., So, J. H., Muth, J. & Dickey, M. D. 3D printing of free standing liquid metal microstructures. *Adv. Mater.* **25,** 5081–5085 (2013).
23. Davis, M. E. & Brewster, M. E. Cyclodextrin-based pharmaceutics: past, present and future. *Nat. Rev. Drug Discov.* **3,** 1023–1035 (2004).
24. Aruffo, A., Stamenkovic, I., Melnick, M., Underhill, C. B. & Seed, B. CD44 is the principal cell surface receptor for hyaluronate. *Cell* **61,** 1303–1313 (1990).
25. Maeda, H., Wu, J., Sawa, T., Matsumura, Y. & Hori, K. Tumor vascular permeability and the EPR effect in macromolecular therapeutics: a review. *J. Control. Release* **65,** 271–284 (2000).
26. Lu, Y., Sun, W. & Gu, Z. Stimuli-responsive nanomaterials for therapeutic protein delivery. *J. Control. Release* **194,** 1–19 (2014).
27. Collery, P., Keppler, B., Madoulet, C. & Desoize, B. Gallium in cancer treatment. *Crit. Rev. Oncol. Hematol.* **42,** 283–296 (2002).
28. Wang, F., Jiang, X., Yang, D. C., Elliott, R. L. & Head, J. F. Doxorubicin-gallium-transferrin conjugate overcomes multidrug resistance: evidence for drug accumulation in the nucleus of drug resistant MCF-7/ADR cells. *Anticancer Res.* **20,** 799–808 (2000).
29. Mo, R., Jiang, T., DiSanto, R., Tai, W. & Gu, Z. ATP-triggered anticancer drug delivery. *Nat. Commun.* **5,** 3364 (2014).
30. Montaseri, H. & Taxol, Z. *Solubility, Stability, and Bioavailability* (Doctoral dissertation, University of Alberta, 1997).
31. Thelen, J., Dickey, M. D. & Ward, T. A study of the production and reversible stability of EGaIn liquid metal microspheres using flow focusing. *Lab Chip.* **12,** 3961–3967 (2012).
32. Vermes, I., Haanen, C., Steffens-Nakken, H. & Reutellingsperger, C. A novel assay for apoptosis Flow cytometric detection of phosphatidylserine expression on early apoptotic cells using fluorescein labelled Annexin V. *J. Immunol. Methods* **184,** 39–51 (1995).
33. Hakem, R. *et al.* Differential requirement for caspase 9 in apoptotic pathways *in vivo*. *Cell* **94,** 339–352 (1998).
34. Zöller, M. CD44: can a cancer-initiating cell profit from an abundantly expressed molecule? *Nat. Rev. Cancer* **11,** 254–267 (2011).
35. Takemura, G. & Fujiwara, H. Doxorubicin-induced cardiomyopathy: from the cardiotoxic mechanisms to management. *Prog. Cardiovasc. Dis.* **49,** 330–352 (2007).
36. Ynag, K. *et al.* *In vivo* pharmacokinetics, long-term biodistribution, and toxicology of PEGylated graphene in mice. *ACS Nano* **5,** 516–522 (2011).
37. Zaias, J., Mineau, M., Cray, C., Yoon, D. & Altman, N. H. Reference values for serum proteins of common laboratory rodent strains. *J. Am. Assoc. Lab. Anim. Sci.* **48,** 387 (2009).

Acknowledgements

This work was supported by NC TraCS, NIH's Clinical and Translational Science Awards (CTSA, 1UL1TR001111) at UNC-CH and the NC State Faculty Research and Professional Development Award to Z.G. M.D.D. and Yi.L. acknowledge support from the NSF CAREER CMMI-0954321 and the Research Triangle NSF MRSEC on Programmable Soft Matter (DMR-1121107). We acknowledge Dr Ran Mo, Dr Tianyue Jiang, Dr Gabriel McKeon, Dr Hong Yuan and Joseph Merrill for their assistance for animal study.

Author contributions

Yu.L. and Z.G. designed the project. Yu.L., Q.H., Yi.L. and C.W. performed the experiments. Yu.L., Z.G., M.D.D. and F.S.L. wrote the paper. All authors analysed and interpreted the data.

Additional information

Competing financial interests: The authors declare no competing financial interests.

6

A plug-and-play approach to antibody-based therapeutics *via* a chemoselective dual click strategy

Antoine Maruani[1], Mark E.B. Smith[1], Enrique Miranda[2], Kerry A. Chester[2], Vijay Chudasama[1] & Stephen Caddick[1]

Although recent methods for the engineering of antibody-drug conjugates (ADCs) have gone some way to addressing the challenging issues of ADC construction, significant hurdles still remain. There is clear demand for the construction of novel ADC platforms that offer greater stability, homogeneity and flexibility. Here we describe a significant step towards a platform for next-generation antibody-based therapeutics by providing constructs that combine site-specific modification, exceptional versatility and high stability, with retention of antibody binding and structure post-modification. The relevance of the work in a biological context is also demonstrated in a cytotoxicity assay and a cell internalization study with HER2-positive and -negative breast cancer cell lines.

[1] Department of Chemistry, University College London, 20 Gordon Street, London WC1H OAJ, UK. [2] UCL Cancer Institute, 72 Huntley Street, London WC1E 6BT, UK. Correspondence and requests for materials should be addressed to V.C. (email: v.chudasama@ucl.ac.uk) or to S.C. (email: s.caddick@ucl.ac.uk).

Antibody–drug conjugates (ADCs) are comprised of antibodies that are armed with highly potent warheads using various conjugation/linker technologies[1–4]. This class of therapeutic combines the directing ability of antibodies (that is, allowing for discrimination between healthy and diseased tissue) with the cell-killing ability of potent cytotoxic drugs. This powerful class of targeted therapy has shown considerable promise in the treatment of various cancers with two US Food and Drug Administration (FDA)-approved ADCs currently on the market (Adcetris and Kadcyla) and over 30 ADCs currently in the clinic[5,6]. However, in order for ADCs to deliver their full potential, sophisticated conjugation technologies to connect the warhead to the antibody and novel strategies and approaches for their construction are required[7,8]. Conjugation to native ADCs is typically achieved through either multiple lysine modification or by functionalization of thiols generated by reduction of interchain disulfide bonds; neither of which is ideal (Fig. 1)[7,8]. Lysine modification is suboptimal as it results in batch-to-batch variability and generates heterogeneous ADCs, which have been shown to have a narrow therapeutic window relative to homogeneous ADCs, therefore having major pharmacokinetic limitations[9,10]. Cysteine modification, following interchain disulfide reduction, results in the permanent loss of structural disulfide bonds, which may reduce the stability of the ADC *in vivo*[7,8]. It also generates heterogeneous mixtures when targeting typical drug-to-antibody ratios. Other approaches using cysteine-based site-directed mutagenesis and unnatural amino acids have also been described[10,11], but similarly have limitations, for example, disulfide scrambling post-reduction and high cost combined with relatively low expression yields, respectively.

Recently, we have described methods for the insertion of small molecules into disulfide bonds in various proteins, engineered antibody single-chain variable fragments, a fragment antigen-binding (Fab) construct to yield site-selectively modified products and a full antibody[12–19]. This has gone some way to addressing the issues of site-selective antibody modification. There is also relevant work in this area by Godwin and coworkers, which highlights an alternative set of reagents for site-selective disulfide bridging of antibodies[20]. However, significant barriers still remain in this rapidly evolving field and there is demand for novel ways of constructing ADCs with stable and versatile linkers with retention of core antibody structure[1–6].

In this manuscript, we describe a significant step towards a platform for next-generation ADCs by providing constructs that combine site-specific native antibody modification, exceptional versatility (*via* a 'dual click' approach), high stability and retention of antibody structure post-modification. The technology, at its core, is based on the insertion of pyridazinediones (PDs) bearing orthogonal 'clickable' handles into native disulfide bonds in antibody fragments and full antibodies, with a view to then carry out two orthogonal transformations to yield multifunctionalized adducts (Fig. 2). This enables the rapid assembly of dual-modified ADCs in a highly convergent manner. The work described herein could pave the way to novel antibody-based therapeutics.

Results

Antibody scaffold, drug and fluorophore selection. To evaluate this chemistry, a suitable antibody system and cytotoxic drug needed to be selected. Trastuzumab (Herceptin), a monoclonal immunoglobulin G1 (IgG1) that targets the internalizing HER2 receptor, has been used successfully in the treatment of HER2+ breast cancer and is the antibody component of a recently FDA-approved ADC therapy for the same indication, trastuzumab emtansine (Kadcyla)[21,22]. Anticancer drug doxorubicin (Dox) has been used as a cytotoxic model payload previously and has a relatively distinctive absorbance maximum at 495 nm to facilitate determination of drug-to-antibody ratios by ultraviolet–visible absorption[12]. As such, Herceptin and Dox were chosen as the antibody and cytotoxic platforms, respectively. To analyse the effectiveness of the 'dual click' approach on a full antibody scaffold, where accurate mass spectrometry analysis is limited, a second light absorbing moiety that absorbs at a distinct wavelength to Dox was needed to enable facile analysis by ultraviolet–visible spectrometry of the loading of each cargo. To this end, a photostable, water-soluble, cyanine-based fluorophore with a maximum absorbance at 646 nm (sulfo-Cy5) was selected.

Choice of linker. In order to deliver a widely applicable and versatile approach to antibody modification, it was rationalized that an exceptionally stable linker bearing multiple modalities that could be introduced *via* conjugation onto native antibodies was required. A suitable scaffold was dibromopyridazinedione (diBrPD) as it has previously been shown to be efficient at inserting into disulfide bonds and the resulting constructs to be exceptionally stable to hydrolysis, even at high temperatures (Fig. 3)[18]. Moreover, their structure is appealing as they are ideally set up for attaching various modalities *via* each nitrogen atom. As we wanted the platform to be versatile and widely applicable, orthogonal 'clickable' handles, one on each nitrogen atom, onto the PD motif were to be attached. To this end, Astra-PD **1**, an alkyne-strained alkyne-pyridazinedione was synthesized (Fig. 3, see Supplementary Information for details on the synthesis).

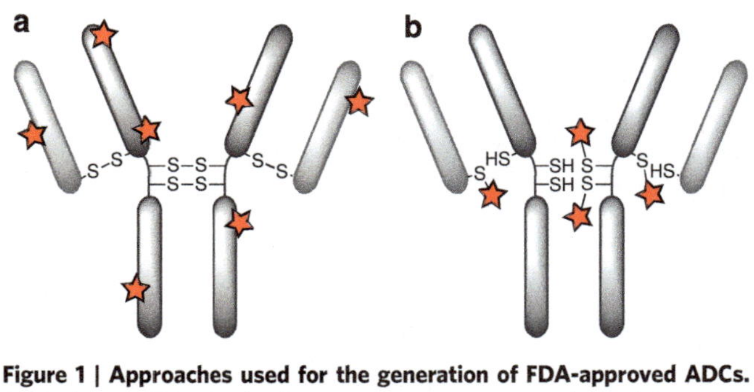

Figure 1 | Approaches used for the generation of FDA-approved ADCs. (**a**) Lysine modification leading to a heterogeneous product mixture, undefined physical and pharmacokinetic properties, and wide distribution of drug-to-antibody ratio. (**b**) Modification of reduced disulfide bonds leading to a heterogeneous product mixture, loss of structural disulfide bonds and a suboptimal therapeutic index.

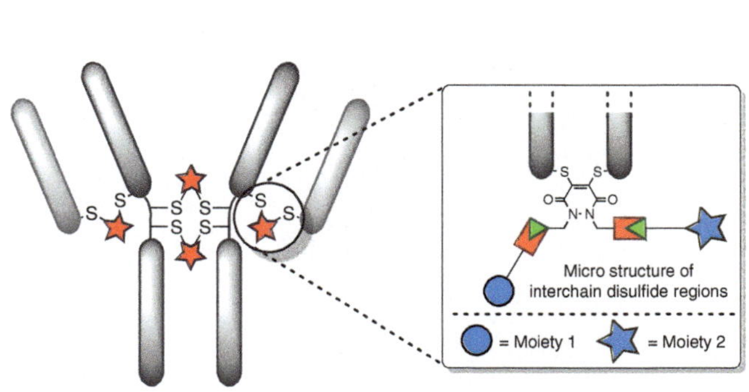

Figure 2 | Functional disulfide re-bridging followed by a dual click approach. Disulfide re-bridging by a pyridazinedione construct yields a site-selectively modified antibody with dual modalities.

Appraisal of dual click strategy on an antibody fragment. With PD-construct **1** in hand, the insertion of this molecule into a Fab fragment of Herceptin; Fab-Her **2**, by simple reduction of the single interchain disulfide bond, followed by functional disulfide re-bridging with the PD construct was carried out (Fig. 4), which afforded exclusive formation of a re-bridged Fab fragment with a PD molecule inserted into the disulfide bond, construct **3**, by mass spectrometry and SDS–PAGE. No further purification was required and a yield in excess of 95% was obtained (Supplementary Figs 11 and 12).

Moreover, the two orthogonal reactive handles could be utilized to introduce distinct functionalities selectively. For example, construct **3** was reacted with PEG$_4$-N$_3$ followed by reaction with sulfo-Cy5-N$_3$, by applying strain-promoted azide–alkyne cycloaddition (SPAAC) and copper(I)-catalysed azide–alkyne cycloaddition (CuAAC) chemistry, respectively, to form species **5** (Fig. 4). These reactions were shown to be clean and high yielding with complete selectivity, as demonstrated by mass spectrometry and SDS–PAGE (Fig. 4 and Supplementary Figs 13 and 14).

Furthermore, model diethyl-PD-modified Fab construct **6** (Fab-Diet) was shown to have: (i) comparable binding activity by enzyme-linked immunosorbent assay (ELISA) to Fab-Her **2**; (ii) excellent stability in blood plasma mimicking conditions for 7 days; (iii) complete stability after 8 months of storage at 4 °C in phosphate-buffered saline (PBS); and stability in (iv) low pH (3.1) and (v) high pH (9.0) conditions at 37 °C over a protracted period (the blood plasma mimicking conditions correspond to the simulated body fluid described by Jalota *et al.*[23] with addition of human serum albumin (600 μM, 40 mg ml^{-1}) and glutathione (20 μM); Fig. 5 and Supplementary Figs 15–20).

There are plentiful opportunities in multimodal imaging and theranostics, which can be made available by the dual labelling technology described by attachment of a fluorophore and a radiolabel or cytotoxic drug, respectively[24,25]. Furthermore, this technology allows for the construction of novel ADC fragment-based products. In terms of therapeutics, antibody fragments offer potential advantages over full antibodies (for example, fragments are smaller and offer greater tissue penetration, and many lack the antibody Fc region and can therefore be expressed in bacterial cell lines as there is no requirement for post-translational glycosylation)[3,26–28]. However, antibody fragments below the *ca.* 60–70 kDa size cut-off for glomerular filtration lack the stability afforded by the antibody Fc region and therefore often have insufficient lifetime in blood serum required to elicit a beneficial therapeutic effect without intensive and frequent therapy[28–33]. It is envisaged that the technology will allow antibody fragment therapeutics to be far more accessible by delivering long lifetime antibody fragment–drug conjugates. This could be achieved by the attachment of an antibody fragment to: (i) a suitable toxic drug; and (ii) a blood serum-stabilizing functionality (for example, PEG, albumin or albumin-binding functionality) to increase lifetime *in vivo*, using the dual labelling technology. Moreover, as the chemistry allows for the controlled stoichiometric addition of each moiety, it will allow for the construction of well-characterized products.

The construction of an antibody fragment functionalized with a life-extension technology and a cytotoxic drug was initiated by choosing Dox as the drug payload and a 20 kDa PEG chain as the

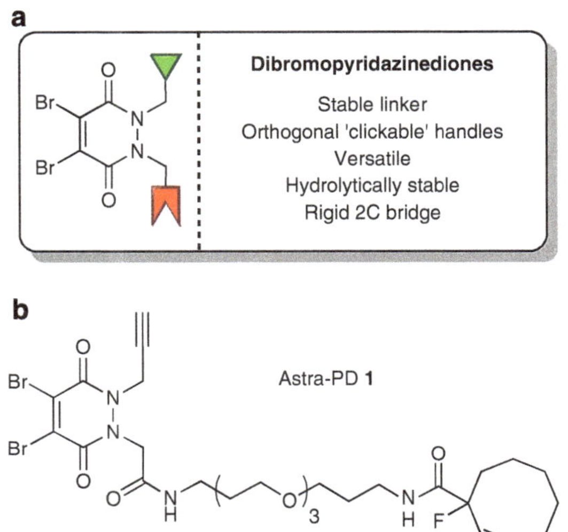

Figure 3 | Properties of dibromopyridazinediones and structure of Astra-PD 1. (**a**) General properties of the pyridazinedione platform. (**b**) Chemical structure of Astra-PD **1** with an alkyne and a strained alkyne handle.

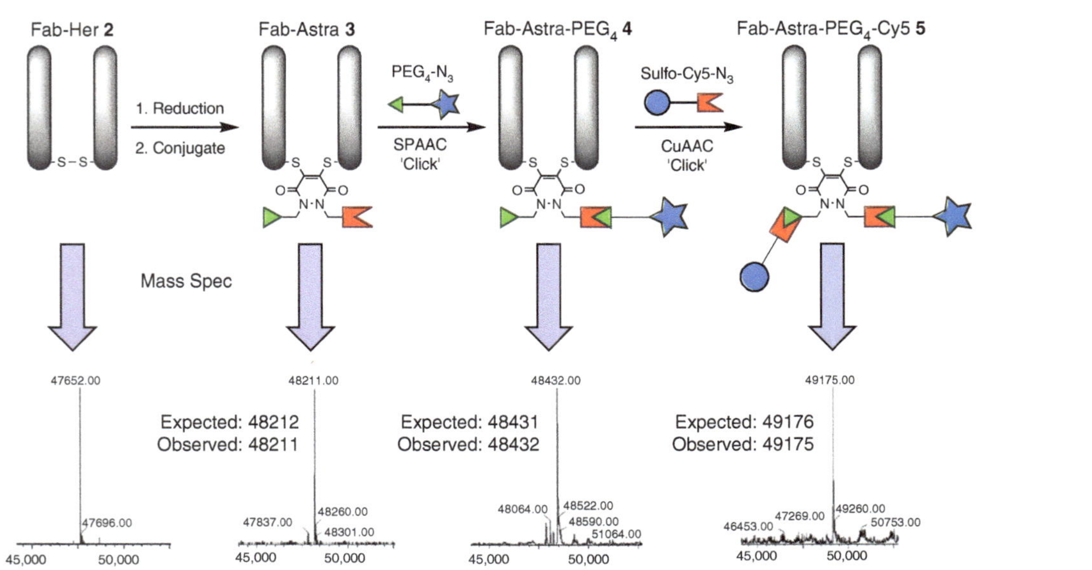

Figure 4 | Sequential modifications from native Fab-Her 2 to afford Fab-Astra-PEG4-Cy5 5. Re-bridging of Fab-Her **2** and regioselective dual modification using SPAAC with PEG$_4$-N$_3$ followed by CuAAC with Sulfo-Cy5-N$_3$ to yield Fab-Astra-PEG$_4$-Cy5 **5** conjugate as a single product.

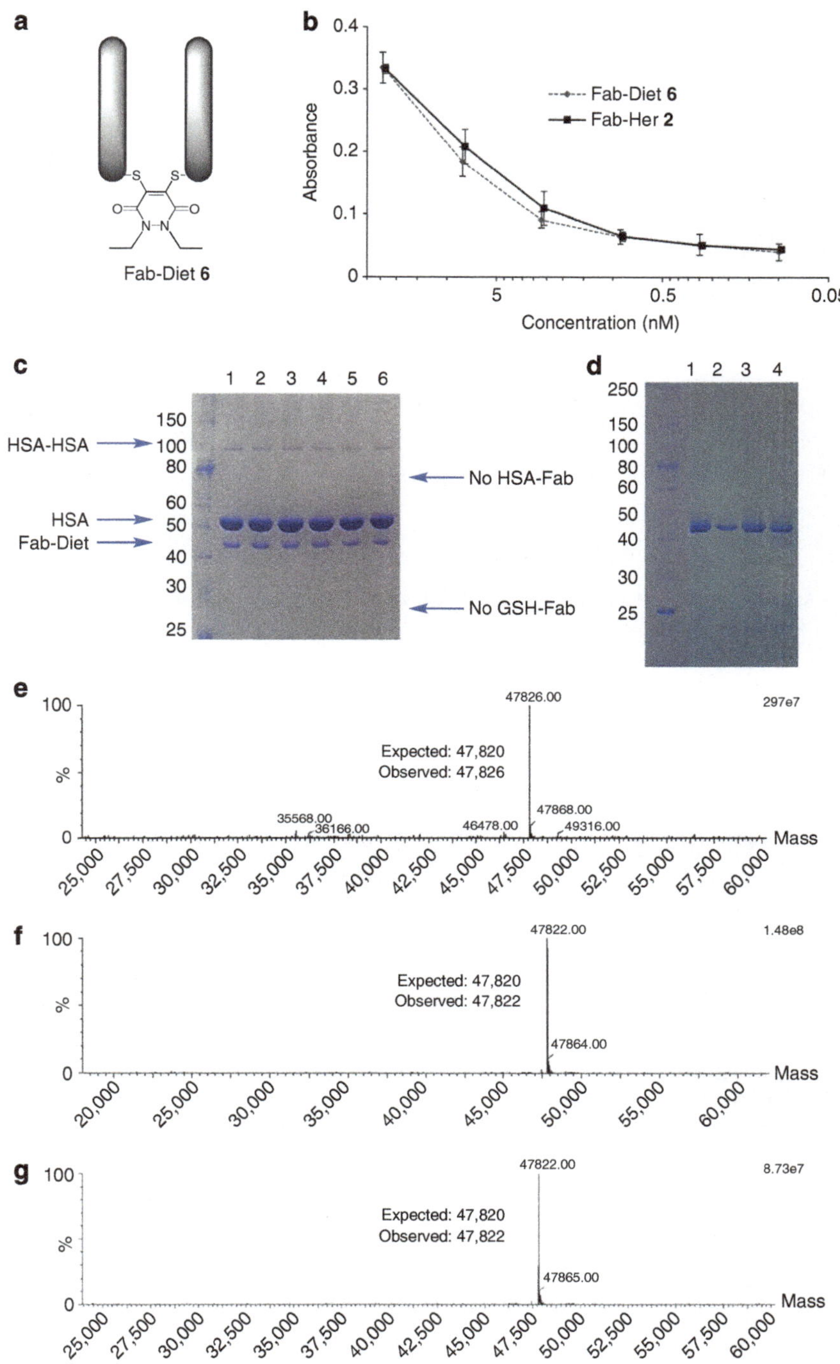

Figure 5 | Activity and stability data for model construct Fab-Diet 6. (**a**) Structure of Fab-Diet **6**. (**b**) Binding activity data for native Fab-Her **2** and Fab-Diet **6** assessed by ELISA. (**c**) SDS-PAGE gel following incubation of Fab-Diet **6** in blood plasma mimicking conditions for 0, 1, 2, 3, 5 and 7 days (lanes 1-6, respectively). (**d**) SDS-PAGE gel of Fab-Diet **6**, Fab-Diet **6** after 8 months of storage at 4 °C in PBS, Fab-Diet **6** after 24 h at 37 °C in buffer pH 3.1 and Fab-Diet **6** after 24 h at 37 °C in buffer pH 9.0 (lanes 1-4, respectively). (**e**) Deconvoluted MS data of Fab-Diet **6** after 8 months of storage at 4 °C in PBS. (**f**) Deconvoluted MS data of Fab-Diet **6** after 24 h at 37 °C in buffer pH 3.1. (**g**) Deconvoluted MS data of Fab-Diet **6** after 24 h at 37 °C in buffer pH 9.0.

lifetime extension modality (as its effects on half-life are well-characterized *in vivo* and it does not significantly compromise the penetration ability of a Fab fragment)[34–36]. As such, Fab-Astra **3** was treated with Dox-N_3 (see Supplementary Information for details on the synthesis) and PEG_{20k}-N_3 in a sequential manner using SPAAC and CuAAC chemistry, respectively, to afford construct **7** by liquid chromatography–mass spectrometry (LC–MS) and SDS–PAGE analysis, without the need for gel-filtration chromatography (Fig. 6 and Supplementary Figs 21 and 22). The formation of this construct with almost complete retention of binding activity by ELISA on HER2 and no unspecific binding observed on EGFR (HER1; Fig. 6 and Supplementary Fig. 23) paves the way for the appraisal of the strategy on delivering antibody fragment-based therapeutics. The retention in binding affinity post-conjugation to a large construct is due to the installation of the PD-bridging moiety at a position that is distal from the binding site, which is another important feature of the chemistry. Moreover, the construction of a wide variety of similar lifetime extension/drug modalities should be facile and allow for rapid evaluation of various combinations

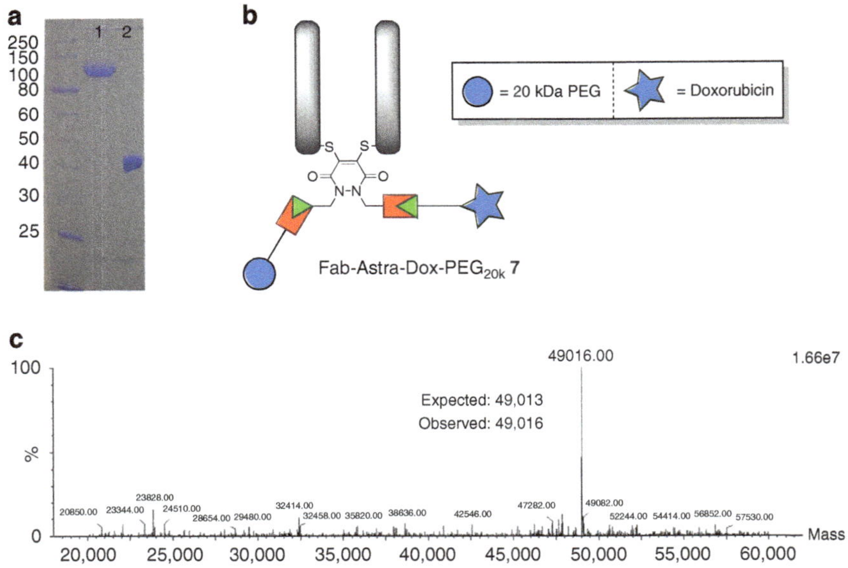

Figure 6 | Construction of Fab-Astra-Dox-PEG$_{20k}$ 7, a fragment-based ADC with an additional lifetime extension modality. (**a**) SDS-PAGE analysis for bioconjugates Fab-Astra-Dox-PEG$_{20k}$ **7** and Fab-Astra-Dox (lanes 1 and 2, respectively). (**b**) Structure of bioconjugate Fab-Astra-Dox-PEG$_{20k}$ **7**. (**c**) Deconvoluted MS data of Fab-Astra-Dox.

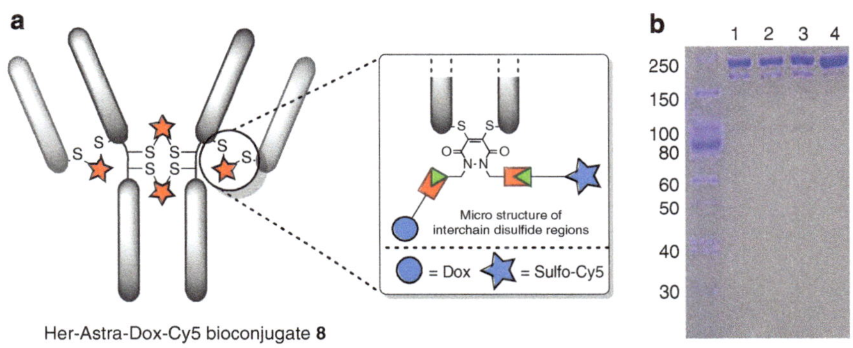

Figure 7 | Construction of Her-Astra-Dox-Cy5 8, a dually modified ADC with both a cytotoxic drug and a fluorophore. (**a**) Structure of fully re-bridged and dually functionalized Herceptin. (**b**) SDS-PAGE analysis of Her-Astra, Her-Astra-Dox, Her-Astra-Dox-Cy5 **8** and Herceptin (lanes 1-4, respectively).

of these modalities. The construction of Fab-Astra-Dox-PEG$_{20k}$ **7** also further demonstrates the flexibility of the chemistry platform to allow for the attachment of large constructs such as a 20-kDa PEG moiety in a facile manner.

Appraisal of dual click strategy on a full antibody. The use of Astra-PD **1** was then applied to the bridging of disulfide bonds in the full antibody system of trastuzumab (Herceptin), a clinically approved IgG1 full antibody comprising four disulfide bonds. Successful application of this method on this scaffold would allow for the construction of a site-specifically dual-functionalized full ADC with retention of core antibody structure.

Initially, Herceptin was modified with PD-construct **1** by reduction of the full antibody interchain disulfide bonds followed by functional re-bridging. Through optimization of temperature and concentration of reagents, a full antibody that was re-bridged in the correct orientation with a 4:1 ratio of PD-to-antibody by SDS–PAGE and ultraviolet–visible spectrometry was obtained (Fig. 7 and Supplementary Figs 24 and 25). This provides the first example of functionally re-bridging the disulfide bonds of a full antibody with almost complete retention of full antibody structure. Following this, the site-selectively modified full antibody was functionalized with a drug (Dox-N_3) and a fluorophore (sulfo-Cy5-N_3) using click chemistry. It was confirmed by ultraviolet-visible spectrometry that each species was attached with a loading of four, and by SDS–PAGE that there was retention of full antibody structure (Fig. 7 and Supplementary Figs 24 and 25). Construct **8** was also stable under a range of pH conditions and in blood plasma mimicking conditions (Supplementary Figs 26 and 27). These results provide proof of concept for the ability to attach multiple modalities in a controlled manner to a full antibody and a new approach for native antibody functionalization[37]. Moreover, retention of binding by ELISA on HER2, and no unspecific binding on EGFR (HER1) or evidence of aggregation over a protracted period could be observed (Fig. 8 and Supplementary Fig. 23).

Internalization study. The effect of the conjugation chemistry on the internalizing properties of Herceptin and Fab-Her **2** was appraised next. For this purpose, Herceptin and Fab-Her **2** were directly conjugated to AlexaFluor488 *via* a PD-linker to form Her-PD-AlexaFluor488 and Fab-PD-AlexaFluor488 (Supplementary Figs 28 and 29). These constructs were then used to treat breast cancer cell lines BT-474 (HER2-positive) and

MDA-MB-468 (HER2-negative). Initially, incubation was performed at 4 °C, a temperature that allowed binding but not internalization; Her-PD-AlexaFluor488 and Fab-PD-AlexaFluor488 bound to BT-474 cells (Fig. 9a) but not to MDA-MB-468 cells (Fig. 9b) indicating the specificity of these constructs to HER2. Internalization of the bound antibodies occurred when the temperature was increased to 37 °C; incubation under these conditions for 1 h resulted in cytoplasmic localization of the labelled constructs in BT-474 cells (Fig. 9a), thus indicating that the chemistry employed does not interfere with the internalizing properties of Herceptin or Fab-Her **2**.

Cytotoxicity assay. Finally, the selectivity and cytotoxicity of constructs Fab-Astra-Dox-PEG_{20k} **7** and Her-Astra-Dox-Cy5 **8** in

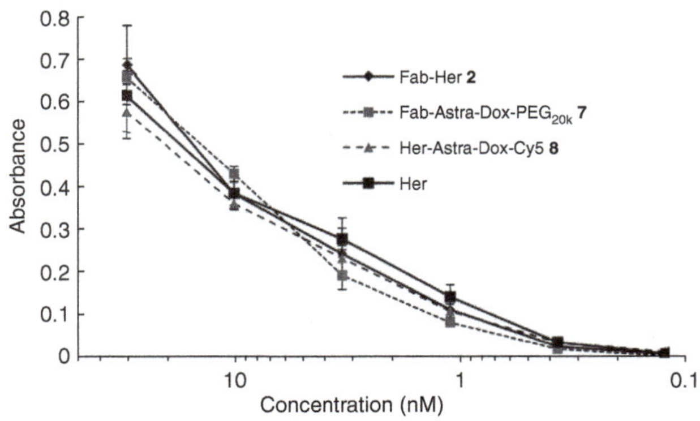

Figure 8 | Binding activity of dually modified constructs by ELISA. Binding activity of Fab-Her **2**, Fab-Astra-Dox-PEG_{20k} **7**, Her-Astra-Dox-Cy5 **8** and unmodified full antibody Herceptin (Her).

in vitro studies, using the same HER2-positive and -negative cell lines in the microscopy studies (that is, BT-474 and MDA-MB-468), were evaluated. Initially, the sensitivity of both cell lines to Dox to assess their suitability in a cytotoxicity assay was appraised. A comparable reduction in cell viability was observed for both cell lines at similar concentrations of the toxic payload ($IC_{50} = 39$ and 10 nM for BT-474 and MDA-MB-468, respectively) (Fig. 10a), thus paving the way for toxicity studies with Dox conjugates **7** and **8**. Notably, BT-474 cell viability was reduced significantly when incubated with the conjugates ($IC_{50} = 2.8$ and 2.1 µM for conjugates **7** and **8**, respectively), especially when compared with the controls of native Herceptin and Fab-Her **2** where the reduction in cell survival was minimal at high concentrations (Fig. 10b). In addition, highlighting the targeted delivery aspect of the method, no toxicity was observed in analogous studies with MDA-MB-468 cells incubated with either conjugate **7** or **8** (Fig. 10c). This is consistent with the observations by microscopy, that is, Herceptin and Fab-Her conjugates do not appear to bind to these cells, and hence Dox could not be internalized and thereby cause a toxic effect. These results highlight the selectivity of conjugates **7** and **8** over Dox alone, and indicate that the toxic drug in conjugates **7** and **8** is delivered by an HER2-dependent internalization mechanism.

Not only does this chemistry have the potential to make a significant contribution to the current ADC field, it is also flexible and has far reaching scope. Indeed, from a full antibody construct modified with Astra-PD **1**, it is wholly feasible that entirely new branches of strategies for ADC development will ensue. For example, the attachment of drugs that operate by an orthogonal mechanism of action, which is a concept that is gathering increasing momentum[38–41].

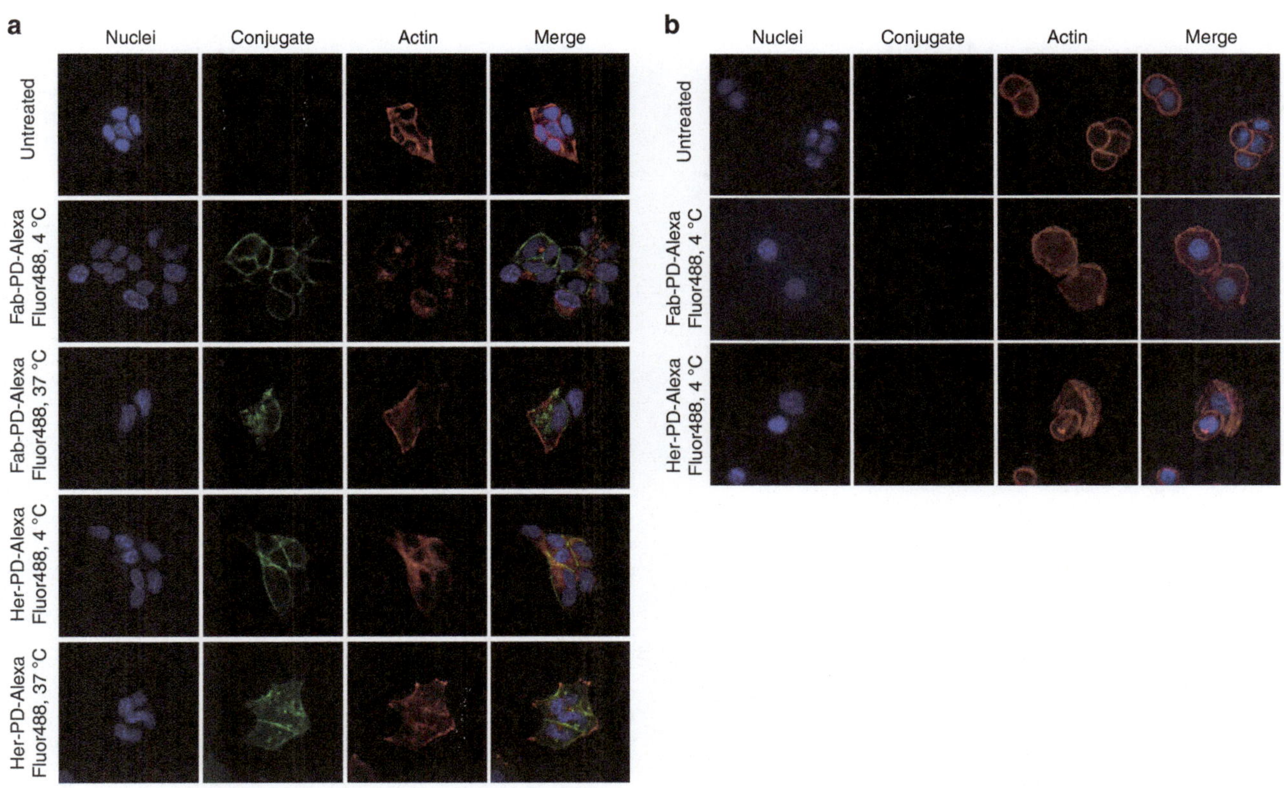

Figure 9 | Internalization study of fragment-based and full length antibody-based bioconjugates. (**a**) BT-474 breast cancer cells were treated at 4 °C with Her-PD-AlexaFluor488 and Fab-PD-AlexaFluor488 (green); both constructs bound at this temperature and were found inside the cells once internalization was allowed by incubating at 37 °C for 1 h. (**b**) No signal from fluorophore-conjugated constructs Her-PD-AlexaFluor488 and Fab-PD-AlexaFluor488 was detected when allowed to bind to the MDA-MB-468 breast cancer cells.

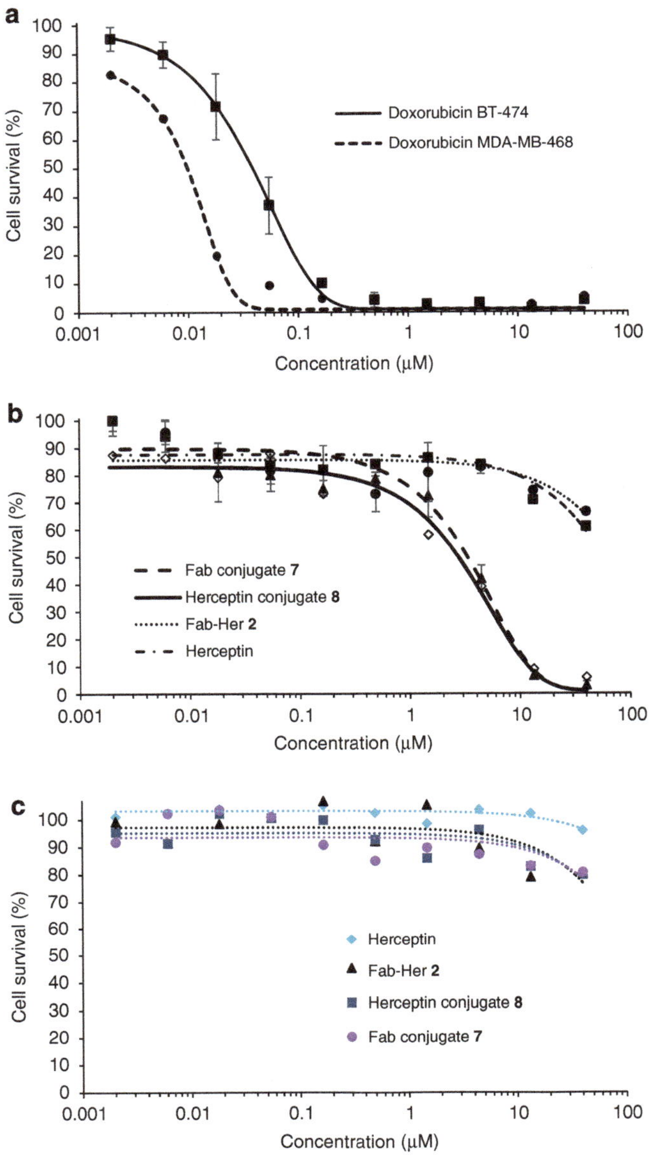

Figure 10 | Inhibition of cell proliferation in cancer cell lines with different levels of HER2 expression. **(a)** BT-474 (HER2-positive) and MDA-MB-468 (HER2-negative): Dox alone; $IC_{50}=39$ and 10 nM for BT-474 and MDA-MB-468, respectively. **(b)** BT-474: Fab-Astra-Dox-PEG_{20k} **7**, Her-Astra-Dox-Cy5 **8** (at similar concentration in Dox) in comparison with Herceptin and Fab-Her **2**; $IC_{50}=2.8$ and 2.1 μM for conjugates **7** and **8,** respectively **(c)** MDA-MB-468: Fab-Astra-Dox-PEG_{20k} **7**, Her-Astra-Dox-Cy5 **8** (at similar concentration in Dox) in comparison with Herceptin and Fab-Her **2**.

Discussion

In conclusion, by the introduction of PD molecules bearing orthogonal 'clickable' handles into the native disulfide bonds of antibodies, we could demonstrate: (i) the opportunity to synthesize multifunctionalized Fab fragments, which should be useful in a variety of applications (for example, imaging and theranostics); (ii) the potential of the technology to make a novel Fab-based ADC drug candidate; (iii) a new approach to a multifunctionalized, homogeneous full antibody to open the door to next-generation full antibody-based drug conjugate therapeutics; (iv) selective internalization into HER2-positive cells with conjugates of Herceptin and Fab-Her **2** that were directly conjugated to AlexaFluor488 *via* a PD-linker; and (v) selective cell killing over Dox alone for conjugates **7** and **8** on appropriate HER2 cell lines. Moreover, the 'plug-and-play' approach has broader applications in fields outside the scope of ADC technologies (for example, dual warhead antibiotics)[39–41].

Methods

General experimental. All reagents were purchased from Sigma-Aldrich, Promega, Molecular Probes, AlfaAesar, Sino Biological, Invitrogen, UCLH or Lumiprobe and were used as received. Where described below, petrol refers to petroleum ether (40–60 °C). All reactions were monitored by thin-layer chromatography on pre-coated SIL G/UV254 silica gel plates (254 μm) purchased from VWR. Flash column chromatography was carried out with Kiesegel 60 M 0.04/0.063 mm (200–400 mesh) silica gel. ^{1}H and ^{13}C NMR spectra were recorded at ambient temperature on a Bruker Avance 500 instrument operating at a frequency of 500 MHz for ^{1}H and 125 MHz for ^{13}C and a Bruker Avance 600 instrument operating at a frequency of 600 MHz for ^{1}H and 150 MHz for ^{13}C in $CDCl_3$ or CD_3OD (as indicated below). The chemical shifts (δ) for ^{1}H and ^{13}C are quoted relative to residual signals of the solvent on the p.p.m. scale. ^{1}H NMR peaks are reported as singlet (s), doublet (d), triplet (t), quartet (q), quint. (quintet), sext. (sextet), oct. (octet), m (multiplet), br (broad), dd (doublet of doublet), dt (doublet of triplets), ABq (AB quartet). Coupling constants (J values) are reported in hertz (Hz) and are H-H coupling constants unless otherwise stated. Signal multiplicities in ^{13}C NMR were determined using the distortionless enhancement by phase transfer spectral editing technique. Infrared spectra were obtained on a Perkin Elmer Spectrum 100 FTIR Spectrometer operating in ATR mode with frequencies given in reciprocal centimetres (cm^{-1}). Melting points were measured with a Gallenkamp apparatus and are uncorrected. Mass spectra of small molecule organic compounds were obtained on a VG70-SE mass spectrometer.

Protein LC-MS. LC–MS was performed on protein samples using a Waters Acquity uPLC connected to Waters Acquity Single Quad Detector. Column: Hypersil Gold C4, 1.9 μm, 2.1 × 50 mm; wavelength: 254 nm; mobile phase: 95:5 water (0.1% formic acid):MeCN (0.1% formic acid); gradient over 4 min (to 5:95 water (0.1% formic acid):MeCN (0.1% formic acid); flow rate: 0.6 ml min^{-1}; MS mode: ES +; scan range: $m/z=250$–2,000; scan time: 0.25 s. Data obtained in continuum mode. The electrospray source of the MS was operated with a capillary voltage of 3.5 kV and a cone voltage of 50 V. Nitrogen was used as the nebulizer and desolvation gas at a total flow of 600 l h^{-1}. Ion series were generated by integration of the ultraviolet-absorbance (at 254 nm) chromatogram over the 1.4–2.0 min range. Total mass spectra for protein samples were reconstructed from the ion series using the MaxEnt 1 algorithm pre-installed on MassLynx software.

Ultraviolet-visible spectroscopy. Ultraviolet-visible spectra were recorded on a Varian Cary 100 Bio UV-Visible spectrophotometer, operating at room temperature. Sample buffer was used as blank for baseline correction. Calculation of molecule over antibody ratio, r, follows the formula below with $\varepsilon_{280}=215{,}380\ M^{-1}\ cm^{-1}$ for Herceptin mAb, $\varepsilon_{280}=68{,}590\ M^{-1}\ cm^{-1}$ for Herceptin Fab, $\varepsilon_{345}=9{,}100\ M^{-1}\ cm^{-1}$ for Astra-PD, $\varepsilon_{495}=8{,}030\ M^{-1}\ cm^{-1}$ for Dox, $\varepsilon_{646}=271{,}000\ M^{-1}\ cm^{-1}$, and 0.724, 0.28 and 0.13 as a correction factor (CF) for Dox, Astra-PD and sulfo-Cy5, respectively, for the absorbance at 280 nm.

$$r=\frac{A_\lambda/\varepsilon_\lambda}{\left(A_{280}-\sum_\lambda CF_\lambda \times A_\lambda\right)/\varepsilon_{280}}$$

With A_λ the absorbance of a molecule at the wavelength λ in nm, and ε_λ, the corresponding extinction coefficient.

SDS-PAGE gels. Non-reducing glycine-SDS–PAGE at 12% acrylamide gels were performed following standard laboratory procedures. A 4% stacking gel was used and a broad-range molecular weight (MW) marker (10–250 kDa, BioLabs) was co-run to estimate protein weights. Samples (3–5 μl at ~10 μM in total mAb) were mixed with loading buffer (1–2 μl, composition for 6 × SDS: 1 g SDS, 3 ml glycerol, 6 ml 0.5 M Tris buffer pH 6.8, 2 mg R-250 dye) and heated at 75 °C for 3 min. The gel was run at constant current (30–35 mA) for 40 min in 1 × SDS running buffer. All gels were stained with Coomassie. Gel photographs were taken with a Wiko-Stairway device.

Chemical biology. Conjugation of Astra-PD **1** onto Fab-Her **2**: To a solution of Fab-Her **2** (1 eq) in borate buffer (25 mM sodium borate, 25 mM NaCl, 0.5 mM EDTA, pH 8.0) was added TCEP (3 eq) and the reaction mixture was incubated at 37 °C for 90 min. After this time, was added a solution of Astra-PD **1** in DMF (5 eq) and the reaction mixture was incubated at 37 °C for 1 h. The excess reagents were then removed by repeated diafiltration into fresh buffer using VivaSpin sample concentrators (GE Healthcare, 10,000 molecular weight cut-off (MWCO)).

Conjugation of Astra-PD **1** onto Herceptin: To a solution of Herceptin (1 eq) in borate buffer (25 mM sodium borate, 25 mM NaCl, 0.5 mM EDTA, pH 8.0) was added TCEP (10 eq) and Astra-PD **1** (20 eq) in DMF and the reaction mixture was incubated at 4 °C for 6 h. The excess reagents were then removed by repeated diafiltration into fresh buffer using VivaSpin sample concentrators (GE Healthcare, 10,000 MWCO).

SPAAC click conditions: To a solution of an antibody or antibody fragment (1 eq) modified with Astra-PD **1** in PBS or borate buffer was added azide cargo (5 eq per re-bridged Astra-PD molecule) and the reaction mixture was incubated at 37 °C for 4 h. The excess reagents were then removed by repeated diafiltration into fresh buffer using VivaSpin sample concentrators (GE Healthcare, 10,000 MWCO).

CuAAC click conditions: To a solution of an antibody or antibody fragment (1 eq) modified with mono-functionalized-Astra-PD **1** in PBS was added tris(3-hydroxypropyltriazolylmethyl)amine (1.25 mM), $CuSO_4$ (250 µM) followed by the addition of azide cargo (3 eq for Herceptin-Fab-Astra-PD and 10 eq for Herceptin-Astra-PD) and sodium ascorbate (final concentration 5 mM), and the reaction mixture was incubated at 37 °C for 16 h. The excess reagents were then removed by repeated diafiltration into fresh PBS with 2 mM EDTA (to remove residual copper ions) using VivaSpin sample concentrators (GE Healthcare, 10,000 MWCO).

Cell lines. Breast cancer cell lines BT-474 and MDA-MB-468 were purchased from ATCC. BT-474 cells were maintained at 37 °C, 5% CO_2 in Hybricare Medium (ATCC) complemented with 10% foetal calf serum (Labtech International, Ringmer, UK). MDA-MB-468 cells were maintained at 37 °C, 5% CO_2 in Dulbecco's modified Eagle's medium complemented with 10% foetal calf serum and 2 mM L-glutamine (PAA Laboratories, UK).

Internalization analysis by confocal microscopy. Cells on coverslips at 70% confluency were incubated with AlexaFluor488-conjugated constructs at 10 µg ml^{-1} for 1 h at 4 °C. Cells were extensively washed with PBS to remove unbound antibodies and incubated at 37 °C in growth media. Internalization was allowed for 1 h, followed by extensive washing and fixation with 4% formaldehyde for 10 min at 4 °C. Coverslips were then blocked with 5% goat serum in 0.3% Triton-X100 (Sigma). Actin was detected with phalloidin-568 (Invitrogen) and Hoechst trihydrochloride (Invitrogen) was used to stain cell nuclei. Coverslips were mounted on slides using ProLong Gold Antifade (Invitrogen) and examined using Perkin Elmer Spinning Disc Confocal microscope and Volocity Visualization software.

Toxicity assays. Cells were seeded in 96-well plates at 10^4 cells/well and allowed to attach for 24 h. Serial dilution of Herceptin, Fab-Her **2**, Dox, and conjugates **7** and **8** were added to the cells at concentrations ranging from 40 to 0 µM in complete growth medium. After 96 h, cell viability was measured using the CellTiter 96 Aqueous Non-radioactive cell proliferation assay (Promega) following manufacturer's instructions. Cell viability was plotted as percentage of untreated cells.

Ellman's test. Herceptin and Fab-Her **2**, respectively; To a solution of trastuzumab (50 µl, 40 µM, 1 eq) in phosphate buffer (100 mM sodium phosphate, 1 mM EDTA, pH 8.0), was added TCEP (final concentration 400 µM, 10 eq) and the reaction mixture was incubated at 37 °C for 2 h. The excess reagents were then removed by repeated diafiltration into fresh buffer using VivaSpin sample concentrators (GE Healthcare, 10,000 MWCO). The concentration of protein was then adjusted to 40 µM. Following this, a solution of Ellman's reagent (5,5′-dithio-*bis*-(2-nitrobenzoic acid)) in phosphate buffer (100 mM sodium phosphate, 1 mM EDTA, pH 8.0; final concentration 1.6 mM, 40 eq) was added, and the reaction mixture was incubated at 21 °C for 30 min. Analysis by ultraviolet–visible spectrometry after a tenfold dilution revealed an absorption at 412 nm of 0.439 that corresponds to 7.8 accessible sulfhydryl groups ($\varepsilon_{412} = 14{,}150\ M^{-1}\,cm^{-1}$). A similar experiment without TCEP reduction before incubation with Ellman's reagent showed no absorbance at 412 nm.

To a solution of Fab-Her **2** (50 µl, 20 µM, 1 eq) in phosphate buffer (100 mM sodium phosphate, 1 mM EDTA, pH 8.0), was added TCEP (final concentration 100 µM, 5 eq) and the reaction mixture was incubated at 37 °C for 2 h. The excess reagents were then removed by repeated diafiltration into fresh buffer using VivaSpin sample concentrators (GE Healthcare, 10,000 MWCO). The concentration of protein was then adjusted to 20 µM. Following this, a solution of Ellman's reagent (5,5′-dithio-*bis*-(2-nitrobenzoic acid)) in phosphate buffer (100 mM sodium phosphate, 1 mM EDTA, pH 8.0; final concentration 200 µM, 10 eq) was added, and the reaction mixture was incubated at 21 °C for 30 min. Analysis by ultraviolet–visible spectrometry revealed an absorption at 412 nm of 0.557 that corresponds to two accessible sulfhydryl groups ($\varepsilon_{412} = 14{,}150\ M^{-1}\,cm^{-1}$). A similar experiment without TCEP reduction before incubation with Ellman's reagent showed no absorbance at 412 nm.

References

1. Sapra, P., Hooper, A. T., O'Donnell, C. J. & Gerber, H.-P. Investigational antibody drug conjugates for solid tumors. *Expert Opin. Investig. Drugs* **20,** 1131–1149 (2011).
2. Alley, S. C., Okeley, N. M. & Senter, P. D. Antibody-drug conjugates: targeted drug delivery for cancer. *Curr. Opin. Chem. Biol.* **14,** 529–537 (2010).
3. Hess, C., Venetz, D. & Neri, D. Emerging classes of armed antibody therapeutics against cancer. *Med. Chem. Comm.* **5,** 408–431 (2014).
4. Adair, J. R., Howard, P. W., Hartley, J. A., Williams, D. G. & Chester, K. A. Antibody-drug conjugates—a perfect synergy. *Expert Opin. Biol. Ther.* **12,** 1191–1206 (2012).
5. Mullard, A. Maturing antibody-drug conjugate pipeline hits 30. *Nat. Rev. Drug Discov.* **12,** 329–332 (2013).
6. Sassoon, I. & Blanc, V. Antibody-drug conjugate (ADC) clinical pipeline: a review. *Methods Mol. Biol.* **1045,** 1–27 (2013).
7. Flygare, J. A., Pillow, T. H. & Aristoff, P. Antibody-drug conjugates for the treatment of cancer. *Chem. Biol. Drug Des.* **81,** 113–121 (2013).
8. Ducry, L. & Stump, B. Antibody-drug conjugates: linking cytotoxic payloads to monoclonal antibodies. *Bioconjug. Chem.* **21,** 5–13 (2010).
9. Junutula, J. R. *et al.* Engineered thio-trastuzumab-DM1 conjugate with an improved therapeutic index to target human epidermal growth factor receptor 2-positive breast cancer. *Clin. Cancer Res.* **16,** 4769–4778 (2010).
10. Junutula, J. R. *et al.* Site-specific conjugation of a cytotoxic drug to an antibody improves the therapeutic index. *Nat. Biotechnol.* **26,** 925–932 (2008).
11. Zimmerman, E. S. *et al.* Production of site-specific antibody-drug conjugates using optimized non-natural amino acids in a cell-free expression system. *Bioconjug. Chem.* **25,** 351–361 (2014).
12. Castañeda, L. *et al.* Acid-cleavable thiomaleamic acid linker for homogeneous antibody-drug conjugation. *Chem. Commun.* **49,** 8187–8189 (2013).
13. Schumacher, F. F. *et al.* Homogeneous antibody fragment conjugation by disulfide bridging introduces 'spinostics'. *Sci. Rep.* **3,** 1525 (2013).
14. Ryan, C. P. *et al.* Tunable reagents for multi-functional bioconjugation: reversible or permanent chemical modification of proteins and peptides by control of maleimide hydrolysis. *Chem. Commun.* **47,** 5452–5454 (2011).
15. Schumacher, F. F. *et al.* Next generation maleimides enable the controlled assembly of antibody-drug conjugates *via* native disulfide bond bridging. *Org. Biomol. Chem.* **12,** 7261–7269 (2014).
16. Schumacher, F. F. *et al.* In situ maleimide bridging of disulfides and a new approach to protein PEGylation. *Bioconjug. Chem.* **22,** 132–136 (2011).
17. Smith, M. E. B. *et al.* Protein modification, bioconjugation, and disulfide bridging using bromomaleimides. *J. Am. Chem. Soc.* **132,** 1960–1965 (2010).
18. Chudasama, V. *et al.* Bromopyridazinedione-mediated protein and peptide bioconjugation. *Chem. Commun.* **47,** 8781–8783 (2011).
19. Maruani, A. *et al.* A mild TCEP-based para-azidobenzyl cleavage strategy to transform reversible cysteine thiol labelling reagents into irreversible conjugates. *Chem. Commun.* **51,** 5279–5282 (2015).
20. Badescu, G. *et al.* Bridging disulfides for stable and defined antibody drug conjugates. *Bioconjug. Chem.* **25,** 1124–1136 (2014).
21. Hudis, C. A. Trastuzumab–mechanism of action and use in clinical practice. *N. Engl. J. Med.* **357,** 39–51 (2007).
22. Verma, S. *et al.* Trastuzumab emtansine for HER2-positive advanced breast cancer. *N. Engl. J. Med.* **367,** 1783–1791 (2012).
23. Jalota, S., Bhaduri, S. B. & Tas, A. C. Using a synthetic body fluid (SBF) solution of 27 mM HCO_3^- to make bone substitutes more osteointegrative. *Mater. Sci. Eng. C* **28,** 129–140 (2008).
24. Thorp-Greenwood, F. L. & Coogan, M. P. Multimodal radio- (PET/SPECT) and fluorescence imaging agents based on metallo-radioisotopes: current applications and prospects for development of new agents. *Dalton Trans.* **40,** 6129–6143 (2011).
25. Crawley, N., Thompson, M. & Romaschin, A. Theranostics in the growing field of personalized medicine: an analytical chemistry perspective. *Anal. Chem.* **86,** 130–160 (2014).
26. Schmidt, M. M. & Wittrup, K. D. A modeling analysis of the effects of molecular size and binding affinity on tumor targeting. *Mol. Cancer Ther.* **8,** 2861–2871 (2009).
27. Yokota, T., Milenic, D. E., Whitlow, M. & Schlom, J. Rapid tumor penetration of a single-chain Fv and comparison with other immunoglobulin forms. *Cancer Res.* **52,** 3402–3408 (1992).
28. Thurber, G. M., Schmidt, M. M. & Wittrup, K. D. Antibody tumor penetration: transport opposed by systemic and antigen-mediated clearance. *Adv. Drug Deliv. Rev.* **60,** 1421–1434 (2008).
29. Holliger, P. & Hudson, P. J. Engineered antibody fragments and the rise of single domains. *Nat. Biotechnol.* **23,** 1126–1136 (2005).
30. Milenic, D. E. *et al.* Construction, binding properties, metabolism, and tumor targeting of a single-chain Fv derived from the pancarcinoma monoclonal antibody CC49. *Cancer Res.* **51,** 6363–6371 (1991).
31. Williams, L. E. *et al.* Numerical selection of optimal tumor imaging agents with application to engineered antibodies. *Cancer Biother. Radiopharm.* **16,** 25–35 (2001).
32. Holt, L. J., Herring, C., Jespers, L. S., Woolven, B. P. & Tomlinson, I. M. Domain antibodies: proteins for therapy. *Trends Biotechnol.* **21,** 484–490 (2003).
33. Deen, W. M., Lazzara, M. J. & Myers, B. D. Structural determinants of glomerular permeability. *Am. J. Physiol. Renal Physiol.* **281,** 579–596 (2001).

34. Chapman, A. P. *et al.* Therapeutic antibody fragments with prolonged in vivo half-lives. *Nat. Biotechnol.* **17,** 780–783 (1999).
35. Yang, K. *et al.* Tailoring structure-function and pharmacokinetic properties of single-chain Fv proteins by site-specific PEGylation. *Protein Eng.* **16,** 761–770 (2003).
36. Fishburn, C. S. The pharmacology of PEGylation: balancing PD with PK to generate novel therapeutics. *J. Pharm. Sci.* **97,** 4167–4183 (2008).
37. Brocchini, S. *et al.* Disulfide bridge based PEGylation of proteins. *Adv. Drug Deliv. Rev.* **60,** 3–12 (2008).
38. Das, M., Jain, R., Agrawal, A. K., Thanki, K. & Jain, S. Macromolecular bipill of gemcitabine and methotrexate facilitates tumor-specific dual drug therapy with higher benefit-to-risk ratio. *Bioconjug. Chem.* **25,** 501–509 (2014).
39. Thiyagarajan, D., Goswami, S., Kar, C., Das, G. & Ramesh, A. A prospective antibacterial for drug-resistant pathogens: a dual warhead amphiphile designed to track interactions and kill pathogenic bacteria by membrane damage and cellular DNA cleavage. *Chem. Commun.* **50,** 7434–7436 (2014).
40. Silver, L. L. Challenges of antibacterial discovery. *Clin. Microbiol. Rev.* **24,** 71–109 (2011).
41. O'Connell, K. M. G. *et al.* Combating multidrug-resistant bacteria: current strategies for the discovery of novel antibacterials. *Angew. Chem. Int. Ed.* **52,** 10706–10733 (2013).

Acknowledgements

We gratefully acknowledge UCL and UCLB for support of our programme, and Dr James Baker for productive discussions.

Authors contributions

V.C. conceived and designed the project; V.C., A.M., M.E.B.S. and S.C. conceived and designed the chemistry and chemical biology experiments; A.M., V.C., E.M. and K.C. conceived and designed the biology experiments; A.M. and V.C. performed the chemistry and chemical biology experiments; A.M. and E.M. performed the biology experiments; V.C. and A.M. analysed the data; V.C., A.M. and S.C. co-wrote the paper.

Additional information

Competing financial interests: The authors declare no competing financial interests.

7

Collaboration between primitive cell membranes and soluble catalysts

Katarzyna P. Adamala[1,*], Aaron E. Engelhart[1,*] & Jack W. Szostak[1]

One widely held model of early life suggests primitive cells consisted of simple RNA-based catalysts within lipid compartments. One possible selective advantage conferred by an encapsulated catalyst is stabilization of the compartment, resulting from catalyst-promoted synthesis of key membrane components. Here we show model protocell vesicles containing an encapsulated enzyme that promotes the synthesis of simple fatty acid derivatives become stabilized to Mg^{2+}, which is required for ribozyme activity and RNA synthesis. Thus, protocells capable of such catalytic transformations would have enjoyed a selective advantage over other protocells in high Mg^{2+} environments. The synthetic transformation requires both the catalyst and vesicles that solubilize the water-insoluble precursor lipid. We suggest that similar modified lipids could have played a key role in early life, and that primitive lipid membranes and encapsulated catalysts, such as ribozymes, may have acted in conjunction with each other, enabling otherwise-impossible chemical transformations within primordial cells.

[1]Howard Hughes Medical Institute and Department of Molecular Biology and Center for Computational and Integrative Biology, Massachusetts General Hospital, 185 Cambridge Street, Boston, Massachusetts 02114, USA. * These authors contributed equally to this work. Correspondence and requests for materials should be addressed to J.W.S. (email: szostak@molbio.mgh.harvard.edu).

A widely believed model of early life contends that early cellular life was comprised of catalysts and nutrients enclosed within membrane vesicles, much like contemporary cells. The membranes and catalysts found in such a scenario, however, would likely have differed from those observed in extant life, due to the absence of sophisticated protein enzyme catalysts. As a result, simpler alternatives to contemporary biomolecules may have been common in early life. In particular, due to the relative ease of synthesis of single-chain amphiphiles, fatty acid-based membranes, in lieu of the diacylphospholipid membranes largely employed by life today, represent a plausible alternative[1–4]. Similarly, due to the self-templating capability of nucleic acids, RNA (or a closely related molecule) represents a highly attractive potential polymer for primitive catalysts[5–7].

Thus, the first protocells could have been comprised of fatty acid membranes and ribozyme catalysts, in contrast to the diacylphospholipid membranes and protein catalysts observed in most contemporary life[5,8]. Unfortunately, these substitutions result in another, seemingly paradoxical, problem. Known ribozymes require moderate to high concentrations of divalent cations (typically on the order of 10^{-3} to 10^{-2} M)[9,10]. Vesicles composed only of fatty acids, however, are disrupted in the presence of such concentrations of divalent cations due to fatty acid precipitation[11,12]. Previously, we demonstrated a potential solution to this problem by employing mixed vesicles containing both a fatty acid and its glycerol ester, which are robust to low millimolar concentrations of Mg^{2+}, thus enabling functional encapsulated ribozymes[11].

While the glycerol ester of a fatty acid represents a partial structural analogue to contemporary diacylphospholipids, esters of carboxylic acids are hydrolytically unstable, and they are subject to ester-amide exchange with free amines. Indeed, ribozymes and simple dipeptide catalysts exist that are capable of promoting the synthesis of amide linkages (including naturally found ones, such as the peptidyl transferase center of the ribosome)[13–16]. Similarly, the nonenzymatic synthesis of amide linkages by ester-amide exchange with a starting polyester was recently demonstrated[17]. Taken together, these results led us to consider whether an amide analogue of a monoacylglycerol, synthesized by ester-amide exchange, could serve as a membrane component that would confer greater robustness to Mg^{2+} than pure fatty acid vesicles.

Here we show that (1) such a lipid can serve as a membrane component in vesicles, rendering these vesicles robust to millimolar concentrations of magnesium. Thus, (2) these vesicles can support catalysis by an encapsulated ribozyme. Finally, (3) lipid vesicles solubilize a precursor to this lipid amide, which enables an encapsulated catalyst (the protein enzyme α-chymotrypsin, used here as a model substitution for an as-yet undescribed primitive acyltransferase ribozyme) to produce this lipid amide within vesicles. Vesicles containing the catalyst (an enzyme in these experiments) are thereby rendered more robust to Mg^{2+} than those of the same composition lacking the encapsulated catalyst. Furthermore, (4) this conversion is not possible in the absence of vesicles, due to the low solubility of the precursor lipid. In light of these results, we suggest that amide analogues of glycerol esters could have played a role in stabilizing the membranes of early protocells. We also speculate that primitive catalysts and membranes could have worked in concert to enable a wider range of chemical transformations than would otherwise have been possible.

Results

Fatty acid derivatives stabilize oleic acid vesicles to Mg^{2+}. We first sought to examine whether a fatty acid glycerol ester (glycerol monooleate, GMO) and its amide analogue (*N*-(2,3-dihydroxypropyl)oleamide, GMOA) were similarly effective in stabilizing oleic acid (OA) vesicles (Fig. 1a). As expected, we observed that pure oleic acid (OA) vesicles (50 mM) containing either an encapsulated small molecule dye (5 mM calcein) or an encapsulated oligonucleotide (2 μM fluorescein-labelled DNA 9mer) readily leaked in the presence of 10 mM Mg^{2+}, with only 21% retention of dye and 37% retention of oligonucleotide after 24 h (Fig. 1d,g). The presence of 15 mol% GMO or GMOA, in each case, stabilized vesicles substantially to magnesium: OA/GMO vesicles retained 85% of dye and 92% of oligonucleotide after 24 h (Fig. 1b,e), while OA/GMOA vesicles gave comparable performance, retaining 76% of dye and 93% of oligonucleotide after 24 h (Fig. 1c,f). Given that oligonucleotide permeability of fatty acid liposomes is highly length dependent, with longer oligonucleotides being markedly better retained than shorter ones[18], these vesicles are anticipated to retain longer oligonucleotides (for example, ribozymes, as employed below) as well or better than the short 9 nucleotide oligomers employed in the leakage study described.

Vesicles containing GMO(A) support ribozyme catalysis. Retention of encapsulated oligonucleotides in the presence of divalent cations is a critical feature of a protocell membrane. This, alone, is not sufficient, however, as most ribozymes require free (that is, unchelated) Mg^{2+}. Having verified that GMOA was as effective as GMO in enabling OA vesicles to retain their contents in the presence of Mg^{2+}, we next investigated whether OA/GMO and OA/GMOA vesicles are (1) permeable to Mg^{2+} and (2) contain free internal Mg^{2+} when incubated in solutions containing Mg^{2+} ions. To do so, we employed mag-fura-2, a vesicle-impermeable dye which gives a ratiometric fluorescence response to free Mg^{2+} (ref. 11). We monitored the transit of Mg^{2+} across the membrane of OA/GMO and OA/GMOA vesicles containing mag-fura-2, using a calibration curve of mag-fura-2 with free Mg^{2+} (Supplementary Figs 1 and 2). For both vesicle compositions, Mg^{2+} equilibration across the membrane was rapid and complete in ca. 2 min. Consistent with the longer acyl chain length, this equilibration took somewhat longer than the 20 s required for the shorter chain myristoleic acid/glycerol monomyristoleate (MA/GMM) vesicles previously studied[11].

Having ascertained that Mg^{2+} transit across OA/GMOA membranes was rapid and free Mg^{2+} was present within these vesicles, we examined the functionality of an encapsulated hammerhead ribozyme[19,20]. This ribozyme exhibits Mg^{2+}-promoted catalytic function. Consistent with the presence of free Mg^{2+} within these vesicles, a reaction containing 1 μM of each ribozyme component and 15 mM $MgCl_2$ exhibited a rate constant of 0.24 h^{-1} in OA/GMO vesicles and 0.25 h^{-1} in OA/GMOA vesicles—comparable between lipid systems and somewhat lower than the vesicle-free rate constant of 0.50 h^{-1} (Fig. 2).

Vesicles with a GMOA synthesis catalyst are Mg^{2+} stabilized. As a model of a prebiotic liposomal system, we encapsulated α-chymotrypsin within vesicles comprised of oleic acid and ethyl oleate, in the presence of (±)-3-amino-1,2-propanediol. As this enzyme is known to convert hydrophobic ethyl esters to amides[21], we reasoned that the α-chymotrypsin-catalyzed formation of GMOA from the ethyl oleate precursor would result in vesicles that were more robust to Mg^{2+}. The conversion of ethyl oleate to GMOA in this system was efficient, with ca. 70% yield over 72 h (Fig. 3), corresponding to ca. 10 mol% GMOA. We

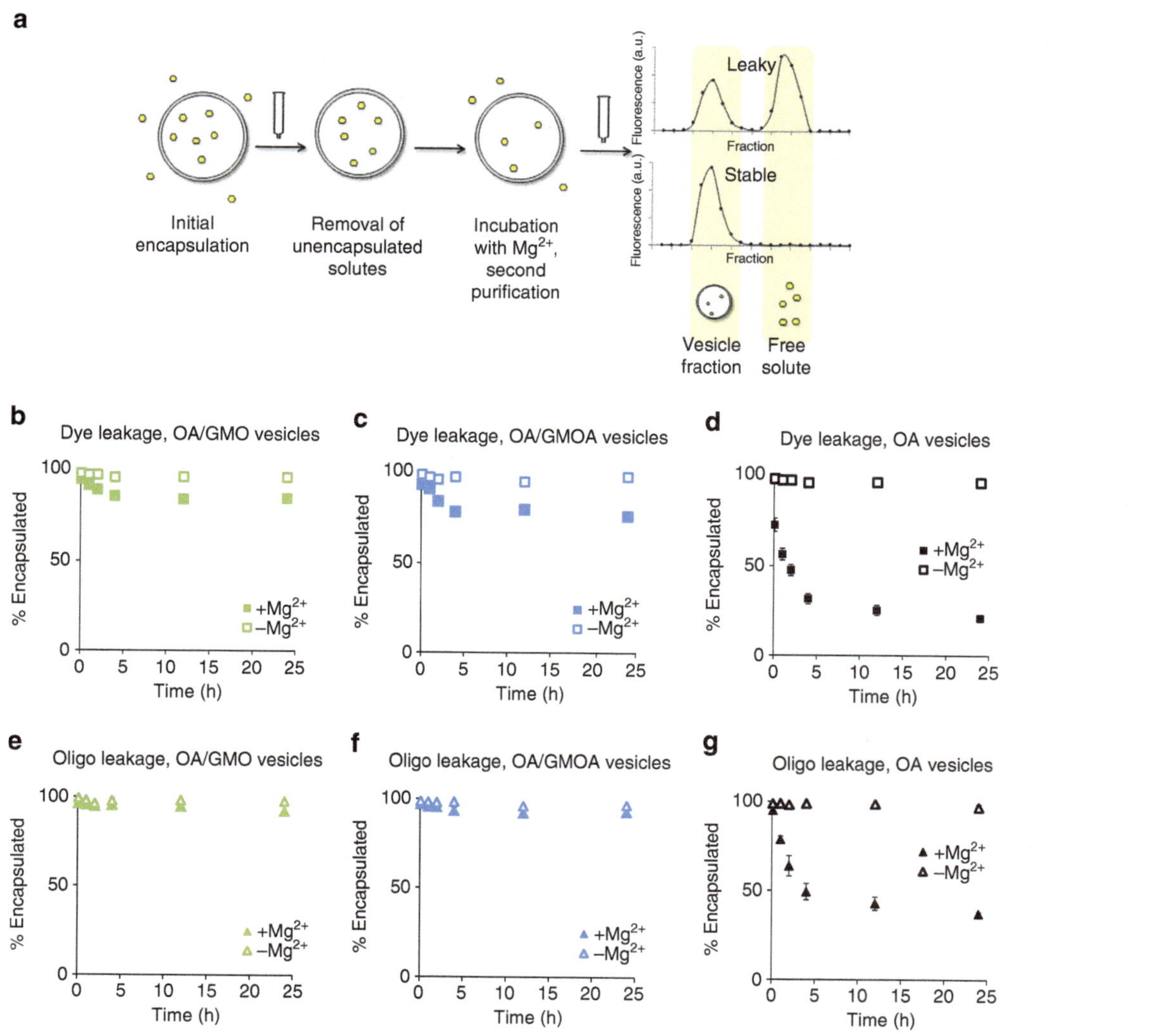

Figure 1 | Vesicle leakage measurements. Magnesium-induced leakage of small molecules and oligonucleotides from vesicles of different membrane composition was assayed by encapsulation of the analyte of interest, removal of unencapsulated solutes, addition of magnesium (if applicable), followed by a second purification to quantitate leakage. General scheme of the vesicle stability experiment. (**a**) Columns between panels represent a size-exclusion chromatography step, in which vesicles are separated from unencapsulated solutes. Size-exclusion chromatograms shown here are intended to be schematic in nature only. Leakage of the encapsulated small molecule (calcein (**b-d**)) and oligonucleotide (**e-g**). Squares: small molecule leakage; triangles: oligonucleotide leakage. Green markers: OA/GMO vesicles; blue markers: OA/GMOA vesicles; black markers: OA vesicles. Error bars are s.e.m. ($n = 4$); error bars that are not visible are sufficiently small they are obscured by markers. All reactions contained vesicles comprised of 50 mM total lipid (50 mM OA, or 42.5 mM OA, 7.5 mM (15 mol%) GMO or GMOA), and 0.2 M Na-glycinamide, pH 8.5 as buffer. Mg^{2+}-containing samples were 15 mM in Mg^{2+}. Calcein, when used, was present at 2.5 mM. Fluorescein-labelled oligonucleotide, when used, was present at 2 μM. Reactions were performed at uncontrolled room temperature (20 ± 2 °C).

tested these post-incubation vesicles for robustness to Mg^{2+}, and observed that they were significantly more effective at retaining their contents, with 68% retention of calcein and 87% retention of oligonucleotide over 24 h of incubation with Mg^{2+} (cf. 19% dye retention and 52% oligonucleotide retention for vesicles of the same composition without encapsulated α-chymotrypsin). These results were similar to those observed in vesicles formed with nonenzymatically synthesized GMOA (Fig. 1, example purification traces are shown in Supplementary Figs 3 and 4).

Vesicles are required for enzymatic GMOA synthesis. In addition to the catalytic impact of α-chymotrypsin, we observed that the vesicles, themselves, are promoters of the transformation of ethyl oleate to GMOA. When we attempted to perform the same enzymatic reaction in aqueous solution lacking liposomes, essentially no conversion occurred, at least in part due to the negligible solubility of ethyl oleate (Fig. 3b). Thus, we suggest that the vesicle membrane itself is a phase-transfer catalyst of this transformation, due to its ability to solubilize ethyl oleate, as well as potential favourable interactions between cationic (±)-3-amino-1,2-propanediol and the negatively charged fatty acid bilayer.

Discussion

We have shown that α-chymotrypsin can act as a catalyst within model protocell vesicles comprised of OA and ethyl oleate, enabling the *in situ* generation of the lipid GMOA. In turn, GMOA is incorporated into the vesicle membranes, thus rendering these vesicles robust to the presence of divalent

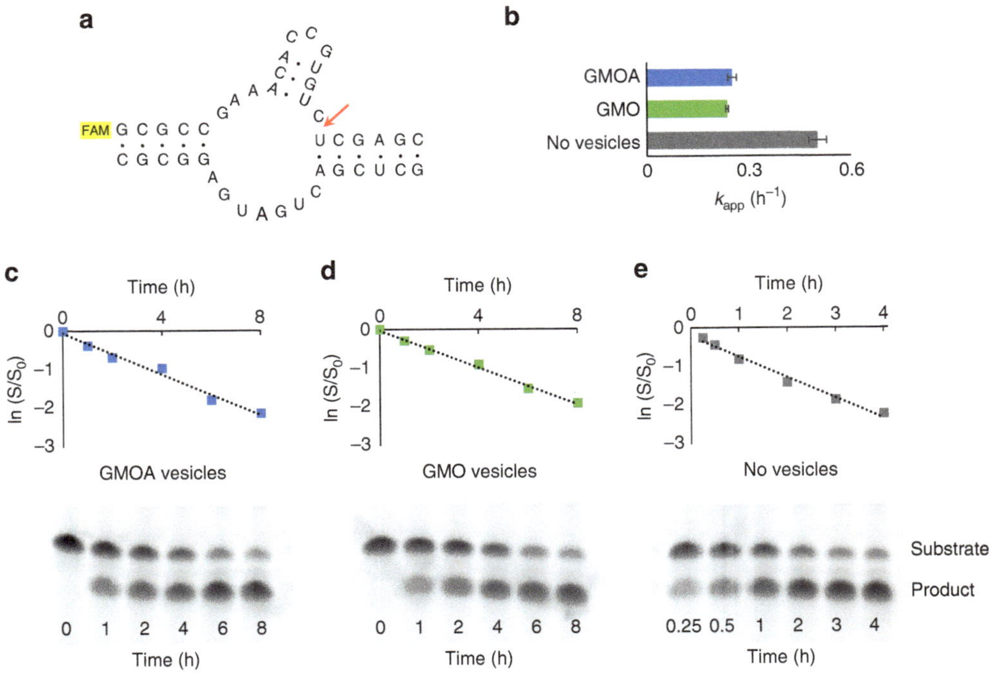

Figure 2 | OA/GMOA and OA/GMO vesicles support ribozyme activity. Fluorescein-labelled hammerhead ribozymes (**a**) were encapsulated within OA/GMOA- and OA/GMO-containing vesicles. Cleavage rates for this ribozyme were measured in both types of vesicles as well as in the absence of vesicles. At 15 mM Mg^{2+}, OA/GMOA vesicles gave similar ribozyme cleavage rates to OA/GMO-containing vesicles (**b-e**) Green markers: OA/GMO vesicles, $k_{app} = 0.24\,h^{-1}$ (s.e.m. = 0.004); blue markers: OA/GMOA vesicles, $k_{app} = 0.25\,h^{-1}$ (s.e.m. = 0.013); grey markers: no vesicles, $k_{app} = 0.50\,h^{-1}$ (s.e.m. = 0.026). Representative polyacrylamide gel electrophoresis (PAGE) analyses of hammerhead ribozyme reactions for each set of vesicle conditions are shown below each graph. Error bars on **b** are s.e.m. ($n = 3$). Solid lines on **c**,**d** and **e** are linear fits; for all, $R^2 > 0.98$. All reactions contained vesicles comprised of 50 mM total lipid (50 mM OA, or 42.5 mM OA, 7.5 mM (15 mol%) GMO or GMOA), and 0.2 M Na-glycinamide, pH 8.5 as buffer, RNA concentration was 1 μM of each hammerhead strand. Mg^{2+} was present at 15 mM. Reactions were performed at uncontrolled room temperature (20 ± 2 °C).

cations—an absolute requirement for an RNA-based form of primitive metabolism[9,22,23]. The high level of retention of oligonucleotides enabled by this system, while simultaneously allowing for some small molecule permeability, would have been a highly adaptive trait for a primitive cell, enabling greater retention of catalysts (such as ribozymes) than otherwise possible, while allowing for permeation of nutrients (such as nucleotide monomers)[24,25].

These results amount to a primitive adaptive behaviour for a compartment, driven by an encapsulated catalyst. While α-chymotrypsin, the catalyst employed in this work, is not a prebiotic enzyme, we note that several ribozymes exist that catalyse the formation of amide linkages, suggesting that the transformations performed in this work are within the catalytic repertoire of RNA[26–28]. Furthermore, given the enhanced hydrolytic stability of amides relative to esters, compounds such as GMOA would have been more stable in the presence of divalent metal ions and more robust to heating and wet/dry cycles on the early earth, as was recently observed in the conversion of esters to depsipeptides[17].

Our observation that the presence of vesicles is essential for the conversion of ethyl oleate to GMOA is fully consistent with previous suggestions that fatty acid membranes themselves could have functioned as reaction promoters in early life[29]. Indeed, before the advent of coded protein synthesis, it is likely that such interactions were crucial. Contemporary life employs proteins that fold with exquisite specificity to form hydrophobic cores capable of binding water-insoluble substrates. It would be more difficult for the highly charged RNA polymer to have played such a role. Fatty acid membranes could have played a key role in solubilizing such molecules and enabling them to undergo chemical transformations that were otherwise inaccessible in aqueous solution. We speculate that scenarios in early life in which membranes and soluble catalysts collaborated could have been critical to early life. Ribozyme catalysts could have enabled reactions of higher specificity than a simple self-assembled membrane bilayer, and the lipid bilayer could have solubilized components that otherwise would not be available for reaction in bulk aqueous solution.

Here, we have shown that GMOA, the amide analogue of GMO, is as effective as GMO in stabilizing lipid vesicles against divalent ion-mediated lysis and precipitation, and vesicles containing either lipid can support ribozyme catalysis. The comparative hydrolytic stability of GMOA relative to GMO leads us to speculate that GMOA (and related amino-sugar-based fatty acid amides) could have served as membrane-stabilizing lipids before the emergence of coded protein catalysts.

Recently, several other groups have observed that lipid assemblies can bind biopolymer building blocks[30] and modulate DNA aptamer specificity[31]. Our results here demonstrate that the functional repertoire of lipids extends yet further, into the realm of enabling soluble catalysts to promote reactions that are otherwise disfavoured or impossible in aqueous solution. Our observation that both a catalyst and a fatty acid-based membrane acting in concert were required for the aqueous synthesis of GMOA suggests that lipids and catalysts may have coevolved in early life, with soluble fatty acids serving to bring

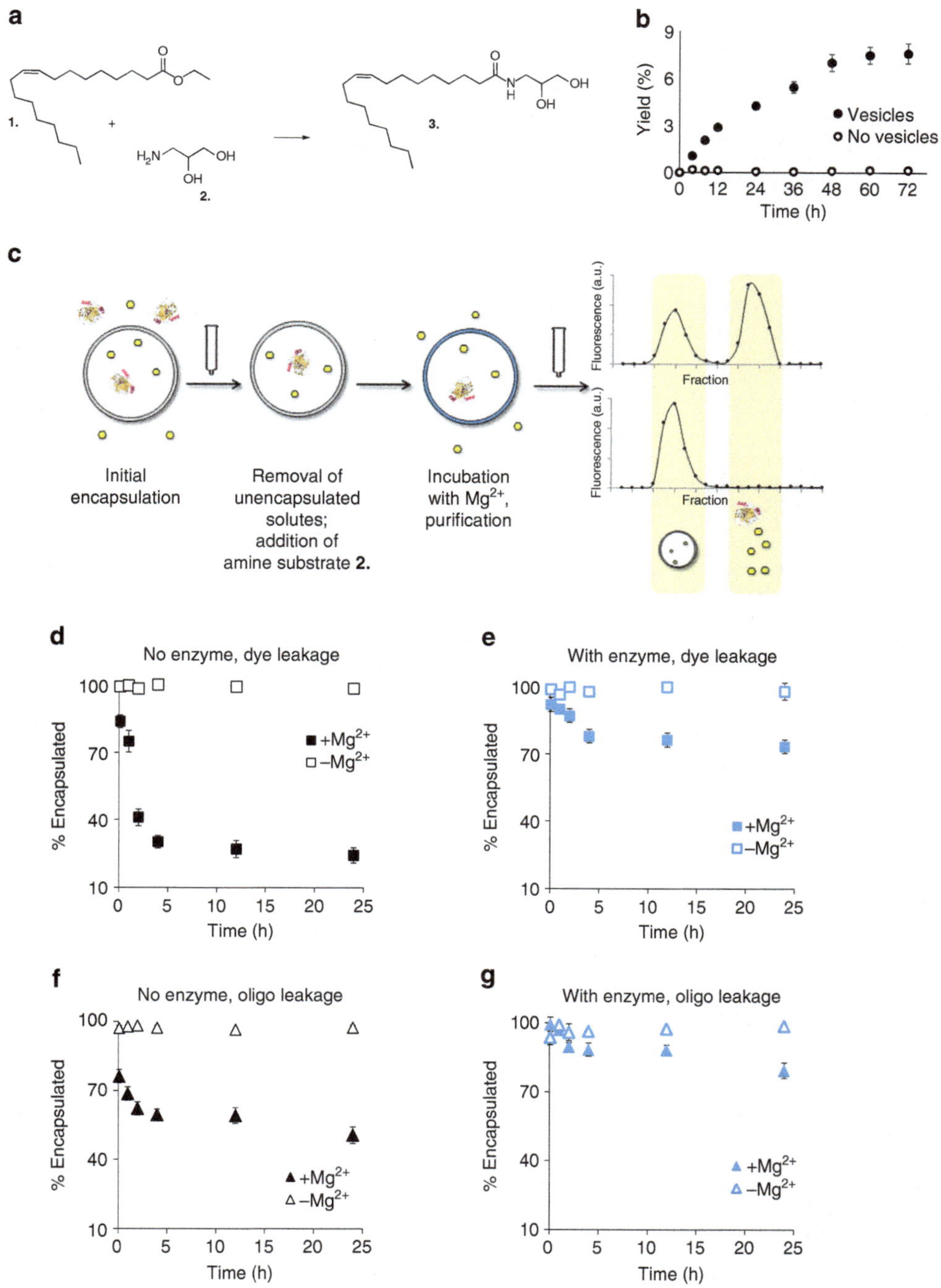

Figure 3 | Encapsulated α-chymotrypsin generates GMOA *in situ* and renders vesicles robust to Mg^{2+}. General scheme of the experiment (**a**). Vesicles containing ethyl oleate within the membrane and α-chymotrypsin within their lumen, in the presence of (±)-3-amino-1,2-propanediol, convert these substrates to GMOA. α-chymotrypsin-catalyzed conversion of ethyl oleate to GMOA requires the presence of vesicles, which solubilize the precursor lipid. (**b**) OA/GMOA vesicles, derived from α-chymotrypsin-catalyzed conversion of ethyl oleate after 72 h incubation, are stabilized to Mg^{2+}-induced leakage. (**d**–**g**) Analysis of vesicles by repurification following Mg^{2+} exposure (general scheme of the assay on **c**; size-exclusion chromatograms shown here are intended to be schematic in nature only) revealed that *in situ* generated GMOA dramatically reduced Mg^{2+}-induced leakage of small molecule dye and oligonucleotides from α-chymotrypsin-containing vesicles. (**d**–**g**) Leakage of the encapsulated small molecule (calcein; **d**,**e**) and oligonucleotide (**f**,**g**). Squares: small molecule leakage; triangles: oligonucleotide leakage. Blue markers: vesicles with the enzyme, black markers: vesicles without the enzyme. Error bars are s.e.m. ($n = 4$); error bars that are not visible are sufficiently small they are obscured by markers. All reactions contained vesicles comprised of 50 mM total lipid (42.5 mM OA, 7.5 mM (15 mol%) EtOA), 25 mM (±)-3-amino-1,2-propandiol, and 0.2 M Na-glycinamide buffer, pH 8.5 as buffer. α-chymotrypsin, when present, was 1 mg ml^{-1} (within vesicles). Reactions were performed at uncontrolled room temperature (20 ± 2 °C).

hydrophobic reactants into solution in primitive encapsulated catalyst systems. As more sophisticated proteins with hydrophobic cores evolved, this crude solubilization role would have been supplanted by the catalysts themselves. In turn, less soluble lipids capable of forming more stable membranes could have supplanted more soluble fatty acids. Early forms of life comprised principally of ribozyme/fatty acid-based systems could, thus, have evolved in concert into the coded protein catalyst/diacylphospholipid system that predominates in life today.

Methods

Suppliers. Unless otherwise stated, all chemicals and enzymes were purchased from Sigma-Aldrich Corp. (St Louis, MO, USA) and used without further purification.

Preparation of vesicles. Oleic acid/oleate (OA) vesicles were prepared by resuspending oleic acid in buffer (0.2 M Na-glycinamide pH 8.5), and then sonicating for about 1 min, followed by five freeze-thaw cycles. After that, samples were typically left tumbling for 24 h and then extruded through a 100 nm filter using a Mini-Extruder system (Avanti Polar Lipids, Inc., Alabaster, AL, USA), as previously described[11]. Glycinamide buffer was used in place of the more common bicine buffer because it is more prebiotically plausible while simultaneously exhibiting a similar pK_a and membrane permeability to bicine. *N*-oleoylglycinamide (NOG), the hypothetical amide condensation product of ethyl oleate and glycinamide, was not observed in reactions (Supplementary Fig. 5). NOG prepared by chemical synthesis could form vesicles in combination with oleic acid, but these vesicles leaked in the presence of magnesium (Supplementary Fig. 6). For vesicles prepared with encapsulated enzyme and dye (or dye-labelled oligonucleotide), the enzyme and dye-containing component were dissolved in buffer prior to addition of lipid, and vesicles were then treated as described above. After extrusion, vesicles were left tumbling to equilibrate for at least 2 h. Vesicles were purified from unencapsulated solutes on a gravity-flow size-exclusion column (Sepharose 4B, wet bead diameter 45–165 μm) with glycinamide buffer as the mobile phase. The vesicles, with an average diameter of 100 nm after extrusion, elute with the void volume of the column, followed by unencapsulated enzyme (30–60 kDa) and calcein dye (< 1 kDa) or oligonucleotide (ca. 3 kDa).

Mixed composition vesicles, made of oleic acid and ethyl oleate (EtOA) or GMO or GMOA or NOG, were prepared by mixing oleic acid with a chloroform or dichloromethane solution of EtOA or GMO or GMOA or NOG, removing the solvent on a rotary evaporator, and then resuspending the mixture in buffer with or without solutes, after which the vesicles were prepared as described above. In each case, the coelution of encapsulated dye or dye-labelled oligonucleotide with the vesicle fraction during gel purification confirmed encapsulation and the presence of vesicles (versus oil droplets, which would not encapsulate solutes). Vesicles were 50 mM total lipid; where a second lipid component was present, it comprised 15 mol% of total lipid (that is, 42.5 mM OA, 7.5 mM EtOA, GMO, GMOA or NOG).

α-Chymotrypsin activity in vesicles. For α-chymotrypsin reactions, vesicles comprised of OA and EtOA with encapsulated α-chymotrypsin and calcein were prepared as described above and incubated in the presence of 25 mM glycerol amine analogue (±)-3-amino-1,2-propanediol. The pH of mixture of calcein and (±)-3-amino-1,2-propanediol in the glycinamide buffer was adjusted to 8.5 before the addition of the enzyme and lipids. The initial concentration of α-chymotrypsin was 1 mg ml^{-1}; vesicles were 50 mM total lipid (42.5 mM OA, 7.5 mM (15 mol%) EtOA). The buffer employed was 0.2 M Na-glycinamide buffer, pH 8.5. In a typical experiment, OA/EtOA vesicles were prepared with α-chymotrypsin in the solution, followed by equilibration, extrusion and removal of unencapsulated solutes. (±)-3-amino-1,2-propanediol was then added and samples were left to tumble at 37 °C for 72 h. The pK_a of the conjugate acid of (±)-3-amino-1,2-propanediol was determined to be 9.18 (Supplementary Fig. 7), indicating that, in bulk solution, 17% of this substance was present as the free base.

The reaction yield was calculated from the ^{13}C NMR spectrum (Varian VXR-400). Samples were prepared as follows: at a given time point, a sample was frozen in liquid nitrogen, then lyophilized; the lipid compounds were redissolved in a small volume of $CDCl_3$. Redissolution in $CDCl_3$ was employed to dissociate vesicles, thus avoiding the extensive line broadening due to the slow tumbling of large vesicles. A representative ^{13}C NMR of the carbonyl region of a GMOA/EtOA mixture and the ^{13}C NMR spectrum of GMOA are given in Supplementary Fig. 5. The yield was calculated by referencing a ratio of the heights of the 178.8 p.p.m. peak corresponding to the ester carbonyl of EtOA and the 174.5 p.p.m. peak corresponding to the amide carbonyl of GMOA to a calibration curve (Supplementary Fig. 8). As both these resonances correspond to carbonyl carbons, we anticipated that intensity variations due to proton decoupling could be neglected. To validate this, control experiments with a $CDCl_3$ solution containing known quantities of EtOA and GMOA were performed; these gave the same ratiometric quantitation results as a $CDCl_3$ extract of the lyophilization residue of an aqueous sample of the same composition.

Synthesis of GMOA. GMOA was synthesized and purified as described previously[32]. Oleic acid (0.159 ml, 0.141 g, 0.5 mmol) was dissolved in 40 ml of dry CH_2Cl_2. *N*-(3-dimethylaminopropyl)-*N*′-ethylcarbodiimide hydrochloride in the measure 0.326 g (2 mmol) was added, followed by 0.279 ml (2 mmol) triethylamine and a catalytic amount of DMAP; after 10 min of vigorous stirring, 0.116 ml (1.5 mmol) of (±)-3-amino-1,2-propanediol was added. The reaction was stirred at RT for 5 h. The solvent was removed by rotary evaporation, and the resulting colourless oil was purified on a silica gel flash column with a hexane:AcOEt gradient (2:8 to 7:3). GMOA in the measure 0.146 g (0.41 mmol, 82% yield) was obtained as a colourless oil which solidified on standing. Electrospray ionization-mass spectrometry (ESI-MS): Calculated *m*/*z* for $C_{21}H_{42}NO_3^+$ [M + H]: 356.3; observed: 356.2. ^{1}H NMR ($CDCl_3$, 400 MHz): 0.87 (t, 3H), 1.26 (m, 20H), 1.59 (m, 2H), 2.15 (m, 6H), 3.38 (m, 2H), 3.59 (m, 2H), 3.70 (br, 2H), 3.98 (m, 1H), 5.41 (m, 2H), 7.21 (br, 1H). ^{13}C NMR ($CDCl_3$, 100 MHz): 14.12, 22.71, 25.78, 28.63, 29.05, 29.11, 29.65, 29.69, 29.74, 29.82, 31.94, 34.12, 36.52, 43.90, 65.06, 71.24, 131.56, 131.60 and 174.51.

Synthesis of NOG. Oleic acid (947 μl, 3 mmol), N, N, N′, N′ - tetramethyl-0-(1H-benzotriazol-1-yl) urcnium hexafluorophosphate (HBTU) (1138 mg, 3 mmol) and diisopropylethylamine (2613 μl, 15 mmol) were stirred in 75 ml dichloromethane for 1 h at room temperature. Glycinamide hydrochloride (365 mg, 3.3 mmol) was then added, and the resulting mixture was stirred overnight at room temperature. The reaction was filtered, and the filtrate was concentrated by rotary evaporation onto 5 g of silica. The reaction was then dry-loaded and purified on a silica gel flash column with a EtOAc:MeOH gradient (1:0 to 9:1). The compound eluted as a single peak comprised of NOG · 3.5 N,N-Diisopropylethylamine DIEA) (by ^{1}H NMR integration), of which 570 mg (0.72 mmol, 24% yield based on overall molar mass of the mixture) was obtained as a white solid. Recrystallization of 150 mg of this mixture from ca. 30 ml ethyl acetate yielded 33 mg of NOG containing no detectable DIEA (51% recovery). Electrospray ionization-mass spectrometry (ESI-MS): Calculated *m*/*z* for $C_{20}H_{38}N_2O_2Na^+$ [M + Na]: 361.5; observed: 361.1. ^{1}H NMR (Methanol-d4, 400 MHz): 0.92 (t, 3H), 1.34 (m, 20H), 1.64 (p, 2H), 2.05 (m, 4H), 2.28 (t, 2H), 3.85 (br s, 2H), 5.36 (m, 2H). ^{13}C NMR ($CDCl_3$, 100 MHz): 14.30, 22.87, 25.76, 27.36, 27.41, 29.31, 29.42, 29.51, 29.71, 29.89, 29.95, 32.09, 36.53, 43.06, 129.90, 130.21, 171.29, 174.07. These spectral data are consistent with those previously reported for this compound[33].

Vesicle leakage after magnesium exposure. Vesicle stability was assessed by measuring the leakage of either an encapsulated small molecule dye (calcein) or a dye-labelled oligonucleotide versus time. After vesicle formation, unencapsulated solute (2.5 mM calcein or 2 μM 5′-d(CCA ATG CGC)-3′-fluorescein) was removed by size-exclusion chromatography (Sepharose 4B), with running buffer containing the same concentration and composition of lipids as the vesicle sample.

In a typical experiment, a 3 ml purified vesicle sample was mixed with 0.2 ml of $MgCl_2$ solution (240 mM, to give the desired final Mg^{2+} concentration of 15 mM) while stirring rapidly to minimize vesicle disruption by transient exposure of vesicles to high magnesium concentrations. We observed that the rapid mixing of magnesium is critical to vesicle stability. If magnesium is added and vesicles are not immediately vigorously vortexed, permeability experiments produce inconsistent results from sample to sample. After Mg^{2+} addition, the sample was allowed to tumble for the given time and repurified on a size-exclusion column, as described above.

Calcein or oligonucleotide label fluorescence was recorded using a SpectraMax Gemini EM plate reader (Molecular Devices). Vesicle samples were diluted to ca. 1 mM lipid to minimize fluorescence artifacts arising from scattering. The total fluorescence of all vesicle and all free dye/oligonucleotide fractions was calculated; leakage is reported as % leakage $= (F_{dye} \times 100\%)/(F_{ves} + F_{dye})$, where F_{ves} is total fluorescence of all vesicle fractions and F_{dye} is total fluorescence of all unencapsulated dye/oligonucleotide fractions collected during the size-exclusion column purification. The initial calcein concentration was 2.5 mM; at this concentration, self-quenching is negligible.

Magnesium exposure experiments presented in the main text were performed in four replicates, independently mixing samples with magnesium and purifying on the column. Reported leakage values are the arithmetic average of four experiments. The reported error bars are s.e.m. ($n = 4$).

Mag-fura-2 assay for free Mg^{2+} measurement. The permeability assays were performed as described previously[11]. Briefly, vesicles were prepared as described above, with mag-fura-2 tetrapotassium salt encapsulated inside vesicles at a concentration of 2.5 mM. Free dye was removed by size-exclusion column chromatography. Vesicles were adjusted to ca. 5 mM total lipid, and Mg^{2+} was added as a stock of $MgCl_2$ to a final concentration of 2.5 mM.

The concentration of free magnesium inside vesicles was calculated by fitting the measured fluorescence ratio of fluorescence emission measured with ($\lambda_{em} = 500$ nm, $\lambda_{ex} = 340$ nm) to that measured with $\lambda_{ex} = 370$ nm, $\lambda_{em} = 500$ nm to a standard calibration curve of known Mg^{2+} (as $MgCl_2$) concentration (Supplementary Fig. 2).

References

1. Gebicki, J. M. & Hicks, M. Ufasomes are stable particles surrounded by unsaturated fatty acid membranes. *Nature* **243,** 232–234 (1973).
2. Hargreaves, W. R. & Deamer, D. W. Liposomes from ionic, single-chain amphiphiles. *Biochemistry* **17,** 3759–3768 (1978).
3. Hanczyc, M. M., Fujikawa, S. M. & Szostak, J. W. Experimental models of primitive cellular compartments: encapsulation, growth and division. *Science* **302,** 618–622 (2003).
4. Deamer, D. & Dworkin, J. in *Prebiotic Chemistry* (ed. Walde, P.) **259,** 1–27 (Springer Berlin Heidelberg, 2005).
5. Ruiz-Mirazo, K., Briones, C. & de la Escosura, A. Prebiotic systems chemistry: new perspectives for the origins of life. *Chem. Rev.* **114,** 285–366 (2014).

6. Gilbert, W. Origin of life: the RNA world. *Nature* **319,** 618–618 (1986).
7. Joyce, G. F. RNA evolution and the origins of life. *Nature* **338,** 217–224 (1989).
8. Szostak, J. W. The eightfold path to non-enzymatic RNA replication. *J. Syst. Chem.* **3,** 2 (2012).
9. Bowman, J. C., Lenz, T. K., Hud, N. V. & Williams, L. D. Cations in charge: magnesium ions in RNA folding and catalysis. *Curr. Opin. Struct. Biol.* **22,** 262–272 (2012).
10. Petrov, A. S., Bowman, J. C., Harvey, S. C. & Williams, L. D. Bidentate RNA-magnesium clamps: on the origin of the special role of magnesium in RNA folding. *RNA* **17,** 291–297 (2011).
11. Chen, I. A., Salehi-Ashtiani, K. & Szostak, J. W. RNA catalysis in model protocell vesicles. *J. Am. Chem. Soc.* **127,** 13213–13219 (2005).
12. Adamala, K. & Szostak, J. W. Nonenzymatic template-directed RNA synthesis inside model protocells. *Science* **342,** 1098–1100 (2013).
13. Nissen, P., Hansen, J., Ban, N., Moore, P. B. & Steitz, T. A. The structural basis of ribosome activity in peptide bond synthesis. *Science* **289,** 920–930 (2000).
14. Cech, T. R. Structural biology. The ribosome is a ribozyme. *Science* **289,** 878–879 (2000).
15. Adamala, K. & Szostak, J. W. Competition between model protocells driven by an encapsulated catalyst. *Nat. Chem.* **5,** 495–501 (2013).
16. Grochmal, A., Prout, L., Makin-Taylor, R., Prohens, R. & Tomas, S. Modulation of reactivity in the cavity of liposomes promotes the formation of peptide bonds. *J. Am. Chem. Soc.* **137,** 12269–12275 (2015).
17. Forsythe, J. G. *et al.* Ester-mediated amide bond formation driven by wet-dry cycles: a possible path to polypeptides on the prebiotic Earth. *Angew. Chem. Int. Ed. Engl.* **54,** 9871–9875 (2015).
18. Mansy, S. S. & Szostak, J. W. Thermostability of model protocell membranes. *Proc. Natl Acad. Sci. USA* **105,** 13351–13355 (2008).
19. Uhlenbeck, O. C. A small catalytic oligoribonucleotide. *Nature* **328,** 596–600 (1987).
20. Adamala, K., Engelhart, A. E. & Szostak, J. W. Generation of functional RNAs from inactive oligonucleotide complexes by non-enzymatic primer extension. *J. Am. Chem. Soc.* **137,** 483–489 (2015).
21. Morihara, K. & Oka, T. alpha-Chymotrypsin as the catalyst for peptide synthesis. *Biochem. J.* **163,** 531–542 (1977).
22. Donghi, D. & Schnabl, J. Multiple roles of metal ions in large ribozymes. *Met. Ions Life Sci.* **9,** 197–234 (2011).
23. Johnson-Buck, A. E., McDowell, S. E. & Walter, N. G. Metal ions: supporting actors in the playbook of small ribozymes. *Met. Ions Life Sci.* **9,** 175–196 (2011).
24. Blain, J. C. & Szostak, J. W. Progress toward synthetic cells. *Annu. Rev. Biochem.* **83,** 615–640 (2014).
25. Mansy, S. S. Membrane transport in primitive cells. *Cold Spring Harb. Perspect. Biol.* **2,** a002188 (2010).
26. Zhang, B. & Cech, T. R. Peptide bond formation by in vitro selected ribozymes. *Nature* **390,** 96–100 (1997).
27. Martin, L. L., Unrau, P. J. & Müller, U. F. RNA synthesis by in vitro selected ribozymes for recreating an RNA world. *Life* **5,** 247–268 (2015).
28. Pressman, A., Blanco, C. & Chen, I. A. The RNA world as a model system to study the origin of life. *Curr. Biol.* **25,** R953–R963 (2015).
29. Cape, J. L., Monnard, P.-A. & Boncella, J. M. Prebiotically relevant mixed fatty acid vesicles support anionic solute encapsulation and photochemically catalyzed trans-membrane charge transport. *Chem. Sci.* **2,** 661 (2011).
30. Black, R. A. *et al.* Nucleobases bind to and stabilize aggregates of a prebiotic amphiphile, providing a viable mechanism for the emergence of protocells. *Proc. Natl. Acad. Sci. U. S. A.* **110,** 13272–13276 (2013).
31. Peterson, A. M., Jahnke, F. M. & Heemstra, J. M. Modulating the substrate selectivity of DNA aptamers using surfactants. *Langmuir* **31,** 11769–11773 (2015).
32. Aneetha, H., O'Dell, D. K., Tan, B., Walker, J. M. & Hurley, T. D. Alcohol dehydrogenase-catalyzed *in vitro* oxidation of anandamide to N-arachidonoyl glycine, a lipid mediator: synthesis of N-acyl glycinals. *Bioorg. Med. Chem. Lett.* **19,** 237–241 (2009).
33. Boger, D. L. Serotonin potentiation by oleamide analogs. Patent US6734204 B1 (2004).

Acknowledgements

We thank Mr Tony Jia, Dr Neha Kamat, Ms Lin Jin, and Dr Christian Hentrich for helpful discussions. This work was supported in part by NASA Exobiology grant NNX07AJ09G to J.W.S. and a grant (290363) from the Simons Foundation to J.W.S. A.E.E. was supported by an appointment to the NASA Postdoctoral Program, administered by Oak Ridge Associated Universities through a contract with NASA and a Tosteson Fellowship from the Massachusetts General Hospital Executive Committee on Research. J.W.S. is an investigator of the Howard Hughes Medical Institute.

Author contributions

K.P.A. and A.E.E. performed the experiments. All authors interpreted the experimental data, contributed ideas, and wrote the manuscript. K.P.A. and A.E.E. contributed equally to this work.

Additional information

Competing financial interests: The authors declare no competing financial interests.

8

Biosynthesis of the mycotoxin tenuazonic acid by a fungal NRPS-PKS hybrid enzyme

Choong-Soo Yun[1,*], Takayuki Motoyama[1,*] & Hiroyuki Osada[1]

Tenuazonic acid (TeA) is a well-known mycotoxin produced by various plant pathogenic fungi. However, its biosynthetic gene has been unknown to date. Here we identify the TeA biosynthetic gene from *Magnaporthe oryzae* by finding two TeA-inducing conditions of a low-producing strain. We demonstrate that TeA is synthesized from isoleucine and acetoacetyl-coenzyme A by TeA synthetase 1 (TAS1). TAS1 is a unique non-ribosomal peptide synthetase and polyketide synthase (NRPS-PKS) hybrid enzyme that begins with an NRPS module. In contrast to other NRPS/PKS hybrid enzymes, the PKS portion of TAS1 has only a ketosynthase (KS) domain and this domain is indispensable for TAS1 activity. Phylogenetic analysis classifies this KS domain as an independent clade close to type I PKS KS domain. We demonstrate that the TAS1 KS domain conducts the final cyclization step for TeA release. These results indicate that TAS1 is a unique type of NRPS-PKS hybrid enzyme.

[1]Chemical Biology Research Group, RIKEN Center for Sustainable Resource Science, 2-1 Hirosawa, Wako-shi, Saitama 351-0198, Japan. * These authors contributed equally to this work. Correspondence and requests for materials should be addressed to H.O. (email: hisyo@riken.jp).

The mycotoxin tenuazonic acid (TeA), (5*S*)-3-acetyl-1,5-dihydro-4-hydroxy-5-[(1*S*)-1-methylpropyl]-2*H*-pyrrol-2-one, is a tetramic acid derivative first isolated from the broth of an *Alternaria tenius* culture[1]. *Alternaria* is a ubiquitous plant pathogenic fungus that causes both spoilage of various food crops and fruits in the field and post-harvest decay[2], and TeA has been detected in various *Alternaria*-contaminated crops, fruits and vegetables[3–5]. In addition, the plant pathogenic fungi *Phoma sorghina* (a pathogen of sorghum) and *Magnaporthe oryzae* (a pathogen of rice) reportedly produce TeA[6,7]. TeA is the most toxic of the *Alternaria* toxins. It inhibits protein biosynthesis on ribosomes by suppressing the release of new protein[8]. It also reportedly shows biological properties including antitumor, antibacterial, antiviral and phytotoxic activities[9–11].

TeA is thought to be a hybrid of an isoleucine and two acetates[12]. Due to its tetramic acid-containing structure (Fig. 1a), TeA is also expected to be a product of a polyketide synthase and non-ribosomal peptide synthetase (PKS–NRPS) hybrid enzyme[13].

Typical fungal secondary metabolite compounds polyketides and non-ribosomal peptides are synthesized by PKSs and NRPSs, respectively. There are two types of fungal PKSs. One is iterative type I PKSs, consisting of multiple catalytic domains that contain ketosynthase (KS), acyltransferase (AT) and acyl carrier protein (ACP) main domains, along with several optional β-keto processing domains, such as β-ketoacyl reductase (KR), dehydratase (DH) and trans-acting enoyl reductase domains in single enzyme[14]. According to the absence or presence of β-keto processing domains, iterative type I PKSs can be subdivided into non-reducing (NR-), partially reducing (PR-) and highly reducing (HR-) PKSs[15,16]. The other type is type III PKSs, consisting of a homodimeric KS[17]. Type II PKSs, which include several individual enzymes[18], have to date only been identified in bacteria.

In addition, fungal PKS–NRPS hybrid enzymes, whose structure consists of an iterative type I PKS followed by a single module NRPS, produce a wide variety of structurally diverse secondary metabolites including polyketide-amino acid hybrids[19–23]. In the typical fungal PKS–NRPS, the PKS portion consists of KS, AT and ACP domains, along with several modifying domains such as KR, DH and methyltransferase domains. The NRPS portion consists of adenylation (A), thiolation (T, also known as PCP for peptidyl carrier protein), condensation (C) and terminal release or cyclization (R, reductase or DKC, Dieckmann cyclization) domains[18]. Similar types of hybrid enzyme have been observed in bacteria[24,25]. In contrast to PKS–NRPS, NRPS–PKS hybrid enzymes (which begin with a NRPS module) also exist; however, to date, this type of enzyme has only been found in bacteria[26–31] (Supplementary Fig. 1).

Herein, we describe the identification of the TeA biosynthetic gene and the functional *in vivo* and *in vitro* analysis of TeA synthetase 1 (TAS1) in *M. oryzae*. TAS1 is a unique type of NRPS–PKS hybrid enzyme of fungal origin. In addition, we discuss the unique domain structure of TAS1 and demonstrate that its KS domain is indispensable for TeA biosynthesis.

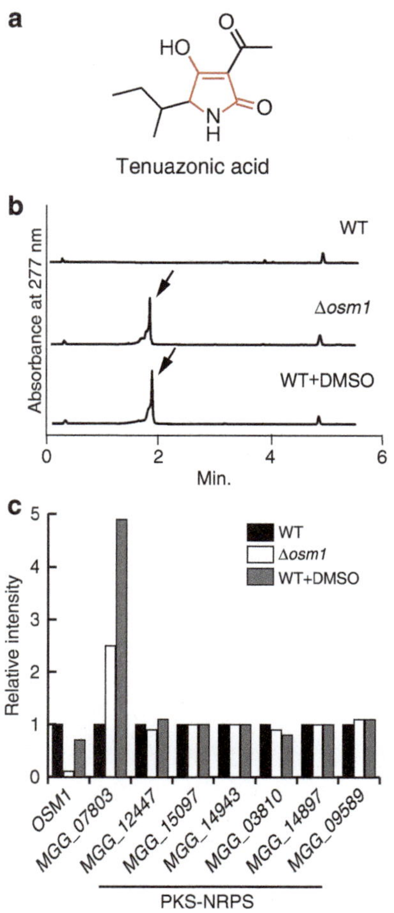

Figure 1 | Tenuazonic acid biosynthetic gene identification. (**a**) Chemical structure of TeA. Red portions indicate the tetramic acid structure. (**b**) UPLC analysis of metabolites extracted from strains of *M. oryzae* Kita 1 (wild type (WT)), *Δosm1* and DMSO-added culture conditions. Each strain was static-cultured for 5 days at 25 °C. Arrows indicate TeA peak. (**c**) Signal intensities of *OSM1*, *MGG_07803* and six PKS-NRPS genes in a microarray analysis with WT (black), *Δosm1* (white) and strains cultivated under DMSO-added culture conditions (grey).

Results

Identification of the TeA biosynthetic gene. Secondary metabolite production in microorganisms is regulated by environmental conditions for adaptation. For example, OSM1, a Hog1-related mitogen-activated protein kinase in *M. oryzae*, is reportedly involved in the response to environmental signals such as hyperosmotic stress[32]. Therefore, we hypothesized that OSM1 is also involved in secondary metabolism regulation in *M. oryzae*. To confirm this hypothesis, we used *M. oryzae* to construct *OSM1* (*MGG_01822*) knockout strains (Δ*osm1*; Supplementary Methods and Supplementary Fig. 2) and analysed their metabolites with ultra performance liquid chromatography/mass spectroscopy (UPLC/MS). The production of one metabolite was induced in the Δ*osm1* strains under static culture conditions (Fig. 1b). We then obtained the ultraviolet pattern and mass spectra of the induced metabolite, which had a molecular mass of 197.2 ($m/z = 196.2$ $[M\text{-}H]^-$) corresponding to the fungal secondary metabolite TeA (Fig. 1a, Supplementary Fig. 3). Although a previous report described the production of TeA in *M. oryzae*, the *M. oryzae* Kita 1 strain used in this study did not produce a detectable amount of TeA under our laboratory culture conditions (Methods section). In addition, while screening compounds that induce the production of secondary metabolites, we found that 1% dimethylsulphoxide (DMSO) also induced TeA production ($14\,mg\,l^{-1}$) in *M. oryzae* under static culture conditions (Fig. 1b). Although a number of studies have detected TeA in plants infected with TeA-producing fungi[3–5], the biosynthetic gene of this compound has been unknown to date. Thus, we undertook this study to identify the biosynthetic gene of TeA in *M. oryzae*. Two TeA-inducing culture conditions enabled us to explore the TeA biosynthetic gene in this organism.

To identify the TeA biosynthetic gene, we performed a custom DNA microarray analysis using total RNA of *M. oryzae* extracted under the two TeA-inducing culture conditions and non-inducing normal culture conditions. TeA is expected to be the product of a PKS–NRPS hybrid enzyme[13], and the genome sequence of *M. oryzae*[33] has six genes encoding such PKS–NRPS enzymes. However, the transcriptional levels of these genes were unchanged under TeA-inducing conditions. Of the remaining secondary metabolism-related genes, only one gene (*MGG_07803*) was found to be commonly upregulated (by more than twofold) under the two TeA-inducing conditions (Fig. 1c, Supplementary Table 1), suggesting that it is involved in TeA biosynthesis.

Confirmation of the TeA biosynthetic gene. To ensure that *MGG_07803* is the biosynthetic gene of TeA, we constructed an *MGG_07803* knockout. The knockout strain lost the capacity to produce TeA under the DMSO-added culture condition (Fig. 2a, Supplementary Fig. 4), confirming that this gene is the TeA biosynthetic gene. Accordingly, we renamed *MGG_07803* to *TAS1*. Because *TAS1* is automatically annotated from the genome sequence of *M. oryzae*, we conducted rapid amplification of complementary DNA (cDNA) ends-polymerase chain reaction with the cDNA of *M. oryzae* that was cultured with 1% DMSO and identified an open reading frame (ORF) region (Supplementary Methods and Supplementary Fig. 5). Using this ORF, we constructed *TAS1* overexpression strains by inserting the *Aspergillus oryzae TEF1* promoter into the upstream of the *TAS1* start codon via homologous recombination. UPLC/MS analysis of extracted metabolites from the resulting strain showed 28 mg l^{-1} TeA-producing capability without DMSO addition (Fig. 2a, Supplementary Fig. 4). This outcome also supported the conclusion that TAS1 is the TeA biosynthetic enzyme. *TAS1* encodes an NRPS–PKS hybrid protein of 1,602 amino acids that consists of C, A and PCP domains in the NRPS portion and a KS domain in the PKS portion (Fig. 2b). The PKS portion of this domain structure is distinct from that of typical fungal PKS–NRPS hybrid enzymes with type I PKSs (Fig. 2b). Intriguingly, the NRPS portion of TAS1 also lacks the terminal release or cyclization domain (R or DKC)[20], required for product release from the enzyme and generally present in fungal PKS–NRPS hybrid enzymes. In contrast, similar to type III PKSs, the PKS portion of TAS1 has only a KS domain. However, it shows very low similarity to known type III PKSs. Moreover, fungal PKS–NRPS hybrid enzymes have been reported but there are no reports of NRPS–PKS type fungal hybrid enzymes, even though bacterial NRPS–PKS enzymes have been identified. Therefore, TAS1 is a novel type of fungal secondary metabolite biosynthetic enzyme.

Substrate specificity of TAS1. On the basis of the structure of TeA, we expected that isoleucine and diketide are necessary for TeA biosynthesis. As TAS1 is a novel fungal NRPS–PKS enzyme with a PKS portion that has low similarity to that of type III PKSs, it is unknown whether this diketide is synthesized by TAS1. If TAS1 can synthesize this diketide, TeA can be synthesized by a single enzyme, TAS1, from isoleucine, acetyl-coenzyme A (CoA) and malonyl-CoA. Alternatively, TAS1 might use acetoacetyl-CoA as a source of diketide. To clarify this point, we prepared a recombinant enzyme and analysed its substrate specificity *in vitro*. The 173-kDa TAS1 protein was overexpressed in a protease-deficient yeast strain BJ5464 using *Saccharomyces cerevisiae* 2 μm episomal vector carrying the genes-encoding TAS1 under the control of the *ADH1* promoter. To obtain the active form of TAS1, the 4′-phosphopantetheinyl transferase (PPTase; MGG_17878) of *M. oryzae*, which catalyses the essential post-translational activation of carrier proteins of NRPS and PKS, was coexpressed as a non-tagged protein. The purified C-terminally His-tagged recombinant TAS1 protein synthesized TeA as a single enzyme from isoleucine with acetoacetyl-CoA, not acetyl-CoA and malonyl-CoA (Fig. 3a,b), which clearly indicated that TAS1 is the TeA biosynthetic enzyme. The k_{cat} was determined to be $(14.9 \pm 0.3) \times 10^{-3}\,min^{-1}$, and the $K_{M(app)}$ was $(44.3 \pm 3.1) \times 10^{-2}$ mM for isoleucine. For acetoacetyl-CoA, the k_{cat} was determined to be $(13.1 \pm 0.2) \times 10^{-3}\,min^{-1}$, and the $K_{M(app)}$ was $(1.2 \pm 0.1) \times 10^{-2}$ mM. For ATP, the k_{cat} was determined to be $(15.3 \pm 0.2) \times 10^{-3}\,min^{-1}$, and the $K_{M(app)}$ was $(8.1 \pm 0.5) \times 10^{-2}$ mM (Supplementary Fig. 6).

Homologue search and phylogenetic analysis. To investigate the distribution of this type of hybrid protein, we conducted a Basic

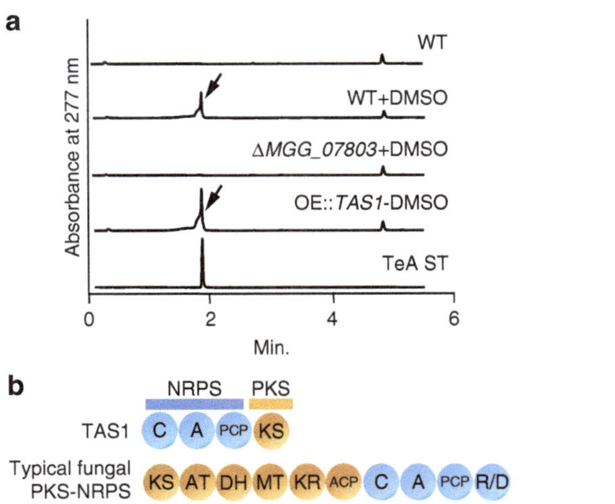

Figure 2 | Confirmation of the TeA biosynthetic gene and domain structure. (**a**) UPLC analysis of metabolites extracted from WT, WT + DMSO, *ΔMGG_07803* + DMSO and a *TAS1* overexpression strain (OE::*TAS1*) without DMSO. Each strain was static-cultured for 5 days at 25 °C. Arrows indicate TeA peak. A 100 μM TeA standard (TeA ST) was also analysed. (**b**) Domain structure of TAS1 and typical fungal PKS-NRPS hybrid enzyme; blue indicate the NRPS portion, and yellow indicates the PKS portion.

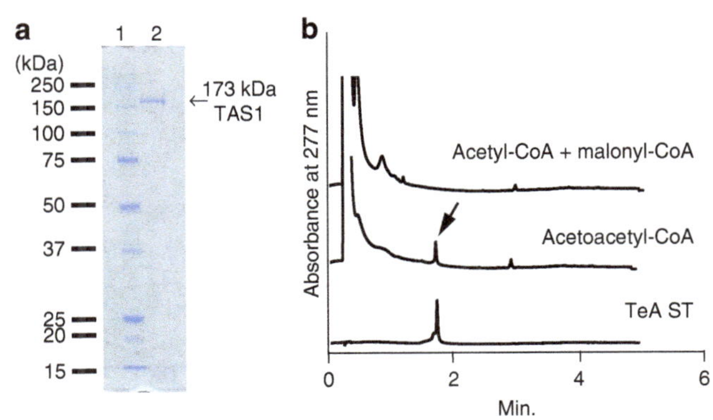

Figure 3 | Recombinant TAS1 expression and substrate specificity. (**a**) Purified C-terminally 8 × His-tagged TAS1 protein was loaded and analysed with a 5-20% sodium dodecyl sulphate-polyacrylamide gel electrophoresis gradient gel. Line 1, molecular mass markers. Line 2, purified TAS1 (500 ng). (**b**) UPLC analysis of the products of the TAS1 enzyme reaction with acetoacetyl-CoA and a mixture of acetyl-CoA and malonyl-CoA. Arrows indicate TeA peak. A 100 μM TeA standard (TeA ST) was also analysed. *In vitro* reaction conditions are described in the Methods section.

Local Alignment Search Tool (BLAST) search for proteins with sequences similar to that of TAS1 via the National Center for Biotechnology Information (http://www.ncbi.nim.nih.gov) and *Alternaria* genomes database (http://alternaria.vbi.vt.edu)[34]. *Alternaria* species are known to be major TeA producers. Homologue proteins that share 50% amino acid identity with TAS1 were found in *Alternaria* genomes database and their domain structures were identical to that of TAS1 (C–A–PCP–KS). TAS1 homologue proteins in *Alternaria* spp. were highly conserved and share over 97% identity. In addition, proteins with domain structures identical to that of TAS1 were also found in other fungal species from the National Center for Biotechnology Information database, which suggests that this type of hybrid enzyme is distributed in Ascomycota and Basidiomycota fungi (Supplementary Table 2).

However, similar protein sequences matched with TAS1 from bacterial sources had domain structures that differed considerably from that of TAS1. These proteins had a typical PKS structures consisted of KS, AT and ACP domains (Supplementary Table 2) at the end of the NRPS. This difference suggests that the C–A–PCP–KS domain structure of the NRPS–PKS hybrid enzyme is specific to fungi. We also conducted phylogenetic analysis of the KS domain in TAS1 with other KS domains from types I, II, III PKSs and PKS–NRPS hybrid proteins including the top five protein sequences from fungal and bacterial sources extracted from the TAS1 BLAST search result (Supplementary Table 2). Despite the single KS domain structure of the PKS portion, the KS domain of TAS1 was not classified as a type III PKS (Fig. 4, Supplementary Fig. 7). In addition, the KS domains from fungal PKS–NRPS hybrid enzymes were classified into one independent clade; however, the KS domain of TAS1 was not classified therein. Interestingly, protein sequences similar to TAS1 in bacteria were classified into an independent bacterial clade of type I PKSs with other proteins that reportedly interact with NRPS modules, including NRPS–PKS hybrid proteins[31]. In contrast, the KS domain of TAS1 formed a new clade with the homologues from fungal sources, which was close to the type I PKSs (Fig. 4, Supplementary Fig. 7). This outcome indicates that TAS1 is a unique type of NRPS–PKS hybrid enzyme.

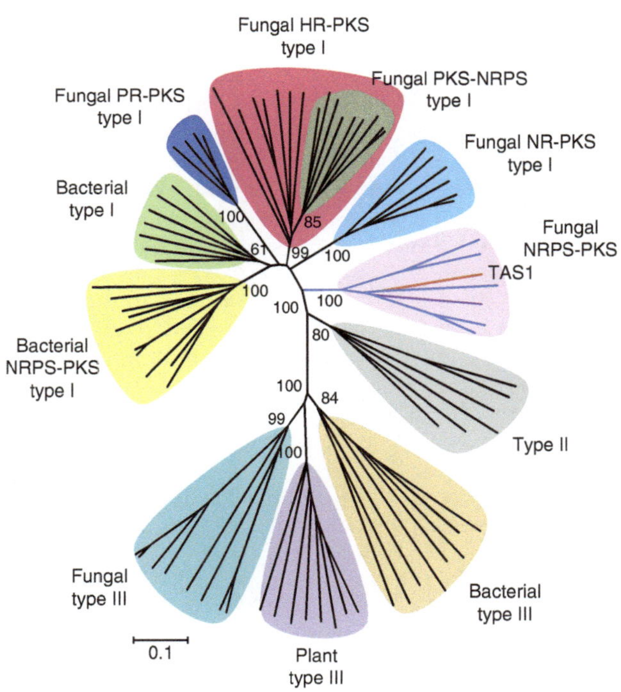

Figure 4 | Phylogenetic analysis of the KS domain in TAS1. Phylogenetic analysis of the TAS1 KS domain and other KS domains. Domains were aligned with MUSCLE, and the tree was constructed with the neighbour-joining method. Scale bar, 0.1 substitutions per site. The TAS1 KS domain is indicted with a red line and purple line indicates TAS1 homologue of *A. alternate*. The KS domains of other TAS1 homologues are indicated with blue lines.

To further analyse TAS1 homologues, we conducted phylogenetic analysis of all KS domains, except for three homologues that have a truncated C-terminal KS domain (Supplementary Table 2), extracted from BLAST search results. All KS domains from TAS1 homologues were also classified into an independent clade different from type I PKS clades (Supplementary Fig. 8). In addition, TAS1 homologues were divided into four sub-clades. Sub-clade A has five KS domains and includes the TAS1 from *M. oryzae* and a TAS1 homologue from *A. alternata*. These two strains were already known to produce TeA. This suggests that strains have a TAS1 homologue in the sub-clade A are TeA producer. Interestingly, sub-clade B and C consists only of homologues from Basidiomycota, whereas sub-clade D mainly contains homologues from fungi that infect insects, except for *Fusarium* spp.

Functional analysis of the TAS1 KS domain. Generally, the KS domain in the PKS enzyme is responsible for the control of polyketide product chain length via ketide extension. However, TAS1 uses acetoacetyl-CoA as a source of diketide to produce TeA, which means that ketide extension to yield the diketide (acetoacetyl-CoA) is unnecessary for TeA synthesis. Thus, we investigated whether the KS domain of TAS1 is necessary for TeA biosynthesis. We constructed the KS domain deleted *TAS1* overexpression strains and found that the constructed strains lost the capability to produce TeA (Fig. 5a, Supplementary Fig. 9). This result showed that the KS domain is necessary for TeA biosynthesis. For further confirmation, we prepared separately expressed recombinant proteins with the C–A–PCP and KS domain portions of TAS1 by using a cell-free protein synthesis system (Supplementary Fig. 10a). Analysis of the TeA biosynthetic capability of these proteins showed that TeA was produced only when the C–A–PCP domains and the KS domain coexisted, which clearly indicated that the KS domain of TAS1 is indispensable for TeA biosynthesis (Fig. 5b). However, the exact role of the KS domain in TeA biosynthesis remains unknown.

From the perspective of the domain structure of TAS1, we proposed a model for TeA biosynthesis and hypothesized that the final cyclization step in TeA production is conducted by the KS domain (Fig. 5c). To support this hypothesis, we synthesized an artificial intermediate of TeA biosynthesis using *N*-acetylcysteamine (SNAC), *N*-acetoacetyl-L-Ile-SNAC (Supplementary Methods and Supplementary Fig. 11), to analyse the cyclization capacity of the KS domain with the KS and C–A–PCP domain proteins synthesized with the cell-free system. The results showed that only the KS domain protein produced TeA, which indicated that this TAS1 domain is responsible for the final cyclization step in TeA production (Fig. 5d). In addition, we confirmed the condensation capacity of activated isoleucine and acetoacetyl-CoA by C–A–PCP or KS domain proteins using L-Ile-SNAC (Supplementary Fig. 11). The results showed that the condensed product (*N*-acetoacetyl-L-Ile-SNAC) is only produced by the C–A–PCP domain protein from L-Ile-SNAC and acetoacetyl-CoA, which indicated that the condensation reaction takes place via the C domain rather than the KS domain in TAS1 (Supplementary Fig. 12).

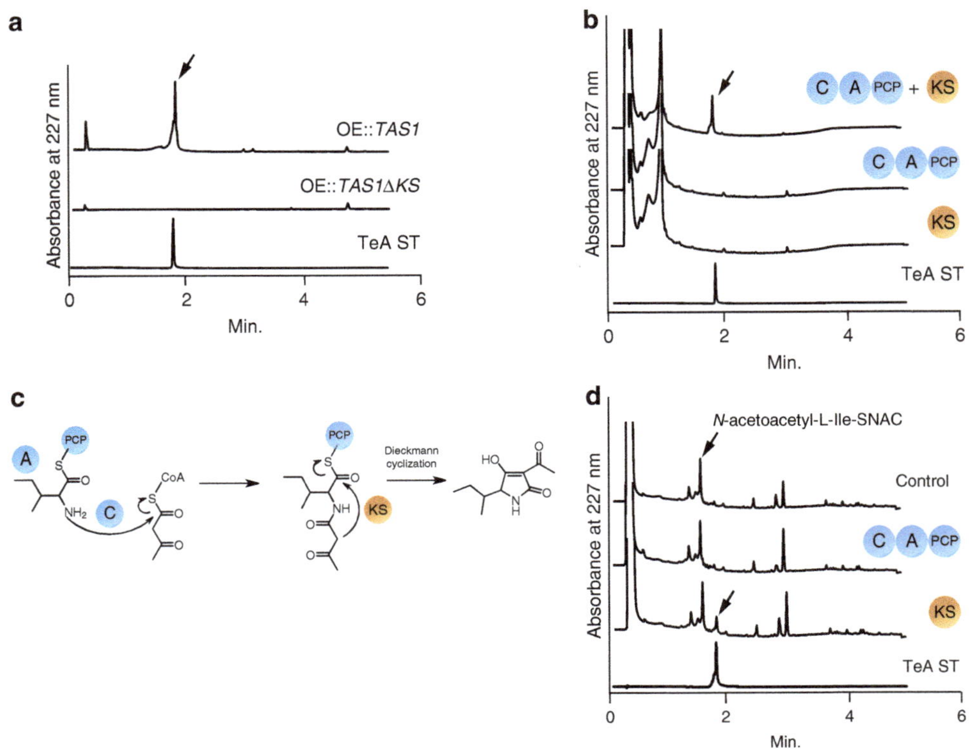

Figure 5 | KS domain analysis. (**a**) UPLC analysis of metabolites extracted from *TAS1* overexpression (OE::*TAS1*) and KS domain-disrupted *TAS1* overexpression (OE::*TAS1ΔKS*) strains. Arrows indicate TeA peak. A 100 µM TeA standard (TeA ST) was also analysed. (**b**) UPLC analysis of enzyme reaction products. C-A-PCP domain proteins, KS domain proteins or both of TAS1 were used as enzymes. A 100 µM TeA ST was also analysed. *In vitro* reaction conditions are described in the Methods section. Arrow indicates TeA peak. (**c**) Proposed model for TeA biosynthesis. (**d**) UPLC analysis of enzyme reaction products. Soluble C-A-PCP domain proteins and KS domain proteins were used to investigate cyclization capability with the artificial intermediate *N*-acetoacetyl-L-Ile-SNAC. A control reaction was conducted with cell-free-synthesized soluble protein without messenger RNA. *In vitro* reaction conditions are described in the Methods section. Arrow indicates TeA peak. SNAC, *N*-acetylcysteamine.

Discussion

The recent sequencing of fungal genomes has shown that fungi are richer in secondary metabolism-related genes than once believed. However, the majority of these genes are silent or quiescent under laboratory culture conditions[35]. During the past decade, successful approaches such as epigenetic regulation, promoter exchange and heterologous expression have been applied to activate these silent gene clusters in fungi[36–39]. In this study, we activated the TeA biosynthetic gene in *M. oryzae* via disruption of *OSM1*, showing the possibility that disruption of an environmental signal response system can influence some of the poorly expressed or silent secondary metabolite genes in *M. oryzae*.

TeA was first reported in 1957 by Rosett *et al.*[1]; however, the TeA biosynthetic gene was not identified in the nearly 60 years that followed despite many reports of TeA detection. The chemical structure of TeA suggested that TeA is synthesized by a PKS and NRPS hybrid enzyme, and fungal PKS–NRPS hybrid enzyme was considered as biosynthetic enzyme of TeA, because reports of NRPS–PKS type fungal enzymes were not forthcoming. The TAS1 that we have identified is the first reported fungal NRPS–PKS enzyme, which may explain why the TeA biosynthetic gene has remained unidentified for so long. Our result also shows the limitations of the *in silico* prediction of polyketide and non-ribosomal peptide biosynthetic pathways from genomic sequence data.

The KS domain of TAS1 is quite different from other KS domains. Phylogenetic position of the KS domain of TAS1 is nearest to PKS type I. However, it does not belong to any clade of fungal NR-, PR- or HR-PKS type I and is clearly separated from the clades of PKS type I KS domains (Supplementary Fig. 7). In contrast to fungal KS domains of PKS type I, the KS domain of TAS1 is non-iterative. Moreover, the conserved catalytic triad residues of the KS domain in TAS1 is different from that of type I PKSs (Cys–His–His) and identical to that of type III PKSs (Cys–His–Asn; Supplementary Fig. 13)[17,40]. The catalytic triads of fungal homologues of TAS1 are also identical to that of TAS1. However, the residues surrounding the catalytic triad of TAS1 are more frequently conserved in the KS domains of type I PKSs than in those of type III PKSs (Supplementary Fig. 13). According to the above data, we suggest that the KS domain of TAS1 can be classified into a new group different from KS domains of type I, II, and III PKSs.

TAS1 consists of an NRPS and a unique type PKS KS domain involved in cyclization and product release. Cyclization and product release generally take place via the thioesterase (TE) or Claisen cyclase domains in fungal type I NR-PKSs. In contrast, there is no fused TE domain in most HR-PKSs and product release and cyclization require an in trans interaction with separate enzymes[41,42]. In addition, the cyclization step of macrocyclic peptides that produced by fungal NRPSs is mainly conducted by a terminal condensation-like domain[43]. Interestingly, there are two product release mechanisms in fungal PKS–NRPS hybrid enzymes[20], which differ from the single PKS or NRPS. The first is the final R domain mediated release mechanism. The R domain is thought to catalyse the reductive release from the sequence similarities to short-chain dehydrogenase/reductase family proteins. The second mechanism is DKC, catalysed by the terminal domain (D). Extensive investigations of tenellin, fusaridione and α-cyclopiazonic acid (fungal PKS–NRPS products) biosynthesis suggested that the formation of tetramic acid derivatives via DKC is the product release mechanism[44–46]. In the proposed model for TeA biosynthesis (Fig. 5c), the final TeA release mechanism is expected to be DKC. However, TAS1 does not contain any of these cyclization and product release related domains. In contrast, type III PKSs have only a KS domain and the elongation/cyclization step for polyketide production is conducted with a single enzyme. In addition, biosynthesis analysis of the polyketide RK-682, which contains tetronic acid (an analogue of tetramic acid), showed that a KS-like enzyme (RkD) responsible for assembly of the tetronic acid moiety by catalysing intermolecular C–C and C–O bond formation[47]. In other polyketides-containing tetronic acid—chlorothricin and quartromicin—biosynthetic pathways also contain these type of KS-like enzymes[48,49]. However, tetramic acid ring formation in TeA occurs via an intramolecular reaction catalysed by a KS domain (Dieckmann cyclization).

To study the peptide cyclization capability of TE domain in bacterial NRPSs, SNAC derivatives were extensively used as artificial intermediates of the enzyme reaction[50,51]. SNAC mimics the enzyme-phosphopantetheine moiety that carries enzymatic intermediates in the cyclization enzyme reaction. However, non-enzymatic self-cyclization and non-reactiveness in SNAC derivatives have also been reported[52,53]. We synthesized *N*-acetoacetyl-L-Ile-SNAC as an intermediate of TeA biosynthesis to investigate the cyclization capability of the KS domain in TAS1; however, synthesized *N*-acetoacetyl-L-Ile-SNAC showed self-cyclization, the rate of which increased with increasing pH. Thus, we conducted the enzyme reaction at pH 6.5 to minimize self-cyclization and demonstrated that the KS domain in TAS1 is responsible for the final step of cyclization in TeA production.

In fungal type I PKSs, the KS domain generally recognizes acetyl-CoA as a starter unit to produce polyketide. However, the TAS1 KS domain recognizes acetoacetyl-L-Ile hybrid to initiate the cyclization reaction to produce TeA. Acetoacetyl-CoA is an intermediate of the mevalonate pathway and some type III PKS enzymes also recognize acetoacetyl-CoA as a starter unit to produce polyketides[54,55].

The Cys–His–Asn catalytic triad that is conserved in the type III PKSs and KS domain of TAS1 may be involved in cyclization reaction and substrate recognition as in type III PKSs. However, at this time we cannot explain the exact cyclization or substrate recognition mechanisms of the TAS1 KS domain for TeA biosynthesis. Further studies such as crystal structure analysis will be required to clarify this mechanism.

In summary, we identified a gene (*TAS1*) for mycotoxin TeA biosynthesis in *M. oryzae*. We also demonstrated that TAS1 biosynthesizes TeA from isoleucine and acetoacetyl-CoA. TAS1 is the first fungal NRPS–PKS hybrid enzyme identified that consists of the unique domain structure described. TeA-producing plant pathogenic fungi are ubiquitous, and TeA-contaminated crops have attracted increasing public health attention in recent years. We expect the results of this study to stimulate additional investigations of TeA biosynthesis and production regulation mechanisms in pathogenic fungi and shed light on TeA contamination. Moreover, because the unique domain structure of the PKS portion of TAS1 was previously unreported, the results of our study will also contribute to research of PKSs.

Methods

Chemicals. Isoleucine, CoA, acetoacetyl-CoA, acetyl-CoA, malonyl-CoA, ATP and TeA were purchased from Sigma-Aldrich (St Louis, MO, USA). Kanamycin (Km), carbenicillin and hygromycin B were purchased from Nacalai (Kyoto, Japan). Blasticidin S was purchased from Funakoshi (Tokyo, Japan). All other reagents were of analytical grade.

Strains and culture conditions. The strains and plasmids used in this study are listed in Supplementary Table 3. The *M. oryzae* strain kita 1, which is pathogenic in rice plants, was used as a wild-type strain. *M. oryzae* was grown on OMA plates (5% oatmeal agar, Difco, Detroit, MI, USA) or PDA plates (3.9% potato dextrose agar, Difco) at 25 °C. For static culture, *M. oryzae* was grown in 100 μl of liquid YG media (0.5% yeast extract, 2% glucose) on a 96-well flat bottom plate (Iwaki, Tokyo, Japan) at 25 °C without agitation for 5 days. *M. oryzae* was transformed with the *Agrobacterium tumefaciens*-mediated transformation method[56,57]. *A. tumefaciens* strain C58 was cultured with a conidial suspension of *M. oryzae* prepared as described previously[58]. Transformants were selected with 500 $\mu g\,ml^{-1}$ hygromycin B or 150 $\mu g\,ml^{-1}$ blasticidin S when culturing *M. oryzae* transformants. *Escherichia coli* DH5α was grown in Lysogeny broth (LB) at 37 °C, and transformation was performed with a standard method[59]. Km or carbenicillin (50 $\mu g\,ml^{-1}$) was added to the *E. coli*-transformant selective medium and solidified with 2% agar for plate culture if needed.

DNA manipulation. DNA isolation and manipulation were performed with a standard method[59]. Genomic DNA was isolated from *M. oryzae* with a DNeasy plant total DNA isolation kit (Qiagen, Valencia, CA, USA). DNA fragments were isolated from agarose with a gel extraction kit (Qiagen). Plasmids were purified with a QIAprep Spin Miniprep kit (Qiagen). PCR was carried out with a DNAEngine Peltier thermal cycler (Bio-Rad, Hercules, CA, USA). DNA amplification via PCR was conducted with KOD-Plus-neo DNA polymerase (Toyobo, Osaka, Japan). DNA fragment ligation and circularization for vector construction was conducted via infusion reaction using an In-Fusion HD cloning system (Clontech, Palo Alto, CA, USA). DNA sequencing was carried out with a BigDye terminator ver3.1 kit (Applied Biosystems, Foster City, CA, USA). Sequencing products were run on an automated 3730 × l capillary DNA analyzer (Applied Biosystems).

***MGG_07803* gene disruption.** *MGG_07803* was disrupted by exchanging a whole-gene ORF with the hygromycin B resistance gene expression unit via homologous recombination. *MGG_07803* gene disruptants were constructed as follows. Two kb of the upstream region of *MGG_07803* was amplified from the genomic DNA of *M. oryzae* via PCR with the primers ΔM07803_UP-F and ΔM07803_UP-R (fragment 1). The hygromycin B resistance gene expression unit was amplified via PCR from pCSN45 (ref. 60) with primers 5HPH and 3HPH (fragment 2). Two kb of the downstream region of *MGG_07803* was amplified from the genomic DNA of *M. oryzae* via PCR with the primers ΔM07803_DN-F and ΔM07803_DN-R (fragment 3). The vector sequence of pBI121 (Clontech) between the right and left borders was amplified from pBI121 via PCR with primers pBI121-RB and pBI121-LB (fragment 4).

All fragments were gel purified and cloned by using the In-Fusion cloning technology to yield pBI-M07803::HPH. The In-Fusion reaction mixtures were used to transform *E. coli* DH5α, and transformants were selected with Km (50 $\mu g\,ml^{-1}$). After the DNA sequences were verified, plasmid DNA containing the correct insert was transformed into *A. tumefaciens*, and the transformants were used for *A. tumefaciens*-mediated transformation. *MGG_07803* disruptants were selected via PCR with primers ΔM07803_CHK_F and ΔM07803_CHK_R that hybridize outside of the upstream and downstream regions of *MGG_07803*, which was used for vector construction. This primer set can amplify 10 kb of the *MGG_07803* ORF, including 2 kb of the up- and downstream regions from wild-type strains, the 5.3-kb hygromycin B resistance gene expression unit, and the up- and downstream regions of *MGG_07803* from disruptants (Supplementary Fig. 4). The primers used are listed in Supplementary Table 4.

***TAS1* gene overexpression.** *TAS1* gene overexpression strains were constructed as follows. Two kb of the upstream region of *TAS1* was amplified from the genomic DNA of *M. oryzae* via PCR using primers OETAS1_UP-F and OETAS1_UP-R (fragment 1). The blasticidin S resistance gene expression unit was amplified from pBF101-Δ*Not*I (ref. 58) via PCR with primers 5HPH and 3HPH (fragment 2). The *TEF1* gene promoter region of *A. oryzae* was amplified from *A. oryzae* genomic DNA via PCR with primers AoTEF1-F and AoTEF1-R (fragment 3). The *TAS1* ORF was amplified from the cDNA of *M. oryzae* that was cultured with 1% DMSO via PCR with primers TAS1-F and TAS1-R (fragment 4). Two kb of the downstream region of the *TAS1* gene was amplified from the genomic DNA of *M. oryzae* via PCR with primers OETAS1_DN-F and OETAS1_DN-R (fragment 5). The vector sequence of pBI121 between the right and left borders was amplified from pBI121 via PCR with primers pBI121-RB and pBI121-LB (fragment 6).

All fragments were gel purified and then cloned with an In-Fusion HD Cloning kit (Clontech) to yield pBI-OE::TAS1. *E. coli* DH5α and *A. tumefaciens* transformation was conducted in the manner described for *MGG_07803* gene disruption. *TAS1* overexpression strains were selected via PCR with primers OETAS1_CHK-F and OETAS1_CHK-R that hybridize outside of the used upstream region and the *TAS1* gene. This primer set can amplify 3.3 kb from wild-type strains and the 5.4-kb blasticidin S resistance gene expression unit containing region from *TAS1* overexpression strains (Supplementary Fig. 4).

KS domain disruption. KS domain-disrupted strains were constructed as follows. Two kb of the upstream region of *TAS1* was amplified from the genomic DNA of *M. oryzae* via PCR with primers ΔKS_UP-F and ΔKS_UP-R (fragment 1). The hygromycin B resistance gene expression unit was amplified from pCSN45 (ref. 60) via PCR with primers 5HPH and 3HPH (fragment 2). The *TEF1* promoter region and the C-A-PCP domains of *TAS1* were amplified from pBI-OE::TAS1 via PCR with primers TEF1 + CAT-F and TEF1 + CAT-R (fragment 3). Two kb of the downstream region of the *TAS1* gene was amplified from the genomic DNA of *M. oryzae* via PCR with primers ΔKS_DN-F and ΔKS_DN-R (fragment 4). The vector sequence of pBI121 between the right and left borders was amplified from pBI121 via PCR with primers pBI121-RB and pBI121-LB (fragment 5).

All fragments were gel purified and cloned by using the In-Fusion cloning technology to yield pBI-TAS1-KS::HPH. *E. coli* DH5α and *A. tumefaciens* transformation was conducted in the manner described for *MGG_07803* gene disruption. The KS domain of *TAS1* disruptants were selected via PCR with primers ΔKS_CHK-F and ΔKS_CHK-R that hybridizes upstream of the KS domain and outside of the downstream region used for vector construction of *TAS1*. This primer set can amplify 4.1 kb from wild-type strains and 2.9 kb from the KS domain disruptants (Supplementary Fig. 9).

Microarray analysis. An RNeasy Plant mini kit (Qiagen) was used to extract total RNA from static-cultured *M. oryzae* (wild type, Δ*osm1* and 1% DMSO-added condition) on day 5 of culture. Double-stranded cDNA synthesis (from 10 μg total RNA) and Cy-3 labelling and hybridization steps were performed by Roche NimbleGen (Madison, WI, USA) according to standard protocols. A custom-designed microarray platform comprising single 60-mer probes designed against 12,825 annotated genes from the genome sequence of *M. oryzae* was used[53]. Five non-overlapping 60-mer probes were designated against various parts of the single target gene, and 64,125 probes were finally designated against the total number of annotated genes. An average transcriptional intensity of five non-overlapping probes in the single gene was used as the transcriptional level of the gene.

Metabolite extraction and analysis. *M. oryzae* culture broth (100 μl) was extracted with four volumes of ethanol, evaporated with an N_2 stream and dissolved in 200 μl methanol to analyse metabolites. UPLC/MS analysis was performed with an Acquity UPLC H-Class system (Waters Alliance, Milford, MA, USA) equipped with a mass spectrometer (API 3200, Applied Biosystems). The UPLC conditions were as follows: column, XTerraMSC_{18} (Waters), 5 μm (2.1 × 150 mm); flow rate, 0.6 ml min^{-1}; solvent A, water-containing 0.05% formic acid; solvent B, acetonitrile. After the extracted sample was injected into a column equilibrated with 5% solvent B, the column was developed with a linear gradient from 5 to 100% solvent B over the course of 3.5 min and kept at 100% solvent B for another 2 min. Mass spectra were collected in the electrospray ionization-positive and electrospray ionization-negative modes.

Protein expression and purification from *S. cerevisiae*. *S. cerevisiae* protease-deficient strain BJ5464 was obtained from the American Type Culture Collection (ATCC 208288) and used as the host cell for TAS1 protein expression. A PCR fragment of 4,809 bp containing the *TAS1* ORF was amplified from the cDNA of *M. oryzae* that was cultured with 1% DMSO via PCR with primers TAS1-F and TAS1-R, and the amplified fragment was inserted into the YEp352 ADH1 vector[61] that amplified with primers YEp352-F and YEp352-R via infusion reaction to construct the N-terminal-8 × His-tagged TAS1 expression plasmid YEp352-TAS1. The nucleotide sequence of the inserted fragment was confirmed. To express TAS1 as an active form, we also amplified the *PPTase* coding sequence (*MGG_17878*, 978 bp) of *M. oryzae* from *M. oryzae* cDNA via PCR with primers MoPPT-F and MoPPT-R and inserted it into YEp352 to construct YEp352-MoPPT. The *ADH1* promoter and terminator containing the *PPTase* coding sequence from YEp352-MoPPT was amplified with primers ADH1-F and ADH1-R, subcloned into the pESC-Trp-modified vector (Stratagene, La Jolla, CA, USA), which was deleted at the 1,960- to 3,302-bp region containing *GAL1* and *GAL10* yeast promoters and terminators by amplification with primers pESC-TRP-F and pESC-TRP-R to obtain pESC-Trp-ADH1-MoPPT for the expression of non-tagged PPTase.

The constructed YEp352-TAS1 and pESC-Trp-ADH1-MoPPT were co-transformed to *S. cerevisiae* strain BJ5464 with the lithium acetate/single-stranded carrier DNA/Polyethylene glycol method[62], and transformants were selected on a synthetic glucose induction (SGI) plate (1 g l^{-1} casamino acids, 7 g l^{-1} yeast nitrogen base, 20 g l^{-1} glucose and 2% agar). Transformed cells were pre-cultured at 30 °C in SGI liquid media for 24 h and then inoculated into 2 l YPD media (2% glucose, 2% bactopeptone and 1% yeast extract) with an initial optical density at 600 nm of 0.015 and cultured for 48 h at 28 °C. The cultured cells were harvested via centrifugation (5,000*g*) for 15 min at 4 °C, and then the collected cells were suspended in 5 ml cold distilled water. Suspended cells were granulized via dripping into liquid N_2 and then homogenized with a Waring mixer with liquid N_2. Fifty grams of homogenized cells were suspended in 100 ml protein extraction buffer (10 mM Tris-HCl (pH 8.0), 1 mM dithiothreitol (DTT), 0.1% TritonX-100 and 15% glycerol). The cell suspension was heavily vortexed for 1 min and sonicated with a Tomy UD-200 sonicator twice on ice for 30 s each with 30-s intervals. Cellular debris was removed via centrifugation (8,000g) for 30 min (Beckman GS-15). The supernatant was diluted in the same volume of 2 × bind buffer (1 mM DTT, 800 mM NaCl, 10 mM imidazole and 15% glycerol), applied to a Ni-nitrilotriacetic acid beads column (2 × 8 cm; Qiagen) equilibrated with equilibration buffer (10 mM Tris-HCl (pH 8.0), 1 mM DTT, 400 mM NaCl and 15% glycerol), and washed with 100 ml wash buffer (10 mM Tris-HCl (pH 8.0), 1 mM DTT, 50 mM NaCl, 5 mM $MgCl_2$, 0.1 mM ethylenediaminetetraacetic acid and 15% glycerol) containing 10 mM imidazole. The column was further washed with 75 ml wash buffer containing 30 mM imidazole.

The His-tag-binding protein was eluted with 60 ml elution buffer (10 mM Tris-HCl (pH 8.0), 1 mM DTT, 50 mM NaCl, 5 mM $MgCl_2$, 0.1 mM ethylenediaminetetraacetic acid and 25% glycerol) containing 200 mM imidazole. The eluted protein was concentrated with an Amicon Ultra-100k (Millipore, Bedford, MA, USA), and then buffer exchange was conducted with buffer A (10 mM Tris-HCl (pH 7.5) and 25% glycerol) using a PD Midi Trap G-25 column (GE Healthcare, Uppsala, Sweden) and applied to a Superdex 200 column (GE Healthcare) for size exclusion. The column was previously equilibrated with running buffer (50 mM Tris-HCl (pH 7.5), 500 mM NaCl and 10% glycerol) at a flow rate of 0.8 ml min^{-1}. Eluted 8 × His-tagged TAS1 was concentrated with the Amicon Ultra-100k (Millipore), and buffer exchange was conducted with buffer A and the PD Midi Trap G-25 column (GE Healthcare). The highly purified His-tagged TAS1 protein was stored at − 80 °C and used for *in vitro* enzyme assay.

Protein expression and purification from *E. coli*. PPTase (MGG_17878) of *M. oryzae* was overexpressed with the *E. coli* BL21(DE3)pLysS strain (Takara). The *PPTase* ORF (978 bp) of *M. oryzae* was amplified from *M. oryzae* cDNA via PCR with primers pET-MoPPT-F and pET-MoPPT-R. The pET19b(+) vector sequence was amplified from pET-H6SDH[63] with primers pET-F and pET-R, and then two fragments were infused to obtain pET-MoPPT for the expression of N-terminally 6x His-tagged PPTase. The constructed pET-MoPPT sequence was confirmed after transformation into *E. coli* BL21(DE3)pLysS. After overnight culture at 37 °C in LB with Km (50 μg ml^{-1}), 2 ml pre-cultured transformed cells was inoculated into 100 ml fresh LB with Km (50 μg ml^{-1}) and cultured at 37 °C for 2 h. Isopropyl-β-D-thiogalactopyranoside was added to a final concentration of 1 mM. After growth for 5 h at 30 °C, the cells were harvested via centrifugation (5,000*g*) for 15 min at 4 °C. The collected cells were suspended in 10 ml protein extraction buffer and sonicated five times on ice for 30 s each with 30-s intervals. Cellular debris was removed via centrifugation (8,000*g* for 30 min). The supernatant was purified on a Ni-nitrilotriacetic acid beads column as described for protein expression and purification from *S. cerevisiae*. Purified PPTase (Supplementary Fig. 10b) was used for the KS domain activity assay.

Protein expression using a wheat germ cell-free system. PCR fragments containing the C–A–PCP domain (1–3,360 nt) or KS domain (3,361–4,809 nt) of *TAS1* were amplified from the cDNA of *M. oryzae* that was cultured with 1% DMSO via PCR with primers pEU_CAT-F and pEU_CAT-R for the C–A–PCP domain and pEU_KS-F and pEU_KS-R for the KS domain. The amplified fragments were inserted into the pEU01 vector (Cell-Free Science, Matsuyama, Japan) amplified with primers pEU01-F and pEU01-R using the In-Fusion HD cloning system (Clontech). The resulting expression plasmids were designated pEU01_C-His_CAT and pEU01_N-His_KS for the synthesis of C-terminally 6 × His-tagged C–A–PCP domains and N-terminally 6 × His-tagged KS domain of TAS1, respectively. Cell-free protein synthesis with wheat embryo extracts was performed with a WEPRO7240 Expression kit (Cell Free Science) using pEU01_C-His_CAT and pEU01_N-His_KS in a 226-μl bilayer format on 96-well flat-bottom plates (Iwaki) according to the manufacturer's instructions. The solubility of the translation product was confirmed with sodium dodecyl sulphate-polyacrylamide gel electrophoresis[64]. The reaction mixture was centrifuged at 13,000*g* for 30 min at 4 °C, and the supernatant was used directly for the TeA biosynthesis assay.

***In vitro* enzyme assay.** A TAS1 activity assay was conducted at 25 °C for 2 h in 100 μl reaction mixture containing 0.5 μM TAS1, 50 mM Tris-HCl (pH 7.5), 10 mM Tris (2-carboxyethyl) phosphine (TCEP), 1 mM ATP, 5% glycerol, 1 mM isoleucine and 1 mM acetoacetyl-CoA. Acetyl CoA (1 mM) and malonyl-CoA (1 mM) were also used instead of 1 mM acetoacetyl-CoA. To determine the kinetic parameters for isoleucine, both of ATP and acetoacetyl-CoA were remained constant at saturating concentrations (2 mM) and the concentration of isoleucine was varied from 0.1 to 3 mM. For acetoacetyl-CoA, both ATP and isoleucine were remained constant at saturating concentrations (2 mM) and the concentration of acetoacetyl-CoA was varied from 0.005 to 1 mM. For ATP, both acetoacetyl-CoA and isoleucine were remained constant at saturating concentrations (2 mM) and the concentration of ATP was varied from 0.01 to 2 mM. The kinetic parameters $K_{M(app)}$ and k_{cat} and their standard errors were determined using non-linear regression to fit the data to the Michaelis–Menten equation using SigmaPlot (Systat software). All reactions were performed in triplicate. A KS domain activity assay was conducted in 260 μl reaction mixture. The first reaction was conducted with 0.8 μM purified MoPPT containing100 μl soluble crude proteins containing C-His-tagged C–A–PCP, N-His-tagged KS protein, or both, 80 μM CoA, 4 mM ATP, 2 mM $MgCl_2$, 8 mM TCEP, 5% glycerol and translation buffer with the WEPRO7240 Expression kit (Cell Free Science) at 30 °C for 1 h. Then, 0.4 mM isoleucine and acetoacetyl-CoA were added and reacted at 25 °C for 5 h. For analysis of the cyclization capability of the KS domain, the enzyme reaction was conducted in 200 μl reaction mixture containing five times-concentrated and buffer (50 mM MES (pH 6.5))-exchanged 100 μl of C-His-tagged C–A–PCP or N-His-tagged KS soluble crude proteins synthesized with the cell-free system, 5% glycerol, 5 mM TCEP and 0.25 mM *N*-acetoacetyl-L-Ile-SNAC at 25 °C for 2 h. For analysis of the condensation capability of the TAS1, the enzyme reaction was conducted in the same manner as cyclization capability analysis, but using 0.25 mM L-Ile-SNAC and acetoacetyl-CoA instead of *N*-acetoacetyl-L-Ile-SNAC.

After the enzyme reaction, the reaction mixtures were extracted with four volumes of ethanol, evaporated with an N_2 stream, and finally dissolved in 200 μl methanol for UPLC/MS analysis of TeA production. After injection of the extracted sample into a column equilibrated with 5% solvent B, the column was developed with a linear gradient from 5 to 100% solvent B over 3.0 min and kept at 100% solvent B for another 2 min. Mass spectra were collected in the electrospray ionization-positive and electrospray ionization-negative modes.

References

1. Rosett, T., Sankhala, R. H., Stickings, C. E., Taylor, M. E. U. & Thomas, R. Biochemistry of microorganisms. CIII. Metabolites of *Alternaria tenuis* auct.: culture filtrate products. *Biochem. J.* **67,** 390–400 (1957).
2. Ostry, V. *Alternaria* mycotoxins: an overview of chemical characterization, producers, toxicity, analysis and occurrence in foodstuffs. *World Mycotoxin J.* **1,** 175–188 (2008).
3. Siegel, D., Rasenko, T., Koch, M. & Nehls, I. Determination of the *Alternaria* mycotoxin tenuazonic acid in cereals by high-performance liquid chromatography-electrospray ionization ion-trap multistage mass spectrometry after derivatization with 2,4-dinitrophenylhydrazine. *J. Chromatogr. A* **1216,** 4582–4588 (2009).
4. Gross, M., Curtui, V., Ackermann, Y., Latif, H. & Usleber, E. Enzyme immunoassay for tenuazonic acid in apple and tomato products. *J. Agric. Food Chem.* **59,** 12317–12322 (2011).
5. Lohrey, L., Marschik, S., Cramer, B. & Humpf, H. U. Large-scale synthesis of isotopically labeled $^{13}C_2$-tenuazonic acid and development of a rapid HPLC-MS/MS method for the analysis of tenuazonic acid in tomato and pepper products. *J. Agric. Food Chem.* **61,** 114–120 (2013).
6. Steyn, P. S. & Rabie, C. Characterization of magnesium and calcium tenuazonate from *Phoma sorghina*. *J. Phytochem.* **15,** 1977–1979 (1976).
7. Umetsu, N., Kaji, J. & Tamari, K. Investigation on the toxin production by several blast fungus strains and isolation of tenuazonic acid as a novel toxin. *Agric. Biol. Chem.* **36,** 859–866 (1972).
8. Shigeru, H. T. & Gordon, C. N. The biological activity of tenuazonic acid. *Biochemistry* **2,** 1132–1137 (1963).
9. Miller, F. A. *et al.* Antiviral activity of tenuazonic acid. *Nature* **200,** 1338–1339 (1963).
10. Gitterman, C. O. Antitumor, cytotoxic, and antibacterial activities of tenuazonic acid and congeneric tetramic acids. *J. Med. Chem.* **8,** 483–486 (1965).
11. Lebrun, M. H. *et al.* Relationships between the structure and the phytotoxicity of the fungal toxin tenuazonic acid. *Phytochemistry* **27,** 77–84 (1988).
12. Stickings, C. E. & Townsend, R. J. Studies in the biochemistry of micro-organisms. 108. Metabolites of *Alternaria tenuis* Auct.: the biosynthesis of tenuazonic acid. *Biochem. J.* **78,** 412–418 (1961).
13. Collemare, J., Billard, A., Böhnert, H. U. & Lebrun, M. H. Biosynthesis of secondary metabolites in the rice blast fungus *Magnaporthe grisea*: the role of hybrid PKS-NRPS in pathogenicity. *Mycol. Res.* **112,** 207–2015 (2008).
14. Fischbach, M. A. & Walsh, C. T. Assembly-line enzymology for polyketide and nonribosomal peptide. Antibiotics: logic, machinery, and mechanisms. *Chem. Rev.* **106,** 3468–3496 (2006).
15. Gallo, A., Ferrara, M. & Perrone, G. Phylogenetic study of polyketide synthases and nonribosomal peptide synthetases involved in the biosynthesis of mycotoxins. *Toxins (Basel)* **5,** 717–742 (2013).
16. Hertweck, C. The biosynthetic logic of polyketide diversity. *Angew. Chem. Int. Ed.* **48,** 4688–4716 (2009).
17. Hashimoto, M., Nonaka, T. & Fujii, I. Fungal type III polyketide synthases. *Nat. Prod. Rep.* **31,** 1306–1317 (2012).
18. Hertweck, C., Luzhetskyy, A., Rebets, Y. & Bechthold, A. Type II polyketide synthase: gaining a deeper insight into enzymatic teamwork. *Nat. Prod. Rep.* **24,** 162–190 (2007).
19. Fisch, K. M. Biosynthesis of natural products by microbial iterative hybrid PKS-NRPS. *RSC Adv.* **3,** 18228–18247 (2013).
20. Boettger, D. & Hertweck, C. Molecular diversity sculpted by fungal PKS-NRPS hybrids. *Chembiochem* **14,** 28–42 (2013).
21. Böhnert, H. U. *et al.* A putative polyketide synthase/peptide synthetase from *Magnaporthe grisea* signals pathogen attack to resistance rice. *Plant Cell* **16,** 2499–2513 (2004).
22. Song, Z., Cox, R. J., Lazarus, C. M. & Simpson, T. J. Fusarin C biosynthesis in *Fusarium moniliforme* and *Fusarium venenatum*. *Chembiochem* **5,** 1196–1203 (2004).
23. Eley, K. L. *et al.* Biosynthesis of the 2-pyridone tenellin in the insect pathogenic fungus *Beauveria bassiana*. *Chembiochem* **8,** 289–297 (2007).
24. Yu, F. *et al.* Structure and biosynthesis of heat-stable antifungal factor (HSAF), a broad-spectrum antimycotic with a novel mode of action. *Antimicrob. Agents Chemother.* **51,** 64–72 (2007).
25. Blodgett, J. A. V. *et al.* Common biosynthetic origins for polycyclic tetramate macrolactams from phylogenetically diverse bacteria. *Proc. Natl Acad. Sci. USA* **107,** 11692–11697 (2010).
26. Müller, S. *et al.* Paenilamicin: structure and biosynthesis of a hybrid nonribosomal peptide/polyketide antibiotic from the bee pathogen *Paenibacillus larvae*. *Angew. Chem. Int. Ed.* **53,** 10821–10825 (2014).
27. Gerc, A. J., Song, L., Challis, G. L., Stanley-Wall, N. R. & Coulthurst, S. J. The insect pathogen *Serratia marcescens* Db10 uses a hybrid non-ribosomal peptide synthetase-polyketide synthase to produce the antibiotic althiomycin. *PLoS ONE* **7,** e44673 (2012).
28. Tang, G.-L., Cheng, Y.-Q. & Shen, B. Leinamycin biosynthesis revealing unprecedented architectural complexity for a hybrid polyketide synthase and nonribosomal peptide synthetase. *Chem. Biol.* **11,** 33–45 (2004).
29. Simunovic, V. *et al.* Myxovirescin A biosynthesis is directed by hybrid polyketide synthases/nonribosomal peptide synthetase, 3-hydroxy-3-methylglutaryl-coa synthases, and trans-acting acyltransferases. *Chembiochem* **7,** 1206–1220 (2006).
30. Silakowski, B. *et al.* New lessons for combinatorial biosynthesis from *Myxobacteria*. The myxothiazol biosynthetic gene cluster of *Stigmatella aurantiaca* DW4/3-1. *J. Biol. Chem.* **274,** 37391–37399 (1999).
31. Du, L., Sánchez, C. & Shen, B. Hybrid peptide-polyketide natural product: biosynthesis and prospects toward engineering novel molecules. *Metab. Eng.* **3,** 78–95 (2001).
32. Dixon, K. P., Xu, J. R., Smirnoff, N. & Talbot, N. J. Independent signaling pathways regulate cellular tumor during hyperosmotic stress and appressorium-mediated plant infection by *Magnaporthe grisea*. *Plant Cell* **11,** 2045–2058 (1999).
33. Dean, R. A. *et al.* The genome sequence of the rice blast fungus *Magnaporthe grisea*. *Nature* **434,** 980–986 (2005).
34. Dang, H. X., Pryor, B., Peever, T. & Lawrence, B. The *Alternaria* genomes database: a comprehensive resource for a fungal genus comprised of saprophytes, plant pathogens, and allergenic species. *BMC Genomics* **16,** 239 (2015).
35. Hertweck, C. Hidden biosynthetic treasures brought to light. *Nat. Chem. Biol.* **5,** 450–452 (2009).
36. Bok, J. W. *et al.* Chromatin-level regulation of cryptic biosynthetic gene clusters in *Aspergillus nidulans*. *Nat. Chem. Biol.* **5,** 462–464 (2009).
37. Chiang, Y. M. *et al.* A gene cluster containing two fungal polyketide synthases encodes the biosynthetic pathway for a polyketide, asperfuranone, in *Aspergillus nidulans*. *J. Am. Chem. Soc.* **131,** 2965–2970 (2009).
38. Wasil, Z. *et al.* One pathway, many compounds: heterologous expression of a fungal biosynthetic pathway reveals its intrinsic potential for diversity. *Chem. Sci.* **4,** 3845–3856 (2013).
39. Lazarus, C. M., Williams, K. & Bailey, A. M. Reconstructing fungal natural product biosynthetic pathways. *Nat. Prod. Rep.* **31,** 1339–1347 (2014).
40. Jez, J. M. *et al.* Structural control of polyketide formation in plant-specific polyketide synthase. *Chem. Biol.* **7,** 919–930 (2000).
41. Korman, T. P. *et al.* Structure and function of an iterative polyketide synthase thioesterase domain catalyzing Claisen cyclization in aflatoxin biosynthesis. *Proc. Natl Acad. Sci. USA* **107,** 6246–6251 (2010).
42. Chooi, Y.-H. & Tang, Y. Navigating the fungal polyketide chemical space: from genes to molecules. *J. Org. Chem.* **77,** 9933–9953 (2012).
43. Gao, X. *et al.* Cyclization of fungal nonribosomal peptides by a terminal condensation-like domain. *Nat. Chem. Biol.* **8,** 823–830 (2012).
44. Halo, L. M. *et al.* Authentic heterologous expression of the tenellin iterative polyketide synthase nonribosomal peptide synthetase requires coexpression with an enoyl reductase. *Chembiochem* **9,** 585–594 (2008).
45. Sims, J. W. & Schmidt, E. W. Thioesterase-like role for fungal PKS-NRPS hybrid reductive domains. *J. Am. Chem. Soc.* **130,** 11149–11155 (2008).
46. Liu, X. & Walsh, C. T. Characterization of cyclo-acetoacetyl-L-tryptophan dimethylallyltransferase in cyclopiazonic acid biosynthesis: substrate promiscuity and site directed mutagenesis studies. *Biochemistry* **48,** 11032–11044 (2009).
47. Sun, Y. *et al.* *In vitro* reconstruction of tetronate RK-682 biosynthesis. *Nat. Chem. Biol.* **6,** 99–101 (2009).
48. Jia, X.-Y. *et al.* Genetic characterization of the chlorothricin gene cluster as a model for spirotetronate antibiotic biosynthesis. *Chem. Biol.* **13,** 575–585 (2006).
49. He, H.-Y. *et al.* Quartromicin biosynthesis: two alternative polyketide chains produced by one polyketide synthase assembly line. *Chem. Biol.* **19,** 1313–1323 (2012).

50. Kohli, R. M., Trauger, J. W., Schwarzer, D., Marahiel, M. A. & Walsh, C. T. Generality of peptide cyclization catalyzed by isolated thioesterase domains of nonribosomal peptide synthetases. *Biochemistry* **40,** 7099–7108 (2001).
51. Trauger, J. W., Kohli, R. M., Mootz, H. D., Marahiel, M. A. & Walsh, C. T. Peptide cyclization catalyzed by the thioesterase domain of tyrocidine synthetase. *Nature* **407,** 215–218 (2000).
52. Xu, W. *et al.* An in vitro system to study cyclopeptide heterophyllin B biosynthesis in the medicinal plant *Pseudostellaria heterophylla. Plant Cell Tissue Organ Cult.* **108,** 137–145 (2012).
53. Sieber, S. A., Walsh, C. T. & Marahiel, M. A. Loading peptidyl-coenzyme A onto peptidyl carrier proteins: a novel approach in characterizing macrocyclization by thioesterase domain. *J. Am. Chem. Soc.* **125,** 10862–10866 (2003).
54. Mori, T. *et al.* Structural basis for the formation of acylalkylpyrones from two β-ketoacyl units by the fungal type III polyketide synthase CsyB. *J. Biol. Chem.* **290,** 5214–5225 (2015).
55. Funa, N., Ohinishi, Y., Ebizuka, Y. & Horinouchi, S. Properties and substrate specificity of RppA, a chalcone synthase-related polyketide synthase in *Streptomyces griseus. J. Biol. Chem.* **277,** 4628–4635 (2002).
56. de Groot, M. J., Bundock, P., Hooykaas, P. J. & Beijersbergen, A. G. Agrobacterium tumefaciens-mediated transformation of filamentous fungi. *Nat. Biotechnol.* **16,** 839–842 (1998).
57. Rho, H. S., Kang, S. & Lee, Y. H. *Agrobacterium* tumefaciens-mediated transformation of the plant pathogenic fungus, *Magnaporthe grisea. Mol. Cells* **12,** 407–411 (2001).
58. Motoyama, T., Imanishi, K. & Yamaguchi, I. cDNA cloning, expression, and mutagenesis of scytalone dehydratase needed for pathogenicity of the rice blast fungus, *Pyricularia oryzae. Biosci. Biotechnol. Biochem.* **62,** 564–566 (1998).
59. Sambrook, J., Fritsch, E. F. & Maniatis, T. *Molecular Cloning: A Laboratory Manual* (Cold Spring Harbor Laboratory Press, Cold Spring Harbor, 2001).
60. Motoyama, T. *et al.* A two-component histidine kinase of the rice blast fungus is involved in osmotic stress response and fungicide action. *Fungal Genet. Biol.* **42,** 200–212 (2005).
61. Hill, J. E., Myers, A. M., Koerner, T. J. & Tzagoloff, A. Yeast/*E.coli* shuttle vectors with multiple unique restriction site. *Yeast* **2,** 163–167 (1986).
62. Gietz, R. D. & Woods, R. A. Transformation of yeast by lithium acetate/single-stranded carrier DNA/polyethylene glycol method. *Methods Enzymol.* **350,** 87–96 (2002).
63. Yamada, N. *et al.* Enzymatic characterization of scytalone dehydratase Val75Met variant found in melanin biosynthesis dehydratase inhibitor (MBI-D) resistant strains of the rice blast fungus. *Biosci. Biotechnol. Biochem.* **68,** 615–621 (2004).
64. Laemmli, U. K. Cleavage of structural proteins during the assembly of the head of bacteriophage T4. *Nature* **227,** 680–685 (1970).

Acknowledgements

The authors wish to thank Dr T. Shimizu for the synthesis of artificial intermediate. This study was supported in part by Grants-in-Aid from the Ministry of Education, Culture, Sports, Science and Technology of Japan and by the Science and Technology Research Promotion Program for Agriculture, Forestry, Fisheries and Food Industry.

Author contributions

C.-S.Y. wrote the manuscript. C.-S.Y. and T.M. performed biochemical and genetic experiments and bioinformatic analysis. C.-S.Y. and T.M. designed the research protocol. H.O. integrated all research.

Additional information

Competing financial interests: The authors declare no competing financial interests.

9

Light-driven oxidation of polysaccharides by photosynthetic pigments and a metalloenzyme

D. Cannella[1,*], K.B. Möllers[2,*], N.-U. Frigaard[2], P.E. Jensen[3], M.J. Bjerrum[4], K.S. Johansen[5] & C. Felby[1]

Oxidative processes are essential for the degradation of plant biomass. A class of powerful and widely distributed oxidative enzymes, the lytic polysaccharide monooxygenases (LPMOs), oxidize the most recalcitrant polysaccharides and require extracellular electron donors. Here we investigated the effect of using excited photosynthetic pigments as electron donors. LPMOs combined with pigments and reducing agents were exposed to light, which resulted in a never before seen 100-fold increase in catalytic activity. In addition, LPMO substrate specificity was broadened to include both cellulose and hemicellulose. LPMO enzymes and pigment derivatives common in the environment of plant-degrading organisms thus form a highly reactive and stable light-driven system increasing the turnover rate and versatility of LPMOs. This light-driven system may find applications in biotechnology and chemical processing.

[1] Department of Geoscience and Natural Resource Management, University of Copenhagen, Rolighedsvej 23, Frederiksberg C, 1958 Copenhagen, Denmark. [2] Marine Biological Section, Department of Biology, University of Copenhagen, Strandpromenaden 5, 3000 Helsingør, Denmark. [3] Copenhagen Plant Science Centre, Department of Plant and Environmental Sciences, University of Copenhagen, Thorvaldsensvej 40, Frederiksberg C, DK-1871 Copenhagen, Denmark. [4] Department of Chemistry, University of Copenhagen, Universitetsparken 5, 2100 Copenhagen, Denmark. [5] Division of Industrial Biotechnology, Department of Chemical and Biological Engineering, Chalmers University of Technology, Kemivägen 10, SE-412 96 Göteborg, Sweden. * These authors contributed equally to this work. Correspondence and requests for materials should be addressed to D.C. (email: dac@ign.ku.dk) or to C.F. (email: cf@ign.ku.dk).

The oxidation of plant biomass by enzyme catalysed reactions is a basic mechanism in the global carbon cycle and of increasing importance and value for industrial applications. The major plant cell wall-derived polymers cellulose, hemicellulose and lignin are oxidized by enzymes, which often are metalloenzymes.

One class of powerful oxidative enzymes is the extracellular copper-based lytic polysaccharide monooxygenases (LPMOs), which are monooxygenases capable of oxidizing the most recalcitrant polysaccharides. When combined with hydrolases there is a strong synergistic effect on biomass degradation[1–4]. LPMOs are found in fungi, bacteria and viruses, spanning a number of terrestrial and aquatic ecosystems and are classified in the carbohydrate active enzyme database as auxiliary activity (AA) enzymes class 9, 10, 11 and 13 (ref. 5). The enzymes play an essential role, yet not well-understood, in the turnover of organic matter. Some organisms have multiple genes for LPMO enzymes, and especially in plant cell wall-degrading fungi, the AA9 family is highly present with some species containing >30 AA9-encoding genes[5]. All identified substrates are naturally abundant polysaccharides including cellulose, hemicellulose, starch and chitin[6], in which the glycosidic bonds are oxidized. The products are oxidized oligosaccharides and their non-oxidized counterparts. Specifically for glucans the oxidation of the pyranose ring can take place at either the C1 position producing aldonic acids, or at the C4 position producing 4-ketoaldose (gemdiols). Independently from the C1 and/or C4 type of oxidation, one molecule of dioxygen is needed for the LPMO catalytic cycle: one oxygen atom is inserted in the carbohydrate chain for breaking the glycosidic bond, and the second atom is released from the reaction as part of a water molecule[1].

The LPMOs have a mononuclear type 2 copper active site, with a t-shaped coordination sphere termed the histidine brace, where two histidines provide three nitrogen ligands, two from N-His and one from the terminal amine. Moreover, LPMOs require an extracellular electron donor to complete their catalytic cycle, and both proteins and plant-derived molecules have been found to serve as donors[1,7,8]. However, the specific types of electron donors, as well as the mechanism of electron transfer are still unresolved. Ascorbic acid has often been used as electron donor experimentally, and is also present in non-lignified plant biomass, but other donors such as lignin have been described[9–11]. In addition, the enzyme cellobiose dehydrogenase (CDH) functions as a donor with the electron transfer presumably taking place from the haem b-containing cytochrome domain of CDH[8,12,13]. However, as CDH is absent in many LPMO-expressing organisms, it is not an absolute prerequisite for all LPMO activity in nature[14].

The apparent promiscuity of LPMOs with respect to the source of electrons, prompted us to investigate the effect of one of the most potent electron donors in nature namely chlorophylls, which are essential components of the photosynthetic light-harvesting and reaction centre complexes. The energy of electromagnetically excited chlorophyll *a* is 1.85 eV and 1.33 eV for the singlet and triplet state, which corresponds to reduction potentials of -1.07 V and -0.55 V for the oxidized chlorophyll to excited state chlorophyll in solution, respectively[15]. Thus, a simple photosystem was set up for light-induced electron transfer consisting of thylakoids or chlorophyllin as electron-donating pigments, an electron accepting LPMO enzyme with an estimated redox potential of $+0.22$ V (ref. 15) and a reductant ascorbic acid, reducing the oxidized pigment. The redox potential of ascorbic acid at pH 7 is $+0.35$ V in its radical form and $+0.06$ V when oxidized to dehydroascorbate. To excite the pigments, the samples were exposed to light of different wavelengths and in different time intervals.

Previous examples of light-driven enzymatic processes are hybrid systems of carbon monoxide dehydrogenase molecules with CdS nanocrystals for light-driven reduction of CO_2 (ref. 16) and a hydrogenase and TiO_2 nanoparticle for light-driven H_2 generation[17]. However, here we combine abundant biological parts: chlorophyll pigments and LPMO enzymes, and report that the catalytic activity of LPMOs can be driven by light and in some cases increased up to 100-fold by light-induced electron transfer. Using both sunlight and LED light sources of different colours, isolated thylakoids from photosynthetic organisms, as well as the chlorophyll derivative chlorophyllin increase the reactivity of three different LPMOs. In addition, we show how different reductants can be applied and suggest how light-induced oxidation may be an active mechanism in the global carbon cycle.

Results

Light boosted LPMO activity: oligosaccharide products. The activity of the LPMOs was detected in an assay using phosphoric acid-swollen cellulose (PASC) as substrate, and the released oligomeric products were detected and quantified by high-performance anion exchange chromatography (HPAEC) after incubation. The gluconic acid on the C1 oxidized oligosaccharides was quantified and used as a measure of enzyme activity. The assay for quantification was based on a complete hydrolysis of the oxidized PASC with a cellulase mixture and subsequent detection of gluconic acid by HPAEC.

Initially, PASC was exposed to LPMO from *Thielavia terrestris* and suspended thylakoid membranes from a cyanobacterium (*Synechococcus* sp. PCC 7002) in the presence of ascorbic acid and sunlight. The light response resulted in a 20-fold increase in release of oligosaccharides as compared with incubation in darkness or incubation under conventional conditions with ascorbic acid and light (without the thylakoids, Fig. 1a). Replacing the cyanobacterial thylakoids with membranes prepared from a higher plant (*Arabidopsis thaliana*) produced similar results (Supplementary Fig. 1).

As LPMOs require dioxygen, we wanted to rule out that dioxygen produced by the thylakoid membranes was the cause of the observed activity enhancement, and therefore the water-soluble chlorophyll derivative chlorophyllin was tested in the same assay. This resulted in an equally pronounced light-driven accumulation of oxidized products (Fig. 1b), showing that chlorophyll pigments on light excitation transfer electrons to the LPMO, regardless of whether the chlorophyll pigment is membrane-bound (thylakoids) or soluble (chlorophyllin).

The light-driven oxidation was expanded to experiments with two LPMOs from *Thermobifida fusca* and *Thermoascus aurantiacus* (Supplementary Figs 2 and 3). For all enzymes tested, both C1 and C4 oxidation increased in the light-driven system. Not only the swollen PASC cellulose but also crystalline cellulose in the form of Avicel was tested as a substrate, and it was similarly oxidized by the light-driven system (Supplementary Fig. 4).

To investigate the light response of the combined pigments and LPMO-catalysed cellulose oxidation, experiments were performed using blue, red, green and sunlight with the same range of intensity. In accordance with the absorption spectra of chlorophyll pigments: sunlight, blue and red light induced the highest levels of cellulose oxidation (Fig. 1c). A light response experiment with alternating darkness and sunlight further proved that light activated the oxidation of cellulose (Fig. 1d).

The level of oxidized products from a 3-h light-induced reaction (Fig. 1c) measured by the amount of glucose units oxidized to gluconic acid, shows that ~10% of the cellulose was oxidized using photo-excited pigments and sunlight, compared with 0.5% without pigments and only ascorbic acid and sunlight.

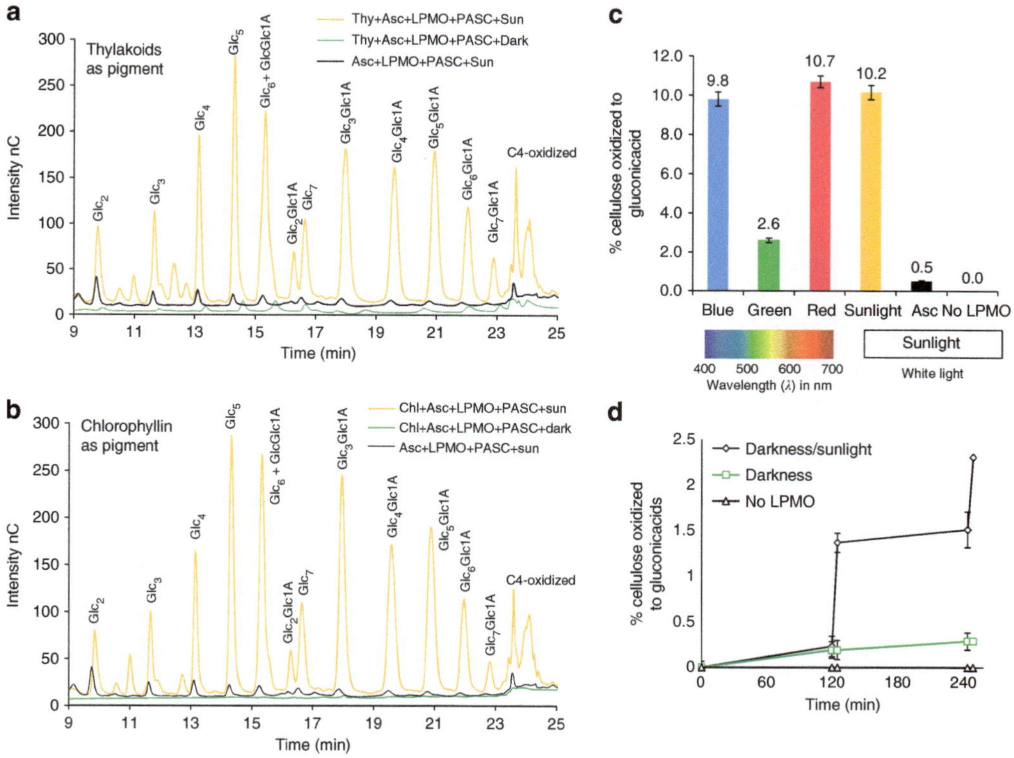

Figure 1 | Cellulose oxidation by *Tt*LPMO9E combined with light-induced electron transfer. The released oligosaccharides from PASC were measured by HPAEC showing non-oxidized (blue) and oxidized (magenta) reaction products. Each chromatogram is average of three replicate experiments. (**a**) Thy + Asc + LPMO + PASC + Sun: Thylakoids, ascorbic acid and LPMO with PASC in sunlight for 3 h (150–200 μmol photons s^{-1} m^{-1}), yellow line. Control 1 Thy + Asc + LPMO + PASC + Dark: same conditions but in darkness, green line. Control 2: Asc + LPMO + PASC + Sun: same conditions but without thylakoids, black line. (**b**) Same experiments as in **a** but with chlorophyllin (Chl) as pigment. Peak annotations of native oligosaccharides (blue) were done using the pure compounds as standard: Glc_2, cellobiose; Glc_3, cellotriose; Glc_4, cellotetraose; Glc_5, cellopentaose; Glc_6, cellohexaose; Glc_7, celloheptaose; oxidized oligosaccharides (magenta) were assigned by comparing with literature chromatograms performed with identical separation conditions as done by Westereng[37]: GlcGlc1A, cellobionic acid; Glc_2Glc1A, cellotrionic acid; Glc_3Glc1A, cellotetraonic acid; Glc_4Glc1A, cellopentaoinic acid; Glc_5Glc1A, cellohexaoinic acid; Glc_6Glc1A, celloeptaonic acid; Glc_7Glc1A, cellooctaonic acid. On the *y*-axis is reported the intensity of the signal in nC (nano Coulomb) without further adjustments. (**c**) PASC oxidation with LPMO (*Tt*AA9E), chlorophyllin and ascorbic acid in response to blue, green, red light (150 μmol photons s^{-1} m^{-1}) and sunlight (150–200 μmol photons s^{-1} m^{-1}). Oxidation was measured with HPAEC quantifying the gluconic acid (oxidation of the C1 position at the pyranose ring). Asc: PASC and ascorbic acid in sunlight (black bar). No LPMO: PASC, chlorophyllin and ascorbic acid in sunlight. (**d**) Chlorophyllin, ascorbic acid, LPMO and PASC incubation was run for two cycles of darkness for 2 h and sunlight for 5 min, or always in darkness (4 h 10 min), black line with diamonds and green light with square, respectively. Negative control experiment missing LPMO was run in parallel with the darkness/sunlight cycles. The percentage of oxidized cellulose was measured by quantification of gluconic acid formed (C1 oxidation).

No products were detected without the enzyme (Fig. 1c,d). The levels of cellulose oxidation catalysed by light-induced electron transfer and LPMOs are to the best of our knowledge higher than any previously reported result[1,9].

Control experiments were performed to investigate the light response of thylakoids and chlorophyllin when incubated with LPMO and without any reductant: the pigments alone were able to donate electrons to LPMO but were subject to photobleaching and degradation (Supplementary Figs 5 and 6 and Supplementary Discussion). Non-enzymatic reactions from light-induced reactive oxygen species, which potentially can break down cellulose, were also investigated. This was done by combining the pigment preparations and ascorbic acid, but without LPMO enzyme added. These experiments did not result in oxidation of cellulose and confirmed that the applied pigments or ascorbic acid caused cellulose oxidation only when combined with LPMO (Fig. 1c,d, and Supplementary Fig. 7).

Oxygen consumption. As LPMOs require oxygen for their catalytic cycle, oxygen measurements were used to monitor the reaction catalysed by the *T. terrestris* LPMO. The results show that oxygen is consumed by the photosystem in a light-dependent manner. In Fig. 2a, the oxygen consumption is shown during incubation in light or darkness: from 0 to 900 s the reactions were carried out in darkness, while from 900 to 2400 s white light was applied. Without the enzyme (blue line) there was no oxygen consumption observed. Without chlorophyllin but with LPMO and ascorbic acid (magenta line) the oxygen consumption was very low, and there was no response to light.

For the complete photosystem composed of reductant, pigment and enzyme (turquoise line) there was the same oxygen consumption in darkness as seen for the incubation without chlorophyllin (magenta line), but when the light was switched on (at 900 s) a rapid consumption of oxygen was observed. The oxygen consumption correlated with the release of gluconic acid. At the end of the reaction after 2,400 s, 0.7% of the PASC cellulose was oxidized at the C1 positions to gluconic acid, corresponding to a stoichiometric ratio of 1:0.8 between dioxygen consumption and C1 product formation. As the C4 oxidation products are not included, it is reasonable to assume a 1:1 ratio between dioxygen and oxidation products. Without the pigment

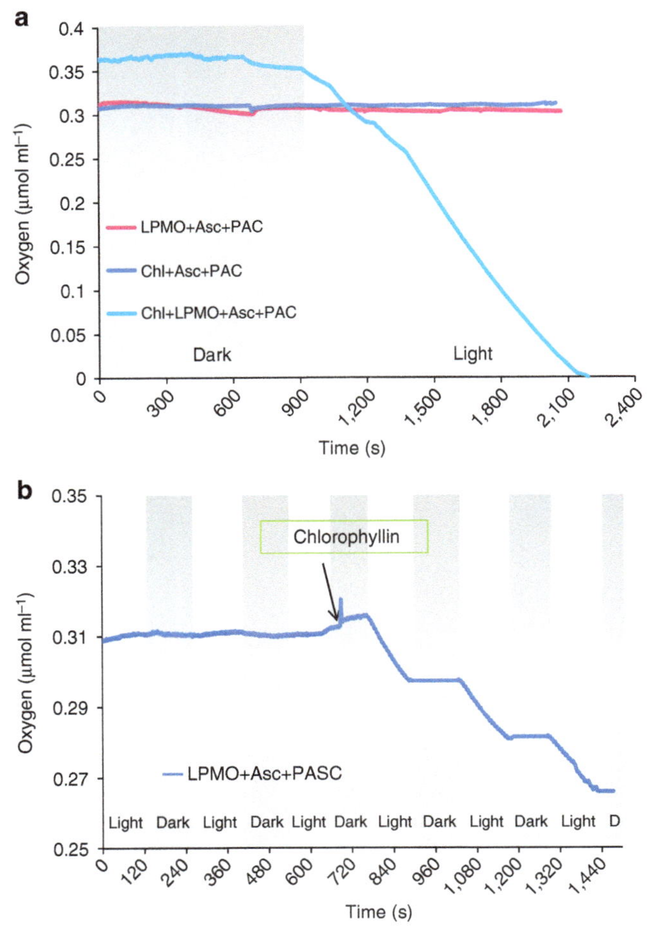

Figure 2 | Oxygen consumption during cellulose oxidation with light or in darkness. (**a**) Oxygen consumption with different combinations of the photosystem: blue line shows chlorophyllin, ascorbic acid and PASC (without *T. terrestris* LPMO); magenta line shows *T. terrestris* LPMO, ascorbic acid and PASC; the turquoise line shows the whole photosystem with chlorophyllin, ascorbic acid, *T. terrestris* LPMO and PASC. Light is turned on at 900 s. (**b**) Alternating cycles of light and darkness: at the beginning of the incubation only *T. terrestris* LPMO, ascorbic acid and PASC are present, then chlorophyllin is added after 700 s.

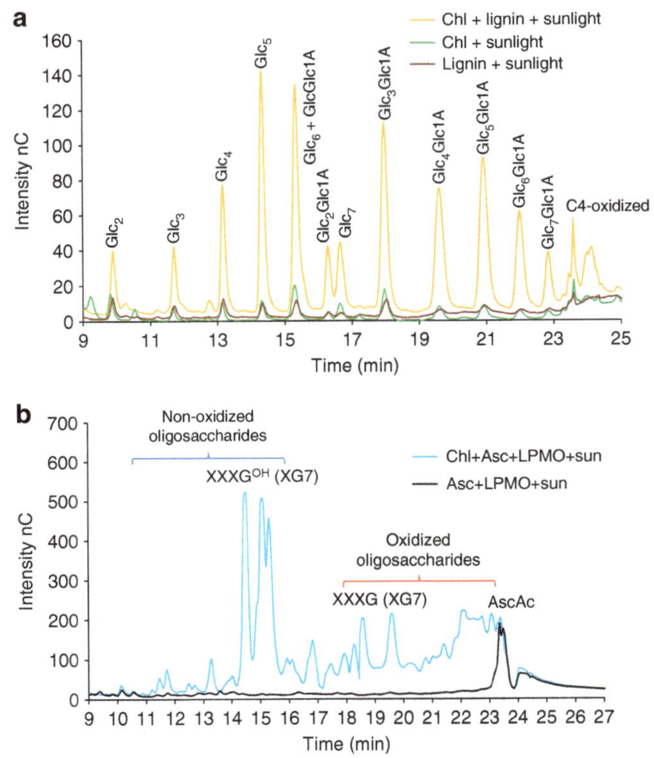

Figure 3 | HPAEC analysis of light-induced polysaccharide oxidation with *Tt*LPMO9E. (**a**) Lignin as reductant: LPMO oxidative activity on cellulose (PASC) when incubated with chlorophyllin and organosolv lignin as reductant in 3 h of sunlight. Controls with only chlorophyllin or organosolv lignin in sunlight are included. Each chromatogram is average of three replicate experiments. For details on peak annotation see Fig. 1 or Supplementary Fig. 1. (**b**) Xyloglucan as substrate: LPMO oxidative activity on xyloglucan (1% w/w) when incubated with chlorophyllin and ascorbic acid and exposed to 3 h of sunlight. Pronounced is the presence of oxidized species detectable from minute 17 to 24. A control experiment with LPMO and ascorbic acid shows no degradation of the xyloglucan. The xyloglucan is a heteropolymer of xylose, glucose, galactose and arabinose. The division of peaks in oxidized and non-oxidized is derived from the model heptamer xylooligosaccharide $XXXG^{OH}$ in its reduced form made of four units of glucose of which three are substituted with one xylose. While XXXG is the same, but oxidized in the reducing end of the last subunit in the glucan backbone[19,20].

(magenta line), the oxidation products were below the detection limit.

The photosystem reactivity could be controlled by switching the light on and off (Fig. 2b). At the start of the experiment, only LPMO, ascorbic acid and PASC were present. After five cycles of light and darkness there was no sign of oxygen consumption. After 700 s, chlorophyllin was added, followed by a clear oxygen consumption in light and no oxygen consumption in darkness (Fig. 2b).

Non-cellulosic substrate and lignin as reductants. Ascorbic acid is a naturally available reductant for pigments only in non-lignified plant tissue. To look for alternative natural reductants, we replaced ascorbic acid with extracted lignin. Indeed, insoluble lignin extracted from wheat straw could replace ascorbic acid in the light-driven assay with chlorophyllin (Fig. 3a). This points to how light-induced electron transfer may also be functional in degradation of lignified biomass. The functionality of lignin as a multiple electron donor to the pigments may be ascribed to its electrons in de-localized and overlapping π-orbitals[18].

To our surprise, the specificity of the *T. terrestris* LPMO changed when used in the light-induced electron-transfer assays. This particular LPMO is thought to be strictly cellulose specific, but when combined with light-induced electron transfer from pigments, the enzyme also oxidized the hemicellulose xyloglucan (Fig. 3b). Oxidation of both cellulose and hemicellulose has been reported for LPMOs from *Neurospora crassa*[19] and from *Podospora anserine*[20]. The degradation pattern of xyloglucan in Fig. 3b is similar to a xyloglucan degradation pattern previously reported[20].

Discussion

The light-induced boosting of PASC oxidation presented here clearly shows the pivotal role of the LPMO enzyme for the light-induced oxidation of carbohydrates. We suggest a model in which light excitation of the pigments generates a strong reductant that efficiently reduces the copper ion in the LPMO enzyme by donating an electron as shown in Fig. 4. Subsequently the

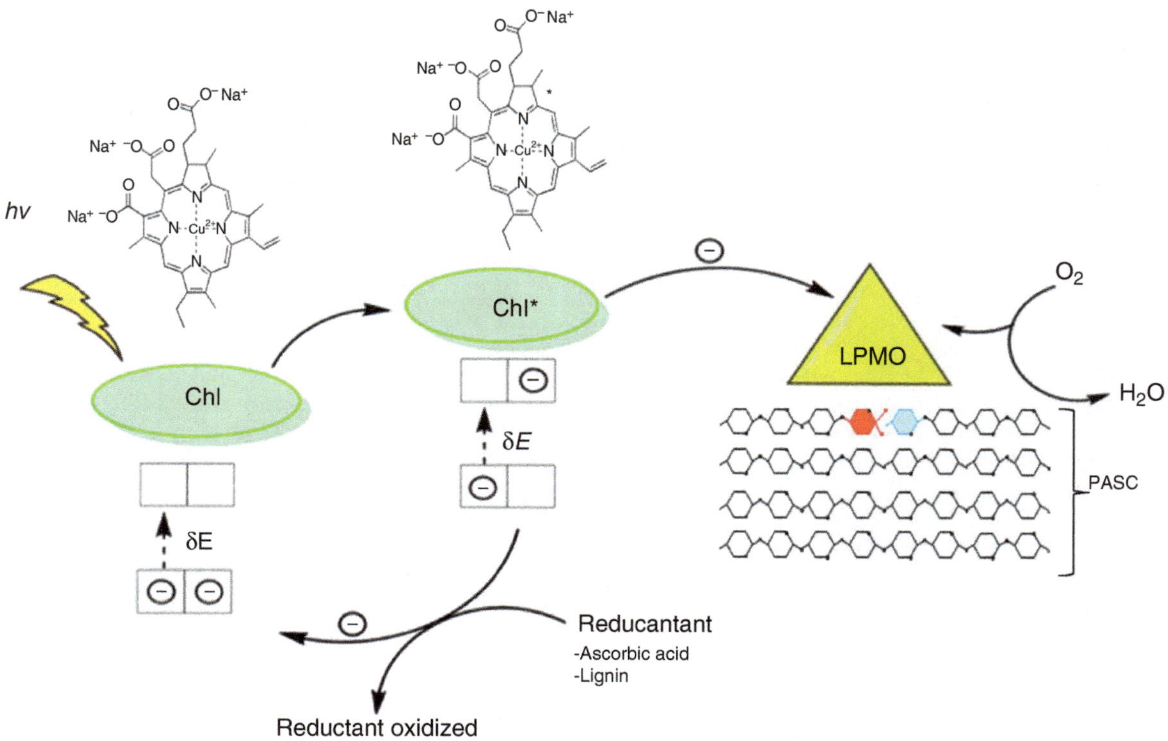

Figure 4 | Proposed mechanism for light-induced electron transfer to LPMO. Light excites the pigment, which in its excited state transfers an electron to the LPMO enzyme. The excited electron reduces the copper in the LPMO-active site, which then activates oxygen and oxidizes the polysaccharide. The oxidized pigment is reduced by acquisition of an electron from a reductant such as ascorbic acid or lignin, thus now available for a new cycle of excitation. During the monooxygenase reaction of LPMO, dioxygen is split, and one oxygen atom is incorporated into the substrate, and the other is reduced to water. The oxidation of a PASC cellulose chain is shown as a red monomer of glucose oxidized in the C1 position to an aldonic acid oligosaccharide, whereas the blue monomer will be released as a non-oxidized oligosaccharide.

oxidized form of the pigments is reduced by a reductant with lower redox potential, in this case ascorbic acid or lignin.

The initial observation of light-induced boosting of PASC oxidation raised the question if non-enzymatic Fenton-reactions were induced by light. Exposure of chlorophyllin or thylakoid preparations, which contain magnesium, copper and zinc, could by themselves generate hydroxyl radicals, that potentially are powerful oxidizers[21]. However, the extensive control experiments examining all the system components revealed no formation of oxidation products from such reactions (Supplementary Fig. 7), also there was no detectable oxygen consumption with an enzyme-free control (Fig. 2a).

Not only the level of oxidized products, but also the reactivity increased with light-induced electron transfer. Using ascorbic acid or CDH as electron donors, an end-level of ~2% oxidized cellulose is reached after 24–36 h (refs 9,20). With the light-induced electron-transfer system, the level of LPMO-oxidized cellulose was 2.3% after two cycles of 5 min (Fig. 1d, Supplementary Fig. 8 and Supplementary Discussion). It is thus possible to achieve the same level of oxidation in less than a hundredth of the time required without light-induced electron transfer. These data are also supported by the LPMO turnover frequency determined to be $0.25\,s^{-1}$ based on the oxygen consumption when incubated with the photosystem at 25 °C (Fig. 2a) and $0.55\,s^{-1}$ at 50 °C based on the accumulation of oxidized product after two cycles of 5 min light exposure (Fig. 1d). Previously published turnover values for a *N. crassa* LPMO by measuring reducing ends of oxidized PASC at 50 °C were $0.01–0.04\,s^{-1}$ (ref. 22).

Based on our results, we conclude that the catalytic activity for LPMOs combined with the photosystem for light-induced electron transfer results in up to 2 orders of magnitude higher level of catalytic activity than previously reported.

The results presented support the proposed mechanism and synergy of light-induced electron donation for the LPMOs (Fig. 4). But how are the electrons donated to the active site, directly or via long-range transfer through the protein? Tan *et al.*[12] suggested a direct electron transfer to the LPMO-active site from CDH via a haem-binding cytochrome complex. Another possibility is through protein long-range electron transfer. This allows the enzyme to be adsorbed to its substrate during the electron transfer, and a possible pathway for electron transfer from CDH has been suggested[3]. In the LPMO AA9 family we found, using molecular modelling, other possible electron-transfer pathways from the surface to the active site, potentially capable of transferring electrons from the excited pigments. Specifically for the *Ta*LPMO9A a putative pathway goes from a histidine (His87) surface ligand through 12 covalent bonds to the Cu(II), equivalent to a tunnelling length of 16.8 Å with a calculated pathway coupling decay value of 2.2×10^{-3}. The histidine is situated on the side of LPMO and thus free of cellulose that is binding to the LPMO during catalysis. Similar pathways were located in other AA9 LPMOs all located very much in the same position but with different ligands at the surface (Supplementary Fig. 9). The exact mechanism for the suggested electron transfer is so far unresolved, and the existence of parallel paths cannot be ruled out.

As the light-induced electron transfer reported caused a tremendous increase in products, as well as a change in enzyme specificity, the energy of the transferred electrons seems to alter the cycle or state of the active site. As suggested above, the high-energy photoelectron from pigments could be donated by long-range electron transfer[23]. Possible targets could be [LPMOCu(II)OO•] and/or [LPMO–Cu(II)-OOH] (− OOH = deprotonated hydrogen peroxide), intermediates in the catalytic cycle of LPMO proposed by Philips *et al.*[8], and later studied by quantum mechanical calculations[24,25]. The products from the latter reduction could via a Fenton-type reaction form OH•, OH–, Cu(II) (or alternatively Cu(III) and 2 OH–) capable of performing the observed polysaccharide oxidation[26–28].

Abiotic light-driven degradation of biomass is well-described[29], here we have identified a biotic mechanism for light-driven degradation. Components common in the natural environments of plant-degrading organisms (that is, chlorophylls and breakdown products hereof, other porphyrins and lignin) can together with LPMOs form a highly reactive system capable of using light energy to boost the degradation of plant cell walls. In light-exposed environments, microorganisms can use electromagnetic radiation to increase the release of glucose and other metabolizable carbohydrates. Our results show that LPMOs can use photo-excited electrons for their catalytic cycle, and the wide distribution of LPMOs in nature suggests that biotic photodegradation of cell wall materials can be a widespread mechanism in different ecosystems. However, the findings of this work need to be extended to prove such mechanisms, and further work may find inspiration from the observations of light-enhanced enzymatic activity in forest floor litter[30].

The observed light-driven enzymatic oxidation performed by LPMO consumes oxygen, oxidizes carbon and produces water. Looking ahead, this process may find applications in technology areas for chemical processing such as conversion of recalcitrant biomass to fuels and chemicals. We speculate that the light-induced oxidation can be applied in a broader application scheme for oxidation of C–H bonds.

Methods

Enzymes. Purified *T. terrestris* LPMO (*Tt*LPMO9E, previously *Tt*GH61E) and *T. auranticus* (*Ta*LPMO9A) were donated from Novozymes A/S (Denmark). The enzymes are produced by expression in a host organism and subsequently purified. *T. fusca* AA10 (TfLPMO10A) cloned and expressed in *Escherichia coli* was purchased from Nzytech Ltd (Portugal). All LPMOs were free of any residual cellulase or hemicellulose activities. Commercial cellulase mixtures Celluclast 1.5l and Novozym 188 were obtained from Novozymes A/S. The Celluclast 1.5l mixture had a protein content of 127 mg g^{-1}, containing 62 filter paper units (FPU) per g, and a β-glucosidase activity of 15 U per g. Novozym 188 had a protein content of 220 mg g^{-1}, containing a β-glucosidase activity of 231 U per g.

Chemicals, materials and substrates. Ascorbic acid was obtained from Sigma-Aldrich. Stock solutions of 100 mM were made in water and kept at − 20 °C in the dark. Avicel microcrystalline cellulose was obtained from Sigma-Aldrich. Xyloglucan product code P-XYGLN was obtained from Megazymes Ltd, Ireland. Chlorophyllin product code C-100.000-WS-P produced by extraction of *Festuca arundinacae* was obtained from Chr. Hansen, Hørsholm, Denmark.

Preparation of PASC microcrystalline cellulose substrate. Avicel (microcrystalline cellulose, Sigma-Aldrich PH101) was swollen with phosphoric acid to generate PASC as described by Wood[31] with a few modifications: 4 g of Avicel were suspended in 100 ml of phosphoric acid (85% w/v) at 40 °C and magnetically stirred for 1 h. The mixture was poured into 1.9 l of water at 40 °C and stirred for 1 h. The suspension was washed four times with 2 l H_2O (MilliQ-quality), two times with 2 l of a 1% $NaHCO_3$ solution to reduce acidity, and then three additional times with 2 l H_2O and stored at 4 °C. The final cellulose content of the PASC suspension was determined by enzymatic hydrolysis (24 h, 50 °C), with an enzymatic dosage of 75 FPU per g of cellulose, using Celluclast 1.5L cellulolytic enzymes and Novozym 188 in a 5:1 ratio; the released glucose was measured according to the protocol described in section 'Enzymes' giving an estimated cellulose content of 1.5% w/v. The average degree of polymerization (DP) of Avicel-derived PASC was determined by measuring the total number of reducing ends[32] and comparing this with the total amount of monomeric glucose, giving a DP of 52.

Organosolv lignin extraction. The lignin fraction was prepared from wheat straw (*Triticum aestivum* L.). The straw was ball milled for 20 min and the cellulose and hemicellulose fractions were removed by hydrolytic enzymes (Celluclast 1.5L and, Novozym 188 in a 5:1 v/v) at a final dosage of 75 FPU per g of dry lignocellulose substrate for 144 h at 50 °C followed by washing with MilliQ water at a ratio of 1 l for 1 g of dry material. The amount of residual carbohydrates was < 2% in the final material. Finally, the residual material was suspended in an aqueous ethanol solution (50:50 water/ethanol) at a 5:1 liquid/solid ratio and heated at 220 °C in a 1 l Parr reactor for 80 min. After heating, the lignin residue was filtered at 75 °C. Solubilized lignin was precipitated by adding water at three times the original amount and recovered by filtration. The insoluble lignin fraction was dried at 40 °C and ground with a pestle and mortar.

Preparation of chlorophyllin. A stock solution of 12% w/v (166 mM) chlorophyllin was prepared by dissolving the powder in water (MilliQ-quality). The chlorophyllin was kept in darkness while stored at 4 °C. Prior to each experiment, an aliquot of the stock solution was incubated in darkness for 2 h at room temperature.

Preparation of cyanobacterial thylakoid suspensions. Thylakoid suspensions, containing light-harvesting antennae (phycobilisomes) and thylakoid membranes, were prepared from the cyanobacterium *Synechococcus* sp. PCC 7002 (referred to as *Synechococcus*) grown in medium A containing 2 g $NaNO_3$ per l as previously described[33]. *Synechococcus* cells were harvested in 50 ml volumes by centrifugation (5,000*g* for 5 min) and subsequently resuspended in 1 ml thylakoid washing buffer (pH 6.35) as described[34]. The resolved pellet was transferred to a microfuge tube containing 500 µl glass beads (glass beads for cell disruption, 0.1–0.25 mm diameter, Retsch Technology GMBH, Haan, Germany) followed by cell disruptive sonication (Amplitude 50, 3 min processing time, 5 s on/off cycle). The cell extract was then centrifuged (12,000*g*, 4 °C, 20 min) and the pellet, containing unbroken cells and cell walls, was discarded. An additional centrifugation (40,000*g*, 4 °C, 30 min) separated the light-harvesting antennae (phycobilisomes, supernatant) from the thylakoid membranes (pellet). Absorption spectra of the thylakoid suspensions in the supernatant were performed with a UV1800 spectrophotometer (Shimadzu, Kyoto, Japan) and the chlorophyll *a* content was calculated as described[35]. The thylakoid suspensions were then used to conduct the light-induced electron-transfer experiments with an average concentration of 0.21 ± 0.06 mg Chl per ml. Putative contaminations derived from cytoplasmic membranes in the thylakoid suspension have no impact on light-induced electron transfer to LPMOs, this was tested with disrupted *E. coli* K-12 cells

Preparation of plant thylakoid membranes. Plant thylakoid membranes were extracted from *A. thaliana* (L.) Heynh. Ecotype Columbia. Plants were grown in compost in a controlled environment *Arabidopsis* chambers (Percival AR-60 I, Boone, IA) at a photosynthetic flux of 130–150 µmol of photons per m^2 per s, 20 °C and 70% humidity. Leaves from ~25 plants were pooled and homogenized using a blender fitted with razor blades in ice cold buffer containing 20 mM Tricine (pH 7.5), 10 mM NaCl, 5 mM $MgCl_2$, 0.4 M sucrose, 5 mg ml^{-1} bovine serum albumin (BSA) and 100 mM sodium ascorbate. The homogenate was immediately filtered through two layers of nylon mesh (31 µm pore size), after which the filtrate was centrifuged (6,000*g*, 4 °C, 15 min). The pellet was resuspended in 5 mM Tricine (pH 7.9) to lyse the chloroplasts. Following lysis, the thylakoids were collected by centrifugation (17,200*g*, 4 °C, 10 min). The pellet was resuspended in a small volume of homogenization buffer without ascorbate and BSA but with 20% glycerol (v/v). Total chlorophyll (Chl) and Chl *a*/*b* ratio were determined in 80% acetone according to the study by Lichtenthaler[36]. The final concentration of the thylakoid membranes applied in the light-induced electron-transfer experiments was 3.57 mg Chl per ml (Chl a/b ratio = 3.0).

Elemental analysis inductively coupled plasma (ICP). Chlorophyllin and thylakoid suspensions were analysed for the presence of metals on an Aurora Elite ICP-MS system from Bruker. Samples of chlorophyllin 12% (w/w) and extracted thylakoid suspensions 2% (w/w) were diluted to 500–1,000 p.p.b. dry matter in 1% nitric acid. The samples were analysed for ^{24}Mg, ^{25}Mg, ^{26}Mg ^{63}Cu, ^{65}Cu, ^{66}Zn, ^{67}Zn and ^{68}Zn. Chlorophyllin contained 0.41% Mg, 2.0% Cu and 0.05% Zn relative to total dry matter. Extracted thylakoid suspensions contained 1.8% Mg, 0.007% Cu and 0.0007% Zn relative to total dry matter.

Quantification of glucose and cellobiose by HPLC. The quantification of D-glucose and D-cellobiose was done using an Ultimate 3000 high-performance liquid chromatography (HPLC, Dionex, Germering, Germany) equipped with refractive index detector (Shodex, Japan) and ultraviolet detector at 210 nm (Dionex). The separation was performed in a Phenomenex Rezex ROA column at 80 °C with 5 mM H_2SO_4 as eluent at a flow rate of 0.8 ml min^{-1}.

Measurement of oligosaccharide and gluconic acid by HPAEC. The samples were prepared as follows: 200 µl were centrifuged at 14,000*g* for 2 min and 100 µl of the supernatant was inserted in the HPLC conical vial, without any further light exposure (wrapped in aluminium foil). HPAEC was run on an ICS5000 system (two different machines available in the laboratory), equipped with a PAD detector

(Dionex, Sunnyvale, CA, USA) with a CarboPac PA1 column (2 × 50 mm guard column followed by a 2 × 250 mm analytical column) and operated at a flow of 0.25 ml min^{-1}, at 30 °C. Chromatographic analysis of aldonic acids separation was conducted as described by Westering *et al.*[37]. The elution involved a linear gradient from 100% *A*:0% *B* to 90% *A*:10% *B* (10 min), followed by an exponential gradient to 70% *A*:30% *B* (15 min) and last an exponential gradient to 100% *B* (5 min). After that a linear gradient was applied for 15 min at the initial conditions 100% *A*:0% *B* (eluent *A* = 0.1 M NaOH, *B* = 0.1 M NaOH and 1 M NaOAc)[37]. The peaks of the cellulose oligomers in the chromatograms are assigned according to Westereng *et al.*[37]. The chromatograms shown in the main paper, and in Supplementary Figs 1–7, are average of three independent experiments.

Quantification of cellulose oxidation for TtLPMO9E. To quantify the total oxidation of cellulose (as shown in Fig. 1c), the residual PASC together with the supernatant containing oligosaccharides were digested with a commercial cellulase mixture, Celluclast supplemented with beta-glucosidase Novozym 188 in a 5:1 v/v ratio (both lacking LPMO activity) and applied based on Celluclast at 75 f.p.u. per g cellulose for 5 h at 50 °C in darkness. These conditions ensured complete hydrolysation of the PASC material. The LPMO reaction was stopped prior to cellulases treatment by boiling the vials/samples for 10 min. The hydrolysates were then analysed by HPLC to quantify the D-glucose and in the HPAEC for quantification of gluconic acid (marker of C1 oxidizing activity of LPMO). A minor C4 component (annotated C4-oxidized after minute 24) appeared in the oxidized products; it was impossible to quantify the amount of the monomeric 4-keto aldose sugar (C4-oxidized glucose) as due to tautomerization there are no standards available.

Standard conditions for enzymatic reactions and LIET. The light-induced electron-transfer (LIET) system was composed of PASC incubated with an LPMO enzyme, a light-absorbing component (chlorophyllin, cyanobacterial thylakoid suspensions or plant thylakoid membranes) and a reductant, that is, ascorbic acid. All experiments presented in the paper, as well as in the Supplementary Information were conducted in triplicates, and repeated on two different ICS5000 HAPEC systems available in the laboratory.

The standard reaction mixture composition for PASC oxidation reactions was: 0.75% w/v PASC, 2 mM ascorbic acid, 100 mM of citrate-phosphate buffer (pH 6.3) and 0.05 mg ml^{-1} LPMO in a 200 µl reaction volume. For the *Tt*LPMO9E with an estimated molecular weight of 22.5 KDa this equals 2.2 nmol of added enzyme. Chlorophyllin was added in a 1:100 dilution out of a 12% stock solution thus a final concentration of 1.6 mM. Freshly prepared cyanobacterial thylakoid suspensions and plant thylakoid membranes were diluted into the reaction mixture 1:2 and 1:20, respectively. The described composition applies for reactions using the AA9 enzymes (*Tt*LPMO9E and *Ta*LPMO9A) and is referred as standard conditions in the rest of the Methods section. For reactions with the AA10 enzyme (*Tf*LPMO10A), a phosphate buffer at pH 7.8 (20 mM) was used and 0.20 mg ml^{-1} of the enzyme preparation was used.

The standard conditions for the light-induced electron-transfer system for cellulose oxidation are defined as: 3 h incubation time, irradiance intensity for sunlight was 150–200 µmol photons per m^2 per s, for blue, red and green light sources 150 µmol of photons per m^2 per s (custom made LEDs for blue light 440 nm, red light 625 nm and green light 528 nm, with a spectral width 18 nm, 18 nm and 33 nm respectively). The irradiance intensities reported were measured at the bottom of the closed 2 ml Eppendorf test tubes, and the light probe (a Spherical Micro Quantum Sensor US-SQS/Lund) was submerged into the reaction mixture to account for light-scattering effects caused by PASC cellulose or other insoluble particles. Incubations were carried out in 2 ml Eppendorf polypropylene microfuge tubes in a thermomixer at 1,000 r.p.m. and 50 °C (Eppendorf thermomixer, Eppendorf, Hamburg, Germany). The thermomixer was modified to carry a transparent microfuge tube rack allowing light penetration through the walls of the microfuge tubes. All experimental preparations were carried out in dim green light at 0 µmol of photons per m^2 per s (Phillips TL-D 36 W colour green Lumen 3600, 540 nm). The degree of PASC oxidation was determined by quantification of oxidized glucose units being the gluconic acid. Optimization of reaction conditions in terms of time of incubation and reactant dosages are shown in Supplementary Fig. 8.

In addition to the standard reactions described above, multiple controls were performed as follows: darkness controls were wrapped in aluminium foil; LPMO activity controls were without photopigments or without light exposure and were incubated with ascorbic acid and PASC; non-enzymatic oxidation controls of chlorophyllin, thylakoid suspensions, lignin and ascorbic acid were individually incubated with PASC and without LPMO enzymes under standard conditions (see above) to assess the possibility of spontaneous cellulose oxidation caused by light exposure (Fenton-type reactions).

Sequential darkness or sunlight exposure. To verify the role of light as an inducer of cellulose oxidation via chlorophyllin, darkness and sunlight were sequentially applied to the same samples (results in Fig. 1d). The experiment was performed using the standard conditions for light-induced electron transfer (see above). The reaction mixture contained PASC, chlorophyllin, ascorbic acid and *Tt*LPMO9E with a lower dosage, using only 4 µg (instead of 10 µg) for a lower level of product release. The reaction mixtures were exposed to darkness for 2 h followed by sunlight exposure for 5 min. This cycle was subsequently repeated. After each change in light/darkness exposure, an aliquot was removed from the vials for further analysis. All aliquots were analysed by high-performance anion exchange chromatography (HPAEC) to measure the release of oligosaccharides and additionally treated with cellulase to quantify the monomeric glucose and gluconic acid (see above). An LPMO activity control, applying the same LPMO dosage was incubated for 24 h and afterwards treated with cellulase to quantify the monomeric glucose and gluconic acid (see above).

Oxygen consumption measurements. The oxygen consumption was measured with a Chlorolab 2 System (Hansatech, England), with an oxygen sensor mounted at the bottom of a sealed reaction chamber. The 1 ml reaction chamber was stirred by a magnet and was equipped with a window for light exposure. Reactants could be added via a capillary cannula at the top, and no oxygen exchange with the external environment was observed. The measurements are given as oxygen concentration in µmol per ml. White LED light (4100 K) at 150 µmol photons per m^2 per s was applied. The reactants were added at the beginning according to the dosages explained in the standard conditions, but at a lower temperature of 25 °C. Note that there is no headspace in the reaction chamber, why much less oxygen is available compared with the conditions described in standard conditions for enzymatic reactions. The combinations tested were: complete photosystem: LPMO (*Tt*LPMO9E) + chlorophyllin + Asc + PASC; the negative control experiments: chlorophyllin + Asc + PASC (no enzyme present); and the LPMO + Asc + PASC (no pigment present). From 0 to 900 s these reactions were carried out in darkness, while from 900 to 2,400 s white light was supplied (results shown in Fig. 2a).

The complete photosystem (LPMO + chlorophyllin + Asc + PASC) was measured with alternating cycles of light/darkness without or with the pigment. The conditions were identical to standard (see above) with the difference that *Tt*LPMO9E was dosed at only 0.01 mg ml^{-1} to lower the rate of oxygen consumption, and a temperature of 25 °C. *Tt*LPMO9E, ascorbic acid and PASC were first added. After five cycles of light and darkness (2 min each), chlorophyllin was added through the cannula, and three cycles of light and darkness were alternated (results shown in Fig. 2b).

Lignin as reductant. Ascorbic acid, generally used as a reductant in the standard conditions for the enzymatic hydrolysis, was replaced with organosolv-extracted lignin to investigate its putative reducing capacity. The organosolv lignin was suspended in citrate-phosphate buffer (pH 6.3) and was added to a final concentration of 5 mg ml^{-1} (equal to a molar concentration of 25 mM based on lignin monomers) to the reaction mixture and incubated under standard conditions (see above) with *Tt*LPMO9E. Control experiments were performed as described above, replacing ascorbic acid with lignin.

Light-induced oxidation of xyloglucans. Xyloglucan 1% w/w was used as a substrate, replacing PASC and incubated with chlorophyllin, ascorbic acid and *Tt*LPMO9E, the others parameters were according to standard conditions (see above). A release of oligosaccharides was detected using HAPEC. The observed pattern was a mixture of cello-oligosaccharides and xylogluco-oligosaccharides, which could not be fully separated, as seen for other published works[19,20], therefore the peaks are separated into non-oxidized and oxidized products only.

Light-induced oxidation of crystalline cellulose. Avicel was used as a cellulose substrate at 1% w/w instead of PASC and all other conditions were same as standard conditions (see above).

Protein structure modelling. The PDB structures from the RCSB PDB Protein data bank were adapted for the HARLEM Molecular Modelling Package program (Kurnikov *et al*, available at http://harlem.chem.cmu.edu/). The PDB structures were (after adding hydrogens) analysed in HARLEM for possible LRET pathways from the surface to the metal ion[21].

References

1. Vaaje-Koldstad, G. *et al.* An oxidative enzyme boosting the enzymatic conversion of recalcitrant polysaccharides. *Science* **330,** 219–222 (2010).
2. Harris, P. V. *et al.* Stimulation of lignocellulosic biomass hydrolysis by proteins of glycoside hydrolase family 61: structure and function of a large, enigmatic family. *Biochemistry* **49,** 3305–3316 (2010).
3. Vaaje-Kolstad, G., Westereng, B. & Eijsink, V. G. H. Novel enzymes for the degradation of cellulose. *Biotechnol. Biofuels* **5,** 45 (2012).
4. Beeson, W. T., Vu, V. V., Span, E. A., Phillips, C. M. & Marletta, M. A. Cellulose degradation by polysaccharide monooxygenases. *Annu. Rev. Biochem.* **84,** 923–946 (2015).
5. Busk, P. & Lange, L. Classification of fungal and bacterial lytic polysaccharide monooxygenases. *BMC Genomics* **16,** 368 (2015).
6. Leggio, L. L. *et al.* Structure and boosting activity of a starch-degrading lytic polysaccharide monooxygenases. *Nat. Commun.* **6,** 5961 (2015).

7. Quinlan, J. R. *et al.* Insights into the oxidative degradation of cellulose by a copper metalloenzyme that exploits biomass components. *Proc. Natl Acad. Sci. USA* **108,** 1579–1584 (2011).
8. Phillips, C. M., Beeson, W. T., Cate, J. H. & Marletta, M. A. Cellobiose dehydrogenase and a copper-dependent polysaccharide monooxygenase potentiate cellulose degradation by Neurospora crassa. *ACS Chem. Biol.* **6,** 1399–1406 (2011).
9. Cannella, D., Hsieh, C. W. C., Felby, C. & Joergensen, H. Production and effect of aldonic acids during enzymatic hydrolysis of lignocellulose at high dry matter content. *Biotechnol. Biofuels* **5,** 26 (2012).
10. Rodrigues-Zuniga, U. F. *et al.* Lignocellulose pretreatment technologies affect the level of enzymatic cellulose oxidation by LPMO. *Green Chem.* **17,** 2896–2903 (2015).
11. Westereng, B. *et al.* Enzymatic cellulose oxidation is linked to lignin by long-range electron transfer. *Sci. Rep.* **5,** 18561 (2015).
12. Tan, T. C. *et al.* Structural basis for cellobiose dehydrogenase action during oxidative cellulose degradation. *Nat. Commun.* **6,** 7542 (2015).
13. Langston, J. A. *et al.* Oxidoreductive cellulose depolymerization by the enzymes cellobiose dehydrogenase and glycoside hydrolase 61. *Appl. Environ. Microbiol.* **77,** 7007–7015 (2011).
14. Hemsworth, G. R., Johnston, E. M., Davies, G. J. & Walton, P. H. Lytic polysaccharide monooxygenases in biomass conversion. *Trends Biotechnol.* **33,** 747–761 (2015).
15. Seeley, G. R. The energetics of electron transfer reactions of chlorophyll and other compounds. *Photochem. Photobiol.* **27,** 639–654 (1978).
16. Chaudhary, Y. S. *et al.* Visible light-driven CO_2 reduction by enzyme coupled CdS nanocrystals. *Chem. Commun. (Camb)* **48,** 58–60 (2012).
17. Caputo, A., Wang, L., Beranek, R. & Reisner, E. Carbon nitride–TiO_2 hybrid modified with hydrogenase for visible light driven hydrogen production. *Chem. Sci.* **6,** 5690–5694 (2015).
18. Elder, T. & Fort, R. C. in *Lignin and Lignans: Advances in Chemistry.* (eds Heitner, C. Dimmel, D. R. & Schmidt, J. A.) 321–347 (CRC Press, 2010).
19. Agger, J. W. *et al.* Discovery of LPMO activity on hemicelluloses shows the importance of oxidative processes in plant cell wall degradation. *Proc. Natl Acad. Sci. USA* **111,** 6287–6292 (2014).
20. Bennati-Granier, C. *et al.* Substrate specificity and regioselectivity of fungal AA9 lytic polysaccharide monooxygenases secreted by *Podospora anserine. Biotechnol. Biofuels* **8,** 90 (2015).
21. Van der Zee, J., Krootjes, B. B., Chignell, C. F., Dubbelman, T. M. & Van Steveninck, J. Hydroxyl radical generation by a light-dependent Fenton reaction. *Free Radic. Biol. Med.* **14,** 105–113 (1993).
22. Borisova, A. S. *et al.* Characterization of a lytic polysaccharide monooxygenase with broad substrate specificity. *J. Biol. Chem.* **290,** 22955–22969 (2015).
23. Onuchic, J. N., Beratan, D. N., Winkler, J. R. & Gray, H. B. Pathway analysis of protein electron-transfer reactions. *Annu. Rev. Biophys. Biomol. Struct.* **21,** 349–377 (1992).
24. Kim, S., Ståhlberg, J., Sandgren, M., Paton, R. S. & Beckham, G. T. Quantum mechanical calculations suggest that lytic polysaccharide monooxygenases use a copper-oxyl, oxygen-rebound mechanism. *Proc. Natl Acad. Sci. USA* **111,** 149–154 (2013).
25. Kjaergaard, C. H. *et al.* Spectroscopic and computational insight into the activation of O_2 by the mononuclear Cu center in polysaccharide monooxygenases. *Proc. Natl Acad. Sci. USA* **111,** 8797–8802 (2014).
26. Sutton, H. C. & Winterbourn, C. C. On the participation of higher oxidation states of iron and copper in Fenton reactions. *Free Radic. Biol. Med.* **6,** 53–60 (1989).
27. Pham, N. A., Xing, G., Miller, C. J. & Waite, T. D.T. Fenton-like copper redox chemistry revisited: Hydrogen peroxide and superoxide mediation of copper-catalysed oxidant production. *J. Catal.* **301,** 54–64 (2013).
28. Dhar, D. & Tolman, W. B. Hydrogen atom abstraction from hydrocarbons by a copper(III)-hydroxide complex. *J. Am. Chem. Soc.* **137,** 1322–1329 (2015).
29. Austin, A. T. & Ballare, C. L. Dual role of lignin in plant litter decomposition in terrestrial ecosystems. *Proc. Natl Acad. Sci. USA* **107,** 4618–4622 (2010).
30. Baker, N. R. & Allison, T. Ultraviolet photodegradation facilitates microbial litter decomposition in a Mediterranean climate. *Ecology* **96,** 1994–2003 (2015).
31. Wood, T. M. Preparation of crystalline, amorphous and dyed cellulase substrates. *Method Enzymol.* **160,** 19–25 (1988).
32. Horn, S. J. & Eijsink, V. G. H. A reliable reducing end assay for chito-oligosaccharides. *Carbohydr. Polym.* **56,** 35–39 (2004).
33. Möllers, K. B., Cannella, D., Jørgensen, H. & Frigaard, N. U. Cyanobacterial biomass as carbohydrate and nutrient feedstock for bioethanol production by yeast fermentation. *Biotechnol. Biofuels* **7,** 64 (2014).
34. Cardona, T. *et al.* Isolation and characterization of thylakoid membranes from the filamentous cyanobacterium Nostoc punctiforme. *Physiol. Plant.* **131,** 622–634 (2007).
35. Holm-Hansen, O. & Riemann, B. Chlorophyll *a* determination: improvements and methodology. *Oikos* **30,** 438–447 (1978).
36. Lichtenthaler, H. K. Chlorophylls and carotenoids: pigments of photosynthetic biomembranes. *Methods Enzymol.* **148,** 350–382 (1987).
37. Westereng, B. *et al.* Efficient separation of oxidized cello-oligosaccharides generated by cellulose degrading lytic polysaccharide monooxygenases. *J. Chrom. A* **1271,** 144–152 (2013).

Acknowledgements

We gratefully acknowledge the donation of enzymes from Novozymes A/S and financial support from the following projects: The lignin-cellulose electron shuttling mechanism funded by The Danish Council for Independent Research (D.C.), The VILLUM Center of Excellence Plant Plasticity (P.E.J.), bioSYNergy funded by the UCPH Excellence Programme for Interdisciplinary Research (P.E.J.), Plant Power: Light-Driven Synthesis of Complex Terpenoids Using Cytochromes P450 (12-131834) funded by the Danish Council for Strategic Research, Programme Commission on Strategic Growth Technologies (P.E.J. and N.U.F.). Controlling metal-catalysed protein oxidation in biology and medicine funded by The Danish Council for Independent Research-Technology and Production Sciences (M.J.B.), Nordic Energy Research project AquaFEED-Conversion of solar energy to infrastructure-ready transport fuels using aquatic photobiological organisms (K.B.M. and N.U.F.).

Authors contributions

D.C., K.B.M. and C.F. planned and participated in the experiments; P.E.J. and N.U.-F. participated in planning of the experiments; M.J.B. planned and performed the protein modelling. K.S.J together with all authors analysed the data and wrote the manuscript.

Additional information

Competing financial interests: The authors declare no competing financial interests.

10

Supramolecular amplification of amyloid self-assembly by iodination

Arianna Bertolani[1], Lisa Pirrie[1,2], Loic Stefan[1], Nikolay Houbenov[3], Johannes S. Haataja[3], Luca Catalano[1], Giancarlo Terraneo[1], Gabriele Giancane[4], Ludovico Valli[4], Roberto Milani[2], Olli Ikkala[3], Giuseppe Resnati[1] & Pierangelo Metrangolo[1,2]

Amyloid supramolecular assemblies have found widespread exploitation as ordered nanomaterials in a range of applications from materials science to biotechnology. New strategies are, however, required for understanding and promoting mature fibril formation from simple monomer motifs through easy and scalable processes. Noncovalent interactions are key to forming and holding the amyloid structure together. On the other hand, the halogen bond has never been used purposefully to achieve control over amyloid self-assembly. Here we show that single atom replacement of hydrogen with iodine, a halogen-bond donor, in the human calcitonin-derived amyloidogenic fragment DFNKF results in a super-gelator peptide, which forms a strong and shape-persistent hydrogel at 30-fold lower concentration than the wild-type pentapeptide. This is remarkable for such a modest perturbation in structure. Iodination of aromatic amino acids may thus develop as a general strategy for the design of new hydrogels from unprotected peptides and without using organic solvents.

[1] Laboratory of Nanostructured Fluorinated Materials (NFMLab), Department of Chemistry, Materials, and Chemical Engineering 'Giulio Natta', Politecnico di Milano, Via Mancinelli 7, Milano I-20131, Italy. [2] VTT-Technical Research Centre of Finland, PO Box 1000, Espoo FI-02044, Finland. [3] Department of Applied Physics, Aalto University, PO Box 15100, Espoo FI-02150, Finland. [4] Dipartimento Beni Culturali, Università del Salento, Lecce I-73100, Italy. Correspondence and requests for materials should be addressed to P.M. (email: pierangelo.metrangolo@polimi.it).

Amyloid fibrils are a class of nanomaterials that can be assembled from a wide variety of peptides and proteins and have an array of functional and pathological roles in nature[1]. In the latter, the self-assembly process can be triggered by peptide mutations and leads to the formation of insoluble fibrillar aggregates that are associated with Alzheimer's, Parkinson's, Creutzfeldt-Jakob and Huntington's diseases[2–6]. Amyloid fibrils also perform a number of physiological roles including being constituents of protective envelopes of fish and insect eggs, essential amphipathic materials of fungi and bacteria and a vital constituent of spider silk[7]. This wide variety of physiological functions of amyloids has inspired many nanocomposites thanks to the robustness of these fibrillar materials, which are comparable to dragline silk and collagen, as well as biomaterial applications like tissue engineering[8–11]. In this context, it is important that the amyloidogenic peptide/protein self-assembles in a tunable and controlled manner to provide fibrils of well-defined architecture[12].

Amyloid fibrils are characterized by their cross-β-sheet structures whereby bundles of β-strands form highly ordered filaments that run perpendicular to the fibre axis[11,13]. The stability of amyloid structures can be attributed to noncovalent interactions, namely hydrogen bonds, π–π stacking and hydrophobic interactions occurring between both side chain and backbone atoms, which hold the β-sheets and strands together[11]. On the basis of the assumption that noncovalent interactions are key to the formation of the amyloid structure, and since halogen bonds have been recently appreciated as new tools for supramolecular engineering[14], we decided to study the impact of introducing halogen-bond donor groups, for example, iodine atoms, into an already known amyloidogenic motif to study the potential of halogen bonding in controlling and promoting amyloid self-assembly.

Although many modifications of amyloidogenic sequences have been utilized to tune their self-assembly behaviour[12], halogenation has rarely been pursued and only on single amino acids or on a dipeptide[15–17]. The advantage of a strategy based on the introduction of halogen atoms on amyloidogenic motifs lies in the fact that halogenation is a minimal structural modification, which, on the other hand, may induce a large difference in the peptide supramolecular behaviour as a consequence of the rich variety of noncovalent interactions given by halogen atoms[18]. It has been shown previously that incorporation of single halogen substituents on the aromatic side chain of amino acids enhances their self-assembly into amyloid-like fibrils that promote hydrogelation in aqueous solvents[19]. In particular, fluorinated amino acids were found to function better than chlorinated and brominated analogues in terms of assembly kinetics and rigidity of the obtained hydrogels. This has been explained by the significant electronic effects of fluorine and the subsequent perturbation of the energetics of the π–π stacking interactions, although hydrophobic and steric effects could not be discounted[20]. Conversely, there is only one report on the fibrillar nanostructures given by a diiodinated dipeptide[17] but, to the best of our knowledge, no detailed study on the effect of iodination on promoting/inhibiting the self-assembly of amyloidogenic peptides. Under the hypothesis that introducing an iodine atom on the *p*-position of the benzyl side chain of an aromatic amino acid may result in the formation of halogen bonding thus promoting fibrils' formation, we studied the structure of *p*-iodo-phenylalanine (*p*-I-Phe) and the self-assembly behaviour of a series of iodinated derivatives of a known amyloidogenic core sequence.

Results

Single crystal X-ray structure of *p*-iodo-phenylalanine. The halogen bond[21] is the strong, specific and directional interaction involving halogen atoms as electrophilic species[22]. Increasing in strength with halogen atom polarizability, halogen bonds are most efficient with iodine substituents. This rather new interaction is virtually unexplored in the field of amino acid and peptide self-assembly. For this reason, to evaluate the possibility of exploiting the halogen bond in the self-assembly of amyloid peptides/proteins, we have first studied the single crystal X-ray structure of the amino acid *p*-I-Phe.

The crystal structure of *p*-I-Phe (Supplementary Fig. 1 and Supplementary Table 1) highlights the amphoteric behaviour of the iodine atom, which functions as Lewis acid along the extension of the C–I bond and as Lewis base at the belt perpendicular to it (Supplementary Fig. 2a). In fact, the crystal packing of this iodinated amino acid is characterized by two strands of amino acids running along the *b* crystallographic axis and connected through hydrogen bonds (Supplementary Fig. 2b). These strands are then laterally assembled into two-dimensional supramolecular sheets thanks to type II iodine···iodine contacts[23], that is, halogen bonds (I···I distance 3.7517(4) Å, I···I···I angle 89.66°), occurring at the strand surface (Supplementary Fig. 2c). Importantly, no sign of π–π stacking is observed, while C–H···π interactions involving the benzyl hydrogens contribute to aromatic side-chain ordering. Despite the fact that inferring structural information on the amyloid state from high-resolution X-ray diffraction analysis of small-molecule models is difficult, we introduced the *p*-I-Phe residue into the amyloidogenic core sequence DFNKF[24] (NH_2-Asp-Phe-Asn-Lys-Phe-COOH, Fig. 1), with the purpose of exploiting similar

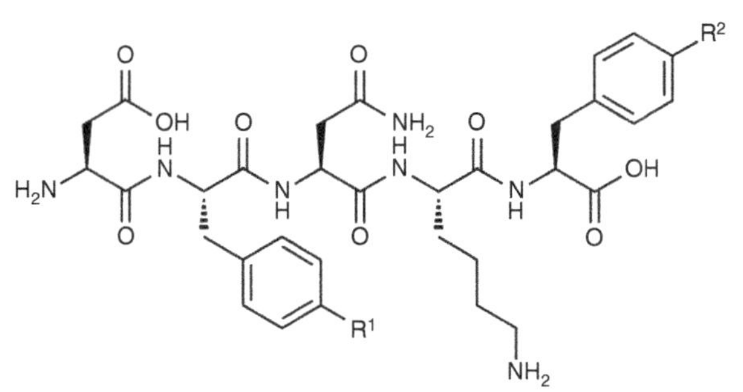

	R^1	R^2
DFNKF	H	H
DFNKF(I)	H	I
DF(I)NKF	I	H
DF(I)NKF(I)	I	I
DF(Br)NKF(Br)	Br	Br
DF(Cl)NKF(Cl)	Cl	Cl

hCT_{15-19} NH_2 – Asp – Phe – Asn – Lys – Phe – COOH
1 2 3 4 5

Figure 1 | Molecular formulae of the peptides used in this study. The natural peptide DFNKF was modified by halogenation on the phenylalanine residues in the para position. Modifications are shown in the peptide sequence in brackets where F(I) denotes 4-iodophenylalanine, F(Br) denotes 4-bromophenylalanine and F(Cl) denotes 4-chlorophenylalanine.

halogen bonds at the fibril strand surface to promote fibril–fibril lamination effects.

DFNKF is a short segment (residues 15–19) of the human calcitonin (hCT) hormone[25], an amyloidogenic protein with 32 amino acids, whose amyloid formation is related to medullary carcinoma of the thyroid[24]. This particular pentapeptide is widely used for investigating the nanoscopic arrangement of a fibril complex since DFNKF forms well-ordered fibrils similar to the aggregates of hCT[24,25]. For this reason, we selected DFNKF as a simple model compound for fibril formation and obtained its halogenated derivatives by introducing I, Br and Cl atoms at the *p*-position of the phenylalanine (Phe) benzene ring. To study the effect on fibrillation of the introduction of halogen atoms on specific positions of the pentapeptide, we studied the 5-*p*-iodo-Phe derivative DFNKF(I), the 2-*p*-iodo-Phe derivative DF(I)NKF, the 2,5-bis-*p*-iodo-Phe derivative DF(I)NKF(I), as well as its bis-brominated and bis-chlorinated analogues DF(Br)NKF(Br) and DF(Cl)NKF(Cl), respectively (Fig. 1). Importantly, all of the studied peptides carry free amino (N) and carboxyl (C) termini.

Hydrogelation properties of halogenated DFNKF derivatives. Hydrogelation is highly indicative of fibrillation, therefore all peptides were in the first instance assessed for their ability to form hydrogels[26,27]. All of the studied peptides were found to self-assemble into hydrogels over a minimum gelation concentration. Only the hydrogel of the wild-type peptide DFNKF has been previously reported, though of its *N*-acetylated form at the N terminus[28]. All the reported halogenated peptides are novel types of gelators and, importantly, formed gels (Fig. 2a and Supplementary Table 2) at concentrations much lower than that of their wild-type parent DFNKF, that is, 75 mM, with the super-gelator peptide DF(I)NKF that showed a minimum gelation concentration of $<0.2\%$ (2.5 mM; 30-fold lower than DFNKF).

Attempts to study the kinetics of fibril formation using Thioflavin T (ThT) were unsuccessful since no increase in fluorescence was observed upon mixing ThT with peptide solutions. This is well in accordance with the literature since previous studies have demonstrated that DFNKF does not bind this fibril marker dye[29].

Subsequently, a working concentration of 15 mM was chosen where all halogenated peptides formed clear and elastic gels within a reasonable time (≤ 12 h). At this concentration, the speed of hydrogel formation followed the order DF(I)NKF(I) > DF(I)NKF ≥ DF(Br)NKF(Br) > DF(Cl)NKF(Cl) ≥ DFNKF(I), whereas DFNKF did not form a gel even after 30 days (Supplementary Table 3). The diiodinated peptide formed the most homogeneous and shape-persistent gel (Fig. 2b), whereas peptides DF(Cl)NKF(Cl) and DFNKF(I) showed phase separation of the fibrous aggregates from the aqueous phase. The obtained hydrogels were all thermoreversible with gel–sol transition temperatures ranging from 65 °C (DFNKF(I) to 119 °C (DF(I)NKF(I); Fig. 2c). In particular, the observed transition temperatures followed the same order of the gel forming kinetics reported above. These results clearly show that, in general, halogenation of the DFNKF model peptide greatly promotes hydrogel formation, with the diiodinated peptide being the most efficient. This efficiency decreases with the polarizability of the halogen atom, that is, DF(I)NKF(I) > DF(Br)NKF(Br) > DF(CI)NKF(CI), and also depends on the specific position of the halogen atom in the peptide scaffold, that is, the peptide iodinated at the Phe in the second position is more efficient than the one iodinated in the fifth.

Characterization of the different halogenated peptide hydrogels was done by oscillatory rheology (ring-cast method) using the 15 mM concentration. At this low concentration, the wild-type peptide DFNKF only formed a viscous fluid and was not studied further. The diiodinated peptide DF(I)NKF(I) was confirmed to form the stiffest gel, which is reflected in its high elastic modulus ($G' > 10^4$ Pa and $G' \gg G''$; Fig. 2d and Supplementary Fig. 3) and by the fact that the gel assumes the well-defined shape of the mould (Fig. 2b). The G' value for DF(I)NKF(I) is comparable to

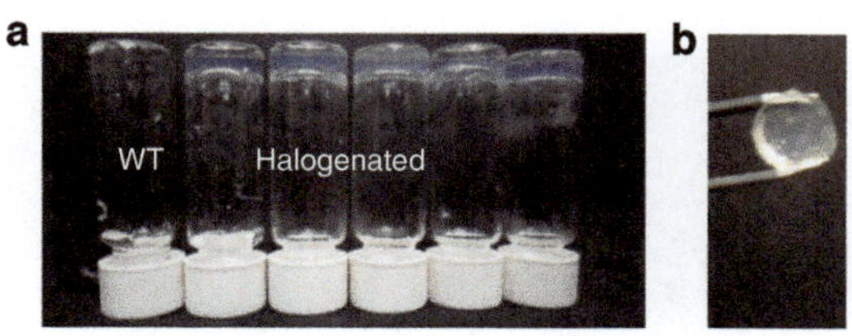

c

Peptide	Transition *T* (°C)
DFNKF(I)	64.7–69.0
DF(I)NKF	88.2–92.7
DF(I)NKF(I)	114.0–119.0
DF(Br)NKF(Br)	92.1–96.4
DF(Cl)NKF(Cl)	78.1–82.4

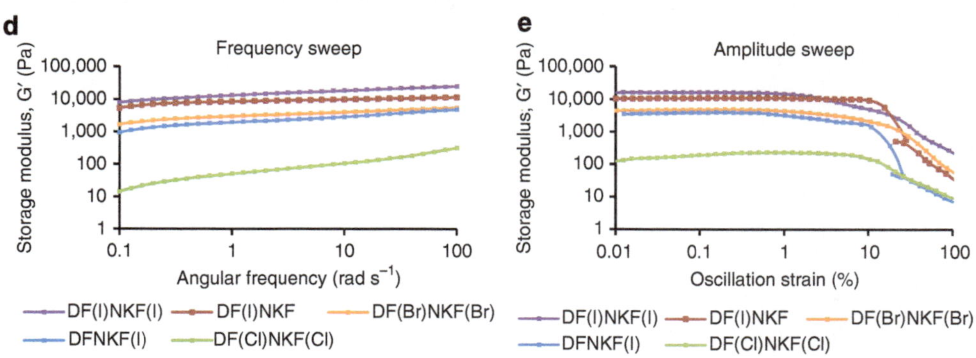

Figure 2 | Characterization of hydrogels formed by halogenated DFNKF derivatives (**a**) Photograph of vials containing 15 mM peptide solutions upon aging 12 h at RT After this period, all halogenated peptides formed clear hydrogels whereas the wild-type peptide DFNKF (WT) remained as a viscous fluid. (**b**) The hydrogel formed by DF(I)NKF(I) was the most shape-persistent of all gels. The gel was prepared using the ring-cast method at 15 mM concentration and imaged after 48 h. (**c**) Gel-sol transition temperatures of 15 mM halogenated peptide hydrogels formed in 48 h. (**d**) Rheological characterization of the gels by frequency sweep studies whereby the storage modulus (G') was recorded as a function of angular frequency (ω). The peptide DF(I)NKF(I) was determined to be the strongest gel with $G' > 10^4$ Pa. (**e**) Amplitude sweep studies of peptide hydrogels showing G' as a function of oscillation strain (γ). All halogenated hydrogels show similar linear viscoelastic region (LVR) profiles. Peptide hydrogels at a concentration of 15 mM were utilized for the rheological measurements after 48 h from preparation. For the loss moduli, see the Supplementary Fig. 3.

that of Ac-LIVAGD, a hexamer that forms one of the best performing hydrogels in terms of ease of hydrogel formation and strength[30,31]. Interestingly, the trend of G′ values parallels those of gel formation efficiency and thermal stability whereby gels of peptides DF(I)NKF(I), DF(I)NKF and DF(Br)NKF(Br) are the stiffest and DFNKF(I) and DF(Cl)NKF(Cl) formed the weakest ones.

Morphology and microarchitecture of the self-assembled fibrillar networks constituting the halogenated hydrogels were evaluated by confocal microscopy (Fig. 3)[32]. The network of the hydrogel of DF(I)NKF(I) appeared to be very filamentous with micrometre-long fibrils assembled in bundles with a helical sense, which propagate along the same direction. These micrometre-long fibrils become fewer and thinner on going from DF(I)NKF(I) to DF(I)NKF and DF(Br)NKF(Br), and disappear in the hydrogels of DF(CI)NKF(CI) and DFNKF(I) where only entangled matrices of small fibrils and aggregates were observed. These observations fit well with the rheology data where the strongest gels result from the entanglement of the longest fibrils. In agreement with the confocal microscopy study, the hydrogel formed by the peptide DF(I)NKF(I) showed an intertwined network of long fibrils in its transmission electron microscopy (TEM) image (Fig. 3f).

The nanostructure of the halogenated fibrils was revealed by atomic force microscopy (AFM) (Fig. 4). The peptides were first imaged after incubation in dilute aqueous conditions for 9 days (see also Supplementary Fig. 4) and Fig. 4a–c highlight the most strongly fibrillating peptides DF(I)NKF(I), DF(I)NKF and DF(Br)NKF(Br). While only small spherical aggregates were observed for the wild-type peptide DFNKF (Supplementary Fig. 4f), all of the halogenated peptides showed pronounced fibrillar structures. Large differences were, however, observed between the monoiodinated peptides DFNKF(I) and DF(I)NKF: Peptide DFNKF(I) (Supplementary Fig. 4e) showed only small protofibrils and aggregates, whereas the peptide DF(I)NKF (Fig. 4b and Supplementary Fig. 4c) showed large twisted fibrils similar to those of the diiodinated peptide DF(I)NKF(I) (Fig. 4a and Supplementary Fig. 4a), which, in turn, had the most complex structures with twisted ribbon-like fibrils tens of nanometre wide and micrometre long. These results highlight the importance of the specific position of the iodine atom in the peptide structure and therefore the subsequent specific interactions, which result in different amyloid fibril topographies. That the fibrillation and twisting is not due to drying is excluded by high-resolution cryo-TEM and electron tomography (ET) studies. This is exemplified for the peptide DF(I)NKF(I) corresponding to the strongest hydrogel. It shows fibrils with lateral sizes of 20–25 nm and helical twisting with periodicities between 250 and 140 nm (Fig. 4h and Supplementary Fig. 4b). Such periodicities correspond to the different maturation steps of amyloid fibrils as previously reported for β-lactoglobulin fibrils[33]. ET allowed three-dimensional visualization of the twisting morphology of a DF(I)NKF(I) fibril imaged in solution (Fig. 4i).

Amyloid structure of fibrils of halogenated peptides. There are three criteria that define a protein aggregate as an amyloid fibril: green birefringence upon staining with Congo Red, fibrillar morphology and β-sheet secondary structure[34]. Therefore, Congo red staining was carried out for all the peptides (at the same 15 mM concentration) and strong green-gold birefringence was observed under polarized light, which is a reporter for long-range cross-β structure (Supplementary Fig. 5).

Further evidence of amyloid nature was obtained by Fourier transform infrared (FT-IR) spectroscopy. The FT-IR spectra of both the solution and the gel state in D_2O were recorded (Supplementary Fig. 6). A solution of the wild-type peptide DFNKF showed no significant amide I' band, however, using a concentration at which it forms a gel (75 mM), this band became visible. This peak at around 1,630 cm^{-1} can be attributed to the peptide bond carbonyl upon formation of the amyloid β-sheet structure and subsequent gelation[35]. The spectra of the gel state

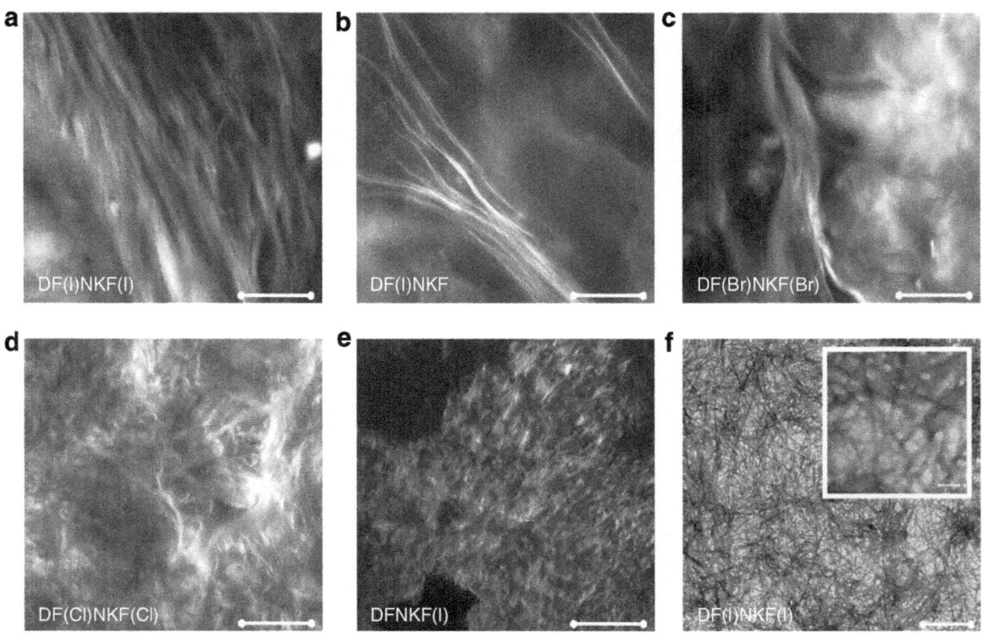

Figure 3 | Fibrillar morphologies of halogenated hydrogels. (**a**–**e**) Confocal microscopy images of peptide hydrogels at a concentration of 15 mM upon aging 48 h at RT and after staining with Rhodamine B (scale bar, 100 μm). Panels **a**, **b** and **c** show bundles of long twisted fibrils belonging to hydrogels of peptides DF(I)NKF(I), DF(I)NKF and DF(Br)NKF(Br), respectively. Panels **d** and **e** show much smaller fibrils of hydrogels of peptides DF(Cl)NKF(Cl) and DFNKF(I), which intertwine and form a matrix-type structure. (**f**) TEM image of the dried hydrogel of peptide DF(I)NKF(I) showing an intertwined network of long fibrils (scale bars, 0.5 μm and 5 nm).

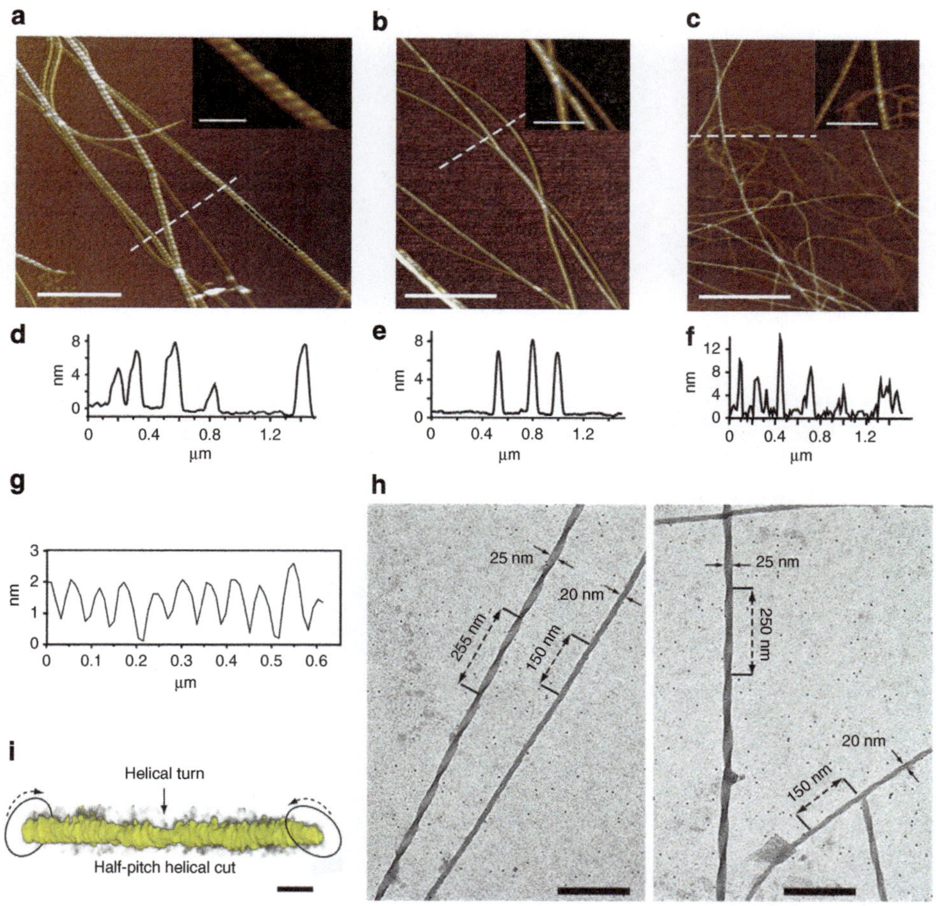

Figure 4 | Nanofibrillation of the halogenated pentapeptides. AFM image close-up insets of (**a**) peptide DF(I)NKF(I), (**b**) peptide DF(I)NKF, and (**c**) peptide DF(Br)NKF(Br) evaporated on mica substrates after 9 days incubation in aqueous solutions (scale bar, 1 µm). (**d**-**f**) Height profiles of fibrils from **a**, **b** and **c** (cross-section lines highlighted in white). (**g**) Cross-sectional analysis on the top of a fibril segment in **a** showing a helicoidal profile along the longitudinal direction of the fibrils (highlighted in black). (**h**) Cryo-TEM images of the dried hydrogel of DF(I)NKF, showing a network of long fibrils (scale bar, 200 nm). (**i**) *In situ* electron tomography of half-pitch helicoidal peptide DF(I)NKF(I) vitrified from aqueous solution. The twisted fibril morphology is highlighted by arrows. Electron tomography reconstructions were collected upon tilting of samples up to ±69° (scale bar, 25 nm).

of the halogenated peptides showed this same amide I' band consistent with the formation of amyloid fibrillar species comprising β-sheet elements. These data confirm that all the studied halogenated derivatives of DFNKF form fibrils of the same amyloid nature to their wild-type parent. Interestingly, this 1,630 cm^{-1} peak is red-shifted in the spectra of DF(I)NKF(I) and DF(I)NKF, while it is not in the one of DFNKF(I) (Supplementary Fig. 6). This red-shift indicates a reduced electron density on the peptide bond and is consistent with the involvement of the carbonyl oxygen in halogen bonding[36].

Raman spectroscopy is a powerful tool to investigate the occurrence of the halogen bond[37]. For this reason, we followed the changes experienced by the C–I stretching band at around 167 cm^{-1} upon formation of the gel. Interestingly, compared with the bulk powders of the starting materials, in the dried gels of the iodinated peptides DF(I)NKF(I) and DF(I)NKF, this band shifted to lower frequency (see Supplementary Fig. 7). This shift is perfectly consistent with the weakening of the C–I bond as a consequence of its involvement in halogen bonding upon formation of the hydrogels[37].

Co-assembly of wild-type peptide with diiodinated analogue. Bromination and chlorination of proteins *in vivo* has been related to a series of oxidative stress-related diseases such as cystic fibrosis[38], atherosclerotic intima[39], sepsis[40] and asthma, among others[41]. A direct correlation exists between the role of oxidative stress and protein fibrillation[42]. However, how halogenation affects protein structure, folding and functioning is not yet known. To study to what extent a partial degree of halogenation may promote the amplified formation of amyloid fibrils, we studied the co-assembly[43] of the strongly self-assembling diiodinated peptide DF(I)NKF(I) with the wild-type peptide DFNKF. Mixing DFNKF and DF(I)NKF(I) at concentrations at which normally neither form gels alone (Fig. 5a), 15 and 0.75 mM (5%), respectively, resulted in a mixed hydrogel within 18 h.

Characterization of the physical properties of the mixed hydrogel was performed by oscillatory rheology (ring-cast method) where the mixed hydrogel was found to be weaker than the hydrogel containing only DF(I)NKF(I) (Fig. 5b versus Fig. 2d). It was also observed that increasing the amount of the strong gelator DF(I)NKF(I) led to an increase in the elastic modulus value, whereas the gel containing 10% was stronger than the one containing 5% of DF(I)NKF(I). Interestingly, circular dichroism (CD) analysis of the mixed hydrogels showed the same structural features observed in the CD spectra of the pure iodinated peptides (see Supplementary Figs 8 and 9).

The resulting fibril morphologies in the mixed hydrogels were studied by AFM (Fig. 5c,d). A 15 mM sample of the wild-type peptide DFNKF showed only a mixture of intertwined aggregates and small fibrils, which is consistent with the observation that no gelation is observed at this concentration since the fibrils appear

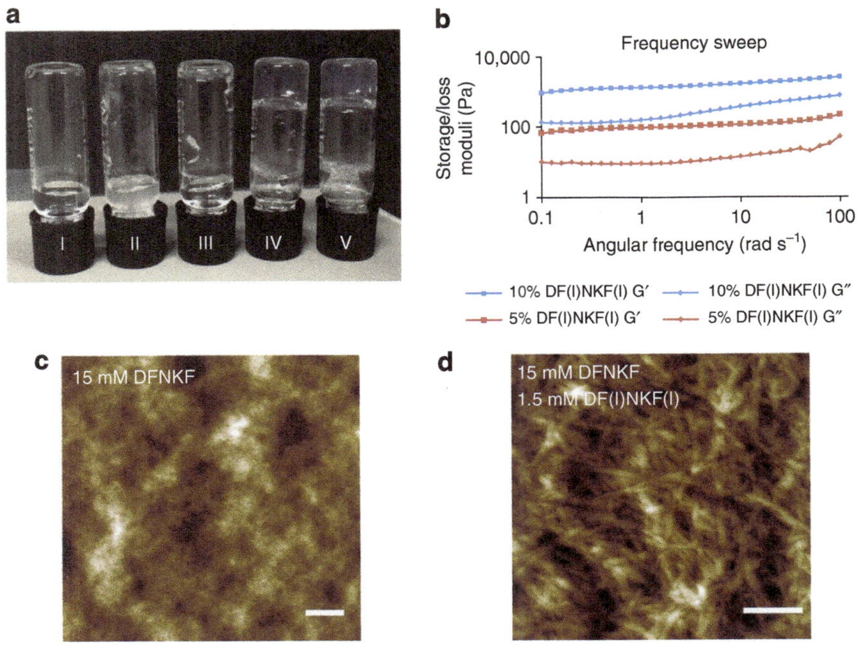

Figure 5 | Mixed hydrogels formed by co-assembly of DFNKF and DF(I)NKF(I). (**a**) Photograph of vials after 48 h containing (I) 15 mM DFNKF, which did not form hydrogel; (II) 1.5 mM DF(I)NKF(I), which did not form hydrogel and some precipitation was observed; (III) 15 mM DFNKF:0.15 mM DF(I)NKF(I), which did not form hydrogel; (IV) 15 mM DFNKF:0.75 mM DF(I)NKF(I), which formed hydrogel within 48 h; (V) 15 mM DFNKF:1.5 mM DF(I)NKF(I), which formed hydrogel within 48 h. (**b**) Rheological characterization of mixed hydrogels by frequency sweep studies whereby the storage concentration of 15 mM after 48 h were utilized for the rheological measurements. (**c**) AFM of a 15 mM solution of wild-type peptide DFNKF showed small fibrils and aggregates unable to network sufficiently to form a hydrogel. (**d**) AFM of a sample of the mixed hydrogel containing 15 mM DFNKF and 1.5 mM DF(I)NKF(I) showed larger and more entangled fibrils (scale bar, 400 nm).

too small to crosslink and form a network (Fig. 5c). On the addition of 10% DF(I)NKF(I), the resulting fibrils were much longer and intertwined to form a fibrous network resulting in the hydrogelation of the wild-type peptide DFNKF solution (Fig. 5d and Supplementary Fig. 10a). The weakness of the resulting mixed hydrogel compared with that of DF(I)NKF(I) alone can be explained by the relative size and network formation of the fibrils (Supplementary Fig. 10b). This result may have important implications in a biological setting since only small amounts of brominated or chlorinated peptides may be required to amplify the supramolecular self-assembly of an amyloid fibril.

Role of halogen bonding in hydrogels' self-assembly. Although no direct structural evidence of halogen bonding between our peptide monomers has been obtained, yet, the reported results, IR and Raman in particular, provided corroborative evidence that iodinated peptides DF(I)NKF and DF(I)NKF(I) potentially use halogen bonds to efficiently self-assemble into mature amyloid fibrils. In fact, the particular position of the iodinated residue in the peptide sequence greatly influences the efficiency of its self-assembly process, thus suggesting the involvement of the iodine atom in a specific noncovalent interaction (Fig. 6a), rather than nonspecific hydrophobic and/or steric effects. This is further supported by considerations of halogen atom polarizability since the dibrominated and dichlorinated derivatives DF(Br)NKF(Br) and DF(Cl)NKF(Cl) showed less tendency to form fibrils than their diiodinated analogue DF(I)NKF(I). This result allows interactions involving halogen atoms as hydrogen bond acceptors to be ruled out as Cl and Br would give stronger interactions than I, which was not observed. Electrostatic interactions involving the ionizable side chains of the peptides, Asp and Lys, may also be ruled out as the formation of these halogenated hydrogels is largely pH-independent (see Supplementary Table 4). As a

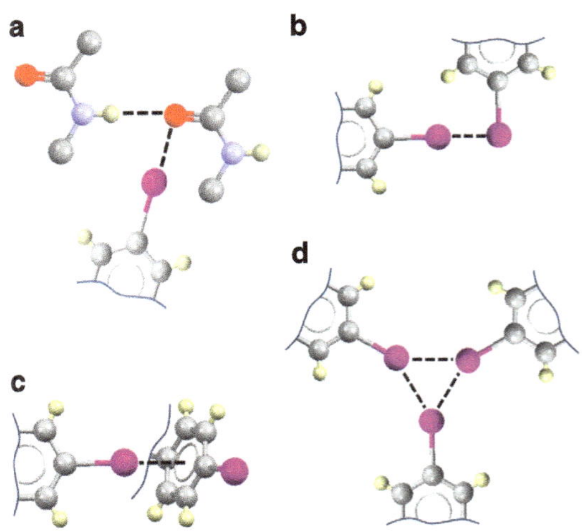

Figure 6 | Potential halogen bonds involving *p*-iodo-phenylalanine. IR and Raman spectroscopies are more consistent with the occurrence of type **a** interaction, however, other interactions reported cannot be discounted. (**a**) Orthogonal hydrogen and halogen bonds involving the carbonyl oxygen atom of the peptide bond[36]. (**b**) Type II iodine···iodine contacts[23]. (**c**) Iodine···π interactions[49]. (**d**) Triangular iodine synthon[50].

confirmation of the minor role played by electrostatic attraction in the self-assembly of the reported halogenated peptides, increasing the ionic strength of the peptide solutions resulted in accelerated formation of the hydrogels (see Supplementary Table 5). It is also unlikely that aromatic π–π interactions are playing an important role in the self-assembly of the halogenated hydrogels, since this would not explain the polarizability dependence that we observed. Finally, hydrophobic interactions could not justify the observed trends in hydrogel properties because

DF(Br)NKF(Br) and DF(Cl)NKF(Cl) are both more hydrophobic than DF(I)NKF (see Supplementary Table 6), which instead forms the second stiffest gel. Moreover, the two monoiodinated peptides, despite having similar hydrophobicity, display completely different properties. For the same reasons discussed above, electrostatic and hydrophobic interactions also should not play an important role in the interfibrillar association observed in the halogenated gels.

Discussion

In summary, we have demonstrated that iodination of a short amyloidogenic core sequence strongly promotes its fibril formation ability and affects the structure of formed fibrils. In particular, the 2-*p*-iodo-Phe derivative DF(I)NKF was demonstrated to be 30-fold more efficient in forming hydrogels than the wild-type pentapeptide DFNKF. This is remarkable for such a modest perturbation in structure. Preliminary results indicate that the strong effect of iodination on peptide self-assembly is general and not limited to the core sequence reported here. In fact, iodination of the amyloidogenic core sequence KLVFF (residues 16–20 of Aβ) as well as of the full-length hCT similarly promotes fibril formation efficiency and self-assembly (data not shown). Iodination of aromatic amino acids may thus develop as a general strategy for the design of new hydrogels starting from unprotected peptides and without the use of organic solvents[44,45].

The results reported in this paper are relevant in the field of the programmed synthesis of amyloid supramolecular assemblies as well as in the context of amyloid-dependent diseases. In the former field, we have demonstrated that single atom mutations of amyloidogenic core sequences by replacing hydrogen with iodine atoms greatly promote fibril formation and self-assembly. Methods promoting the formation of mature fibrils are advantageous for use, among others, as scaffolds for tissue engineering, controlled drug release, surgical reconstruction applications, microfluidic devices, biosensors and bioswitches[7–11]. As far as amyloid-dependent diseases are concerned, our results suggest that oxidative stress-induced halogenation of proteins might potentially be the triggering point for the transformation of a natively folded protein into a halogen-bonded and fibrillar malfunctioning form. Research in this direction is currently being carried out in our laboratories and will be reported elsewhere.

Methods

Reagents. Congo red, Rhodamine B, NaCl and D_2O were purchased from Sigma-Aldrich and used without further purification. Peptides with confirmed amino acid analysis (purity ≥98%), were purchased from Biopeptek (Malvern, USA). The integrity of all peptides was confirmed by ion spray mass spectrometry and the purity was determined by reverse phase high-pressure liquid chromatography.

Hydrogel preparation. Hydrogels were prepared by dissolving each peptide (15 mM for halogenated peptides and 75 mM for the wild-type peptide DFNKF) in deionized water (18.2 Mω cm) or in D_2O (≥99.9 atom % deuterium). The glass vials containing the 500 µl solutions were sealed, sonicated for 20 s, heated using a heat gun until complete dissolution of the peptides, before slow cooling to room temperature (RT) All the samples were stored at RT for 48 h before analysis.

Preparation of peptide solutions. Peptide solutions (40 µM) were freshly prepared in deionized water (18.2 Mω cm), sonicated for 20 s, and gently warmed to reach 90 °C before filtration through a 0.22 µm Millipore filter. The peptide solutions were stored in sealed vials at RT for varying time points before analysis.

X-ray crystallography. The *p*-iodo-phenylalanine was suspended in water at RT and kept in an open vial under a hood. After 6 months, colourless crystalline needles appeared. Suitable crystals for XRD measurements were analysed without further manipulation. The crystals were measured using Mo-Kα radiation on a Bruker KAPPA APEX II diffractometer with a Bruker KRYOFLEX low temperature device. The crystal structure was solved by direct method and refined against F^2 using SHELXL97[46]. Packing diagrams were generated using Mercury[47]. The non-hydrogen atoms were refined anisotropically and hydrogen atoms were fixed geometrically and refined isotropically (CCDC deposition number: 1049844).

Mixed gels preparation. Mixed hydrogels were prepared using the same method as previously described. Briefly, peptides were weighed into a glass vial and MilliQ water added. The glass vial was sealed and sonicated for 20 s before heating with a heat gun until complete solubilization of the peptides. The vials were allowed then to cool slowly to RT and stored at RT for 7 days before analysis.

Thermal stability. The vials containing the hydrogels at 15 mM were inverted and fixed at the bottom of an oil bath with stirring. The temperature was kept at 25 °C for 10 min to equilibrate the system and then gradually increased from 25 to 140 °C (at 1 °C min^{-1}). The temperature at which the gels break is reported as a range: the initial temperature corresponds to the fall of the first drop and the final one to the complete breakdown of the gel.

Rheology. Rheology experiments were performed using a TA instrument ARG2 Rheometer. A 20 mm stainless steel, parallel-plate geometry was used with a gap distance of 1,000 µm. Oscillatory frequency sweep studies were performed for a range of 0.1–100 rad s^{-1}, using a 0.5% strain. Oscillatory amplitude sweep studies were conducted from 0.01 to 100% strain with an angular frequency of 1 rad s^{-1}. The ring-cast method was used for hydrogel preparation at a peptide concentration of 15 mM. The peptide solution was sonicated for 20 s in a sealed glass vial before heating to 90 °C to afford complete dissolution of the peptide. After cooling, the solutions were transferred into ring casts of 22 mm diameter and placed in tightly sealed tissue-culture dishes for 48 h. All the measurements were repeated a minimum of three times.

Circular dichroism spectroscopy. All the CD experiments were carried out in deionized water (18.2 MΩ cm) in a 0.01 mm detachable quartz cuvette, using a JASCO J-815 CD spectrometer. Acquisitions were performed between 190 and 300 nm with a 0.1 nm data pitch, 1 nm bandwidth, 100 nm min^{-1} scanning speed and 1 s response time. All the spectra are an average of 10 scans and were corrected from a reference solution, comprising deionized water (18.2 MΩ.cm) alone. Raw data (θ, in millidegree) were subsequently converted to mean residue ellipticity ([θ] in deg cm^2 $dmol^{-1}$) for the sake of comparison, in accordance with the following formulae:

$$[\theta] = \frac{\theta}{10 \times l \times c \times (n-1)}, \quad (1)$$

where θ is the observed ellipticity in millidegree, *c* is the concentration of the sample in mol l^{-1}, $(n-1)$ is the number of peptide bonds and *l* is the path length of the cuvette in centimetre.

Confocal microscopy. Hydrogels were imaged using a Zeiss LSM 710 microscope with a He/Ne laser (λ_{ex} = 543 nm). The fluorescent dye, Rhodamine B, was incorporated into an aged hydrogel (48 h) scaffold by the addition of 10 µl of the dye solution (0.1% w/v). Following complete absorption of the dye, the sample was excited at 543 nm and emitted light recorded using the E570LP emission filter.

Congo red staining. All the samples were monitored for green birefringence using an Olympus BX50 polarizing microscope with a SensiCam PCO camera used to display and enhance images. An 80% ethanol solution saturated with NaCl and Congo red was freshly prepared before each measurement. A piece of each peptide hydrogel was placed on a glass microscope, allowed to air dry and then stained with Congo red solution. Subsequently, excess Congo red solution was blotted off the slide and the samples were analysed using both bright and polarized light.

Infrared spectroscopy (FT-IR). Infrared spectra were recorded at RT using a Nicolet iS50 FT-IR spectrometer equipped with a DTGS detector. Peptides were analysed as solutions (after heating at 100 °C to break any pre-formed fibrils) or gels at 15 mM in D_2O (for DFNKF 75 mM concentration was used). Spectra represent an average of 64 scans recorded in a single-beam mode with a 4 cm^{-1} resolution and corrected for the background. The second-derivative analyses of the spectra were performed using the Nicolet FTIR software, Omnic 9.0, with a 13-point and third polynomial order Savitzky and Golay function. Second-derivative spectra generated negative bands as compared with the original spectra, thus for comparison all the second-derivative spectra were multiplied by −1.

Raman spectroscopy. Raman spectra were acquired at RT by using a Horiba Xplora MicroRaman instrument equipped with an Olympus BX-41 Microscope. An excitation wavelength of 785 nm was used. Laser power was attenuated by neutral density filters with a final power density of ~0.07 mW μm^{-2}. The low wavenumber detection limit is 140 cm^{-1}. Each spectrum was acquired with an exposure time of 5 s over 35 cycles. The raw data were first corrected from the baseline, using the JASCO Nicolet FTIR software, Omnic 9.0, between 140 and 340 cm^{-1}. Obtained data were subsequently normalized, for the sake of comparison, and plotted using Origin Pro 8.

Atomic force microscopy. Peptide solution (10 µl, 40 µM) or diluted gel was deposited onto freshly cleaved mica surface and air dried. AFM characterizations were performed on a Veeco Dimension 5,000 Scanning Probe Microscope with a Nanoscope V controller (Digital Instruments, Inc.). All the samples were prepared on mica substrate and measured without treatment. Al-coated silicon AFM tips (NSC 15/AIBS, MikroMasch, Estonia) with a tip radius of 10 nm were used to probe the surface profiles of the films. Tapping-mode AFM imaging was used according to well-established procedures. All the images were post-treated with NanoScope Analysis 1.5 Software.

Transmission electron microscopy. High-resolution transmission electron microscopy imaging was carried out using JEM-3200Fsc field emission microscope (JEOL) operated at 300 kV in bright-field mode with Omega-type Zero-loss energy filter. The images were acquired with ULTRASCAN 4,000 CCD camera (GATAN) and with GATAN DIGITAL MICROGRAPH software, while the specimen temperature was maintained at − 187 °C. Dried TEM and ET samples were prepared by placing 4 µl of 40 µM DF(I)NKF(I) solution on 200 mesh carbon only grid (CFT200-Cu) and the excess removed with filter paper.

Electron tomography. For image alignment purposes, the TEM grids were dipped in 11-mercapto-1-undecanol ligand-coated gold particle solution ($d = 3$–10 nm) before sample deposition. ET tilt series were acquired with the SERIALEM-software package between tilt angles of ± 69°. Prealignment of tilt image series was done with IMOD and the fine alignment and cropping with JPEGANIM software package[48]. The images were binned twice to reduce noise and computation time. Maximum entropy method reconstruction scheme was carried out with MEM software package[48] on Linux cluster with regularization parameter value of $\delta = 5.0 \times 10^{-2}$. Data visualization, volumetric graphics and analyses were performed with the UCSF CHIMERA package. The tomogram was filtered with CHIMERA's Gaussian filter with 1.5 voxel width.

References

1. Knowles, T. P. J. & Buehler, M. J. Nanomechanics of functional and pathological amyloid materials. *Nat. Nanotechnol.* **6,** 469–479 (2011).
2. Selkoe, D. J. Alzheimer's disease: genes, proteins, and therapy. *Physiol. Rev.* **81,** 741–766 (2001).
3. Dobson, C. M. Protein folding and misfolding. *Nature* **426,** 884–890 (2003).
4. Selkoe, D. J. Folding proteins in fatal ways. *Nature* **426,** 900–904 (2003).
5. Chiti, F. & Dobson, C. M. Protein misfolding, functional amyloid and human disease. *Annu. Rev. Biochem.* **75,** 333–366 (2006).
6. Kreplak, L. & Aebi, U. From the polymorphism of amyloid fibrils to their assembly mechanism and cytotoxicity. *Adv. Protein Chem.* **73,** 217–233 (2006).
7. Hauser, C. A. E., Maurer-Stroh, S. & Martins, I. C. Amyloid-based nanosensors and nanodevices. *Chem. Soc. Rev.* **43,** 5326–5345 (2014).
8. Adamcik, J. & Mezzenga, R. Fibrils from a polymer physics perspective. *Macromolecules* **45,** 1137–1150 (2012).
9. Li, C. & Mezzenga, R. The interplay between carbon nanomaterials and amyloid fibrils in bio-nanotechnology. *Nanoscale* **5,** 6207–6218 (2013).
10. Mankar, S., Anoop, A., Sen, S. & Maji, S. K. Nanomaterials: amyloids reflect their brighter side. *Nano Rev.* **2,** 6032–6043 (2011).
11. Cherny, I. & Gazit, E. Amyloids: not only pathological agents but also ordered nanomaterials. *Angew. Chem. Int. Ed.* **47,** 4062–4069 (2008).
12. Griffith Jones, O. & Mezzenga, R. Inhibiting, promoting, and preserving stability of functional protein fibrils. *Soft Matter* **8,** 876–895 (2012).
13. Petkova, A. T. *et al.* A structural model for Alzheimer's β-amyloid fibrils based on experimental constraints from solid state NMR. *Proc. Natl Acad. Sci. USA* **99,** 16742–16747 (2002).
14. Metrangolo, P. & Resnati, G. Halogen versus hydrogen. *Science* **321,** 918–919 (2008).
15. Ryan, D. M., Anderson, S. B., Senguen, F. T., Youngman, R. E. & Nilsson, B. L. Self-assembly and hydrogelation promoted by F5-phenylalanine. *Soft Matter* **6,** 475–479 (2010).
16. Wang, Y., Zhang, Z., Xu, L., Li, X. & Chen, H. Hydrogels of halogenated Fmoc-short peptides for potential application in tissue engineering. *Colloids Surf. B Biointerfaces* **104,** 163–168 (2013).
17. Reches, M. & Gazit, E. Designed aromatic homo-dipeptides: formation of ordered nanostructures and potential nanotechnological applications. *Phys. Biol.* **3,** S10–S19 (2006).
18. Metrangolo, P., Pilati, T. & Resnati, G. Halogen bonding and other noncovalent interactions involving halogens: a terminology issue. *Cryst. Eng. Comm.* **8,** 946–947 (2006).
19. Ryan, D. M., Doran, T. M., Anderson, S. B. & Nilsson, B. L. Effect of C-terminal modification on the self-assembly and hydrogelation of fluorinated Fmoc-Phe derivatives. *Langmuir* **27,** 4029–4039 (2011).
20. Ryan, D. M., Anderson, S. B. & Nilsson, B. L. The influence of side-chain halogenation on the self-assembly and hydrogelation of Fmoc-phenylalanine derivatives. *Soft Matter* **6,** 3220–3231 (2010).
21. Desiraju, G. R. *et al.* Definition of the halogen bond (IUPAC Recommendations 2013). *Pure Appl. Chem.* **85,** 1711–1713 (2013).
22. Metrangolo, P. & Resnati, G. Tracing iodine. *Nat. Chem.* **3,** 260 (2011).
23. Metrangolo, P. & Resnati, G. Type II halogen··· halogen contacts are halogen bonds. *IUCrJ* **1,** 5–7 (2014).
24. Berger, G., Berger, N., Guillaud, M. H., Trouillas, J. & Vauzelle, J. L. Calcitonin-like immunoreactivity of amyloid fibrils in medullary thyroid carcinomas. An immunoelectron microscope study. *Virchows Arch. A Pathol. Anat. Histopathol.* **412,** 543–551 (1988).
25. Arvinte, T., Cudd, A. & Drake, A. F. The structure and mechanism of formation of human calcitonin fibrils. *J. Biol. Chem.* **268,** 6415–6422 (1993).
26. Steed, J. W. Supramolecular gel chemistry: developments over the last decade. *Chem. Commun.* **47,** 1379–1383 (2011).
27. Meazza, L. *et al.* Halogen-bonding-triggered supramolecular gel formation. *Nat. Chem.* **5,** 42–47 (2013).
28. Lakshmanana, A. *et al.* Aliphatic peptides show similar self-assembly to amyloid core sequences, challenging the importance of aromatic interactions in amyloidosis. *Proc. Natl Acad. Sci. USA* **110,** 519–524 (2013).
29. Shtainfeld, A., Sheynis, T. & Jelinek, R. Specific mutations alter fibrillation kinetics, fiber morphologies, and membrane interactions of pentapeptides derived from human calcitonin. *Biochemistry* **49,** 5299–5307 (2010).
30. Hauser, C. A. E. *et al.* Natural tri- to hexapeptides self-assemble in water to amyloid β-type fiber aggregates by unexpected α-helical intermediate structures. *Proc. Natl Acad. Sci. USA* **108,** 1361–1366 (2011).
31. Mishra, A. *et al.* Ultrasmall natural peptides self-assemble to strong temperature-resistant helical fibers in scaffolds suitable for tissue engineering. *Nano Today* **6,** 232–239 (2011).
32. Smith, D. K. In *Supramolecular Chemistry: From Molecules to Nanomaterials* (John Wiley and Sons, Inc., 1999).
33. Adamcik, J. *et al.* Understanding amyloid aggregation by statistical analysis of atomic force microscopy images. *Nat. Nanotechnol.* **5,** 423–428 (2010).
34. Nilsson, M. R. Techniques to study amyloid fibril formation in vitro. *Methods* **34,** 151–160 (2004).
35. Shivu, B. *et al.* Distinct β-sheet structure in protein aggregates determined by ATR-FTIR spectroscopy. *Biochemistry* **52,** 5176–5183 (2013).
36. Vasylyeva, V. *et al.* Orthogonal halogen and hydrogen bonds involving a peptide bond model. *Cryst. Eng. Comm.* **16,** 8102–8105 (2014).
37. Messina, M. T. *et al.* Infrared and Raman analyses of the halogen-bonded non-covalent adducts formed by α-ϖ-diiodoperfluoroalkanes with DABCO and other electron donors. *J. Mol. Struct.* **524,** 87–94 (2000).
38. Vliet, V. D. *et al.* Myeloperoxidase and protein oxidation in cystic fibrosis. *Am. J. Physiol. Lung Cell. Mol. Physiol.* **279,** L537–L546 (2000).
39. Hazen, S. L. 3-chlorotyrosine, a specific marker of myeloperoxidase? Catalyzed oxidation is markedly elevated in low density lipoprotein isolated from human atherosclerotic intima. *J. Clin. Invest.* **99,** 2075–2075 (1997).
40. Gaut, J. P. Neutrophils employ the myeloperoxidase system to generate antimicrobial brominating and chlorinating oxidants during sepsis. *Proc. Natl Acad. Sci. USA* **98,** 11961–11961 (2001).
41. Wu, W. *et al.* Eosinophils generate brominating oxidants in allergen-induced asthma. *J. Clin. Invest.* **105,** 1455–1463 (2000).
42. Mazzulli, J. R., Hodara, R., Lind, S. & Ischiropoulos, H. Oxidative stress and protein deposition diseases in protein misfolding, aggregation, and conformational diseases. *Prot. Rev.* **4,** 123–133 (2006).
43. Ryan, D. M., Doran, T. M. & Nilsson, B. L. Complementary π-π interactions induce multicomponent coassembly into functional fibrils. *Langmuir.* **27,** 11145–11156 (2011).
44. Gazit, E. Searching sequence space. *Nat. Chem.* **7,** 14–15 (2015).
45. Frederix, P. W. J. M. *et al.* Exploring the sequence space for (tri-)peptide self-assembly to design and discover new hydrogels. *Nat. Chem.* **7,** 30–37 (2015).
46. Sheldrick, G. M. A short history of SHELX. *Acta Crystallogr. Sect. A* **64,** 112–122 (2008).
47. Macrae, C. F. *et al.* Mercury CSD 2.0 - new features for the visualization and investigation of crystal structures. *J. Appl. Cryst.* **41,** 466–470 (2008).
48. Engelhardt, P. Electron tomography of chromosome structure. *Encyclopedia of analytical chemistry* (2006).
49. Sirimulla, S., Bailey, J. B., Vegesna, R. & Narayan, M. Halogen interactions in protein–ligand complexes: implications of halogen bonding for rational drug design. *J. Chem. Inf. Model.* **53,** 2781–2791 (2013).
50. Lu, Y. *et al.* Triangular halogen trimers. A DFT study of the structure, cooperativity, and vibrational properties. *J. Phys. Chem. A* **109,** 11956–11961 (2005).

Acknowledgements

The European Research Council is acknowledged for the Starting Grant ERC-2012-StG_20111012 FOLDHALO (Grant Agreement Number 307108) to P.M. Dr Claudia Pigliacelli (Politecnico di Milano) is acknowledged for help with the TEM experiment of Fig. 3f.

Author contributions

A.B. and L.P. performed most of the peptide self-assembly work. L.S. performed the IR spectroscopy and ionic strength-pH-dependence experiments. N.H. measured the samples by AFM. J.S.H. measured the samples by TEM. L.C. and G.T. performed the single crystal X-ray analysis. G.G. and L.V. performed the Raman spectroscopy. P.M., R.M., O.I., and G.R. conceived the experiments and contributed to results discussion. All the authors contributed to manuscript writing.

Additional information

Accession codes: The X-ray crystallographic coordinates for the structure reported in this study have been deposited at the Cambridge Crystallographic Data Centre (CCDC), under deposition number 1049844. These data can be obtained free of charge from The Cambridge Crystallographic Data Centre via www.ccdc.cam.ac.uk/data_request/cif.

Competing financial interests: The authors declare no competing financial interests.

11

De novo branching cascades for structural and functional diversity in small molecules

Miguel Garcia-Castro[1], Lea Kremer[1,2], Christopher D. Reinkemeier[1], Christian Unkelbach[2], Carsten Strohmann[2], Slava Ziegler[1], Claude Ostermann[3] & Kamal Kumar[1,2]

The limited structural diversity that a compound library represents severely restrains the discovery of bioactive small molecules for medicinal chemistry and chemical biology research, and thus calls for developing new divergent synthetic approaches to structurally diverse and complex scaffolds. Here we present a *de novo* branching cascades approach wherein simple primary substrates follow different cascade reactions to create various distinct molecular frameworks in a scaffold diversity phase. Later, the scaffold elaboration phase introduces further complexity to the scaffolds by creating a number of chiral centres and incorporating new hetero- or carbocyclic rings. Thus, employing *N*-phenyl hydroxylamine, dimethyl acetylenedicarboxylate and allene ester as primary substrates, a compound collection of sixty one molecules representing seventeen different scaffolds is built up that delivers a potent tubulin inhibitor, as well as inhibitors of the Hedgehog signalling pathway. This work highlights the immense potential of cascade reactions to deliver compound libraries enriched in structural and functional diversity.

[1] Max-Planck-Institut für Molekulare Physiologie, Abteilung Chemische Biologie, Otto-Hahn-Strasse 11, 44227 Dortmund, Germany. [2] Technische Universität Dortmund, Fakultät für Chemie und Chemische Biologie, Otto-Hahn-Strasse 6, 44227 Dortmund, Germany. [3] Compound Management and Screening Center (COMAS), Max-Planck-Institut für Molekulare Physiologie, Otto-Hahn-Strasse 11, 44227 Dortmund, Germany. Correspondence and requests for materials should be addressed to K.K. (email: kamal.kumar@mpi-dortmund.mpg.de).

Structural diversity has a profound impact on the performance of a compound collection exposed to biological screenings[1–3]. Probe and drug discovery research, therefore, beseech quality-based compound libraries rich in structural diversity[4,5]. The latter in turn is primarily determined by the number of diverse scaffolds or chemotypes that represent a compound library[6–8]. Consequently, new synthetic challenges have emerged aiming at divergent access to structurally distinct and complex scaffolds[9–11]. Different approaches have been developed to incorporate structural diversity to a compound collection, for instance, in the build-couple-pair strategies[12,13], folding[14] and branching pathways[15–17], structural variations either in the building blocks or the reacting partners of common substrates derive the formation of new scaffolds. Synthesis of natural product scaffold-based[18–21] compound libraries either employs accessible complex natural products or their derivatives for generating new and complex scaffold structures or build up compound libraries around privileged scaffolds[1,22]. However, most of the above mentioned synthesis designs require carefully functionalized substrates, as well as their reacting partners and often deliver structural diversity in a compound library at the cost of tedious multistep synthesis protocol. Nature displays it's amazing ability to assemble a limited pool of simple building blocks into structurally and functionally diverse natural products[23,24]. For instance, terpenes that represent a large class of natural products are formally generated from only one biosynthetic unit[25]. Infact, terpenes and sesquiterpenes represent interesting examples of *de novo* biosynthesis designs where compound collections are generated in two important phases; the first cyclase phase builds up scaffolds from simple acyclic substrates with the help of cyclase enzymes and the next phase elaborates these scaffolds, for instance, with oxidative modifications, to generate a number of diversely functionalized molecules (Fig. 1a)[25,26].

In contrast to divergent biosynthetic designs, laboratory syntheses targeting diverse scaffolds either choose the substrates already equipped with the ring systems desired in different products or follow multistep and tedious synthetic validations for every target scaffold before producing a library of molecules. Cascade or domino reaction sequences are highly efficient synthetic tools that rapidly build up molecular complexity[27,28]. However, their potential in scaffold diversity synthesis remains underexplored. The 'branching cascades' strategy is a scaffold diversity synthesis approach wherein common precursors follow different cascade or domino reactions and provide structurally distinct scaffolds for library synthesis[29]. Inspired by the *de novo* biogenesis of secondary metabolites, we envisioned that the *de novo* branching cascades could be designed to transform simple acyclic substrates into appreciable scaffold diversity (Fig. 1b). In the absence of enzyme weaponry that nature exploits in biogenesis of diverse natural products, the *de novo* scaffold diversity synthesis with cascade reactions is a formidable synthetic challenge that seeks careful reaction design. Herein we report our efforts towards scaffold diversity synthesis with a *de novo* branching cascades strategy.

Taking cognizance of the biosynthetic library design, we planned to first establish a scaffold diversity phase (SD phase) by transforming simple acyclic substrates into diverse ring systems via branching cascades. While the same set of simple substrates generates different scaffolds, molecular properties like molecular weight and clogP of the resulting distinct scaffolds remain within a threshold limit and can be controlled by small structural variations in the substrates. Therefore, there is a large scope to further modify the generated scaffolds and their appended functionalities in subsequent scaffold elaboration phase (SE phase) and deliver a compound collection rich in structural diversity and molecular complexity (Fig. 1b). To establish the principle, a *de novo* cascade reaction design is chosen in which simple acyclic substrates, *N*-phenylhydroxylamine, acetylenedicarboxylates and allene esters undergo branching cascades reactions under different reaction conditions to provide seven distinct scaffolds in SD phase. Four of these scaffolds are further

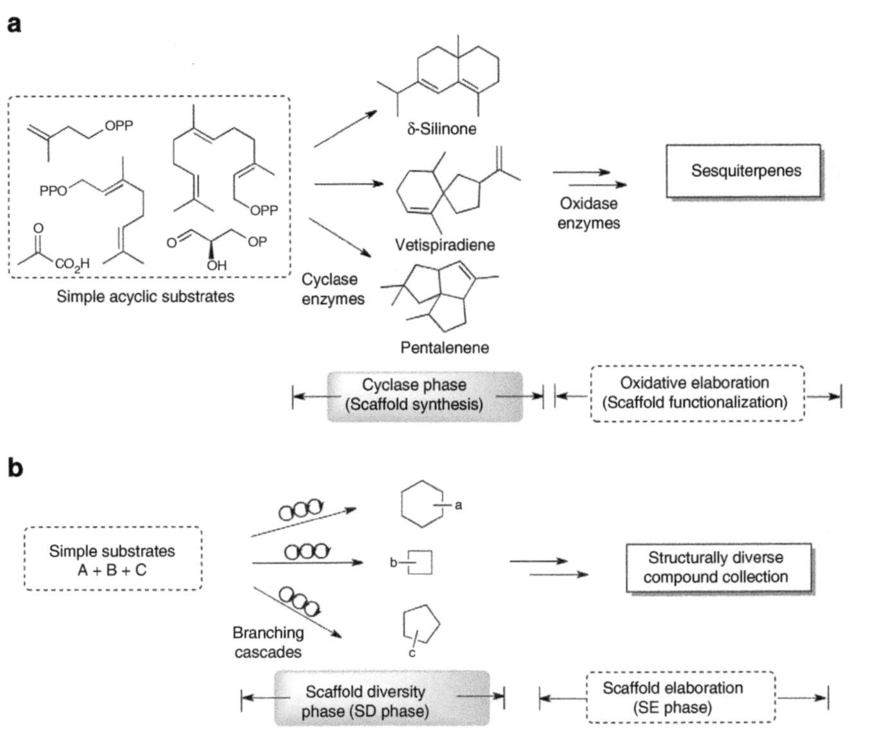

Figure 1 | Nature's divergent biogenesis strategy inspires synthetic planning to scaffold diversity. (**a**) Biogenesis of diverse terpenes from simple substrates; (**b**) a *de novo* branching cascades strategy to build a structurally diverse compound collection.

elaborated in SE phase to yield a collection of 61 molecules spread over 17 molecular frameworks. Cell-based screenings identify potent and novel bioactive molecules, representing four different structural classes, as inhibitors of tubulin polymerization or hedgehog signalling, and thereby validate the notion that functional diversity is bequeathed to a compound collection by scaffold diversity.

Results

Planning of the *de novo* branching cascades approach. Targeting a compound collection of diverse azaheterocycles and employing simple substrates lacking any azaring system, we focused on a cascade reaction sequence that could provide a suitable point of divergence, that is, an intermediate that could be transformed into distinct chemotypes (azaheterocycles) by merely modulating the reaction conditions. *N*-phenyl nitrones (**1**) undergo [3 + 2] cycloadditions with allenic esters (**2**) and the corresponding isoxazolidines (**3**) follow a pericyclic rearrangement leading to azepinones (**5**, Fig. 2)[30–34]. Azepinone **5** beholds different reactive functionalities like a secondary amine and a β-ketoester that can be exploited in branching cascade approach to build distinct azaheterocyclic frameworks. Moreover, similar azepinones are known to rearrange and yield vinyl indoles (**6**)[35]. Cascade reaction sequences, thus, can be designed to explore various reactive functionalities in either the azepinone **5** or vinyl indole **6** to generate new and distinct scaffolds (Fig. 2).

Scaffold diversity phase. With this planning, reaction conditions were optimized (Supplementary Table 1) for an *in situ* generation of the nitrone (**10**) by addition of phenyl hydroxylamine (**7**) to dimethyl acetylenedicarboxylate (DMAD) (**8**), and followed by a [3 + 2] cycloaddition with allene ester (**2**), leading to azepinone **11** (*de novo* cascade I, Fig. 3a). DMAD facilitates *in situ* generation of nitrone[36,37] **10** and provides reactive functionalities to build new ring systems in SD phase, as well as to elaborate scaffolds later in SE phase. Figure 3 depicts building of SD phase in a *de novo* branching cascades approach from primary substrates (indicated by blue arrows), as well as from **11** (indicated by red arrows). Formation of scaffolds **13–17** from azepinone **11** not only confirmed its intermediacy in different cascade reaction sequences, but also provided an efficient access to scaffolds **14** and **17**, which were rather difficult to access directly from primary substrates (Fig. 3b).

Azepinone **11** itself is an interesting scaffold rich in sp^3 character. Using differently substituted allene esters (**2**), the first cascade reaction sequence (cascade I) under optimized reaction conditions (Supplementary Table 1) delivered highly substituted azepinones (**11a–o**). Except for the α-substituted allene esters, which provided **11n–o** (R^4 = Me and prenyl) with two consecutive quaternary centres in low yields, azepinones **11a–m** were obtained in moderate to high yields (40–76%; Fig. 3b; Supplementary Fig. 1). Interestingly, just changing the reaction condition in cascade I from 80 °C to room temperature provided a different scaffold **12**, that is, a dihydroisoxazole (cascade II; Fig. 3b; Supplementary Fig. 1).

To find suitable reaction conditions that could transform the primary substrates into diverse scaffolds via azepinone as intermediate, **11a** (R^1–R^2, R^4 = H, R^3 = Et) was screened against different Lewis and Brønsted acids. While Lewis acids like $BF_3.OEt_2$ or TMSOTf provided only fragmentation products of the azepinone, $AlCl_3$ yielded an inseparable complex mixture of products (Supplementary Table 2). In a successful case (cascade III), catalytic trifluoroacetic acid (TFA) transformed azepinone **11a** into allyl indole (**13a**; Fig. 3b) at high temperature (100 °C). After slight modification in this reaction condition, we were able to synthesize allyl indole **13** directly from primary substrates (cascade IV; Fig. 3b). Allyl indole **13** supports an electron-poor olefin appended to an electron-rich indole, and therefore is a potential building block for higher-order polycyclic indoles. This opportunity was realized by overnight heating of azepinones **11** in toluene with 20 mol% of TFA that would have led to allyl indole (**13**), and followed by treatment with sodium hydroxide (NaOH) at room temperature in dioxane that provided us benzo[*b*]indolizine molecules (**14**) embodying a naturally occurring scaffold[38,39] in high yields (cascade V; Fig. 3b). In a separate reaction, treatment of allyl indoles **13** with 1 M NaOH also yielded benzo[*b*]indolizines **14** in excellent yields (80–94%). However, attempts to develop a *de novo* access to **14** directly from primary substrates (**2** + **7** + **8**) did not succeed in this case.

Azepinone **11** contains more than one nucleophilic site in proximity to different carbonyl functions. Therefore, various modes of cyclization and/or ring-distortion reactions can be realized in the presence of a suitable base, leading to different heterocyclic systems. With this hope, we resorted to a reaction screen of azepinone **11a** employing various bases and in different solvents (Supplementary Table 3). The reaction screening revealed various levels of reaction control, that is, effects of different reagents, solvents and temperature in directing substrates to follow a preferred reaction sequence among many possible pathways. For instance, overnight stirring of **11a** with potassium carbonate in dimethylformamide (DMF) at room temperature provided a novel scaffold **15** as the only product in low yield (< 15%). Increasing the temperature (100 °C) led to the formation of **15** along with another scaffold **16** in

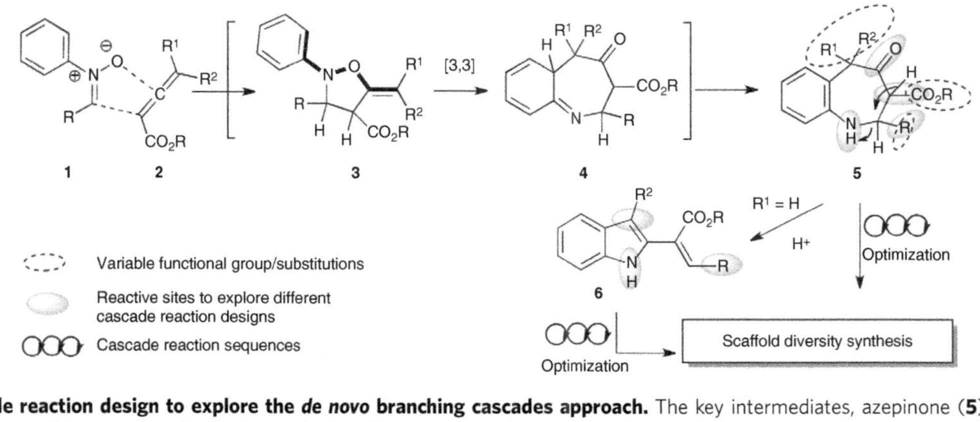

Figure 2 | A cascade reaction design to explore the *de novo* branching cascades approach. The key intermediates, azepinone (**5**) and vinyl indole (**6**) that could be transformed into structurally distinct molecular frameworks under cascade reaction conditions.

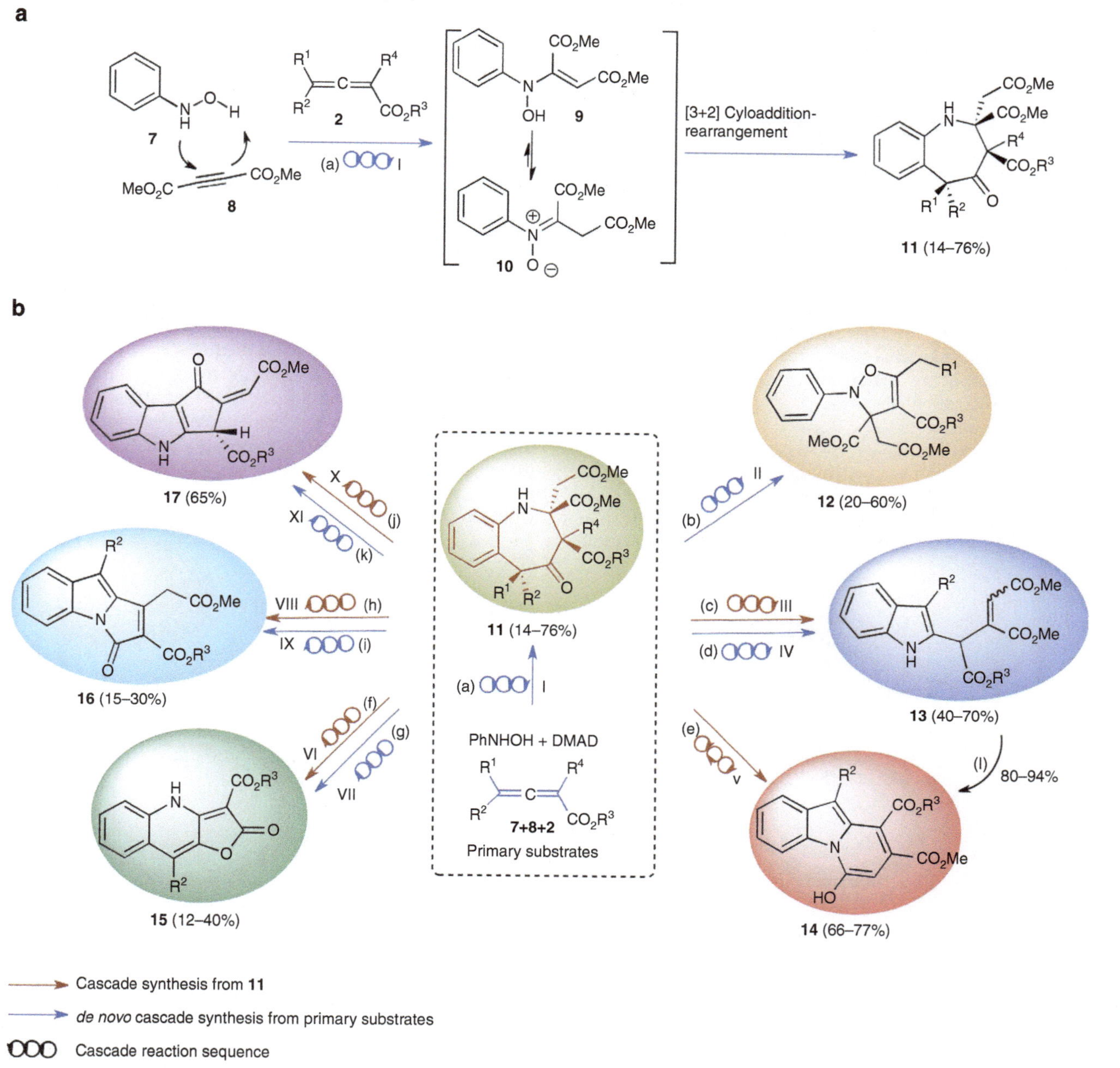

Figure 3 | Building scaffold diversity phase (SD phase) by *de novo* branching cascades. (**a**) Cascade synthesis of azepinone **11** from primary substrates. (**b**) SD phase; *de novo* branching cascades from primary substrates (blue arrows) and branching cascades from azepinone (red arrows) leading to diverse scaffolds, for details see Supplementary Methods. (a) **7** + **8** in MeCN, 10 min, room temperature (RT), then **2**, 80 °C, 8 h, 14-76%; (b) **7** + **8** in MeCN, 10 min, RT, then **2**, RT, 8 h, 20-60%; (c) TFA (20 mol%) in toluene, 100 °C, 6 h, 60-89%; (d) **7** + **8** in MeCN, 10 min, RT, then **2**, 80 °C, 8 h, then TFA (20 mol%) in toluene, 100 °C, 6 h, 40-70%; (e) TFA (20 mol%) in toluene, 100 °C, 6 h, then NaOH(aq) 1 M, 1,4-dioxane, RT, 6 h and neutralization with HCl (3 M) until pH = 6-7, 66-77%; (f) K_2CO_3, DMF, 100 °C, 6 h, 25-76%; (g) **7** + **8** in DMF, 10 min, RT, then **2**, 80 °C, 8 h, then K_2CO_3, 100 °C, 6 h, 12-40%; (h) KOAc, EtOH, 60 °C, 6 h, 69% or K_2CO_3, DMF, 100 °C, 6 h, 15-32%; (i) **7** + **8** in DMF, 10 min, RT, then **2**, 80 °C, 8 h, then K_2CO_3, 100 °C, 6 h, 15-30%; (j) NaH, DMF, 0 °C, 30 min, 65%; (k) **7** + **8** in DMF, 10 min, RT, then **2**, 80 °C, 8 h, then KOAc, EtOH, 60 °C, 4%; and (**l**) NaOH(aq) 1 M, 1,4-dioxane, RT, 6 h and neutralization with HCl 3 M until pH = 6-7, 80-94%.

appreciable yields. While some other bases attempted in the reaction screening either led to incomplete reactions or heavy decomposition (Supplementary Table 3), this reaction condition was adapted in *de novo* cascade synthesis from primary substrates delivering substituted scaffolds **15** and **16** in acceptable yields (cascades VII and IX; Fig. 3b; Supplementary Table 3).

To our delight, treatment of azepinone **11a** ((R^1–R^2, R^4 = H, R^3 = Et) with NaH in DMF at 0 °C (decomposition at room temperature) yielded another novel scaffold **17** in high yield (65%, cascade X). Compound **17** embodies an indole-fused cyclopentanone supporting an exocyclic olefin (Fig. 3b) that apparently infers C3-indole cyclization to one of the ester function derived from DMAD. Unfortunately, the corresponding *de novo* synthesis of **17** from primary substrates was very low yielding (4% via cascade XI) and could not be improved. Overall, the *de novo* branching cascades successfully delivered five scaffolds (**11–13** and **15–16**) and two (**14** and **17**) were obtained from azepinone **11**. A compound collection of 46 molecules was thus generated in the SD phase (Supplementary Fig. 1).

The proposed reaction mechanisms for the *de novo* cascades I, II and IV leading to the formation of azepinones **11**, isoxazolidines (**12**) or allyl indoles (**13**, Fig. 4a), respectively, are supported by various reported cycloaddition reactions between *N*-phenyl nitrones and allenes[30–32]. Notably, *de novo* cascade II to **12** does not pass through azepinone and the initial [3 + 2] cycloadducts **18** (Fig. 4a) preferred energetically favoured olefinic isomerization to **12** over a sigmatropic rearrangement to azepinone **11** that requires higher temperature. Formation of allyl indoles (**13**) apparently occurs via isomerization of vinyl indoles (**20**) formed from **11** under optimized reaction conditions (Fig. 4a). Benzo[*b*]indolizines (**14**) were synthesized under the reaction conditions that first generate an allyl indole **13**, which cyclizes on treatment with a base to yield **14** (cascade V; Fig. 4a). Although we also expected the formation of scaffold **21** by *N*-indole cyclization to C2-ester in allyl indoles **13**, however, no trace of **21** was detected (Fig. 4a).

Structural features of **17** (formed under basic reaction conditions from **11**) were suggestive of a C3-indole cyclization to one of the ester moieties in allyl indole **13**. However, treatment of **13** with NaH in a separate reaction did not yield indolocyclopentanone **17** (Fig. 4a). We assume that treatment of **11** with NaH generates a benzylic anion (**23**) via vinylic aminal **22** that adds to one of the esters regioselectively before aromatizing to yield **17** (Fig. 4a, cascade XI).

Formation of scaffolds **15** and **16** (for X-ray analysis see Supplementary Figs 9–10) was rather unexpected. For the synthesis of **15** (cascade VII; Fig. 4b), we propose that azepinone **11** undergoes a retro-Michael reaction generating an anilide **24** that adds to highly electron-poor Michael acceptor in **24** to give an enolate intermediate **25**. Addition of enolate to ethyl ester moiety generates a cyclopropane that opens concomitantly leading to formation of enolate **27** that cyclizes to form tricyclic lactone **28**. Electrocyclic ring opening in the latter forms the benzylic vinylogous amide **29** that enolizes under basic reaction conditions and cyclizes to provide the scaffold **15**.

Scaffold **21** (expected but not formed) and **16** are structurally similar but support the ester and methyl acetate functions (arising from allene ester and DMAD, respectively) on different positions (Fig. 4a,b). Mechanistically, we assume that addition of enolate in **30** to the methyl ester on the quaternary centre generates the intermediate **31** that undergoes 1,2-acyl shift to yield anionic malenoate **32**. Ketone formation followed by concomitant cyclopropane and azepane ring opening forms anilide **33** that condenses with keto moiety to form vinyl indole **34**. *N*-indole cyclization in **34** then generates the final adducts **16** (cascade IX; Fig. 4b).

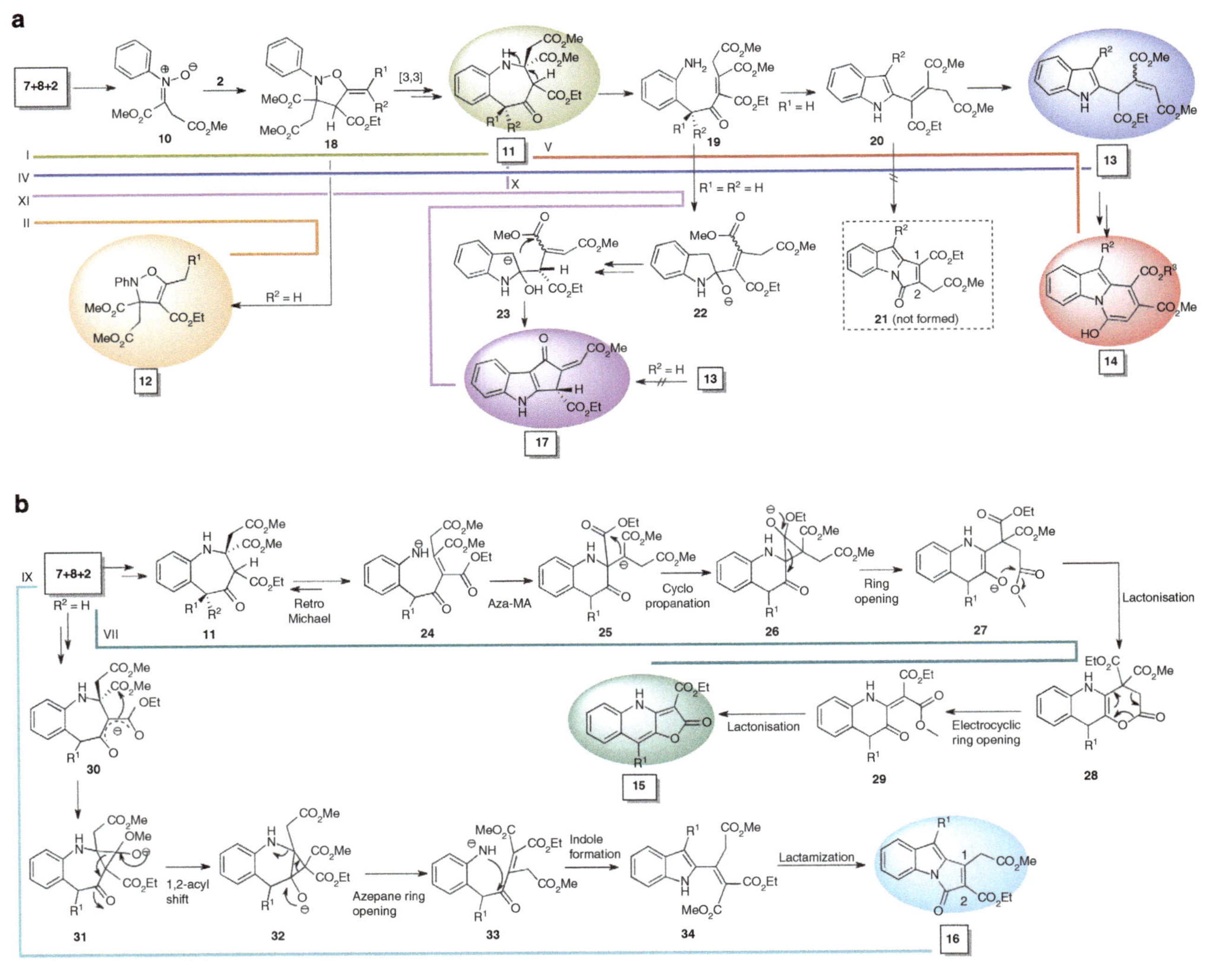

Figure 4 | Proposed cascade routes to diverse scaffolds in SD phase. (**a**) Mechanistic proposal for the cascade synthesis of scaffolds **11-14**, **17** and (**b**) for **15-16**. Compound numbers shown in boxes depict isolated molecules. Differently coloured thick bars display different cascade reaction sequences leading to diverse scaffolds in SD phase.

Scaffold elaboration phase. The *de novo* cascade reactions were designed to adorn scaffolds in SD phase with functional groups that can be exploited to accomplish further complex scaffolds in terms of embodying ring systems, percentage of sp^3 carbons and number of chiral centres, and thus furnishing greater structural diversity to the library. To provide a proof of this principle, we chose four scaffolds generated by *de novo* cascade design, including structurally flat scaffolds (**15–16**) for structural elaboration in SE phase (Fig. 5). Following simple one-step transformations, azepinone **11** delivered further complex benzazepane scaffolds **35–37**. Treatment of **11** with ammonium fluoride yielded azepinamine **35**. To our delight, two different chemo- and diastereoselective reductions of phenacetone in **11** provided hydroxyl azepane **36** supporting three consecutive chiral centres in high yield and a tricyclic benzazepane lactone **37** that was isolated in low yield as single diastereoisomers (Fig. 5).

Birch reduction transformed the novel but flat scaffold **15** into sp^3-rich framework **38** in high yields (Supplementary Fig. 1). Another flat but novel scaffold **16** was transformed into six distinct scaffolds **39–44** carrying greater sp^3 character and number of chiral centres (Fig. 5). A stereoselective palladium catalysed hydrogenation of **16** yielded single diastereoisomer of 1,2-dihydro-3*H*-pyrrolo[1,2-*a*]indol-3-one (**39**). Employing higher pressure of the hydrogen in this reaction provided, along with **39**, another sp^3-rich scaffold, tetrahydro-3*H*-pyrrolo[1,2-*a*]-indol-3-one (**40**), although as a mixture of diastereomers (dr ~ 5:1, Fig. 5). Oxidation of **16** with oxone-generated scaffold **41** presumably via an intermediary epoxide that opened up under basic condition affording *Z*-**41** as single diastereomer in high yield (Fig. 5, see Supplementary Fig. 53 for the nuclear overhauser effect spectroscopy (NOESY) experiment to confirm relative stereochemistry in **41**).

Scaffold **16** beholds a tetrasubstituted electron-poor olefin that might be explored in different cycloaddition/annulation reactions and could generate two consecutive quaternary centres. However, tetrasubstituted olefins could provide steric resistance to reacting partners. Gratifyingly, scaffold **16** proved to be a nice dipolarophile for various dipolar cycloaddition reactions. A dipolar cycloaddition of **16** with an *in situ* generated azomethine ylide from *N*-methoxymethyl-*N*-(trimethylsilylmethyl)benzylamine[40] led to tetracyclic indole scaffold **42** in high yields (71–81%) and with excellent stereoselectivity. Another *in situ* generated dipole, the nitrile iminoester[41], in a low conversion reaction with **16** afforded cycloadduct **43** in a complete regio- and stereoselective manner. Phosphine-catalysed [3 + 2] cycloaddition of **16** with the zwitterion generated from allene ester (**2**, R^1–R^2, R^4 = H, R^3 = Et) went smoothly and in a regio- and stereoselective manner to provide adduct **44** in moderate yield (Fig. 5)[42]. Notably, this

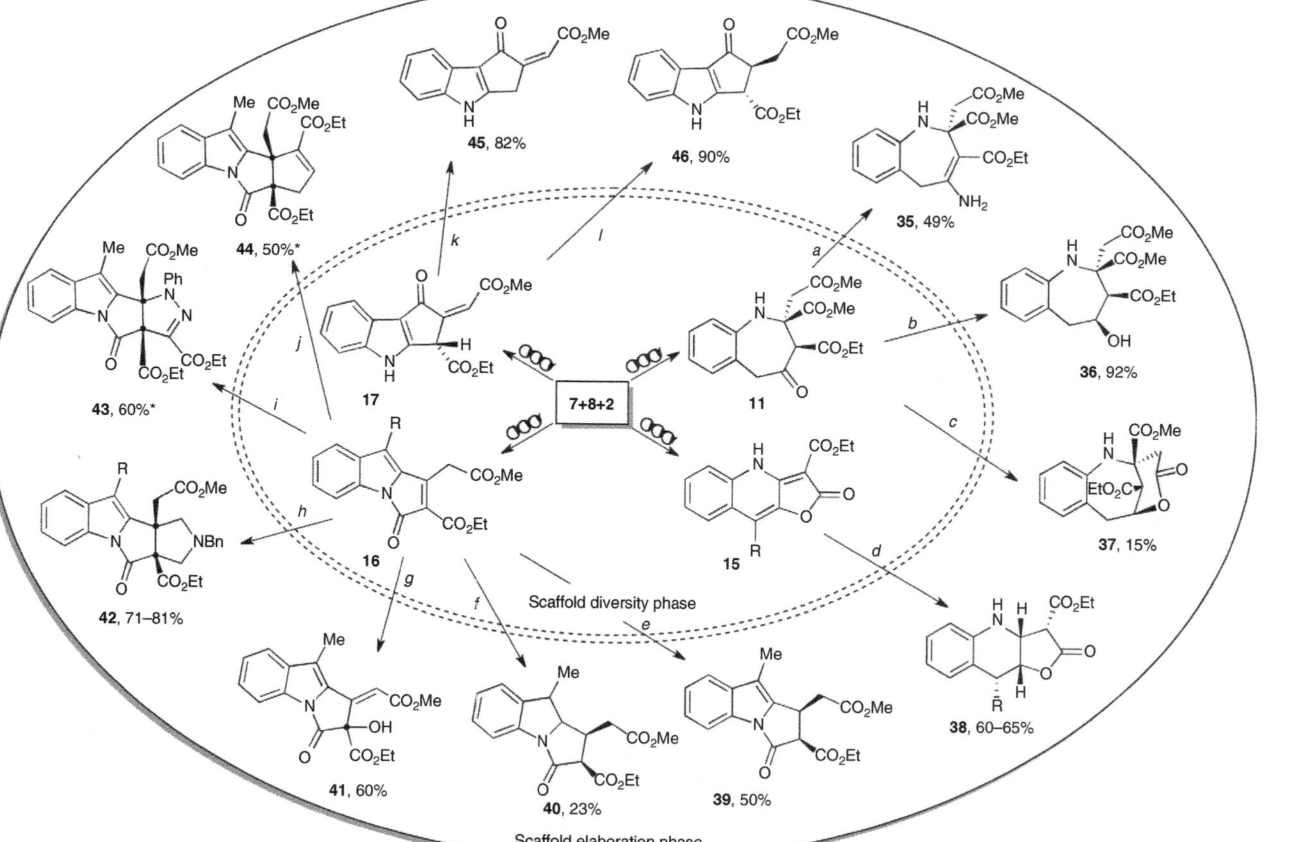

Figure 5 | SD and SE phases in *de novo* branching cascades strategy. For details, see Supplementary Methods. (a) NH_4F, MeOH, room temperature (RT), 8 h, 49%; (b) BH_3-morpholine, toluene, RT, 6 h, 92% (single diastereomer); (c) DIBAL-H 1.0 M, DCM, from −78 °C to 0 °C, 1 h, 15% (single diastereomer); (d) Na, $NH_{3(liq)}$, THF, −78 °C, 15 min, 60–65% (dr = 2.3-4.0:1); (e) 1 bar H_2, MeOH, RT, Pd-C, 50% of **39** (single diastereomer); (f) 8 bar H_2, MeOH, RT, Pd-C, 34% of **39** and 23% of **40** (dr = 4.9:1); (g) Oxone, K_2CO_3, MeCN-H_2O, RT, 1 h, 60% (single diastereomer); (h) *N*-Methoxymethyl-*N*-(trimethylsilylmethyl)benzylamine, TFA (1.2 eqv.), toluene, RT, 8 h, 71–81% (single diastereomer for **42a**); (i) Hydrazonoyl chloride, TEA (2.0 eqv.), toluene, 70 °C, 8 h, 60% (*brsm, single diastereomer); (j) allene ester (**2a**), tris(4-methoxyphenyl)-phosphine (40 mol%), toluene, RT, 8 h, 50% (*brsm, single diastereomer); (k) TEA(3.5 eqv.), toluene, reflux, 8 h, 82% (*E*:*Z* = 1:4); and (l) 1 bar H_2, Pd-C, MeOH, RT, 12 h, 90% (single diastereomer). Dots in **41–44** mark the new quaternary centres generated from scaffold **16**.

phosphine-catalysed transformation is among the rare cases of dipolar annulations of allene-derived zwitterions with tetrasubstituted olefins[43,44]. All cycloadducts formed in above cases, that is, **42–44** behold two consecutive quaternary chiral centres. Elaboration of the fourth scaffold **17** involved a base-mediated selective decarboxylation to yield scaffold **45** and a Pd-catalysed hydrogenation to yield scaffold **46** as a single diastereoisomer in excellent yields (Fig. 5). Functional groups like esters or amines present in the scaffolds from SD and SE phases can be utilized to generate a suitable number of compounds adequately representing each scaffold in the compound library.

Scaffold diversity and molecular properties analysis. Scaffold diversity synthesis outlined in Figures 3 and 5 provided a collection of 61 molecules based on 17 distinct molecular frameworks (Fig. 6a; Supplementary Fig. 1). Unlike the conventional multistep synthesis of the complex molecules, each distinct scaffold in SD and SE phases was a one-step product that could be easily purified by silica gel column chromatography providing molecules in sufficient amounts for a wide range of biological screenings (see Supplementary Methods). Some of the cascade reactions in SD phase were performed at 1–2-g scale providing enough amounts of scaffolds for further elaborations. For instance, *de novo* cascades I, VII and IX provided differently substituted **11**, **15** and **16** in appreciable yields (18–76%) at 1–2-g scale. The fact that no combinatorial synthesis step such as amide synthesis, reductive aminations or coupling reactions etc. were applied to the scaffolds, the potential to build a large and structurally diverse library with this approach is remarkably high.

Although the molecular architectures of the 17 scaffolds in this compound collection are clearly distinctive, the scaffold diversity was quantified by applying similarity metrics to 17 Bemis–Murcko frameworks (Fig. 6a)[45]. To this end, Tanimoto coefficients[46] were generated using extended connectivity fingerprints (ECFP_6, for further details see Supplementary Fig. 2)[47]. Figure 6b clearly depicts that most of the 17 scaffolds are largely distinct from one another (from 1.0 to 0, 1.0 for being the same scaffold). Moreover, the compound collection is represented by ring systems found in natural products, drugs and some unprecedented azaheterocycles, and thus covers biologically relevant as well as novel chemical space.

Substitutions on substrates directly influence the physical properties of library members. In the *de novo* branching cascade approach, simple acyclic and low molecular weight substrates

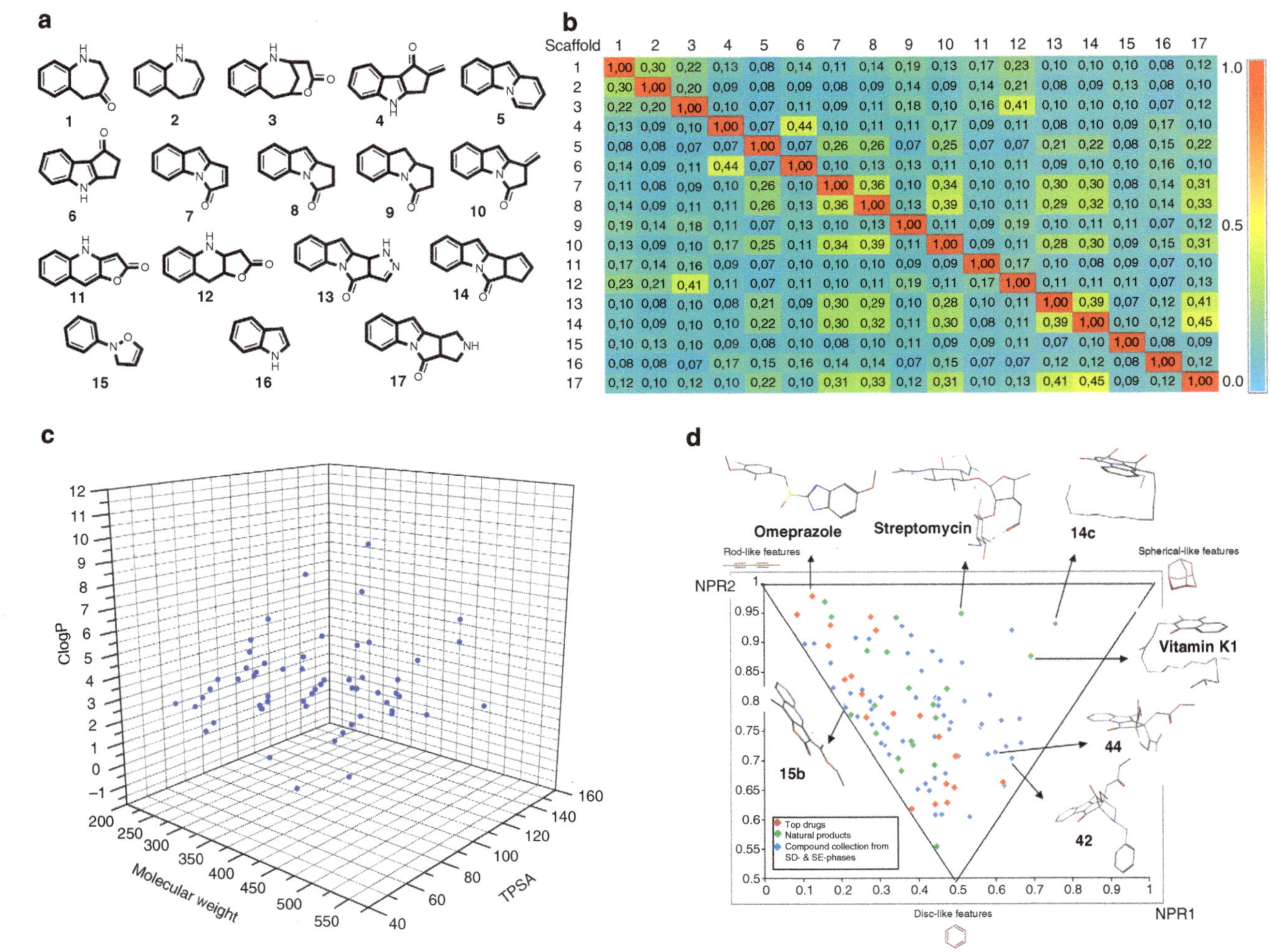

Scaffold	1	2	3	4	5	6	7	8	9	10	11	12	13	14	15	16	17
1	1,00	0,30	0,22	0,13	0,08	0,14	0,11	0,14	0,19	0,13	0,17	0,23	0,10	0,10	0,10	0,08	0,12
2	0,30	1,00	0,20	0,09	0,08	0,09	0,08	0,09	0,14	0,09	0,14	0,21	0,08	0,09	0,13	0,08	0,10
3	0,22	0,20	1,00	0,10	0,07	0,11	0,09	0,11	0,18	0,10	0,16	0,41	0,10	0,10	0,10	0,07	0,12
4	0,13	0,09	0,10	1,00	0,07	0,44	0,10	0,11	0,11	0,17	0,09	0,11	0,08	0,10	0,09	0,17	0,10
5	0,08	0,08	0,07	0,07	1,00	0,07	0,26	0,26	0,07	0,25	0,07	0,07	0,21	0,22	0,08	0,15	0,22
6	0,14	0,09	0,11	0,44	0,07	1,00	0,10	0,13	0,13	0,11	0,10	0,11	0,09	0,10	0,10	0,16	0,10
7	0,11	0,08	0,09	0,10	0,26	0,10	1,00	0,36	0,10	0,34	0,10	0,10	0,30	0,30	0,08	0,14	0,31
8	0,14	0,09	0,11	0,11	0,26	0,13	0,36	1,00	0,13	0,39	0,10	0,11	0,29	0,32	0,10	0,14	0,33
9	0,19	0,14	0,18	0,11	0,07	0,13	0,10	0,13	1,00	0,11	0,09	0,19	0,10	0,11	0,11	0,07	0,12
10	0,13	0,09	0,10	0,17	0,25	0,11	0,34	0,39	0,11	1,00	0,09	0,11	0,28	0,30	0,09	0,15	0,31
11	0,17	0,14	0,16	0,09	0,07	0,10	0,10	0,10	0,09	0,09	1,00	0,17	0,10	0,08	0,09	0,07	0,10
12	0,23	0,21	0,41	0,11	0,07	0,11	0,10	0,11	0,19	0,11	0,17	1,00	0,11	0,11	0,11	0,07	0,13
13	0,10	0,08	0,10	0,08	0,21	0,09	0,30	0,29	0,10	0,28	0,10	0,11	1,00	0,39	0,07	0,12	0,41
14	0,10	0,09	0,10	0,10	0,22	0,10	0,30	0,32	0,11	0,30	0,08	0,11	0,39	1,00	0,10	0,12	0,45
15	0,10	0,13	0,10	0,09	0,08	0,10	0,08	0,10	0,11	0,09	0,09	0,11	0,07	0,10	1,00	0,08	0,09
16	0,08	0,08	0,07	0,17	0,15	0,16	0,14	0,14	0,07	0,15	0,07	0,07	0,12	0,12	0,08	1,00	0,12
17	0,12	0,10	0,12	0,10	0,22	0,10	0,31	0,33	0,12	0,31	0,10	0,13	0,41	0,45	0,09	0,12	1,00

Figure 6 | Diversity and molecular properties analysis of compounds created in SD and SE phases by *de novo* branching cascades approach. (**a**) Structures of 17 diverse scaffolds that make up the core frameworks in the compound collection. (**b**) Relative tanimoto similarity coefficients for 17 scaffolds (1.0 represents perfect similarity and 0.0 totally distinct scaffolds), for a tanimoto-matrix analysis with different connectivity (see Supplementary Fig. 2). (**c**) Molecular properties of the compounds generated in SD and SE phases. (**d**) Principal moment of inertia (PMI) plot. The molecular shape of the compounds generated in SD and SE phases with branching cascade approach (blue squares) and comparison with 20 natural products (green square) and 20 drugs (red squares).

provide scaffold diversity, and thereby keep an inherent check over physical properties of the library members. A major portion of the generated library possesses molecular properties within Lipinski's limits[48] as far as molecular weight, polar surface area and clog*P* values are concerned. The only molecules that were generated from allene esters supporting lipophilic alkyl chains at γ-position displayed the expected deviations (Fig. 6c; Supplementary Table 4). In branching cascades approach, some of the reaction sequences were apparently directed by low-energy pathways leading to thermodynamically stable aromatic and flat products. Introduction of greater sp^3 character to the library members enhances the chance of a bioactive molecule to specifically interact with a protein target. Therefore, structural elaboration of the flat scaffolds in SE phase is highly significant. We observed that transition from SD phase to SE phase on average reduced the clog*P* value by 16% and enhanced the fraction of sp^3-hybridized carbons (Fsp3) value by >50% (Supplementary Fig. 3)[49], thereby supporting the elaboration of primary scaffolds into more complex frameworks that deliver bioactive molecules (Supplementary Fig. 4).

Principal moment of inertia (PMI) analysis of the lowest-energy conformations of library members, selected natural products and available drugs (Fig. 6d; Supplementary Fig. 5) was performed to compare their three-dimensional shape diversity[50]. The selected natural products and drugs include a number of well-known substances (for example, Taxol, Penicillin and etc.), as well as molecules embodying scaffolds similar to those presented in the SD and SE phases. The library members appeared to cover a large area from the central towards the rod-disc side of the triangle quite in a similar manner to biologically active natural products and drug molecules (Fig. 6d; for details see Supplementary Figs 6 and 7 and Supplementary Tables 5–7).

Biological evaluation. Scaffold diversity in a compound collection is expected to bestow the molecules' ability to modulate different biological functions, and thereby sets a platform for medicinal chemistry and probe discovery research projects. To investigate this possibility, the compound collection was screened in two cell-based assays. In the first case, molecules were subjected to a high-content screen that monitors changes in cytoskeleton and DNA in the human cervical carcinoma HeLa cell line[51]. Treatment of cells with compounds at a concentration of 30 μM for 24 h and subsequent staining of DNA, actin filaments and microtubules revealed structurally similar molecules **17** and **45** causing the cells to round up as if they were entering mitosis (Supplementary Figs 11 and 12). The phenotype caused by **17** was similar to that of nocodazole, a known microtubule destabilizer. Mitotic accumulation induced by **17** was further confirmed by a concentration-dependent increase in the percentage of mitotic HeLa cells that were stained for the mitotic marker phospho-histone H3 (see Fig. 7a and Supplementary Fig. 13). Treatment with **17** and **45** for 48 h reduced the viability of HeLa cells with similar half-maximal inhibitory concentrations (IC_{50}) of 3.87 ± 0.01 μM and 3.86 ± 0.77 μM, respectively (Supplementary Fig. 14). Live-cell imaging of HeLa cells treated with 2.5 μM **17** demonstrated that cells were arrested in mitosis for several hours before undergoing apoptosis as detected by membrane blebbing and cell shrinkage (Supplementary Movies 1–3).

A closer look at the influence of **17** (10 μM), which is a more functionalized analogue of **45**, on the cytoskeleton in interphase cells revealed a disorganization of the microtubule network already after 2 h of treatment (Fig. 7b). In contrast to control cells wherein microtubules emerged from the microtubule organizing centre (MTOC) near the nucleus (visible as the most intense tubulin staining) and extend towards the cell periphery, microtubules in **17** (10 μM)-treated HeLa cells did not converge in the MTOC and their radial organization was distorted. Microtubules are highly dynamic structures that are important for the maintenance of cell shape, inner cellular transport and cell division[52], and remain an attractive target for anticancer treatment[52,53].

The influence of **17** on microtubule dynamics was investigated *in vitro* using porcine brain tubulin. A concentration-dependent inhibition of tubulin polymerization was monitored by means of the increase of 4′,6-diamidino-2-phenylindole (DAPI) fluorescence on binding to microtubules (Fig. 7c)[54]. **17** also inhibited the polymerization of microtubules in HeLa cells after cold treatment (Fig. 7d; Supplementary Fig. 15). Microtubules reversibly disintegrate at low temperatures and their repolymerization can be monitored after rewarming cells to 37 °C. While in dimethylsulphoxide (DMSO)-treated cells, microtubules started to repolymerize 2 min after rewarming, and nearly complete reconstitution of the mictrotubule cytoskeleton was observed after 10 min, in HeLa cells treated with 10 μM solution of **17**, no regrowth of microtubules was detected even 10 min after rewarming. Thus, **17** is a novel microtubule destabilizing small molecule[55].

Among the three well-characterized binding sites in tubulin, tubulin destabilizers bind to either colchicine or vinca alkaloid-binding sites. Binding of **17** to these binding sites was assessed by means of competition experiments. On binding to tubulin, the intrinsic fluorescence of colchicine increases and displacement of colchicine by small molecules leads to the decrease in fluorescence as detected for nocodozole (Supplementary Fig. 16a)[56]. Unfortunately, due to autofluorescence of **17**, it was not possible to determine a putative influence of the compound on the binding of colchicine to tubulin (Supplementary Fig. 16b). However, **17** but not colchicine could displace BODIPY-FL-vinblastine from tubulin in a concentration-dependent manner with a half-maximal effective concentration (EC_{50}) of 0.67 ± 1.51 μM (Fig. 7e; Supplementary Fig. 17), and thus most likely binds to the vinca site in tubulin.

Molecules from SD and SE phases were also subjected to a screen that monitors modulation of Hedgehog signalling. The Hedgehog pathway plays a fundamental role during animal embryonic and post-embryonic development by regulating proliferation, migration and differentiation[53,57]. In adults, the pathway is silenced and can be reactivated for tissue repair and regeneration[58]. Moreover, the Hedgehog pathway can be involved in tumorigenesis since aberrant Hedgehog signalling is detected in various cancers[53,58]. Therefore, small-molecule modulators of the Hedgehog pathway are highly desired for drug discovery and chemical biology investigations. To find inhibitors of Hedgehog signalling, we employed the pluripotent mesenchymal C3H10T1/2 cells that undergo osteogenic differentiation on activation with Hedgehog ligands or purmorphamine, which in turn is characterized by the expression of alkaline phosphatase (AP) and can be used to monitor Hedgehog signalling[59,60]. Screening of the compound collection identified **11b** (dr ~ 5:2:1), **15g** and **16c** as inhibitors of Hedgehog signalling, which dose-dependently decreased AP activity in C3H10T1/2 cells (without influencing cell viability) with IC_{50} of 0.79, 0.84 and 0.16 μM, respectively (Fig. 8a,b). The high-performance liquid chromatography purified major diastereomer of azepinone **11b** (for purification and structural assignment, see Supplementary Methods) was found to be more potent than the two different isomeric mixtures of **11b** employed in the assay (Supplementary Fig. 18). Binding of lipophilic Hedgehog proteins to the transmembrane protein patched 1 (Ptch-1) triggers a signalling cascade by relieving Ptch-1-induced repression of the transmembrane protein Smoothened (Smo). This results in the activation of glioma-

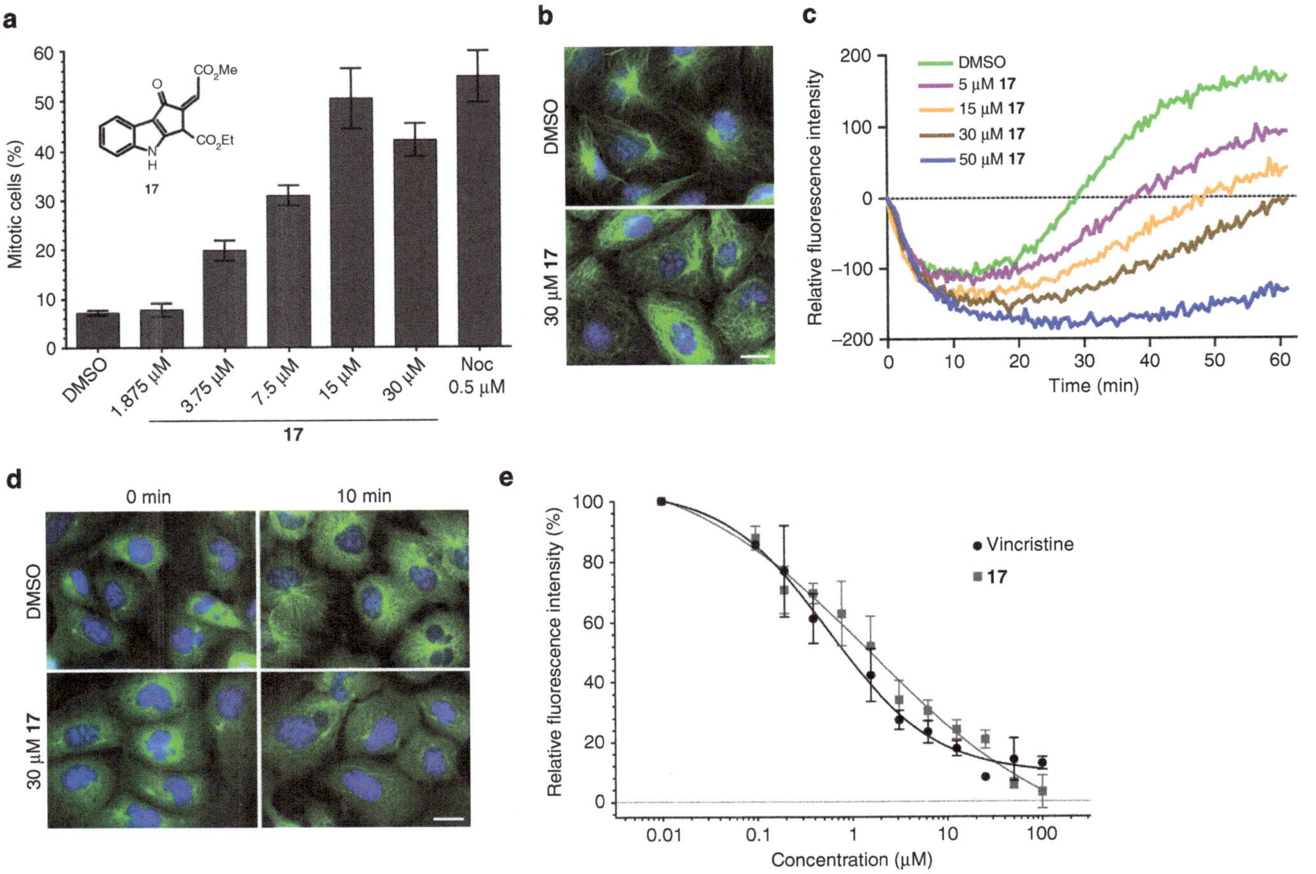

Figure 7 | Influence of 17 on HeLa cells and microtubules. (**a**) **17** induces mitotic arrest. HeLa cells were treated for 24 h with different concentrations of **17** or DMSO and nocodazole (Noc) as controls. Cells were then fixed and stained for DNA, the mitotic marker phospho-histone H3 and tubulin. High-content analysis was performed to determine the percentage of mitotic cells using the MetaMorph software. Data are shown as mean values ($n = 3$) ± s.d. (**b**) **17** impairs the microtubule cytoskeleton in HeLa cells. Cells were treated with 10 μM **17** or DMSO for 2 h. After fixation, cells were stained for tubulin and DNA using an anti-tubulin antibody coupled to FITC (green) and DAPI (blue), respectively. Scale bar, 20 μm. (**c**) **17** inhibits tubulin polymerization *in vitro*. Tubulin polymerization was monitored by means of increase of fluorescence intensity of DAPI on binding to microtubules at ex/em 340/460 nm. Data are representative of three independent experiments. (**d**) **17** inhibits the regrowth of microtubules in HeLa cells. Cells were treated with 10 μM **17** and DMSO for 2 h before incubation on ice for 1 h. Cells were rewarmed at 37 °C for given time intervals and then fixed and stained as described in **b**. Scale bar, 20 μm. Pictures shown are representative of three biological replicates. (**e**) **17** displaces BODIPY-FL-vinblastine from tubulin. Porcine brain tubulin was incubated with BODIPY-FL-vinblastine and different concentrations of **17** or vincristine as a control for 40 min at 25 °C. Fluorescence intensity was then monitored at ex/em 470/514 nm. Decrease in fluorescence indicates competition with BODIPY-FL-vinblastine for binding to the vinca-binding site. Data are shown as mean values ($n = 3$) ± s.d. and were normalized to DMSO.

associated oncogene homologues (Gli)-dependent transcription of Hedgehog pathway-specific target genes[61]. **11b**, **15g** and **16c** suppressed the expression of the Hedgehog target gene Ptch-1 (ref. 62) in NIH/3T3 cells on stimulation with purmorphamine (Fig. 8c). Furthermore, all three molecules inhibited the expression of a Gli-responsive luciferase reporter gene in Shh-LIGHT2 cells[63] (Fig. 8d). Several Hedgehog inhibitors operate by binding to and inhibiting Smo[64]. However, **11b**, **15g** and **16c** failed to displace BODIPY-cyclopamine from Smo, and thus most likely do not bind to this receptor (Fig. 8e; Supplementary Fig. 19). As a result, three new structural classes of Hedgehog inhibitors were discovered providing vital starting points in medicinal chemistry research that targets Hedgehog inhibition-based therapeutics. To obtain an acceptable structure activity relationship (SAR), screening of a larger set of molecules is required. Therefore, a conclusive SAR for the Hedgehog inhibition by molecules based on scaffolds **15** and **16** could not be realized (see Supplementary Fig. 4 for results from the primary screen). However, the results of the primary high-throughput screening for Hedgehog inhibition with benazepinones (**11**) indicate that the benzylic substitution significantly modulates the bioactivity and prefers methyl and ethyl groups over the bulkier ones (Supplementary Fig. 4). Although **11e** (with an ethyl-β-ketoester and an ethyl group on benzylic carbon) appeared to be active in the primary screening, it displayed an $IC_{50} > 10$ μM for the inhibition of purmorphamine-induced osteogenesis (data not shown).

Discussion

Recent analyses of large data sets of synthetic compounds have indicated that a major part of them is presented by a small percentage of scaffolds and the same is true for drug molecules[65]. The redundancy in the scaffolds representing compound libraries that are used in the discovery research severely restrains the biological scope of the small molecules. Synthetic designs leading to significant scaffold diversity are expected to yield highly useful novel small-molecule candidates for drug and probe discovery research. The *de novo* branching cascades strategy imbibes inspiration from biogenesis of natural products, employs simple

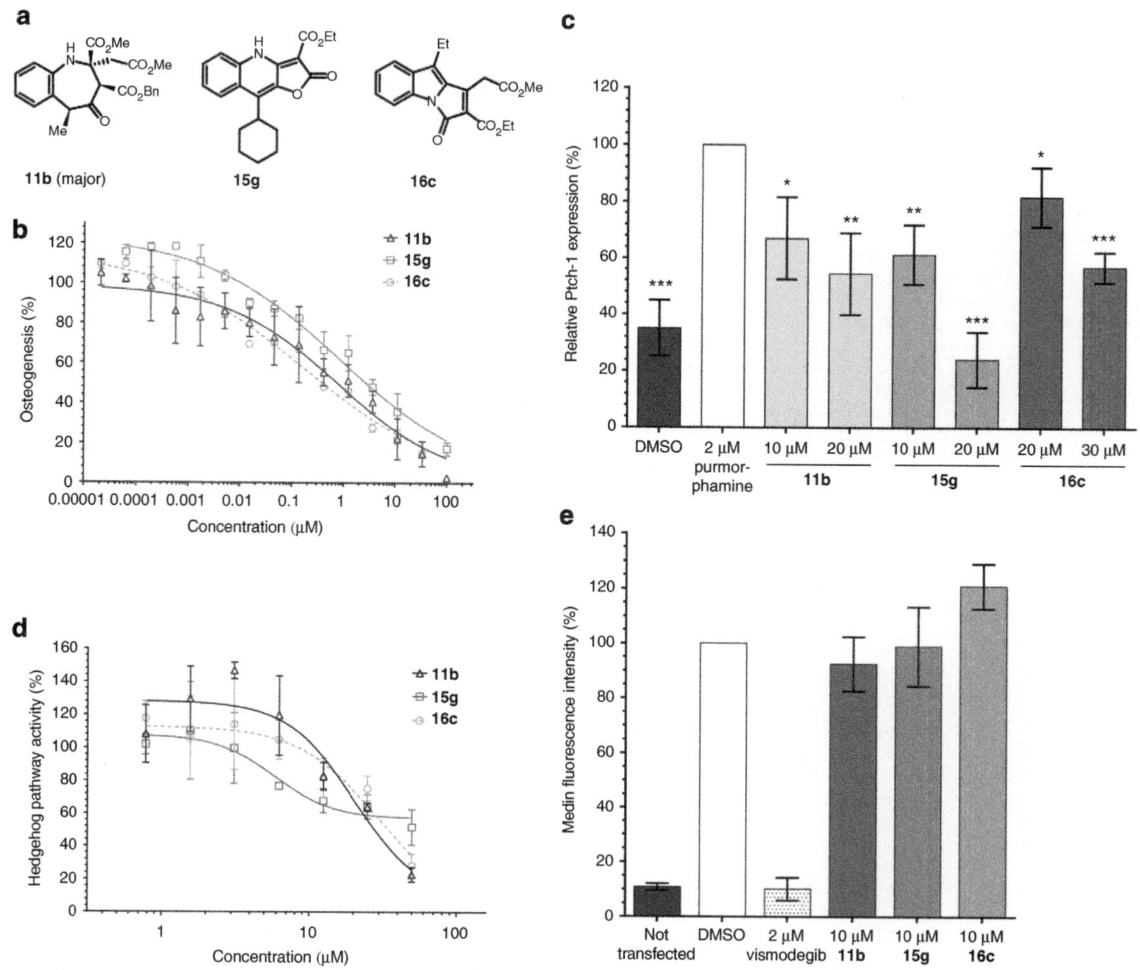

Figure 8 | Influence of selected compounds on Hedgehog signalling. (**a**) Chemical structures of **11b** (structure for the major diastereomer is shown), **15g** and **16c**. (**b**) Influence of **11b**, **15g** and **16c** on purmorphamine-induced osteogenesis in C3H10T1/2 cells as determined by the activity of alkaline phosphatase. C3H/10T1/2 cells were treated for 96 h with 1.5 μM purmorphamine and different concentrations of the compounds or DMSO as control. Activity of alkaline phosphatase was determined using a luminescent readout. Nonlinear regression was performed using a four parameter fit. Data are mean values ($n = 3$) ± s.d. and were normalized to purmorphamine-treated cells. (**c**) Influence of **11b**, **15g** and **16c** on the relative expression of Ptch-1. NIH/3T3 cells were incubated with 2 μM purmorphamine and different concentrations of the compounds or DMSO as control for 48 h. Following cDNA preparation, the relative expression levels of Ptch-1 and Gapdh were determined by means of quantitative PCR. Data are mean values ($n = 3$) ± s.d. and were normalized to purmorphamine-treated cells (*$P < 0.05$, **$P < 0.01$ and ***$P < 0.001$). (**d**) Influence of **11b**, **15g** and **16c** on Gli-mediated reporter gene expression. Shh-LIGHT2 cells were treated with 4 μM purmorphamine and different concentrations of the compounds or DMSO as control for 48 h. Luciferase activity was determined as a measure of Hedgehog pathway activity. Data are mean values ($n = 3$) ± s.d. and normalized to cells treated with purmorphamine. (**e**) Influence of the compounds on the binding of BODIPY-cyclopamine to Smo. HEK293T cells were transfected with a Smo-expression construct. Two days later, cells were treated with the compounds or DMSO as control in the presence of 5 nM BODIPY-cyclopamine for 5 h. Cells were then subjected to flow cytometric analysis to detect Smo-bound BODIPY-cyclopamine. The graph shows the median BODIPY-cyclopamine fluorescence intensity on treatment with the compounds. Data are mean values ($n = 3$) ± s.d.

substrates in the reaction designs to generate suitably functionalized scaffolds via cascade reactions and leads to a compound collection rich in scaffold diversity. With just three simple substrates, that is, phenylhydroxylamine, DMAD and allene ester, a collection of 61 molecules represented by 17 scaffolds was generated without employing any combinatorial synthesis step. The structurally diverse compound collection in turn delivered functionally diverse small molecules as potent inhibitors of the tubulin cytoskeleton or the Hedgehog signalling pathway, thus paving the way for their further biological applications.

We believe that endeavours to develop divergent access to novel chemical space get more exciting and challenging when driven alongside by chemists' desire to explore novel chemical reactivity. In fact, many reactive intermediates reported in various cascade or domino reactions including multi-component reactions might be explored in scaffold diversity synthesis and in unravelling new chemical transformations of broader synthetic applications. This work highlights on the one hand, the immense potential of cascade reactions in building structural diversity and molecular complexity from simple substrates and on the other hand validates the notion that functional diversity of a compound collection is a direct consequence of its scaffold diversity.

Methods

Chemical synthesis. Compounds were synthesized according to the procedures specified in Supplementary Methods. X-ray crystallographic data and images are reported in Supplementary Figs 8–10 and Supplementary Tables 8–20. For ^{1}H, ^{13}C and two-dimensional nuclear magnetic resonance spectra of compounds see Supplementary Figs 20–59).

Scaffold diversity assesment. Two different heat maps were generated to assess the scaffold similarity. The calculation was performed using the software PipelinePilot 9.0.2 from the company Accelrys. Calculation based on the ECFP_4 and ECFP_6 (extended connectivity feature-based fingerprint on four and six bonds, respectively; Supplementary Fig. 2). Molecular properties were calculated using ChemBioDraw 12.0 software (Supplementary Table 4). Representative values of Fsp3 and clogP for SD phase and SE phase molecules are shown in Supplementary Fig. 3.

PMI calculations. We compared the molecular shape diversity of our library with established reference sets of 20 top-selling brand name drugs and 20 diverse natural products (Supplementary Figs 6 and 7).

PMI were calculated using Molecular Operating Environment, MOE software package, after minimization of energy of each molecule using a MMFF94x force field with the generalized Born solvation model; eps = r, cutoff [8,10] and gradient = 0.1 RMS Kcal mol^{-1} A^{-2}. The PMI and related calculations are performed in units of daltons (AMU) and angstroms. The stochastic conformational search algorithm in the MOE software package was used to generate three-dimensional conformers for each compound. Sampling and minimization parameters were implemented as follows: stochastic search limit: 7; refinement conformation limit: 300; stochastic search failure limit: 100; stochastic search iteration limit: 1,000; energy minimization iteration limit: 200; and energy minimization gradient test: 0.01; only the conformer with the lowest energy was retained for PMI calculations in each conformational sampling run (Supplementary Fig. 5; Supplementary Tables 5–7).

Normalized PMI ratios (I1/I3 and I2/I3) of these conformers were obtained from MOE and then plotted on a triangular graph, with the coordinates (0,1), (0.5,0.5) and (1,1) representing a perfect rod, disc and sphere, respectively, based on the report in ref. 50.

Phenotypic screen. HeLa cells were obtained from DSMZ GmbH, Germany and were seeded in black clear bottom 96-well microtiter plates. After incubation overnight, cells were treated with the compounds for 24 h at 30 μM. Cells were fixed with 3.7% formaldehyde in Tris-buffered saline (TBS) and permeabilized with 0.1% Triton X-100 in TBS for 15 min each before blocking using 2% bovine serum albumin (BSA) in TBS/0.1% Tween-20 (TBS-T). Cells were then stained for actin, tubulin and DNA with phalloidin coupled to tetramethylrhodamine, anti-α-tubulin antibody coupled to fluorescein isothiocyanate (FITC) and DAPI, as well as anti-phospho-histone H3 (phospho S10) coupled to AlexaFluor594 (in case of quantification of mitotic cells). Image acquisition was performed on an automated microscope Axiovert M200 (Carl Zeiss, Germany) at 20 × magnification using MetaMorph 7.7.8.0 software (Molecular Devices, USA).

Immunocytochemistry. After seeding and treatment with compounds, cells were fixed with 3.7% formaldehyde and permeabilized with 0.1% Triton X-100 in TBS. Samples were then blocked with 2% BSA in TBS-T before staining for tubulin and DNA with an anti-α-tubulin antibody coupled to FITC and DAPI, respectively. Axiovert Observer Z1 or Axiovert M200 microscopes (Carl Zeiss) were used for image acquisition.

Fluorescence-based tubulin polymerization assay. 10 μM Porcine α/β-tubulin (> 99% pure, cytoskeleton, USA, in 80 mM Na-PIPES pH 6.9, 1 mM $MgCl_2$, 1 mM EGTA and 0.88 mM Na-glutamate) was dissolved in general tubulin buffer containing 16.67% tubulin glycerol buffer, 1 mM GTP and 0.01 mg/ml DAPI on ice. The compound or DMSO were added and fluorescence was measured at 37 °C using the Infinite® M200 plate reader (Tecan, Austria) with excitation/emission wavelengths of 340/460 nm.

Microtubule regrowth assay. HeLa cells were seeded on cover slips and incubated overnight followed by treatment with 10 μM **17** or DMSO as a control for 2 h. Depolymerization of microtubules was achieved by cold treatment at 4 °C for 1 h. Afterwards, the microtubule cytoskeleton was allowed to repolymerize by placing cells to 37 °C. Cells were fixed in 3.7% formaldehyde in TBS at given intervals before or after rewarming. Microtubules were visualized with an anti-α-tubulin antibody coupled to FITC. DAPI was used to stain DNA. Axiovert Observer microscope Z1 (Carl Zeiss) was used for image acquisition.

Colchicine competition assay. A 1:2 dilution series of the compound or nocodazole as a control was prepared on ice using a master mix containing 5 mM tubulin (dissolved in general tubulin buffer), 1 mM GTP, 50 μM colchicine, 16.9% v/v tubulin glycerol buffer and 76.6% v/v TR-FRET buffer. After incubation for 40 min at 25 °C, fluorescence intensity was measured in black 96-well plates at ex/em 365/435 nm using the Infinite M200 plate reader (Tecan). Blank values were subtracted from all sample values. Values were normalized to the DMSO control.

Vinblastine competition assay. A 1:2 dilution series of the compound or vincristine as a control was prepared on ice using a master mix containing 5 mM tubulin (dissolved in general tubulin buffer), 1 mM GTP, 5 μM BODIPY-FL-vinblastine (Invitrogen, Germany), 16.9% v/v tubulin glycerol buffer and 76.6% v/v TR-FRET buffer. After incubation for 40 min at 25 °C, fluorescence intensity was measured in black 96-well plates at ex/em 470/514 nm using the Infinite M200 plate reader (Tecan). Blank values were subtracted from all sample values. Values were normalized to the DMSO control.

Osteogenesis. C3H10T1/2 cells were obtained from ATCC, USA. Eight hundred C3H/10T1/2 cells were seeded per well in white 384-well plates. On the next day, cells were treated with 1.5 μM purmorphamine and different concentrations of the compounds or DMSO as a control. After 96 h, the luminogenic AP substrate CDP-Star (Roche) was added to the wells to detect AP activity. One hour after addition of CDP-Star, luminescence was measured on an Infinite M200 plate reader (Tecan). Nonlinear regression was performed using a four parameter fit (GraphPad Prism 6, GraphPad Software, La Jolla, California, USA).

Quantitative PCR. NIH/3T3 cells were obtained from DSMZ GmbH and were seeded in 24-well plates (2×10^4 cells per well). After incubation overnight, cells were treated with 2 μM purmorphamine and the compounds or DMSO as a control for 48 h. Complementary DNA (cDNA) was prepared using the FastLane Cell cDNA Kit (Qiagen) following the manufacturer's instructions. The relative messenger RNA amount of the Hedgehog target gene Ptch-1 and the housekeeping gene Gapdh (glyceraldehyde-3-phosphate dehydrogenase) was assessed using the QuantiFast SYBR Green PCR Kit (Qiagen) and the following primers: 5′-CAGTG CCAGCCTCGTC-3′ and 5′-CAATCTCCACTTTG-CCACTG-3′ for Gapdh; and 5′-CTCTGGAGCAGATTTCCAAGG-3′ and 5′-TGCCGCAGTTCTTTTGA ATG-3′ for Ptch-1 (ref. 62) The SYBR Green signal was detected with an iQ5 Real-Time PCR Detection System (Bio-Rad, Germany). Expression levels of Ptch-1 were normalized to Gapdh and were related to the expression level of purmorphamine-treated cells. Significance was determined using the unpaired *t*-test using the GraphPad Prism 6 software (San Diego, USA). Differences were considered statistically significant at $P < 0.05$, confidence interval: 95%.

Gli-mediated reporter gene assay. For detection of the Gli-mediated reporter gene expression, the reporter cell line Shh-LIGHT2 was employed. Shh-LIGHT2 cells are NIH/3T3 cells, stably transfected with a Gli-responsive firefly luciferase reporter plasmid and a pRL-TK vector for constitutive expression of Renilla luciferase[61,63]. Shh-LIGHT2 cells (3.0×10^4) were seeded per well in 96-well plates. After incubation overnight, cells were treated with 4 μM purmorphamine and the compounds or DMSO as a control for 48 h. Luciferase expression and activity were detected by means of the Dual-Luciferase Reporter Assay System (Promega) using the Infinite M200 plate reader (Tecan). Nonlinear regression was performed using a four parameter fit (GraphPad Prism 6, GraphPad Software).

Smoothened-binding assay. Flow cytometric analysis of BODIPY-cyclopamine-labelled cells was performed as described in ref. 66. Briefly, 2.5×10^5 HEK293T cells were seeded per well in 6-well plates. After incubation overnight, the cells were transfected with a Smo-expression construct (pGEN-mSmo, a gift from Philip Beachy, Addgene no. 37673)[63] using Fugene HD (Promega). Two days after transfection, cells were treated with the compounds or DMSO in DMEM containing 0.5% FBS and 5 nM BODIPY-cyclopamine (Carbosynth Limited) for 5 h at 37 °C. Cells were then detached and diluted in DMEM containing 0.5% FBS before centrifugation at 129 RCF for 5 min at room temperature. Cells were washed twice in ice-cold PBS and were finally collected by centrifugation at 129 RCF for 5 min at 4 °C. Cells were resuspended in ice-cold PBS and subjected to flow cytometric analysis employing the BD LSR II Flow Cytometer (laser line: 488 nm, emission filter: 530/30) to detect BODIPY. Data were analysed with the FlowJo software, version 7.6.5 (Tree Star Inc., USA) and the Flowing software, version 2.5.1 (by Perttu Terho, University of Turku, Finland/ Turku Bioimaging).

References

1. Shelat, A. A. & Guy, R. K. Scaffold composition and biological relevance of screening libraries. *Nat. Chem. Biol.* **3,** 442–446 (2007).
2. Dandapani, S. & Marcaurelle, L. A. Accessing new chemical space for 'undruggable' targets. *Nat. Chem. Biol.* **6,** 861–863 (2010).
3. Ibbeson, B. M. *et al.* Diversity-oriented synthesis as a tool for identifying new modulators of mitosis. *Nat. Commun.* **5,** 3155 (2014).
4. Villar, H. O. & Hansen, M. R. Design of chemical libraries for screening. *Expert Opin. Drug Discov.* **4,** 1215–1220 (2009).
5. Burke, M. D., Berger, E. M. & Schreiber, S. L. Generating diverse skeletons of small molecules combinatorially. *Science* **302,** 613–618 (2003).
6. Lee, M. L. & Schneider, G. Scaffold architecture and pharmacophoric properties of natural products and trade drugs: application in the design of natural product-based combinatorial libraries. *J. Comb. Chem.* **3,** 284–289 (2001).
7. Renner, S. *et al.* Bioactivity-guided mapping and navigation of chemical space. *Nat. Chem. Biol.* **5,** 585–592 (2009).
8. Ertl, P., Jelfs, S., Muhlbacher, J., Schuffenhauer, A. & Selzer, P. Quest for the rings. In silico exploration of ring universe to identify novel bioactive heteroaromatic scaffolds. *J. Med. Chem.* **49,** 4568–4573 (2006).

9. Tan, D. S. Diversity-oriented synthesis: exploring the intersections between chemistry and biology. *Nat. Chem. Biol.* **1,** 74–84 (2005).
10. Kumar, K. & Waldmann, H. Synthesis of natural product inspired compound collections. *Angew. Chem. Int. Ed. Engl.* **48,** 3224–3242 (2009).
11. Burke, M. D. & Schreiber, S. L. A planning strategy for diversity-oriented synthesis. *Angew. Chem. Int. Ed. Engl.* **43,** 46–58 (2004).
12. Marcaurelle, L. A. *et al.* An aldol-based build/couple/pair strategy for the synthesis of medium- and large-sized rings: discovery of macrocyclic histone deacetylase inhibitors. *J. Am. Chem. Soc.* **132,** 16962–16976 (2010).
13. Morton, D., Leach, S., Cordier, C., Warriner, S. & Nelson, A. Synthesis of natural-product-like molecules with over eighty distinct scaffolds. *Angew. Chem. Int. Ed. Engl.* **48,** 104–109 (2009).
14. Oguri, H. & Schreiber, S. L. Skeletal diversity via a folding pathway: Synthesis of indole alkaloid-like skeletons. *Org. Lett.* **7,** 47–50 (2005).
15. Kwon, O., Park, S. B. & Schreiber, S. L. Skeletal diversity via a branched pathway: Efficient synthesis of 29 400 discrete, polycyclic compounds and their arraying into stock solutions. *J. Am. Chem. Soc.* **124,** 13402–13404 (2002).
16. Kumagai, N., Muncipinto, G. & Schreiber, S. L. Short synthesis of skeletally and stereochemically diverse small molecules by coupling Petasis condensation reactions to cyclization reactions. *Angew. Chem. Int. Ed. Engl.* **45,** 3635–3638 (2006).
17. Robbins, D. *et al.* Synthesis of natural-product-like scaffolds in unprecedented efficiency via a 12-fold branching pathway. *Chem. Sci.* **2,** 2232–2235 (2011).
18. Morrison, K. C. & Hergenrother, P. J. Natural products as starting points for the synthesis of complex and diverse compounds. *Nat. Prod. Rep.* **31,** 6–14 (2014).
19. Huigens, R. W. A ring-distortion strategy to construct stereochemically complex and structurally diverse compounds from natural products. *Nat. Chem.* **5,** 195–202 (2013).
20. Beckmann, H. S. G. *et al.* A strategy for the diversity-oriented synthesis of macrocyclic scaffolds using multidimensional coupling. *Nat. Chem.* **5,** 861–867 (2013).
21. Cordier, C., Morton, D., Murrison, S., Nelson, A. & O'Leary-Steele, C. Natural products as an inspiration in the diversity-oriented synthesis of bioactive compound libraries. *Nat. Prod. Rep.* **25,** 719–737 (2008).
22. Newman, D. J. & Cragg, G. M. Natural product scaffolds as leads to drugs. *Future Med. Chem.* **1,** 1415–1427 (2009).
23. Firn, R. D. & Jones, C. G. Natural products—a simple model to explain chemical diversity. *Nat. Prod. Rep.* **20,** 382–391 (2003).
24. Fischbach, M. A. & Clardy, J. One pathway, many products. *Nat. Chem. Biol.* **3,** 353–355 (2007).
25. Maimone, T. J. & Baran, P. S. Modern synthetic efforts toward biologically active terpenes. *Nat. Chem. Biol.* **3,** 396–407 (2007).
26. Razzak, M. & De Brabander, J. K. Lessons and revelations from biomimetic syntheses. *Nat. Chem. Biol.* **7,** 865–875 (2011).
27. Tietze, L. F. Domino reactions in organic synthesis. *Chem. Rev.* **96,** 115–136 (1996).
28. Nicolaou, K. C., David, J. E. & Paul, G. B. Cascade reactions in total synthesis. *Angew. Chem. Int. Ed. Engl.* **45,** 7134–7186 (2006).
29. Liu, W., Khedkar, V., Baskar, B., Schurmann, M. & Kumar, K. Branching cascades: a concise synthetic strategy targeting diverse and complex molecular frameworks. *Angew. Chem. Int. Ed. Engl.* **50,** 6900–6905 (2011).
30. Padwa, A., Kline, D. N. & Norman, B. H. Synthesis of the benzazepin-4-one ring-system via dipolar cyclo-addition of n-phenylnitrones with activated allenes. *J. Org. Chem.* **54,** 810–817 (1989).
31. Padwa, A., Bullock, W. H., Kline, D. N. & Perumattam, J. Heterocyclic synthesis via the reaction of nitrones and hydroxylamines with substituted allenes. *J. Org. Chem.* **54,** 2862–2869 (1989).
32. Wilkens, J., Kuhling, A. & Blechert, S. Hetero-cope rearrangements.6. short and stereoselective syntheses of 2-vinylindoles by a tandem-process. *Tetrahedron* **43,** 3237–3246 (1987).
33. Mo, D. L., Wink, D. J. & Anderson, L. L. Solvent-controlled bifurcated cascade process for the selective preparation of dihydrocarbazoles or dihydropyridoindoles. *Chemistry* **20,** 13217–13225 (2014).
34. Pecak, W. H., Son, J., Burnstine, A. J. & Anderson, L. L. Synthesis of 1,4-enamino ketones by [3,3]-rearrangements of dialkenylhydroxylamines. *Org. Lett.* **16,** 3440–3443 (2014).
35. Wirth, T. & Blechert, S. Synthesis of 2,3-disubstituted indoles. *Synlett* 717–718 (1994).
36. Nguyen, T. B., Martel, A., Dhal, R. & Dujardin, G. N-benzyl aspartate nitrones: unprecedented single-step synthesis and [3 + 2] cycloaddition reactions with alkenes. *Org. Lett.* **10,** 4493–4496 (2008).
37. Zhang, X. *et al.* Asymmetric synthesis of α,α-disubstituted amino acids by cycloaddition of (e)-ketonitrones with vinyl ethers. *Org. Lett.* **16,** 1936–1939 (2014).
38. Devkota, K. P. *et al.* Compounds from Simarouba berteroana which inhibit proliferation of NF1-defective cancer cells. *Phytochem. Lett.* **7,** 42–45 (2014).
39. Cebrian-Torrejon, G. *et al.* Alkaloids from Rutaceae: activities of canthin-6-one alkaloids and synthetic analogues on glioblastoma stems cells. *Medchemcomm* **3,** 771–774 (2012).
40. Srihari, P., Yaragorla, S. R., Basu, D. & Chandrasekhar, S. Tris(pentafluorophenyl)borane-catalyzed synthesis of N-benzyl pyrrolidines. *Synthesis* 2646–2648 (2006).
41. Singh, A., Loomer, A. L. & Roth, G. P. Synthesis of oxindolyl pyrazolines and 3-amino oxindole building blocks via a nitrile imine [3 + 2] cycloaddition strategy. *Org. Lett.* **14,** 5266–5269 (2012).
42. Dakas, P. Y., Parga, J. A., Hoing, S., Scholer, H. R., Sterneckert, J. & Kumar, K. *et al.* Discovery of neuritogenic compound classes inspired by natural products. *Angew. Chem. Int. Ed. Engl.* **52,** 9576–9581 (2013).
43. Fan, Y. C. & Kwon, O. Advances in nucleophilic phosphine catalysis of alkenes, allenes, alkynes, and MBHADs. *Chem. Commun.* **49,** 11588–11619 (2013).
44. Lu, Z., Zheng, S. Q., Zhang, X. M. & Lu, X. Y. An unexpected phosphine-catalyzed [3 + 2] annulation. Synthesis of highly functionalized cyclopentenes. *Org. Lett.* **10,** 3267–3270 (2008).
45. Langdon, S. R., Brown, N. & Blagg, J. Scaffold diversity of exemplified medicinal chemistry space. *J. Chem. Inf. Model.* **51,** 2174–2185 (2011).
46. Rogers, D. J. & Tanimoto, T. T. A computer program for classifying plants. *Science* **132,** 1115–1118 (1960).
47. Rogers, D. & Hahn, M. Extended-connectivity fingerprints. *J. Chem. Inf. Model.* **50,** 742–754 (2010).
48. Lipinski, C. A., Lombardo, F., Dominy, B. W. & Feeney, P. J. Experimental and computational approaches to estimate solubility and permeability in drug discovery and development settings. *Adv. Drug Deliv. Rev.* **23,** 3–25 (1997).
49. Yan, A. X. & Gasteiger, J. Prediction of aqueous solubility of organic compounds by topological descriptors. *Qsar Comb. Sci.* **22,** 821–829 (2003).
50. Sauer, W. H. B. & Schwarz, M. K. Molecular shape diversity of combinatorial libraries: a prerequisite for broad bioactivity. *J. Chem. Inf. Comp. Sci.* **43,** 987–1003 (2003).
51. Duckert, H. *et al.* Natural product-inspired cascade synthesis yields modulators of centrosome integrity. *Nat. Chem. Biol.* **8,** 179–184 (2012).
52. Dumontet, C. & Jordan, M. A. Microtubule-binding agents: a dynamic field of cancer therapeutics. *Nat. Rev. Drug Dis.* **9,** 790–803 (2010).
53. Atwood, S. X., Chang, A. L. S. & Oro, A. E. Hedgehog pathway inhibition and the race against tumor evolution. *J. Cell. Biol.* **199,** 193–197 (2012).
54. Kawaratani, Y. *et al.* New microtubule polymerization inhibitors comprising a nitrooxymethylphenyl group. *Bioorg. Med. Chem.* **19,** 3995–4003 (2011).
55. Kavallaris, M. Microtubules and resistance to tubulin-binding agents. *Nat. Rev. Cancer* **10,** 194–204 (2010).
56. Bhattacharyya, B. & Wolff, J. Promotion of fluorescence upon binding of colchicine to tubulin. *Proc. Natl Acad. Sci. USA* **71,** 2627–2631 (1974).
57. Mullor, J. L., Sanchez, P. & Altaba, A. R. Pathways and consequences: Hedgehog signaling in human disease. *Trends Cell Biol.* **12,** 562–569 (2002).
58. McMillan, R. & Matsui, W. Molecular pathways: the hedgehog signaling pathway in cancer. *Clin. Cancer Res.* **18,** 4883–4888 (2012).
59. Nakamura, T. *et al.* Induction of osteogenic differentiation by hedgehog proteins. *Biochem. Bioph. Res. Commun.* **237,** 465–469 (1997).
60. Wu, X., Walker, J., Zhang, J., Ding, S. & Schultz, P. G. Purmorphamine induces osteogenesis by activation of the hedgehog signaling pathway. *Chem. Biol.* **11,** 1229–1238 (2004).
61. Sasaki, H., Hui, C. C., Nakafuku, M. & Kondoh, H. A binding site for Gli proteins is essential for HNF-3 beta floor plate enhancer activity in transgenics and can respond to Shh in vitro. *Development* **124,** 1313–1322 (1997).
62. Lipinski, R. J., Gipp, J. J., Zhang, J., Doles, J. D. & Bushman, W. Unique and complimentary activities of the Gli transcription factors in Hedgehog signaling. *Exp. Cell Res.* **312,** 1925–1938 (2006).
63. Taipale, J. *et al.* Effects of oncogenic mutations in Smoothened and Patched can be reversed by cyclopamine. *Nature* **406,** 1005–1009 (2000).
64. Amakye, D., Jagani, Z. & Dorsch, M. Unraveling the therapeutic potential of the Hedgehog pathway in cancer. *Nat. Med.* **19,** 1410–1422 (2013).
65. Lipkus, A. H. *et al.* Structural diversity of organic chemistry. A scaffold analysis of the CAS Registry. *J. Org. Chem.* **73,** 4443–4451 (2008).
66. Chen, J. K., Taipale, J., Cooper, M. K. & Beachy, P. A. Inhibition of Hedgehog signaling by direct binding of cyclopamine to Smoothened. *Gene Dev.* **16,** 2743–2748 (2002).

Acknowledgements

This work was supported by research funds from the Max Planck Society. We thank Prof. Dr H. Waldmann (MPI, Dortmund) for his support and encouragement.

Author contributions

M.G.-C. designed and carried out the synthesis experiments and PMI analysis. L.K. and C.D.R. performed the biological and biochemical experiments. C.U. and C.S. performed the X-ray analysis of single crystals. C.O. performed the cheminformatics analysis to calculate the scaffold diversity. K.K. and S.Z. designed the experiments and supervised the project. All authors discussed the results and commented on the manuscript. K.K., S.Z. and L.K. wrote the manuscript.

Additional information

Accession codes: The X-ray crystallographic coordinates for structures reported in this article have been deposited at the Cambridge Crytallographic Data Center, under deposition numbers CCDC 943382, 943383 and 943385. These data can be obtained free of charge from Cambridge Crytallographic Data Centre via www.ccdc.cam.ac.uk/data_request/cif.

Competing financial interests: The authors declare no competing financial interests.

12

A squalene synthase-like enzyme initiates production of tetraterpenoid hydrocarbons in *Botryococcus braunii* Race L

Hem R. Thapa[1], Mandar T. Naik[1,2], Shigeru Okada[3,4], Kentaro Takada[3,4], István Molnár[5], Yuquan Xu[5,6] & Timothy P. Devarenne[1]

The green microalga *Botryococcus braunii* is considered a promising biofuel feedstock producer due to its prodigious accumulation of hydrocarbon oils that can be converted into fuels. *B. braunii* Race L produces the C_{40} tetraterpenoid hydrocarbon lycopadiene via an uncharacterized biosynthetic pathway. Structural similarities suggest this pathway follows a biosynthetic mechanism analogous to that of C_{30} squalene. Confirming this hypothesis, the current study identifies C_{20} geranylgeranyl diphosphate (GGPP) as a precursor for lycopaoctaene biosynthesis, the first committed intermediate in the production of lycopadiene. Two *squalene synthase* (*SS*)-like complementary DNAs are identified in race L with one encoding a true SS and the other encoding an enzyme with lycopaoctaene synthase (LOS) activity. Interestingly, LOS uses alternative C_{15} and C_{20} prenyl diphosphate substrates to produce combinatorial hybrid hydrocarbons, but almost exclusively uses GGPP *in vivo*. This discovery highlights how SS enzyme diversification results in the production of specialized tetraterpenoid oils in race L of *B. braunii*.

[1] Department of Biochemistry and Biophysics, Texas A&M University, College Station, Texas 77843, USA. [2] Biomolecular NMR Laboratory, Department of Biochemistry and Biophysics, Texas A&M University, College Station, Texas 77843, USA. [3] Laboratory of Aquatic Natural Products Chemistry, Graduate School of Agricultural and Life Sciences, The University of Tokyo, Yayoi, Bunkyo, Tokyo 113-8657, Japan. [4] Japan Science and Technology Agency-Core Research for Evolutional Science and Technology (CREST), Gobancho, Chiyoda, Tokyo 102-0076, Japan. [5] Natural Products Center, School of Natural Resources and the Environment, The University of Arizona, Tucson, Arizona 85739, USA. [6] Biotechnology Research Institute, The Chinese Academy of Agricultural Sciences, Beijing 100081, China. Correspondence and requests for materials should be addressed to T.P.D. (email: tpd8@tamu.edu).

Microalgae are a promising next-generation source of feedstocks for biofuel production with the potential to serve as a practical alternative to petroleum-based transportation fuels[1]. Depending on the microalgal species, the oils produced vary from triacylglycerols to hydrocarbons[2,3]. Hydrocarbon-based fuels are preferred over other biofuels, as they are highly compatible with existing petroleum infrastructures and possess superior fuel properties[4,5]. The colony-forming green microalga *B. braunii* is an exciting candidate for biofuel feedstock production, as it produces up to 61% of its dry weight as liquid hydrocarbon oils[6]. These hydrocarbons are produced inside the cells of the colony, seen as intracellular oil bodies and secreted into the colony extracellular matrix where the majority of the hydrocarbons are stored[6] (Fig. 1a). Most importantly, catalytic hydrocracking of hydrocarbons from this alga results in petroleum-equivalent fuels of gasoline, kerosene and diesel[7]. Intriguingly, geologic evidence also shows a direct contribution of this alga to the formation of currently used fossil fuel deposits around the globe[8–12]. Despite the aforementioned advantages of *B. braunii*, its use for biofuel feedstock production is hindered by a slow growth rate and the lack of transformation systems to achieve targeted genetic modification[3]. Thus, the identification of *B. braunii* hydrocarbon biosynthetic pathways and associated genes/enzymes can provide options for metabolically engineering these pathways into heterologous hosts with better growth characteristics and the ability to be genetically manipulated. This would then allow the development of improved versions of hydrocarbon biosynthetic enzymes, to direct production towards the most commercially desirable products[13].

There are three different races of *B. braunii* based on the hydrocarbons synthesized. Race A produces fatty acid-derived C_{23}–C_{33} alkadienes and alkatrienes. Races B and L produce isoprenoid-derived hydrocarbons: methylsqualenes and C_{30}–C_{37} botryococcene triterpenoids in race B and the C_{40} tetraterpenoid lycopadiene, the focus of this study, in race L[6]. Green algae have been shown to possess only the plastid-localized methyl erythritol phosphate pathway to supply isoprenoid precursors for terpene production[14,15], and thus lycopadiene is predicted to be generated from C_5 precursors via this route. However, the exact mechanism of lycopadiene biosynthesis from C_{20} prenyl diphosphate intermediates has been a mystery and two possible biosynthetic routes have been suggested (Fig. 1b)[16]. The first entails C_{20} geranylgeranyl diphosphate (GGPP) reduction by GGPP reductase to produce C_{20} phytyl diphosphate (PPP; Fig. 1b). Two molecules of PPP would then undergo head-to-head condensation (1-1′ linkage) to produce lycopadiene (Fig. 1b). The second possibility is the head-to-head condensation of two GGPP molecules to produce lycopaoctaene, followed by stepwise enzymatic reduction to produce lycopadiene (Fig. 1b).

Using either proposed route, the condensation of PPP or GGPP is predicted to proceed in a reaction mechanism similar to that carried out by the enzyme squalene synthase (SS), which forms C_{30} squalene, a precursor required for sterol biosynthesis in eukaryotes[17] and hopanoids in some prokaryotes[18]. SS enzymes catalyse a two-step reaction. First, the condensation of two C_{15} farnesyl diphosphate (FPP) molecules yields the cyclopropyl intermediate presqualene diphosphate (PSPP; Fig. 1c). Second, PSPP undergoes NADPH-dependent reductive rearrangement to form squalene with a 1-1′ linkage between the two FPP molecules (Fig. 1c)[19]. Herein we report the elucidation of the first committed step in the lycopadiene hydrocarbon biosynthetic pathway in *B. braunii* Race L, which is catalysed by a new SS-like (SSL) enzyme.

Results

L race hydrocarbon content and related enzyme activity. Previous studies reported *trans,trans*-lycopadiene as the predominant hydrocarbon (98% of total hydrocarbons) produced by race L, with a small amount of lycopatriene also documented[16,20,21]. However, the lycopatriene structure was not determined. Thus, we first set out to ascertain and refine the hydrocarbon profile of *B. braunii* Race L, to provide a baseline for our biosynthetic models. Analysis of purified hydrocarbons by gas chromatography–mass spectrometry (GC–MS) showed that lycopadiene accounts for 95% of the total hydrocarbon pool (Fig. 2a (I)), with six other minor compounds constituting the remaining 5%: lycopatriene, lycopatetraene, lycopapentaene, lycopapentaene isomer, lycopahexaene and a $C_{35}H_{64}$ molecule (Fig. 2a (II–VII) and Supplementary Figs 1–4). Ozonolysis experiments suggested lycopatriene and lycopapentaene share an identical reduced C_{20} moiety with lycopadiene (Supplementary Figs 5–7). Nuclear magnetic resonance (NMR) spectroscopy was used to confirm identity and structure, including double bond positions, of each molecule (Fig. 2a, Supplementary Table 1 and Supplementary Figs 8–15). This is the first report of lycopatetraene, lycopapentaene and lycopahexaene hydrocarbons from *B. braunii* race L. The lack of detectable amounts of lycopaoctaene and lycopaheptaene in the total hydrocarbon fraction could be due to rapid conversion of these metabolites into more highly saturated homologues. Interestingly, the unique $C_{35}H_{64}$ isoprenoid hydrocarbon detected contains seven isoprene units and thus is assigned to the recently named C_{35} terpene class 'sesquarterpenes'[22].

To gain more insight into the mechanism of lycopadiene biosynthesis and assess the validity of the two possible lycopadiene biosynthetic routes (Fig. 1b), a cell-free *in vitro* enzyme assay was developed using total, soluble and microsomal protein fractions with ^{3}H-labelled PPP and GGPP substrates for potential incorporation into lycopadiene and lycopaoctaene, respectively. Squalene production by SS enzyme activity with ^{3}H-labelled FPP as substrate was used as a control[23]. Lycopadiene production from PPP was not detected in any protein fraction tested, whereas lycopaoctaene biosynthesis from GGPP was easily observed, mainly in the microsomal fraction (Fig. 2b). This indicates a lycopaoctaene synthase (LOS) activity is localized to a membrane system, possibly the endoplasmic reticulum (ER) as is seen for SS[24]. Tellingly, LOS activity in these assays was detected when using the same cofactor and divalent metal cation as those used for SS activity measurement. These results suggest direct lycopadiene biosynthesis from PPP is not a significant contributor to C_{40} hydrocarbon production in race L, whereas conversion of GGPP into lycopaoctaene occurs readily, as proposed in the second hypothetical route (Fig. 1b). Correspondingly, LOS enzyme activity directly correlated with lycopadiene accumulation over the growth cycle, suggesting this activity is related to hydrocarbon biosynthesis (Supplementary Fig. 16). Detection of LOS activity from an algal homogenate in an assay similar to that of SS supports the notion that an LOS enzyme may be similar to a typical SS enzyme.

Identification of an LOS enzyme. To identify the gene(s) responsible for lycopaoctaene biosynthesis, we generated and computationally screened an L race transcriptomic database for *SSL* sequences. Two *SSL* cDNAs were identified and named based on the function of their encoded proteins as detailed below: *SS* from race L (*LSS*) and *LOS*. Both the LSS and LOS proteins contain all five conserved activity domains, the transmembrane domain and the NADPH-binding residues found in typical SS

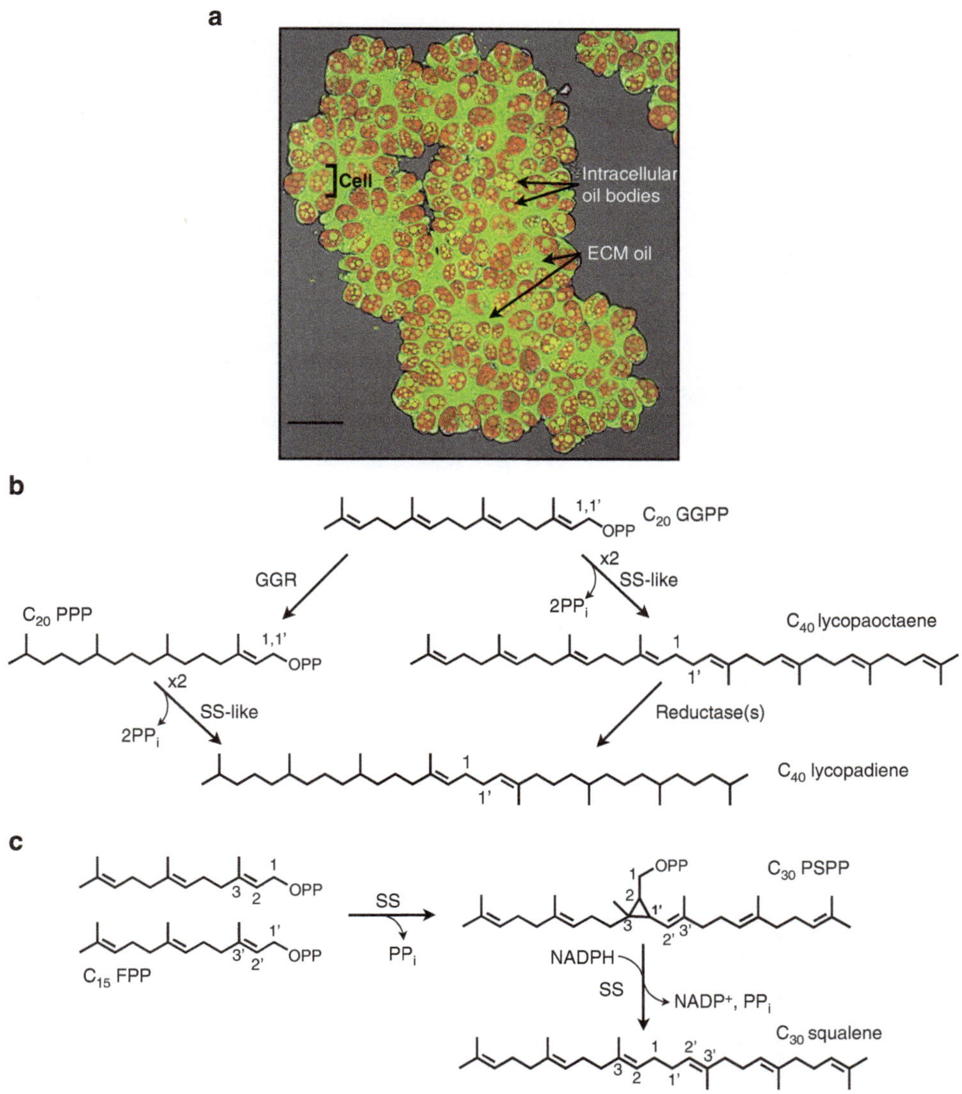

Figure 1 | Background information on lycopadiene biosynthesis in *B. braunii* race L. (**a**) Confocal microscopy image of an L race colony of *B. braunii*. Shown are hydrocarbon oils stained using the neutral lipid-binding stain Nile red (false coloured green) and chlorophyll autofluorescence (red). Indicated are intracellular hydrocarbon oil bodies, hydrocarbon oil in the extracellular matrix (ECM) and an individual cell as defined by chlorophyll autofluorescence from the single chloroplast in each cell. Image is an overlay of Nile red signal, chlorophyll autofluorescence signal and a bright-field image. Scale bar, 20 μm. (**b**) Two possible pathways for lycopadiene biosynthesis. Possibility 1: GGPP reduction to PPP followed by condensation of two PPP molecules, to directly produce lycopadiene. Possibility 2: condensation of two molecules of GGPP to form lycopaoctaene, which would then be reduced to lycopadiene. (**c**) Two-step reaction catalysed by single SS enzyme: two molecules of FPP are condensed to squalene through the PSPP intermediate. In **b** and **c**, enzyme names are coloured blue.

enzymes (Supplementary Figs 17 and 18)[17]. Comparison of the encoded amino acid sequences showed 57.5 and 49.2% identity to SS from race B (BSS)[25] for LSS and LOS, respectively (Supplementary Fig. 17). In addition, comparison of LOS with other SS and SSL enzymes showed 52.5% identity to LSS, 47.3% identity to the *Solanum lycopersicum* (tomato) SS (*Sl*SS), 44.6, 61.7 and 44.1% identity to the SSL-1, -2 and -3 from the B race of *B. braunii*, and 33.9% identity to the human SS (HSS; Supplementary Fig. 18). Several studies[17,26–28], including the recent determination of the crystal structure of HSS[29], have elucidated the catalytic mechanism of the SS reaction. These studies have shown the FLAP domain, domains I–V and the JK loop of HSS form the SS active site[29]. Furthermore, several HSS residues (Ser51, Sr53, Tyr171 and Gln212), two DxxED motifs in domains II and IV, and the NADPH-binding residues were determined to be key for catalytic activity (see Supplementary Fig. 18)[29]. All of these catalytic residues and motifs are also conserved in both the LSS and LOS enzymes (Supplementary Fig. 18), making it difficult at this time to determine which domains/residues are specific for LOS activity in a comparison of LSS and LOS.

For initial characterization of the enzymatic activity of LSS and LOS, both proteins were expressed in *Escherichia coli* and the purified recombinant proteins tested for enzyme activity. LSS was shown to yield squalene as its sole product in an *in vitro* assay with FPP as the substrate, as expected (Fig. 3a,b (I)). However, purified recombinant LOS incubated with GGPP in an *in vitro* assay showed lycopaoctaene as the sole reaction product (Fig. 3a,b (II)). As LOS may have arisen from an SS paralogue that evolved to accept GGPP as substrate for lycopaoctaene production, we considered that LOS may have retained the ability to use FPP to produce squalene. Indeed, squalene production was detected

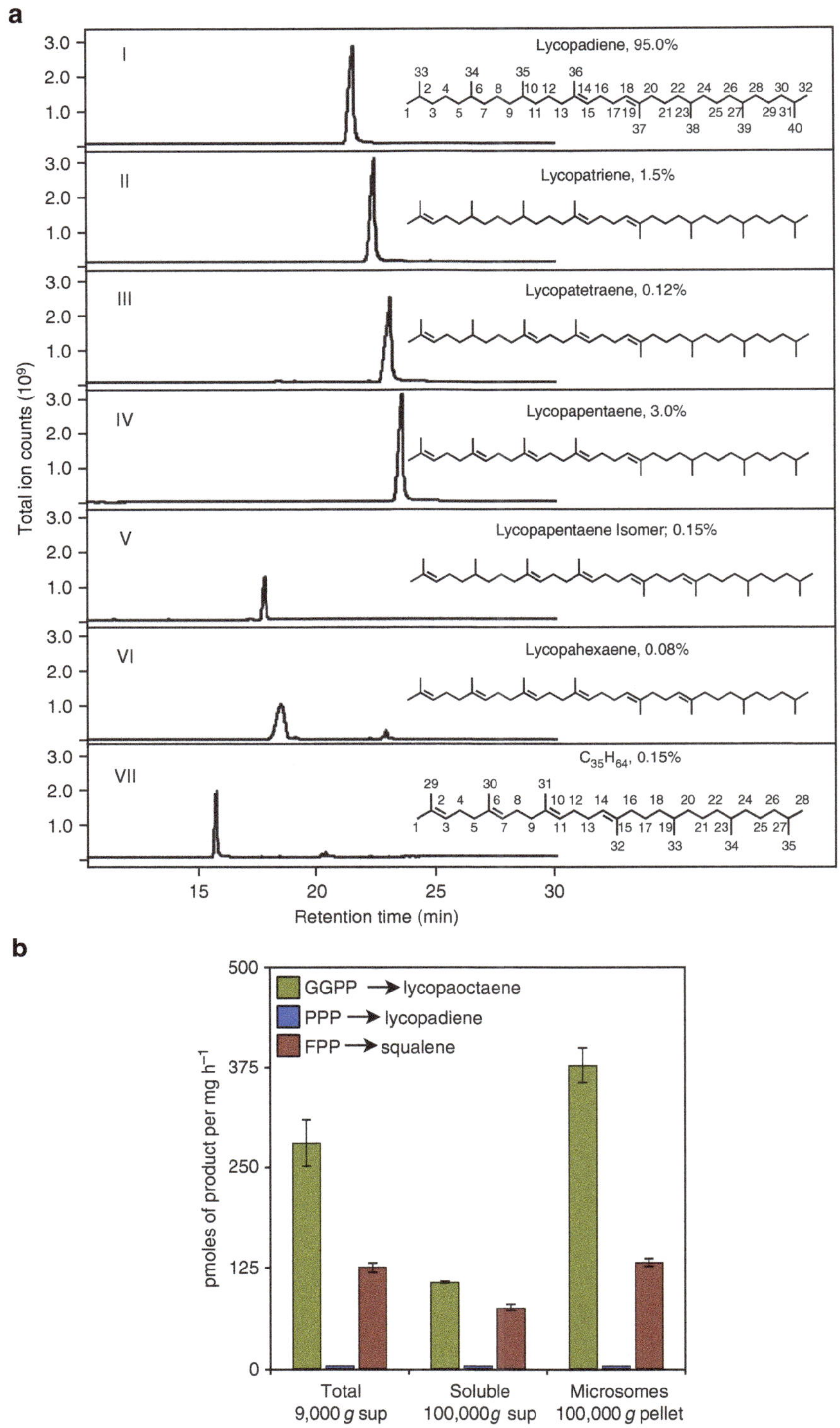

Figure 2 | Hydrocarbon composition and hydrocarbon-related enzyme activities in *B. braunii* race L. (**a**) Representative GC-MS profiles of purified hydrocarbons and their corresponding names and structures (I-VII). Structures of hydrocarbons were determined by ^{1}H- and ^{13}C-NMR. Lycopadiene carbon numbering (I) is used for numbering all C_{40} hydrocarbons of race L. Percentage next to the name of each hydrocarbon corresponds to its average amount present in total hydrocarbon pool from three independent culture cycles ($n = 3$). (**b**) Radioactive enzyme assays were conducted using different protein fractions of race L cell lysates to test the two possible lycopadiene biosynthetic pathways. Enzyme activities were calculated based on the incorporation of ^{3}H-GGPP into lycopaoctaene, ^{3}H-PPP into lycopadiene and ^{3}H-FPP into squalene. SS enzyme activity is used as a positive control. Values shown are the mean ± s.e. obtained from three independent experiments ($n = 3$).

when LOS was incubated with FPP in an *in vitro* reaction (Fig. 3a,b (III)). More surprising, LOS incubation with PPP yielded lycopadiene, albeit at levels much lower than lycopaoctaene or squalene (Fig. 3a,b (IV)). Similar *in vitro* incubations of LSS with GGPP or PPP did not result in lycopaoctaene or lycopadiene production (see below). As LOS can use three different substrates, steady-state kinetic experiments were performed with FPP, GGPP and PPP, to analyse LOS

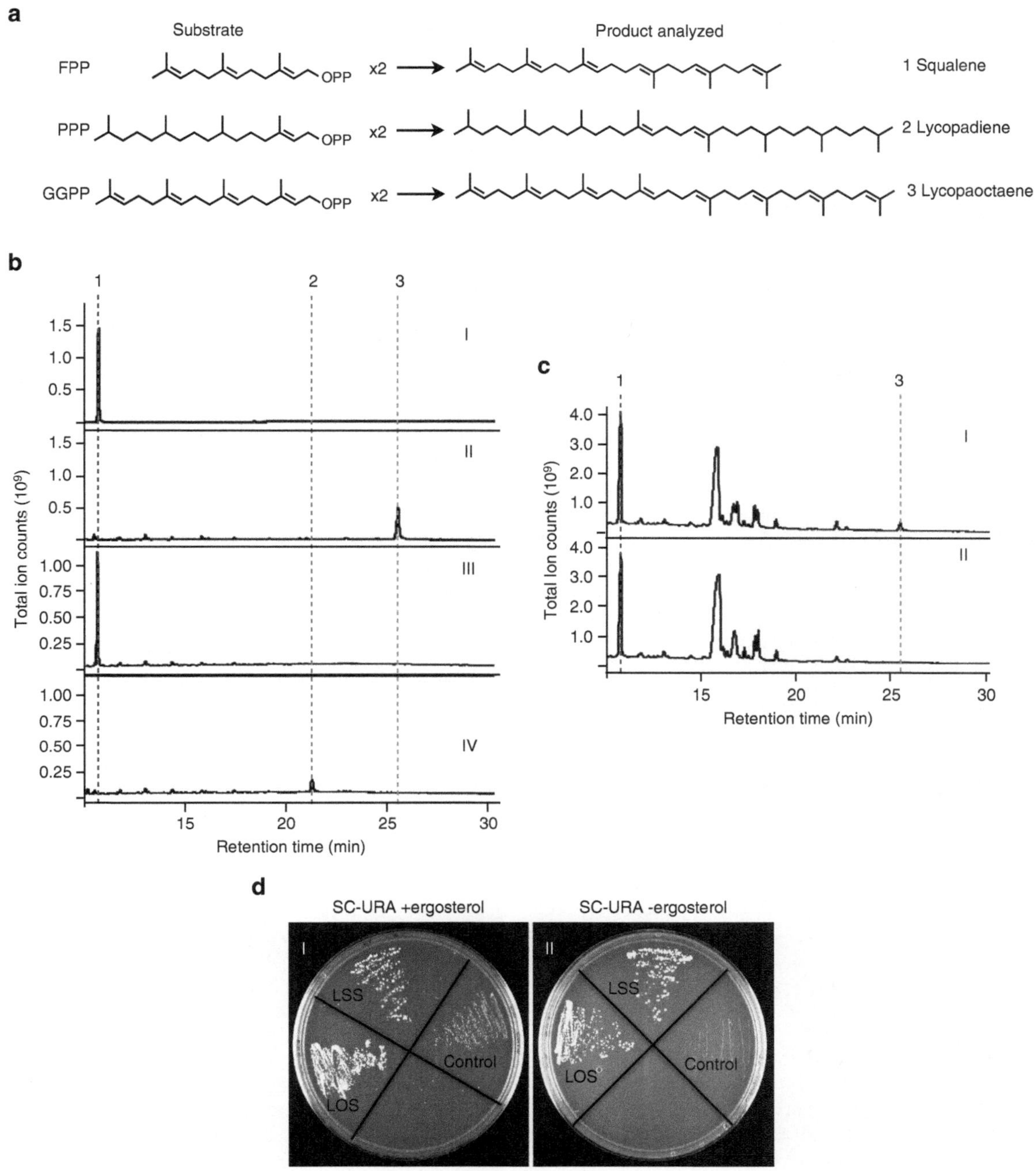

Figure 3 | Functional characterization of SSL enzymes from race L *in vitro* and *in vivo*. (**a**) Diagram of reactions showing substrates and products for the analysis in **b** and **c**. (**b**) GC-MS profiles of LSS and LOS *in vitro* reaction products: I, LSS products using FPP as substrate to produce squalene (1); II, LOS products using GGPP as substrate to produce lycopaoctaene (3); III, LOS products using FPP as substrate to produce squalene (1); and IV, LOS products using PPP as substrate to produce lycopadiene (2). (**c**) GC-MS profiles of LOS yeast *in vivo* reaction products: I, *n*-hexane extractable metabolites of yeast expressing *At*GGPPS11 and LOS, to make lycopaoctaene (3); or II, yeast expressing *At*GGPPS11 alone. (**d**) Expression of LSS and LOS in the *SS* (*ERG9*) yeast knockout strain: I, in the presence of ergosterol, or II, in the absence of ergosterol, to test restoration of ergosterol prototrophy. The control carries the expression vector without a transgene. Data shown in **b**,**c** and **d** are representatives from three independent experiments ($n = 3$).

substrate specificity. The LOS enzyme shows higher substrate affinity (K_m) for GGPP compared with that of FPP and PPP, and the turnover number (k_{cat}) and catalytic efficiency (k_{cat}/K_m) for PPP are an order of magnitude less than for FPP and GGPP (Table 1 and Supplementary Fig. 19).

To further characterize LOS, it was coexpressed in yeast with *Arabidopsis thaliana* GGPP synthase-11 (*At*GGPPS11)[30], as unlike FPP, yeast GGPP biosynthesis is limited and is considered a bottleneck for GGPP-derived isoprenoid production[31]. Coexpression of LOS and *At*GGPPS11 in yeast resulted in lycopaoctaene production (Fig. 3c (I)), which was undetectable when *At*GGPPS11 was expressed without LOS (Fig. 3c (II)). Furthermore, when expressed in a yeast *SS* knockout strain, LOS restored ergosterol prototrophy, indicating its ability to produce squalene *in vivo* (Fig. 3d (I,II)). LSS was also able to restore ergosterol prototrophy as expected (Fig. 3d (I,II)). Taken together, the results in Fig. 3 suggest LSS is a true SS enzyme, whereas LOS appears to be a promiscuous SSL enzyme with broader substrate chain length and saturation specificity.

Table 1 | LOS steady-state kinetic parameters for GGPP, FPP and PPP substrates.

	GGPP	FPP	PPP
K_m (mM)	0.07 ± 0.02	0.13 ± 0.02	0.11 ± 0.01
k_{cat} (s^{-1})	1.14×10^{-2}	2.05×10^{-2}	1.16×10^{-3}
k_{cat}/K_m ($\mu M^{-1} s^{-1}$)	1.65×10^{-4}	1.62×10^{-4}	1.03×10^{-5}

FPP, farnesyl diphosphate, GGPP, geranylgeranyl diphosphate; LOS, lycopaoctaene synthase; PPP, phytyl diphosphate.
K_m values shown are the mean ± s.e. obtained from three independent measurements ($n = 3$).

Previous studies have reported the production of lycopaoctaene (a.k.a. lycopersene) *in vitro* from protein extracts during studies conducted to decipher the carotenoid biosynthetic pathway[32–34]; however, subsequent studies determined lycopaoctaene could not be an intermediate in carotenoid biosynthesis[35,36], raising the possibility that lycopaoctaene production in these studies was an *in vitro* artefact. A more relevant study reported the ability of purified yeast SS to use GGPP for lycopaoctaene production *in vitro*, at levels much lower than native squalene production[32]. Importantly, we did not detect any lycopaoctaene production when *At*GGPPS11 was expressed alone in wild-type yeast (Fig. 3c (II)), suggesting that yeast SS does not use GGPP *in vivo* under the conditions employed. Similarly, the ability of LOS to catalyse the conversion of PPP to lycopadiene may not have biological significance as we did not detect lycopadiene production using PPP as the substrate *ex vivo* in race L cell lysates (Fig. 2b) and the efficiency of LOS to use PPP as a substrate is quite low (Table 1). Together, these results suggest that the relevant *in vivo* route to lycopadiene production is unlikely to involve the condensation of two PPP molecules by LOS, but rather, as we contend, the condensation of two GGPP molecules to form lycopaoctaene, for eventual conversion to lycopadiene.

LOS is promiscuous towards prenyl substrates. The LOS enzyme was further characterized using combinations of FPP, GGPP and PPP as substrates in *in vitro* reactions. Interestingly, in the presence of FPP and GGPP, LOS produced significant amounts of squalene and a $C_{35}H_{58}$ molecule, and lesser amounts of lycopaoctaene (Fig. 4a (I)). The $C_{35}H_{58}$ molecule is a chimera produced from head-to-head condensation of one FPP molecule and one GGPP molecule, and its identity was confirmed by GC–MS (Supplementary Fig. 20). Next, when LOS was supplied with FPP and PPP, squalene production predominated with small amounts of $C_{35}H_{64}$ and lycopadiene (Fig. 4a (II)). $C_{35}H_{64}$, also a chimeric hydrocarbon, is produced as a result of head-to-head condensation of one molecule each of FPP and PPP. Finally, LOS incubation with GGPP and PPP produced lycopadiene and lycopapentaene as minor products and lycopaoctaene as the major product (Fig. 4a (III)).

To the best of our knowledge, this is the first report where a wild-type, non-mutated eukaryotic SS or SSL enzyme has been shown to be able to use three naturally occurring prenyl diphosphate substrates to yield hydrocarbon products in all possible combinations. To support this notion, several SS and SSL enzymes were tested for their ability to use FPP, GGPP and PPP as substrates (Fig. 4b). The enzymes tested included LOS and three typical SS enzymes: LSS, BSS and *Sl*SS. In addition, enzymes SSL-1, SSL-2 and SSL-3 from race B were chosen because of their SSL activities. SSL-1 uses FPP to produce PSPP, which is then converted by SSL-2 to squalene, or by SSL-3 to C_{30}-botryococcene as the major product and squalene as a minor product[37]. As shown in Fig. 4b, the typical SS enzymes display SS activity as their main catalytic function, with BSS and *Sl*SS also using PPP to produce minute amounts of lycopadiene. In contrast, the B race SSL enzymes have limited substrate flexibility, allowing SSL-2 and SSL-1 plus SSL-2 to use GGPP to generate minor amounts of lycopaoctaene (Fig. 4b). However, LOS is the only enzyme tested that is able to use all three substrates and to produce significant amounts of lycopaoctaene, squalene and lycopadiene (Fig. 4b).

The LOS reaction uses a cyclopropyl intermediate. We next conducted enzyme assays to determine whether the LOS reaction mechanism with GGPP as substrate is similar to that of SS, that is, uses a PSPP-like cyclopropyl diphosphate intermediate, which is termed prelycopaoctaene diphosphate (PLPP; Fig. 5a). First, assays were conducted with or without a dinucleotide reducing agent, which would be required to convert PLPP to lycopaoctaene (Fig. 5a). LOS successfully used both NADH and NADPH as reducing agents for lycopaoctaene production, with preference for NADPH (Fig. 5b). In the absence of a dinucleotide reducing agent, LOS activity was lost (Fig. 5b), suggesting the presence of the PLPP reaction intermediate. This result is consistent with previous reports where yeast SS used GGPP *in vitro* to produce lycopaoctaene; however, without NADH or NADPH only the reaction intermediate PLPP accumulated[32]. Next, we showed that LOS is strongly inhibited by squalestatin (Fig. 5c), a potent SS inhibitor that mimics PSPP binding[23,38–40]. Finally, we identified PLPP as a reaction intermediate by conducting a GGPP-based LOS assay in the absence of NADPH, followed by acid phosphatase treatment to convert PLPP to prelycopaoctaene alcohol (PLOH). Analysis by GC–MS showed the presence of PLOH (Fig. 5d) with a fragmentation pattern consistent with previous reports of PLOH (Supplementary Fig. 21)[32]. Taken together, these results are consistent with a two-step LOS reaction to produce lycopaoctaene. By analogy to SS, LOS catalyses the condensation of two GGPP units in the first half reaction to form the cyclopropylcarbinyl diphosphate intermediate PLPP, with concomitant release of one molecule of inorganic pyrophosphate (Fig. 5a). In the second half reaction, the PLPP cyclopropyl ring is cleaved and rearranged to form a 1-1′ linkage and further reduction by NADPH forms lycopaoctaene (Fig. 5a).

Discussion

Our studies identified the first committed step in lycopadiene biosynthesis in race L of *B. braunii* and describe a new SSL enzyme, LOS, which carries out this reaction. Importantly, this enzyme may be used to engineer the effective production of hydrocarbon biofuel feedstocks in other photosynthetic or heterotrophic organisms in the future. This novel enzyme produces the tetraterpenoid lycopaoctaene from GGPP in a reaction analogous to that of SS. Lycopaoctaene would then undergo sequential reduction by a yet-to-be identified reductase enzyme(s) to form lycopahexaene, lycopapentaene, lycopatetraene, lycopatriene and finally lycopadiene as the major hydrocarbon product of the pathway (Fig. 6). Furthermore, we show that the wild-type LOS enzyme displays remarkable substrate flexibility and can use at least three different, naturally occurring C_{15} and C_{20} prenyl diphosphate substrates *in vivo* and *in vitro* for the combinatorial biosynthesis of hydrocarbons

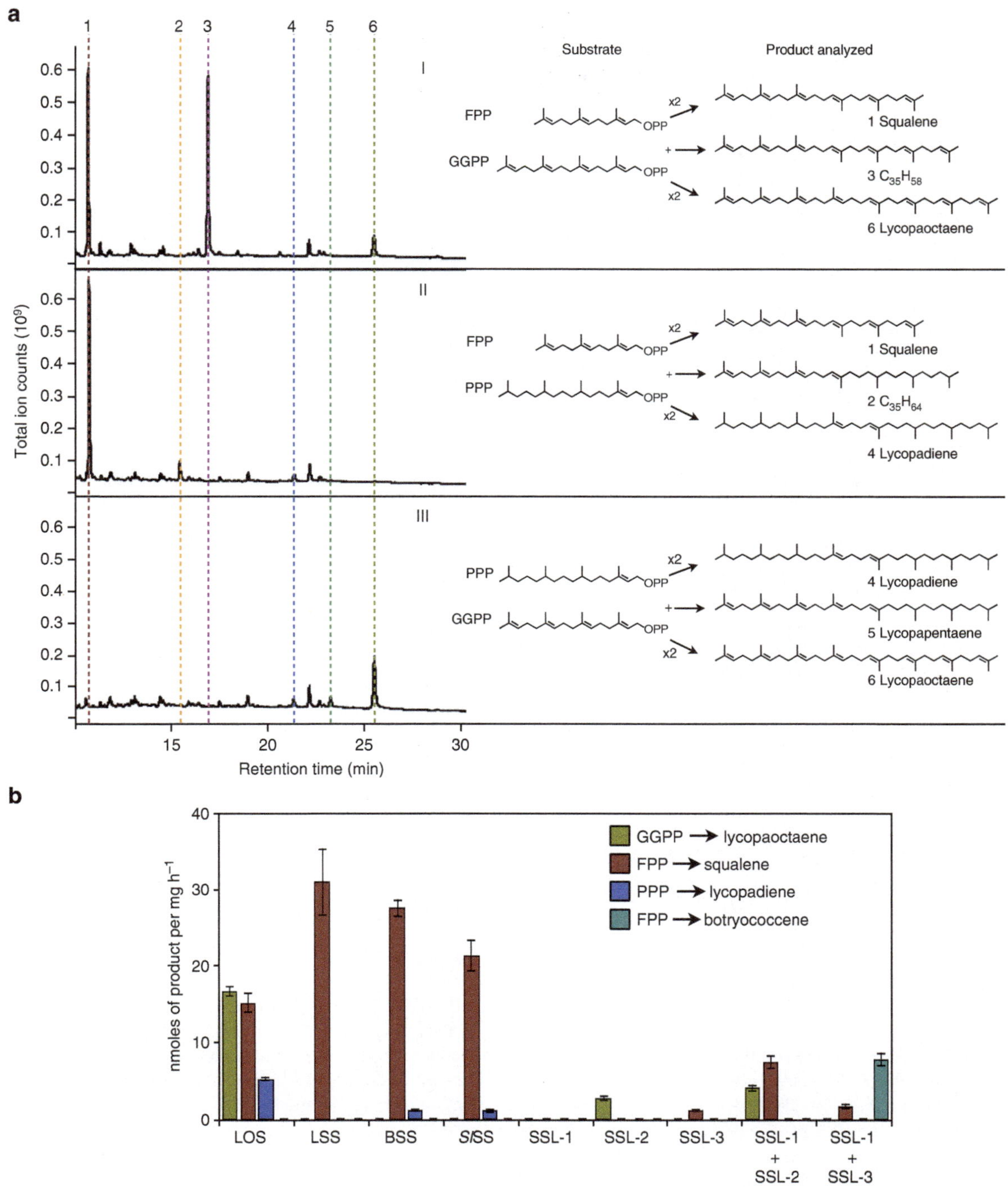

Figure 4 | Characterization of substrate use for SS and SSL enzymes. (**a**) Diagram of reactions showing substrates and products, and GC–MS profiles of LOS enzyme assay products with: I, FPP and GGPP substrates to produce squalene (1), $C_{35}H_{58}$ (3) and lycopaoctaene (6); II, FPP and PPP substrates to make squalene (1), $C_{35}H_{64}$ (2) and lycopadiene (4); and III, GGPP and PPP substrates to produce lycopadiene (4), lycopapentaene (5) and lycopaoctaene (6). Data shown are representatives from three independent experiments ($n=3$). (**b**) Radioactive enzyme assays using ^{3}H substrates GGPP, FPP and PPP with selected SS or SSL enzymes. SSL-1 produces PSPP from FPP substrate; however, PSPP was not analysed for in this assay. BSS, race B SS; LOS, lycopaoctaene synthase; LSS, race L SS; *Sl*SS, SS from *S. lycopersicum* (tomato), and SSL-1, SSL-2, SSL-3 are SSL enzymes from the B race of *B. braunii*. Values shown are the mean ± s.e. obtained from three independent measurements ($n=3$).

belonging to three terpene classes: triterpenes, sesquarterpenes and tetraterpenes.

When used under artificial reaction conditions, other SSL enzymes have been described to show relaxed substrate flexibility and to yield products of varying lengths. One such enzyme is CrtM, a dehydrosqualene synthase from *Staphylococcus aureus* that is involved in C_{30} carotenoid biosynthesis[41–43]. In *S. aureus* and also when overexpressed in *E. coli*, CrtM uses two FPP molecules to produce dehydrosqualene, a C_{30} squalene-like molecule with a *cis* double bond at the 1-1′ linkage of the two FPP molecules[44]. Expression of a GGPP synthase in *E. coli* also expressing CrtM led to the depletion of FPP and the overproduction of GGPP[41]. Under these conditions, CrtM was found to generate its native C_{30} product, a C_{35} homologue and C_{40} phytoene, indicating the ability of CrtM to accept GGPP as a substrate when the natural FPP substrate is limiting[41]. CrtM

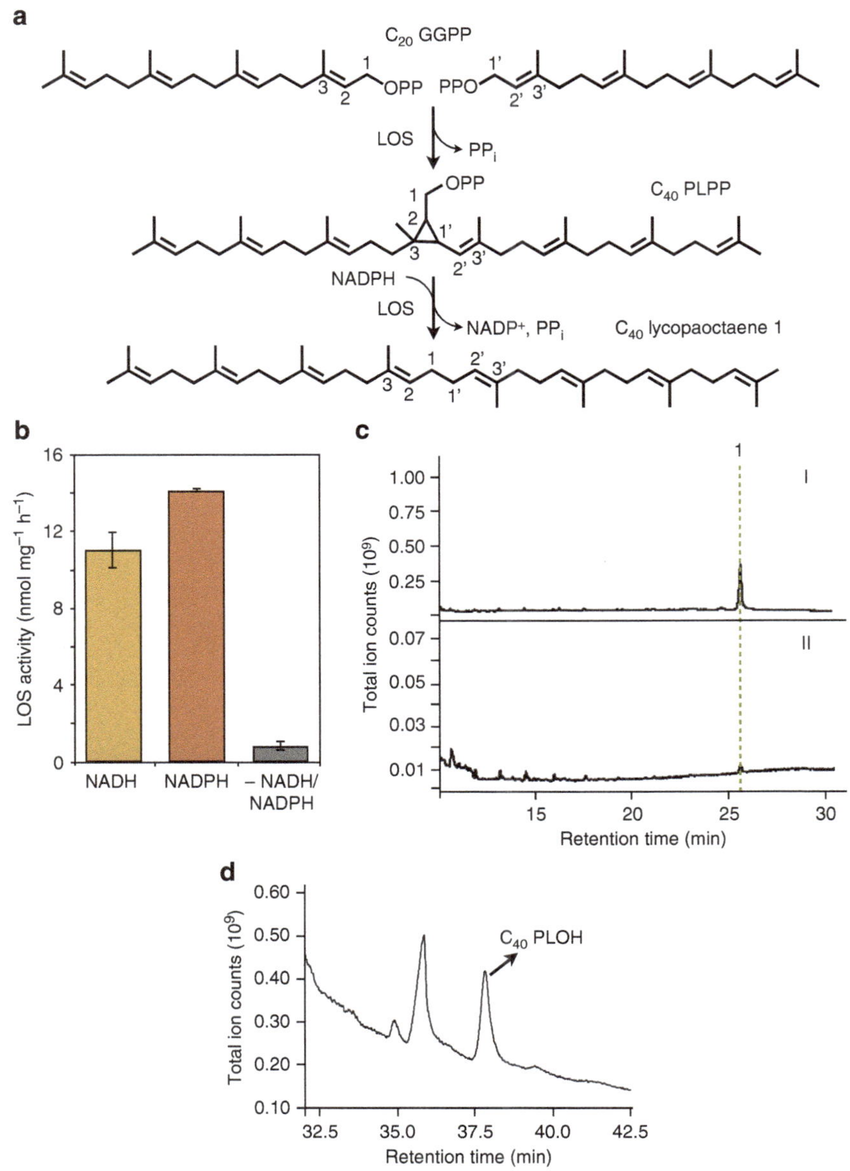

Figure 5 | Characterization of the LOS reaction intermediate. (**a**) Proposed LOS reaction mechanism to produce the PLPP intermediate and lycopaoctaene final product. (**b**) Radioactive LOS enzyme assays conducted with ^{3}H-GGPP substrate with or without NADH or NADPH reducing agents. Values shown are the mean ± s.e. obtained from three independent measurements ($n=3$). (**c**) GC-MS profiles of LOS enzyme assay with GGPP as substrate to produce lycopaoctaene (1): I, in the absence, or II, in the presence of 20 nM squalestatin. (**d**) GC-MS profile of products from LOS enzyme assay with GGPP substrate in the absence of NADPH to produce PLPP, which was hydrolysed by acid phosphatase to yield PLOH for GC-MS analysis. Data shown in **c** and **d** are representatives from three different experiments ($n=3$).

was further engineered through mutations to accept C_{25} farnesylgeranyl diphosphate, yielding C_{40}, C_{45} and C_{50} homologues of dehydrosqualene in the presence of FPP, farnesylgeranyl diphosphate and GGPP[42,43]. In another study, SSs from yeast, pig and rat have been shown to accept non-natural derivatives of FPP, to generate several unnatural terpene products *in vitro*[45,46]. In addition, yeast SS has been shown to yield alternative products *in vitro* from FPP as a substrate under various non-physiological reaction conditions, such as in the absence of NADPH with extended incubation times or in the presence of an unreactive NADPH analogue[47–49]. Irrespective of these examples, the promiscuity of LOS is still remarkable, considering that the wild-type LOS enzyme is able to accept three naturally occurring prenyl diphosphate substrates and produce the chimeric hydrocarbons described here without the need for mutations or altered, non-physiological reaction conditions. Such remarkable intrinsic substrate promiscuity has not been documented for any other eukaryotic SS or SSL enzyme up until now.

Ours and other studies[37,50] raise the question about how SSL enzymes arise. LSS and LOS from *B. braunii* share >50% sequence identity and both contain the important catalytic residues and conserved domains of SS enzymes (Supplementary Fig. 18), but they still have different substrate preferences and product formation. Thus, we hypothesize that an ancient SS gene may have undergone gene duplication in race L. The paralogue that gave rise to LSS maintained the ability to produce squalene as its sole product to support sterol biosynthesis for primary metabolism. However, the paralogue that yielded LOS developed the use of alternative substrates for hydrocarbon oil

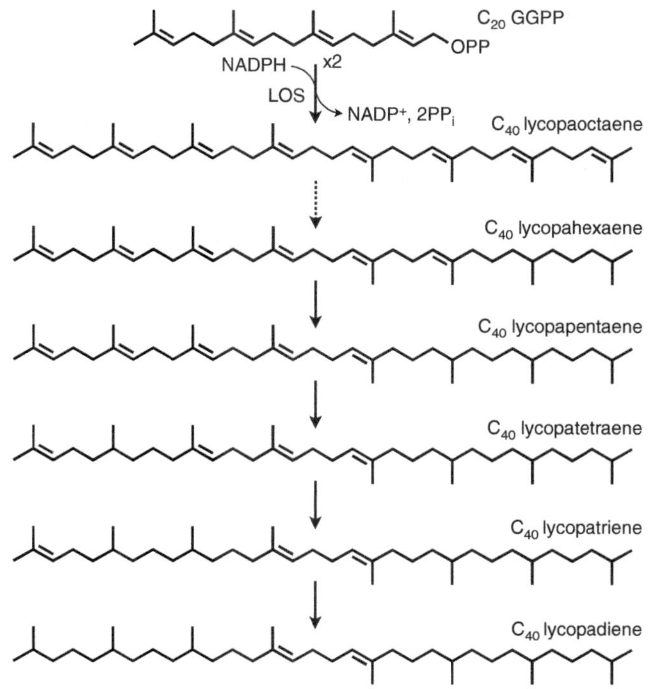

Figure 6 | Proposed pathway for lycopadiene biosynthesis. LOS uses GGPP in a two-step reaction to produce lycopaoctaene, which is further reduced by a yet unknown enzyme(s) to produce lycopahexaene, lycopapentaene, lycopatetraene, lycopatriene and finally lycopadiene.

production, while also preserving SS activity. A similar scenario has been suggested as a possible mechanism for generating new secondary metabolites that may provide a fitness benefit to the host organism[42,51]. For *B. braunii*, it has been proposed that hydrocarbon oils and their derivatives increase the buoyancy of colonies for increased exposure to sunlight[50,52,53]. Similar gene diversifications have also been proposed in the B race of *B. braunii* for botryococcene production, and in some bacteria for squalene generation. For botryococcene production, *SS* duplications gave rise to the *SSL-1* and *SSL-3* genes/enzymes responsible for botryococcene production, while retaining a residual ability to produce squalene[37]. For bacteria, a three-enzyme system for squalene production arose from successive gene duplication events[50].

LOS produces a variety of terpenes *in vitro* when offered a mixture of substrates and may likely also do so *in vivo*. In the *in vitro* mixed substrate assays with LOS (Fig. 4a), a $C_{35}H_{64}$ molecule was observed from FPP and PPP condensation. This product has the exact same GC retention time and mass spectrum as the $C_{35}H_{64}$ hydrocarbon detected in the total hydrocarbon pool (Fig. 2a (VII) and Fig. 4a (II)). This suggests that LOS is also promiscuous in *B. braunii* cells, but at best has limited access to the FPP and PPP pools, as the $C_{35}H_{64}$ product only comprises 0.15% of the total hydrocarbons (Fig. 2a (VII)). The FPP–GGPP condensation product $C_{35}H_{58}$ seen in these *in vitro* assays (Fig. 4a (I)) is not detected in the total hydrocarbon pool, possibly because this molecule is reduced to $C_{35}H_{64}$ *in vivo*. It should be noted that two previous studies described several C_{35} squalene-like isoprenoids from *Methanococcus jannaschii* and *Thermococcus barophilus*, including the $C_{35}H_{58}$ and $C_{35}H_{64}$ molecules identified here[54,55], although the enzymatic basis for the biosynthesis of these products remains uncharacterized. In addition, the ability of LOS to produce lycopapentaene as an observed *in vitro* condensation product of GGPP and PPP (Fig. 4a (III)) may indicate the existence of an alternative *in vivo* biosynthetic route to this molecule and may be partially responsible for the relative prevalence of lycopapentaene (3% of total hydrocarbons; Fig. 2a (IV)) among the minor hydrocarbons of race L. This would suggest that LOS can also use PPP *in vivo*. However, LOS does not appear to use PPP as a single substrate for lycopadiene production *in vivo*, as we were not able to detect lycopadiene synthesis activity in cell fractions using PPP as substrate (Fig. 2b).

The promiscuous LOS described in this work is nevertheless used by *B. braunii* Race L to produce lycopaoctaene preferentially over squalene, as shown by the large amounts of hydrocarbon oils produced by this alga and the absence of detectable amounts of squalene or non-sterol squalene derivatives in the total hydrocarbon pool. In contrast, accumulation of the non-sterol squalene derivatives methylsqualenes in race B are thought to arise from squalene produced by the additional enzyme SSL-2 (ref. 37). Thus, detection of non-sterol squalene derivatives in race L would be expected if LOS was producing squalene *in vivo*.

We hypothesize that the observed preference of LOS for C_{40} lycopaoctaene biosynthesis over C_{30} squalene production *in vivo* may result from an increased flux from GGPP over FPP due to metabolic channelling, to form a biosynthetic metabolon specific for lycopadiene production. Formation of such metabolons is common in plants and is used for channelling substrates and intermediates required for the biosynthesis of different classes of plant natural products, including isoprenoids[56–58]. When expressed in yeast, LOS may produce squalene presumably due to a lack of this proposed metabolon, thus giving LOS access to the FPP pool. We therefore conclude that LOS is responsible for synthesizing lycopaoctaene as the first committed step towards lycopadiene hydrocarbon biosynthesis in race L of *B. braunii*.

Methods

Reagents. [1-^{3}H]-FPP (specific activity, 18.2 Ci mmol^{-1}) was purchased from PerkinElmer. [1-^{3}H]-GGPP (specific activity, 20.0 Ci mmol^{-1}), [1-^{3}H]-PPP (specific activity, 20.0 Ci mmol^{-1}) and non-radiolabelled PPP were purchased from American Radiolabeled Chemicals. When required, specific activities of radiolabelled chemicals were adjusted with non-labelled chemicals. Solvents and chemicals were purchased from VWR. All other reagents were purchased from Sigma unless otherwise noted.

Culturing of *B. braunii*. *B. braunii*, race L, Songkla Nakarin strain[16] was obtained from Algobank-Caen Microalgal Culture Collection, University of Caen Basse-Normandie, France, and grown in a modified Chu 13 medium, pH 7.5 (ref. 59) at 22 °C under continuous aeration of filter-sterilized air enriched with 2.5% CO_2. The concentrations of chemicals in modified Chu 13 medium were as follows: KNO_3 (0.4 g l^{-1}), $MgSO_4{\bullet}7H_2O$ (0.1 g l^{-1}), K_2HPO_4 (0.052 g l^{-1}), $CaCl_2{\bullet}2H_2O$ (0.054 g l^{-1}), FeNa EDTA (0.01 g l^{-1}), H_3BO_4 (2.86 mg l^{-1}), $MnSO_4{\bullet}H_2O$ (1.54 mg l^{-1}), $ZnSO_4{\bullet}7H_2O$ (0.22 mg l^{-1}), $CuSO_4{\bullet}5H_2O$ (0.08 mg l^{-1}), $NaMoO_4{\bullet}2H_2O$ (0.06 mg l^{-1}) and $CoSO_4{\bullet}7H_2O$ (0.09 mg l^{-1}). The cultures were grown under a 12:12 h light:dark cycle with a light intensity of 120 μE m^{-2} s^{-1}. Algae cells were subcultured by inoculation of 100 ml of a floating 6-week-old culture to 750 ml of fresh medium.

Hydrocarbon staining and microscopy. Nile Red (Sigma) staining of L race hydrocarbons was done by treating 500 μl of *B. braunii* L race colonies in medium at early phase (5 day) density with 2.5 μl of a stock solution of Nile Red dissolved in acetone (0.15 mg ml^{-1}) for a final concentration of Nile red of 0.375 g ml^{-1} and acetone of 0.25%. Samples were kept in the dark and incubated at room temperature for 15 min before imaging. Confocal microscopy images were obtained using an Olympus IX83-FV1200 inverted confocal microscope with spectral detection unit using an Olympus UPLSAPO ×60 oil-immersion objective (numerical aperture = 1.35). The confocal aperture diameter was set to 1 Airy disk unit and Z-series images were acquired with voxel dimensions of 101 × 101 × 490 (nm; *x*-*y*-*z*, respectively). Nile Red and chlorophyll were excited using a 488-nm laser and detected using 540–590 nm and 655–755 nm barrier filter settings, respectively. Bright-field images were acquired via the transmitted detector using the 488-nm laser. Maximum intensity projections and brightness/contrast adjustments were completed using the Olympus FV-ASW software version 4.

Hydrocarbon purification. Hydrocarbons were purified as described previously[60,61] and as follows. Algal cells were harvested by vacuum filtration using a 10-μm nylon-mesh filter, freeze-dried, extracted twice with *n*-hexane for 2 h, to recover extracellular hydrocarbons, and then twice with $CHCl_3$:MeOH (2:1) for 12 h stirring at room temperature, to recover intracellular hydrocarbons. Both extracts were concentrated separately using a rotary evaporator and resuspended in a small amount of *n*-hexane before running through separate gravity-fed silica gel columns with *n*-hexane as solvent. The *n*-hexane eluent before the pigment front was collected as the hydrocarbon fraction. Both extracellular and intracellular hydrocarbon fractions were combined and evaporated to dryness, to recover the total hydrocarbon pool of the algae. Individual hydrocarbon molecules were purified by injecting the total hydrocarbon sample onto an HPLC Develosil 60 silica column (20 mm × 250 mm) using *n*-hexane as a mobile phase at a flow rate of 6 ml min^{-1} and detection at 210 nm. An aliquot of the purified hydrocarbons were analysed by GC–MS for purity analysis.

GC-MS conditions. Enzyme assay products, purified hydrocarbons and yeast extracts were analysed by GC–MS (Bruker 436-GC-SCION SQ Premium) using a 5% Phenyl BR-5 ms capillary column (30 m × 0.25 mm, film thickness: 0.25 μm) in electron ionization (70 eV) mode. Initial oven temperature was 220 °C, held for 1 min, then increased to 280 °C at the rate of 5 °C min^{-1} and then ramped to 300 °C at a rate of 2 °C min^{-1} and held for 20 min. Helium was used as a carrier gas at a flow rate of 2.58 ml min^{-1}. Temperatures of injection port, interface and ion source were 280 °C, 250 °C and 200 °C, respectively.

Hydrocarbon NMR analysis. One-dimensional NMR spectra of race L hydrocarbons molecules suffer from severe spectral overlap and the data interpretation is further complicated by degenerate chemical shifts as a result of symmetry planes in the lycopadiene structure. Consequently, hetero-nuclear two-dimensional (2D) experiments were used for unambiguous structural characterization of each entity. NMR experiments were performed at 25 °C on purified samples dissolved in deuterated chloroform using a 500 MHz Bruker Avance III HD spectrometer equipped with an inverse detection TXI probe. We have relied on 2D heteronuclear experiments acquired at natural abundance of ^{13}C nuclei, to deduce structures of these molecules. Multiplicity-edited 2D heteronuclear single quantum coherence spectrum was acquired for each compound and the resonances were assigned using a combination of 2D heteronuclear 2 bond correlation, 2D heteronuclear multiple bond correlation and 2D heteronuclear single quantum coherence–total correaltion spectroscopy experiments. The ^{13}C chemical shifts obtained for lycopadiene match those previously obtained[16].

Spectral data for hydrocarbons analysed. Lycopadiene: ^1H NMR (500 MHz, $CDCl_3$): δ 0.87, 0.89, 1.08, 1.09, 1.16, 1.2, 1.29, 1.39, 1.4, 1.54, 1.61, 1.96, 2.04, 2.17, 5.15; ^{13}C NMR (125 MHz, $CDCl_3$): δ 15.9, 19.7, 22.7, 24.5, 24.7, 25.4, 28.0, 28.3, 31.2, 32.8, 36.8, 37.4, 39.4, 40.0, 124.0, 135.9; MS (*m/z*): [M^+] calcd. for $C_{40}H_{78}$, 558.61; found, 558.70. Lycopatriene: ^1H NMR (500 MHz, $CDCl_3$): δ 0.87, 0.89, 1.08, 1.09, 1.13, 1.16, 1.2, 1.29, 1.29, 1.39, 1.4, 1.54, 1.61, 1.63, 1.71, 1.96, 1.98, 2.04, 2.16, 5.12, 5.15; ^{13}C NMR (125 MHz, $CDCl_3$): δ 15.9, 17.6, 19.7, 22.7, 24.5, 24.7, 24.7, 25.4, 25.6, 25.7, 28.0, 28.3, 30.8, 32.7, 36.8, 37.1, 37.4, 39.4, 40.0, 124.1, 125.1, 131.1, 135.9; MS (*m/z*): [M^+] calcd. for $C_{40}H_{76}$, 556.59; found, 556.70. Lycopatetraene: ^1H NMR (500 MHz, $CDCl_3$): δ 0.87, 0.89, 1.08, 1.13, 1.16, 1.2, 1.29, 1.39, 1.4, 1.54, 1.61, 1.63, 1.71, 1.95, 1.96, 1.99, 2.04, 2.16, 5.12, 5.15; ^{13}C NMR (125 MHz, $CDCl_3$): δ 15.9, 17.6, 19.7, 22.7, 24.5, 24.7, 25.4, 25.5, 25.7, 28.0, 28.3, 30.8, 32.7, 36.6, 37.1, 37.4, 39.3, 39.4, 40.0, 124.1, 125.2, 131.1, 135.9; MS (*m/z*): [M^+] calcd. for $C_{40}H_{74}$, 554.58; found, 554.60. Lycopapentaene: ^1H NMR (500 MHz, $CDCl_3$): δ 0.87, 0.89, 1.08, 1.16, 1.2, 1.29, 1.39, 1.4, 1.55, 1.62, 1.63, 1.71, 2, 2.04, 2.09, 2.23, 5.15; ^{13}C NMR (125 MHz, $CDCl_3$): δ 16.0, 17.7, 19.8, 22.7, 24.5, 24.7, 25.4, 25.7, 26.7, 28.0, 28.3, 31.0, 32.8, 36.6, 37.4, 39.4, 39.8, 124.2, 131.1, 135.9; MS (*m/z*): [M^+] calcd. for $C_{40}H_{72}$, 552.56; found, 552.70. Lycopapentaene isomer: ^1H NMR (500 MHz, $CDCl_3$): δ 0.87, 0.89, 1.08, 1.13, 1.16, 1.2, 1.29, 1.4, 1.55, 1.62, 1.63, 1.71, 1.99, 2, 2.04, 2.09, 5.1, 5.15; ^{13}C NMR (125 MHz, $CDCl_3$): δ 16.0, 17.7, 19.8, 22.7, 24.5, 24.7, 25.5, 25.7, 26.7, 28.0, 28.3, 32.7, 36.6, 37.1, 37.4, 39.4, 39.8, 124.2, 124.8, 131.1, 135.9; MS (*m/z*): [M^+] calcd. for $C_{40}H_{72}$, 552.56; found, 552.90. Lycopahexaene: ^1H NMR (500 MHz, $CDCl_3$): δ 0.87, 0.89, 1.08, 1.16, 1.2, 1.29, 1.4, 1.55, 1.61, 1.63, 1.71, 1.96, 1.99, 2.04, 2.06, 5.15; ^{13}C NMR (125 MHz, $CDCl_3$): δ 16.0, 17.7, 19.8, 22.7, 24.5, 24.7, 25.7, 26.8, 28.0, 28.3, 32.8, 36.6, 37.4, 39.4, 39.8, 124.3, 131.1, 135.9; MS (*m/z*): [M^+] calcd. for $C_{40}H_{70}$, 550.55; found, 550.60. Lycopaoctaene: ^1H NMR (500 MHz, $CDCl_3$): δ 1.62, 1.63, 1.71, 2, 2.03, 2.04, 5.15; ^{13}C NMR (125 MHz, $CDCl_3$): δ 16.0, 17.7, 25.7, 27.2, 28.3, 39.8, 124.3, 131.1, 135.9; MS (*m/z*): [M^+] calcd. for $C_{40}H_{66}$, 546.52; found, 546.70. $C_{35}H_{64}$: ^1H NMR (500 MHz, $CDCl_3$): δ 0.83, 0.86, 1.06, 1.13, 1.19, 1.21, 1.24, 1.32, 1.36, 1.52, 1.61, 1.68, 1.98, 2.01, 2.06, 5.13; ^{13}C NMR (125 MHz, $CDCl_3$): δ 16.3, 17.9, 20.0, 22.9, 24.4, 24.8, 25.1, 26.0, 26.3, 27.2, 28.2, 33.0, 37.7, 39.6, 40.0, 124.8, 131.4, 135.1; MS (*m/z*): [M^+] calcd. for $C_{35}H_{64}$, 484.50; found, 484.70.

Ozonolysis experiments. Owing to the limited amount of minor hydrocarbons in the total L race hydrocarbon pool, ozonolysis experiments were conducted only on those hydrocarbons that could be purified to obtain at least 10 mg: lycopadiene, lycopatriene and lycopapentaene. These hydrocarbons were separately dissolved in dichloromethane and submitted to ozone cleavage for 5 min at −78 °C. Each product from reductive cleavage of the resulting ozonide was directly subjected to GC–MS (electron ionization) analyses without purification. GC–MS analysis was carried out using a GCMS-QP2010 Ultra spectrometer (Shimadzu, Kyoto, Japan) equipped with a capillary column (InertCap 1MS, GL Science; 60 m × 0.25 mm, film thickness: 0.25 μm). The column temperature was programmed as follows: 50 °C for 1 min, raised at 10 °C min^{-1} from 50 to 220 °C, then at 2 °C min^{-1} from 220 to 260 °C and held for 22 min at the final temperature. Helium was used as a carrier gas at a flow rate of 41.2 cm s^{-1}. Temperatures of injection port, interface and ion source were 260, 250 and 200 °C, respectively.

Preparation of algal cell lysate. Algae from the rapid growth phase were harvested, snap frozen with liquid nitrogen and stored at −80 °C for future use. In a typical preparation, 200 mg of frozen algae was added to eight individual tubes containing 0.8 ml extraction buffer (50 mM MOPS pH 6.8, 20 mM $MgCl_2$, 5 mM β-mercaptoethanol, 5 mM EGTA and 20% (v/v) glycerol) with 200 μl of stainless steel metal beads (mixture of 0.9–2 mm diameter beads) and then homogenized using a Bullet Blender Storm 24 (Next Advance, Inc., USA) for 10 min at a speed of 10. The algal homogenates were centrifuged at 9,000 *g* for 10 min at 4 °C. The 9,000 *g* supernatants were pooled and then centrifuged at 100,000 *g* for 1 h to obtain the soluble fraction. The pellet was washed with 4 ml extraction buffer and centrifuged again at 100,000 *g*. This process was repeated twice and the pellet was resuspended in 0.5 ml extraction buffer to obtain a microsomal fraction. An aliquot of each protein fraction was used for enzyme assays.

Enzyme assays. Radioactive enzyme assays were conducted in a 50-μl total reaction volume as described previously[23,37] and as follows. Reactions were initiated by adding 10 μg algal lysate or 1 μg of recombinant enzyme purified from *E. coli* to a reaction buffer containing 50 mM MOPS pH 6.8, 2.5 mM β-mercaptoethanol, 20 mM $MgCl_2$, 2 mM NADPH or NADH and 10 μM [1-^3H]-prenyl-PP substrate (specific activity = 0.25 Ci mmol^{-1}). Enzyme assays were incubated at 37 °C for 60 min and terminated by adding 60 μl of *n*-hexane, followed by brief vortexing and centrifugation. Thirty microlitres of the organic layer was spotted onto silica gel 60 thin-layer chromatography plates along with authentic standards of squalene (R_f = 0.17), C_{30}-botryococcene (R_f = 0.2), lycopaoctaene (R_f = 0.09) and lycopadiene (R_f = 0.5), and then developed with *n*-hexane. Hydrocarbon standards were visualized by iodine vapour and spots corresponding to authentic standards were scraped and analysed by liquid scintillation counter, to determine radioactive incorporation into the respective hydrocarbons. SS and botryococcene synthase assays were conducted using ^3H-FPP, whereas LOS and lycopadiene synthase assays were conducted using ^3H-GGPP and ^3H-PPP, respectively. For GC–MS analysis of reaction products, enzyme assays were conducted with non-radiolabelled substrates in glass vials by scaling the total reaction volume and contents to 500 μl. Two separate 500 μl reactions were each extracted twice with 1 ml *n*-hexane, the extracts pooled, dried under a stream of nitrogen gas, resuspended in 100 μl *n*-hexane and a 5-μl aliquot analysed by GC–MS.

Hydrocarbon accumulation and LOS activity over growth cycle. Several flasks of race L culture were grown over a 39- day culture period. Twenty-five millilitres of culture was collected at indicated times on pre-weighed Whatman GF/C filters by vacuum filtration. The filters with algae were dried in an 80 °C oven and total hydrocarbons were extracted using the protocol described above. Similarly, 500 ml of algae cells were also harvested at indicated times by filtering through a 10-μm nylon mesh and the samples used to determine LOS enzyme activity over the culture period. LOS enzyme assays were conducted using 10 μg of 9,000 *g* supernatant.

RNA isolation. Total RNA was isolated as previously described[62] and as follows. For *B. braunii*, liquid nitrogen frozen samples were pulverized using a Tissuelyser II (Qiagen, Valencia, CA). Approximately 200 mg of this frozen pulverized *B. braunii* tissue, or frozen leaf tissues of *A. thaliana* and *S. lycopersicum*, were ground in a mortar and pestle with liquid nitrogen. The samples were then added to 1 ml of TRIzol (Life Technologies, Grand Island, NY) and the total RNA was isolated following the manufacturer's instructions. The RNA from *A. thaliana* and *S. lycopersicum* were used directly for gene cloning (see below). For the *B. braunii* RNA, contaminating polysaccharides were removed before the use for gene cloning as follows. The RNA pellet was resuspended in 0.5 ml 2 M LiCl, incubated for 5 min and centrifuged at 12,000 *g* for 15 min at 4 °C to pellet the total RNA. The polysaccharides remain in the supernatant. This process was repeated until the size of RNA pellet remained constant. The RNA pellet was then dissolved in 0.5 ml 1 × Tris-EDTA (TE), extracted with an equal volume of phenol/chloroform/isoamyl alcohol mix (25:4:1) and centrifuged at 12,000 *g* for 15 min at 4 °C. The aqueous supernatant phase was removed and extracted with 0.5 ml of chloroform followed by centrifugation at 12,000 *g* for 15 min at 4 °C. The RNA was precipitated from the aqueous supernatant by adding 0.1 volume 3 M sodium acetate and 2.5 volumes of 100% ethanol, incubated for 20 min at −20 °C and centrifuged at 12,000 *g* for 15 min at 4 °C. The RNA pellet was washed twice with 0.5 ml 70%

ethanol, centrifuged at 12,000 *g* for 15 min at 4 °C, the pellet dried using a speedvac and the RNA resuspended in 50 μl of RNAase-free water.

***LSS* and *LOS* cloning.** RNAseq analysis on the L race of *B. braunii* was carried out as previously reported[62] and as follows. RNA isolated from days 0, 3, 7, 14, 21 and 28 over the 4-week culture cycle were kept separate, paired-end libraries were prepared from each RNA sample and each sample was sequenced under the Illumina platform. A transcriptome contig library was created using the Trinity software suite. *SSL* sequences were computationally screened using this transcriptomic database with *BSS* as a query. Two *SSL* cDNAs, *LSS* and *LOS*, were identified. Total RNA was extracted from a day-3 race L culture and first-strand cDNA was prepared using the SuperScript III first-strand synthesis kit (Invitrogen). Primers specific to each cDNA based on the transcriptome sequence were used to amplify the PCR product from first-strand cDNA using GoTaq DNA polymerase mix (Promega) followed by cloning into the pGEM-T vector (Promega). Gene-specific primers were as follows: For *LSS*, forward primer 5′-ATGGGGAAGCTACAGGAGGTTTTGAAGC-3′ and reverse primer 5′-TCAGGCAAGGCCGCCGCGAAG-3′; for *LOS*, forward primer 5′-ATGAAGTACACAGATTTCCTTGCGC-3′ and reverse primer 5′-TCACACAGTCTTGAGGGCGAG-3′.

***AtGGPPS11* and *SlSS* cloning.** The cDNA sequences for *AtGGPPS11* (*At*4g36810)[30] and *SlSS* were obtained from the NCBI nucleotide database. Total RNA isolated from the leaves of *A. thaliana* and *S. lycopersicum* were used for first-strand cDNA synthesis and PCR products corresponding to each cDNA were amplified using gene-specific primers followed by cloning into the pGEM-T vector. Gene-specific primers used for PCR amplification were as follows: for *AtGGPPS11*, forward primer 5′-ATGGCTTCAGTGACTCTAGGTTC-3′ and reverse primer 5′-TCAGTTCTGTCTATAGGCAATG-3′; for *SlSS*, forward primer 5′-ATGGGAACATTGAGGGCA-3′ and reverse primer 5′-CTAAGACCGGCTGCCAAAAAGTTG-3′.

Protein expression and purification. Except for *SSL-1* and *SSL-3*, DNA sequences encoding the predicted transmembrane domain at the carboxy terminus of each SS or SSL protein were deleted and then cloned to pET28a, to encode for amino terminal 6 × His-tagged proteins. DNA templates of *BSS*, *SSL-1*, *SSL-2* and *SSL-3* were received from previous studies[25,37]. The pET28a expression constructs were made using appropriate restriction sites; $LOS^{\Delta391-444}$ with NheI and HindIII; $LSS^{\Delta399-462}$, $SlSS^{\Delta387-411}$, $BSS^{\Delta399-461}$, *SSL-1* and *SSL-3* with NdeI and HindIII; and $SSL\text{-}2^{\Delta392-465}$ with NdeI and SalI. The expression constructs were transformed into *E. coli* BL21(DE3), grown at 37 °C to OD_{600} = 0.8 and protein expression for each gene induced by adding 1 mM isopropylthio-β-D-galactoside. The induced cultures were then grown for an additional 6 h at 25 °C.

His-tagged proteins were purified at 4 °C by standard procedures. In a typical purification, pellets from a 100-ml culture were resuspended in 10 ml extraction buffer (50 mM sodium phosphate buffer pH 7.8, 300 mM NaCl, 10 mM imidazole, 1 × general protease inhibitor cocktail (Sigma), 1 mM $MgCl_2$ and 1% glycerol (v/v)), and then sonicated four times for 15 s at 70% maximum power with 2 min interval between each sonication. The sonicated samples were centrifuged at 16,000 *g* for 10 min at 4 °C. The supernatants were applied to a gravity-fed column containing Ni-NTA agarose (Qiagen) and the His-tagged proteins were purified according to the manufacturer's recommendations. The wash buffer (50 mM sodium phosphate buffer pH 7.8, 300 mM NaCl, 60 mM imidazole, 1 mM $MgCl_2$ and 1% glycerol (v/v)) and the elution buffer (50 mM sodium phosphate buffer pH 7.8, 300 mM NaCl, 400 mM imidazole, 1 mM $MgCl_2$ and 1% glycerol (v/v)) were used for protein purification. The eluted fractions were dialysed using storage buffer (300 mM NaCl, 20 mM Tris-HCl pH 7.5, 5 mM dithiothreitol and 2 mM $MgCl_2$), concentrated with an Amicon Ultra centrifugal filter (0.5 ml, 30 kDa cutoff; EMD Millipore) to desired protein concentration, an equal amount of 100% glycerol added and stored at − 20 °C for 1–2 months without a loss of enzyme activity.

LOS steady-state kinetic experiments. Michaelis–Menten enzyme kinetics experiments with the LOS enzyme were set up using the protocol described previously[63] and as follows. Pilot experiments were initially conducted using radioactive enzyme assays, to determine the reaction conditions where reaction velocity is linear (< 10% turnover) with respect to enzyme concentration. For steady-state kinetics experiments, enzyme assays (50 μl total reaction volume) were conducted with 100 nM of purified LOS enzyme at the indicated concentration of ^{3}H-GGPP, ^{3}H-FPP or ^{3}H-PPP for a set time interval. Kinetics for the NADPH cofactor were not determined and this factor was held constant at 2 mM in all assays. The reaction velocities were plotted against substrate concentrations to generate Michaelis–Menten curves and the kinetics parameters of LOS enzyme for GGPP, FPP and PPP were determined by analysing the data using GraphPad Prism 6 software.

Yeast expression. The DNA sequence encoding the 56 N-terminal amino acids for the plastid-targeting signal of *AtGGPPS11* were deleted resulting in $AtGGPPS11^{\Delta57}$, which was cloned into pESC-TRP using BamHI and SalI restriction sites for expression under the inducible GAL1 promoter. For coexpression studies, the DNA sequences encoding the predicted C-terminal transmembrane domain of *LOS* was deleted resulting in $LOS^{\Delta391-444}$, which was cloned into the second multiple cloning site of $AtGGPPS11^{\Delta57}$:pESC-TRP using EcoRI and SpeI restriction sites for expression under the inducible GAL10 promoter. The yeast expression constructs $AtGGPPS11^{\Delta57}$:pESC-TRP and $AtGGPPS11^{\Delta57} + LOS^{\Delta391-444}$:pESC-TRP were introduced into yeast strain CKY457 (*MAT***a**, leu2Δ1, *ura3-52*, *trp1Δ63*, *his3Δ200* and *lys2-128∂*) via lithium acetate transformation followed by selection on yeast synthetic dropout medium (SC-TRP). Positive transformants were grown at 30 °C in 150 ml SC-TRP media to mid-log phase, induced with 2% final galactose concentration and grown for an additional 130 h. Yeast cells were harvested, freeze-dried and extracted with *n*-hexane for 2 h by stirring at room temperature. The organic extracts were centrifuged at 1,000 *g*, the supernatant dried using a rotary evaporator, resuspended in 500 μl *n*-hexane and a 5-μl aliquot analysed by GC–MS to evaluate for the production of lycopaoctaene.

Purification of lycopaoctaene standard. A 3-l culture of yeast strain CKY457 expressing $AtGGPPS11^{\Delta57} + LOS^{\Delta391-444}$:pESC-TRP was grown and the metabolites were extracted with *n*-hexane using the protocol described above. The organic extracts were applied to a silica gel gravity-fed column and metabolites eluted sequentially with two different solvent systems, *n*-hexane and 1% diethyl ether in *n*-hexane. The 1% diethyl ether in *n*-hexane eluent fraction containing lycopaoctaene and other metabolites was concentrated and resuspended in a small volume of *n*-hexane before injecting the samples onto a 10 mm × 250 mm Cosomil $5C_{18}$-AR-II HPLC column. Lycopaoctaene was purified by reversed phase HPLC using methanol:acetone (60:40) as a mobile phase at a flow rate of 3 ml min^{-1} and detection at 210 nm. Identity of lycopaoctaene molecule was confirmed by GC–MS (Supplementary Fig. 3) and NMR (Supplementary Table 1 and Supplementary Fig. 15).

LSS and LOS complementation of yeast *SS* knockout strain. The yeast strain ZX 178-08 (*MAT***a**, *his3*, *leu2*, *met15*, *ura3*, *SUE* and *erg9Δ::HPH*), which has a knockout of the endogenous *SS* gene (*ERG9*), was used for complementation experiments[64]. The expression vector XURA used in our study was made by replacing the GAL1 and GAL10 promoters in pESC-URA vector backbone with constitutive promoters TEF1 and GPD, respectively[65]. Full-length *LSS* was cloned into XURA using Not1 and SpeI restriction sites. Full-length *LOS* was cloned into EcoRI and SpeI sites of XURA vector. The expression constructs *LSS*:XURA, *LOS*:XURA and XURA were transformed into ZX 178-08 yeast strain via lithium acetate transformation followed by selection on yeast synthetic dropout medium supplemented with ergosterol synthetic complete-Uracil (SCE URA). The positive transformants were further characterized by streaking on selection media with or without ergosterol, to test the ability of individual genes to restore ergosterol prototrophy of ZX 178-08 yeast.

GC–MS analysis of PLOH. The enzyme assay for PLPP was conducted following the protocol described previously for analysis of PSPP[37,50] and as follows. The PLPP assay contained 50 mM MOPS (pH 6.8), 2.5 mM β-mercaptoethanol, 20 mM $MgCl_2$, 180 μM GGPP and 750 μg of purified LOS enzyme in a 2.5-ml total reaction volume. The reaction mixture was incubated at 37 °C for 2 h and terminated by snap freezing with liquid nitrogen followed by lyophilization. For hydrolysis of PLPP to PLOH, the white residue obtained after lyophilization was resuspended in 2 ml phosphatase solution (20% 1-propanol, 100 mM sodium acetate pH 4.7, 0.1% Triton-X-100 and 50 units of sweet potato acid phosphatase) and then incubated for 16 h in a 28 °C shaker. The dephosphorylation reaction mixture was extracted three times with 4 ml of methyl tertiary butyl ether (MTBE), the extracts pooled, dried under a stream of nitrogen gas, resuspended in 100 μl *n*-hexane and a 5-μl aliquot analysed by GC–MS as under the conditions described above with the following differences. Chemical ionization using methane gas at 20 psi was employed for analysis of the PLOH molecule, as electronic ionization did not result in a molecular ion of PLOH. Initial oven temperature was 70 °C, held for 1 min, then increased to 200 °C at the rate of 8 °C min^{-1} and then ramped to 300 °C at a rate of 20 °C min^{-1} and held for 20.75 min. Temperatures of injection port, interface and ion source were 260 °C, 250 °C and 200 °C, respectively.

References

1. Moody, J. W., McGinty, C. M. & Quinn, J. C. Global evaluation of biofuel potential from microalgae. *Proc. Natl Acad. Sci. USA* **111**, 8691–8696 (2014).
2. Chisti, Y. Constraints to commercialization of algal fuels. *J. Biotechnol.* **167**, 201–214 (2013).
3. Banerjee, A., Sharma, R., Chisti, Y. & Banerjee, U. C. *Botryococcus braunii*: a renewable source of hydrocarbons and other chemicals. *Crit. Rev. Biotechnol.* **22**, 245–279 (2002).
4. Yoshida, M., Tanabe, Y., Yonezawa, N. & Watanabe, M. M. Energy innovation potential of oleaginous microalgae. *Biofuels* **3**, 761–781 (2012).
5. Lee, S. Y., Kim, H. M. & Cheon, S. Metabolic engineering for the production of hydrocarbon fuels. *Curr. Opin. Biotechnol.* **33**, 15–22 (2015).
6. Metzger, P. & Largeau, C. *Botryococcus braunii*: a rich source for hydrocarbons and related ether lipids. *Appl. Microbiol. Biotechnol.* **66**, 486–496 (2005).

7. Hillen, L. W., Pollard, G., Wake, L. V. & White, N. Hydrocracking of the oils of *Botryococcus braunii* to transport fuels. *Biotechnol. Bioeng.* **24,** 193–205 (1982).
8. Moldowan, J. M. & Seifert, W. K. First discovery of botryococcane in petroleum. *J. Chem. Soc. Chem. Commun.* **19,** 912–914 (1980).
9. Mckirdy, D. M., Cox, R. E., Volkman, J. K. & Howell, V. J. Botryococcane in a new class of Australian non-marine crude oils. *Nature* **320,** 57–59 (1986).
10. Lichtfouse, E., Derenne, S., Mariotti, A. & Largeau, C. Possible algal origin of long chain odd *n*-alkanes in immature sediments as revealed by distributions and carbon isotope ratios. *Org. Geochem.* **22,** 1023–1027 (1994).
11. Glikson, M., Lindsay, K. & Saxby, J. *Botryococcus*-A planktonic green alga, the source of petroleum through the ages: transmission electron microscopical studies of oil shales and petroleum source rocks. *Org. Geochem.* **14,** 14 (1989).
12. Adam, P., Schaeffer, P. & Albrecht, P. C40 monoaromatic lycopene derivatives as indicators of the contribution of the alga *Botryococcus braunii* race L to the organic matter of Messel oil shale (Eocene, Germany). *Org. Geochem.* **37,** 584–596 (2006).
13. Khan, N. E., Nybo, E. S., Chappell, J. & Curtis, W. R. Triterpene hydrocarbon production engineered into a metabolically versatile host—*Rhodobacter capsulatus*. *Biotechnol. Bioeng.* **112,** 1523–1532 (2015).
14. Rohmer, M. The discovery of a mevalonate-independent pathway for isoprenoid biosynthesis in bacteria, algae and higher plants. *Nat. Prod. Rep.* **16,** 565–574 (1999).
15. Sato, Y., Ito, Y., Okada, S., Murakami, M. & Abe, H. Biosynthesis of the triterpenoids, botryococcenes and tetramethylsqualene in the B race of *Botryococcus braunii* via the non-mevalonate pathway. *Tetrahedron Lett.* **44,** 7035–7037 (2003).
16. Metzger, P. & Casadevall, E. Lycopadiene, a tetraterpenoid hydrocarbon from new strains of the green-alga *Botryococcus braunii*. *Tetrahedron Lett.* **28,** 3931–3934 (1987).
17. Robinson, G. W., Tsay, Y. H., Kienzle, B. K., Smith-Monroy, C. A. & Bishop, R. W. Conservation between human and fungal squalene synthetases: similarities in structure, function, and regulation. *Mol. Cell. Biol.* **13,** 12 (1993).
18. Rohmer, M. Hopanoids. *Handbook of Hydrocarbon and Lipid Microbiology* 133–142 (Springer, 2010).
19. Poulter, C. D. Biosynthesis of non-head-to-tail terpenes—formation of 1'-1 and 1'-3 linkages. *Acc. Chem. Res.* **23,** 70–77 (1990).
20. Metzger, P., Pouet, Y. & Summons, R. Chemtotaxonomic evidence for the similarity between *Botryococcus braunii* L race and *Botryococcus neglectus*. *Phytochemistry* **44,** 1071–1075 (1997).
21. Zhang, Z., Metzger, P. & Sachs, J. P. Biomarker evidence for the co-occurrence of three races (A, B and L) of *Botryococcus braunii* in El Junco Lake, Galápagos. *Org. Geochem.* **38,** 20 (2007).
22. Sato, T. *et al.* Sesquarterpenes (C35 terpenes) biosynthesized via the cyclization of a linear C35 isoprenoid by a tetraprenyl-beta-curcumene synthase and a tetraprenyl-beta-curcumene cyclase: identification of a new terpene cyclase. *J. Am. Chem. Soc.* **133,** 9734–9737 (2011).
23. Okada, S., Devarenne, T. P., Murakami, M., Abe, H. & Chappell, J. Characterization of botryococcene synthase enzyme activity, a squalene synthase-like activity from the green microalga *Botryococcus braunii*, Race B. *Arch. Biochem. Biophys.* **422,** 110–118 (2004).
24. Stamellos, K. D. *et al.* Subcellular localization of squalene synthase in rat hepatic cells. Biochemical and immunochemical evidence. *J. Biol. Chem.* **268,** 12825–12836 (1993).
25. Okada, S., Devarenne, T. P. & Chappell, J. Molecular characterization of squalene synthase from the green microalga *Botryococcus braunii*, Race B. *Arch. Biochem. Biophys.* **373,** 307–317 (2000).
26. Gu, P., Ishii, Y., Spencer, T. A. & Shechter, I. Function-structure studies and identification of three enzyme domains involved in the catalytic activity in rat hepatic squalene synthase. *J. Biol. Chem.* **273,** 12515–12525 (1998).
27. Pandit, J. *et al.* Crystal structure of human squalene synthase. A key enzyme in cholesterol biosynthesis. *J. Biol. Chem.* **275,** 30610–30617 (2000).
28. Liu, C. I., Jeng, W. Y., Chang, W. J., Ko, T. P. & Wang, A. H. Binding modes of zaragozic acid A to human squalene synthase and staphylococcal dehydrosqualene synthase. *J. Biol. Chem.* **287,** 18750–18757 (2012).
29. Liu, C. I. *et al.* Structural insights into the catalytic mechanism of human squalene synthase. *Acta Crystallogr. D* **70,** 231–241 (2014).
30. Beck, G. *et al.* Characterization of the GGPP synthase gene family in *Arabidopsis thaliana*. *Plant Mol. Biol.* **82,** 393–416 (2013).
31. Ignea, C. *et al.* Efficient diterpene production in yeast by engineering Erg20p into a geranylgeranyl diphosphate synthase. *Metab. Eng.* **27,** 65–75 (2015).
32. Qureshi, A. A., Barnes, F. J., Semmler, E. J. & Porter, J. W. Biosynthesis of prelycopersene pyrophosphate and lycopersene by squalene synthetase. *J. Biol. Chem.* **248,** 2755–2767 (1973).
33. Barnes, F. J., Qureshi, A. A., Semmler, E. J. & Porter, J. W. Prelycopersene pyrophosphate and lycopersene. Intermediates in carotene biosynthesis. *J. Biol. Chem.* **248,** 2768–2773 (1973).
34. Kushwaha, S. C. & Kates, M. Isolation and identification of 'bacteriorhodopsin' and minor C_{40}-carotenoids in *Halobacterium cutirubrum*. *Biochim. Biophys. Acta* **316,** 235–243 (1973).
35. Gregonis, D. & Rilling, H. The stereochemistry of trans-phytoene synthesis. Some observations on lycopersene as a carotene precursor and a mechanism for the synthesis of *cis*- and *trans*-phytoene. *Biochemistry* **13,** 1538–1542 (1974).
36. Kushwaha, S. C., Kates, M. & Porter, J. W. Enzymatic synthesis of C_{40} carotenes by cell-free preparation from *Halobacterium cutirubrum*. *Can. J. Biochem.* **54,** 816–823 (1976).
37. Niehaus, T. D. *et al.* Identification of unique mechanisms for triterpene biosynthesis in *Botryococcus braunii*. *Proc. Natl Acad. Sci. USA* **108,** 12260–12265 (2011).
38. Baxter, A. *et al.* Squalestatin 1, a potent inhibitor of squalene synthase, which lowers serum cholesterol *in vivo*. *J. Biol. Chem.* **267,** 11705–11708 (1992).
39. Bergstrom, J. D. *et al.* Zaragozic Acids—a family of fungal metabolites that are picomolar competitive inhibitors of squalene synthase. *Proc. Natl Acad. Sci. USA* **90,** 80–84 (1993).
40. Lindsey, S. & Harwood, H. J. Inhibition of mammalian squalene synthetase-activity by zaragozic acid a is a result of competitive-inhibition followed by mechanism-based irreversible inactivation. *J. Biol. Chem.* **270,** 9083–9096 (1995).
41. Umeno, D. & Arnold, F. H. A C35 carotenoid biosynthetic pathway. *Appl. Environ. Microbiol.* **69,** 3573–3579 (2003).
42. Umeno, D. & Arnold, F. H. Evolution of a pathway to novel long-chain carotenoids. *J. Bacteriol.* **186,** 1531–1536 (2004).
43. Tobias, A. V. & Arnold, F. H. Biosynthesis of novel carotenoid families based on unnatural carbon backbones: a model for diversification of natural product pathways. *Biochim. Biophys. Acta* **1761,** 235–246 (2006).
44. Wieland, B. *et al.* Genetic and biochemical analyses of the biosynthesis of the yellow carotenoid 4,4'-diaponeurosporene of *Staphylococcus aureus*. *J. Bacteriol.* **176,** 7719–7726 (1994).
45. Ortiz de Montellano, P. R., Wei, J. S., Vinson, W. A., Castillo, R. & Boparai, A. S. Substrate selectivity of squalene synthetase. *Biochemistry* **16,** 2680–2685 (1977).
46. Koyama, T., Ogura, K. & Seto, S. Substrate specificity of squalene synthetase. *Biochim. Biophys. Acta* **617,** 218–224 (1980).
47. Jarstfer, M. B., Zhang, D. L. & Poulter, C. D. Recombinant squalene synthase. Synthesis of non-head-to-tail isoprenoids in the absence of NADPH. *J. Am. Chem. Soc.* **124,** 8834–8845 (2002).
48. Blagg, B. S., Jarstfer, M. B., Rogers, D. H. & Poulter, C. D. Recombinant squalene synthase. A mechanism for the rearrangement of presqualene diphosphate to squalene. *J. Am. Chem. Soc.* **124,** 8846–8853 (2002).
49. Zhang, D. L. & Poulter, C. D. Biosynthesis of non-head-to-tail isoprenoids—synthesis of 1'-1-structures and 1'-3-structures by recombinant yeast squalene synthase. *J. Am. Chem. Soc.* **117,** 1641–1642 (1995).
50. Pan, J. J. *et al.* Biosynthesis of squalene from farnesyl diphosphate in bacteria: three steps catalyzed by three enzymes. *ACS Cent. Sci.* **1,** 77–82 (2015).
51. Vining, L. C. Roles of secondary metabolites from microbes. *Ciba Found. Symp.* **171,** 184–194 (1992).
52. Weiss, T. L. *et al.* Colony organization in the green alga *Botryococcus braunii* (Race B) is specified by a complex extracellular matrix. *Eukaryot. Cell* **11,** 1424–1440 (2012).
53. Niehaus, T. D. *et al.* Functional identification of triterpene methyltransferases from *Botryococcus braunii* race B. *J. Biol. Chem.* **287,** 8163–8173 (2012).
54. Manquin, B. P., Morgan, J. A., Ju, J., Muller-Spath, T. & Clark, D. S. Production of C35 isoprenoids depends on H2 availability during cultivation of the hyperthermophile *Methanococcus jannaschii*. *Extremophiles* **8,** 13–21 (2004).
55. Cario, A., Grossi, V., Schaeffer, P. & Oger, P. M. Membrane homeoviscous adaptation in the piezo-hyperthermophilic archaeon *Thermococcus barophilus*. *Front. Microbiol.* **6,** 1152 (2015).
56. Jorgensen, K. *et al.* Metabolon formation and metabolic channeling in the biosynthesis of plant natural products. *Curr. Opin. Plant Biol.* **8,** 280–291 (2005).
57. Field, B. & Osbourn, A. E. Metabolic diversification--independent assembly of operon-like gene clusters in different plants. *Science* **320,** 543–547 (2008).
58. Nutzmann, H. W. & Osbourn, A. Gene clustering in plant specialized metabolism. *Curr. Opin. Biotechnol.* **26,** 91–99 (2014).
59. Chu, S. P. The influence of the mineral composition of the medium on the growth of planktonic algae: part I. Methods and culture media. *J. Ecol.* **30,** 284–325 (1942).
60. Metzger, P., Allard, B., Casadevall, E., Berkaloff, C. & Coute, A. Structure and chemistry of a new chemical race of *Botryococcus braunii* (Chlorophyceae) that produces lycopadiene, a tetraterpenoid hydrocarbon. *J. Phycol.* **26,** 258–266 (1990).

61. Weiss, T. L. *et al.* Raman spectroscopy analysis of botryococcene hydrocarbons from the green microalga *Botryococcus braunii*. *J. Biol. Chem.* **285,** 32458–32466 (2010).
62. Molnar, I. *et al.* Bio-crude transcriptomics: gene discovery and metabolic network reconstruction for the biosynthesis of the terpenome of the hydrocarbon oil-producing green alga, *Botryococcus braunii* race B (Showa). *BMC Genomics* **13,** 576 (2012).
63. Garrett, S. R., Morris, R. J. & O'Maille, P. E. Steady-state kinetic characterization of sesquiterpene synthases by gas chromatography-mass spectroscopy. *Methods Enzymol.* **515,** 3–19 (2012).
64. Zhuang, X. & Chappell, J. Building terpene production platforms in yeast. *Biotechnol. Bioeng.* **112,** 1854–1864 (2015).
65. Zhuang, X. Engineering novel Terpene Production Platforms in the Yeast Saccharomyces cerevisae. *Theses and Dissertations–Plant and Soil Sciences* **Paper 17,** 1–189 (2013).

Acknowledgements

This work was supported by grants NSF-EFRI-PSBR 1240478 and DOE-ARPA-E-PETRO DE-AR0000203 to T.P.D.; by contract DE-EE0003046 awarded to the National Alliance for Advanced Biofuels and Bioproducts (NAABB) from the U.S. Department of Energy, of which T.P.D. and I.M. are members; and by a grant from the Japan Science and Technology (JST) agency, Core Research for Evolutionary Science and Technology (CREST) programme to S.O. We thank Joe Chappell, University of Kentucky, for yeast strain ZX 178-08 and XURA vector; Eric Bridenbaugh of Olympus Scientific Solutions Americas, for assistance with microscopy; and the following people at Texas A&M University: Craig Kaplan for yeast strain CKY457, Larry Dangott for HPLC use, Dan Browne for help with bioinformatics, and Indranil Malik and Ben Kaster for helpful comments.

Author contributions

H.R.T., M.T.N., S.O. and T.P.D. conceived and designed the experiments. H.R.T., M.T.N., S.O., K.T. and I.M. performed experiments. H.R.T., M.T.N., S.O., I.M., Y.X. and T.P.D. analysed data. H.R.T., S.O., I.M. and T.P.D. co-wrote the paper.

Additional information

Accessions codes: New DNA sequences described in this study have been deposited in GenBank: *LSS* accession #KT388100, *LOS* accession #KT388101 and *SlSS* accession #NM_001247787.

Competing financial interests: The authors declare no competing financial interests.

13

Chemically related 4,5-linked aminoglycoside antibiotics drive subunit rotation in opposite directions

Michael R. Wasserman[1], Arto Pulk[2,3], Zhou Zhou[1], Roger B. Altman[1], John C. Zinder[4], Keith D. Green[5], Sylvie Garneau-Tsodikova[5], Jamie H. Doudna Cate[2,3] & Scott C. Blanchard[1,4]

Dynamic remodelling of intersubunit bridge B2, a conserved RNA domain of the bacterial ribosome connecting helices 44 (h44) and 69 (H69) of the small and large subunit, respectively, impacts translation by controlling intersubunit rotation. Here we show that aminoglycosides chemically related to neomycin—paromomycin, ribostamycin and neamine—each bind to sites within h44 and H69 to perturb bridge B2 and affect subunit rotation. Neomycin and paromomycin, which only differ by their ring-I 6′-polar group, drive subunit rotation in opposite directions. This suggests that their distinct actions hinge on the 6′-substituent and the drug's net positive charge. By solving the crystal structure of the paromomycin-ribosome complex, we observe specific contacts between the apical tip of H69 and the 6′-hydroxyl on paromomycin from within the drug's canonical h44-binding site. These results indicate that aminoglycoside actions must be framed in the context of bridge B2 and their regulation of subunit rotation.

[1] Department of Physiology and Biophysics, Weill Cornell Medical College, New York, New York 10065, USA. [2] Department of Molecular and Cell Biology, University of California at Berkeley, Berkeley, California 94720, USA. [3] Department of Chemistry, University of California at Berkeley, Berkeley, California 94720, USA. [4] Tri-Institutional Training Program in Chemical Biology, Weill Cornell Medical College, Rockefeller University, Memorial Sloan-Kettering Cancer Center, New York, New York 10065, USA. [5] Department of Pharmaceutical Sciences, University of Kentucky College of Pharmacy, Lexington, Kentucky 40536, USA. Correspondence and requests for materials should be addressed to J.H.D.C. (email: jcate@lbl.gov) or to S.C.B. (email: scb2005@med.cornell.edu).

Aminoglycoside antibiotics are potent, broad-spectrum bactericidal agents that inhibit cell growth by targeting functional centres within the bacterial ribosome to alter global aspects of the translation mechanism[1–5]. However, despite their clinical effectiveness, aminoglycosides are commonly restricted to topical applications. Oral administration is typically a last resort to fight multi-drug-resistant pathogens due to their propensities to elicit adverse side effects in patients, including oto- and nephrotoxicities[6]. The molecular origins of aminoglycoside toxicities and their bactericidal activities are currently a matter of significant debate.

Initiatives spanning the past half century indicate that aminoglycosides with a 2-deoxystreptamine ring target the conserved helix 44 (h44) messenger RNA (mRNA) 'decoding site' region of ribosomal RNA (rRNA) within the small (30S) subunit of the bacterial ribosome[7,8], reducing translational fidelity by promoting the promiscuous incorporation of near- and non-cognate aminoacyl-transfer RNAs (tRNAs) (miscoding)[9,10]. Investigations of RNA oligonucleotide model systems[2], ribosomal subunits[11] and intact ribosomes[8,12–14] suggest that aminoglycosides do so by binding the major groove of h44 to shift the position and dynamics of two universally conserved residues (A1492 and A1493) responsible for the recognition of the mRNA codon-aminoacyl–tRNA (aa–tRNA) complex[2,15].

While these investigations have greatly advanced our understanding of the molecular determinants of ribosome binding, they fall short of elucidating how the aminoglycosides affect each stage of the translation mechanism—initiation, elongation, termination and recycling[3,5,16–19]. They are also unable to fully explain how subtle chemical distinctions between structurally related aminoglycosides give rise to unique inhibition and resistance profiles[20–22]. Furthermore, a miscoding mechanism of action is at odds with observations that bacterial strains harbouring error-prone ribosomes are viable[23,24] and evidence that specific aminoglycosides inhibit protein synthesis while exhibiting little, to no, miscoding[25]. A comprehensive understanding of aminoglycoside action on the ribosome must provide a rationale for these observations, as well as data suggesting that 2-deoxystreptamine aminoglycosides may operate through more than one functional site on the bacterial ribosome[18,21,26].

Crystal structures of intact 70S *Escherichia coli* (*E. coli*) ribosomes have recently revealed that aminoglycosides bind to sites in both the small and large (50S) subunit of the bacterial ribosome[1,19,27]. We have shown that neomycin, the most potent and toxic member of the 4,5-linked aminoglycoside class[21] (Fig. 1a), promotes translational miscoding and shutdown by simultaneously binding to the h44-decoding site and the major groove of H69 within the 30S and 50S subunits, respectively[1] (Fig. 1b). These conserved rRNA elements, which comprise the central intersubunit bridge B2, undergo marked structural rearrangements during small-subunit rotation[1], a process underpinning multiple aspects of the translation cycle[19,28–31]. It is not presently known whether, and to what extent, chemically related 4,5-linked aminoglycosides within the neomycin class impact the dynamics of subunit rotation.

Here, we present a detailed examination of structure–function relationships in chemically related 4,5-linked aminoglycosides to reveal that neomycin and paromomycin, which differ by only a single functional group, bind h44 and H69 to drive subunit rotation in opposite directions. These data, which include an analysis of both ribosome- and drug-resistance mutations, indicate that the distinct impacts of aminoglycosides on the collective motions of bridge B2 play a central role in their mechanisms of action. By solving the structure of the

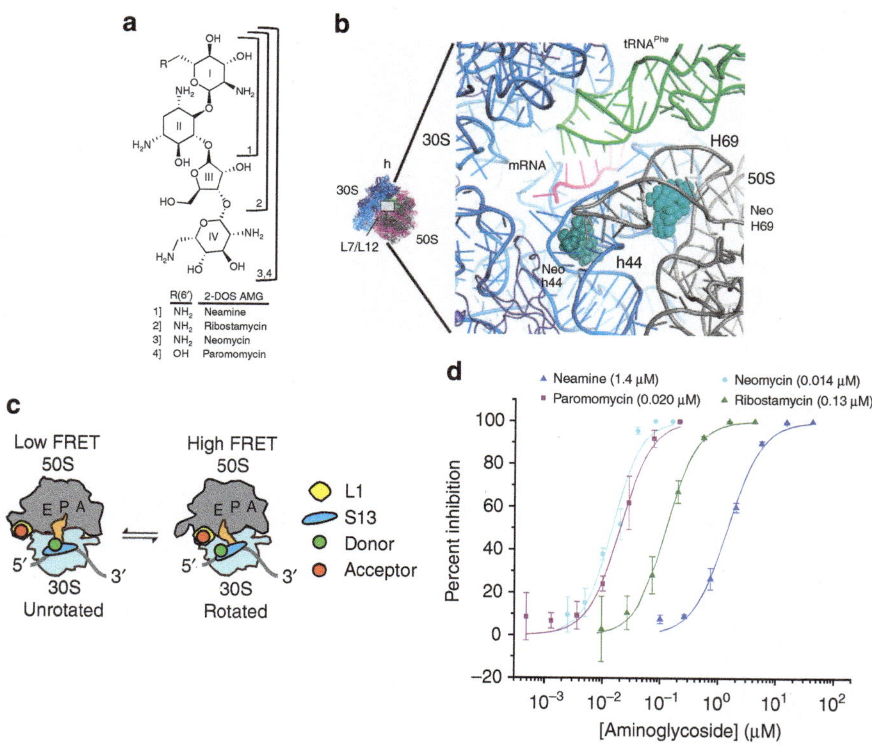

Figure 1 | Investigation of the neomycin family of aminoglycosides. **(a)** 4,5-linked 2-deoxystreptamine (2-DOS) aminoglycosides in the neomycin family. **(b)** Overview of neomycin-binding sites in the small-subunit decoding site (h44, dark green) and in the large subunit H69 (light blue). 16S rRNA (30S) is shown in blue, 23S rRNA (50S) in grey, P-site tRNA in green and mRNA in magenta. **(c)** Cartoon illustrating the ribosome labelling strategy used for monitoring intersubunit rotation via single-molecule FRET, which has previously been shown to be affected by neomycin[1]. **(d)** Aminoglycoside-induced inhibition of *in vitro* translation. IC_{50} values are indicated. Experiments were performed in triplicate and the mean ± s.d. is plotted.

paromomycin–ribosome complex to 3.1 Å resolution, we show that such distinctions hinge on interactions of paromomycin with the apical tip of H69 via its 6′-hydroxyl moiety from within its canonical h44-binding site. These findings provide compelling evidence that mechanisms describing the potency and selectivity of this important class of anti-infective agents must include their capacity to alter the dynamic properties of bridge B2 via interactions with H69 of the large subunit.

Results

We investigated the structural and dynamic impacts of 4,5-linked aminoglycoside antibiotic (Fig. 1a) binding to bridge B2 (Fig. 1b) within the intact *E. coli* 70S ribosome via single-molecule FRET (smFRET) by site-specifically labelling ribosomal proteins S13 and L1 within the small and large subunits with donor and acceptor fluorophores, respectively[1] (Fig. 1c). To enable quantitative smFRET measurements of subunit rotation dynamics[32–35], we employed intra-molecularly stabilized donor and acceptor fluorophores, which exhibit markedly enhanced brightness and photostability[36] (see the Methods section). Using this system, smFRET data could be acquired at ~2.6-fold faster time resolution than previously reported at an improved signal-to-noise ratio, where a total of ~500 photons could be detected for each individual ribosome complex per imaging frame.

Coupled *in vitro* transcription–translation assays using purified translation components (see Methods) showed that ribosomes site-specifically labelled on proteins S13 and L1 were fully functional in processive translation reactions (Supplementary Fig. 1). This system was also employed to validate that neomycin and the chemically related 4,5-linked aminoglycosides paromomycin, ribostamycin and neamine (Fig. 1a) inhibit the function of our purified 70S *E. coli* ribosomes at concentrations consistent with those reported using S30 or S100 extracts[37,38]. Notably, neomycin and paromomycin exhibited roughly equivalent half-maximal inhibitory concentration (IC_{50}) values (14–20 nM), whereas ribostamycin and neamine exhibited substantially reduced potencies (130 and 1,400 nM, respectively) (Fig. 1d). At face value, these data suggest that neomycin and paromomycin operate through similar mechanisms, where rings III and IV contribute as much as 100-fold to the overall affinity of the drug–ribosome interaction[7,39].

Native ribosomes exhibit spontaneous subunit rotation. To explore the impact of distinct 4,5-linked aminoglycosides on the structure and dynamics of intact ribosomes, site-specifically labelled 70S complexes bearing deacylated $tRNA^{fMet}$ in the P-site were imaged in the absence and presence of specific concentrations of 4,5-linked aminoglycosides. Such complexes spontaneously interconverted between high-FRET (0.67 ± 0.01) and low-FRET (0.26 ± 0.01) states on the sub-second timescale. Although these sites of labelling are in highly mobile elements of the ribosome—the L1 stalk[33,34,40–42] and head domain[43,44]—dynamic FRET events principally reflect reversible transitions between 'rotated'/'hybrid' and 'unrotated'/'classical' ribosome configurations in which the S13 and L1 proteins are proximal and distal to each other, respectively[1] (Figs 1c and 2a; Supplementary Fig. 2). This interpretation was further corroborated by imaging the process of subunit rotation from two distinct structural perspectives (Supplementary Figs 3 and 4).

Hidden Markov modelling-based idealization[45] of the S13–L1 data obtained using the QuB software package (see the Methods section) also revealed evidence of a third, intermediate-FRET (0.45 ± 0.01) state that appeared to be on path between low- and high-FRET configurations (Fig. 2a; Supplementary Fig. 5). The existence and kinetic linkage of these FRET states was corroborated by generating transition density plots from the idealized data[46], which revealed six well-defined peaks symmetrically distributed with respect to the diagonal axis (Supplementary Fig. 5). Such findings suggest that at least one transient intermediate state of subunit rotation is transited on the path between unrotated and fully rotated ribosome configurations[1].

Steady-state actions of neomycin on intact 70S ribosomes. Ribosome complexes bearing deacylated $tRNA^{fMet}$ in the P-site exhibited a pronounced bimodal response to the addition of neomycin[1] (Fig. 2b; Supplementary Fig. 6a). Consistent with h44 binding[1,18], low neomycin concentrations (up to 100 nM) increased the occupancy of the low-FRET, unrotated ribosome configuration by ~1.4- to 1.6-fold (Fig. 2b; Supplementary Fig. 6a). The improved time resolution and signal-to-noise ratio afforded by the use of intra-molecularly stabilized fluorophores also revealed that neomycin concomitantly stabilized the intermediate-FRET state by ~2.5-fold. At concentrations above 100 nM, the intermediate-FRET, partially rotated ribosome configuration was preferentially stabilized, accompanied by an approximately fivefold decrease in the global rate of intersubunit dynamics (0.7 versus 4.9 s^{-1} or 5.4 s^{-1}) (Fig. 2b). These effects have been specifically attributed to neomycin's interaction with the major groove of H69, which blocks the axial compression of H69 required for subunit rotation[1]. Hence, the predominant impact of neomycin binding to both h44 and H69 is to destabilize the unrotated ribosome configuration in favour of intermediate states of subunit rotation. Having established how ribosomes quantitatively respond to neomycin, we next set out to examine whether these changes in dynamics were altered by known drug-resistance modifications.

Consistent with previous investigations of resistance modifications to neomycin[21,47], we observed increased IC_{50} values for both the ring-I-modified, 6′-*N*-acetyl-neomycin (0.75 μM) and the ring II-modified 3-*N*-acetyl-neomycin (34 μM) (Supplementary Fig. 7; Supplementary Table 1). In agreement with the mode of interaction between neomycin and the ribosome observed crystallographically[1] in which rings I and II of the drug recognize chemical features within the major groove of both h44 and H69, subunit rotation dynamics were only weakly affected (1.4–2-fold) by 6′-*N*-acetyl-neomycin and largely unaffected by 3-*N*-acetyl-neomycin (see the Methods section; Supplementary Fig. 8). 6′-*N*-acetyl-neomycin modestly stabilized (twofold) both low- and intermediate-FRET states at concentrations above 1 μM (Supplementary Fig. 8). By contrast, 3-*N*-acetyl-neomycin exhibited little, to no, detectable impact on the FRET-state distribution even at the highest concentrations tested (10 μM; Supplementary Fig. 8).

The contributions of rings III and IV of neomycin to the observed impact on subunit rotation dynamics and processive translation were examined through analogous investigations of the closely related 4,5-linked aminoglycosides, ribostamycin (rings I–III) and neamine (rings I and II) (Fig. 1a). Consistent with previous literature[7,20,37] and the crystallographically observed modes of interaction of neomycin with h44 and H69 (ref. 1), both ribostamycin and neamine exhibited substantially reduced impacts on subunit rotation dynamics and correspondingly higher IC_{50} values in coupled *in vitro* transcription–translation reactions (Supplementary Table 1). At the highest concentrations tested (10 μM), both drugs elicited low- and intermediate-FRET stabilization (Supplementary Fig. 6). Collectively, these findings reveal that rings I and II are necessary and sufficient to mediate drug binding to h44 and H69.

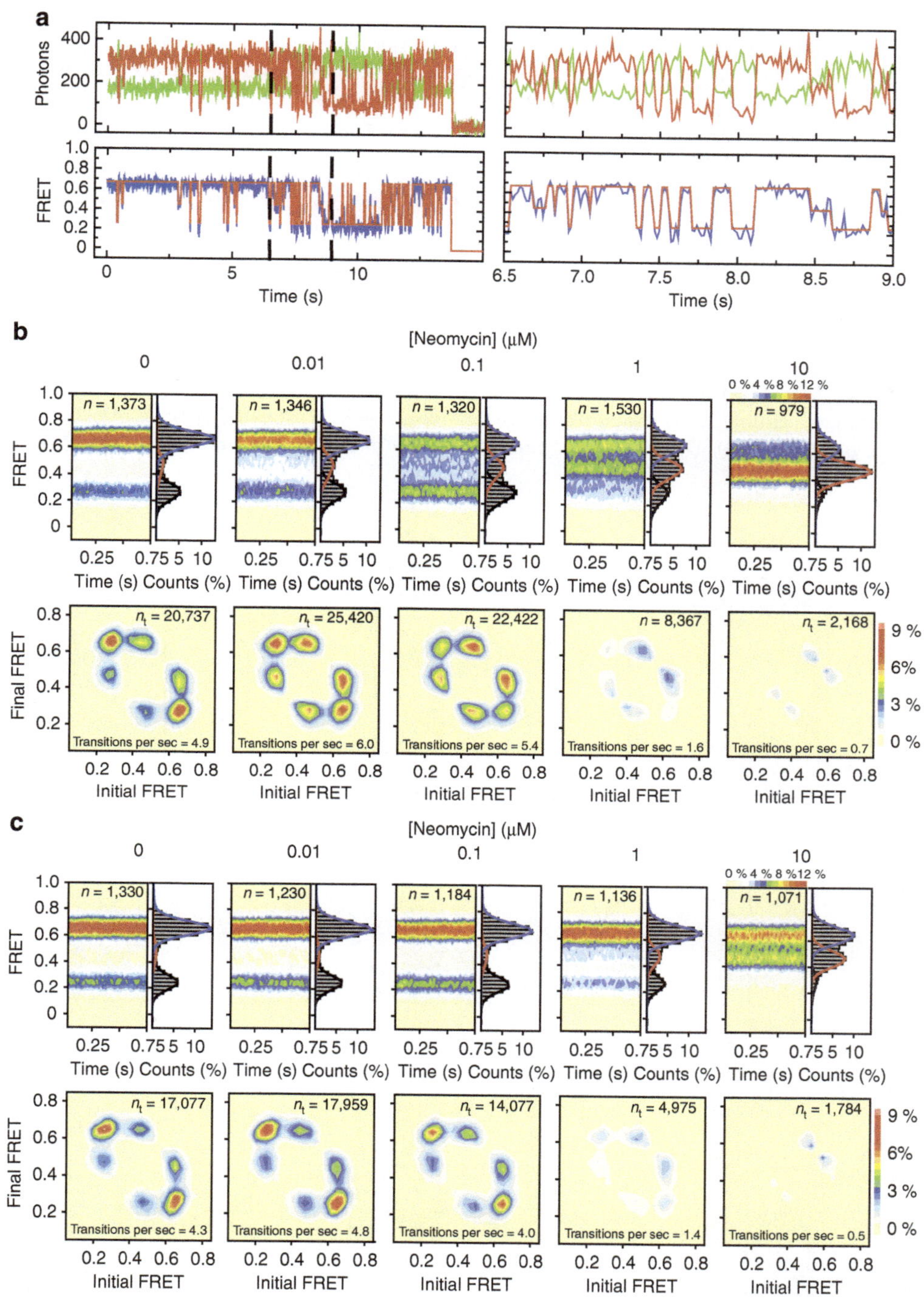

Figure 2 | Neomycin-induced effects on intersubunit rotation in wild-type and A1408G aminoglycoside-resistant ribosomes. (**a**) (Left) Single-molecule fluorescence (donor—green; acceptor—red) and FRET (blue) trajectories illustrating typical conformational changes in ribosomes labelled as shown in Fig. 1c imaged in the absence of drug. FRET idealization is overlaid in red. (Right) A zoomed-in view highlights the transient nature of the intermediate-FRET state. (**b,c**) (top panels) smFRET trajectories summed into FRET histograms reveal the population behaviours across a range of neomycin concentrations in (**b**) wild-type and (**c**) A1408G ribosomes. (Bottom panels) Initial and final FRET values for each transition summed into two-dimensional histograms (transition density plots). Experiments were performed in triplicate on three separate days.

The h44- and H69-binding sites are functionally linked. Clinical isolates resistant to the 4,5-linked aminoglycosides exhibit rRNA mutations that cluster in the h44-decoding site[2]. To investigate the impact of an h44 mutation on drug binding to the spatially separated h44- and H69-binding sites (Fig. 1b), identical experiments were performed on ribosome complexes bearing an A1408G point mutation that markedly reduces aminoglycoside binding to h44 (refs 7,20) (Fig. 1b; see the Methods section) and the IC_{50} of *in vitro* transcription–translation (42 versus 0.014 μM) (Supplementary Fig. 7; Supplementary Table 1). Ribosomes bearing the A1408G mutation exhibited globally similar dynamic behaviours to wild-type ribosomes (Fig. 2b,c), and as anticipated from reduced binding at the h44-decoding site, the low-FRET, unrotated-state population did not increase in the presence of neomycin (Fig. 2c). A modest increase in intermediate-FRET state occupancy was evidenced only at elevated (>1 μM) neomycin concentrations that correlated with a global dampening of intersubunit rotation dynamics (Fig. 2c; Supplementary Fig. 9a). These distinctions are consistent with a functional linkage between the spatially separated neomycin-binding sites, where abrogation of drug binding at h44 concomitantly reduces the affinity and/or accessibility of the H69-binding site.

Paromomycin stabilizes an unrotated ribosome configuration. To gain a deeper understanding of the contribution of ring I to 4,5-linked aminoglycoside interactions with the intact ribosome, we next set out to investigate the impact of paromomycin on wild-type and A1408G ribosomes. Paromomycin differs chemically from neomycin at a single position on ring 1: in neomycin, the 6′-position is a primary amine; in paromomycin, the 6′-position is a hydroxyl group (Fig. 1a). Despite their chemical similarities and their nearly indistinguishable IC_{50}'s for the wild-type ribosome (Fig. 1d), neomycin and paromomycin exhibit unique sensitivities to resistance mutations in the decoding site[7,18,20–22]. Such distinctions have been attributed to subtle differences in their modes of interaction with specific bases within the h44 major groove[38]. Consistent with earlier studies, we observed that the A1408G mutation conferred substantially lower levels of resistance to paromomycin (IC_{50} = 1.6 μM) than for neomycin (IC_{50} = 42 μM) (Supplementary Fig. 7; Supplementary Table 1).

Strikingly, and in line with its distinct impacts on the translation mechanism[21], paromomycin was observed to preferentially stabilize the low-FRET, unrotated ribosome configuration at all concentrations tested (0.01–10 μM) (Fig. 3a; Supplementary Fig. 6). At concentrations up to 1 μM, such effects were notably correlated with a marked (~100%) increase in subunit rotation dynamics (Fig. 3a, lower panels). Hence, despite

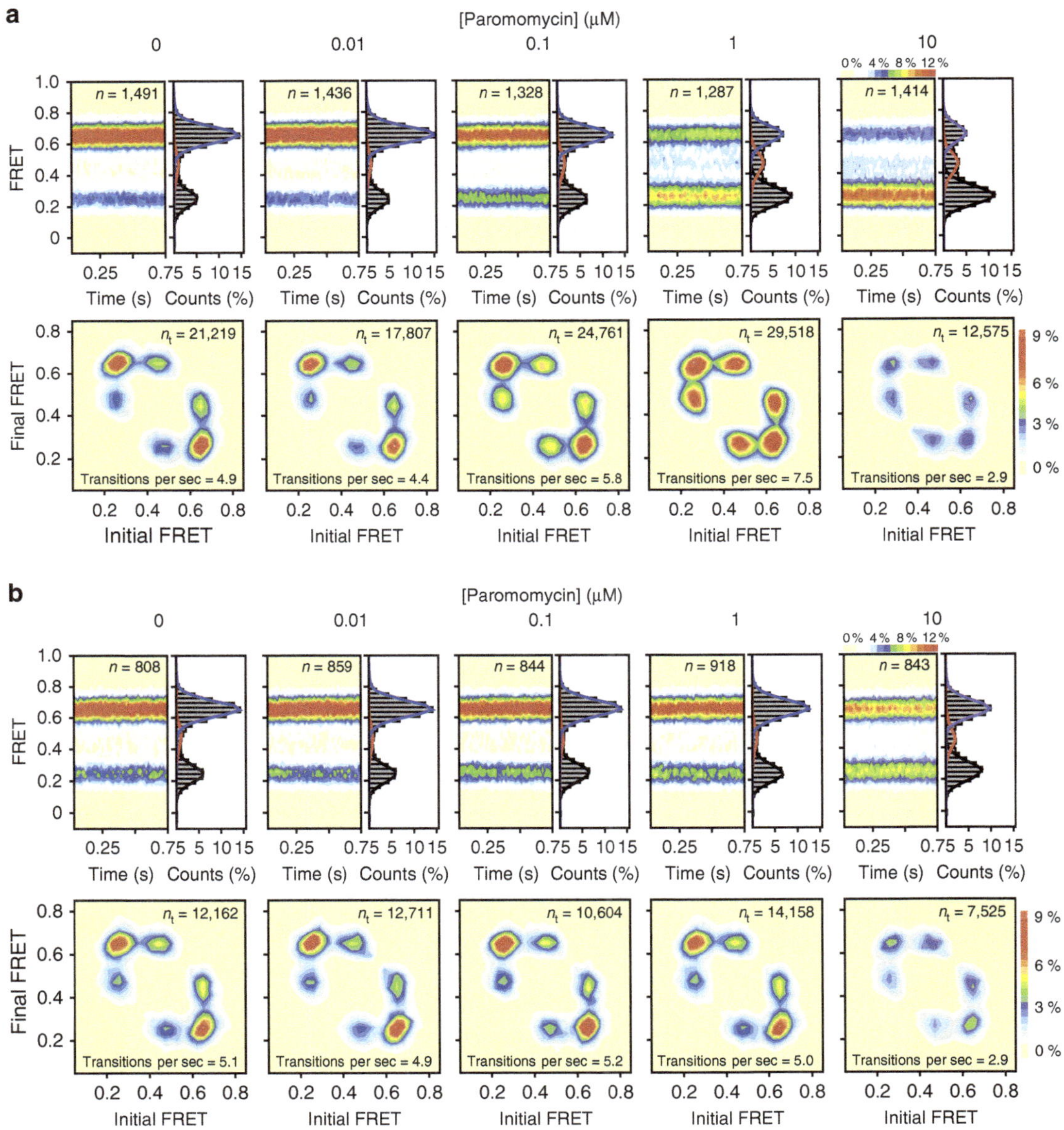

Figure 3 | Paromomycin effects on intersubunit rotation in wild-type and A1408G aminoglycoside-resistant ribosomes. (a,b) (Top panels) smFRET trajectories summed into FRET histograms reveal the population behaviours across a range of paromomycin concentrations in **(a)** wild-type and **(b)** A1408G ribosomes. (Bottom panels) Initial and final FRET values for each transition summed into two-dimensional histograms (transition density plots). Experiments were performed in triplicate on three separate days.

paromomycin's close chemical relation, its impacts are starkly opposite to those of neomycin. At concentrations above 1 μM, however, evidence of a modest stabilization of the intermediate-FRET state was observed; subunit rotation dynamics were also suppressed (Fig. 3a; Supplementary Fig. 6). Ribosomes bearing the A1408G mutation were largely unaffected by paromomycin up to a concentration of 1 μM, above which clear evidence of low- and intermediate-FRET state stabilization emerged, together with a concomitant dampening of intersubunit rotation dynamics (Fig. 3b; Supplementary Fig. 9b). Such findings, together with data obtained at lower pH (Supplementary Fig. 10), suggest that paromomycin's distinct impacts on subunit rotation hinge on alterations in the drug's interactions at both h44- and H69-binding sites and that these changes arise via the 6′-hydroxyl group and the drug's net positive charge.

Paromomycin contacts H69 from within the h44-binding site. To ascertain whether paromomycin's distinct impacts on subunit rotation dynamics stem from unique modes of interaction with h44 and/or H69, we solved the crystal structure of the paromomycin–70S ribosome complex at a resolution of 3.1 Å (Table 1) under conditions enforcing drug protonation (below neutral pH) and intermediate-FRET state stabilization[1] (see the Methods section; Supplementary Fig. 10). As observed for neomycin[1], unbiased $F_{obs} - F_{calc}$ difference electron density maps revealed clear, positive electron density for paromomycin in the major grooves of both h44 and H69. To a first approximation, paromomycin's modes of interaction with both sites appeared indistinguishable from those of neomycin (Fig. 4a), corroborating the placement and orientation of both 4,5-linked aminoglycosides within H69. The global conformations of both the neomycin- and paromomycin-bound ribosome were also indistinguishable. As observed for neomycin, and as suggested by the intermediate-FRET state value observed in the presence of both neomycin and paromomycin, the large and small subunits adopt a partially rotated configuration in which deacylated P-site tRNA is bound in an intermediate hybrid state (P/pe)[1] (Fig. 4b; Supplementary Fig. 11).

However, one potentially critical distinction was observed between the neomycin- and paromomycin-bound ribosome crystal structures. In the presence of paromomycin, strong electron density was observed for the universally conserved A1913 residue located at the apical tip of the large subunit H69, whereas this was not observed in the neomycin-bound structure (Fig. 5a; Supplementary Fig. 12). These data suggest that the A1913 residue remains dynamic when neomycin is bound[1], but is relatively static when paromomycin is bound. Inspection of the paromomycin data revealed the 6′-hydroxyl moiety of the h44-bound paromomycin molecule to be in hydrogen bonding distance (~2.9 Å) of the A1913 Watson–Crick face (Fig. 5b). This finding, together with the observation that subunit rotation dynamics are largely unaffected by a 6′-deoxy-paromomycin analogue[48] (Supplementary Fig. 13), indicates that the contact between the 6′-substituent in paromomycin and the apical tip of H69 is a critical determinant that drives the preferential stabilization of the unrotated ribosome configuration. Given that residue A1913 is observed to directly contact the anticodon loop region of tRNA bound within the A-site-decoding region[13,49], the role of this contact on the aminoglycoside-induced miscoding mechanism warrants further exploration.

Table 1 | Data collection and refinement statistics.

Data collection	
Space group	$P2_12_12_1$
Cell dimensions	
a, b, c (Å)	212.10, 435.24, 614.44
Resolution (Å)*	70-3.1 (3.2-3.1)
CC1/2*	99.6 (42.5)
$I/\sigma I$*	6.53 (0.68)
R_{meas}*	17.8 (106.9)
Completeness (%)*	86.3 (54.7)
Redundancy*,†	4.9 (1.7)
Refinement	
Resolution (Å)	70-3.0
Number of reflections	909,645
R_{work}/R_{free}	0.2141/0.2816
Number of atoms	
Protein/RNA	291,292
Ligand/ion	848
Water	1,729
B-factors	
Protein/RNA	64.3
Ligand/ion	61.6
Water	62.3
R.m.s.d.	
Bond lengths (Å)	0.009
Bond angles (°)	1.29

r.m.s.d., root mean squared deviation.
*Values in parentheses are for the highest-resolution shell.
†Data were measured from 16 crystals.

Discussion

Functional investigations aimed at quantifying the impact of small-molecule inhibitors on the mechanism of protein synthesis face significant challenges when examined in the context of full-length protein production. This is due to the fact that mechanisms of action may be masked by additive or synergistic effects on multiple steps of protein synthesis. Inhibition of functional protein production may arise through impacts on any aspect of the system including translation, transcription, protein folding or by promoting protein inactivation or non-specific aggregation. Our examination of the 4,5-linked aminoglycosides in reconstituted protein translation reactions using purified components revealed direct evidence of hidden and complex modes of inhibition, as each of the 4,5-linked aminoglycosides examined, with the exception of neamine, exhibited IC_{50} values (14–130 nM) that were lower (2–18-fold) than the ribosome concentration in the reaction (250 nM). Moreover, both neomycin and paromomycin were observed to exhibit nearly indistinguishable IC_{50} values (Fig. 1d; Supplementary Table 1) despite their apparently distinct impacts on the dynamics of subunit rotation (Figs 2 and 3). By contrast, isolated structural investigations provide only snapshots of a drug's impact on its target. Indeed, our comparison of neomycin- and paromomycin-bound ribosome structures revealed just a single obvious distinction: the position and dynamics of residue A1913 (Fig. 5; Supplementary Fig. 12). These findings provide a compelling argument that single-molecule imaging methods offer a means to bridge these distinct approaches to generate insights into dynamic structural processes directly related to function.

The process of subunit rotation underpins essential aspects of the translation mechanism and entails large-scale, functional remodelling events within intersubunit bridge B2 (refs 1,31). In line with cryo-electron microscopy investigations of the bacterial ribosome[50], we find that the subunit rotation process is rapid and

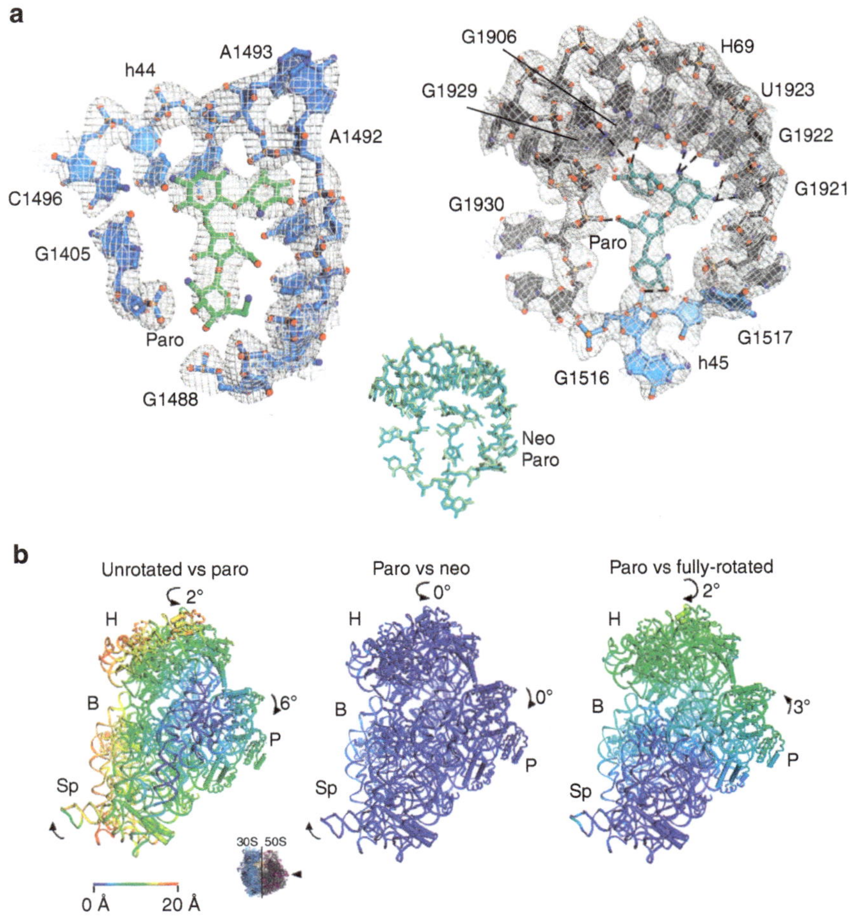

Figure 4 | Crystal structure of the paromomycin-bound partially rotated ribosome. (**a**) Paromomycin binding within the (left) h44-decoding site and (right) within H69. 16S rRNA (blue), 23S rRNA H69 (grey) and paromomycin (h44—green; H69—teal) are shown, along with a ($2F_{obs} - F_{calc}$) electron density map, calculated in PHENIX and contoured at 1.4 s.d. from the mean. Paromomycin and H69 rRNA contacts <3.5 Å are shown as dashed lines. (Inset) Paromomycin contacts within H69 are indistinguishable from those formed by neomycin (r.m.s.d. = 0.412 Å). (**b**) Paromomycin induces global rearrangements of the 70S ribosome that are indistinguishable from those stabilized by neomycin. (Inset) View of the 30S subunit from the perspective of the 50S subunit. (Left) Difference in the vector shifts between equivalent RNA phosphorus atoms and protein Cα atoms in the unrotated compared with the partially rotated paromomycin-bound ribosome; (middle) superposition of the partially rotated neomycin-bound and partially rotated paromomycin-bound ribosomes. (Right) The fully rotated compared with the partially rotated paromomycin-bound ribosome. The vectors are colour coded as indicated in the scale. Ribosomes were superimposed using the 50S subunit as the frame of reference. 30S head domain: H; 30S body: B; 30S platform: P; 30S spur: Sp.

reversible under the present experimental conditions (ca. $5\,s^{-1}$; see the Methods section). We have leveraged this inherent metastability and the synergies of smFRET and X-ray crystallography to gain a deeper understanding of the actions of chemically related aminoglycosides on the intact 70S ribosome. These data reveal that both neomycin and paromomycin bind the major grooves of both h44 and H69, central features of the bridge B2 element. Drug binding at these structurally distinct, functionally linked, sites impacts the nature and rate of intersubunit rotation dynamics in opposing ways. Binding to h44 stabilizes an unrotated ribosome configuration, whereas drug binding to H69 stabilizes an intermediate state of subunit rotation. Simultaneous binding at both sites secures the ribosome in an intermediate state of rotation that is incompetent for basal translation elongation reactions[1]. Together with the present investigations of chemically related 4,5-linked aminoglycosides and resistance modifications to both neomycin and the ribosome, we conclude that binding at both the h44 and H69 sites within bridge B2 contributes to the impact of 4,5-linked aminoglycoside antibiotics on protein synthesis.

The present observation of functional linkages between the h44- and H69-binding sites is likely a natural consequence of the collective conformational changes that occur within bridge B2 during subunit rotation. This conclusion warrants further examination, particularly as it relates to aminoglycoside mechanisms of action and toxicity[6,16]. While h44 sequences vary substantially between bacterial and human cells, the corresponding sequences of H69 are more closely related[51]. The extent of H69 sequence conservation and its functional role in the process of subunit rotation may explain why some aminoglycosides impinge upon the translation machinery of the host cells and why aminoglycoside-resistance mutations have yet to be reported within H69. Our own efforts to site-specifically alter the H69 sequence to confer aminoglycoside resistance have only led to modest reductions in neomycin binding when four nucleotides in the helical region of H69 were swapped to the corresponding human sequence. Notably, these mutations imparted a marked increase in subunit rotation dynamics and modestly stabilized the high-FRET, rotated ribosome configuration (Supplementary Fig. 14; see the Methods section).

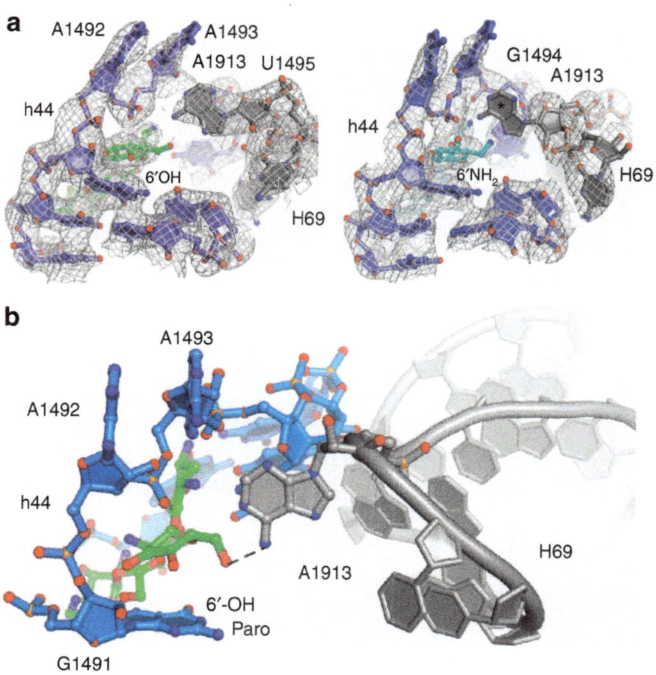

Figure 5 | Paromomycin contacts the universally conserved A1913 residue of H69 from its canonical h44 site of binding. **(a)** The decoding site region of the paromomycin- (left) and neomycin- (right) bound ribosome (unrotated) exhibit strong and weak electron density for residue A1913 of H69, respectively. The 16S rRNA (blue), 23S rRNA (grey), h44-bound paromomycin (green) and h44-bound neomycin (light blue) are overlaid with feature-enhanced electron density maps[69], calculated in PHENIX and contoured at 1.4 s.d. from the mean. The temperature factors for residue A1913 are as follows. Unrotated paromomycin-bound: ~ 90 Å^2; unrotated neomycin-bound: ~ 430 Å^2. The lack of density for A1913 in the neomycin-bound structure is highlighted with an asterisk. **(b)** Paromomycin (green) bound within the canonical small-subunit h44-decoding site (blue) contacts the apical tip of H69 via paromomycin ring I (6′-O) and the universally conserved base A1913 (N6) of H69 (grey).

While more significant H69 mutations may be expected to confer greater aminoglycoside resistance, they are likely to be lethal when made in isolation. To best address this issue, a comprehensive, targeted screen for resistance mutations will need to be performed taking into account the consideration that bridge B2 exhibits collective conformational transitions that regulate the nature and rate of intersubunit rotation.

A molecular rationale that explains how the A1408G-resistance mutation in the h44-decoding site reduces the apparent impact of aminoglycoside binding to H69 (Figs 2c and 3b; Supplementary Fig. 9) is not immediately obvious. The insight provided by the paromomycin-bound ribosome crystal structure, that aminoglycosides located within the h44 major groove can directly contact the apical tip of H69, suggests that mutations that reduce h44 affinity may simultaneously alter the local geometry of bridge B2 in a manner that indirectly affects drug interactions with the H69 major groove. A reduced affinity for H69 may also arise from global changes in the accessibility of the H69-binding site related to compression of its major groove during subunit rotation[1,31]. For instance, the H69-binding site is sterically occluded when the ribosome adopts a fully rotated configuration and may only be optimally formed upon partial subunit rotation. In this view, small changes in the underlying energy landscape of subunit rotation may conformationally mask the H69 major groove by altering the effective concentration of the transient H69-binding site. Considerations of this kind may indeed contribute to aminoglycoside resistance in eukaryotes, as the 80S ribosome preferentially adopts rotated states[52,53].

Preliminary efforts to examine kinetic features of neomycin and paromomycin binding to the ribosome reveal that their on- and off-rates to both h44 and H69 must also be taken into consideration (Fig. 6). Neomycin rapidly and simultaneously binds the major grooves of both h44 and H69 to stabilize the partially rotated ribosome, whereas following buffer exchange, the distribution of states exhibited resemble those observed in the context of the A1408G mutation, in which residual binding at H69 persists (Figs 2c and 6a). By contrast, paromomycin appears to quickly bind (ca. <1 s) and preferentially stabilize the unrotated ribosome configuration while dissociating relatively slowly from its h44-binding site upon buffer exchange (Fig. 6b). These observations suggest that aminoglycosides reversibly bind the h44-decoding site[18], whereas binding may be more avid to H69. Thus, distinct aminoglycoside actions in translation must include considerations about the disparities in the kinetic parameters of h44 and H69 binding. A quantitative understanding of aminoglycoside actions on the ribosome should therefore include considerations of the ribosome and drug concentrations, the dynamics of subunit rotation and the drug's net charge. This latter point is particularly crucial, given that the cell's proton gradient has been reported to dissipate upon aminoglycoside treatment[16], an outcome that is expected to increase the avidity of drug binding to bridge B2.

The capacity to image aminoglycoside activities on intact ribosomes greatly extends our knowledge of the mechanisms of action of specific aminoglycosides beyond that which could be achieved using model oligonucleotide systems and offers a powerful approach for exploring drug actions on defined ribosome complexes. However, the present investigations, together with observations that conformational processes in the ribosome are context dependent[18], suggest that the impact of aminoglycosides may be non-uniform in translation. Thus, targeted investigations probing the many distinct ribosome complexes transited during processive translation reactions must be further explored. A complete understanding of aminoglycoside actions in the cell may ultimately depend on the marriage of technological developments affording increased imaging throughput with those providing global snapshots of ribosome positions on specific mRNA species during active translation[54].

Methods

Reagents. Neomycin trisulfate salt hydrate (Sigma-Aldrich), paromomycin sulfate salt (Sigma-Aldrich), ribostamycin sulfate salt (Sigma-Aldrich), neamine hydrochloride (Toronto Research Chemicals) and puromycin dihydrochloride (Sigma-Aldrich) were used at the purity stated by the commercial suppliers ($\geq 97\%$). Chemoenzymatic preparation of acetylated forms of neomycin is described below. 6′-deoxy-paromomycin pentaacetate salt was synthesized by the Vasella group (ETH Zurich) as described[48] and was a kind gift from David Crich (Wayne State University). All smFRET experiments were performed in Tris-polymix buffer (pH 7.5) containing 50 mM Tris acetate, pH 7.5, 5 mM $Mg(OAc)_2$, 100 mM KCl, 5 mM NH_4OAc, 0.5 mM $CaCl_2$, 0.1 mM EDTA, 5 mM putrescine and 1 mM spermidine.

Generation of site-specifically labelled ribosome complexes. 30S subunits enzymatically labelled on S13 (N-Sfp) with LD550 (Lumidyne Technologies) and 50S subunits labelled with LD650 on L1 (T202C) were prepared and purified as previously described[1,40]. To label ribosomal protein L5 on native ribosomes, L5 was PCR cloned from *E. coli* strain K12 genomic DNA into the pPROEX HTb vector with a TEV-protease-cleavable histidine $(His)_6$ tag and a 12-residue peptide encoding the A1 epitope for the AcpS phosphopantetheinyl transferase reaction[55] (amino-acid sequence, GDSLDMLEWSLM) fused at the N terminus (N-AcpS). After plasmid shuffling into an *E. coli* ΔL5 knockout strain[56], cells were cultured and ribosomes were harvested, cleaved and labelled *in situ* as performed for S13-labelled subunits[1]. Cy5-fMet-tRNA^{fMet} (Cy5-s^{4}U8) was purified[57] and labelled[58] as previously described.

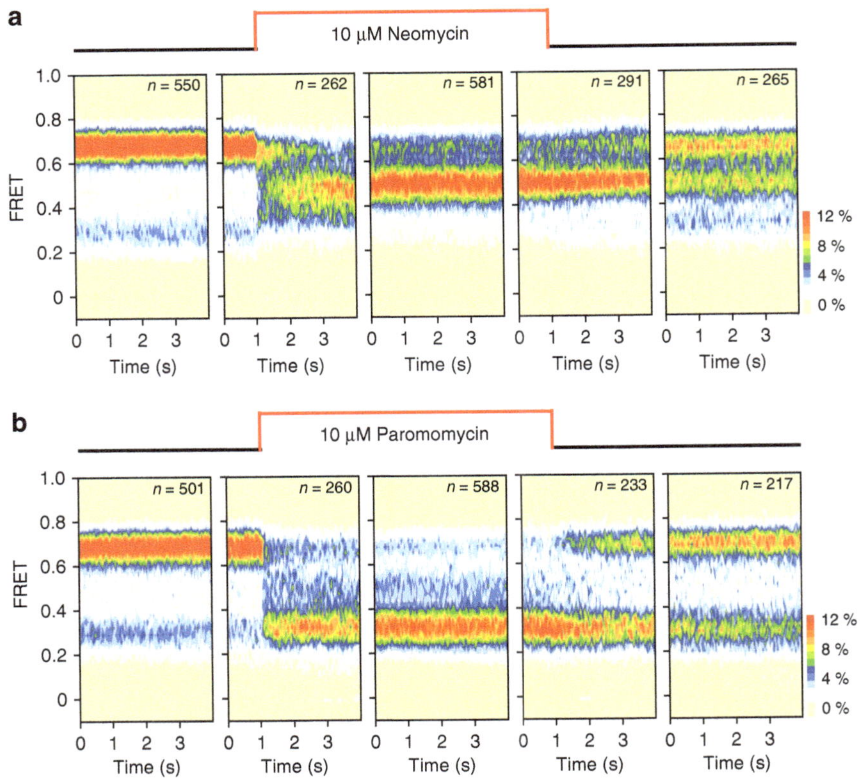

Figure 6 | Neomycin and paromomycin rinse off of h44 but not H69. Aminoglycoside real-time deliveries and rinse-outs were performed on wild-type ribosomes. For both (**a**) neomycin and (**b**) paromomycin, five population FRET histograms are shown representing the chronological order of the experiment. Left to right: (1) no drug; (2) real-time delivery of 10 µM aminoglycoside; (3) aminoglycoside equilibrium; (4) real-time rinse-out; (5) 90 s following rinse-out.

The A1408G point mutation in h44 of 16S rRNA was introduced into *in vivo* assembled ribosomes containing N-Sfp S13 using the previously described MS2 tag protocol[59] (pSpurMS2). Similarly, the H69 mutations (C1908G, C1909A, G1921U and G1922C) in 23S rRNA were introduced into *in vivo* assembled ΔL1 ribosomes[40] using the MS2 tag protocol[59] (p278MS2). To form intact 70S ribosome complexes, LD550-S13 30S and LD650-L1 50S subunits were heat activated at 42° C for 10 min in Tris-polymix Mg^{2+} buffer. Ribosomes were then initiated with fMet-$tRNA^{fMet}$ as previously described[60].

Acquisition and analysis of subunit rotation dynamics. smFRET experiments were performed using a prism-based total internal reflection microscope[45,60] at 25 °C. Ribosome complexes programmed with biotinylated mRNA and bearing P-site fMet-$tRNA^{fMet}$ were surface immobilized via a biotin–streptavidin bridge within polyethylene glycol (PEG)-passivated, streptavidin-coated quartz microfluidic chambers[45,60] and imaged in Tris-polymix buffer with 5 mM Mg^{2+}, supplemented with an oxygen scavenging system[61] and triplet-state quenching compounds[62] (1 mM cyclooctatetraene, 1 mM nitrobenzyl alcohol and 1 mM Trolox). Prior to experiments, P-site fMet-$tRNA^{fMet}$ was deacylated by incubation with 2 mM puromycin, a mimetic substrate of the peptidyltransferase centre that liberates nascent peptide chains[63], for 10 min at 25 °C in Tris-polymix buffer at pH 8.5. Aminoglycoside titrations were executed at pH 7.5, unless otherwise noted.

smFRET data were acquired by directly exciting the LD550 fluorophore linked to ribosomal protein S13 at 532 nm (LaserQuantum), while the LD550 and LD650 intensities were simultaneously recorded in Metamorph (Molecular Devices) at a 15-ms integration time. Using this approach, fluorescence and FRET trajectories could be obtained from hundreds of surface-immobilized, fluorescently labelled ribosome complexes simultaneously, where an average of 500 photons were collected for each ribosome complex per image frame. FRET traces were calculated as: $FRET = I_{LD650}/(I_{LD550} + I_{LD650})$, where I_{LD550} and I_{LD650} are the instantaneous donor and acceptor fluorescence intensities, respectively. FRETing molecules were selected for analysis using custom-made analytical software implemented in MATLAB (MathWorks) using the following criteria: a single catastrophic photobleaching event, at least 8:1 signal-to-background-noise ratio and 6:1 signal-to-signal-noise ratio, less than four donor fluorophore blinking events, a correlation coefficient between donor/acceptor <0.5 and a lifetime of at least 50 frames (750 ms) in any FRET state ≥ 0.15. FRET trajectories were idealized to a four-state hidden Markov model (Supplementary Fig. 5) using the segmental k-means algorithm[45]. Occupancies in each FRET state were calculated from the idealized dwell times in each state divided by the total dwell time in all nonzero FRET states, and were plotted in Origin (OriginLab).

Prior to the puromycin reaction, ribosome complexes bearing fMet-$tRNA^{fMet}$ in the P-site predominantly ($\sim 80\%$) occupied a stable, low-FRET (0.26 ± 0.01; mean-fitted FRET value ± s.d. of the Gaussian centre over three biological replicates) state (Supplementary Fig. 2a). A battery of published biological controls[1], together with high-resolution ribosome structures[1,31], suggest that this FRET value corresponds to an 'unrotated'/'classical' ribosome configuration in which the S13 and L1 proteins are distal to each other (Fig. 1c). A small fraction ($\sim 20\%$) of ribosomes exhibited high-FRET (0.67 ± 0.01) (Supplementary Fig. 2a). This FRET value is consistent with a 'rotated'/'hybrid' configuration in which the S13–L1 inter-protein distance is substantially reduced (~ 40 Å closer) (Fig. 1c). Consistent with this representing a subpopulation of ribosomes bearing deacylated P-site tRNA, all ribosomes were observed to exhibit a predominantly high-FRET state upon puromycin treatment (Supplementary Fig. 2b). Visual inspection of individual fluorescence and FRET trajectories obtained from puromycin-released complexes revealed time-dependent, anti-correlated fluctuations in donor and acceptor fluorescence intensities corresponding to low- to high-FRET state transitions (Fig. 2a). Consistent with the notion that P-site tRNA deacylation 'unlocks' the ribosome to enable the process of subunit rotation[42], ribosome complexes bearing deacylated $tRNA^{fMet}$ in the P-site spontaneously interconverted between unrotated and rotated configurations on the sub-second timescale. P-site tRNA deacylation increased the average transition rate between all three FRET states by $\sim$50-fold (~ 4.9 versus $\sim 0.1\ s^{-1}$) (Supplementary Fig. 5e). These data indicate that the observed motions predominantly report on transitions between unrotated, partially rotated and fully rotated configurations of the ribosome, rather than fast independent motions of the L1 stalk alone. Hence, we find that the process of subunit rotation in native ribosome complexes can occur at rates that are roughly an order of magnitude faster than EF-G-catalysed translocation[57] and the rates previously reported through investigations of *in vitro* reconstituted ribosome particles[32] or ribosomes hybridized to fluorescent oligonucleotides[35]. This metastability provided an ideal platform for exploring whether and to what extent chemically related aminoglycosides impact the ribosome's energy landscape upon binding.

Aminoglycoside association and dissociation experiments. The neomycin and paromomycin rinse-in/out experiments presented were performed using stopped-flow instrumentation on ribosomes labelled at L1–S13 and imaged as above, except the integration time was lowered to 40 ms to allow for a longer imaging window.

Purification of 70S ribosomes for *in vitro* translation. Tight-coupled 70S ribosomes of a homogenous rRNA sequence were purified from strain SQ171 (Quan S., Skovgaard O. and Squires C.L., unpublished results) expressing the wild-type or A1408G rrnB operon using a previously described protocol[60] with a slight modification. Prior to storage, ribosomes were buffer exchanged into 20 mM HEPES-KOH pH 7.8, 30 mM KCl, 10 mM $Mg(OAc)_2$ and 1 mM TCEP-HCl using a 100-kDa MWCO Microcon (Millipore), and 1 molar equivalent of purified *E. coli* S1 protein was added. The ribosome/S1 mixture was then incubated at 37 °C for 10 min and cooled to 4 °C over ~5 min. The solution was then exchanged into the same buffer and concentrated to [70S] > 10 μM before flash freezing and storing in liquid nitrogen.

Construction of the DNA template for *in vitro* translation. The DNA template used for *in vitro* protein synthesis reactions was a PCR product made by amplifying the *Photinus pyralis* luciferase gene. The primers were designed such that a T7 promoter followed by a bacterial Shine–Dalgarno sequence was added prior to the start codon and a T7 terminator sequence was added after the stop codon, as per the manufacturer's recommendations (New England BioLabs).The primer sequences are listed below, where the firefly luciferase sequence (lower case letters), the T7 promoter (underlined), the Shine–Dalgarno sequence (bold) and the stop codon (italics) are indicated. Forward: 5′-GCGAATTAATACGACTCACTATAGG GCTTAAGTATA**AGGAGG**AAAAAATatggaagacgccaaaaacat-3′. Reverse: 5′-AAA CCCCTCCGTTTAGAGAGGGGTTATGCTAG*TTA*cacggcgatctttccgccct-3′. Prior to use in coupled *in vitro* transcription–translation reactions, the PCR product was purified using the QIAquick PCR Purfication Kit (Qiagen) and resuspended in 0.5× TE buffer to a final concentration of 50 ng μl^{-1}.

***In vitro* protein synthesis reactions.** Solution A and Factor Mix were purchased from New England BioLabs as part of the PURExpress ΔRibosomes *In Vitro* Protein Synthesis Kit. Individual reactions contained 2 μl Solution A, 0.6 μl Factor Mix, 250 nM or 100 nM tight-coupled 70S ribosomes and 5 nM DNA template in a total volume of 5 μl. For the IC_{50} experiments, 4 μl of a reaction mix lacking aminoglycosides was added to 1 μl of aminoglycoside dilution in 0.2 ml thin-walled tubes on ice. For the ribosome activity assay, 4.1 μl of reaction mix was added to 0.9 μl of ribosomes or buffer on ice. Synthesis was initiated by incubation at 37 °C in a thermocycler (95 °C top) and terminated after 30 min by lowering the temperature to 4 °C for 10 min.

Luciferase activity assay. Luciferase activity was quantified using the Luciferase Assay System (Promega). Fifty microlitre of luciferase assay reagent was added to the completed reactions, which were then transferred to white 96-well plates (Corning, 360 μl well volume) for luminescence quantification by an Infinite M1000 Pro microplate reader (Tecan). To determine IC_{50} of aminoglycosides, luciferase activity assays were performed in triplicate across a series of drug concentrations. A no-drug control (100% activity) was used to normalize luminescence measurements. Activity percentages were then converted to percent inhibition by subtraction from 100%. These values were then plotted on a semi-log scale versus aminoglycoside concentrations. To estimate IC_{50} values, plots were fit to the Hill equation: $(100 \times [\text{aminoglycoside}]^n)/((IC_{50})^n + [\text{aminoglycoside}]^n)$.

Chemoenzymatic preparation of 3- and 6′-*N*-acetyl-neomycin. The enzymes used to generate the *N*-acetyl-neomycin derivatives, AAC(3)-IV and AAC(6′)/APH(2″) were purified as previously reported[64]. Reaction mixtures (5 ml) containing trisulfate salt hydrate (5 mM, Sigma-Aldrich), acetyl coenzyme A lithium salt (7.5 mM, Sigma-Aldrich), ammonium bicarbonate (pH 7.0, 50 mM) and appropriate enzyme (0.1 mg ml^{-1} for AAC(6′)/APH(2″) and 0.03 mg ml^{-1} for AAC(3)-IV) were incubated with shaking at 37 °C (for AAC(6′)/APH(2″)) or 25 °C (for AAC(3)-IV). The reaction progress was monitored by thin-layer chromatography (TLC) (Merck, SiO_2 gel 60 F_{254}) using a 3:2/MeOH:NH_4OH eluent system and staining with a cerium-molybdate stain ($(NH_4)_2Ce(NO_3)_6$ (5 g), $(NH_4)_6Mo_7O_{24} \bullet 4H_2O$ (120 g), H_2SO_4 (80 ml) and H_2O (720 ml)). Reactions were generally complete after 24 h of incubation. Once the enzymatic conversion was completed, the enzyme was precipitated by addition of an equal volume (5 ml) of ice-cold MeOH. The mixture was cooled at −20 °C for at least 10 min and centrifuged (5,500 r.p.m., 10 min, 4 °C, Sorvall RC6 Plus centrifuge, rotor F21S-8x50) to pellet the protein. To extract any *N*-acetylated neomycin that could be associated with the protein, the protein pellet was resuspended in 1 ml of cold H_2O, 1 ml of ice-cold MeOH was added and the precipitation process was repeated two more times. The 14 ml of combined solvent was removed *in vacuo* and the residue was resuspended in 1 ml of H_2O. *N*-acetylated neomycin products were purified away from the CoA side-product on an Amberlite CG-50 (100–200 mesh: ammonium form, column packed = 1 × 10 cm). The residual solution was loaded onto the ion-exchange column and washed with 15 ml H_2O. The compounds were eluted stepwise (0.4% increments of 3 ml each for AAC(6′)/APH(2″) and 0.2% increments of 3 ml each for AAC(3)-IV) using a gradient of NH_4OH (0–2%). The size of each fraction collected was 1.5 ml. To facilitate loading of the fractions onto the TLC plate for visualization, the fractions were diluted with 1.5 ml of MeOH. Fractions containing desired *N*-acetylated neomycin, as determined by TLC, were combined and the solvent was removed *in vacuo*. Both 6′-*N*-acetyl-neomycin and 3-*N*-acetyl-neomycin eluted between 0.4 and 0.8% NH_4OH. Reactions yielded 13.1 mg of 3-*N*-acetyl-neomycin (80%) and 6.4 mg of 6′-*N*-acetyl-neomycin (39%). Note: the yield of the 6′-*N*-acetyl-neomycin is lower as it was utilized to optimize the purification protocol.

Ribosome purification for crystallization. Ribosome purification from *E. coli* strain MRE600 was prepared as previously described[65]. Ribosome complexes with ribosome recycling factor (RRF) were made by incubating 2 μM ribosomes, 8 μM mRNA (5′-GGCAAGGAGGUAAAAUUCUACAAA-3′ Thermo Scientific) and 4 μM $tRNA^{Phe}$ for 15 min at 37 °C. Next, 8 μM RRF was added and incubated for 15 min at 37 °C. Ribosomes were crystallized at 18 °C using microbatch 96-well plates and buffers containing 4.0–5.0% 2-methyl-2,4-pentanediol, 4.1–4.5% PEG 8000, 4.0 mM $MgCl_2$, 380 mM NH_4Cl, 5.7 mM putrescine, 5.0 mM spermidine, 10 mM Tris, 40 mM MES, pH 6.5–7.0 and 0.25 mM EDTA.

Data collection and processing. Ribosome crystals were stepwise cryo-protected to the final conditions containing 7.0% 2-methyl-2,4-pentanediol, 7.0% PEG 8000 and 24% PEG 400, pH 4.8, to allow cryo-cooling of the crystals to liquid nitrogen temperatures. During the last cryo-protection step (PEG 400 24%), paromomycin sulfate salt (Sigma-Aldrich) was added at a 200-μM concentration to the cryo-protection buffer, and crystals were incubated at 4 °C with paromomycin. After 2 h of incubation, crystals were frozen with liquid nitrogen. Diffraction data were measured from crystals cooled to 100 K using 0.1–0.3° oscillations at the Advanced Light Source (beamlines 8.3.1 and 12.3.1), each of which is equipped with an ADSC Q315 area detector. Data were reduced using XDS[66], yielding the statistics shown in Table 1.

Molecular replacement and structure refinement. Molecular replacement and structure refinement was carried out in PHENIX[67] platform, and previously published/deposited coordinates of neomycin-bound ribosomes (PDB ID: 4GAQ 4GAR 4GAS 4GAU) were used as a reference model[1]. Neomycin molecules were removed from the molecular replacement model to reduce possible model bias during refinement. Corrections to the resulting model were carried out in Coot[68]. The resulting structural models were then refined using rounds of manual rebuilding in Coot. Electron density maps were generated from the PHENIX output directly. RNA rebuilding concentrated on the tRNAs, H69 and h44 (neomycin-/paromomycin-binding sites) and paromomycin structures were inserted and refined using Coot and PHENIX. Paromomycin occupancies were determined by matching paromomycin atomic displacement parameters to the mean value of neighbouring rRNA, and then the group occupancy of each paromomycin was refined in PHENIX as previously described[1]. As with neomycin[1], additional paromomycin-binding sites were found near A482 (H24), G551 (H25), G1157 (H41), U1240 (H46), A1858 (H68) and C2674 (H95) in 23S rRNA, and near C658 (h22), G902 (h27) and U1420 (h44) in 16S rRNA. Maps comparing bridge B2a in neomycin-bound (PDB ID 4V9C) and paromomycin-bound ribosome structures used feature-enhanced maps generated by phenix.fem in the PHENIX-1.9-1692 release, with default parameters plus the combined-omit map algorithm[69].

Superpositions. Superpositions were carried out in PyMOL using the 'pair_fit' command[70]. Disordered or flexible regions of 23S rRNA were not used (for example, the L1 stalk, the L7/L12 stalk, H38 and H69) in the superpositions. Superpositions were performed using ribose C1′ positions or phosphorus atoms in nucleotides. The angles of rotation of the 30S subunit domains were calculated essentially as described previously[1,31]. Angles given for the rotation of the head domain were calculated from 30S subunit structures superimposed by means of their platform domains. A rotation of 0° is defined as centering the head domain over the 30S P-site, as seen in the structure of the unrotated ribosome[31]. Superpositions of P/P, P/pe and P/E tRNAs used the C1′ atoms of nucleotides 31–39 in the anticodon stem loop[31]. The comparisons of tRNA bending angles used the glycosidic bond of position 31 near the end of the anticodon stem loop and the glycosidic bond of nucleotide 63 in superimposed tRNAs[31]. The calculated bending angles were 24° for P/P tRNA compared with P/pe tRNA, and 14° for P/pe tRNA compared with P/E tRNA. For the H69-bound neomycin and paromomycin rRNA comparison, 23S rRNA positions 1906–1909, 1919–1931 were used.

Figure preparation. All structure figures were made using the program PyMOL[70].

References

1. Wang, L. *et al.* Allosteric control of the ribosome by small-molecule antibiotics. *Nat. Struct. Mol. Biol.* **19,** 957–963 (2012).
2. Puglisi, J. D. *et al.* in *The Ribosome: Structure, Function, Antibiotics and Cellular Interactions.* (eds Garrett, R. A., Douthwaite, S. R., Liljas, A., Matheson, A. T., Moore, P. B. & Noller, H. F.) (ASM Press, 2000).

3. Cabanas, M. J., Vazquez, D. & Modolell, J. Inhibition of ribosomal translocation by aminoglycoside antibiotics. *Biochem. Biophys. Res. Commun.* **83,** 991–997 (1978).
4. Davies, J., Gorini, L. & Davis, B. D. Misreading of RNA codewords induced by aminoglycoside antibiotics. *Mol. Pharmacol.* **1,** 93–106 (1965).
5. Peske, F., Savelsbergh, A., Katunin, V. I., Rodnina, M. V. & Wintermeyer, W. Conformational changes of the small ribosomal subunit during elongation factor G-dependent tRNA-mRNA translocation. *J. Mol. Biol.* **343,** 1183–1194 (2004).
6. Becker, B. & Cooper, M. A. Aminoglycoside antibiotics in the 21st century. *ACS Chem. Biol.* **8,** 105–115 (2013).
7. Recht, M. I., Douthwaite, S. & Puglisi, J. D. Basis for prokaryotic specificity of action of aminoglycoside antibiotics. *EMBO J.* **18,** 3133–3138 (1999).
8. Moazed, D. & Noller, H. F. Interaction of antibiotics with functional sites in 16S ribosomal RNA. *Nature* **327,** 389–394 (1987).
9. Ogle, J. M. & Ramakrishnan, V. Structural insights into translational fidelity. *Annu. Rev. Biochem.* **74,** 129–177 (2005).
10. Rodnina, M. V. & Wintermeyer, W. Fidelity of aminoacyl-tRNA selection on the ribosome: kinetic and structural mechanisms. *Annu. Rev. Biochem.* **70,** 415–435 (2001).
11. Carter, A. P. *et al.* Functional insights from the structure of the 30S ribosomal subunit and its interactions with antibiotics. *Nature* **407,** 340–348 (2000).
12. Vaiana, A. C. & Sanbonmatsu, K. Y. Stochastic gating and drug-ribosome interactions. *J. Mol. Biol.* **386,** 648–661 (2009).
13. Selmer, M. *et al.* Structure of the 70S ribosome complexed with mRNA and tRNA. *Science* **313,** 1935–1942 (2006).
14. Jerinic, O. & Joseph, S. Conformational changes in the ribosome induced by translational miscoding agents. *J. Mol. Biol.* **304,** 707–713 (2000).
15. Ogle, J. M., Murphy, F. V., Tarry, M. J. & Ramakrishnan, V. Selection of tRNA by the ribosome requires a transition from an open to a closed form. *Cell* **111,** 721–732 (2002).
16. Davis, B. D. Mechanism of bactericidal action of aminoglycosides. *Microbiol. Rev.* **51,** 341–350 (1987).
17. Burke, J. F. & Mogg, A. E. Suppression of a nonsense mutation in mammalian cells in vivo by the aminoglycoside antibiotics G-418 and paromomycin. *Nucleic Acids Res.* **13,** 6265–6272 (1985).
18. Feldman, M. B., Terry, D. S., Altman, R. B. & Blanchard, S. C. Aminoglycoside activity observed on single pre-translocation ribosome complexes. *Nat. Chem. Biol.* **6,** 54–62 (2010).
19. Borovinskaya, M. A. *et al.* Structural basis for aminoglycoside inhibition of bacterial ribosome recycling. *Nat. Struct. Mol. Biol.* **14,** 727–732 (2007).
20. Hobbie, S. N. *et al.* Binding of neomycin-class aminoglycoside antibiotics to mutant ribosomes with alterations in the A site of 16S rRNA. *Antimicrob. Agents Chemother.* **50,** 1489–1496 (2006).
21. Benveniste, R. & Davies, J. Structure-activity relationships among the aminoglycoside antibiotics: role of hydroxyl and amino groups. *Antimicrob. Agents Chemother.* **4,** 402–409 (1973).
22. Pfister, P. *et al.* Mutagenesis of 16S rRNA C1409-G1491 base-pair differentiates between 6'OH and 6'NH3+ aminoglycosides. *J. Mol. Biol.* **346,** 467–475 (2005).
23. Jorgensen, F. & Kurland, C. G. Death rates of bacterial mutants. *FEMS Microbiol. Lett.* **40,** 43–46 (1987).
24. O'Connor, M., Goringer, H. U. & Dahlberg, A. E. A ribosomal ambiguity mutation in the 530 loop of E. coli 16S rRNA. *Nucleic Acids Res.* **20,** 4221–4227 (1992).
25. Perzynski, S., Cannon, M., Cundliffe, E., Chahwala, S. B. & Davies, J. Effects of apramycin, a novel aminoglycoside antibiotic on bacterial protein synthesis. *Eur. J. Biochem.* **99,** 623–628 (1979).
26. Misumi, M., Nishimura, T., Komai, T. & Tanaka, N. Interaction of kanamycin and related antibiotics with the large subunit of ribosomes and the inhibition of translocation. *Biochem. Biophys. Res. Commun.* **84,** 358–365 (1978).
27. Zhou, J., Lancaster, L., Donohue, J. P. & Noller, H. F. How the ribosome hands the A-site tRNA to the P site during EF-G-catalyzed translocation. *Science* **345,** 1188–1191 (2014).
28. Maivali, U. & Remme, J. Definition of bases in 23S rRNA essential for ribosomal subunit association. *RNA* **10,** 600–604 (2004).
29. Horan, L. H. & Noller, H. F. Intersubunit movement is required for ribosomal translocation. *Proc. Natl Acad. Sci. USA* **104,** 4881–4885 (2007).
30. Agrawal, R. K. *et al.* Visualization of ribosome-recycling factor on the Escherichia coli 70S ribosome: functional implications. *Proc. Natl Acad. Sci. USA* **101,** 8900–8905 (2004).
31. Dunkle, J. A. *et al.* Structures of the bacterial ribosome in classical and hybrid states of tRNA binding. *Science* **332,** 981–984 (2011).
32. Cornish, P. V., Ermolenko, D. N., Noller, H. F. & Ha, T. Spontaneous intersubunit rotation in single ribosomes. *Mol. Cell* **30,** 578–588 (2008).
33. Fei, J. *et al.* Allosteric collaboration between elongation factor G and the ribosomal L1 stalk directs tRNA movements during translation. *Proc. Natl Acad. Sci. USA* **106,** 15702–15707 (2009).
34. Fei, J., Kosuri, P., MacDougall, D. D. & Gonzalez, Jr. R. L. Coupling of ribosomal L1 stalk and tRNA dynamics during translation elongation. *Mol. Cell* **30,** 348–359 (2008).
35. Marshall, R. A., Dorywalska, M. & Puglisi, J. D. Irreversible chemical steps control intersubunit dynamics during translation. *Proc. Natl Acad. Sci. USA* **105,** 15364–15369 (2008).
36. Zheng, Q. *et al.* Ultra-stable organic fluorophores for single-molecule research. *Chem. Soc. Rev.* **43,** 1044–1056 (2014).
37. Greenberg, W. A. *et al.* Design and synthesis of new aminoglycoside antibiotics containing neamine as an optimal core structure: Correlation of antibiotic activity with in vitro inhibition of translation. *J. Am. Chem. Soc.* **121,** 6527–6541 (1999).
38. Perez-Fernandez, D. *et al.* 4'-O-substitutions determine selectivity of aminoglycoside antibiotics. *Nat. Commun.* **5,** 3112 (2014).
39. Kaul, M. & Pilch, D. S. Thermodynamics of aminoglycoside-rRNA recognition: the binding of neomycin-class aminoglycosides to the A site of 16S rRNA. *Biochemistry* **41,** 7695–7706 (2002).
40. Munro, J. B., Altman, R. B., Tung, C. S., Sanbonmatsu, K. Y. & Blanchard, S. C. A fast dynamic mode of the EF-G-bound ribosome. *EMBO J.* **29,** 770–781 (2010).
41. Cornish, P. V. *et al.* Following movement of the L1 stalk between three functional states in single ribosomes. *Proc. Natl Acad. Sci. USA* **106,** 2571–2576 (2009).
42. Valle, M. *et al.* Locking and unlocking of ribosomal motions. *Cell* **114,** 123–134 (2003).
43. Guo, Z. & Noller, H. F. Rotation of the head of the 30S ribosomal subunit during mRNA translocation. *Proc. Natl Acad. Sci. USA* **109,** 20391–20394 (2012).
44. Ratje, A. H. *et al.* Head swivel on the ribosome facilitates translocation by means of intra-subunit tRNA hybrid sites. *Nature* **468,** 713–716 (2010).
45. Munro, J. B., Altman, R. B., O'Connor, N. & Blanchard, S. C. Identification of two distinct hybrid state intermediates on the ribosome. *Mol. Cell* **25,** 505–517 (2007).
46. McKinney, S. A., Joo, C. & Ha, T. Analysis of single-molecule FRET trajectories using hidden Markov modeling. *Biophys. J.* **91,** 1941–1951 (2006).
47. Biddlecome, S. *et al.* Enzymatic modification of aminoglycoside antibiotics: a new 3-N-acetylating enzyme from a Pseudomonas aeruginosa isolate. *Antimicrob. Agents Chemother.* **9,** 951–955 (1976).
48. Pathak, R., Bottger, E. C. & Vasella, A. Design and synthesis of aminoglycoside antibiotics to selectively target 16S ribosomal RNA position 1408. *Helv. Chim. Acta* **88,** 2967–2985 (2005).
49. Jenner, L., Demeshkina, N., Yusupova, G. & Yusupov, M. Structural rearrangements of the ribosome at the tRNA proofreading step. *Nat. Struct. Mol. Biol.* **17,** 1072–1078 (2010).
50. Fischer, N., Konevega, A. L., Wintermeyer, W., Rodnina, M. V. & Stark, H. Ribosome dynamics and tRNA movement by time-resolved electron cryomicroscopy. *Nature* **466,** 329–333 (2010).
51. Scheunemann, A. E., Graham, W. D., Vendeix, F. A. & Agris, P. F. Binding of aminoglycoside antibiotics to helix 69 of 23S rRNA. *Nucleic Acids Res.* **38,** 3094–3105 (2010).
52. Ben-Shem, A. *et al.* The structure of the eukaryotic ribosome at 3.0A resolution. *Science* **334,** 1524–1529 (2011).
53. Budkevich, T. *et al.* Structure and dynamics of the mammalian ribosomal pretranslocation complex. *Mol. Cell* **44,** 214–224 (2011).
54. Ingolia, N. T., Ghaemmaghami, S., Newman, J. R. & Weissman, J. S. Genome-wide analysis in vivo of translation with nucleotide resolution using ribosome profiling. *Science* **324,** 218–223 (2009).
55. Zhou, Z. *et al.* Genetically encoded short peptide tags for orthogonal protein labeling by Sfp and AcpS phosphopantetheinyl transferases. *ACS Chem. Biol.* **2,** 337–346 (2007).
56. Shoji, S., Dambacher, C. M., Shajani, Z., Williamson, J. R. & Schultz, P. G. Systematic chromosomal deletion of bacterial ribosomal protein genes. *J. Mol. Biol.* **413,** 751–761 (2011).
57. Wang, L., Altman, R. B. & Blanchard, S. C. Insights into the molecular determinants of EF-G catalyzed translocation. *RNA* **17,** 2189–2200 (2011).
58. Watson, B. S. *et al.* Macromolecular arrangement in the aminoacyl-tRNA.elongation factor Tu.GTP ternary complex. A fluorescence energy transfer study. *Biochemistry* **34,** 7904–7912 (1995).
59. Youngman, E. M. & Green, R. Affinity purification of in vivo-assembled ribosomes for in vitro biochemical analysis. *Methods* **36,** 305–312 (2005).
60. Blanchard, S. C., Kim, H. D., Gonzalez, Jr. R. L., Puglisi, J. D. & Chu, S. tRNA dynamics on the ribosome during translation. *Proc. Natl Acad. Sci. USA* **101,** 12893–12898 (2004).

61. Aitken, C. E., Marshall, R. A. & Puglisi, J. D. An oxygen scavenging system for improvement of dye stability in single-molecule fluorescence experiments. *Biophys. J.* **94,** 1826–1835 (2008).
62. Dave, R., Terry, D. S., Munro, J. B. & Blanchard, S. C. Mitigating unwanted photophysical processes for improved single-molecule fluorescence imaging. *Biophys. J.* **96,** 2371–2381 (2009).
63. Traut, R. R. & Monro, R. E. The puromycin reaction and its relation to protein synthesis. *J. Mol. Biol.* **10,** 63–72 (1964).
64. Green, K. D., Chen, W., Houghton, J. L., Fridman, M. & Garneau-Tsodikova, S. Exploring the substrate promiscuity of drug-modifying enzymes for the chemoenzymatic generation of N-acylated aminoglycosides. *Chembiochem* **11,** 119–126 (2010).
65. Blaha, G. *et al.* Preparation of functional ribosomal complexes and effect of buffer conditions on tRNA positions observed by cryoelectron microscopy. *Methods Enzymol.* **317,** 292–309 (2000).
66. Kabsch, W. Xds. *Acta Crystallogr. D Biol. Crystallogr.* **66,** 125–132 (2010).
67. Adams, P. D. *et al.* PHENIX: a comprehensive Python-based system for macromolecular structure solution. *Acta Crystallogr. D Biol. Crystallogr.* **66,** 213–221 (2010).
68. Emsley, P., Lohkamp, B., Scott, W. G. & Cowtan, K. Features and development of Coot. *Acta Crystallogr. D Biol. Crystallogr.* **66,** 486–501 (2010).
69. Afonine, P. V. *et al.* FEM: feature-enhanced map. *Acta Crystallogr. D Biol. Crystallogr.* **71,** 646–666 (2015).
70. Schrödinger LLC. The PyMOL Molecular Graphics System. Version 1.3 (2002).

Acknowledgements

We thank R. Green (Johns Hopkins University) for providing the S13 knockout strain, P. Schultz for providing the L5 knockout strain (Scripps Research Institute), T. Suzuki (University of Tokyo) for providing the pKK3535 ribosome plasmids, D. Crish (Wayne State University) for providing 6′-deoxy-paromomycin and M. O'Connor (University of Missouri-Kansas City) for helpful discussions throughout the course of this work. We also acknowledge helpful discussions and insights provided by all members of the Blanchard and Cate laboratories and J. Headd (Lawrence Berkeley National Laboratory) for help with PHENIX refinement. This work was supported by the US National Institutes of Health (2R01GM079238 to S.C.B., 1R01GM65050 to J.H.D.C., 1R01AI090048 to S.G.-T. and National Cancer Institute grant CA92584 for the SIBYLS and 8.3.1 beamlines at the Advanced Light Source (ALS), Lawrence Berkeley National Laboratory), the Human Frontiers in Science Program (RGY0088), the National Science Foundation (0644129) and the US Department of Energy (DE-AC0376SF00098 for the SIBYLS and 8.3.1 beamlines at the ALS, Lawrence Berkeley National Laboratory).

Author contributions

M.R.W. prepared dye-labelled ribosomes, performed the smFRET imaging and analysed the results. A.P. crystallized, collected, processed and refined X-ray data. J.C.Z. performed *in vitro* transcription/translation assays. K.D.G. and S.G.-T. prepared and purified 3- and 6′-*N*-acetyl-neomycin. Z.Z. and R.B.A. synthesized the next-generation fluorophores used for smFRET imaging. M.R.W. and A.P. made the figures. J.H.D.C. and S.C.B. designed the study. All authors discussed the results and contributed to the writing of the manuscript.

Additional information

Accession codes: The structures reported here have been deposited in the Protein Data Bank with codes 4WOI, 4WON, 4WOM and 4WOO.

Competing financial interests: S.C.B and R.B.A have an equity interest in Lumidyne Technologies. The remaining authors declare no competing financial interests.

14

Triaminopyrimidine is a fast-killing and long-acting antimalarial clinical candidate

Shahul Hameed P.[1,*], Suresh Solapure[1,*], Vikas Patil[1], Philipp P. Henrich[2], Pamela A. Magistrado[3], Sowmya Bharath[1], Kannan Murugan[1], Pavithra Viswanath[1], Jayashree Puttur[1], Abhishek Srivastava[1], Eknath Bellale[1], Vijender Panduga[1], Gajanan Shanbag[1], Disha Awasthy[1], Sudhir Landge[1], Sapna Morayya[1], Krishna Koushik[1], Ramanatha Saralaya[1], Anandkumar Raichurkar[1], Nikhil Rautela[1], Nilanjana Roy Choudhury[1], Anisha Ambady[1], Radha Nandishaiah[1], Jitendar Reddy[1], K.R. Prabhakar[1], Sreenivasaiah Menasinakai[1], Suresh Rudrapatna[1], Monalisa Chatterji[1], María Belén Jiménez-Díaz[4], María Santos Martínez[4], Laura María Sanz[4], Olivia Coburn-Flynn[2], David A. Fidock[2,5], Amanda K. Lukens[3], Dyann F. Wirth[3], Balachandra Bandodkar[1], Kakoli Mukherjee[1], Robert E. McLaughlin[6], David Waterson[7], Lyn Rosenbrier-Ribeiro[8], Kevin Hickling[8], V. Balasubramanian[1], Peter Warner[1], Vinayak Hosagrahara[1], Adam Dudley[1], Pravin S. Iyer[1], Shridhar Narayanan[1], Stefan Kavanagh[8] & Vasan K. Sambandamurthy[1,†]

The widespread emergence of *Plasmodium falciparum* (*Pf*) strains resistant to frontline agents has fuelled the search for fast-acting agents with novel mechanism of action. Here, we report the discovery and optimization of novel antimalarial compounds, the triaminopyrimidines (TAPs), which emerged from a phenotypic screen against the blood stages of *Pf*. The clinical candidate (compound **12**) is efficacious in a mouse model of *Pf* malaria with an ED_{99} $<30\,mg\,kg^{-1}$ and displays good *in vivo* safety margins in guinea pigs and rats. With a predicted half-life of 36 h in humans, a single dose of 260 mg might be sufficient to maintain therapeutic blood concentration for 4-5 days. Whole-genome sequencing of resistant mutants implicates the vacuolar ATP synthase as a genetic determinant of resistance to TAPs. Our studies highlight the potential of TAPs for single-dose treatment of *Pf* malaria in combination with other agents in clinical development.

[1] Department of Innovative Medicines, AstraZeneca India Pvt. Ltd., Bellary Road, Hebbal, Bangalore 560024, India. [2] Department of Microbiology and Immunology, Columbia University Medical Center, New York, New York 10032, USA. [3] Harvard School of Public Health, 665 Huntington Avenue, Boston, Massachusetts 02115, USA. [4] Tres Cantos Medicines Development Campus. Diseases of Developing World (DDW), GlaxoSmithKline, Severo Ochoa 2, Tres Cantos, Madrid 28760, Spain. [5] Division of Infectious Diseases, Department of Medicine, Columbia University Medical Center, New York, New York 10032, USA. [6] AstraZeneca Infection Innovative Medicines, 35 Gatehouse Drive, Waltham, Massachusetts 02451, USA. [7] Medicines for Malaria Venture, International Center Cointrin, Geneva 1215, Switzerland. [8] AstraZeneca, Alderley Park, Cheshire SK10 4TF, UK. * These authors contributed equally to this work. † Present address: Mazumdar Shaw Center for Translational Research (MSCTR), Bangalore, India. Correspondence and requests for materials should be addressed to S.K. (email: stefan.kavanagh@astrazeneca.com) or to V.K.S. (email: vasan2005@gmail.com).

Malaria accounted for an estimated 584,000 deaths in 2013, especially among children and pregnant women[1]. The widespread emergence and dissemination of *Plasmodium falciparum* (*Pf*) strains resistant to conventional antimalarial drugs has intensified the efforts to discover and develop novel, structurally diverse drugs against multidrug-resistant *Plasmodium* strains. This situation is further aggravated by the emergence of *Pf* strains resistant to artemisinin derivatives in parts of South East Asia[2–4]. Even more worrisome is the emergence of *Pf* strains resistant to most available antimalarial medicines in areas along the Cambodia–Thailand border[1].

A global strategy aimed at administering a combination of novel agents that target the asexual and sexual forms of the parasite is likely to prevent the emergence of resistance. The goal of single-exposure radical cure and prophylaxis can be accomplished through suitable combination of partner drugs with fast-killing and/or long half-life attributes[5]. The advent of high-throughput whole-cell screening against *Pf* has led to the progression of several novel antimalarial agents into preclinical and clinical development[6–10].

The present study describes the discovery and optimization of a novel antimalarial series belonging to the triaminopyrimidines (TAPs) class with the potential for long half-life in humans and displaying potent activity against a panel of clinical strains harbouring resistance to known antimalarial drugs as well as agents in clinical development. The TAPs kill *Pf* parasites rapidly, and the emergence of spontaneous resistance under *in vitro* conditions to this chemical class is rare. We present extensive pharmacokinetic (PK) and preclinical safety data that supports the progression of this novel antimalarial agent towards clinical development to treat malaria.

Results

Optimization of TAPs as potent antimalarial agents. The TAPs series emerged from a high-throughput screen (HTS) of 500,000 compounds from the AstraZeneca corporate library against the asexual blood stage of *Pf* using a high-content image-based approach[10]. The synthetic scheme for representative compounds (**7**, **8**, **9** and **12**) is shown in Fig. 1. The procedure for synthesis of compounds **1-6**, **10**, **11** using a scheme similar to Fig. 1 is provided in Supplementary Note 1. The initial hit provided multiple diversification points for a robust lead optimization campaign. A schematic of the medicinal chemistry optimization and structure activity relationship (SAR) of TAPs are shown in Fig. 2a and Table 1. The SAR analysis revealed essentiality of the basic ring substitution at the C-5 position of the pyrimidine core for *Pf* potency. The initial hit (compound **1**) exhibited cardiac ion channel inhibition (hERG, IC_{50}–3.7 μM) and poor solubility (10 μM). Replacement of the phenyl ring with a 4-pyridyl ring system at the 2-amino position improved solubility (830 μM) and reduced the hERG liability (IC_{50} 85 μM). Compounds with 2-pyridyl modification at the 2-amino position of pyrimidine core improved potency and solubility at the cost of potent activity against hERG and acetylcholine esterase (AChE; compounds **3-6**; Table 1). Compound **7**, with a 2-aminopyrazole modification reduced the potency against hERG (IC_{50} > 33 μM) and AChE (IC_{50}–66 μM) with a concomitant, four-fold loss in *Pf* potency. The observed hERG and AChE selectivity improvement with **7** could be attributed to the polar nature of the pyrazole group and its role in reducing the pKa of the core pyrimidine ring. The introduction of a cyclopropyl group at the 4-position of pyridine in **7** resulted in compound **9** with a > 10-fold improvement in *Pf* potency with an excellent hERG and AChE selectivity, albeit with suboptimal bioavailability (Table 1). Compound **12** with *N*-methylpiperazine at the C-5 position (*N*-methyl analogue of **9**)

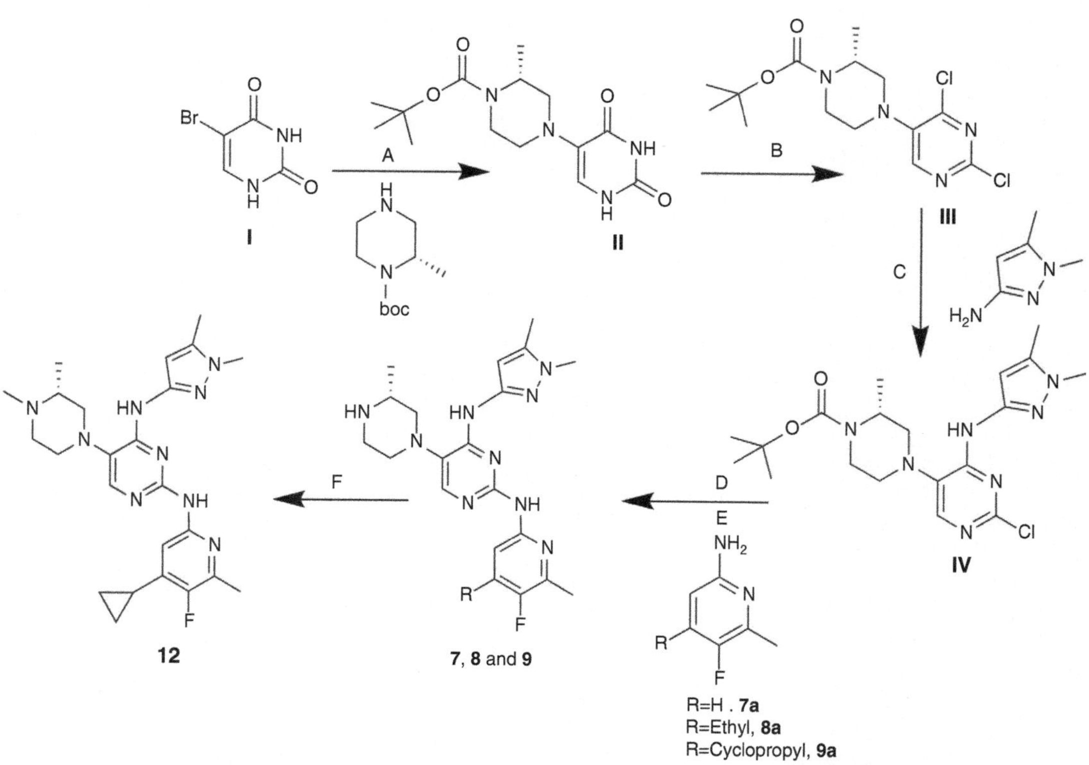

Figure 1 | Synthetic schemes and reaction conditions for compounds 7, 8, 9 and 12. (A) Pyridine, microwave, 150 °C, 45 min. (B) (i) $POCl_3$, reflux, 6 h (ii) sodium carbonate, di-tert-butyl dicarbonate, room temperature, 16 h. (C) N,N-Diisopropylethylamine (DIPEA), ethanol, microwave, 110 °C, 1 h. (D) (i) Potassium tert-butoxide, 2,2′-bis(diphenylphosphino)-1,1′-binaphthyl (BINAP), $pd_2(dba)_3$, toluene, reflux, 12 h. (E) HCl (4 N) in dioxane, 15–30 min. (F) Compound **9**, DIPEA, dichloromethane, formaldehyde (HCHO), sodium cyanoborohydride, 15 min.

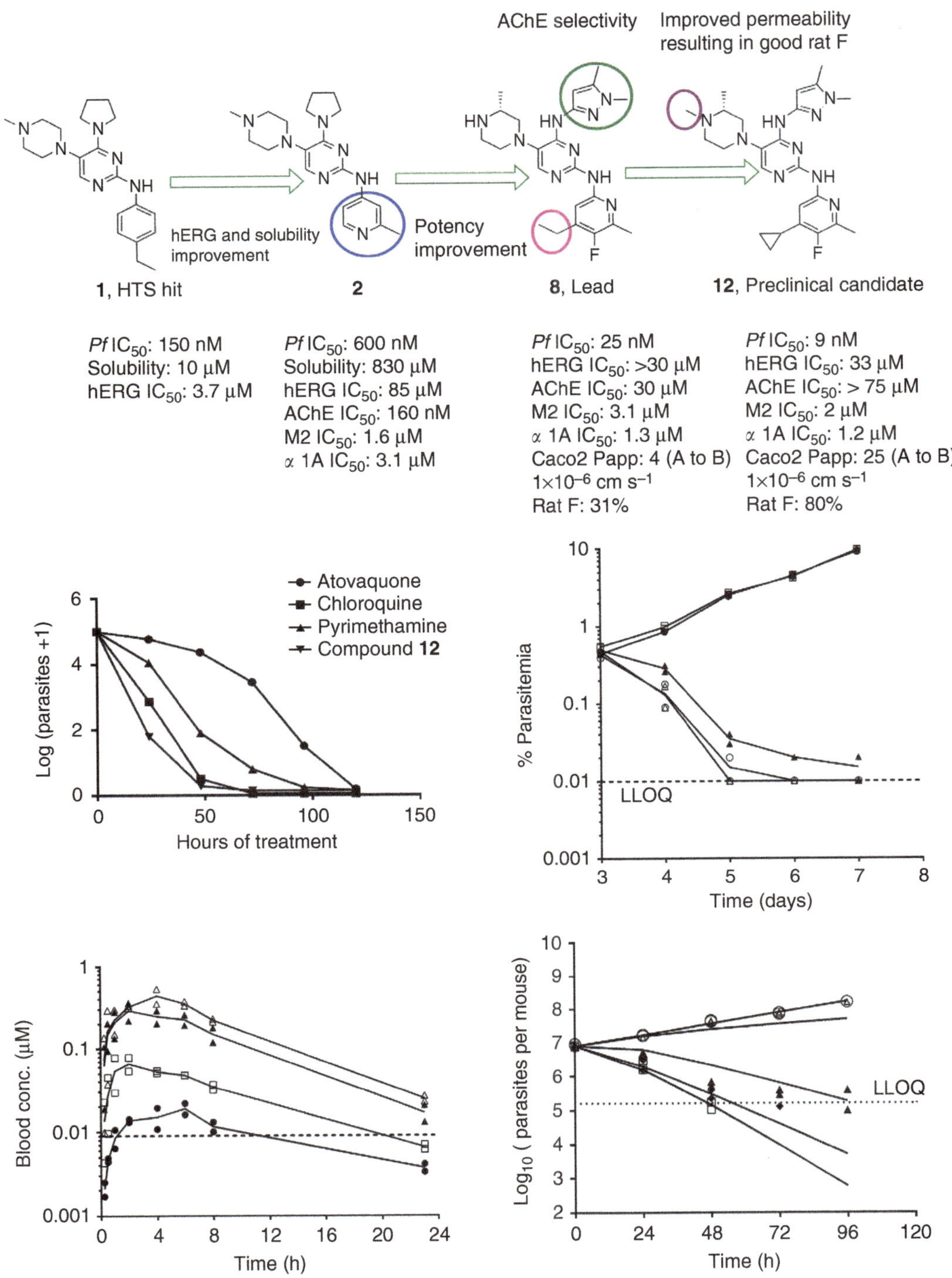

Figure 2 | Evolution of clinical candidate compound 12 and its *in vitro* and *in vivo* killing kinetics profile. (**a**) Schematic of medicinal chemistry optimization and identification of clinical candidate **12**. (**b**) *In vitro* parasite reduction ratio (PRR) for compound **12**. The graph shows change in the number of viable parasites over time after exposure of *Pf* 3D7A to atovaquone (●), chloroquine (■), pyrimethamine (▲) or compound 12 (▼) at a concentration (conc.) equal to 10 times their respective IC_{50}s. (**c**) Percentage parasitemia in peripheral blood of mice infected with *Pf*3D7$^{0087/N9}$ ($n = 2$ per group) after treatment with compound **12** at 10 (□), 20 (▲), 40 (○) or 80 (Δ) mg kg^{-1} or with vehicle (●). Dotted line indicates the lower limit of quantification (LLOQ) for % parasitemia estimation. (**d**) Blood concentration versus time profile for compound **12** in infected mice is depicted after the first dose of 10 (●), 20 (□), 40 (▲) or 80 (Δ) mg kg^{-1}. Dotted line indicates the *Pf* IC_{50}. Two mice were used per dose group. (**e**) Predicted (lines) and observed (symbols) change in the total parasite burden in infected mice in the *Pf*/SCID model after treatment with compound **12** at 10 (Δ), 20 (▲), 40 (♦) or 80 (□) mg kg^{-1} or with vehicle (○). SCID, severe combined immunodeficient.

showed the best potency (9 nM) combined with a >1,000-fold selectivity against hERG and AChE (Table 1). In addition, **12** displayed improved bioavailability (80%) as compared with its demethylated analogue, **9** owing to its improved Caco2 permeability (Fig. 2a). Compound **12** exhibiting desirable PK properties was evaluated for *in vivo* efficacy and safety. To assess the potential for cross-resistance, representative TAPs were screened against a panel of clinical isolates and *in vitro*-generated mutant strains displaying varying mechanisms of resistance to antimalarial agents currently in use and under

Table 1 | Structure activity relationship and cytotoxicity selectivity index for compounds 1-12.

Compound number	Structure	*Pf* IC_{50} NF54 (nM)	*Pf* IC_{50} K1 (nM)	THP-1 IC_{50} (μM)	Solubility (μM)	hERG IC_{50} (μM)	AChE IC_{50} (μM)
1	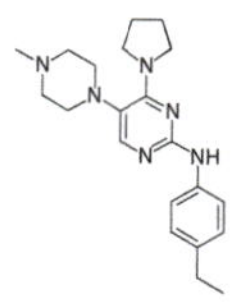	150	110	10	10	3.7	>100
2	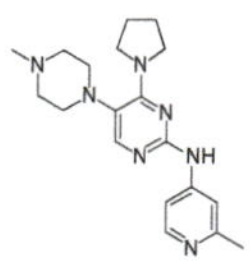	600	560	25	830	85	0.16
3	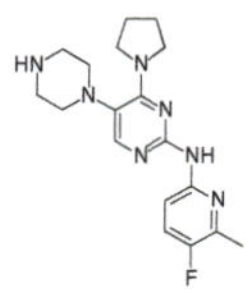	30	70	35	680	6.7	ND
4	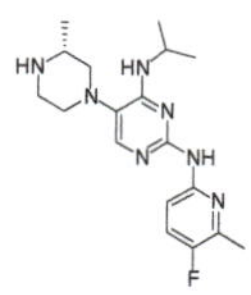	50	100	35	840	16	1.7
5	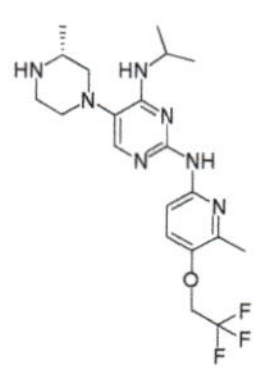	35	50	12	650	4.2	2.5
6	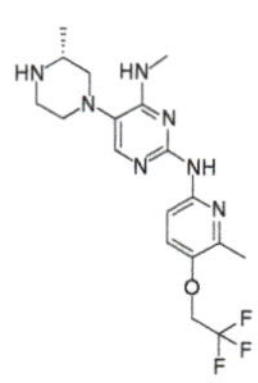	15	15	10	420	4.8	1
7	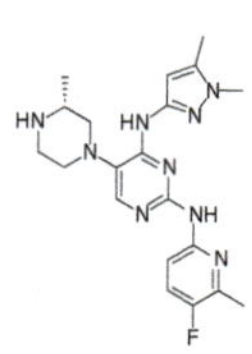	190	350	>50	100	>33	66
8	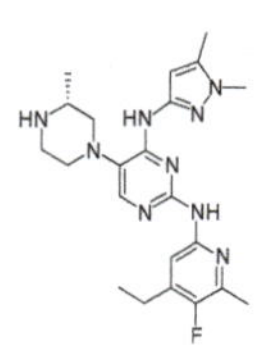	25	49	30	>1,000	>30	30
9	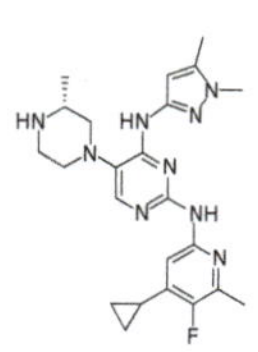	14	19	10	>1,000	19	26

Table 1 (Continued)

Compound number	Structure	*Pf* IC_{50} NF54 (nM)	*Pf* IC_{50} K1 (nM)	THP-1 IC_{50} (μM)	Solubility (μM)	hERG IC_{50} (μM)	AChE IC_{50} (μM)
10	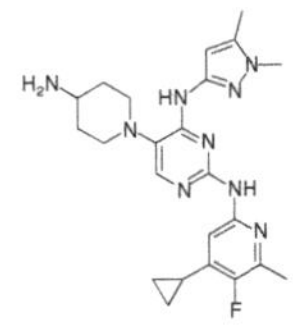	15	40	8	ND	34	32
11	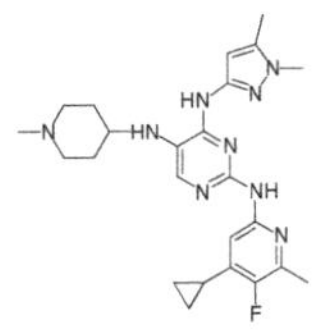	5	8	9	>1,000	15	44
12	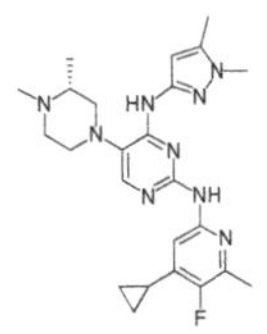	9	15	>50	>1,000	36	75

AChE, acetylcholine esterase; ND, not determined; *Pf*, *Plasmodium falciparum*.
IC_{50}s against NF54 and K1 strains of *Pf* were determined using a SYBR Green based assay. The cytotoxicity was assessed in the mammalian THP-1 cell line using an Alamar blue assay.

development, respectively, (Supplementary Tables 1–3). TAPs retained their IC_{50} against these strains, thereby suggesting a novel mechanism of action.

TAPs lack appreciable activity against the sexual forms or liver stages of the parasite (IC_{50} > 1 μM). TAPs were found to be inactive in the *P. berghei* liver schizont assay (IC_{50} > 10μM) as well as the *P. cynomolgi* liver schizont and hypnozoite assay (IC_{50} > 10 μM). Compound **12** was tested for its *in vitro* activity in the *Pf* male and female dual gamete assay and was found to be inactive (16.5 ± 6.3% inhibition at 1 μM).

Pharmacokinetics-pharmacodynamics (PK-PD) of TAPs. Faster reduction in the blood parasite burden is essential to provide quick relief from clinical symptoms and to minimize the risk of emergence of drug resistance[5]. Compound **12** produced a > 4-log kill after 48 h of exposure in the *in vitro* parasite reduction ratio (PRR) assay[11], an effect similar to chloroquine (Fig. 2b).

In the *Pf*/SCID model[12], **12** cleared *Pf* parasites to below detection limit following 4 days of daily treatment with 20 mg kg^{-1} dose administered through the oral route. A maximum kill rate was observed at 40 mg kg^{-1} (Fig. 2c). Blood C_{min} (0.04 μM) observed at this dose (Fig. 2d) was considered as the minimum parasiticidal concentration (MPC) for the human dose prediction. Blood samples collected from mice in the efficacy study were analysed for the presence of parent as well as its active metabolite, compound **9** (*Pf* IC_{50} 14 nM). Compound **9** was evaluated in the *Pf*/SCID model to understand its PK-PD relationship (Supplementary Note 2, Supplementary Figs 4 and 5, and Supplementary Tables 4 and 5). Kill rate constant (K_{kill}) and blood concentration required for half-maximal kill rate (EC_{50}) were estimated by fitting a sigmoidal E_{max} model[13–16]. Efficacy data for **12** was modelled using the PK-PD model for parent plus metabolite assuming an additive effect[14] (Supplementary Fig. 6 and Supplementary Table 6). The predicted and observed time course of parasitemia following treatment with **12** in the *Pf*/SCID model is shown in Fig. 2e. The blood levels of **9** as a metabolite did not contribute significantly to the observed efficacy of **12** (Supplementary Fig. 7). In addition, the formation of **9** in humans is predicted to be lower than in mice (Supplementary Note 3, Supplementary Figs 8 and 9, and Supplementary Table 7). Therefore, the *in vivo* activity of the metabolite was not considered for the human dose prediction. Compound **12** showed a good *in vitro*–*in vivo* correlation (IVIVC) between the intrinsic clearance (Cl_{int}) in rat and dog hepatocytes and blood clearance (CL) observed *in vivo* in rat and dog, respectively (Supplementary Note 4). Compound **12** displayed a high volume of distribution (V_{ss}) of 10–20 l kg^{-1}, low CL (< 35% liver blood flow) and long half-life (11–13 h in an intravenous (i.v.) PK and 14–16 h in an oral PK study) in rat and dog (Supplementary Figs 10 and 11 and Supplementary Tables 8 and 9). Human CL was predicted by *in vitro*–*in vivo* extrapolation from the *in vitro* Cl_{int} in human hepatocytes and by rat and dog allometry (Supplementary Note 5 and Supplementary Table 10). Human V_{ss} was predicted by rat and dog allometry (Supplementary Table 11). The predicted half-life of **12** in human is ~36 h. The predicted half-life of 36 h may not be long enough as compared with many known antimalarial drugs. However, it is sufficiently long for a fast-killing compound such as **12**, as compared with a fast-killing drug such as artemisinin with a half-life of 1 h in humans.

The predicted human exposure after a single dose of 260 mg is likely to sustain the maximum kill rate observed in the *Pf*/SCID model for 4–6 days to achieve > 3-log reduction in parasitemia (Supplementary Note 6, Supplementary Figs 12 and 13, and Supplementary Table 12). A patient with clinical symptoms of uncomplicated malaria may have 1–2% parasitemia in blood, which is equivalent to 11–12 log parasites in total, assuming 13 log erythrocytes in human blood[17]. At a dose of 260 mg, compound **12** is predicted to maintain a constant kill rate over two life cycles (96 h) of the parasite to yield a 3 to 4-log reduction in the total parasite load. Therefore, a single-dose treatment with ~500 mg might result in >8-log reduction in the parasite load. In a combination-based malaria therapy that is clinically practiced, an additive or synergistic effect of the partner drug with compound **12** is predicted to cure patients following a single-dose treatment.

Genetic basis of resistance to TAPs. To gain insight into the molecular target, mutant selection was attempted with several TAPs against *Pf* Dd2 strain. The frequency of spontaneous resistance emergence under *in vitro* drug selection with TAPs was very low (<1 in 10^{10} asexual blood stage parasites). However, after ~200 days of increasing *in vitro* drug pressure, resistant mutants to **6** were obtained and cloned by limiting dilution. Five clones from two independent selections displayed a three- to six-fold increase in IC_{50} for **6** and **12** compared with the parental Dd2 line, while retaining susceptibility to other antimalarial agents (Supplementary Table 13). These findings provided compelling evidence that compound **6** and **12** act through the same genetic mechanism. Whole-genome sequencing revealed a novel G29V mutation in the vacuolar ATP synthase (V-type H^+-ATPase) subunit D (PF13_0227) in all the clones, suggesting the involvement of this protein in conferring resistance to TAPs (Supplementary Table 14). T-coffee alignment of PF13_0227 with *Saccharomyces cerevisiae*, *Homo sapiens* and *Thermus thermophilus* shows the highest homology to subunit D (Supplementary Fig. 14). This gene is highly conserved in the natural parasite population; no non-synonymous mutations have been observed in 1,931 sequenced isolates from South/South East Asia (52%) and Africa (48%; available from the Pf3K Project, accessed as of 19 January 2015), suggesting that the gene is under mutational constraint, and that resistance to TAPs might not develop readily in the field.

This V-type H^+-ATPase has been localized in the infected host and *Pf* plasma membrane and digestive vacuole, serving as a major route for ATP-dependent H^+ efflux and pH regulation[18–21]. The V-type H^+-ATPase is a large multisubunit complex composed of a proton-translocation domain (V_0) and an ATP-hydrolytic domain (V_1), where the subunit D is located[19]. Subunit D is part of the central stalk connecting the V_1 and V_0 domains, implicating a role in reversible dissociation of the V_1V_0 complex to conserve ATP or coupling of proton transport with ATP hydrolysis[19]. As demonstrated by specific inhibitors of V-type H^+ ATPase, namely bafilomycin A_1 (ref. 22) and concanamycin A[23], inhibition of this pump is likely to result in parasite death due to physiological disturbances. However, specific inhibitors

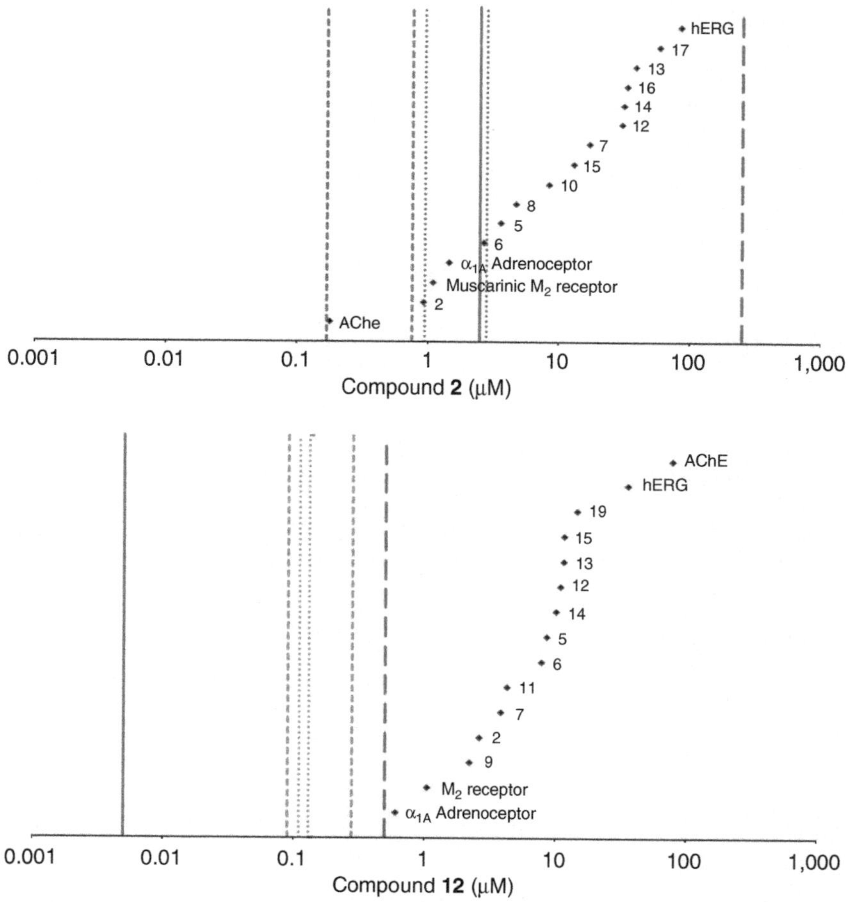

Figure 3 | Improvement in safety margins results in the nomination of compound 12 as a clinical candidate. (**a**) Secondary pharmacology selectivity plot for compound **2**. Selectivity ratio for AChE, α_{1A} and M_2 (IC_{50} against targets (♦)/blood C_{min} at ED_{90} in the *Pf*/SCID model (solid vertical line) was <1. Free plasma concentration range achieved in rat (dotted lines; C_{min} and C_{max}) and guinea pig (dashed line; C_{min} and C_{max}) *in vivo* illustrates targets covered during toxicity studies, and that selectivity against all targets was less than our target of >100-fold (long dash). (**b**) Secondary pharmacology selectivity plot for compound **12**. Blood C_{min} at ED_{90} in the *Pf*/SCID model (solid line) is significantly lower than that of compound **2**. Success in achieving desired *in vitro* selectivity illustrated as all off-target potencies (♦) sit to the right of our selectivity target: >100-fold (long dash). Free plasma concentration range achieved in rat (dotted lines; C_{min} and C_{max}) and guinea pig (dashed line; C_{min} and C_{max}) *in vivo* illustrates reduced pharmacological coverage during toxicity studies compared with compound **2**, which, in turn, translated into an improved safety margin for compound **12**.

Table 2 | Summary of the findings from the preclinical *in vivo* toxicology studies.

Study	Dose (mg kg^{-1} per day)	A—plasma C_{max}(μM) free	Summary of clinical signs and effects	B—predicted mean human plasma C_{max} (μM) free	Safety margin (A/B)-free C_{max}
3-Day rat tolerability	75 (Oral)	0.11	NOEL	0.0087	12.6
	150 (Oral)	0.13	NOEL	0.0087	15
Acute anaesthetized guinea pig CVS model	10 (i.v.)	0.09	NOEL	0.0087	10.3
	30 (i.v.)	0.28	Contractility and QTcB effects	0.0087	32.2

i.v., intravenous; NOEL, no observed effect level; C_{max}, maximum concentration reached in plasma; cardiovascular (CVS); QTcB, heart rate-corrected QT interval.
Safety margins are based upon the predicted human plasma C_{max} at 260 mg dose.

have so far failed to show potency *in vivo*[24]. TAPs being efficacious *in vivo* provide a novel therapeutic approach to treat malaria by targeting this essential proton pump. A very low rate of spontaneous resistance to TAPs coupled with the lack of point mutations in the V-type H^+-ATPase gene from a diverse collection of global isolates confirm the lack of pre-existing mutations to this important gene in *Pf*.

To test whether the identified mutation in the V-type H^+-ATPase subunit D protein affects the ionic homeostasis of the parasite's vacuole, we determined the size of vacuoles relative to overall parasite size from microscopic Giemsa-stained images. Results showed vacuolar size ratios of the parent (Dd2-B2) and compound **6**-resistant mutant strains F2-D6 and F4-D9 cloned from two independent selections (Supplementary Fig. 15). Both mutants showed a relative increase of the digestive vacuole size of 23–28%, implying that the acquired mutation may impair vacuolar maintenance. Further investigations are necessary to characterize the physiological effects on the vacuolar pH, drug accumulation and ion homeostasis of drug-exposed *Pf* parasites.

Safety studies of TAPs in rats and guinea pigs. To assess potential for *in vivo* toxicity mediated *via* secondary pharmacological effects, we profiled TAPs through a diverse panel of radioligand binding, enzyme activity and cellular functional assays covering 25 targets, plus hERG (Supplementary Table 15). The findings from the *in vivo* safety studies are described in Supplementary Note 7. The early lead, **2**, was active against several secondary targets (Fig. 3a). *In vivo* toxicity was observed for **2** in the rat and guinea pig models (Supplementary Table 16). However, only a handful of these targets were likely to be pharmacologically occupied at the systemic exposures achieved during those studies, leading to a hypothesis that the adverse events observed *in vivo* could be attributed to perturbation of AChE, α_{1A} adrenoceptor (antagonism) or muscarinic M_2 receptor (antagonism). We established SAR during lead optimization, resulting in good *in vitro* selectivity against hERG and AChE, with moderate selectivity against α_{1A} adrenoceptor/M_2 receptor (Fig. 2a). Despite the overall off-target profiles for **2** and **12** being similar (Fig. 3a,b), our success in reducing potency against our primary safety targets coupled with improvements in *Pf* potency (>60-fold) and reduction in free fraction (10-fold) translated into good safety margins for **12** *in vivo* in rat toxicity and guinea pig cardiovascular studies (Table 2 and Supplementary Fig. 16).

Discussion

A phenotypic high-throughput screen of the AstraZeneca corporate library against the asexual blood stage of *Pf* resulted in the identification of TAPs as a novel class of antimalarial agents with multiple diversification points for lead optimization. TAPs display excellent *in vitro Pf* potency (*Pf* IC_{50} in the range of 5–30 nM) with good solubility and low *in vitro* CL_{int} during the lead identification phase. However, the series had secondary pharmacology-related liabilities, for example, AChE inhibition along with hERG inhibition. By using a diverse panel of radioligand binding enzyme and cellular functional assays covering a total of 85 targets, we established significant improvement in the *Pf* potency (IC_{50} < 10 nM), with a concomitant improvement in off-target selectivity (*in vitro*) and good bioavailability (80%). An improved *in vivo* efficacy and selectivity profile of compound **12** translated to good safety margins in a 3-day rat toxicology study. No treatment-related effects in biochemical parameters, clinical pathology or histopathology were observed following dosing with **12** in the rats. These data support the progression of **12** into further preclinical toxicology studies, including second species cardiovascular telemetry. In conclusion, compound **12** with a novel mechanism of action, low frequency of resistance emergence, long half-life *in vivo* and ability to kill parasites rapidly is a potential clinical candidate for malaria.

Methods

Synthetic scheme for representative TAPs (compounds 7, 8, 9 and 12). All reagents, starting materials and solvents described in the procedure were commercially available and used without further purification. Analytical thin-layer chromatography was performed on SiO_2 plates on alumina and the purity of all final derivatives for biological testing was confirmed to be >95% using the following conditions: a Shimadzu HPLC (high-performance liquid chromatography) instrument with a Hamilton reverse-phase column (HxSil, C18, 3 μm, 2.1 mm × 50 mm (H2)). Eluent A: 5% CH_3CN in H_2O and eluent B: 90% CH_3CN in H_2O. A flow rate of 0.2 ml min^{-1} was used with ultraviolet detection at 254 and 214 nm. The structure of the intermediates and end products was confirmed by proton (^1H), carbon (^{13}C) nuclear magnetic resonance (NMR) and mass spectroscopy. Proton magnetic resonance spectra were determined in dimethylsulphoxide (DMSO) d_6 unless otherwise stated, using Bruker DRX 300 or Bruker DRX-400 spectrometers, operating at 300 or 400 MHz, respectively. Splitting patterns are indicated as follows: s, singlet; d, doublet; t, triplet; m, multiplet; and br, broad peak. Liquid Chromatography-Mass Spectrometer (LC-MS) data was acquired using Agilent LCMS VL series. Source: Electron Spray Ionization (ESI), coupled with an Agilent 1100 series HPLC system and an Agilent 1100 series Photodiode Array (PDA) as the front end. high-resolution mass spectrometry (HRMS) data was acquired using an Agilent 6520, Quadrupole-Time of flight tandem mass spectrometry (MS/MS) coupled with an Agilent 1200 series HPLC system.

Generic synthetic route for the synthesis of key compounds (**7**, **8**, **9** and **12**) are illustrated in Fig. 1. The synthesis of compounds **1–6**, **10** and **11** are provided in Supplementary Note 1 using chemical scheme analogues to the one illustrated in Fig. 1 in the supplementary section.

A mixture of 5-bromouracil (90 g, 471 mmol, Aldrich) and tert-butyl (R)-2-methyl-4l2-piperazine-1-carboxylate (141 g, 706.8 mmol, Aldrich) were taken in pyridine (200 ml) to give a white suspension. The suspension was vortexed and subjected to microwave irradiation for 45 min at 150 °C to give a clear brown solution. Solvent was removed under vacuum and the residue was then triturated with ethyl acetate. The resultant suspension was filtered out and dried under vacuum to get (R)-tert-butyl-4-(2, 4-dioxo-1, 2, 3, 4-tetrahydropyrimidin-5-yl)-2-methylpiperazine-1-carboxylate (**I**, 39.6 g, 44%) as an off-white solid. Yield: 44%, purity: >95% by HPLC (ultraviolet at 220 and 254 nm). ^1H NMR (300 MHz, DMSO-d_6) δ 11.10 (s, 1H), 10.51 (s., 1H), 6.73 (d, J=4.7 Hz, 1H), 4.12 (s., H), 3.72 (d, J=13.2 Hz, 1H), 3.22–2.93 (m, 3H), 2.42 (dd, J=11.3, 3.6 Hz, 1H), 2.30 (d, J=2.8 Hz, 1H), 1.40 (s, 9H),1.19 (d, J=6.8 Hz, 3H). ^{13}C-NMR (126 MHz, DMSO-d_6) δ 161.20, 154.21, 150.73, 127.09, 126.62, 79.22, 54.63, 50.12, 46.88, 40.82. 39.15,

28.52 and 15.94, HRMS, electrospray ionization (HRMS (ESI)): m/z calculated for $C_{14}H_{22}N_4O_4$ +H [M+H]: 308.1532. Found: 308.3420.

Synthesis of (R)-rert-butyl 4-(2, 4-dichloropyrimidin-5-yl)-2-methylpiperazine-1-carboxylate (**II**), R)-tert-butyl 4-(2, 4-dioxo-1, 2, 3, 4-tetrahydropyrimidin-5-yl)-2-methylpiperazine-1-carboxylate (**II**, 22 g, 71.62 mmol) was taken in phosphorus oxychloride (750 ml), to give a brown suspension. The resulting reaction mixture was refluxed for 6 h and the excess phosphorus oxychloride was distilled out under reduced pressure. The remaining oil was diluted with THF (250 ml) under ice cold condition. Subsequently, the reaction mixture was basified to pH 8 with Na_2CO_3. To this di-tert-butyl dicarbonate (22.17 ml, 96.41 mmol, Aldrich) was added and stirred at room temperature for 16 h. The reaction mixture was diluted with methanol and filtered it off to remove excess salt. The solvent was removed under vacuum and residue was diluted with water and extracted with ethyl acetate. The combined organic layers were dried over sodium sulphate and then removed under reduced pressure. The crude residue was purified by flash chromatography on silica using ethyl acetate and hexane as eluents (1:4) to obtain pure solid of (R)-rert-butyl 4-(2, 4-dichloropyrimidin-5-yl)-2-methylpiperazine-1-carboxylate (**II**, 23.00 g, 93%). Yield: 93%, Purity: >95% by HPLC (ultraviolet at 220 and 254 nm). ^1H NMR (300 MHz, $CDCl_3$) δ 8.11 (s, 1H), 4.33 (s., 1H), 3.94 (d, $J=13.9$ Hz, 1 H), 3.29–3.15 (m, 3H), 2.93–2.70 (m, 2H) 1.47–1.37 (m, 9H), 1.31 (d, $J=6.8$ Hz, 3H). ^{13}C-NMR (126 MHz, DMSO-d_6) δ 155.65, 154.11, 152.94, 151.49, 143.31, 79.52, 54.82, 50.18, 28.50 and 15.54, HRMS (ESI): m/z calculated for $C_{14}H_{20}Cl_2N_4O_2$ +H [M+H]: 347.1011. Found: 347.1140.

Synthesis of (R)-tert-butyl 4-(2-chloro-4-[(1, 5-dimethyl-1H-pyrazol-3-yl)amino)pyrimidin-5-yl]-2-methylpiperazine-1-carboxylate (**III**). To a mixture of (R)-tert-butyl 4-(2, 4-dichloropyrimidin-5-yl)-2-methylpiperazine-1-carboxylate (II, 6.92 g, 20 mmol) and 1,5-dimethyl-1H-pyrazol-3-amine (2.2 g, 20 mmol, Matrix scientific) in ethanol (100 ml) was added to DIPEA (5.30 ml, 30 mmol) and the reaction mixture was subjected to microwave irradiation at 110 °C for 1 h. The solvent was removed under vacuum and then ice cold water was added to the residue. The precipitated solid was filtered out and dried under vacuum. The crude product purified by flash chromatography on silica using dichloro methane and methanol as eluents (9:1) to obtain pure solid of (R)-tert-butyl 4-(2-chloro-4-((1,5-dimethyl-1H-pyrazol-3-yl)amino)pyrimidin-5-yl)-2-methylpiperazine-1-carboxylate (2.61 g). Yield: 31%, Purity: >95% by HPLC (ultraviolet at 220 and 254 nm). ^1H NMR (300 MHz, MeOD) δ 7.99 (s, 1H), 6.61 (s., 1H), 4.39 (m, 1H), 3.99 (d, $J=13.9$ Hz, 1 H), 3.73(s, 3H), 3.4–3.50 (m, 1H), 3.0–3.15 (m, 2H) 2.89–2.92 (m, 2H), 2.70–2,77(m, 1H), 2.33(s, 3H), 1.37–1.48 (m, 12H). ^{13}C-NMR (126 MHz, MeOD) δ 156.24, 155.17, 154.88, 146.93, 144.96, 140.28, 130.91, 97.11, 80.04, 55.56, 51.50, 34.43, 27.26, 14.26 and 9.82, HRMS (ESI)): m/z calculated for $C_{19}H_{28}ClN_7O_2$+H [M+H]: 422.1193. Found: 422.2010.

Synthesis of (R)-N4-(1, 5-dimethyl-1H-pyrazol-3-yl)-N2-(5-fluoro-6-methylpyridin-2-yl)-5-(3-methylpiperazin-1-yl)pyrimidine-2,4-diamine (**7**). To a solution of (R)-tert-butyl 4-(2-chloro-4-((1,5-dimethyl-1H-pyrazol-3-yl)amino)pyrimidin-5-yl)-2-methylpiperazine-1-carboxylate (**III**, 1.14 g, 2.71 mmol) dissolved in toluene (40 ml), were added 2-amino-5-fluoro-6-methylpyridine (0.513 g, 4.07 mmol, Aldrich), potassium tert-butoxide (0.46 g, 4.076 m mol) and degassed for 10 min. Then, BINAP (0.010 g, 0.16 mmol) and $Pd_2(dba)_3$(0.004 g, 0.00543 mol) were added and again degassed for 10 min. The reaction was then refluxed at 110 °C for 12 h. The reaction mixture was filtered through a celite bed and the organic layer was washed with brine, dried over anhydrous sodium sulphate and concentrated in vacuum. The crude residue was purified by flash chromatography on neutral alumina with gradient elution of 0.5–1.0% methanol in dichloromethane to get (R)-tert-butyl 4-(4-((1,5-dimethyl-1H-pyrazol-3-yl)amino)-2-((5-fluoro-6-methylpyridin-2-yl)amino)pyrimidin-5-yl)-2-methylpiperazine-1-carboxylate as an off-white solid. Then, the N-boc-protected precursor was subjected to boc group de-protection using 4 N HCl in dioxane (0.3 ml, Aldrich) at 10 °C with stirring for 15 min. The reaction mixture was concentrated under vacuum afforded the title compound **7** as light brown solid (450 mg). Yield: 31%, purity: >95% by HPLC (ultraviolet at 220 and 254 nm). ^1H NMR (300 MHz, DMSO-d_6) δ 11.51 (s, 1H), 10.25 (s, 1H) 9.52 (s, 1H) 9.10 (s, 1H) 8.11 (s, 1H) 7.83 (s, 1H) 7.28 (s, 1H) 6.68 (s, 1H) 3.53–3.44 (m, 2 H) 3.38 (s, 3 H) 3.17–3.07 (m, 3H) 3.05–2.93 (m, 1H) 2.85–2.76 (m, 1H) 2.56 (d, $J=2.8$ Hz, 3H) 2.31 (s, 3H) 1.28 (d, $J=5.6$ Hz, 3H. ^{13}C-NMR (126 MHz, DMSO-d_6) δ 157.22, 149.36, 147.53, 143.39, 142.96, 139.26, 133.33, 126.63, 125.59, 112.59, 99.92, 47.79, 35.69, 17.08, 15.44 and 10.9, HRMS (ESI): m/z calculated for $C_{20}H_{26}FN_9$+H [M+H]: 412.2295. Found: 412.2367.

The synthesis of 4-ethyl-5-fluoro-6-methylpyridin-2-amine (**8a**) and 4-cyclopropyl-5-fluoro-6-methylpyridin-2-amine (**9a**) was carried out as described earlier[10].

4-Ethyl-5-fluoro-6-methylpyridin-2-amine (8a). Yield: 22%, purity: >95% by HPLC (ultraviolet at 220 and 254 nm). ^1H NMR (300 MHz, DMSO-d_6) δ 7.36 (s, 1 H), 2.63–2.81 (m, 2 H), 2.26 (s, 3 H) 1.18 (t, $J=7.54$ Hz, 3 H)). ^{13}C-NMR (126 MHz, MeOD) δ 155.11, 151.24, 148.89, 140.41, 106.74, 21.60, 15.04 and 12.65, HRMS (ESI): m/z calculated for $C_8H_{11}FN_2$+H [M+H]: 155.0906. Found: 155.0830.

4-Cyclopropyl-5-fluoro-6-methylpyridin-2-amine (9a), Yield: 22%, purity: >95% by HPLC (ultraviolet at 220 and 254 nm). ^1H NMR (300 MHz, DMSO-d_6) δ 14.19 (s., 1H), 7.49 (s., 2H), 6.33 (d, $J=5.8$ Hz, 1H) 2.37 (d, $J=3.0$ Hz, 3H) 2.18–1.97 (m, 1H) 1.27–1.14 (m, 2H) 0.92–0.77 (m, 2H). ^{13}C-NMR (126 MHz, MeOD) δ 153.78, 152.05, 149.93, 147.61, 104.59, 11. 21, 10.58 and 9.16, HRMS (ESI): m/z calculated for $C_9H_{11}FN_2$+H [M+H]: 167.0906. Found: 167. 0960.

Synthesis of (R)-N4-(1,5-dimethyl-1H-pyrazol-3-yl)-N2-(4-ethyl-5-fluoro-6-methylpyridin-2-yl)-5-(3-methylpiperazin-1-yl)pyrimidine-2,4-diamine dihydrochloride (8) is illustrated in Fig. 1 and synthesis procedure followed analogues to compound **7** using (R)-tert-butyl 4-(2-chloro-4-((1,5-dimethyl-1H-pyrazol-3-yl)amino)pyrimidin-5-yl)-2-methylpiperazine-1-carboxylate (III) and 4-ethyl-5-fluoro-6-methylpyridin-2-amine (8a). Yield: 22%, purity: >95% by HPLC (ultraviolet at 220 and 254 nm). ^1H NMR (300 MHz, DMSO-d_6) δ δ11.52 (s, 1H)10.28 (s, 1H) 9.82 (s, 1H) 9.40 (br. s., 1H) 8.13 (s, 1H) 7.18 (d, $J=4.3$ Hz, 1H) 6.74 (s, 1H) 3.76–3.60 (m, 4H) 3.54–3.24 (m, 4H) 3.22–2.96 (m, 3H) 2.85 (d, $J=11.5$ Hz, 1H) 2.67 (q, $J=7.7$ Hz, 2H) 2.54 (d, $J=3.0$ Hz, 3H) 2.31 (s, 3H) 1.29 (d, $J=6.0$ Hz, 3H) 1.24–1.15 (m, 3H); ^{13}C-NMR (126 MHz, MeOD) δ 156.24, 155.17, 154.88, 146.93, 144.96, 140.28, 130.91, 97.11, 80.04, 55.56, 51.10, 34.43, 27.65, 14.26 and 9.82, HRMS (ESI): m/z calculated for $C_{22}H_{30}FN_9$+H [M+H]: 440.2608. Found: 440.2500.

Synthesis of N2-(4-cyclopropyl-5-fluoro-6-methylpyridin-2-yl)-N4-(1, 5-dimethyl-1H-pyrazol-3-yl)-5-(3-methylpiperazin-1-yl)pyrimidine-2,4-diamine dihydrochloride (9) is illustrated in Fig. 1 and the synthesis procedure followed analogues to compound **7** using (R)-tert-butyl4-(2-chloro-4-((1,5-dimethyl-1H-pyrazol-3-yl)amino)pyrimidin-5-yl)-2-methylpiperazine-1-carboxylate (III) and 4-cyclopropyl-5-fluoro-6-methylpyridin-2-amine (9a). Yield: 25%, purity: >95% by HPLC (ultraviolet at 220 and 254 nm). ^1H NMR (300 MHz, DMSO-d_6) δ 13.93 (s., 1H)11.39 (s, 1H) 10.29 (s, 1H) 9.81 (s., 1H) 9.38 (s., 1H) 8.12 (s, 1H) 6.89–6.64 (m, 2H) 3.42 (s., 1H) 3.80–3.56 (m, 4H) 3.38–3.25 (m, 1H) 3.18–2.95 (m, 3H) 2.89–2.76 (m, 1H) 2.54 (d, $J=3.2$ Hz, 3H) 2.31 (s, 3H) 2.13 (t, $J=4.7$ Hz, 1H) 1.29 (d, $J=6.2$ Hz, 3H) 1.22–1.15 (m, 2H) 0.82–0.73 (m, 2H). ^{13}C-NMR (126 MHz, DMO-d_6) δ 154.49, 154.03, 152.15, 150.22, 148.03, 146.16. 144.60, 141.40, 139.51, 138.12, 124.82, 105.50, 96.42, 58.92, 51.62, 34.58, 18.69. 16.55, 10.30 and 8.05, HRMS (ESI): m/z calculated for $C_{23}H_{30}FN_9$ +H [M+H]: 452.2608. Found: 452.2683.

Synthesis of (R)-N2-(4-cyclopropyl-5-fluoro-6-methylpyridin-2-yl)-N4-(1, 5-dimethyl-1H-pyrazol-3-yl)-5-(3, 4-dimethylpiperazin-1-yl)pyrimidine-2,4-diamine (**12**). (R)-N2-(4-cyclopropyl-5-fluoro-6-methylpyridin-2-yl)-N4-(1,5-dimethyl-1H-pyrazol-3-yl)-5-(3-methylpiperazin-1-yl)pyrimidine-2,4-diamine hydrochloride (compound **9**, 190 mg, 0.42 mmol) was taken in dichloromethane (2 ml) to give a yellow suspension. To this Hunig's Base (0.184 ml, 1.05 mmol) was added and the suspension turned clear. After 10 min of stirring, reaction mixture turned into a white suspension and then it was concentrated to dryness. Resultant residue was dissolved in ethanol (absolute, 99.5%) (3 ml), and formaldehyde (0.042 ml, 0.63 mmol) was added and stirred for 10 min. To this clear solution, sodium cyanoborohydride (26.4 mg, 0.42 mmol) was added in one portion to get a white suspension. The reaction mixture was concentrated and the crude product was purified through reverse-phase chromatography to get the pure off-white solid of (R)-N2-(4-cyclopropyl-5-fluoro-6-methylpyridin-2-yl)-N4-(1, 5-dimethyl-1H-pyrazol-3-yl)-5-(3,4-dimethylpiperazin-1-yl)pyrimidine-2,4-diamine (80 mg, 40.8%). Yield: 40.8%, purity: >95% by HPLC (ultraviolet at 220 and 254 nm). ^1H NMR (300 MHz, DMSO-d_6) δ 9.26 (s,1H), 8.03 (s, 1H) 8.00 (s, 1H) 7.67 (d, $J=5.1$ Hz, 1H) 6.83 (s, 1H) 3.33 (s, 3H) 2.96–2.73 (m, 4H) 2.75–2.50 (m, 1H) 2.38–2.30 (m, 4H) 2.23 (s, 7H) 2.10–1.96 (m, 1H),1.08–1.02 (m, 2H) 1.00 (d, $J=6.2$ Hz, 3H) 0.78–0.67 (m, 2H). ^{13}C-NMR (126 MHz, DMO-d_6) δ 155.30, 154.67, 152.10, 150.93, 148.98, 146.81. 145.29, 141.95, 140.31, 138.81, 124.91, 106.20, 97.07, 58.78, 51.87, 42.16, 35.28, 17.23. 10.99 and 8.77, HRMS (ESI): m/z calculated for $C_{24}H_{32}FN_9$+H [M+H]: 466.2765. Found: 466. 2838. Traces of LC-MS, HRMS, ^1H NMR and ^{13}C-NMR of compound **12** are shown in Supplementary Figs 1–3.

Assessment of cross-resistance to antimalarial agents. To eliminate overlapping mode of action with known targets, cross-resistance testing of representative TAPs against drug-resistant lines with well-characterized mutations in *Pf* dihydroorotate dehydrogenase[25], *Pf* heat-shock protein 90, *Pf* chloroquine resistance transporter[25] or *Pf* cytochrome *b* reduction site[26] was performed. Mutations in these genes contribute to resistance to antimalarial agents under development, namely Genz-669178, geldanamycin, IDI-3783 or IDI-5918, respectively. To assess cross-resistance, the IC_{50} or *Pf* growth inhibition was measured using the SYBR green growth assay—the standard non-radioactive, high-throughput 72-h assay that uses a nucleic acid-binding fluorescent dye[27,28]. Assays were carried out in 384-well format with three technical replicates within an assay day and biological replicates carried out on three different assay days to assign a final EC_{50}.

Determination of efficacy in the *Pf*/SCID mouse model. All animal studies were ethically reviewed and carried out in accordance with European Directive 86/609/EEC and the GSK Policy on the Care, Welfare and Treatment of Animals. During this experiment, female mice were used from the batch NSGJ 01/1 (The Jackson Laboratory, USA). The therapeutic efficacy of compound **12** against *Pf*3D7$^{0087/N9}$ was studied using a '4-day test' reported earlier[12]. Briefly, NODscidIL2Rγ^{null} mice engrafted with human erythrocytes were infected with 20×10^6 *Pf*-infected erythrocytes. Infections were performed by i.v. inoculation. All mice were randomly assigned to their corresponding treatment. The treatment started at day 3 and finished at day 6 after infection. The compound was administered orally as a suspension in 0.5% w/v hydroxypropyl methylcellulose (HPMC) with 0.1% v/v Tween 80, using an oral gavage at a maximum dose volume of 20 ml kg^{-1}. Vehicle alone was administered in untreated mice. In all cases, parasitemia was assessed in samples from peripheral blood obtained at days 3–7 after infection. A qualitative analysis of the effect of treatment on *Pf*3D7[0087/N9] was assessed by microscopy and flow cytometry. Fresh samples of peripheral blood from *Pf*-infected mice were stained with TER-119-Phycoerythrine (marker of murine erythrocytes) and SYTO-16 (nucleic acid dye) and then analysed by flow cytometry (FACS Calibur, BD). Microscopy analysis was performed with Giemsa-stained blood smears from samples taken at days 5 and 7 (48 and 96 h after starting treatment, respectively).

PKs in the blood of infected *Pf*/SCID mice. Peripheral blood samples (25 µl) were taken at different times (0.25, 0.5, 1, 2, 4, 6, 8 and 23 h), mixed with 25 µl of H_2O Mili-Q and immediately frozen on dry ice. The frozen samples were stored at − 80 °C until analysis. Vehicle-treated mice experienced the same blood-sampling regimen. Blood samples were processed by liquid–liquid extraction by mixing 10 µl diluted blood with 180 µl AcN:MeOH (80:20; v-v) mixture. Quantitative analysis by Liquid chromatography-tandem mass spectrometry (LC-MS/MS) was performed using UPLC (Waters) and Sciex API4000. The lower limit of quantification in this assay was 0.005 $\mu g\,ml^{-1}$.

***In vitro* generation of drug-resistant *Pf* lines.** In order to identify the mode of action of TAPs, a chemogenomic approach was used whereby resistant parasites are generated by *in vitro* selection under drug pressure and whole-genome sequencing is used to identify the genetic basis of resistance[29]. Generation of *in vitro* resistance to compound **6** was performed as described[26] with the following modifications. Four independent selections (F1-4) each with Dd2 parasites at an initial population of 10^8–10^9 were subjected to drug pressure at 10 × IC_{50} between 2 and 8 days or until all parasites were dead. The drug was then washed off and the parasites were allowed to recover in normal growth media. Four rounds of intermittent drug pressure at 10 × IC_{50} were performed. On the fifth round, the drug pressure was increased to 20 × IC_{50}. Drug resistance was confirmed by measuring IC_{50}s with the SYBR green growth assay as described above. After ~200 days of intermittent drug pressure, drug-resistant lines were cloned from F2 and F4 by limiting dilution for single-cell isolation[30], genomic DNA extracted using the QIAmp DNA Blood Mini Kit (Qiagen, USA) and whole-genome sequenced as described below.

Whole-genome sequencing and analyses. Raw reads were analysed with a customized pipeline based on the GATK toolkit version 2.8.1. In summary, the raw reads where aligned with the BWA-aligner[31], PCR duplicates were removed, Indel regions were realigned and the resulting assemblies were base-recalibrated using the known single-nucleotide polymorphism (SNP) database for Dd2 parasites (PlasmoDB 11.0). Next, a hard filter was applied in order to mark positions of low-quality scores according to the GATK best practices recommendations for haploid genomes[32]. These regions were not excluded from the final analysis. Only SNPs within genes[33] were considered and *Pf* variant genes and surface antigens were excluded from the analysis. The results of all strains were combined to evaluate whether the identified SNPs are conserved within different mutants. High-quality SNPs were called if the SNP was identified by at least 87.5% of all reads from the corresponding position, otherwise the SNPs were considered polymorphic.

Data retrieval. Next-Generation Sequencing (NGS) assembly files of all *Pf* strains used in this study to identify the resistance determinants can be accessed from the Sequence Read Archive at the NCBI (http://www.ncbi.nlm.nih.gov/sra/?term=SRP052918). Please consult the Sequence Read Archive toolkit documentation (http://www.ncbi.nlm.nih.gov/sites/books/NBK158900/) on how to extract the corresponding read data. The reference used for assembly was taken from PlasmoDB.org, version 11.0 (http://plasmodb.org/common/downloads/release-11.0/Pfalciparum3D7/fasta/) and reordered as indicated in the assembly's file header. The MD5 checksum of the reference file is included in the @CO header line.

hERG profiling and assessment of cytotoxicity of TAPs. hERG IC_{50} was determined in an electrophysiology-based hERG assay using IonWorks HT (CHO cells) as described earlier[34].

The THP-1 cytotoxicity screen was based upon determination of fluorescent signal generated by the reduction of non-fluorescent resazurin (7-hydroxy-3H-phenoxazin-3-one 10-oxide) to the fluorescent resorufin (Alamar blue assay). Cellular reduction of resazurin is dependent on a pool of reductase or diaphorase enzymes derived from the mitochondria and cytosol. Therefore, Alamar blue can be used as an oxidation–reduction indicator in cell viability assays for mammalian cells[35,36].

***In vitro* safety pharmacology profiling of TAPs.** An *in vitro* pharmacological profiling panel was specifically designed for the detection of potential high- risk clinical adverse drug reactions. The panel comprised of 26 diverse targets (6 enzymes, 2 transporters, 9 G-protein-coupled receptors (GPCRs), 5 ion channels along with hERG, M2, alpha1A and AchE) and was conduced externally by CEREP, France and was conducted externally by CEREP, France. For assessment of activity against these targets, compounds were tested in an 8-point concentration–response covering half-log units up to a top test concentration of 100 µM, and an IC_{50} or K_i was determined, using either radioligand binding assay or an enzyme activity assay (HTRF, LANCE). Detailed methodology for *in vitro* pharmacological profiling assays provided by CEREP can be found at http://www.cerep.fr, with specific reference to the AChE (cat. no. 0363), Muscarinic M_1 receptor (cat. no. 0091) and Adrenergic α_{1A} receptor (cat. no. 2338) assays. For enzyme and kinase assays, the results were expressed as a percent inhibition of control-specific activity obtained in the presence of the test compound. The IC_{50} values (concentration causing a half-maximal inhibition of control-specific activity) and Hill coefficients (n_H) were determined by nonlinear regression analysis of the inhibition/concentration–response curves using Hill equation curve fitting as described earlier[37].

For radioligand binding assays, the results are expressed as % inhibition of control-specific binding obtained in the presence of the test compounds. The IC_{50} values (concentration causing a half-maximal inhibition of control-specific binding) and Hill coefficients (n_H) were determined by nonlinear regression analysis of the competition curves generated with mean replicate values using Hill equation curve fitting. The inhibition constants (K_i) were subsequently calculated by applying the Cheng Prusoff equation[38].

All secondary pharmacology data for both compounds **2** and **12** from the panel were specifically tailored to this compound series.

Methodology for 3-day rat toxicity study. The study was conducted at AstraZeneca R&D, Alderley Park, Cheshire, UK. All animal studies were undertaken in accordance with the AstraZeneca Bioethics Policy, which requires that work involving animals is carefully considered and justified to ensure that the study is scientifically necessary, and that the principles of the 3Rs (replacement, reduction and refinement) are applied. Animal studies were conducted in compliance with all relevant local and national laws and regulations, and with the principles of the 'Guide for the Care and Use of Laboratory Animals' 8th edition (Institute for Laboratory Animal Research).

Two groups of male rats ($n = 3$, Han Wistar) were obtained from Charles River Laboratories, UK. Animals were group housed for ~1 week before the initiation of dosing. On the first day of dosing, rats were ~11 weeks of age and weighed 250–300 g and received compound **12** via oral gavage at dose levels of 75 or 150 $mg\,kg^{-1}$ for three consecutive days and necropsied 24 h after the final dose. Vehicle: 1% pluronic F127; dose volume: 10 $ml\,kg^{-1}$. In addition to regular monitoring of the animals' general condition, toxicokinetic profiles were obtained on day 1 and clinical chemistry, haematology and histopathology were assessed from all animals on day 4. Female animals were not included in the study as sex differences in either exposure or tolerability were not expected and control animals were excluded from this preliminary study design to reduce animal usage. Body weights were collected during acclimation and before each dose. Food and water consumption were available *ad libitum* but not recorded.

Toxicokinetic profiles were obtained on day 1 from all animals (time points: 0.5, 1, 2, 4, 6 and 24 h post dose) via microsample tail prick (32 µl blood) and were analysed using a qualified analytical procedure. Clinical pathology blood samples were taken at necropsy (24 h after third and final dose) from all animals, and analysed following a routine toxicological parameter list. Urinalysis was not performed. A limited list of key organs were taken at necropsy, processed and examined microscopically to determine whether treatment resulted in any clear pathological effects (heart, kidneys, liver, lungs and gastrointestinal tract).

Results of 3-day rat toxicity study. No abnormalities were observed in the behaviour or general condition of the animals during treatment. Analysis of the clinical pathology and histopathology revealed no findings considered related to treatment with compound **12**. As no findings of significance were observed, these data are not presented here.

Design of anaesthetized guinea pig cardiovascular study. The study was conducted in accordance with the European Union (EU) animal welfare regulation for animal use in experimentation (European Directive 2010/63/EEC) and approved by Biotrial Ethical Committee 'Comité de reflexion Ethique en Expérimentation Animale (CR2EA)'.

Male Dunkin Hartley guinea pigs ($n = 6$ per group; 342–401 g; 4–5 weeks old; Charles River Laboratories, L'Arbresle, Cedex) were anaesthetized with sodium pentobarbitone (60 $mg\,kg^{-1}$ intraperitoneally followed by 6 $mg\,kg^{-1}\,h^{-1}$ i.v. (jugular) for maintenance) and mechanically ventilated. Animals were surgically prepared for the measurement of haemodynamic (blood pressure, heart rate and contractility (including left ventricular pressure)) and electrocardiographic (ECG)-derived parameters (QTcB, PR interval and QRS duration). Briefly, catheters were introduced into a carotid artery and a jugular vein for the measurement of arterial blood pressure and administration of the test item, respectively. Needle electrodes were placed subcutaneously in a lead II position for ECG recording. A fluid-filled catheter was introduced into the left ventricle via the opposite carotid artery for left ventricular pressure measurement.

Animals received two consecutive i.v. infusions of either vehicle (group 1; 40% v/v dimethylacetamide/20% v/v polyethylene glycol 400/17% w/v hydroxyl propyl beta cyclodextrin (HPβCD/kleptose) in water for injection) or compound **12** at 10 or 30 $mg\,kg^{-1}$ (group 2). Each infusion lasted for 15 min at a dose volume of

3 ml kg^{-1} (rate of 0.2 ml kg^{-1} min^{-1}). Animals were monitored for a further 30-min washout period. Blood samples were taken at 5, 10 and 15 min after the start of each infusion and at 5, 10, 20 and 30 min during the washout phase.

Following a stabilization period, haemodynamic and ECG parameters were measured (using HEM software, version 4.3, Notocord Systems) at 10 and 5 min before administration (mean of measurements obtained at these time points = baseline; T0), and 2.5, 5, 10 and 15 min after the start of each i.v. infusion of compound **12** or vehicle and 5, 10, 20 and 30 min after the end of the second infusion. A detailed description of the model, its background and validation has been described earlier[39,40].

All parameters were analysed statistically using SAS software. Baseline values were defined as mean of measurments taken 10 or 5 min before dosing. For each parameter, homogeneity of baseline values (T0) between the two groups were tested using a two-tailed Student's *t*-test. Changes from baseline were assessed using a two-way analysis of variance (group and time) with repeated measurements over the time and group × time interaction.

References

1. WHO. World Malaria Report 2014, http://www.who.int/malaria/publications/world_malaria_report_2014/en/.
2. White, N. J. *et al.* Malaria. *Lancet* **383,** 723–735 (2014).
3. Ariey, F. *et al.* A molecular marker of artemisinin resistant *Plasmodium falciparum* malaria. *Nature* **505,** 50–55 (2014).
4. Ashley, E. A. *et al.* Spread of artemisinin resistance in *Plasmodium falciparum* malaria. *N. Engl. J. Med.* **371,** 411–423 (2014).
5. Burrows, J. N. *et al.* Antimalarial drug discovery—the path towards eradication. *Parasitology* **141,** 128–139 (2014).
6. Lee, J. A. *et al.* Modern phenotypic drug discovery is a viable, neoclassic pharma strategy. *J. Med. Chem.* **55,** 4527–4538 (2012).
7. Rottmann, M. *et al.* Spiroindolones, a potent compound class for the treatment of malaria. *Science* **329,** 1175–1180 (2010).
8. Hameed, S. P. *et al.* Aminoazabenzimidazoles, a novel class of orally active antimalarial agents. *J. Med. Chem.* **57,** 5702–5713 (2014).
9. Jiménez-Díaz, M. B. *et al.* (+)-SJ733, a clinical candidate for malaria that acts through ATP4 to induce rapid host-mediated clearance of Plasmodium. *Proc. Natl Acad. Sci. USA* **111,** E5455–E5462 (2014).
10. Ramachandran, S. *et al.* N-Aryl-2-aminobenzimidazoles: Novel, efficacious, antimalarial lead compounds. *J. Med. Chem.* **57,** 6642–6652 (2014).
11. Le Manach, C. *et al.* Fast *in vitro* methods to determine the speed of action and the stage-specificity of anti-malarials in *Plasmodium falciparum*. *Malar. J.* **16,** 424–430 (2013).
12. Angulo-Barturen, I. *et al.* A murine model of falciparum-malaria by *in vivo* selection of competent strains in non-myelodepleted mice engrafted with human erythrocytes. *PLoS ONE* **3,** e2252 (2008).
13. White, N. J. Pharmacokinetic and pharmacodynamic considerations in antimalarial dose optimization. *Antimicrob. Agents Chemother.* **57,** 5792–5807 (2013).
14. Winter, K. & Hastings, I. M. Development, evaluation, and application of an *in silico* model for antimalarial drug treatment and failure. *Antimicrob. Agents Chemother.* **55,** 3380–3392 (2011).
15. Simpson, J. A. *et al.* Mefloquine pharmacokinetic-pharmacodynamic models: implications for dosing and resistance. *Antimicrob. Agents Chemother.* **44,** 3414–3424 (2000).
16. Fidock, D. A. *et al.* Antimalarial drug discovery: efficacy models for compound screening. *Nat. Rev. Drug Discov.* **3,** 509–520 (2004).
17. Dondorp, A. M. *et al.* Estimation of the total parasite biomass in acute falciparum malaria from plasma PfHRP2. *PLoS Med.* **2,** e204 (2005).
18. Marchesini, N. *et al.* A malaria parasite encoded vacuolar H + -ATPase is targeted to the host erythrocyte. *J. Biol. Chem.* **280,** 36841–36847 (2005).
19. Forgac, M. Vacuolar ATPases: rotary proton pumps in physiology and pathophysiology. *Nat. Rev. Mol. Cell. Biol.* **8,** 917–929 (2007).
20. Saliba, K. J. & Kirk, K. pH regulation in the intracellular malaria parasite, *Plasmodium falciparum* H^+ extrusion via a V-type H^+-ATPase. *J. Biol. Chem.* **274,** 33213–33219 (1999).
21. Saliba, K. J. *et al.* Acidfication of the malaria parasite's digestive vacuole by a H^+-ATPase and a H^+-pyrophosphatase. *J. Biol. Chem.* **278,** 5605–5612 (2003).
22. Isaka, M. *et al.* Potent *in vitro* antimalarial activity of metacycloprodigiosin isolated from *Streptomyces spectabilis* BCC 4785. *Antimicrob. Agents Chemother.* **46,** 1112–1113 (2002).
23. Auparakkitanon, S. & Wilairat, P. Antimalarial activity of concanamycin A alone and in combination with pyronaridine. *Southeast Asian J. Trop. Med. Pub Health* **37,** 619–621 (2006).
24. van Schalkwyk, D. A. *et al.* Inhibition of *Plasmodium falciparum* pH regulation by small molecule indole derivatives results in rapid parasite death. *Biochem. Pharmacol.* **79,** 1291–1299 (2010).
25. Lukens, A. K. *et al.* Harnessing evolutionary fitness in *Plasmodium falciparum* for drug discovery and suppressing resistance. *Proc. Natl Acad. Sci. USA* **111,** 799–804 (2014).
26. Lukens, A. K. *et al.* Diversity-oriented synthesis probe targets *Plasmodium falciparum* cytochrome b ubiquinone reduction site and synergizes with oxidation site inhibitors. *J Infect Dis.* doi:10.1093/infdis/jiu565 (21 October 2014).
27. Bennett, T. N. *et al.* Novel, rapid, and inexpensive cell-based quantification of antimalarial drug efficacy. *Antimicrob. Agents Chemother.* **48,** 1807–1810 (2004).
28. Smilkstein, M. *et al.* Simple and inexpensive fluorescence-based technique for high-throughput antimalarial drug screening. *Antimicrob. Agents Chemother.* **48,** 1803–1806 (2004).
29. Flannery, E. L. *et al.* Using genetic methods to define the targets of compounds with antimalarial activity. *J. Med. Chem.* **56,** 7761–7771 (2013).
30. Rosario, V. Cloning of naturally occurring mixed infections of malaria parasites. *Science* **212,** 1037–1038 (1981).
31. Li, H. & Durbin, R. Fast and accurate short read alignment with Burrows-Wheeler transform. *Bioinformatics* **25,** 1754–1760 (2009).
32. Van der Auwera, G. A. *et al.* From FastQ data to high-confidence variant calls: the genome analysis toolkit best practices pipeline. *Curr. Protoc. Bioinformatics* **11,** 11.10.1–11.10.33 (2013).
33. Gouet, P. *et al.* ESPript/ENDscript: extracting and rendering sequences and 3D information from atomic structures of proteins. *Nucleic Acids Res.* **31,** 3320–3323 (2003).
34. Bridgland-Taylor, M. H. *et al.* Optimisation and validation of a medium-throughput electrophysiology-based hERG assay using IonWorks HT. *J. Pharmacol. Toxicol. Methods* **54,** 189–199 (2006).
35. McMillian, M. K. *et al.* An improved resazurin-based cytotoxicity assay for hepatic cells. *Cell. Biol. Toxicol.* **18,** 157–173 (2002).
36. O'Brien, J. *et al.* Investigation of the Alamar Blue (resazurin) fluorescent dye for the assessment of mammalian cell toxicity. *Eur. J. Biochem.* **267,** 5421–5426 (2000).
37. Hill, A. V. The possible effects of the aggregation of the molecules of haemoglobin on its dissociation curves. *J. Physiol.* **40,** iv–vii (1910).
38. Cheng, Y. & Prusoff, W. Relationship between the inhibition constant (K1) and the concentration of inhibitor which causes 50 per cent inhibition (I50) of an enzymatic reaction. *Biochem. Pharmacol.* **22,** 3099–3108 (1973).
39. Marks, L. *et al.* The role of the anaesthetised guinea-pig in the preclinical cardiac safety evaluation of drug candidate compounds. *Toxicol. Appl. Pharm.* **263,** 171–183 (2012).
40. Mooney, L. *et al.* Optimising conditions for studying the acute effects of drugs on indices of cardiac contractility and on haemodynamics in anaesthetized guinea pigs. *J. Pharmacol. Toxicol. Methods* **66,** 43–51 (2012).

Acknowledgements

We thank the Medicines for Malaria Venture (MMV) for their financial support of this project (grant MMV 09/5400). D.A.F. gratefully acknowledges funding from the National Institutes of Health (R01 AI50234 and AI109023). We express our heartfelt thanks to Dr Tanjore Balganesh, Dr Simon Campbell and Dr Steve Ward for their scientific advice during the course of this project. We express our special thanks to Rajkumar Thimmaiah, Subhash Rajanna and Robert Nanduri for their help with the *in vivo* efficacy studies. We would like to thank the Laboratory Animal Sciences and Drug Metabolism and Pharmacokinetics (DMPK) groups, AstraZeneca, Alderley Park, UK for conducting the 3-day *in vivo* rat toxicology study and bioanalysis, respectively; Biotrial, France for conducting the *in vivo* Guinea Pig cardiovascular study on compound **12**; CEREP, France for performing the secondary pharmacology screening and Charles River laboratories, Edinburgh for performing the Guinea Pig exposure bioanalysis. We thank Eva Lenz and Chad Elmore, DSM Department, Alderley Park, AstraZeneca UK for their help in determining the ^{13}C-NMR structure of compounds. We would like to thank Syngene International for their help in compound synthesis and scale up for *in vivo* studies. We would like to thank Dr Sergio Wittlin, Swiss Tropical and Public Health Institute (SwissTPH) for generating the cross-resistance data for TAPs against a panel of *Pf* isolates with resistance to antimalarial agents in preclinical/clinical development.

Author contributions

S.H.P., S.S., S.K. and V.K.S. designed and directed the study and wrote the manuscript with contributions from all co-authors; S.H.P., V.P., K.M., J.P., E.B., G.S., S.L., K.K., A.R., N.R.C., S.M., S.R., B.B., P.W. and P.S.I., were responsible for medicinal chemistry design, organic synthesis of compounds and data analyses; S.S., S.B., A.S., V.P., R.S., J.R., K.R.P., A.D. and V.H. were responsible for the pharmacokinetics and pharmacodynamics study design and data analysis; P.V., D.A., S.M., N.R., A.A., R.N., M.C., K.M., V.B., S.N. and V.K.S. were responsible for the biological profiling of compounds in various biological assays and data analysis; L.R-R., K.H. and S.K. were responsible for designing the *in vitro* safety pharmacology and *in vivo* safety studies and for data analysis; M.B.J.-D., M.S.M. and L.M.S., were responsible for determining the *in vitro* parasite reduction ratio and evaluating the efficacy parameters in the *Pf*/SCID model; P.A.M., A.K.L. and D.F.W. were responsible for generating cross-resistance and mutant selection data. O.C-F. and D.A.F. were responsible for generating cross-resistance data and mutant selection data; P.P.H. was responsible for measuring the parasite vacuolar size and whole-genome sequence data analysis; D.W. contributed to chemistry design; R.E.M was responsible for generating the whole-genome sequence data of resistant mutants.

Additional information

Accession codes: NGS assembly files of all *Pf* strains used in this study to identify the resistance determinants have been deposited in the NCBI Sequence Read Archive (SRA) with accession code SRP052918.

Competing financial interests: The authors declare no competing financial interests.

15

Luminescent platinum(II) complexes with functionalized N-heterocyclic carbene or diphosphine selectively probe mismatched and abasic DNA

Sin Ki Fung[1,*], Taotao Zou[1,2,*], Bei Cao[1], Tianfeng Chen[3], Wai-Pong To[1], Chen Yang[1,2], Chun-Nam Lok[1] & Chi-Ming Che[1,2]

The selective targeting of mismatched DNA overexpressed in cancer cells is an appealing strategy in designing cancer diagnosis and therapy protocols. Few luminescent probes that specifically detect intracellular mismatched DNA have been reported. Here we used Pt(II) complexes with luminescence sensitive to subtle changes in the local environment and report several Pt(II) complexes that selectively bind to and identify DNA mismatches. We evaluated the complexes' DNA-binding characteristics by ultraviolet/visible absorption titration, isothermal titration calorimetry, nuclear magnetic resonance and quantum mechanics/molecular mechanics calculations. These Pt(II) complexes show up to 15-fold higher emission intensities upon binding to mismatched DNA over matched DNA and can be utilized for both detecting DNA abasic sites and identifying cancer cells and human tissue samples with different levels of mismatch repair. Our work highlights the potential of luminescent Pt(II) complexes to differentiate between normal cells and cancer cells which generally possess more aberrant DNA structures.

[1] State Key Laboratory of Synthetic Chemistry, Institute of Molecular Functional Materials, Chemical Biology Centre, and Department of Chemistry, The University of Hong Kong, Pokfulam Road, Hong Kong, China. [2] HKU Shenzhen Institute of Research and Innovation Shenzhen, Shenzhen 518053, China. [3] Department of Chemistry, Jinan University, Guangzhou 510632, China. * These authors contributed equally to this work. Correspondence and requests for materials should be addressed to C.-M.C. (email: cmche@hku.hk).

The binding interactions between metal complexes and nucleic acids have been the subject of numerous studies in recent decades[1–4]. These interactions are likely to lead to therapeutic (for example, anti-cancer) effects and/or be used for diagnostic (for example, luminescent probes) purposes[5–11]. To date, the selective targeting of nucleic acids in cancer cells by chemotherapeutic metal complexes in a manner that minimizes off-target bindings and hence diminishes side effects remains a great challenge[12,13]. Endeavours in this research area have led to luminescent metal complexes that show emission responses with specificity for specific nucleic acid structures, such as G-quadruplex DNA[14,15], double-stranded RNA[16,17] and mismatched DNA[13,18,19]. The recognition of mismatched DNA is of importance for cancer diagnosis and therapy, because DNA mismatches are associated with oncogenic transformation[20–22]. Furthermore, cancer cells, such as those of colorectal cancer, show a high frequency of DNA mismatches as a result of their deficiency in mismatch repair (MMR)[13,23–26]. Pioneering work by Barton and co-workers has shown that the octahedral d^6 metal complexes $[Rh(bpy)_2chrysi]^{3+}$ (bpy = 2-phenylpyridine, chrysi = 5,6-chrysenequinone diimine)[18,27,28] and Δ-$[Ru(bpy)_2dppz]^{2+}$ (dppz = dipyrido[3,2-*a*:2′,3′-c] phenazine)[19,29,30] are able to selectively bind to mismatched DNA. Other researchers have reported that octahedral cobalt(III) complexes[31], a six-coordinated silicon complex containing a sandwich ruthenium moiety[32], some non-metal compounds[33–41] and others[13] can recognize DNA mismatches with specificity.

The d^8 pincer Pt(II) complexes have been well documented to intercalate into matched DNA[1,2,4,11,42]. With intrinsic emission properties that are highly sensitive to the microenvironment, these types of Pt(II) complexes have been developed as luminescent DNA probes and have been identified as potent therapeutic agents for cancer treatment[14,17,43–45]. However, metal complexes with bare planar coordination geometries likely exhibit non-discriminative intercalative binding interactions with DNA. In the crystal structures of mismatched DNA bound to Rh(III)[18,28] or Ru(II)[19] complexes, the planar extended π-conjugated ligands (with a width of up to 11.3 Å and a length of up to 9.4 Å) favour strong π-stacking interactions with nearby base pairs at mismatched sites, concomitantly with the ejection of mismatched base pairs, whereas the other auxiliary bipyridine ligands promote a groove binding interaction. We therefore hypothesized that a pincer Pt(II) complex coordinated with an out-of-plane bulky ancillary ligand may exhibit selective binding affinity towards mismatched DNA, as discussed in the next section. The bulky ancillary ligand can be designed to hamper the intercalation of the platinum complex with well-matched DNA while exhibiting high binding affinity with mismatched DNA. This difference is possible because the mismatched sites possess larger binding pockets. We have previously described a luminescent $[Pt^{II}(C^\wedge N^\wedge N)(NHC^{2C4})]^+$ ($HC^\wedge N^\wedge N$ = 6-phenyl-2,2′-bipyridine, NHC = *N*-heterocyclic carbene, **1b** in the current report) complex with an NHC ligand almost perpendicular to the C-deprotonated $C^\wedge N^\wedge N$ plane[46]. This complex exhibits a weak binding affinity to matched DNA. Thus, varying the *N*-alkyl/aryl substitution(s) of NHC, $[Pt^{II}(C^\wedge N^\wedge N)(NHC)]^+$ may generate a candidate scaffold to target mismatched DNA via cooperative π-stacking and groove binding interactions.

Here we describe several luminescent, mismatched DNA probes based on the $[Pt^{II}(C^\wedge N^\wedge N)(NHC)]^+$, $[Pt^{II}(N^\wedge C^\wedge N)(NHC)]^+$ and $[Pt^{II}_2(C^\wedge N^\wedge N)_2(\mu\text{-dcpm})]^{2+}$ complexes. The mononuclear Pt(II) complex selectively detects DNA containing CC mismatches, and the dinuclear Pt(II) complex exhibits high selectivity towards several types of DNA mismatches. These luminescent Pt(II) complexes also selectively bind thermodynamically unstable abasic DNA, are able to differentiate cancer cells that have different levels of MMR capacity and can differentiate colon tumour tissue from normal tissue.

Results

Structure and emission responses to CC mismatched DNA. We synthesized and characterized a series of $[Pt^{II}(C^\wedge N^\wedge N)(NHC)]^+$ complexes with different alkyl chains, aromatic groups and hydrophilic alcohol moieties on NHC and/or on C-deprotonated $C^\wedge N^\wedge N$ ligands (Fig. 1; details in Supplementary Information). In addition, we examined the emission responses of complexes that bind mismatched and matched DNA (Supplementary Fig. 1). The most unstable CC mismatched DNA was chosen in the primary screen because metalloinsertors display the highest binding to this type of mismatched DNA[29]. We found that as the steric hindrance of the NHC ligand increased, the binding of the

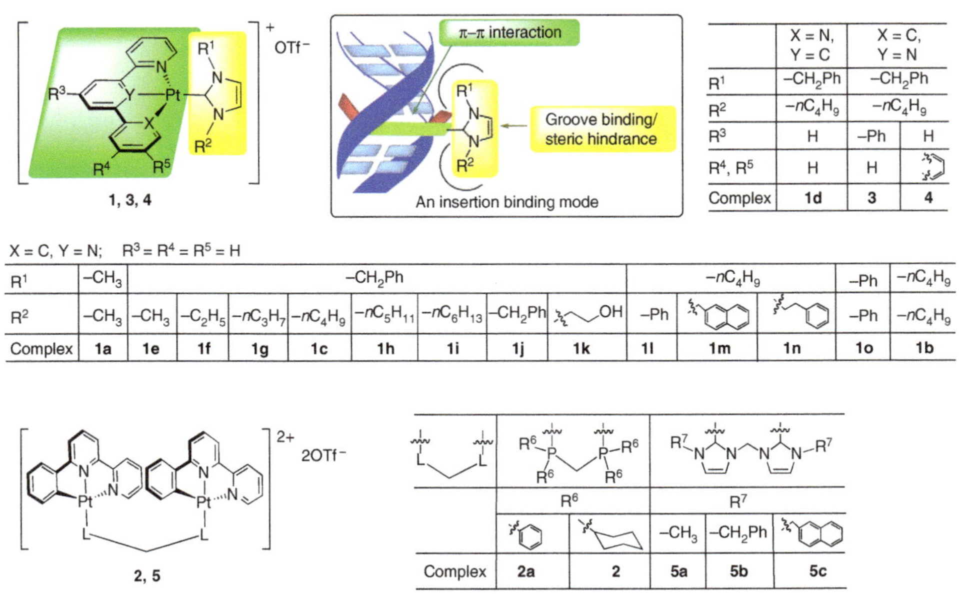

Figure 1 | Chemical structures and proposed binding modes. Platinum(II) complexes **1a–1o**, **2**, **2a**, **3**, **4** and **5a–5c**. The counteranions for all complexes are $CF_3SO_3^-$. The inset shows the molecular design of Pt(II) complexes tested for binding mismatched DNA.

Pt(II) complexes to matched DNA decreased, whereas the binding with CC mismatched DNA remained high for certain bulky NHC ligands (Fig. 2; the DNA sequences in all the experiments are depicted in the Supplementary Table 2). For example, for the $[Pt(C^{\wedge}N^{\wedge}N)(NHC)]^+$ type complexes of **1e** ($-CH_3$), **1f** ($-C_2H_5$), **1g** ($-C_3H_7$), **1c** ($-C_4H_9$), **1h** ($-C_5H_{11}$), **1i** ($-C_6H_{13}$) and **1j** ($-CH_2Ph$), which contain the same N^3-benzyl group but different N^1-substituents on the NHC ligands (Table 1), an increase in the length of the N^1-alkyl was accompanied by a marked decrease in emission enhancement towards matched DNA (I_M), from 38.9-fold to <3-fold. Although increasing the chain length from the N^1-methyl of **1e** to the N^1-ethyl of **1f** led to a decrease in emission enhancement in the case of mismatched DNA (I_{MM}), the emission intensity of **1g**, which bears N^1-propyl, was much stronger than that of **1f**, suggesting that N^1-propyl favours the groove binding of the Pt(II) complex to DNA. This may enhance the tightness of the insertion binding mode, as proposed. Further lengthening the alkyl chain to *n*-butyl, *n*-pentyl and *n*-hexyl gradually decreased the binding with mismatched DNA. Comparing the I_{MM}/I_M ratios, **1c**, containing N^1-(*n*-butyl) and N^3-benzyl substitutions on the NHC (Fig. 1) showed the most significantly different emission responses between CC mismatched DNA and matched DNA. We characterized the structure of this complex by $^1H-^1H$ correlation spectroscopy (COSY) and nuclear Overhauser effect spectroscopy (NOESY) nuclear magnetic resonance (NMR; Supplementary Fig. 2). As shown in Fig. 3a,c, the emission intensity of **1c** ($\lambda_{max} = 535$ nm) in a Tris-buffered solution increased by 26.4-fold in the presence of 1 equivalent of a hairpin DNA oligomer with a CC mismatch[29] (Fig. 3c). However, the emission of **1c** was enhanced only 2.6-fold upon the addition of the same amount of the hairpin DNA, having the cognate sequence with the mismatched bases replaced by matched GC pairs. In another mismatched DNA model using a 17-mer double-stranded DNA (dsDNA) formed by two complementary oligomers[47], the emission intensity of **1c** displayed 27.5-fold and 2.8-fold increases in emission intensity upon the addition of CC mismatched 17-mer dsDNA and matched 17-mer dsDNA, respectively (Supplementary Fig. 3). In the literature, a rhodium-Oregon Green conjugate has been reported to exhibit a ~3.2-fold higher emission intensity in the presence of CC mismatched DNA compared with matched DNA[47]. In a control experiment, the classical DNA intercalator, ethidium bromide (EB), showed increases in emission intensity with hairpin mismatched DNA and matched DNA of 12.7-fold

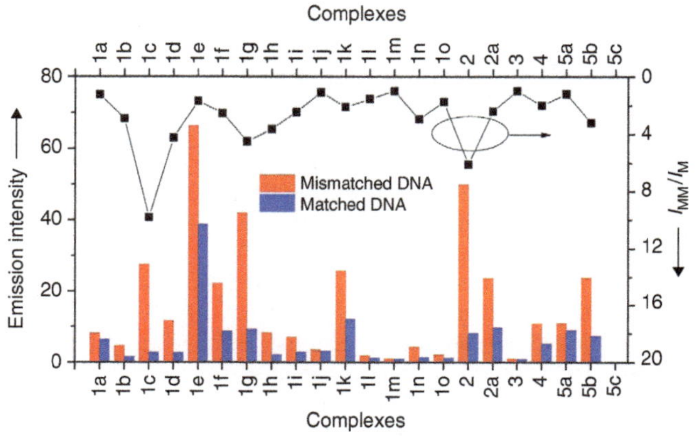

Figure 2 | Emission responses of Pt(II) complexes to matched and mismatched DNA. A summary of emission intensities (at near level off) of different Pt(II) complexes towards mismatched and matched DNA and the emission intensity ratios (I_{MM}/I_M).

and 14.1-fold, respectively, revealing a 0.9-fold difference (Supplementary Fig. 4). Under similar conditions, **1a**, containing a less bulky *N*-methyl substitution on the NHC, displayed similar 8.5-fold and 7.1-fold emission enhancements for hairpin CC mismatched DNA and matched DNA, respectively (1.2-fold difference; Supplementary Fig. 5). This result can be explained by the fact that the less bulky *N*-methyl substituted NHC does not impose enough steric hindrance to diminish the binding of the Pt complex with matched DNA, resulting in similarly high enhancement in emission intensity for both matched and mismatched DNA. Conversely, the more bulky N^1-(*n*-butyl) and N^3-benzyl groups diminished the binding of **1c** with matched DNA, while a much stronger binding affinity with mismatched DNA was still maintained.

Characterizations of bindings to CC mismatched DNA. The following experiments were performed to shed light on the DNA-binding interactions of the novel complexes. In ultraviolet/visible absorption titration experiments, upon the addition of dsDNA containing a CC mismatch to **1c** in a Tris-buffered solution, a hypochromicity of ~34% at $\lambda = 335$ nm, together with isosbestic spectral changes, was observed (Supplementary Fig. 6a); however, only a minor hypochromicity of ~12% was detected at the same wavelength for **1c** when matched dsDNA with a similar sequence was used instead (Supplementary Fig. 6b). Isothermal titration calorimetry (ITC) experiments (Fig. 3b) revealed that the titration of CC mismatched DNA with **1c** resulted in a significant exothermic reaction, revealing binding between the two components, with a dissociation constant (K_d) of 51.5 ± 8.0 µM. In contrast, the reaction between **1c** and matched DNA was a very weak exothermic reaction, with a K_d estimated to be >600 µM. In a similar experiment, the K_d values of the binding of **1a** to mismatched and matched DNA were 66.2 ± 15.7 and 56.4 ± 13.6 µM, respectively (Supplementary Fig. 7a,b). The K_d values of the binding of EB to mismatched DNA and matched DNA were 1.32 ± 0.24 and 1.60 ± 0.28 µM, respectively (Supplementary Fig. 8a,b). ^1H NMR and two-dimensional (2D) NOESY, total correlation spectroscopy (TOCSY) and COSY NMR experiments were performed using a self-complementary DNA oligonucleotide, 5′-$C_1G_2G_3A_4\underline{C_5}T_6C_7C_8G_9$-3′, containing CC mismatches in D_2O (phosphate buffer, pH 6.1; Fig. 4a–d; Supplementary Figs 9 and 10)[48]. In the range of 6.8–8.0 p.p.m., the aromatic ^1H signals of C5 and A4 show substantial shifts, whereas those of the other bases remain nearly unshifted in the presence of **1c** (^1H NMR in Fig. 4a). In addition, in the 2D H1′ × aromatic NOESY spectrum of **1c** with DNA at a 1:1 ratio, a sequential NOESY walk (Fig. 4d), which showed a marked shift at A4 and C5 compared with DNA only (Fig. 4b), was found in combination with cytosine H6/H5 TOCSY (Supplementary Fig. 9) and H1′ × H2′–H2″ TOCSY (Fig. 4c) NMR. At [**1c**]:[DNA] ratios of 0.25:1 and 0.5:1 (Supplementary Fig. 10), a gradual disappearance of C5 and A4 aromatic protons of complex-free DNA and the subsequent appearance of new C5 and A4 aromatic protons of **1c**-bound DNA in the NOESY spectra were found. Thus, both one-dimensional and 2D NMR results indicate the possible binding of **1c** at the mismatched site. Collectively, the more significant the hypochromicity in ultraviolet/visible spectral changes, the higher the amount of heat generation, and the marked shifts of NMR signals at the mismatched site together revealed that **1c** preferentially binds to DNA containing CC mismatches. In the emission quenching experiments using $[Cu(phen)_2]^{2+}$ (phen = 1,10-phenanthroline), a minor-groove-specific quencher, the emission intensity of **1c** in the presence of CC mismatched DNA gradually decreased upon increasing the amount of $[Cu(phen)_2]^{2+}$ (Supplementary

Table 1 | A comparison of structure-emission response relationships of [Pt(C^N^N)(NHC)]CF_3SO_3 complexes containing *N*-benzyl groups towards matched and mismatched DNA.

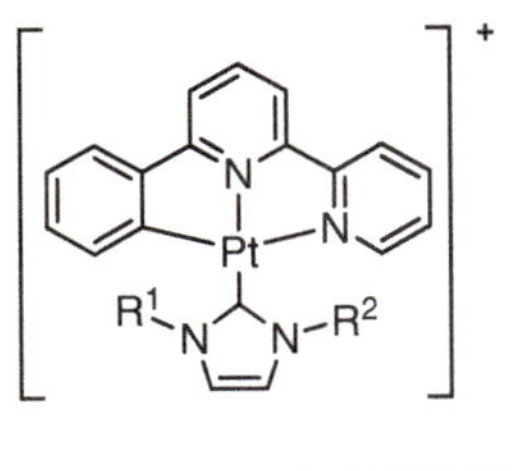

R^1	$-CH_2Ph$						
R^2	$-CH_3$	$-C_2H_5$	$-nC_3H_7$	$-nC_4H_9$	$-nC_5H_{11}$	$-nC_6H_{13}$	$-CH_2Ph$
Complex	**1e**	**1f**	**1g**	**1c**	**1h**	**1i**	**1j**
I_M*	38.9	8.7	9.3	2.6	2.3	2.8	3.2
I_{MM}*	66.2	22.2	42.0	26.4	8.2	7.0	3.5

*I_M and I_{MM} denote the relative emission intensities of the Pt(II) complexes towards matched and mismatched DNA, respectively.

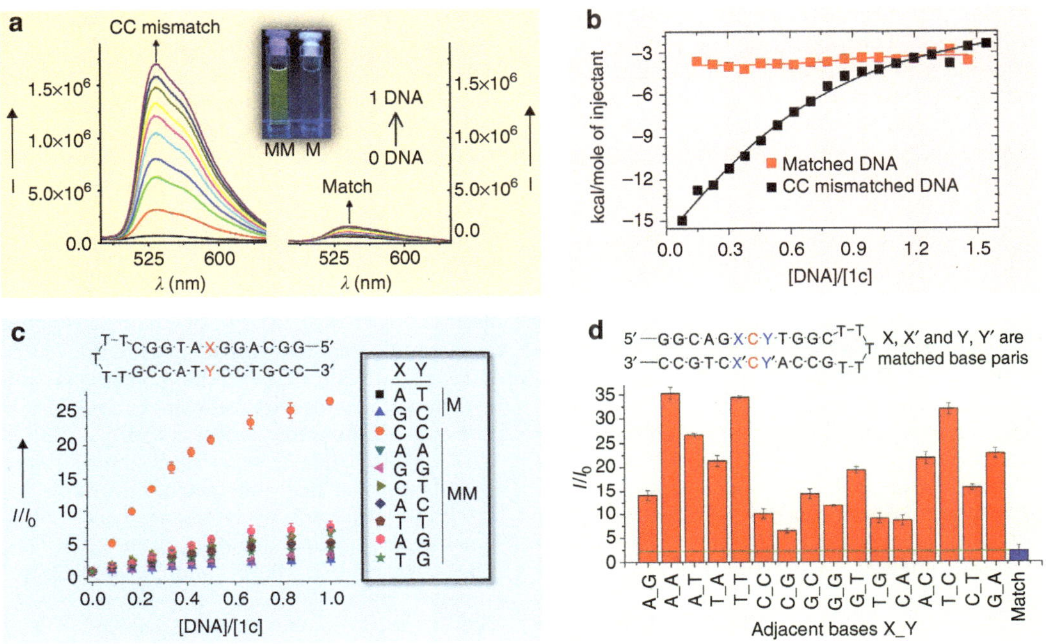

Figure 3 | Binding of 1c with mismatched DNA. (**a**) The emission spectra of complex **1c** (5 μM) in a Tris buffer solution (50 mM NaCl and 2 mM Tris, pH 7.5) after binding to different concentrations of CC mismatched DNA (MM) and matched DNA (M). (**b**) Plot of integrated ITC data for the exothermic interaction between CC mismatched DNA (0.75 mM) and **1c** (0.1 mM). (**c**) Changes in emission intensity (± s.e.m.) at 535 nm of complex **1c** (5 μM) in the Tris buffer solution upon the addition of different types of DNA. M, match; MM, mismatch. (**d**) Relative emission enhancement of **1c** towards CC MM DNA with different adjacent base pairs.

Fig. 11), indicating the binding of the complex through the minor groove[29]. Indeed, a favourable binding mode between **1c** and CC mismatched DNA has also been identified in molecular docking experiments (Supplementary Fig. 12), consistently with the insertion of the Pt(C^N^N) plane between base pairs and the favourable minor-groove binding interaction of the N^1-(*n*-butyl) and N^3-benzyl moieties. Complex **1d**, which contains an N^C^N instead of a C^N^N ligand (Fig. 1), exhibited 4.5-fold higher emission responses towards 1 equivalent of CC mismatched DNA than matched DNA (Supplementary Fig. 13).

The emission responses of **1c** towards all other types of DNA mismatches, including CT, AC, TT, AG, TG, AA and GG mismatches, all of which are more thermodynamically stable than CC mismatches[49], were tested (Fig. 3c). However, no significant difference in emission responses between these mismatched DNA and matched DNA was observed. We also examined the emission responses of **1c** to CC mismatched DNA having 16 different adjacent base pairs (Fig. 3d). Complex **1c** displayed higher emission responses to all the CC mismatched DNA sequences than the cognate matched DNA, with adjacent 5′-<u>A</u>C<u>A</u>-3′ revealing the highest emission enhancement (15.0-fold higher than matched DNA). These experiments suggest that **1c** is a selective probe for CC mismatched DNA.

Targeting thermodynamically more stable DNA mismatches. For recognition of the more stable DNA mismatches, an increase in the binding affinity of the Pt(II) complex with DNA is required. The dinuclear analogues of **1c**, which contain an additional positive charge and a larger molecular dimension, are expected to have higher DNA-binding affinities as a result of stronger ionic bonding interactions and possibly a better geometry to fit into the pocket at the mismatched site. Therefore, we examined the emission responses of a series of doubly charged dinuclear platinum complexes with two Pt^{II}(C^N^N) moieties linked by bridging bis(NHC) or diphosphine ligands (Fig. 1) in the presence of hairpin mismatched and matched DNA. Among the dinuclear Pt(II) complexes, $[Pt^{II}_2(C^\wedge N^\wedge N)_2(\mu\text{-dcpm})]^{2+}$ **2** (Fig. 1; the X-ray crystallographic structure is depicted in

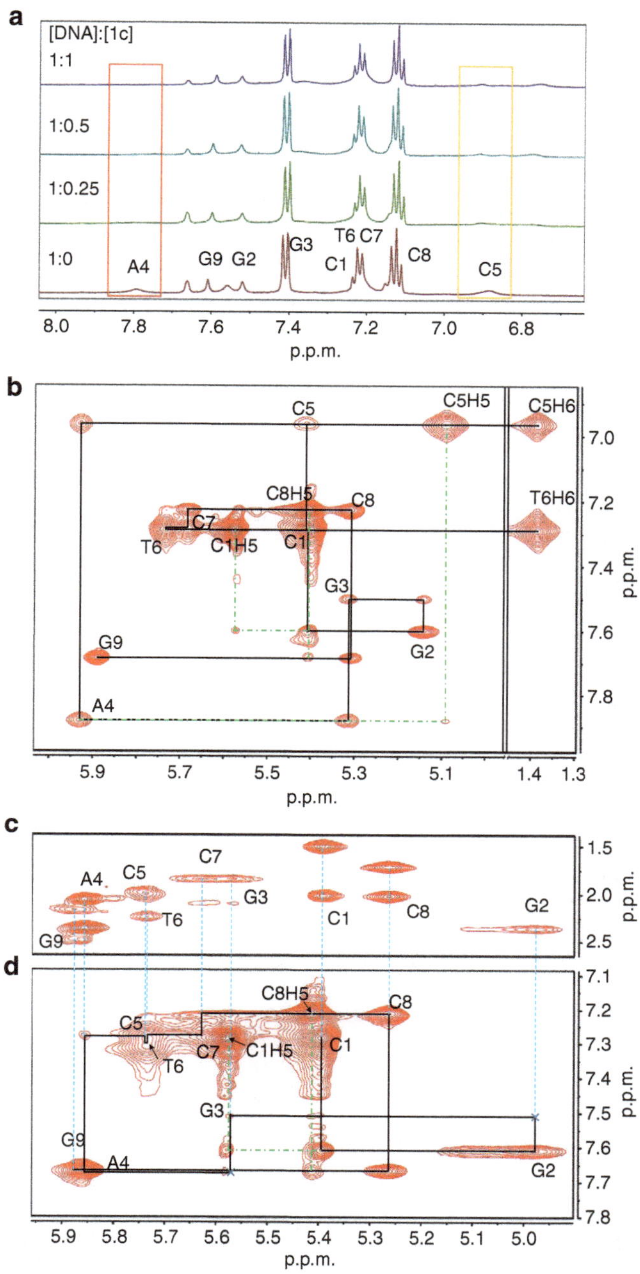

Figure 4 | NMR analysis of the binding of 1c with mismatched DNA. (**a**) ^{1}H NMR (phosphate buffer = 50 mM, pH 6.1, NaCl = 20 mM, in D_2O, 10 °C) spectra of a DNA oligonucleotide in the presence of different amounts of **1c**. (**b**) 2D H1′ × aromatic (left) and T(Me) × aromatic (right) NOESY spectra of a DNA oligonucleotide in phosphate buffer (50 mM in D_2O, containing NaCl = 20 mM, 10 °C); a sequential NOESY walk along the full strand was identified. (**c**) TOCSY spectrum of H1′ × H2′-H2″ and (**d**) NOESY spectrum of H1′ × aromatic of a DNA oligonucleotide in the presence of an equimolar concentration of **1c** in phosphate buffer (50 mM in D_2O, containing NaCl = 20 mM, 10 °C). We also found a sequential NOESY walk that was significantly different at A4 and C5 in comparison with DNA only in **b**, suggestive of interactions of **1c** at a mismatched site. Note that the interactions of C1H5 with G2H8 and C8H5 with G9H8 at the terminal sites in **b** and **d**, and the interaction of C5H5 with A4H8 at the mismatched site in **b** were identified.

Supplementary Fig. 14 and described in Supplementary Table 1, showing an intramolecular Pt-Pt distance of 3.201 Å) was observed to display higher selectivity towards hairpin CC mismatched DNA (Fig. 2). Figure 5a and Supplementary Fig. 15 illustrate that the emission intensity of **2** at $\lambda = 634$ nm was elevated 50-fold upon the addition of only 0.5 equivalents of hairpin DNA containing a CC mismatch but enhanced only 8.2-fold in the presence of the same amount of matched DNA, that is, a 6.1-fold difference in emission responses.

The binding affinity of **2** with DNA was examined by ITC. The interaction of **2** was much more exothermic with CC mismatched DNA than with matched DNA (Supplementary Fig. 16). Notably, the K_d of the binding of **2** to CC mismatched DNA was found to be 7.0 ± 2.1 μM, which is 7.4-fold higher than that of **1c**. The K_d of the binding of **2** with matched DNA was negligible. In addition, the melting temperature (T_m), as determined by the change in absorbance (OD_{260nm}) of DNA was increased by 6.8 °C for mismatched DNA in the presence of **2** at an equimolar concentration (Supplementary Fig. 17a). However, only a 0.5 °C increase in melting temperature was found for matched DNA in the presence of **2** (Supplementary Fig. 17b).

Complex **2** also exhibited higher emission responses to CC mismatched DNA containing 16 different adjacent sequences (Fig. 5b), with a nearby 5′-T̲CT̲-3′ sequence generating the highest fold increase in emission (13.5-fold higher than matched DNA). In addition to its response to CC mismatches, **2** also showed higher emission responses to several other types of mismatched DNA over matched DNA, and the emission increased 4.9-fold for CT, 3.8-fold for AC and 3.4-fold for TT compared with matched DNA but no higher for other mismatches (AG, 1.8-fold; GT, 1.4-fold; AA, 1.4-fold; GG, 0.8-fold; Fig. 5a). Such differences in the emission responses between mismatched and matched DNA partially correlated with the thermodynamic stability of the DNA base pairs, $CC \leq AC \leq CT < AA \leq TT < GA \leq GT < GG$[49], suggesting an insertion binding mode[50]. The emission response to AC was slightly lower than that to CT, and the response to AA was lower than that to TT; these observations are not fully consistent with the trend of thermodynamic stability of mismatched DNA. The lower emission response towards A-containing AC and AA mismatches may have been caused by the weak coordination of the N7 of purine in adenine at the mismatch site to the Pt complex, leading to decrease of emission intensity (note that the pyrimidine N3 in C and T is not accessible to the Pt(II) complexes due to steric hindrance). We performed hybrid quantum mechanics/molecular mechanics (QM/MM) calculations to study the mode of binding between **2** and mismatched DNA. As shown in Fig. 5c, **2** bound to mismatched DNA through an insertion binding mode at the minor groove (this result is also consistent with the results of a $[Cu(phen)_2]^{2+}$ quenching experiment, Supplementary Fig. 18); the C^N^N planes interacted with nearby GC and AT base pairs by π-stacking interactions (inter-planar distance between the C^N^N plane and the base pairs of ~3.2 Å). The QM/MM simulation also indicated that the dinuclear structure better fit the mismatched sites, which have larger pockets (the inter-planar separation between nearby base pairs at mismatched sites can be >9 Å), but not the matched site, which contains less space between base pairs.

We also tested the emission responses of **2** to dsDNA of poly(dC)-poly(dG), poly(dC-dG)$_2$, poly(dA)-poly(dT) and poly(dA-dT)$_2$; single-stranded DNA of oligo(dT)$_{20}$ and oligo(dC)$_{20}$; dsRNA and bovine serum albumin, all of which were observed to have at least 4.5-fold lower emission intensity than that of CC mismatched DNA (Fig. 5d). Notably, the emission intensity of **2** with RNA containing a CC mismatch was 3.3-fold higher than that with matched RNA (Fig. 5d).

Photophysical properties upon binding to mismatched DNA. To better understand the increase in emission upon binding with

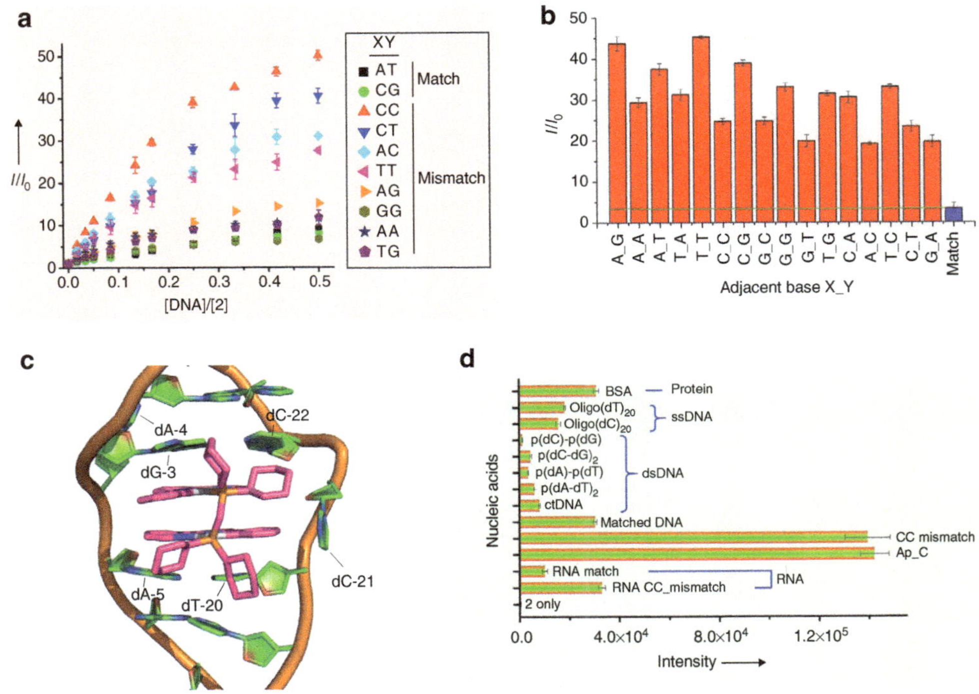

Figure 5 | Binding of 2 with mismatched DNA. (**a**) Changes in emission intensity (± s.e.m.) at 634 nm of **2** (5 μM) in Tris buffer solution (50 mM NaCl, 2 mM Tris, pH 7.5) after the addition of different concentrations of various DNA. (**b**) Relative emission enhancement of **2** towards CC mismatched DNA with different adjacent base pairs. (**c**) QM/MM optimized structure of mismatched DNA bound to **2**. (**d**) Emission responses (intensity ± s.e.m.) of **2** towards different analytes.

mismatched DNA, we measured the emission spectra of the Pt(II) complex (**1c** or **2**) in buffer solutions containing CC mismatched DNA under air or argon. Under Ar, no significant changes in emission intensity were found compared with that under air conditions (Supplementary Fig. 19). The two Pt(II) complexes were weakly emissive in buffer solutions, but their emission intensities were significantly higher in acetonitrile, even under air (11-fold for **1c**, 9-fold for **2**, Supplementary Fig. 20a,b; the emission intensity was found to increase with decreasing donor strength of various solvents, as shown in Supplementary Fig. 20c). Thus, the increased emission could not be accounted for by the shielding of oxygen quenching upon mismatched DNA binding to the Pt(II) complexes. In addition, variations of pH from 5 to 11 did not evoke a significant change in emission intensity (Supplementary Fig. 21). We propose that the shielding of quenching by a buffer solution and/or a more ordered structure of the Pt(II) complex after mismatched DNA binding could account for the emission enhancement[51,52].

We used time-resolved emission spectroscopy to examine the binding between the Pt complexes and DNA. As shown in Supplementary Fig. 22, the emission of **2** in the presence of CC mismatches had a lifetime of 2.3 μs, whereas in the presence of matched DNA, the lifetime is composed of a major portion of short decay (< 100 ns) and a small amount of 1.2-μs decay. The difference in emission intensity, $I_{mismatch}/I_{match}$, was observed to increase with increasing decay time, from 2.3-fold at $t = 0$ to 11.8-fold at $t = 10$ μs (Supplementary Fig. 23). Therefore, time-resolved emission spectroscopy may be a useful strategy to achieve higher sensitivity than steady-state emission spectroscopy for the detection of mismatched DNA using luminescent Pt(II) complexes.

Targeting abasic DNA. The abasic (AP) site, in which a purine or a pyrimidine base is missing in a nucleotide, represents another type of thermodynamically unstable DNA defect[50]. In the presence of a 27-mer dsDNA oligomer containing an abasic site[50] (AP_C with a complementary base of C), the emission intensity of an equimolar amount of **1c** at $\lambda_{max} = 535$ nm exhibited a 32.2-fold increase (Supplementary Fig. 24). This elevation in emission was even higher than that observed for the binding of **1c** with CC mismatched DNA with similar sequences (26.4-fold). For comparison, only a 4.7-fold increase in emission was detected when **1c** was incubated with normal 27-mer dsDNA with a cognate sequence. In the presence of other abasic sites of AP_A, AP_T and AP_G, the emission intensities of **1c** were increased by 14.8, 21.1 and 14.4 times, respectively. Complex **2** showed 17.0–29.3-fold increases in emission intensity in the presence of different abasic DNA sequences, but only a 6.1-fold increase in emission intensity with normal DNA (Supplementary Fig. 25). We also measured the effects of **2** on the T_m of DNA bearing an abasic site (Supplementary Fig. 26). The melting temperature of abasic DNA in the presence of **2** was increased by 5.3 °C, whereas the T_m of the similar DNA sequence without an abasic site was not affected, indicating a high selectivity of **2** towards abasic DNA over matched DNA. Thus, these Pt(II) complexes could be used to probe DNA abasic sites. Such a property is also found in Rh(III)-based metalloinsertors, in which the metal complex binds to thermodynamically unstable DNA structures through an insertion binding mode[50].

Detection of genomic DNA base pair defects. The human colorectal carcinoma cell line HCT116 is known to be MMR deficient, owing to mutations in the MMR-associated *hMLH1* gene, whereas HCT116 cells containing a transferred, normal chromosome 3 (denoted as HCT116N) restore MMR activity[53]. We found that **2** showed significantly stronger emission at $\lambda_{max} = 634$ nm in the presence of genomic DNA isolated from HCT116 cells compared with that from HCT116N cells ($P < 0.01$, $n = 3$; Fig. 6a; Supplementary Fig. 27). We then measured and

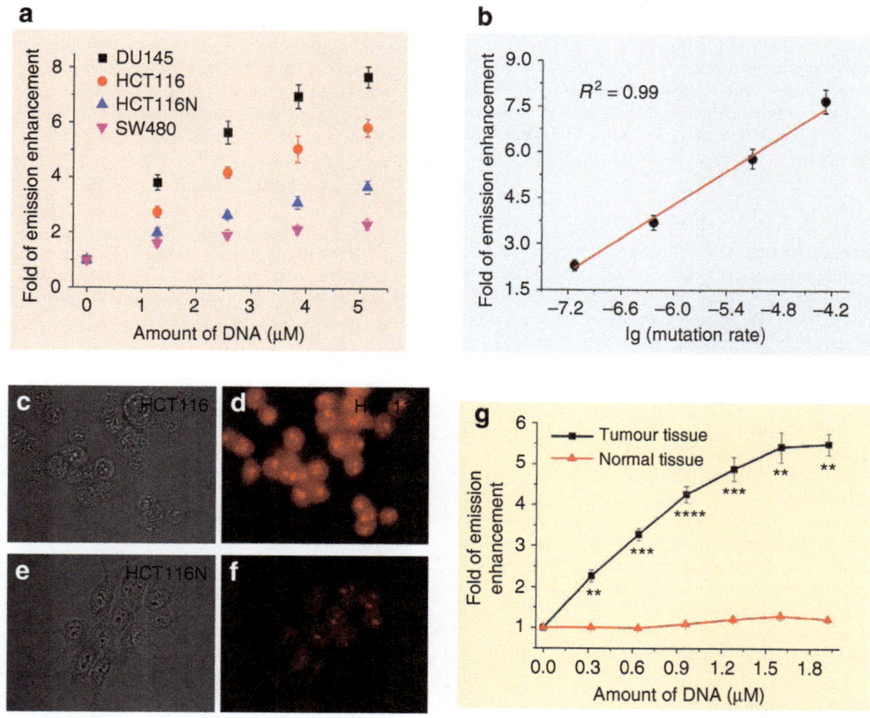

Figure 6 | Detection of mismatched DNA in cancer cells. (**a**) Changes in emission intensity (± s.e.m.) of **2** (5 μM) at 634 nm in an aqueous buffer solution (50 mM NaCl and 2 mM Tris, pH 7.5) after binding to genomic DNA extracted from different cancer cell lines. (**b**) Plot of fold of increased emission versus reported mutation rates in different cancer cell lines, lg (mutation rate); a linear fit is shown. Fluorescence microscopy studies of HCT116 (**d**) and HCT116N (**f**) cells pretreated with digitonin and stained with 10 μM **2**; (**c,e**) bright-field images of HCT116 and HCT116N cells, respectively. (**g**) Plot of emission responses (fold of increase in emission ± s.e.m.) of **2** versus different concentrations of DNA extracted from colon tumour tissue or adjacent normal tissue. All the concentrations of genomic DNA are shown as base pair concentrations.

compared the emission responses of **2** with several types of cancer cells with different mutation rates. The human cancer cell lines DU145, HCT116, HCT116N and SW480 have mutation rates (mutations per cell per generation) in the *hprt* gene locus of 529×10^{-7}, 77.5×10^{-7}, 5.9×10^{-7} and 0.75×10^{-7}, respectively[54], and the relative emission intensities of **2** (5 μM) at 5.1 μM of DNA are 7.7, 5.8, 3.7 and 2.3, respectively (Fig. 6a shows an increase of emission intensity at 0–5.1 μM of genomic DNA from different cancer cells). Notably, the plot of the fold of emission increase versus the lg (mutation rate) yields a linear fit with $R^2 = 0.99$, as shown in Fig. 6b, indicating a strong correlation between emission response and mutation rate.

We next tested whether **2** could be used to selectively detect mismatched DNA in permeabilized cell samples by fluorescence microscopy. Because the Pt(II) complexes are lipophilic and tend to accumulate in cytoplasmic compartments, the HCT116 and HCT116N cancer cells were permeabilized by digitonin and then washed to minimize the fluorescence background due to interactions of the Pt(II) complexes with cytoplasmic contents. As shown in Fig. 6c–f, after labelling with **2**, the emission signal of the HCT116 cells was significantly stronger than that of the HCT116N cells. In addition, complex **2** emitted significantly stronger signals in cancerous HCT116 cells compared with those in the immortalized normal human colon mucosal epithelial cell line NCM460 (ref. 55; Supplementary Fig. 28); complex **2** also consistently exhibited significantly lower emission responses in the presence of the DNA isolated from NCM460 cells compared with that from HCT116 cells (Supplementary Fig. 27). The differences in emission intensity were not caused by variations in cellular uptake of Pt(II) complexes as shown by inductively coupled plasma mass spectrometer measurements (Supplementary Fig. 29). For comparison, no significant differences in the emission properties were found between the two cell lines labelled with the general DNA intercalator EB (Supplementary Fig. 30).

We further examined the ability of the Pt(II) complex to detect mismatched DNA of primary human tumours of colon cancer which is frequently characterized by MMR deficiency. DNA samples were extracted from colorectal adenocarcinoma (well to moderately differentiated) and the surrounding normal tissues of a patient (see details in the Methods section). The results of the emission responses of **2** (5 μM) to the DNA from the tumour and normal tissues are shown in Fig. 6g. In the presence of 0.3–1.9 μM of DNA extracted from normal colon tissues, weak emission responses of up to 1.3-fold of the initial emission intensity were detected; in contrast, the emission intensity increased in a concentration-dependent manner in the presence of 0–1.6 μM of DNA extracted from colon tumour tissues, showing up to a 5.5-fold increase in emission at 1.9 μM. Thus, there was a difference between the emission responses of **2** to the DNA of human colon tumour tissue and the normal adjacent tissue.

In conclusion, we have identified two classes of luminescent Pt(II) complexes that can be used for the detection of DNA mismatches and abasic DNA sites. These Pt(II) complexes are able to differentiate between cells with different levels of MMR activity. Importantly, the differential emission responses of the Pt(II) complexes to DNA from human colon cancer tissues and normal colon tissue suggest that the Pt(II) complexes are promising candidates for tumour diagnosis via simple emission spectroscopy measurements of DNA extracts. The selective targeting of DNA mismatches also lends the complexes therapeutic potential for cancer treatment[21]; and they may be used as scaffolds in the design of new anti-cancer agents for the treatment of MMR-deficient cancers.

Methods

General. The synthesis and characterization of the platinum complexes and the experimental details of X-ray crystallographic analysis, emission titration experiments, ultraviolet/visible absorption titration experiments, ITC, ^{1}H NMR titration experiments, melting temperature experiments and molecular docking model experiments are provided in the Supplementary Methods. The nucleic acid sequences for the different experiments are listed in the Supplementary Information.

Quantum mechanics/molecular mechanics. The structures of mismatched DNA and complex **2** were taken from the literature (PDB ID: 2O11[56], unit Å) and the X-ray crystal analysis (CCDC 1061095), respectively. In general, molecular docking studies were performed using ICM-pro software to obtain the initial structure of DNA–complex **2** adducts[57]; subsequently, the binding between the DNA and complex was further optimized using the QM/MM approach and NWChem software[58]. The quantum part (QM) included complex **2** (total 133 atoms), and the rest of the system was modelled at the MM level. The QM region was treated at the density functional theory level using the M06L method[59]. The 6-31G* Pople basis set[60] was employed for C, H, N and P atoms, and for the Pt atom, we applied the LANL2DZ[61] basis set with effective core potential. The mismatched DNA was described with the AMBER parm99 force field[62]. Van der Waals parameters for the Pt atom were taken from the literature[63]. The docked structure was solvated in a 65-Å cubic box with 8,830 classical SPC/E water molecules. Twenty-two-positive (Na^+) counterions were added to neutralize the charges in the system. The structure was optimized using the BFGS algorithm for the QM part and the steepest descent algorithm for the MM part. The optimization of these two regions (QM and MM) was alternated until self-consistency was reached.

DNA extraction from cancer cells or tumour tissues. HCT116 and HCT116N cell lines were generously provided by Thomas A. Kunkel (NIH, USA). The human tissue samples were obtained from Jinan University, Guangzhou, China, and the experimental uses were authorized by the Human Experimentation Ethics Committee of Jinan University. Genomic DNA from cells or tissues was isolated using DNAzol Reagent (Thermo Fisher) according to the manufacturer's protocol. For DNA extraction from cell lines, 1 ml of DNAzol Reagent was added to $1–3 \times 10^7$ cells grown on 10-cm^2 culture plates. Then, the cells were lysed by agitating the culture plate and gently pipetting the lysate into a microfuge tube. For the tissue sample, 1 ml of DNAzol Reagent was added to $\sim$50 mg of tissues. Then, the tissues were homogenized by applying three strokes in the homogenizer. The homogenate was sedimented for 10 min at 10,000g at 4 °C. Following centrifugation, the resulting viscous supernatant was transferred to a fresh tube, mixed with 0.5 ml of 100% ethanol by inversion and stored at room temperature for 1–3 min. Then, the DNA precipitate was washed twice with 1.0 ml of 75% ethanol. The DNA was air-dried and dissolved in 200 µl of 8 mM NaOH, and the pH was adjusted to 7.5 by the addition of HEPES. The emission spectrum of a solution of complex **1c** or **2** (5 µM) in buffer solution (50 mM NaCl and 2 mM Tris, pH 7.5) was recorded and titrated with increasing amounts of the isolated DNA. The emission spectra were recorded after equilibration for 3 min in each addition.

Fluorescence microscopy examination. Cells (HCT116 or NCM460) were seeded in glass bottom culture dishes (35 mm) at a density of 1.0×10^5 ml^{-1}. After 12 h, the dishes were placed on ice, washed twice with 1 ml of cold PBS and then incubated in 1 ml of permeabilization buffer (20 mM HEPES, pH 7.5, 110 mM potassium acetate, 5 mM magnesium sulphate and 250 mM sucrose) containing 40 µg ml^{-1} digitonin for 30 min. After removal of the digitonin solution, the dishes were washed three times with cold permeabilization buffer for 1, 5 and 10 min. The cells were then incubated with complex 2 (10 µM) in transport buffer (20 mM HEPES, pH 7.3, 110 mM potassium acetate, 5 mM sodium acetate, 2 mM magnesium sulphate and 250 mM sucrose) for 15 min. Cells were directly subjected to fluorescence imaging without removing the transport buffer. For co-culture experiments, HCT116 cells were first seeded in culture dishes for 12 h. Next, the dishes were treated with a non-toxic concentration of Hoechst 33342 (1 µg ml^{-1}) for 30 min and washed twice with PBS. Then, NCM460 cells were seeded in the same dishes for another 12 h. Finally, the co-cultured cells were incubated in permeabilization buffer, stained with complex **2** and subjected to fluorescence imaging as aforementioned.

References

1. Erkkila, K. E., Odom, D. T. & Barton, J. K. Recognition and reaction of metallointercalators with DNA. *Chem. Rev.* **99,** 2777–2796 (1999).
2. Metcalfe, C. & Thomas Jim, A. Kinetically inert transition metal complexes that reversibly bind to DNA. *Chem. Soc. Rev.* **32,** 215–224 (2003).
3. Haas, K. L. & Franz, K. J. Application of metal coordination chemistry to explore and manipulate cell biology. *Chem. Rev.* **109,** 4921–4960 (2009).
4. Liu, H.-K. & Sadler, P. J. Metal complexes as DNA intercalators. *Acc. Chem. Res.* **44,** 349–359 (2011).
5. Demas, J. N. & DeGraff, B. A. Applications of luminescent transition platinum group metal complexes to sensor technology and molecular probes. *Coord. Chem. Rev.* **211,** 317–351 (2001).
6. Ma, D.-L., He, H.-Z., Leung, K.-H., Chen, D. S.-H. & Leung, C.-H. Bioactive Luminescent Transition-Metal Complexes for Biomedical Applications. *Angew. Chem. Int. Ed.* **52,** 7666–7682 (2013).
7. Zeglis, B. M., Pierre, V. C. & Barton, J. K. Metallo-intercalators and metallo-insertors. *Chem. Commun.* 4565–4579 (2007).
8. Richards, A. D. & Rodger, A. Synthetic metallomolecules as agents for the control of DNA structure. *Chem. Soc. Rev.* **36,** 471–483 (2007).
9. Lo, K. K. W. in *Photofunctional Transition Metals Complexes*. Vol. 123 (ed. Yam, V. W. W.) 204–245 (Springer-Verlag, 2007).
10. Gao, F., Chao, H. & Ji, L.-N. DNA binding, photocleavage, and topoisomerase inhibition of functionalized ruthenium(II)-polypyridine complexes. *Chem. Biodivers.* **5,** 1962–1979 (2008).
11. Cummings, S. D. Platinum complexes of terpyridine: interaction and reactivity with biomolecules. *Coord. Chem. Rev.* **253,** 1495–1516 (2009).
12. Komor, A. C. & Barton, J. K. The path for metal complexes to a DNA target. *Chem. Commun.* **49,** 3617–3630 (2013).
13. Granzhan, A., Kotera, N. & Teulade-Fichou, M.-P. Finding needles in a basestack: recognition of mismatched base pairs in DNA by small molecules. *Chem. Soc. Rev.* **43,** 3630–3665 (2014).
14. Ma, D.-L., Che, C.-M. & Yan, S.-C. Platinum(II) complexes with dipyridophenazine ligands as human telomerase inhibitors and luminescent probes for G-quadruplex DNA. *J. Am. Chem. Soc.* **131,** 1835–1846 (2009).
15. Murat, P., Singh, Y. & Defrancq, E. Methods for investigating G-quadruplex DNA/ligand interactions. *Chem. Soc. Rev.* **40,** 5293–5307 (2011).
16. Carlson, C. B., Stephens, O. M. & Beal, P. A. Recognition of double stranded RNA by proteins and small molecules. *Biopolymers* **70,** 86–102 (2003).
17. Zou, T. *et al.* Luminescent cyclometalated platinum(II) complex forms emissive intercalating adducts with double-stranded DNA and RNA: differential emissions and anticancer activities. *Angew. Chem. Int. Ed.* **53,** 10119–10123 (2014).
18. Pierre, V. C., Kaiser, J. T. & Barton, J. K. Insights into finding a mismatch through the structure of a mispaired DNA bound by a rhodium intercalator. *Proc. Natl Acad. Sci. USA* **104,** 429–434 (2007).
19. Song, H., Kaiser, J. T. & Barton, J. K. Crystal structure of Δ-[Ru(bpy)$_2$dppz]$^{2+}$ bound to mismatched DNA reveals side-by-side metalloinsertion and intercalation. *Nat. Chem.* **4,** 615–620 (2012).
20. Kinsella, T. J. Coordination of DNA mismatch repair and base excision repair processing of chemotherapy and radiation damage for targeting resistant cancers. *Clin. Cancer Res.* **15,** 1853–1859 (2009).
21. Martin, S. A., Lord, C. J. & Ashworth, A. Therapeutic targeting of the DNA mismatch repair pathway. *Clin. Cancer Res.* **16,** 5107–5113 (2010).
22. Bridge, G., Rashid, S. & Martin, S. A. DNA mismatch repair and oxidative DNA damage: implications for cancer biology and treatment. *Cancers* **6,** 1597–1614 (2014).
23. Lothe, R. A. Microsatellite instability in human solid tumors. *Mol. Med. Today* **3,** 61–68 (1997).
24. Hoeijmakers, J. H. J. Genome maintenance mechanisms for preventing cancer. *Nature* **411,** 366–374 (2001).
25. Loeb, L. A., Loeb, K. R. & Anderson, J. P. Multiple mutations and cancer. *Proc. Natl Acad. Sci. USA* **100,** 776–781 (2003).
26. Iyer, R. R., Pluciennik, A., Burdett, V. & Modrich, P. L. DNA mismatch repair: functions and mechanisms. *Chem. Rev.* **106,** 302–323 (2006).
27. Jackson, B. A., Alekseyev, V. Y. & Barton, J. K. A versatile mismatch recognition agent: specific cleavage of a plasmid DNA at a single base mispair. *Biochemistry* **38,** 4655–4662 (1999).
28. Zeglis, B. M., Pierre, V. C., Kaiser, J. T. & Barton, J. K. A bulky rhodium complex bound to an adenosine-adenosine DNA mismatch: general architecture of the metalloinsertion binding mode. *Biochemistry* **48,** 4247–4253 (2009).
29. Lim, M. H., Song, H., Olmon, E. D., Dervan, E. E. & Barton, J. K. Sensitivity of Ru(bpy)$_2$dppz^{2+} luminescence to DNA defects. *Inorg. Chem.* **48,** 5392–5397 (2009).
30. McConnell, A. J. *et al.* Luminescent properties of ruthenium(II) complexes with sterically expansive ligands bound to DNA defects. *Inorg. Chem.* **51,** 12511–12520 (2012).
31. Chen, H., Yang, P., Yuan, C. & Pu, X. Study on the binding of base-mismatched oligonucleotide d(GCGAGC)$_2$ by cobalt(III) complexes. *Eur. J. Inorg. Chem.* **2005,** 3141–3148 (2005).
32. Fu, C., Harms, K. & Zhang, L. Meggers, E. DNA mismatch recognition by a hexacoordinate silicon sandwich-ruthenium hybrid complex. *Organometallics* **33,** 3219–3222 (2014).
33. Yang, X. L. *et al.* Imidazole-imidazole pair as a minor groove recognition motif for T:G mismatched base pairs. *Nucleic Acids Res.* **27,** 4183–4190 (1999).
34. Nakatani, K., Sando, S., Kumasawa, H., Kikuchi, J. & Saito, I. Recognition of guanine-guanine mismatches by the dimeric form of 2-amino-1,8-naphthyridine. *J. Am. Chem. Soc.* **123,** 12650–12657 (2001).

35. David, A. *et al.* DNA mismatch-specific base flipping by a bisacridine macrocycle. *Chembiochem.* **4,** 1326–1331 (2003).
36. Chin, K. H., Chen, F. M. & Chou, S. H. Solution structure of the ActD-5′-CCGTT$_3$GTGG-3′ complex: drug interaction with tandem G · T mismatches and hairpin loop backbone. *Nucleic Acids Res.* **31,** 2622–2629 (2003).
37. Nakatani, K. *et al.* Small-molecule ligand induces nucleotide flipping in (CAG)n trinucleotide repeats. *Nat. Chem. Biol.* **1,** 39–43 (2005).
38. Ono, A. *et al.* Specific interactions between silver(I) ions and cytosine-cytosine pairs in DNA duplexes. *Chem. Commun.* 4825–4827 (2008).
39. Jourdan, M., Granzhan, A., Guillot, R., Dumy, P. & Teulade-Fichou, M.-P. Double threading through DNA: NMR structural study of a bis-naphthalene macrocycle bound to a thymine-thymine mismatch. *Nucleic Acids Res.* **40,** 5115–5128 (2012).
40. Takada, T., Ashida, A., Nakamura, M. & Yamana, K. Cationic perylenediimide as a specific fluorescent binder to mismatch containing DNA. *Biorg. Med. Chem.* **21,** 6011–6014 (2013).
41. Lo, Y.-S., Tseng, W.-H., Chuang, C.-Y. & Hou, M.-H. The structural basis of actinomycin D-binding induces nucleotide flipping out, a sharp bend and a left-handed twist in CGG triplet repeats. *Nucleic Acids Res.* **41,** 4284–4294 (2013).
42. Jennette, K. W., Lippard, S. J., Vassiliades, G. A. & Bauer, W. R. Metallointercalation reagents. 2-Hydroxyethanethiolato(2,2′,2″-terpyridine)-platinum(II) monocation binds strongly to DNA by intercalation. *Proc. Natl Acad. Sci. USA* **71,** 3839–3843 (1974).
43. Che, C.-M., Yang, M., Wong, K.-H., Chan, H.-L. & Lam, W. Platinum(II) complexes of dipyridophenazine as metallointercalators for DNA and potent cytotoxic agents against carcinoma cell lines. *Chem. Eur. J.* **5,** 3350–3356 (1999).
44. Wang, P. *et al.* Specific blocking of CREB/DNA binding by cyclometalated platinum(II) complexes. *Angew. Chem. Int. Ed.* **50,** 2554–2558 (2011).
45. Frezza, M. *et al.* *In vitro* and *in vivo* antitumor activities and DNA binding mode of five coordinated cyclometalated organoplatinum(II) complexes containing biphosphine ligands. *J. Med. Chem.* **54,** 6166–6176 (2011).
46. Sun, R. W.-Y. *et al.* Luminescent cyclometalated platinum(II) complexes containing N-Heterocyclic carbene ligands with potent *in vitro* and *in vivo* anti-cancer properties accumulate in cytoplasmic structures of cancer cells. *Chem. Sci.* **2,** 728–736 (2011).
47. Zeglis, B. M. & Barton, J. K. A mismatch-selective bifunctional rhodium-oregon green conjugate: a fluorescent probe for mismatched DNA. *J. Am. Chem. Soc.* **128,** 5654–5655 (2006).
48. Cordier, C., Pierre, V. C. & Barton, J. K. Insertion of a bulky rhodium complex into a DNA cytosine-cytosine mismatch: an NMR solution study. *J. Am. Chem. Soc.* **129,** 12287–12295 (2007).
49. SantaLucia, J. & Hicks, D. The thermodynamics of DNA structural motifs. *Annu. Rev. Biophys. Biomol. Struct.* **33,** 415–440 (2004).
50. Zeglis, B. M., Boland, J. A. & Barton, J. K. Targeting abasic sites and single base bulges in DNA with metalloinsertors. *J. Am. Chem. Soc.* **130,** 7530–7531 (2008).
51. Tong, G. S.-M. & Che, C.-M. Emissive or nonemissive? A theoretical analysis of the phosphorescence efficiencies of cyclometalated platinum(II) complexes. *Chem. Eur. J.* **15,** 7225–7237 (2009).
52. Lo, K. K.-W., Choi, A. W.-T. & Law, W. H.-T. Applications of luminescent inorganic and organometallic transition metal complexes as biomolecular and cellular probes. *Dalton Trans.* **41,** 6021–6047 (2012).
53. Koi, M. *et al.* Human chromosome 3 corrects mismatch repair deficiency and microsatellite instability and reduces *N*-Methyl-*N*′-nitro-*N*-nitrosoguanidine tolerance in colon tumor cells with homozygous *hMLH1* mutation. *Cancer Res.* **54,** 4308–4312 (1994).
54. Glaab, W. E. & Tindall, K. R. Mutation rate at the hprt locus in human cancer cell lines with specific mismatch repair-gene defects. *Carcinogenesis* **18,** 1–8 (1997).
55. Moyer, M., Manzano, L., Merriman, R., Stauffer, J. & Tanzer, L. NCM460, a normal human colon mucosal epithelial cell line. *In Vitro Cell. Dev. Biol. Anim.* **32,** 315–317 (1996).
56. Pierre, V. C., Kaiser, J. T. & Barton, J. K. Insights into finding a mismatch through the structure of a mispaired DNA bound by a rhodium intercalator. *Proc. Natl Acad. Sci. USA* **104,** 429–434 (2007).
57. Neves, M. A. C., Totrov, M. & Abagyan, R. Docking and scoring with ICM: the benchmarking results and strategies for improvement. *J. Comput. Aided Mol. Des.* **26,** 675–686 (2012).
58. Valiev, M. *et al.* NWChem: A comprehensive and scalable open-source solution for large scale molecular simulations. *Comput. Phys. Commun.* **181,** 1477–1489 (2010).
59. Zhao, Y. & Truhlar, D. G. A new local density functional for main-group thermochemistry, transition metal bonding, thermochemical kinetics, and noncovalent interactions. *J. Chem. Phys.* **125,** 194101 (2006).
60. Hariharan, P. C. & Pople, J. A. The influence of polarization functions on molecular orbital hydrogenation energies. *Theor. Chim. Acta* **28,** 213–222 (1973).
61. Hay, P. J. & Wadt, W. R. Ab initio effective core potentials for molecular calculations. Potentials for K to Au including the outermost core orbitals. *J. Chem. Phys.* **82,** 299–310 (1985).
62. Cornell, W. D. *et al.* A second generation force field for the simulation of proteins, nucleic acids, and organic molecules. *J. Am. Chem. Soc.* **117,** 5179–5197 (1995).
63. Hambley, T. W. van der Waals radii of Pt(II) and Pd(II) in molecular mechanics models and an analysis of their relevance to the description of axial M · · · H(− C), M · · · H(− N), M · · · S, and M · · · M (M = Pd(II) or Pt(II)) interactions. *Inorg. Chem.* **37,** 3767–3774 (1998).

Acknowledgements

This work was supported by the University Grants Committee of the HKSAR of China (Area of Excellence Scheme AoE/P-03/08), the National Key Basic Research Program of China (2013CB834802), the Innovation and Technology Fund (ITF-Tier 2 ITS/130/14FP) and Special Equipment Grant of UGC (SEG_HKU02) for MS analysis. We thank Thomas A. Kunkel (NIH, USA) for providing the colon cancer cell lines, and we are grateful to Kong Hung Sze and Siu-Cheong Yan for help with 2D NMR experiments.

Author contributions

T.Z. and C.-M.C. designed the research. S.K.F., B.C., T.C., W.-P.T. and C.Y. performed the research. S.K.F., T.Z., C.-N.L and C.-M.C. analysed the data, and S.K.F., T.Z. and C.-M.C. wrote the paper.

Additional information

Competing financial interests: The authors declare no competing financial interests.

16

Deoxynybomycins inhibit mutant DNA gyrase and rescue mice infected with fluoroquinolone-resistant bacteria

Elizabeth I. Parkinson[1], Joseph S. Bair[1], Bradley A. Nakamura[1], Hyang Y. Lee[1], Hani I. Kuttab[1], Emma H. Southgate[1], Stéphane Lezmi[2], Gee W. Lau[2] & Paul J. Hergenrother[1]

Fluoroquinolones are one of the most commonly prescribed classes of antibiotics, but fluoroquinolone resistance (FQR) is widespread and increasing. Deoxynybomycin (DNM) is a natural-product antibiotic with an unusual mechanism of action, inhibiting the mutant DNA gyrase that confers FQR. Unfortunately, isolation of DNM is difficult and DNM is insoluble in aqueous solutions, making it a poor candidate for development. Here we describe a facile chemical route to produce DNM and its derivatives. These compounds possess excellent activity against FQR methicillin-resistant *Staphylococcus aureus* and vancomycin-resistant *Enterococci* clinical isolates and inhibit mutant DNA gyrase *in-vitro*. Bacteria that develop resistance to DNM are re-sensitized to fluoroquinolones, suggesting that resistance that emerges to DNM would be treatable. Using a DNM derivative, the first *in-vivo* efficacy of the nybomycin class is demonstrated in a mouse infection model. Overall, the data presented suggest the promise of DNM derivatives for the treatment of FQR infections.

[1] Roger Adams Laboratory, Department of Chemistry, University of Illinois at Urbana-Champaign, 600 South Mathews Avenue, Urbana, Illinois 61801, USA. [2] College of Veterinary Medicine, Veterinary Medicine Basic Sciences Building, University of Illinois at Urbana-Champaign, 2001 South Lincoln Avenue, Urbana, Illinois 61802, USA. Correspondence and requests for materials should be addressed to P.J.H. (email: hergenro@illinois.edu).

Fluoroquinolones (FQs) were introduced into the clinic in the early 1980s and since then have become one of the most widely prescribed classes of antibiotics[1–3]. Although early FQs were primarily used to treat Gram-negative infections, later generation FQs were also commonly employed against infections caused by Gram-positive pathogens[1,4]. FQs are prescribed for severe or antibiotic-resistant urinary tract infections, respiratory tract infections[2,4], gonoccocal infections, tuberculosis and as a prophylactic for anthrax[5]. FQs act by inhibiting bacterial type IIA topoisomerases, specifically DNA gyrase (composed of GyrA and B subunits) and topoisomerase IV (composed of ParC and E subunits). These enzymes catalyse the introduction of negative supercoils and the decatenation of interlinked chromosomes, respectively[6–8].

Although FQs have demonstrated great utility in the clinic, their widespread use has resulted in significant resistance[4]. Nearly all vancomycin-resistant enterococcus (VRE) and methicillin-resistant *S. aureus* (MRSA) are also resistant to FQs[9]; thus, FQs can no longer be used to treat such infections. FQs are commonly prescribed for *Neiserria gonorrhoeae* and *Pseudomonas aeruginosa* infections, but FQ resistance (FQR) is now observed in a substantial fraction of these isolates, necessitating other treatments[10]. Target-site mutation is the major contributor to FQR[1,4], with high-level resistance observed in bacteria possessing key mutations in both GyrA and ParC[4]. VRE and MRSA both harbour these target-site mutations, with point mutations in the quinolone resistance-determining region (QRDR) of the GyrA subunit of DNA gyrase and the ParC subunit of topoisomerase IV. These mutations alter residues important for the binding of FQs, resulting in an approximately tenfold decrease in binding affinity[11,12]. Nearly 100% of MRSA substitute Ser84 of GyrA with Leu[13–21]. Similarly, nearly all FQR VRE substitute Ser83 of GyrA with Ile, Arg or Tyr[22–26].

Nybomycin (NM) is a natural product first identified from a culture of a streptomycete isolated from a Missouri soil sample and found to have antibacterial activity[27,28]. During efforts to determine its structure, Rinehart and Renfroe[29] synthesized a related compound, deoxynybomycin (DNM, Fig. 1a), which was later found to be a natural product and to have more potent activity than NM against a range of bacteria[29,30]. More recently, DNM was found to have activity against FQR MRSA with the S84L mutation in GyrA of DNA gyrase[15]. However, isolation of NM and DNM from natural sources is non-trivial [31] and the only reported total synthesis of DNM is very low yielding[32,33]. In addition, the low solubility of DNM in any solvent other than concentrated acid presents challenges for its biological evaluation and limits its potential *in-vivo*.

Described herein is an efficient total synthesis of DNM and modifications of this route are used to construct the first DNM derivatives. DNM and several of the derivatives show outstanding antibacterial potency and selectivity against FQR MRSA and VRE clinical isolates. DNM and its derivatives inhibit the mutant DNA gyrase responsible for FQR and resistance to DNM and derivatives results in re-sensitization to FQs, suggesting a resistance cycling that could be useful in the clinic. Finally, using a DNM derivative with superior solubility and pharmacokinetic properties, the first *in-vivo* activity of this class of compounds is demonstrated.

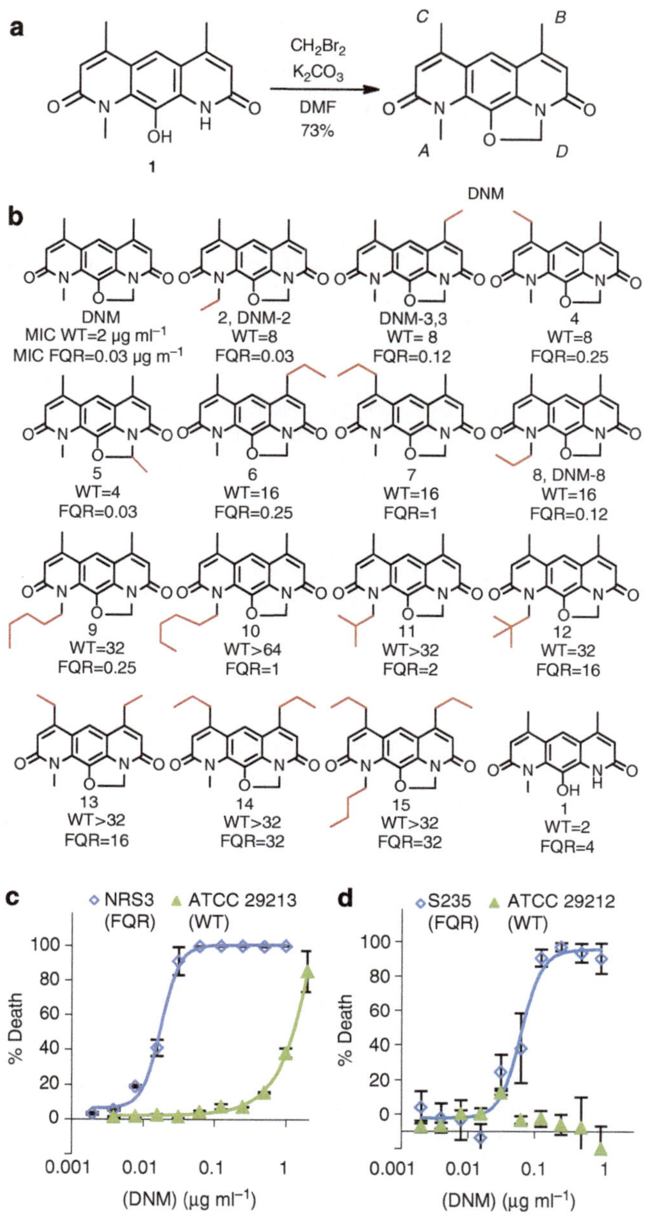

Figure 1 | Synthesis and antibacterial activity of DNM and derivatives. (**a**) Final step in the synthesis of DNM. The letters A, B, C and D around the structure of DNM denote sites of derivitization. (**b**) Structure of DNM and derivatives, and their activities against WT *S. aureus* (ATCC 29213, WT) and FQR *S. aureus* (NRS3, FQR). Activity is from three independent replicates of the microdilution broth assay and is reported as the MIC in μg ml^{-1}. (**c**) Dose-response curves for FQS *S. aureus* (ATCC 29213) and FQR *S. aureus* (NRS3) treated with DNM. Data shown are from three independent replicates ± s.e.m. (**d**) Dose-response curves for FQS *Enterococcus* (ATCC 29212) and FQR *Enterococcus* (S235) treated with DNM. Data shown are from three independent replicates ± s.e.m.

Results

Total synthesis of DNM and construction of derivatives. Owing to the documented difficulty of isolating DNM from natural sources[31], we aimed to develop an efficient, modular and flexible synthesis of DNM that could also be used to construct derivatives. Previously, we reported a synthesis of the natural product deoxynyboquinone (DNQ) that relies on a mixed Suzuki cross-coupling followed by a palladium-catalysed ring closing and deprotection to give diazaanthracenol **1** (Fig. 1a)[34]. To construct DNQ, **1** is oxidized to give the desired quinone[34]. We found that **1** could be converted to DNM in a single step by insertion of the methylene bridge in a reaction inspired by Rinehart's degradation studies and by bridge insertions in similar systems[32,33,35]. Reaction of **1** with dibromomethane gave DNM in a 73% yield (Fig. 1a). Through this route, DNM was obtained in seven steps with an overall yield of 11%, an improvement over the only other reported total synthesis (0.84% overall yield)[32,33].

This flexible synthetic route also allowed for rapid generation of a variety of derivatives that have not been found as natural products. We hypothesized that addition of alkyl chains would disrupt π-stacking between DNM molecules, thus increasing both aqueous and organic solubility, similar to what was observed with DNQ derivatives[36]. By changing the iodoamides used in the Suzuki cross-coupling (see General protocol A in Supplementary Methods), three compounds were synthesized that substituted ethyl for methyl at positions *A*, *B* and *C* (compounds **2**, **3**, and **4**, respectively, Fig. 1b). The derivative with a methyl substitution at *D* was generated by using 1,1-dibromoethane in place of dibromomethane in the final step, to provide compound **5** (Fig. 1b). Other compounds with single sites of derivatization (**6–12**) and multiple sites of derivatization (**13–15**) were also constructed. Full synthetic routes along with experimental details and characterization data can be found in the Supplementary Methods. Compounds with small alkyl appendages have markedly improved solubility (3- to 13-fold) in pH 7.4 PBS relative to DNM (Supplementary Table 1) and all compounds synthesized have improved dimethyl sulfoxide (DMSO) solubility compared with the parent compound (Supplementary Table 1).

Evaluation of DNM and derivatives against FQR MRSA and VRE. DNM was evaluated against both FQ-sensitive *S. aureus* (ATCC 29213) and FQR MRSA (NRS3, which has GyrA S84L and ParC S80F). DNM showed modest activity against the FQ-sensitive (FQS) strain 29213. However, DNM showed excellent activity against the FQR NRS3 (Minimum Inhibitory Concentration, MIC = 0.03 μg ml^{-1}, Fig 1b,c). This MIC compares favourably with standard-of-care treatments for Gram-positive infections including vancomycin (MIC for NRS3 = 8 μg ml^{-1}), daptomycin (MIC for NRS3 = 8 μg ml^{-1}) and linezolid (MIC for NRS3 = 0.5 μg ml^{-1}). The sensitivity of FQR VRE was also explored. DNM had no detectable activity against FQS *Enterococcus* (ATCC 29212, MIC > 1.0 μg ml^{-1}), but it potently inhibited the growth of FQR VRE (clinical isolate S235, which has GyrA S83I and ParC S80I, MIC = 0.125 μg ml^{-1}, Fig. 1d). DNM was also evaluated against a panel of Gram-negative bacteria (Supplementary Table 2). It showed no detectable activity against wild-type (WT) or FQR *P. aeruginosa* or *Acinetobacter baumannii*. Moderate activity was seen with a DNM derivative against a permeabilized strain of *Escherichia coli*, suggesting that DNM and its derivatives are unable to penetrate Gram-negative bacteria.

DNM derivatives were evaluated against both FQS *S. aureus* (ATCC 29213) and FQR MRSA (NRS3), and their MIC values are listed in Fig. 1b. Similar to DNM, most derivatives showed significantly enhanced activity against FQR NRS3 compared with FQS 29213. In general, compounds with a single methyl addition retained good activity against NRS3 (**2–5**). Further substitution at *B* was relatively well tolerated (**6**), while substitution at *C* was generally less well tolerated (**7**). Compounds possessing longer chains at *A* generally retained potency (**8–10**). However, compounds with bulky substitutions at *A* (**11–12**), multiple substitutions (**13–15**), or without the methylene bridge (**1**) were markedly less active.

Activity of DNM or derivatives against FQR clinical isolates. DNM and two of the most potent derivatives (DNM-2 and DNM-8) were evaluated against a panel of MRSA and VRE clinical isolates (Figs 2a,b). As shown in Fig. 2, all MRSA and VRE strains were sensitive to these compounds and resistant to ciprofloxacin (CIP). To understand this selectivity, the QRDRs of GyrA and ParC for many of these isolates were sequenced

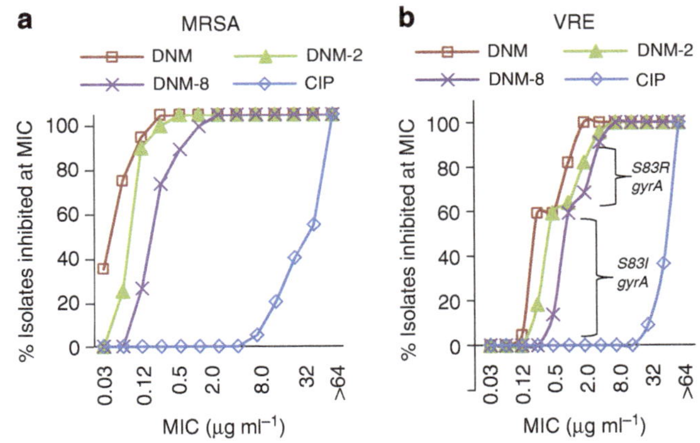

Figure 2 | Sensitivity of MRSA and VRE clinical isolates to DNM, DNM-2, DNM-8 and CIP. **(a)** The percentage of MRSA clinical isolates (n = 21) with an MIC at or lower than the concentration shown. **(b)** The percentage of VRE clinical isolates (n = 22) with an MIC at or lower than the concentration shown.

(Supplementary Tables 3 and 4). Although different substitution patterns were found for MRSA ParC, all sequenced strains have the same mutation in GyrA (S84L), consistent with the notion that this mutation sensitizes bacteria to DNM. Similar to the MRSA isolates, the VRE isolates have many different substitutions in ParC, which do not appear to correlate with sensitivity. Unlike the MRSA isolates, the majority of VRE isolates have two different substitutions for GyrA (S83I or S83R). The sensitivity of these strains is affected by this substitution with VRE harbouring the S83I mutation being very sensitive to DNM (MIC = 0.125–1 μg ml^{-1}) and those with the S83R mutation being less sensitive (MIC ≥ 1 μg ml^{-1}). The activity of the DNM-2 and DNM-8 against these panels of clinical isolates closely mirrors that of DNM (Fig. 2a,b). Full details of the sensitivity of each strain can be found in Supplementary Tables 3 and 4.

Inhibition of mutant DNA gyrase by DNM and derivatives. To further investigate the importance of the GyrA mutation for DNM activity, the ability of DNM, DNM-2 and CIP to inhibit DNA gyrase was determined by using an *in-vitro* DNA cleavage assay. In this assay, DNA gyrase is coincubated with supercoiled DNA and the compound of interest. Inhibition at the cleavage complex of DNA gyrase leads to an increase in either doubly nicked linear (L) DNA (for example, inhibition by CIP[37]) or singly nicked open circular (OC) DNA (for example, inhibition by GSK299423 (ref. 38)). GSK299423 is a recently discovered DNA gyrase inhibitor that is hypothesized to stabilize the DNA–enzyme complex, either pre-cleavage or after the formation of a single-strand break resulting in a buildup of OC DNA[38]. Similar to previous studies, we found that CIP potently inhibits WT DNA gyrase with a greater than sevenfold increase in L DNA being observed at concentrations as low as 0.68 μM (Fig. 3a, full gels in Supplementary Fig. 1, quantification in Supplementary Fig. 2). In addition, in a time-course assay, inhibition of WT DNA gyrase by CIP resulted in a time-dependent buildup of L DNA (Fig. 3b, full gels in Supplementary Fig. 3). Alternatively, when either DNM or DNM-2 was incubated with WT DNA gyrase, neither showed similar increases in L or OC DNA, suggesting that these compounds are poor inhibitors of WT DNA gyrase (Fig. 3a). In addition, during the time-course study with these compounds, buildup of L DNA was only observed at later time points and to a smaller degree (Fig. 3b and Supplementary Fig. 2).

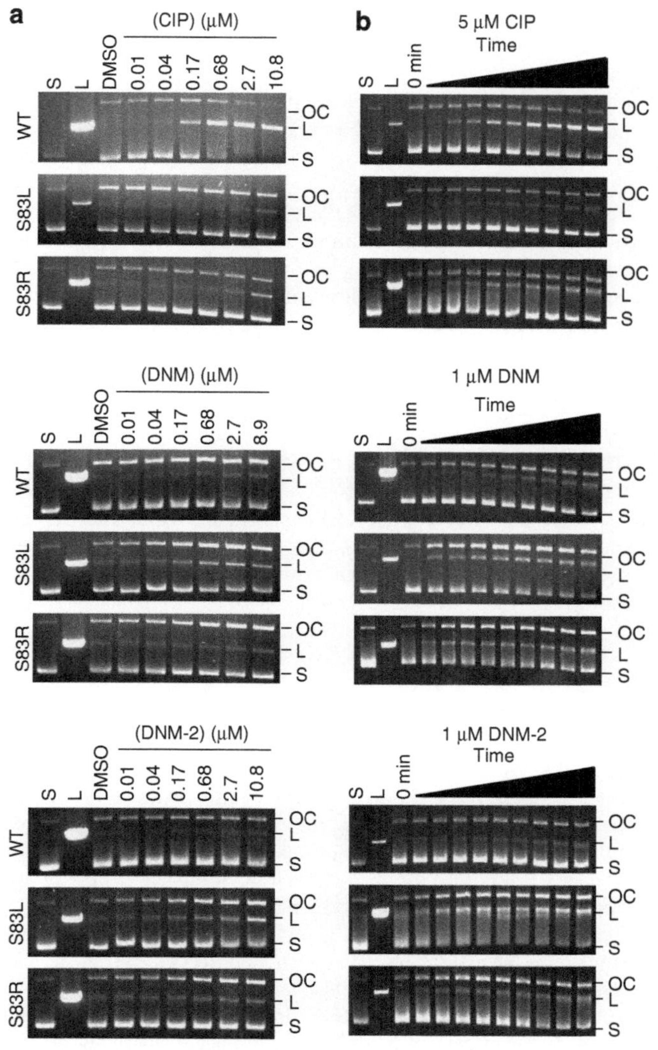

Figure 3 | Inhibition of WT and mutant DNA gyrase. (**a**) DNA cleavage assay with WT, S83L and S83R *E. coli* DNA gyrase in the presence of increasing concentrations of CIP, DNM and DNM-2. Concentrations were 0.01, 0.04, 0.017, 0.68, 2.7 and 10.8 μM, except for DNM, which was 8.9 μM for the highest concentration. S, supercoiled; L, linear; OC, open circular or nicked DNA. (**b**) Time course of DNA cleavage with WT, S83L and S83R *E. coli* DNA gyrase in the presence of 5 μM CIP, 1 μM DNM and 1 μM DNM-2. Time points were 0, 1, 3, 5, 10, 15, 20, 30, 60 and 90 min. All gels are representative data from at least three independent experiments.

The ability of CIP, DNM and DNM-2 to inhibit S83L or S83R DNA gyrase was then determined. CIP was much less effective at inhibiting either S83L or S83R DNA gyrase compared with WT, with only small increases in L DNA being observed (Fig. 3a). In addition, minimal change was observed with 5 μM CIP at up to 1.5 h (Fig. 3b). Time-course studies performed with an increased concentration of CIP (200 μM) and S83L DNA gyrase revealed a similar pattern of inhibition to that of WT DNA gyrase, suggesting that the residual inhibition goes through a similar mechanism (Supplementary Fig. 4a). DNM and DNM-2 induce only small increases in L DNA with S83L DNA gyrase (Fig. 3a). Instead, DNM inhibition of S83L DNA gyrase led to a significant buildup of OC DNA at 0.17 μM ($P<0.05$), with a similar increase observed for DNM-2 (Fig. 3a and Supplementary Fig. 2). This buildup does not diminish over time (Fig. 3b), suggestive of a mode of inhibition more similar to GSK 299423 than to CIP. Increasing concentrations of DNM-2 to 200 μM and increasing the time up to 2 h confirmed that this OC buildup is not a fleeting event as occurs with CIP (Supplementary Fig. 4b). DNM or DNM-2 inhibition of S83R also led to a buildup of OC DNA similar to that seen with S83L DNA gyrase only at a slightly higher concentration or longer time points, consistent with the activity of these compounds against VRE with the S83R DNA gyrase. Overall, these results are consistent with the clinical isolate data, supporting the critical importance of a mutant DNA gyrase for sensitizing bacteria to DNM. Finally, to determine the selectivity of DNM and derivatives for bacterial DNA gyrase, a decatenation assay with human topoisomerase II was performed. Although doxorubicin inhibited human topoisomerase II at concentrations as low as 3 μM, DNM-2 showed no significant inhibition at concentrations up to 30 μM (Supplementary Fig. 5).

Development of resistance to DNM. To explore the development of resistance to both CIP and DNM in *S. aureus*, resistant strains of ATCC 29213 were generated. Consistent with previous reports[39,40], high-level resistance to CIP was generally not achieved in a single step. Instead, low-level resistance (CIP MIC = 4–8 μg ml^{-1}) was usually achieved with the first step and corresponded to a mutation in ParC (for example, E84K or S80F; Fig. 4a and Supplementary Fig. 6). Unsurprisingly, these low-level resistant strains that do not have the S84L mutation in GyrA were not sensitive to DNM or DNM-2. Development of high-level CIP resistance (CIP MIC = 16–64 μg ml^{-1}) similar to what is often seen in clinical isolates[13,15,17] was observed at the second step and corresponded to an S84L mutation in GyrA. These high-level CIP-resistant strains were extremely sensitive to DNM (MIC = 0.03–0.06 μg ml^{-1}) and DNM-2 (MIC = 0.06–0.12 μg ml^{-1}). These FQR bacteria were then exposed to DNM in an effort to create DNM-resistant isolates. The development of DNM resistance in high-level CIP-resistant strains was a rare event, with resistance frequencies ranging from 1×10^{-10} to 7×10^{-10} (Supplementary Fig. 6). When these strains were found, they showed dramatically improved sensitivity to CIP (MIC = 0.25–8.0 μg ml^{-1}). All these strains had reverted to WT GyrA (Ser84), with the more CIP-sensitive strains also having WT ParC and the less CIP-sensitive strains retaining ParC mutations (Supplementary Fig. 6). This complete cycle of complementary resistance/sensitivity of CIP and DNM is shown in Fig. 4a and the complete list of resistant strains generated and the sequences of their QRDR is shown in Supplementary Fig. 6.

Resistance development on co-treatment with CIP and DNM-2 was then explored (Supplementary Fig. 7). A low-level CIP-resistant strain (29213-C1) was used in these studies. On treatment with either CIP or DNM-2, resistant colonies were observed. However, no colonies were observed on co-treatment (resistance frequency $<1.0\times10^{-10}$).

***In-vivo* efficacy of DNM-2.** As a prelude to exploring *in-vivo* efficacy, the toxicity and pharmacokinetic profile of DNM and key derivatives was evaluated. Treatment of red blood cells (RBCs) with DNM and key derivatives indicated that none of these compounds induce haemolysis (Supplementary Fig. 8a). In addition, DNM-2 demonstrated no significant DNA intercalation at concentrations up to 30 μM (Supplementary Fig. 8b). This data combined with previously published data showing that DNM is non-toxic to normal (that is, non-cancerous) cell lines[41] suggests that these compounds would probably be well tolerated *in-vivo*. Treatment of mice with increasing concentrations of DNM, DNM-2 and DNM-3 showed that all three compounds were well tolerated up to the highest dose evaluated (50 mg kg^{-1} by oral gavage). Pharmacokinetic studies were next performed on DNM, DNM-2 and DNM-3. Although DNM itself showed very low

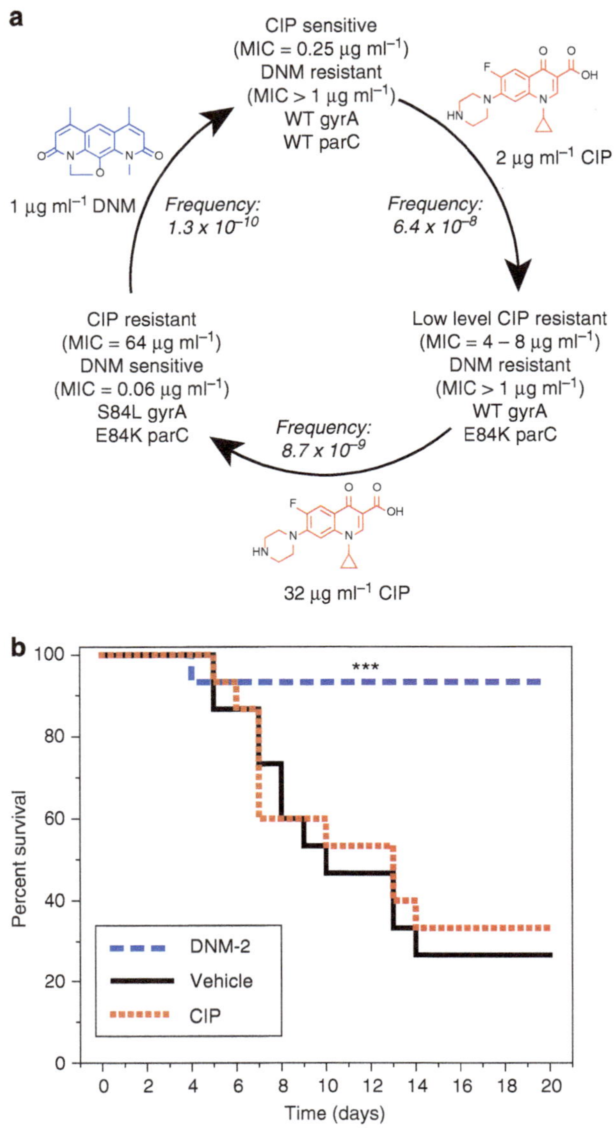

Figure 4 | Resistance mechanisms and *in-vivo* activity. (**a**) Representative data of the resistance cycle observed when bacteria are sequentially treated with CIP then DNM. Each strain generated is listed with the MIC of CIP (red), DNM (blue), as well as mutation status of the QRDR of GyrA (green) and ParC (black) shown below. The selection pressure used in each step is shown over the arrow along with the mutation frequency. (**b**) Kaplan-Meier curves showing the survival rates of mice infected with MRSA (NRS3, FQR). The mice received vehicle alone, 50 mg kg^{-1} CIP, or 50 mg kg^{-1} DNM-2 by oral gavage once daily for 10 days; $n = 15$ for each group. ***$P < 0.005$ versus vehicle and CIP, log rank survival test.

serum exposure ($C_{max} < 0.20$ µM or 0.060 µg ml^{-1}) after a 50-mg kg^{-1} oral dose, DNM-2 showed good bioavailability with a peak serum concentration of 42.6 µM (12.8 µg ml^{-1}) and an area under the curve of 44 h µg ml^{-1} (Supplementary Fig. 9 and Supplementary Table 5). DNM-3 showed an intermediate level of bioavailability with a peak serum concentration of 4.3 µM (1.26 µg ml^{-1}) and an area under the curve of 4 h µg ml^{-1}. The bioavailability of these compounds mirrors the aqueous solubility, suggesting that at least for this limited set of compounds aqueous solubility could be a reasonable predictor of oral bioavailability.

To explore the effect of sustained treatments *in-vivo*, DNM-2 was administered to mice once daily for 10 days (via oral gavage at 50 mg kg^{-1}) and markers of haematological and non-haematological toxicity were examined. No clinically significant evidence for myelosuppression, renal injury or hepatic toxicity was identified (Supplementary Table 6). Kidney, brain, lung, liver, spleen, heart, and stomach sample examinations also did not reveal any evidence of toxicity. In small intestine sections, mild intestinal dilation associated with villi atrophy were noted (Supplementary Fig. 10). Increased vacuolation of white and brown adipocytes with a minimal increase in triglyceride levels were noted as well. These changes were probably indirectly related to the drug and possibly due to the antibiotic effects on the intestinal flora. As none of the mice showed any clinical symptoms, these changes were considered of minimal significance. With this indication that DNM-2 offered good exposure on oral dosing with no observable toxicity, an *in-vivo* model of mouse sepsis was conducted. Mice were infected with FQR MRSA (NRS3) via tail vein injection. Mice were treated with CIP (50 mg kg^{-1}, oral gavage), DNM-2 (50 mg kg^{-1}, oral gavage) or vehicle control once daily for 10 days. As shown by the Kaplan–Meier survival curve in Fig. 4b, mice treated with DNM-2 showed a significant survival difference relative to both CIP and vehicle-treated control ($P < 0.005$, Fig. 4b).

Discussion

Using a synthetic route building on a palladium-catalysed mixed cross-coupling and a methylene bridge insertion, hundreds of milligrams of the natural-product DNM were prepared as described herein. The modular nature of the synthesis also allows access to non-natural DNM derivatives, many of which display similar antibacterial efficacy but with significantly better solubility properties than the parent. Specifically, small alkyl appendages greatly improve aqueous solubility: DNM has aqueous solubility of only 9 µM compared with 121 µM for DNM-2. This improved solubility is probably because of the ability of the short alkyl chains to break up π-stacking similar to what was observed with derivatives of DNQ[36,42]. Longer or more alkyl chains do not display similar increases in aqueous solubility probably because of the increased hydrophobicity of these compounds. With these compounds in hand, a structure–activity relationship was established and derivatives were found to have comparable potencies to the parent against FQR MRSA and VRE in cell culture and against mutated DNA gyrase *in-vitro*.

The major mechanism of FQR for bacteria involves the mutation of FQ targets DNA gyrase and topoisomerase IV. Although nearly all FQR bacteria found to date have such mutations, the exact mutation can vary based on the bacterium. For MRSA, nearly 100% have the S84L mutation in DNA gyrase [13–21]. Similarly, *Bacillus anthracis*, *E. coli* and *A. baumannii* also have the analogous serine mutated to leucine[43–45]. In VRE the serine is mutated to multiple different residues (Ile, Arg and Tyr), while in *Streptococcus pneumoniae*, *Klebsiella pneumonia* and *N. gonorrhoeae* this Ser is often changed to either Phe or Tyr[22–26,46–48]. *P. aeruginosa* differs in that it naturally has a Thr instead of the Ser. However, *P. aeruginosa* is similar to VRE in that the Thr is mutated to an Ile in FQR strains. We now show that DNM has excellent activity versus Ser→Leu or Ser→Ile mutants with moderate activity in Ser→Arg strains, with these cell culture results correlating with the potencies of the compounds in *in-vitro* assays with DNA gyrase and mutants. The ability of DNM to target these different mutants suggests that this natural product or its derivatives could be broadly applicable against FQR bacteria regardless of the exact nature of the Ser mutation.

FQs inhibit DNA gyrase causing double-stranded breaks that appear as a buildup of L DNA in a DNA gyrase cleavage assay (Fig. 3 and ref. 37). Kampranis and Maxwell[37] have demonstrated that FQs initially stabilize a single phosphotyrosine bond as evidenced by an initial buildup of OC DNA, also observed in Fig. 3b. However, the FQ stabilization of a single-strand break causes an even faster second cleavage event that is also stabilized by FQs, thus explaining the rapid buildup of linear DNA[37]. Another DNA gyrase inhibitor, GSK299423, acts via a different mechanism[38]. Unlike FQs that bind within the two active sites, it binds between the active sites stabilizing either an uncleaved or a single-stranded cleaved DNA. The stabilization induced by GSK299423 differs from that of CIP in that it does not result in a second cleavage event and instead causes a buildup of OC DNA. The more potent activity of DNM against FQR-mutant DNA gyrase suggests that DNM probably binds similarly to FQs, near the mutated residues and thus near the two active sites. Despite this similarity in binding position, the phenotype of DNM in the DNA cleavage assay (that is, the buildup of OC DNA) suggests that its overall mode of inhibition is more similar to that of GSK299423. The mutational status of ParC does not affect sensitivity to DNM, as shown by the data from the clinical isolates and resistance mutants.

In this study, we showed that an entire resistance/sensitization cycle is possible beginning with *S. aureus* (ATCC 29213) that is FQS/DNM resistant. After multiple rounds of selection against CIP, a FQR/DNM-sensitive strain was generated. Next, after a selection round with DNM, a FQS/DNM-resistant strain was found. These results suggest intriguing clinical possibilities for DNM, either alone or in combination with FQs. As surveillance data show the ubiquity of FQR in MRSA and VRE, a DNM compound could be an outstanding therapeutic option for these infections; indeed, a new orally available treatment for MRSA and VRE is a well-recognized clinical need[49] and would be a welcome addition to the antibiotic arsenal. As resistance to DNM emerges, the data predict that such bacteria would be sensitive to FQs. At this point, a diagnostic test could be used to choose between FQ or DNM, or a co-treatment with both a FQ and DNM would be possible. As reported herein, co-resistance was not generated in cell culture, with no colonies being observed on treatment with $4\,\mu g\,ml^{-1}$ CIP and $6\,\mu g\,ml^{-1}$ DNM-2.

Before this study, little to no data existed about the administration of DNM to animals. However, a related but less potent compound NM ($C = CH_2OH$) has been examined in mice. It was found to be well tolerated when dosed either subcutaneously, orally or by intraperitoneal injection[27]. However, it showed no activity in mice infected with various bacteria (*K. pneuomoniae*, *S. aureus* or *Mycobacterium tuberculosis*), leading Brock and Sokolski[27] to suggest that this high tolerability and lack of efficacy is probably a result of the very poor solubility of NM (similar to DNM, it is only soluble in concentrated acid) and thus lack of absorption. We demonstrated that DNM also has a high maximal tolerated dose ($>50\,mg\,kg^{-1}$ oral gavage), but pharmacokinetic analysis indicate that it is not absorbed to any appreciable degree. However, DNM-2, which has improved solubility, was also very well tolerated (maximal tolerated dose $>50\,mg\,kg^{-1}$) and showed favourable pharmacokinetic properties when dosed orally. In addition, in this study we found that orally administered DNM-2 is effective in treating mice infected with MRSA, thus showing the first *in-vivo* efficacy for this class of compounds.

FQR pathogens are now a significant medical problem and the data presented herein reveal the considerable translational potential of DNM derivatives, including the following: (1) a short and efficient synthetic route has been developed that can readily supply large amounts of compound. (2) DNM-2 has outstanding pharmacokinetic (PK) properties with a peak serum concentration ($\sim 50\,\mu M$ when given at $50\,mg\,kg^{-1}$ orally) far exceeding the MIC. (3) DNM-2 is extremely well tolerated in mice, with no signs of toxicity at the dose levels tested. This is consistent with our data showing these compounds do not induce haemolysis or inhibit human topoisomerase II. (4) DNM-2, when given orally, has outstanding efficacy in a mouse model of MRSA infection. An oral drug for MRSA and VRE is a well-recognized clinical need and DNM-2 has tremendous promise in this regard. (5) Finally, resistance to DNM is much more difficult than resistance to CIP, as shown by the resistance frequencies in Fig. 4a. A problem with novel antibacterials is that bacterial resistance typically necessitates the development of a new drug to treat those drug-resistant pathogens. However, when the inevitable resistance to DNM/derivatives does arise clinically, these bacteria will be sensitive to FQs, a widely used and well-understood class of antibiotics.

Methods

Synthesis of DNM and derivatives. Synthesis of the diazaanthracenols of DNM and the derivatives and chemical characterizations are described in the Supplementary Methods. The NMR spectra are shown in Supplementary Note 1.

Aqueous solubility determination. Initially, a small amount of solid compound (generally 0.5–1.5 mg) was measured into a 1.7-ml Eppendorf tube. PBS (pH 7.4) was added to give a maximum final concentration of $1\,mg\,ml^{-1}$ of compound. The compound was vortexed for ~ 30 s before being placed into a bath sonicator (Cole Parmer, ultrasonic cleaner) for 1 h. Longer incubation times (up to 24 h) were performed with select compounds and no difference in solubility was observed; thus, 1 h was used for all subsequent testing. The tubes were vortexed again for 30 s before being centrifuged at maximum speed ($13{,}000 \times g$) for 10 min. The supernatant was then filtered through a 0.22-μm syringe filter (Millipore Millex MP). The filtrate was then analysed by liquid chromatography–mass spectrometry ($\lambda = 254$ nm, electrospray ionization–time-of-flight in positive mode, Agilent Technologies 6230 TOF LC/MS). The filtrate was diluted 1:2 and 1:4 and all three samples ($1\times$, $0.5\times$ and $0.25\times$) were analysed in triplicate. Three independent replicates of each compound were performed. A calibration curve for each compound was generated from 1 to 40 μM by dissolving the compound in DMSO and making dilutions of the stock in DMSO. The calibration curve (measured by ultraviolet absorbance) was linear over this range. The concentration of the samples was calculated based on the calibration curves.

Bacterial strains. MRSA and *P. aeruginosa* isolates were from Cubist Pharmaceuticals (Lexington, MA). VRE isolates were from a previously published collection[50]. *E. coli* strains were obtained either from ATCC or Professor Cari Vanderpool (UIUC). *A. baumannii* isolates were obtained from Dr John Quale[51].

Antibiotic susceptibility tests. Susceptibility testing was performed in triplicate, using the microdilution broth method as outlined by the Clinical and Laboratory Standards Institute[52]. Mueller Hinton (MH) broth was used.

DNA amplification and sequencing analysis. A single colony of *S. aureus* grown on MH agar or a single colony of *Enterococcus* grown on Brain-Heart Infusion (BHI) agar was suspended in 50 μl of the PCR mixture containing the primers (Supplementary Table 7) and PCR master mix (Platinum *Taq*DNA Polymerase, Invitrogen). PCR amplification was performed using an initial denaturation step of 94 °C for 2 min followed by 35 cycles of 94 °C for 30 s, 52 °C (*S. aureus*. or *Enterococcus faecalis*) or 58 °C (*Enteroccocus faecium*) for 30 s, 72 °C for 50 s. PCR products were purified further on a 1% agarose gel and DNA was extracted (QIAquick Gel Extraction Kit, Qiagen). DNA sequencing was performed by the W. M. Keck Center for Comparative and Functional Genomics (UIUC). The NCBI standard nucleotide BLAST database was used to verify the identity of the PCR products and determine mutations within the sequences.

Site-directed mutagenesis. pTRCHisA-GyrA plasmid containing the gene for *E. coli* gyrase A was kindly provided by Professor David Hooper[53]. Primers for mutagenesis were designed using QuikChange Primer Design (Agilent) and their sequences can be found in Supplementary Table 7. Site-directed mutagenesis was carried out with the QuikChange Lightning Site-Directed Mutagenesis Kit (Agilent), according to the manufacturer's instructions, with the modification that NEB Turbo Competent *E. coli* were used as the host strain. All clones were confirmed by sequencing.

DNA gyrase expression. Expression of WT *E. coli* gyrase A and gyrase B was performed as previously described[53]. Briefly, pTRCHisA-GyrA, pTRCHisA-GyrAS83L, pTRCHisA-GyrAS83R, or pTRCHisA-GyrB were introduced into One Shot BL21 Star (DE3) (NEB) by chemical transformation. Transformed cells were selected for on an Luria-Bertani (LB) ampicillin plate. Single colonies from a fresh plate were inoculated into 50 ml of LB with 50 μg ml^{-1} ampicillin and incubated aerobically at 37 °C with shaking at 250 r.p.m. overnight (14–16 h). The overnight culture was then used to inoculate 1 litre LB with 50 μg ml^{-1} ampicillin. The culture was grown aerobically with shaking at 250 r.p.m. until A_{600} reached 0.4–0.6. Protein expression was induced with a final concentration of 0.5 mM of IPTG (isopropyl-β-D-thiogalactoside) at 37 °C with shaking at 250 r.p.m. for 4 h. The culture was harvested by centrifugation at 5,000 × g for 5 min at 4 °C. Cell pellets were frozen at −20 °C, thawed on ice for 30 min and resuspended in TGN_{150} (20 mM Tris-HCl pH 7.5, 10% glycerol, 150 mM NaCl) with 0.5 mg ml^{-1} lysozyme with 2 μg ml^{-1} aprotinin, 1 μg ml^{-1} leupeptin, 1 μg ml^{-1} pepstatin A and 100 μM phenylmethanesulfonylfluoride. Cells were lysed by sonication at 35% amplitude (10 s pulse with 30 s rest, 6 times). The lysate was cleared by centrifugation at 35,000 × g for 30 min at 4 °C. The supernatant was batch loaded onto 1 ml of 1:1 Ni-NTA agarose (Qiagen) at 4 °C for 30 min with inversion. The resin was washed with 20 ml TGN_{150} with 10 mM imidazole followed by 10 ml of wash buffer (20 mM Tris-HCl pH 7.5, 10% glycerol, 300 mM NaCl, 10 mM imidazole) and eluted with TGN_{150} containing imidazole concentrations of 25, 50, 100, 200, 300 and 500 mM. Eluted fractions were assessed by SDS–PAGE, using 4–20% TGX Mini-PROTEAN gels (Bio-Rad). Fractions containing pure protein were pooled and dialysed against TDEN buffer (50 mM Tris-HCl pH 7.5, 5 mM dithiothreitol (DTT), 1 mM EDTA, 150 mM NaCl) overnight at 4 °C, using a Slide-A-Lyzer Dialysis Cassette, 10 000 MWCO (Thermo Scientific) and concentrated to ~0.5–1 ml, using an Amicon Ultra-15 50 K Centrifugal Filter Device. The concentration was determined by Bradford assay (Sigma) using BSA (Thermo Scientific) as the control. Expression of S83L and S83R GyrA was performed identically to expression of the WT GyrA.

DNA gyrase cleavage. DNA gyrase cleavage assays were performed as previously described with minor changes[12,37,54]. Ten micrograms per millilitre supercoiled DNA (pBR322, Inspiralis) was added to buffer (35 mM Tris-HCl pH 7.5, 24 mM KCl, 4 mM $MgCl_2$, 2 mM DTT, 1.8 mM spermidine, 6.5% glycerol, 0.1 mg ml^{-1} albumin) with compound or vehicle. Compound concentrations were 0.01, 0.04, 0.17, 0.68, 2.7 and 10.8 μM, except for DNM, which was 8.9 μM for its highest concentration. DNA gyrase was added to a final concentration of 16 nM gyrA and 32 nM gyrA (giving a final concentration of A_2B_2 of 8 nM) for 25 min at 30 °C. Linear product was revealed by addition of 0.2% SDS and 0.1 μg ml^{-1} proteinase K for 30 min at 37 °C. DNA loading dye (Thermo Scientific) was added to the samples and they were run on 1% agarose gels containing 0.5 μg ml^{-1} ethidium bromide. Gels were imaged on a Molecular Imager Gel Doc XR+ (Biorad) and bands were quantified using ImageJ[4]. Per cent of type of DNA was calculated with total DNA in each lane being 100%. For time-course cleavage assays, the same protocol was followed, except that the initial incubation was for varying times (0, 1, 3, 5, 10, 15, 20, 30, 60, 90, 120 and 180 min) instead of 25 min.

Human topoisomerase decatenation assay. The decatenation assay was performed with the Human Topo II Decatenation Assay Kit (Inspiralis), according to the manufacturer's instructions, with minor modifications. First, a master mix was made containing 2 μl of 10 × assay buffer (500 mM Tris-HCl pH 7.5, 1250 mM NaCl, 100 mM $MgCl_2$, 50 mM DTT, 1,000 μg ml^{-1} albumin), 0.67 μl of 30 mM ATP, 1.34 μl of 0.1 ng μl^{-1} kDNA, and 14.3 μl of nuclease-free water per sample is made. DMSO or 30 × compound is added to a 0.5-ml Eppendorf tube (0.67 μl per tube). The master mix is then added to each tube (18.3 μl per tube). Finally, 1 U of human topoisomerase (1 μl of 1 U μl^{-1} stock) is added to each tube for a final volume of 20 μl. The tubes are then incubated at 37 °C for 30 min. Reactions are stopped by the addition of 20 μl of 24:1 chloroform:isoamyl alcohol and 20 μl of stop dye (40% sucrose, 1 mM EDTA, 100 mM Tris-HCl pH 7.5, 0.5 μg ml^{-1} bromophenol blue). Samples were run on 1% agarose gels containing 0.5 μg ml^{-1} ethidium bromide for 1 h at 110 V or until the dye front was approximately halfway down the gel. Gels were imaged on a Molecular Imager Gel Doc XR+ (Biorad).

Resistant mutant generation. Agar plates (15 cm) were prepared containing MH broth and antibacterial compounds at concentrations detailed in the Supplementary Fig. 6. Forty millilitres of an overnight bacterial culture was centrifuged at 3,000 × g for 10 min and resuspended in 0.4 ml of sterile PBS. Plates were inoculated with 100 μl of bacteria in PBS by spreading with beads. Inoculated plates were then incubated at 37 °C for 72 h and the number of resistant colonies was counted. To determine the number of viable colonies spread onto each plate, dilutions of the overnight culture in sterile PBS were spread onto non-selective MH agar plates and plates were incubated overnight at 37 °C before counting colonies.

***In-vitro* haemolysis assay.** Haemolysis assays were performed as previously described[55]. Briefly, assays were performed using human erythrocytes within 3 days of receipt. One millilitre of human blood purchased from Bioreclamation, Inc. (Hicksville, NY) was centrifuged (10,000 × g, 2 min). The pellet was washed three times with sterile saline (0.9% NaCl in water) by repeated gentle suspension and centrifugation. The pellet was resuspended in RBC buffer (10 mM Na_2HPO_4, 150 mM NaCl, 1 mM $MgCl_2$ pH 7.4). To evaluate haemolytic activity of DNM and derivatives, 1 μl either 3.2 mg ml^{-1} DMSO stock (or the most concentrated stock of the compound available if not soluble at 3.2 mg ml^{-1} in DMSO) was transferred to 0.5 ml Eppendorf tubes containing 19 μl RBC buffer. Negative control tubes contained 1 μl DMSO and 19 μl RBC buffer, and positive control tubes contained 1 μl DMSO and 19 μl sterile deionized water. A suspension of washed erythrocytes (10 μl) was added to each tubes and samples were incubated at 37 °C for 2 h. Samples were centrifuged at 10,000 × g for 2 min and the supernatants from each sample (25 μl) were transferred to a clear, sterile 384-well plate. The absorbance of these supernatants was measured at 540 nm using a SpectraMaxPlus384 absorbance plate reader (Molecular Devices). Per cent haemolysis of each sample was calculated relative to the average absorbance values measured for positive controls.

Intercalation assay. Intercalation assays were performed as previously described[34]. The ability of DNM-2 to intercalate into DNA was measured by an ethidium bromide displacement assay. Herring Sperm DNA (34 μg ml^{-1} final) was premixed with buffer containing ethidium bromide (50 mM Tris base, 100 mM NaCl, 1 mM EDTA, 5 μM EtBr pH 7.5). Ninety-five microlitres of this solution was added to a 96-well plate containing 5 μl of DMSO solutions of compounds. In addition to vehicle controls, wells lacking either DNA or EtBr were also used to ensure that these did not have an effect on fluorescence. Doxorubicin was used as a positive control. The reactions were allowed to incubate for 30 min. Fluorescence was then read on a Gemini microplate reader (Molecular Devices, excitation = 545 nm, emission = 595 nm).

Pharmacokinetic assessment. The animal studies (PK, *in-vivo* toxicity and *in-vivo* efficacy) were carried out in strict accordance with the recommendations in the Guide for the Care and Use of Laboratory Animals of the National Institutes of Health. The protocol was approved by the Institutional Animal Care and Use Committee (IACUC) at the University of Illinois at Urbana-Champaign (Protocol Number: 13406). In these studies, 10- to 12-week-old female C57BL/6 mice purchased from Charles River were used. DNM, DNM-2 and DNM-3 were formulated as slurries at 8.3 mg ml^{-1} in 25% Cremophor RH40/water (v/v). Before beginning the pharmacokinetic assessment, mice were first tested for their ability to tolerate the DNM, DNM-2 and DNM-3 at 50 mg kg^{-1} (p.o.). After establishing that this dose was well tolerated, mice were treated with DNM, DNM-2 or DNM-3 (all 50 mg kg^{-1}) via oral gavage, with three mice per time point (15, 30, 60, 120, 240 and 480 min). At specified time points, mice were killed and blood was collected, centrifuged and the serum was frozen at −80 °C until analysis. The proteins in a 50-μl aliquot of serum were precipitated by the addition of 50 μl of acetonitrile and the sample was centrifuged to remove the proteins. Serum concentrations of DNM, DNM-2 and DNM-3 were determined by HPLC. PK parameters were determined using GraphPad Prism Version 5.00 for Windows.

***In-vivo* toxicity assessment.** The protocol was approved by the IACUC at the University of Illinois at Urbana-Champaign (Protocol Number: 14032). Six-week-old male pathogen-free BALB/c mice were purchased from Taconics Biosciences (Albany, NY). All animals were housed in a pathogen-free environment and received sterile food and water. Mice ($n = 5$) were treated once daily for 10 days, with 50 mg kg^{-1} DNM-2 or vehicle (25% Cremophor RH 40/PBS (v/v)), by oral gavage. Toxicity was assessed as previously described[56]. Specifically, heparinized whole blood was collected for assessment of total white blood cells, neutrophils, lymphocytes, haematocrit, platelets, creatinine, blood urea, nitrogen, albumin, alanine aminotransferase, alkaline phosphatase and total bilirubin. Mice were euthanized by overdosing with ketamine/xylazine and the heart, lung, kidney, liver, spleen, gastrointestinal tract and the brain were collected for histopathology. Tissue samples were fixed 24 h in 10% neutral buffered formalin, processed and paraffin embedded, sectioned (5 μm thickness) and stained with haematoxylin and eosin. All slides were systematically evaluated by a single board-certified veterinary anatomic pathologist (SL) for evidence of acute or chronic inflammation and toxicity. All lesions were characterized, recorded and scored for severity (minimal = 1, mild = 2, moderate = 3, marked = 4, and severe = 5).

***In-vivo* efficacy.** The protocol was approved by the IACUC at the University of Illinois at Urbana-Champaign (Protocol Number: 14032). Six-week-old male pathogen-free BALB/c mice were purchased from Harlan Sprague--Dawley (Indianapolis, IN). All animals were housed in a pathogen-free environment and received sterile food and water. For the inoculation, overnight cultures of *S. aureus* clinical isolate NRS3 were diluted 1:100 into fresh tryptic soy broth and grown for 2 h at 37 °C. Bacteria were washed and resuspended in sterile PBS. The mice were anaesthetized with ketamine and xylazine. The mouse tails were pre-warmed in 45 °C for 5 min before 1.2×10^8 colony-forming units of *S. aureus* in 50 μl of PBS were injected into a tail vein using a 29-gauge needle. This number of bacteria was determined from a series of preliminary studies in which groups of mice were infected with a range of 10^6–10^9 colony-forming units of *S. aureus*. Infected mice (15 mice per group) were then treated once daily for 10 days with 50 mg kg^{-1} DNM-2, 50 mg kg^{-1} CIP or vehicle (25% Cremophor RH 40/PBS (v/v)), by oral gavage. For survival analyses, a Kaplan–Meier Log Rank Survival Test was performed using OriginPro 9 (Northampton, MA).

References

1. Hooper, D. C. Fluoroquinolone resistance among Gram-positive cocci. *Lancet Infect. Dis.* **2,** 530–538 (2002).
2. Linder, J. A., Huang, E. S., Steinman, M. A., Gonzales, R. & Stafford, R. S. Fluoroquinolone prescribing in the United States: 1995 to 2002. *Am. J. Med.* **118,** 259–268 (2005).
3. Hicks, L. A., Taylor, Jr. T. H. & Hunkler, R. J. U.S. outpatient antibiotic prescribing, 2010. *N. Engl. J. Med.* **368,** 1461–1462 (2013).
4. Redgrave, L. S., Sutton, S. B., Webber, M. A. & Piddock, L. J. Fluoroquinolone resistance: mechanisms, impact on bacteria, and role in evolutionary success. *Trends Microbiol.* **22,** 438–445 (2014).
5. Chambers, H. F. in *Basic and Clinical Pharmacology.* (ed Katzung, B. G.) Ch. 46, 763–770 (McGraw-Hill, 2007).
6. Chen, S. H., Chan, N. L. & Hsieh, T. S. New mechanistic and functional insights into DNA topoisomerases. *Annu. Rev. Biochem.* **82,** 139–170 (2013).
7. Pommier, Y. Drugging topoisomerases: lessons and challenges. *ACS Chem. Biol.* **8,** 82–95 (2013).
8. Gellert, M., Mizuuchi, K., O'Dea, M. H. & Nash, H. A. DNA gyrase: an enzyme that introduces superhelical turns into DNA. *Proc. Natl Acad. Sci. USA* **73,** 3872–3876 (1976).
9. Jacoby, G. A. Mechanisms of resistance to quinolones. *Clin. Infect. Dis.* **41**(Suppl 2): S120–S126 (2005).
10. Dalhoff, A. Resistance surveillance studies: a multifaceted problem--the fluoroquinolone example. *Infection* **40,** 239–262 (2012).
11. Aldred, K. J. *et al.* Overcoming target-mediated quinolone resistance in topoisomerase IV by introducing metal-ion-independent drug-enzyme interactions. *ACS Chem. Biol.* **8,** 2660–2668 (2013).
12. Barnard, F. M. & Maxwell, A. Interaction between DNA gyrase and quinolones: effects of alanine mutations at GyrA subunit residues Ser(83) and Asp(87). *Antimicrob. Agents Chemother.* **45,** 1994–2000 (2001).
13. Kwak, Y. G. *et al.* Association of norB overexpression and fluoroquinolone resistance in clinical isolates of *Staphylococcus aureus* from Korea. *J. Antimicrob. Chemother.* **68,** 2766–2772 (2013).
14. Wang, S., Wang, Y., Shen, J., Wu, Y. & Wu, C. Polymorphic mutation frequencies in clinical isolates of *Staphylococcus aureus*: the role of weak mutators in the development of fluoroquinolone resistance. *FEMS. Microbiol. Lett.* **341,** 13–17 (2013).
15. Hiramatsu, K. *et al.* Curing bacteria of antibiotic resistance: reverse antibiotics, a novel class of antibiotics in nature. *Int. J. Antimicrob. Agents* **39,** 478–485 (2012).
16. Baba, K. *et al.* Prevalence and mechanism of antimicrobial resistance in *Staphylococcus aureus* isolates from diseased cattle, swine and chickens in Japan. *J. Vet. Med. Sci.* **74,** 561–565 (2012).
17. Costa, S. S. *et al.* Exploring the contribution of efflux on the resistance to fluoroquinolones in clinical isolates of *Staphylococcus aureus. BMC Microbiol.* **11,** 241 (2011).
18. Sanfilippo, C. M., Hesje, C. K., Haas, W. & Morris, T. W. Topoisomerase mutations that are associated with high-level resistance to earlier fluoroquinolones in *Staphylococcus aureus* have less effect on the antibacterial activity of besifloxacin. *Chemotherapy* **57,** 363–371 (2011).
19. Aligholi, M. *et al.* Phenotypic and genotypic evaluation of fluoroquinolone resistance in clinical isolates of *Staphylococcus aureus* in Tehran. *Med. Sci. Monit.* **17,** PH71–PH74 (2011).
20. Yoon, E. J. *et al.* Extended spectrum of quinolone resistance, even to a potential latter third-generation agent, as a result of a minimum of two GrlA and two GyrA alterations in quinolone-resistant *Staphylococcus aureus. Chemotherapy* **56,** 153–157 (2010).
21. Coskun-Ari, F. F. & Bosgelmez-Tinaz, G. grlA and gyrA mutations and antimicrobial susceptibility in clinical isolates of ciprofloxacin- methicillin-resistant *Staphylococcus aureus. Eur. J. Med. Res.* **13,** 366–370 (2008).
22. Sadowy, E. *et al.* High abundance and diversity of antimicrobial resistance determinants among early vancomycin-resistant *Enterococcus faecium* in Poland. *Eur. J. Clin. Microbiol. Infect. Dis.* **32,** 1193–1203 (2013).
23. Werner, G. *et al.* High-level ciprofloxacin resistance among hospital-adapted *Enterococcus faecium* (CC17). *Int. J. Antimicrob. Agents* **35,** 119–125 (2010).
24. Grohs, P., Houssaye, S., Aubert, A., Gutmann, L. & Varon, E. *In vitro* activities of garenoxacin (BMS-284756) against *Streptococcus pneumoniae*, viridans group streptococci, and *Enterococcus faecalis* compared to those of six other quinolones. *Antimicrob. Agents Chemother.* **47,** 3542–3547 (2003).
25. Tremblay, C. L., Charlebois, A., Masson, L. & Archambault, M. Characterization of hospital-associated lineages of ampicillin-resistant *Enterococcus faecium* from clinical cases in dogs and humans. *Front. Microbiol.* **4,** 245 (2013).
26. Rathnayake, I. U., Hargreaves, M. & Huygens, F. Antibiotic resistance and virulence traits in clinical and environmental *Enterococcus faecalis* and *Enterococcus faecium* isolates. *Syst. Appl. Microbiol.* **35,** 326–333 (2012).
27. Brock, T. D. & Sokolski, W. T. Biological studies on the antibiotic, nybomycin. *Antibiot. Chemother. (Northfield Ill)* **8,** 631–636 (1958).
28. Strelitz, F., Flon, H. & Asheshov, I. N. Nybomycin, a new antibiotic with antiphage and antibacterial properties. *Proc. Natl Acad. Sci. USA* **41,** 620–624 (1955).
29. Rinehart, K. L. & Renfroe, H. B. The structure of nybomycin. *J. Am. Chem. Soc.* **83,** 3729–3731 (1961).
30. Naganawa, H., Wakashiro, T., Yagi, A., Kondo, S. & Takita, T. Deoxynybomycin from a *Streptomyces. J. Antibiot.* **23,** 365–378 (1970).
31. Adelmann, S., Baldhoff, T., Koepcke, B. & Schembecker, G. Selection of operating parameters on the basis of hydrodynamics in centrifugal partition chromatography for the purification of nybomycin derivatives. *J. Chromatogr. A.* **1274,** 54–64 (2013).
32. Forbis, R. M. & Rinehart, K. L. Nybomycin.4. Total synthesis of deoxynybomycin. *J. Am. Chem. Soc.* **92,** 6995–699 (1970).
33. Forbis, R. M. & Rinehart, K. L. Nybomycin.7. Preparative routes to nybomycin and deoxynybomycin. *J. Am. Chem. Soc.* **95,** 5003–5013 (1973).
34. Bair, J. S., Palchaudhuri, R. & Hergenrother, P. J. Chemistry and biology of deoxynyboquinone, a potent inducer of cancer cell death. *J. Am. Chem. Soc.* **132,** 5469–5478 (2010).
35. Angibaud, P. *et al.* 5-imidazolyl-quinolinones, -quinazolinones and -benzo-azepinones as farnesyltransferase inhibitors. *Bioorg. Med. Chem. Lett.* **13,** 1543–1547 (2003).
36. Parkinson, E. I., Bair, J. S., Cismesia, M. & Hergenrother, P. J. Efficient NQO1 substrates are potent and selective anticancer agents. *ACS Chem. Biol.* **8,** 2173–2183 (2013).
37. Kampranis, S. C. & Maxwell, A. The DNA gyrase-quinolone complex. ATP hydrolysis and the mechanism of DNA cleavage. *J. Biol. Chem.* **273,** 22615–22626 (1998).
38. Bax, B. D. *et al.* Type IIA topoisomerase inhibition by a new class of antibacterial agents. *Nature* 466, 935–940, 2010).
39. Ferrero, L., Cameron, B. & Crouzet, J. Analysis of gyrA and grlA mutations in stepwise-selected ciprofloxacin-resistant mutants of *Staphylococcus aureus. Antimicrob. Agents Chemother.* **39,** 1554–1558 (1995).
40. Hori, S., Ohshita, Y., Utsui, Y. & Hiramatsu, K. Sequential acquisition of norfloxacin and ofloxacin resistance by methicillin-resistant and -susceptible *Staphylococcus aureus. Antimicrob. Agents Chemother.* **37,** 2278–2284 (1993).
41. Egawa, K. *et al.* Deoxynybomycin is a selective anti-tumor agent inducing apoptosis and inhibiting topoisomerase I. *Biol. Pharm. Bull.* **23,** 1036–1040 (2000).
42. Li, S. *et al.* Pseudonocardians A-C, new diazaanthraquinone derivatives from a deap-sea actinomycete *Pseudonocardia* sp. SCSIO 01299. *Mar. Drugs* **9,** 1428–1439 (2011).
43. Price, L. B. *et al. In vitro* selection and characterization of *Bacillus anthracis* mutants with high-level resistance to ciprofloxacin. *Antimicrob. Agents Chemother.* **47,** 2362–2365 (2003).
44. Vila, J. *et al.* Association between double mutation in gyrA gene of ciprofloxacin-resistant clinical isolates of *Escherichia coli* and MICs. *Antimicrob. Agents Chemother.* **38,** 2477–2479 (1994).
45. Vila, J., Ruiz, J., Goni, P., Marcos, A. & Jimenez de Anta, T. Mutation in the gyrA gene of quinolone-resistant clinical isolates of *Acinetobacter baumannii. Antimicrob. Agents Chemother.* **39,** 1201–1203 (1995).
46. Vernel-Pauillac, F., Hogan, T. R., Tapsall, J. W. & Goarant, C. Quinolone resistance in *Neisseria gonorrhoeae*: rapid genotyping of quinolone resistance-determining regions in gyrA and parC genes by melting curve analysis predicts susceptibility. *Antimicrob. Agents Chemother.* **53,** 1264–1267 (2009).
47. Deguchi, T. *et al.* Alterations in the GyrA subunit of DNA gyrase and the ParC subunit of topoisomerase IV in quinolone-resistant clinical isolates of *Klebsiella pneumoniae. Antimicrob. Agents Chemother.* **41,** 699–701 (1997).
48. Bast, D. J. *et al.* Fluoroquinolone resistance in clinical isolates of *Streptococcus pneumoniae*: contributions of type II topoisomerase mutations and efflux to levels of resistance. *Antimicrob. Agents Chemother.* **44,** 3049–3054 (2000).
49. Talbot, G. H. *et al.* Bad bugs need drugs: an update on the development pipeline from the Antimicrobial Availability Task Force of the Infectious Diseases Society of America. *Clin. Infect. Dis.* **42,** 657–668 (2006).
50. Moritz, E. M. & Hergenrother, P. J. Toxin-antitoxin systems are ubiquitous and plasmid-encoded in vancomycin-resistant enterococci. *Proc. Natl Acad. Sci. USA* **104,** 311–316 (2007).
51. Bratu, S., Landman, D., Martin, D. A., Georgescu, C. & Quale, J. Correlation of antimicrobial resistance with beta-lactamases, the OmpA-like porin, and efflux pumps in clinical isolates of *Acinetobacter baumannii* endemic to New York City. *Antimicrob. Agents Chemother.* **52,** 2999–3005 (2008).
52. Wikler, M. A. *et al. Methods for Dilution Antimicrobial Susceptibility Tests for Bacteria That Grow Aerobically; Approved Standard* 7 edn, Vol. 26Clinical and Laboratory Standards Institute, 2006).
53. Tran, J. H., Jacoby, G. A. & Hooper, D. C. Interaction of the plasmid-encoded quinolone resistance protein Qnr with *Escherichia coli* DNA gyrase. *Antimicrob. Agents Chemother.* **49,** 118–125 (2005).
54. Fisher, L. M. & Pan, X. S. Methods to assay inhibitors of DNA gyrase and topoisomerase IV activities. *Methods Mol. Med.* **142,** 11–23 (2008).

55. Eibergen, N. R., Im, I., Patel, N. Y. & Hergenrother, P. J. Identification of a novel protein synthesis inhibitor active against Gram-positive bacteria. *Chembiochem* **13,** 490 (2012).
56. Botham, R. C. *et al.* Dual small-molecule targeting of procaspase-3 dramatically enhances zymogen activation and anticancer activity. *J. Am. Chem. Soc.* **136,** 1312–1319 (2014).

Acknowledgements

We thank Cubist Pharmaceuticals for generating the MRSA clinical isolate data (Fig. 2a) and providing DNA from those isolates. We thank Professor David C. Hooper for providing the plasmids pTRC-HisA-GyrA and pTRC-HisA-GyrB. We thank Professor Tim Fan, Professor Levent Dirikolu and Andrew Lee for technical assistance. This work was supported by the University of Illinois (P.J.H.) and the National Institutes of Health (NIH) Grant R01HL090699 (G.W.L.). E.I.P. is a National Science Foundation pre-doctoral fellow and an ACS Medicinal Chemistry pre-doctoral fellow. B.A.N. is a National Defense Science and Engineering pre-doctoral fellow.

Author contributions

P.J.H., E.I.P. and J.S.B. conceived the study. J.S.B. designed the synthesis of DNM and all derivatives, except compound **5**. E.I.P. and E.H.S. designed the synthesis of derivative **5**. E.I.P., B.A.N., H.I.K. and E.H.S. executed the synthesis of the compounds used in this study. E.I.P. designed and executed the biological experiments including MIC determinations, sequencing of the QRDR of clinical isolates, *in vitro* analysis of DNA gyrase inhibition and development of resistant mutants. H.Y.L. performed the pharmacokinetic analysis. G.W.L. designed and executed the mouse model of infection. S.L. performed clinical pathology and histopathology toxicity evaluations. P.J.H. and E.I.P. analysed the data and wrote the manuscript, with the assistance of J.S.B., B.A.N., H.Y.L. S.L., and G.W.L.

Additional information

Competing financial interests: The University of Illinois has filed patents on DNM and its derivatives, with E.I.P., J.S.B and P.J.H. listed as inventors. All other author declare no competing financial interests.

17

Design of a bioactive small molecule that targets r(AUUCU) repeats in spinocerebellar ataxia 10

Wang-Yong Yang[1], Rui Gao[2], Mark Southern[3], Partha S. Sarkar[2] & Matthew D. Disney[1]

RNA is an important target for chemical probes of function and lead therapeutics; however, it is difficult to target with small molecules. One approach to tackle this problem is to identify compounds that target RNA structures and utilize them to multivalently target RNA. Here we show that small molecules can be identified to selectively bind RNA base pairs by probing a library of RNA-focused small molecules. A small molecule that selectively binds AU base pairs informed design of a dimeric compound (**2AU-2**) that targets the pathogenic RNA, expanded r(AUUCU) repeats, that causes spinocerebellar ataxia type 10 (SCA10) in patient-derived cells. Indeed, **2AU-2** (50 nM) ameliorates various aspects of SCA10 pathology including improvement of mitochondrial dysfunction, reduced activation of caspase 3, and reduction of nuclear foci. These studies provide a first-in-class chemical probe to study SCA10 RNA toxicity and potentially define broadly applicable compounds targeting RNA AU base pairs in cells.

[1] Departments of Chemistry and Neuroscience, The Scripps Research Institute, Scripps Florida, Jupiter, Florida 33458, USA. [2] Mitchell Center for Neurodegenerative Disorders, Department of Neurology, Neuroscience and Cell Biology, University of Texas Medical Branch, Galveston, Texas 77555, USA. [3] Informatics Core, The Scripps Research Institute, Scripps Florida, Jupiter, Florida 33458, USA. Correspondence and requests for materials should be addressed to M.D.D. (email: Disney@scripps.edu).

RNA has diverse cellular functions. For example, messenger RNAs (mRNAs) encode protein, microRNAs regulate the lifetime of mRNAs, and the ribosome translates mRNAs into proteins[1,2]. In bacteria, riboswitches control the production of proteins by binding to small molecule metabolites[3,4]. In fact, many non-coding RNAs have been found to play significant roles in cellular biology, and these discoveries expand even further the known functions of RNA[5].

Because of the important cellular functions of RNA under normal conditions, it is not surprising that mutations in RNA can cause disease. Single-nucleotide polymorphisms can give rise to cryptic alternative pre-mRNA splicing sites, leading to production of aberrant, defective proteins, as is the case with β-thalassaemia[6,7]. Expanded RNA repeats can also contribute to disease and can be present in 5′ and 3′ untranslated regions (UTRs; fragile X-associated tremor ataxia syndrome (FXTAS)[8] and myotonic dystrophy type 1 (DM1)[9]), introns (spinocerebellar ataxia type 10 (SCA10)[10] and myotonic dystrophy type 2 (DM2)[11]) or coding regions (Huntington's disease (HD)[12]). Small molecules that target these RNA and inhibit its dysfunction are thus highly desirable.

The bacterial ribosome is the most widely exploited RNA target[13,14]. Ribosomes and ribosomal RNA (rRNA) are privileged targets as (i) ribosomes play essential roles in cellular homeostasis and modulation of ribosome activity can have drastic cellular consequences, and (ii) rRNA comprises about 80–90% of the total RNA content of a cell[15]. Riboswitches are also emerging and established targets of RNA-directed small molecules. Compounds identified to bind to and modulate riboswitches generally mimic the structure of the RNA's natural metabolite[3,16], akin to substrate mimicry to design enzyme inhibitors[17]. Most RNA targets to which a small molecule binder is desired, however, are of low abundance and have no natural metabolite to inform drug design.

To aid RNA-targeting endeavours, our group has developed two-dimensional combinatorial screening (2DCS) to identify optimal (high affinity and selective) RNA motif–small molecule interactions. These interactions are deposited into a database and can be used to design small molecules to target RNAs by comparing motifs in the desired target to the database. This approach has been used to design small molecules targeting the RNAs that cause DM[18–20], FXTAS[21] and HD[22]. All of the interactions that are presently in the RNA motif–small molecule database are between small molecules and RNA loops such as hairpins, bulges and internal loops. Herein, we report small molecules that bind selectively to RNA base pairs. Among a variety of compounds tested, small molecules with benzamidine moieties were identified to bind selectively to AU base pairs. These data were leveraged to design the first bioactive small molecule to target the expanded r(AUUCU) repeat that causes SCA10, an incurable neuromuscular disorder. The compound targets the central AU base pairs in r(AUUCU)^{exp} and its dimeric compound (**2AU-2**) displaces sequestered proteins and improves defects in patient-derived cells. The observation that base pair-binding modules can provide bioactive compounds suggests that many other RNAs can be exploited as targets of small molecules.

Results and Discussion

Binding of RNA-focused small molecules to RNA base pairs. By using chemical similarity searching, small molecules with features that should pre-dispose them for binding RNA[23,24] were collected from both the National Cancer Institute's and The Scripps Research Institute's chemical libraries, including benzimidazole, benzamidine, aniline moieties. Compounds were further restrained to be fluorescent to allow for easy screening of binding events, affording 104 small molecules (Fig. 1a and Supplementary Table 1).

Compounds were screened for binding to different base pairs using four model constructs with stretches of AU and GC base pairs embedded in the stem of a common hairpin loop (Fig. 1b). The four RNAs display AU or GC pairs with different nearest neighbours. AUAU and AAUU RNAs have 5′AU/3′UA or 5′AAUU/3′UUAA stretches, respectively, while GCGC and GGCC have 5′GC/3′CG or 5′GGCC/3′CCGG stretches, respectively. Due to significant differences in the thermodynamic stability of the hairpins, RNAs with AU pairs had 12 base pairs in the stem while RNAs with GC pairs had 8 (ref. 25). The $\Delta G°_{37}$ for AUAU and AAUU is −8.5 and −6.7 kcal mol^{-1}, respectively, while for GCGC and GGCC it is −17.5 and −19.1 kcal mol^{-1}, respectively. If only eight AU base pairs were present, then the free energy drops considerably to −3.7 and −2.5 kcal mol^{-1}, respectively.

In initial compound screens, each small molecule was incubated with the four RNAs and the change in emission was measured (Fig. 1c). Changes in emission were not only analysed for statistical significance for binding to these RNAs in general, but also for binding to the different RNA structures. For this compound collection, 29% of the compounds exhibited a

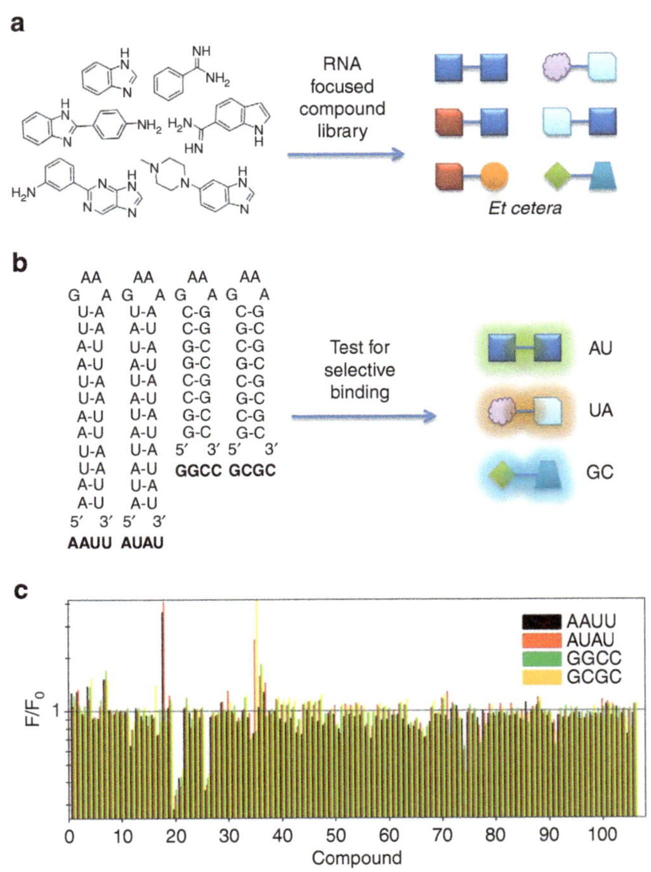

Figure 1 | Overall scheme to identify small molecules that bind to RNA base pairs. (**a**) A small molecule library was collected in which the small molecules have chemotypes present in compounds known to bind RNA. (**b**) The RNA-focused small molecule library was tested for binding to four different RNAs that have different orientations and identities of RNA base pairs. (**c**) Compounds were tested in a fluorescence emission assay for selectively binding to RNAs with different paired elements. Compounds that were selective were further analysed.

change in emission upon binding to any of the RNAs used in this study, and 18% of the compounds showed preferential binding to AU or GC pair-containing duplexes (Supplementary Table 2).

On the basis of the substructures within compounds that bind, a Venn diagram was constructed to correlate chemotypes and binding to AU and GC base pairs (Fig. 2). For example, various functionalized purines bound to RNAs with GC base pairs. Previously, other compounds that bind RNA base pairs have been identified. For example, the Beal group has used threading intercalators with acridine or other aromatic functionalities to target RNA bulges with nearby GC pairs[26]. Thus, it appears that acridine, and related compounds, can provide a GC base-pair-binding module. The group of Tor has shown that ethidium bromide is a 20-fold selective binder to poly r(A)-poly r(U) over poly r(G)–poly r(C)[27].

Each small molecule from the initial screen was tested for saturable binding when incubated with serially diluted RNA. The benzamidine compounds **1** and **2** provided the most robust and saturable change in emission and were found to bind selectively to AU base pairs (Supplementary Figs 1, 2 and 3). Due to the binding of these compounds to AU base pairs and their saturable emission properties, these compounds were selected for further study. This assay, by nature of using emission to study binding, could generate false negatives; however, it allowed us to further characterize positive hits and produce a first-in-class bioactive ligand that targets r(AUUCU), *vide infra*.

Compounds 1 and 2 bind selectively to AU base pairs. The selectivities of compounds **1** and **2** were further assessed by measuring EC_{50}s for all four RNAs. The EC_{50}s for **1** for binding to AAUU, AUAU, GGCC and GCGC are 170, 45, 390 and 240 nM, respectively, while the EC_{50}'s for **2** are 170, 190, 8,550 and 6,950 nM, respectively (Table 1). On the basis of these data, **2** is much more selective for AU base pairs than **1**, with an AU pair selectivity of ~40-fold (Fig. 3). Measurement of the K_d's of **2** to AUAU and AAUU give values of 210 and 320 nM, respectively, and stoichiometries of 2.9 and 3.7, respectively (Supplementary Fig. 4). These data suggests that **2** is a high affinity binder to stretches of AU base pairs and that the compound interacts with between 3 and 4 base pairs in each of the model hairpins. Previous studies have suggested that the aminoglycoside tobramycin recognizes model poly(rI) · poly(rC) RNA duplexes and that the compound interacts with approximately 4 bp, or a similar number of interacting sites that are proposed for **2** (ref. 28). Diphenylfuran amidine was previously reported to bind polyr(A) · polyr(U) duplex via intercalation, as determined from viscosity and circular dichroism studies[29,30]. These observations were further supported by docking studies in which the intercalation of **2** is stabilized through stacking interactions with the positively charged amidine residing in the major groove; minor groove binding could not be accommodated[29,30]. Further, the substituents in this class of compounds affect the thermal stability of polyr(A) · polyr(U), with imidazoline providing the largest enhancement[29,30].

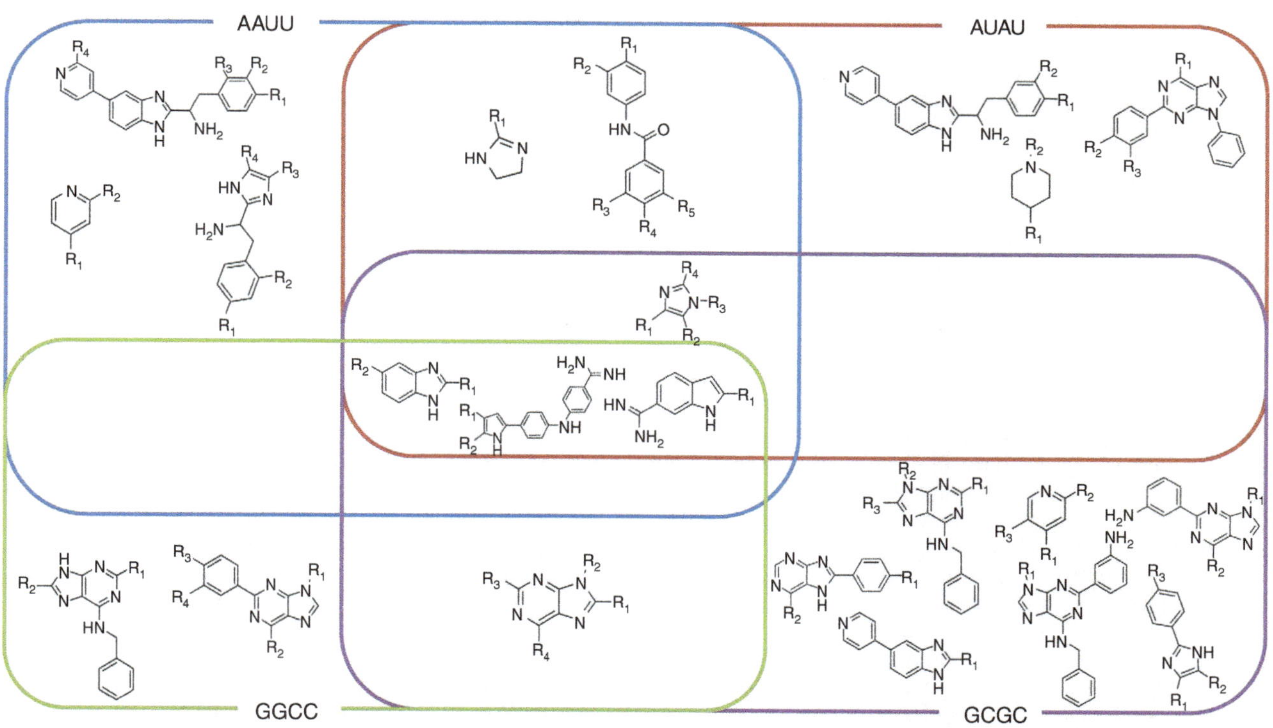

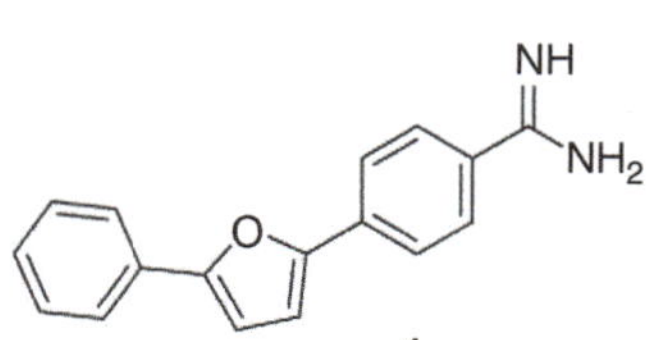

1

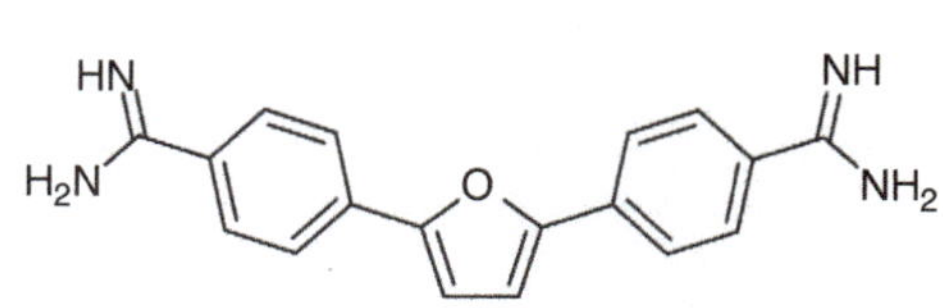

2

Figure 2 | Analysis of compounds identified to bind RNA base pairs. Top, Venn diagram of substructures in compounds that were found to bind to RNA from the fluorescence screening assay showed in Fig. 1c. Data were compiled by using compounds that had a *P* value of <0.001 for binding to the RNA hairpins. Bottom, structures of compounds **1** and **2** that were the most avid for binding to AUAU and AAUU RNA hairpins.

Table 1 | EC_{50}s (nM) and Hill coefficients (shown in parentheses) of compounds binding to the base paired RNAs shown in Fig. 1.

Compound	AAUU	AUAU	GGCC	GCGC
1*	170 ± 10 (3.1 ± 0.79)	45 ± 1 (1.5 ± 0.32)	390 ± 20 (1.4 ± 0.34)	240 ± 10 (0.92 ± 0.34)
2†	170 ± 40 (1.5 ± 0.24)	190 ± 20 (1.6 ± 0.28)	8,550 ± 250 (0.85 ± 0.08)	6,950 ± 250 (0.77 ± 0.16)
AU-azide†	45 ± 6 (1.0 ± 0.17)	31 ± 4 (1.0 ± 0.13)	>6,000	>2,000

*[Compound] = 3 μM.
†[Compound] = 1 μM.

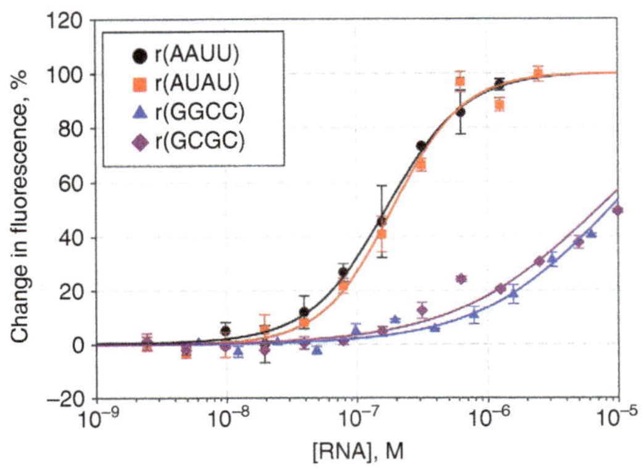

Figure 3 | Binding of compound 2 to four different RNAs Fluorescence binding assays of compound 2 to RNAs with different base pairs show that 2 binds AU base pairs selectively over GC base pairs by >40-fold.

Compounds 1 and 2 bind r(AUUCU) repeats that cause SCA10. The ultimate goal of identifying small molecules that bind RNA motifs *in vitro* is to apply these identified interactions to target an RNA that causes disease. Fortuitously, the expanded repeating r(AUUCU) RNA (r(AUUCU)^{exp}) that causes spinocerebellar ataxia type 10 (SCA10) contains stretches of AU base pairs[10], in particular repeating 5′AU/3′UA base pair steps (Fig. 4b)[31]. SCA10 is an incurable neuromuscular disorder that is mainly found in Latin America[32]. It is a slowly progressive disease that results in poor balance followed by loss of control over upper limbs. Previous experiments have shown that the disease is due to an RNA gain-of-function in which r(AUUCU)^{exp}, located within spliced intron 9 of the ataxin 10 (*ATX10*) mRNA, binds to and sequesters proteins involved in RNA biogenesis such as heterogeneous nuclear ribonucleoprotein K (hnRNP K; Fig. 4a)[33]. Sequestration of hnRNP K causes a host of cellular defects that include the formation of RNA nuclear foci, translocation of protein kinase C-δ (PKCδ) in mitochondria resulting in mitochondrial dysfunction, and activation of caspase 3 and subsequent apoptosis[33]. The binding of the repeats to RNA-binding proteins causes the transcript to be retained in nuclear foci in patient-derived cells and model cellular systems[33].

We assessed the selectivities of compounds **1** and **2** for binding r(AUUCU) repeats over other disease-associated RNA repeats including $r(CUG)_{12}$ (DM1)[9], $r(CGG)_{12}$ (FXTAS and FXS)[8,34], and $r(CCUG)_{12}$ (DM2)[11]. Compound **1** has a similar EC_{50} for all RNA repeats (36–87 nM), indicating that **1** binds to RNA internal loops as well as to AU base pairs with similar binding affinity (Supplementary Fig. 5). Compound **2** selectively recognizes $r(AUUCU)_{11}$ with 5–15-fold selectivity over the other repeats (Table 2 and Supplementary Fig. 6). Compound **2** bound $r(AUUCU)_{11}$ with a K_d of 300 nM and a stoichiometry of ca. 11:1 (**2**:$r(AUUCU)_{11}$). Because $r(AUUCU)_{11}$ contains 10 AU base pairs, it indicates that **2** is binding to each AU base pair in the target. Indeed, a nuclease protection assay revealed that **2** protects each AU base pair in $r(AUUCU)_{11}$ from cleavage (Fig. 5). Interestingly, calculated Hill coefficients indicate that **2** binds to $r(AUUCU)_{11}$ with positive cooperativity ($n = 1.7$; Table 2), suggesting that **2** could be an ideal module to design a polyvalent compound to target multiple adjacent sites in r(AUUCU)^{exp} simultaneously.

An AU base pair-binding module to target r(AUUCU)^{exp}. Previously, our group has developed modularly assembled small molecules that target various repeating RNAs[18,19,35–38]. In this approach, an RNA-binding module or modules that recognize different motifs in an RNA target are displayed on a single molecule. These polyvalent compounds allow for simultaneous recognition of multiple motifs in an RNA target, thereby increasing affinity and selectivity relative to monomeric binders (Fig. 6).

To enable this approach for **2** and r(AUUCU)^{exp}, a derivative of **2**, **AU-azide** (Fig. 6), was synthesized to install an orthogonally reactive group. The azide is used for conjugation onto peptoid polyvalent scaffolds that display alkynes via a Cu-catalysed Huisgen cycloaddition reaction. Dibenzimidate (**3**) was synthesized from the furan via two reaction steps as reported previously[39,40], followed by two amidation reactions to obtain **AU-azide** (Supplementary Fig. 19; see Supplementary Information for chemical synthesis details and compound characterization). After obtaining the desired compound, it was tested for selective recognition of AU over GC base pairs (Table 1 and Supplementary Fig. 7). Interestingly, **AU-azide** has enhanced selectivity of AU base pair than **2**, >45-fold selective for AU over GC base pairs (Table 1). The selectivity of **AU-azide** for $r(AUUCU)_{11}$ over other RNA repeats compared with **2** was also improved (Table 2 and Supplementary Fig. 8).

Development of dimeric compounds to target r(AUUCU)^{exp}. We next synthesized a small library of AU dimers (Fig. 6) by using a previously published approach[19,35–37]. A peptoid backbone was used as a polyvalent scaffold that contains two propargylamine submonomers separated by different distances afforded by varying the number of propylamine spacers inserted into the backbone. The **AU-azides** (RNA-binding modules) were then conjugated by using a Cu-catalysed click reaction (Supplementary Fig. 21). The nomenclature used for these compounds is **2AU-n** where **AU** indicates the RNA-binding module; the number before **AU** indicates valency; and the number after the dash indicates the number of propylamine spacing modules between **AU** RNA-binding modules.

The library of dimeric compounds was then tested for binding to $r(AUUCU)_{11}$ by using a filtre-binding assay (Supplementary Fig. 10). After incubation of radioactively labelled $r(AUUCU)_{11}$ with dimeric compound, the compound-bound RNA was separated from free RNA by using a Dot-Blot apparatus. These studies showed that **2AU-2** bound to the RNA to the greatest extent, 2–20-fold greater than the other dimers. Additional filtre-binding assays were completed with 35-fold excess transfer RNA (tRNA) to gain insight into selectivity. Only a modest

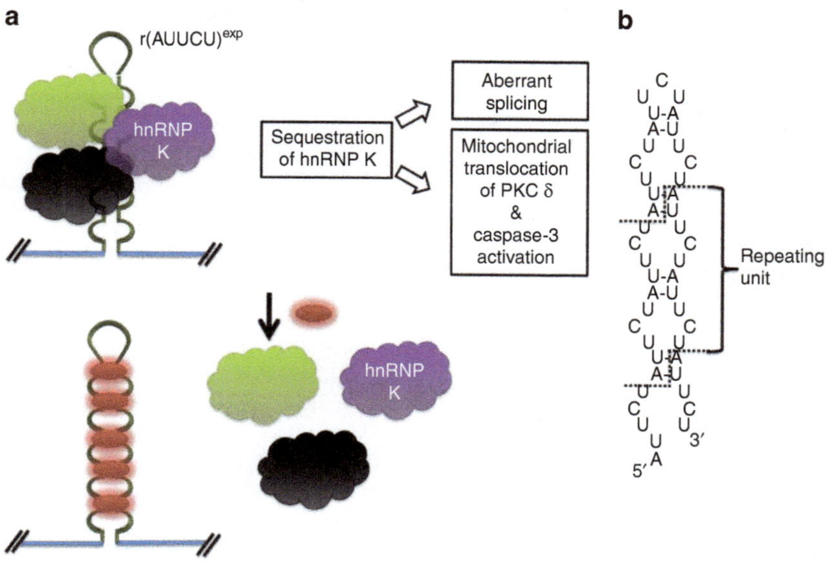

Figure 4 | Schematic of the pathogenic mechanism of SCA10 and a therapeutic approach by targeting the secondary structure of r(AUUCU) repeats. (**a**) r(AUUCU)^{exp}, located within intron 9 of ataxin 10 (*ATX10*) mRNA, sequesters proteins including hnRNP K. Sequestration of hnRNP K results in aberrant splicing of transcripts, mitochondrial translocation of PKCδ and caspase-3 activation, leading to apoptosis (top). A possible therapeutic approach is using small molecules that bind r(AUUCU) repeats and displace sequestered proteins (bottom). (**b**) secondary structure of r(AUUCU) repeats with periodically repeating 5′AU/3′UA base pair steps.

Table 2 | K_ds, Hill coefficients (shown in parentheses) and stoichiometries of 2, AU-azide and 2AU-2 binding to RNA repeats.

RNA repeats	r(AUUCU)$_{11}$		r(CUG)$_{12}$*		r(CGG)$_{12}$*		r(CCUG)$_{12}$*	
Compound	K_d, nM	Stoichiometry	K_d, nM	Stoichiometry	K_d, nM	Stoichiometry	K_d, nM	Stoichiometry
2	300 ± 47 (1.7 ± 0.2)	11 ± 0.5	~4,000	~7	~2,000	~6	~4,000	~11
AU-azide	375 ± 34 (1.3 ± 0.1)	13 ± 0.9	~5,000	~11	~3,000	~9	~5,000	~16
2AU-2	185 ± 1.0 (1.4 ± 0.4)	7 ± 0.1	~4,000	~9	~4,000	~6	~4,000	~6

*The stoichiometry from unsaturated binding curves was estimated from data fit by equation (1). The K_d was calculated by using the estimated stoichiometry.

decrease (~20%) of **2AU-2** binding was observed, suggesting the compound is selective.

To further confirm the results of the filtre binding assays, we developed an assay that evaluates the potency of small molecules for inhibiting protein loading onto r(AUUCU)$_{11}$. It is known that DiGeorge syndrome critical region 8 (DGCR8), a protein that is involved in microRNA biogenesis, binds a wide variety of RNAs[41] and thus could be of potential use as the protein component in this assay. To establish that DGCR8Δ binds r(AUUCU)$_{11}$, a gel mobility shift assay was completed, affording a dissociation constant of 1.6 μM for the RNA–protein interaction (Supplementary Fig. 11). Such affinity is similar to those found between other RNA repeats and proteins, including r(CUG)^{exp}-MBNL1 (muscleblind-like 1), r(CAG)^{exp}-MBNL1, and r(CGG)^{exp}-DGCR8 (refs 20–22).

Time-resolved fluorescence resonance energy transfer (TR-FRET) assays have been developed to screen for inhibitors of each RNA–protein complexes mentioned above[20–22]. Thus, we developed a TR-FRET assay for r(AUUCU)$_{11}$ and DGCR8Δ analogously. Screening results showed that dimers are better inhibitors than the monomer and that **2AU-2** is the most potent inhibitor amongst them (Fig. 7a). At 1.5 μM concentration, **AU-azide** only inhibits ~25% of r(AUUCU)$_{11}$-DGCR8Δ complex formation while **2AU-2** inhibits ~70%. Further, the IC_{50} of **2AU-2** was 2.7-fold less than that of **AU-azide** (Supplementary Fig. 12). Thus, both the TR-FRET and filtre-binding assays establish that **2AU-2** is the most potent binder of r(AUUCU)$_{11}$ and that there is considerable enhancement for the dimer relative to the monomer.

Consistent with results from filtre binding and TR-FRET assays, selective binding of **2AU-2** to r(AUUCU)$_{11}$ over other RNA repeats was observed (Table 2 and Supplementary Fig. 9). The stoichiometry and the K_d of dimer **2AU-2** to r(AUUCU)$_{11}$ was measured. The number of **2AU-2** per r(AUUCU)$_{11}$ was approximately half that of **AU-azide** ($n = 7$ and $n = 13$, respectively), suggesting that each RNA-binding module interacts with each AU pair. The affinity of **2AU-2** for r(AUUCU)$_{11}$ is twofold greater than the monomer and **AU-azide** and **2AU-2** maintained positively cooperative binding to r(AUUCU)$_{11}$. Similar increases in affinity when increasing valency from $n = 1$ to $n = 2$ have been observed with other repeats[19,35]. Furthermore, we studied the binding of **2AU-2** and r(AUUCU)$_{11}$ under molecular crowding conditions that mimic a cellular environment by adding 20% (w/v) PEG 8000. A ninefold enhancement in binding was observed (EC_{50} of 16 ± 4 nM) relative to non-molecular crowding conditions (EC_{50} of 146 ± 8 nM; Supplementary Fig. 13).

The binding of **2AU-2** to a DNA hairpin that contains a stretch of four consecutive AT pairs was studied. Saturable binding was not observed (K_d > 10 μM; Supplementary Fig. 14). Interestingly, **AU-azide** binds to the DNA hairpin (EC_{50} = 116 ± 15 nM) with similar affinity as the AU-rich RNAs (EC_{50} = 111 ± 3 and 113 ± 8 nM for AAUU and AUAU, respectively; Supplementary Figs 14 and 15). Thus, a monomeric binding module with modest selectivity can be reprogrammed to afford highly selective compounds for repetitive sequences by linking the binding modules together.

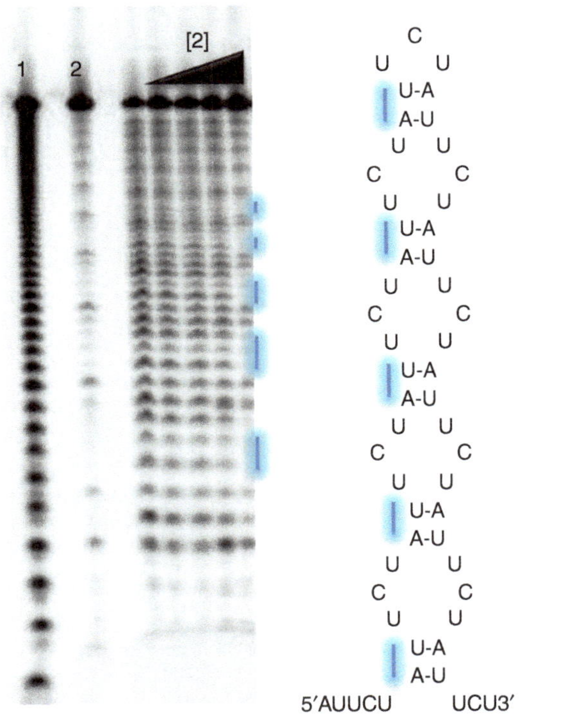

Figure 5 | Compound 2 protects r(AUUCU)$_{11}$ from cleavage by RNase V1. Left, representative gel image of the RNase protection assay. Lane 1, alkaline hydrolysis of r(AUUCU)$_{11}$; lane 2, r(AUUCU)$_{11}$ + No RNase V1; lanes 3-7, r(AUUCU)$_{11}$ + RNase V1 + **2** (0, 0.02, 0.2, 2 and 20 μM). Lines indicate sites of protection. Right, secondary structure of r(AUUCU)$_{11}$. Protected nucleotides are indicated with blue lines.

Recognition of r(AUUCU)$_{500}$ by 2AU-2 in cellular lysates. To investigate whether the designer dimer binds to r(AUUCU)$_{500}$, we developed a method named Chem-Quant-Seq. To enable Chem-Quant-Seq a biotinylated derivative, **2AU-2-Biotin**, was synthesized (Fig. 7b and Supplementary Fig. 23). After incubation of 250 nM of **2AU-2-Biotin** with total RNA isolated from cells that express r(AUUCU)$_{500}$ (ref. 33), bound RNAs were isolated with streptavidin resin. After extensive washing and elution of bound RNAs, the amount of r(AUUCU)$_{500}$ bound to **2AU-2-Biotin** was quantified by using quantitative PCR with reverse transcription (qRT–PCR). Results show that there is a significant enrichment of the r(AUUCU)$_{500}$ in the pulled-down material, as compared to 18S rRNA, showing that the compound indeed recognizes that target in the presence of cellular RNAs (Fig. 7c). To further assess the ability of the compound to recognize r(AUUCU)$_{500}$ over shorter repeats of non-pathogenic length, various concentrations of r(AUUCU)$_{11}$ in excess of r(AUUCU)$_{500}$ were added to the lysate and enrichment of r(AUUCU)$_{500}$ in the pulled-down fraction was quantified. Results from this competition experiment showed that 100-fold excess of r(AUUCU)$_{11}$ was required to decrease enrichment of r(AUUCU)$_{500}$ in the pulled-down fraction (Fig. 7d). Thus, **2AU-2-Biotin** recognizes long repeats over shorter ones. Furthermore, these data also point to cooperative binding of **2AU-2** to r(AUUCU) repeats as being a manner in which longer repeats are preferred over shorter ones. Positive cooperativity for binding to **2AU-2** was also observed *in vitro* (Hill coefficients; Table 2).

We further profiled target selectivity amongst 93 highly abundant transcripts in the pulled-down fraction via qRT-PCR. The RNAs span the diverse biology in the transcriptome, including rRNAs, mRNAs, small RNAs (sRNAs) and tRNAs. The rRNAs included all four rRNA subunits (18S, 28S 5S and 5.8S) and 45S rRNA. The 50 mRNAs were chosen from the most abundant mRNAs in HeLa cells for which there are established qRT–PCR primers[42]. The 17 sRNAs were selected from different structural and functional classes[43]: small nucleolar RNAs (HBII-85, HBII-420, U105 C/D Box snoRNAs, and ACA-16, ACA-44,

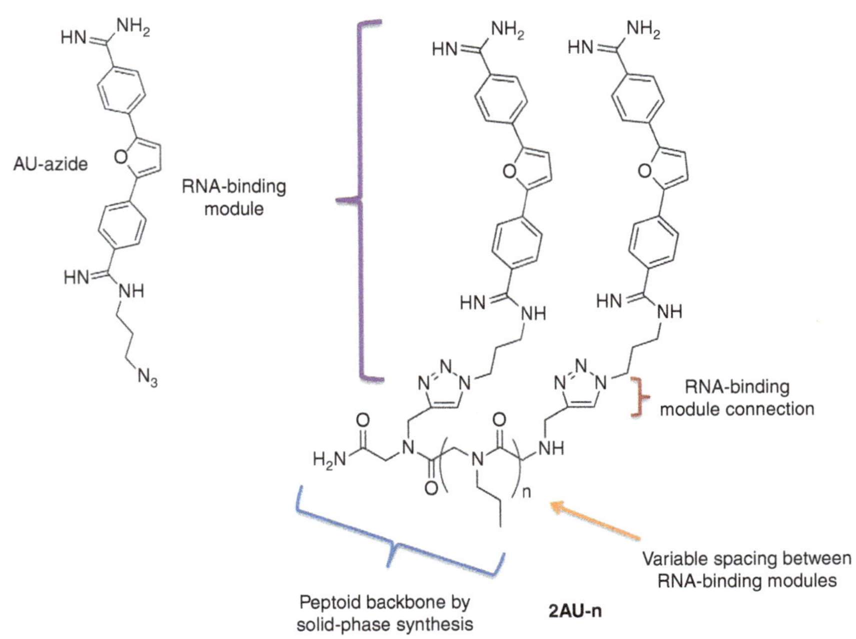

Figure 6 | Design of dimeric compounds to target r(AUUCU)^{exp}. The RNA-binding module, **AU-Azide** (left) was assembled onto various peptoid backbones to afford dimeric compounds (right). The **AU-azides** were separated by different distances by varying the number of spacing modules (n) to generate a series of compounds named **2AU-n**, where **2AU** denotes two RNA-binding modules and n denotes the number of spacing modules.

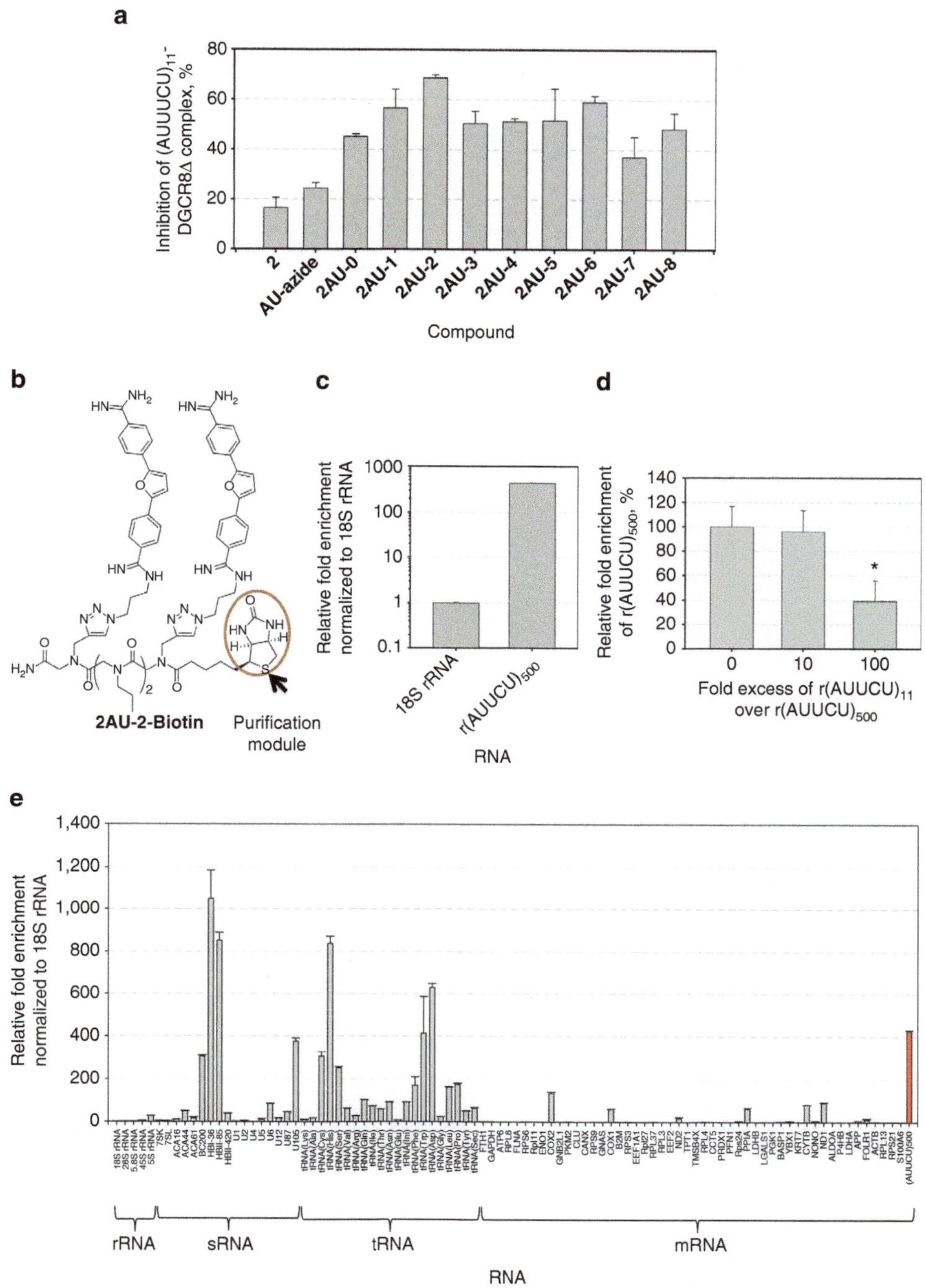

Figure 7 | 2AU-2 binds r(AUUCU) repeats *in vitro* and *in vivo* and inhibits protein binding. (**a**) quantification of inhibition of the r(AUUCU)$_{11}$-DGCR8Δ complex by monomeric and dimeric compounds (1.5 μM). (**b**) structure of **2AU-2-Biotin**. (**c**) relative fold enrichment of r(AUUCU)$_{500}$ pulled down by **2AU-2-Biotin** from total cellular RNAs. (**d**) **2AU-2-Biotin** selectively binds r(AUUCU)$_{500}$ over r(AUUCU)$_{11}$ as determined by a competition experiment. r(AUUCU)$_{500}$ levels were normalized to 18S rRNA. (**e**) Profiling of 93 highly abundant RNAs to determine whether they bind **2AU-2-Biotin** in the context of total cellular RNA.

ACA-61, HBI-36 H/ACA box snoRNAs), small cajal body-specific RNA (U87 scaRNA), small nuclear RNAs (U1, U2, U4, U5, U6 and U12 snRNAs), BC200 RNA, 7SK RNA and 7SL RNA. The 21 tRNAs profiled were randomly selected.

Only a fraction of these RNAs were enriched comparably to r(AUUCU)$_{500}$, showing **2AU-2** it possesses reasonable selectivity for the desired target (Fig. 7e). Of the nine RNAs that show significant enrichment, most are tRNAs. Since **2AU-2** was shown to pull down tRNAs, we studied its effect on translation in two different ways in HeLa cells: (i) by transfecting a plasmid that encodes green fluorescent protein (GFP); and (ii) co-transfecting plasmids that encode GFP and r(AUUCU)$_{500}$. GFP is a commonly used gene reporter in translational studies because of its visually identifiable characteristics[44]. After 24 h incubation, the fluorescence intensity of GFP was measured (Supplementary Fig. 16). Importantly, no change in the expression of GFP was observed in either system after treatment with **2AU-2**, as compared with an untreated control (Supplementary Fig. 16). Thus, **2AU-2** does not affect translation at its active concentrations where it improves SCA10-associated defects, 50 and 100 nM, *vide infra*. Interestingly, 40% of the amino acids in GFP are encoded by seven of the tRNAs pulled down by **2AU-2**. Taken together with the lack of toxicity in healthy and SCA10 patient-derived cells (*vide infra*), these data suggest that binding to tRNAs in a pull-down is not sufficient to elicit a biological effect.

Bioactivity of 2AU-2 in SCA10 patient-derived fibroblasts. Encouraged by these results, the bioactivity of **2AU-2** was assessed by measuring the ability of the compound to improve three downstream disease-associated defects in SCA10 patient-derived fibroblasts[33]. As mentioned above, caspase 3 is abnormally activated in SCA10 fibroblasts by greater than twofold[33]. Thus, the effect of **2AU-2** on the caspase-3 activity in control and SCA10 fibroblasts was measured. When SCA10-affected cells were treated with 50 and 100 nM of **2AU-2** for 48 h, caspase-3 activity was reduced to levels observed in healthy cells (Fig. 8a). For comparison, we also tested the monomer **AU-azide** and two dimers with reduced *in vitro* potencies, **2AU-3 and 2AU-4**. In contrast to **2AU-2**, treatment with 100 nM **AU-azide** reduced levels of over-activated caspase-3 activity by only ~30% while **2AU-3 and 2AU-4** were inactive up to 100 nM dosage despite showing activity *in vitro* (Supplementary Fig. 17). These results suggest that **2AU-3** and **2AU-4**, which have suboptimal distances between binding modules, may bind off-targets in cells. Thus, optimal bioactivity of **2AU-2** is a function of valency and the spacing between RNA-binding modules, which affords affinity and selectivity.

In agreement with the downregulated caspase-3 activity, **2AU-2** also reduced the mitochondrial abundance of PKCδ in SCA10 fibroblasts (Fig. 8c). PKCδ translocated to 70–80% of mitochondria in SCA10 cells, whereas only 10–15% mitochondria included PKCδ after treatment with 50 nM **2AU-2**, similar to levels observed in healthy fibroblasts. Next, the ability of **2AU-2** to disrupt formation of nuclear foci was measured. After 48 h incubation, **2AU-2** (50 nM) diminished ca. 70% of nuclear foci in SCA10 cells (Fig. 8b).

To confirm that improvement of SCA10-associated defects was due to binding of the compound to $r(AAUCU)^{exp}$ and not toxicity, we studied the cytotoxicity of **2AU-2** in healthy and SCA10 fibroblasts by measuring released lactate dehydrogenase (LDH)[45]. No significant toxicity of the compound was observed at its active concentration, 50 nM, in either healthy or SCA10 fibroblasts (Supplementary Fig. 18). Thus, the observed downregulation of caspase-3 activity and reduction of the mitochondrial abundance of PKCδ in SCA10 fibroblasts does not result from compound toxicity. Likewise, these results are consistent with our observation that caspase-3 activity is unchanged in healthy fibroblasts upon **2AU-2** treatment.

Further, the observed bioactivity of **2AU-2** is not due to reduced abundance of the mutant *ATXN10* transcript carrying expanded AUUCU repeats as determined by qRT–PCR analysis (Fig. 8d). That is, the compound works at the RNA level, not at the transcriptional level. Interestingly, previous studies have shown that silencing of the *ATXN10* transcript improves

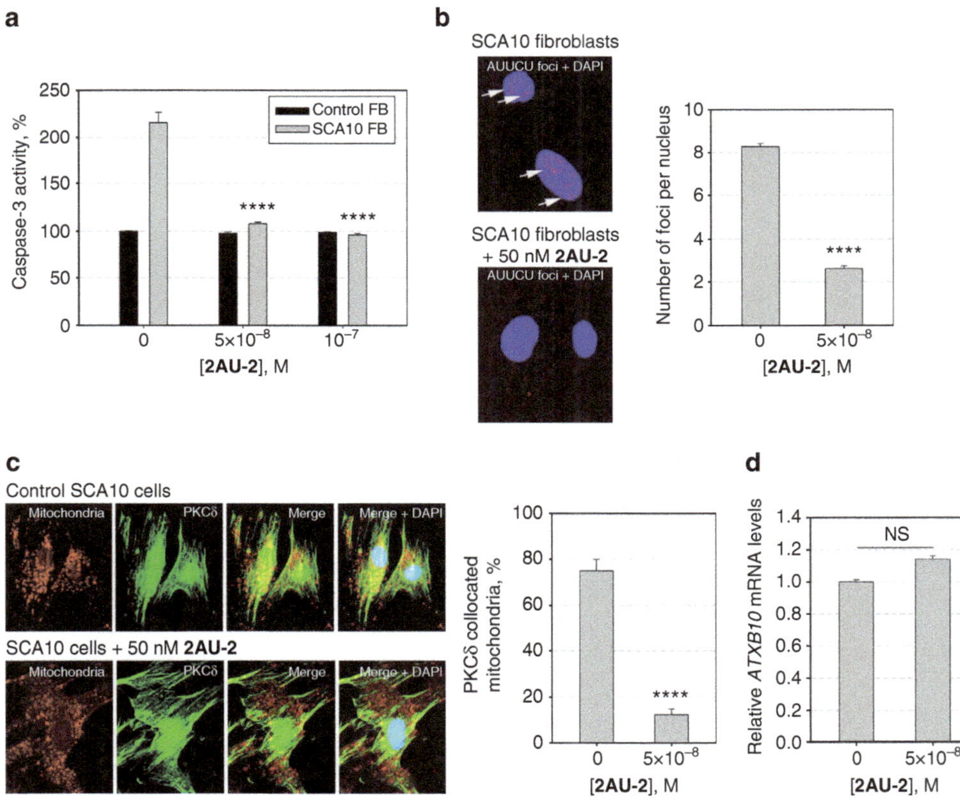

Figure 8 | Studying the bioactivity of 2AU-2 in SCA10 patient-derived fibroblasts. (**a**) Relative caspase-3 activities in normal and SCA10 fibroblasts before and after treatment with compound **2AU-2**. Caspase-3 in SCA10 was reduced to normal levels when cells were treated with 50 nM compound. (**b**) Confocal images showing $r(AUUCU)^{exp}$-containing nuclear foci in SCA10 cells without treatment (left, top) and after treatment with 50 nM **2AU-2** (left, bottom). Quantification of the number of $r(AUUCU)^{exp}$-containing nuclear foci (right) in treated and untreated cells. (**c**) confocal images showing mitochondrial translocation of PKCδ after treatment with 50 nM **2AU-2** (left, bottom) compared with control, untreated SCA10 cells (left, top). Quantification of PKCδ collocated to mitochondria in treated and untreated cells (right). (**d**) Real-time RT-PCR analysis of *ATXN10* mRNA expression levels after treatment with **2AU-2**. The amount of mRNA was normalized relative to *GAPDH* mRNA. Values are reported as the mean ± s.e. ($n = 6$). $^{****}P < 0.0001$ and 'ns' denotes $P > 0.05$, as compared with the untreated sample with a two-tailed Student *t*-test.

SCA10-associated defects[33] and some small molecules that improve microsatellite disease-associated defects work at the transcriptional level[46].

Taken together, **2AU-2** markedly improves defects from hnRNP K sequestration by r(AUUCU)^{exp} in SCA10 patient-derived cells, suggesting that the compound binds to the cellular target, r(AUUCU)^{exp}, and frees sequestered proteins. **2AU-2** is the most potent inhibitor known for a traditional non-covalent binder to repeat expansions.

Conclusion

To identify selective RNA base pair binders, we screened small molecules that have RNA-binding scaffolds. The *bis*-benzamidine compound **2** is a selective AU base pair binder, as it binds AU base pairs 40-fold more strongly than GC base pairs. This compound was applied to target the pentanucleotide r(AUUCU) expansion that causes SCA10. The repeat periodically displays 5′AU/UA3′ base pair steps in its secondary structure. To improve affinity and selectivity, we modularly assembled **2** and determined that the optimal distance between RNA-binding modules was afforded by two propylamine spacing modules, or **2AU-2**. **2AU-2** significantly improves SCA10-associated defects to wild-type levels when patient-derived fibroblasts are treated with 50 nM compound. It is the first bioactive small molecule targeting r(AUUCU)^{exp}. The potent bioactivity of **2AU-2** suggests that base pair-targeting RNA modules could have broad utility to provide bioactive compounds targeting other RNAs in the transcriptome.

Methods

Instrumentation. All pH measurements were performed at room temperature using a Mettler Toledo SG2 pH metre that was standardized at pH 4.0, 7.0, and 10.0. Absorption and emission spectra were measured using SpectraMax M5 plate reader (Molecular Devices, Inc.). Sigma Plot (version 11.0) was used for all curve fitting.

Small molecules. All small modules were procured from the National Cancer Institute (NCI) or The Scripps Research Institute. Emission spectra (excitation: 300 nm, cutoff: 325 nm, emission: 330–600 nm) of all compounds were measured in a 384-well plate (Greiner Low-Volume 784076) to select fluorescent compounds for screening (50 μM compound in 1 × Screening Buffer (8 mM Na_2HPO_4, pH 7.0, 185 mM NaCl, 0.1 mM EDTA)).

Compound purification and analysis. Preparative HPLC was performed using a Waters 1525 Binary HPLC pump equipped with a Waters 2487 dual-absorbance detector system and a Waters Sunfire C18 OBD 5-μm 19 × 150 mm column. Absorbance was monitored at 220 and 345 nm. A gradient of 20–100% methanol in H_2O with 0.1% trifluoroacetic acid (TFA) over 60 min was used for compound purification. Analytical HPLC was performed using a Waters Symmetry C18 5 μm 4.6 × 150 mm column. Compounds were analysed using a gradient of 20–60% MeOH in H_2O with 0.1% TFA over 30 min. All compounds evaluated had ≥95% purity as determined by analytical HPLC. Mass spectrometry was performed with an Applied Biosystems MALDI ToF/ToF Analyzer 4800 Plus and Microflex (Bruker) using an α-hydroxycinnamic acid matrix. See the Supplementary Information for details of compound synthesis and compound characterization.

Oligonucleotide preparation and purification. The RNAs used in fluorescence binding assays, nuclease mapping, and filtre-binding assays were purchased from Dharmacon. The ACE protecting groups were cleaved by using Dharmacon's deprotection buffer by incubating at 60 °C for 30 min. The samples were lyophilized, resuspended in water and gel purified. Concentrations were determined by absorbance using a Beckman Coulter DU800 ultraviolet–visible spectrophotometer at 85 °C. Extinction coefficients (at 260 nm) were calculated using the HyTher server, which uses nearest-neighbour parameters[47–49].

Initial screen for small molecules that bind RNA base pairs. RNAs were folded in 1 × Screening Buffer at 95 °C for 2 min followed by slowly cooling to room temperature on the bench top. A 10-μl aliquot of a 1 μM RNA solution was dispensed into each well of a black 384-well plate (Greiner Low-Volume 784076) using an Aurora Discovery FRD-1B liquid dispenser. A 10-nl aliquot of a 2.5 mM stock of small molecule was pinned into each well using Biomek NXP Laboratory Automation Workstation that was equipped with a 384-pin head. The solution was incubated at room temperature for 30 min. Fluorescence intensity was measured using the maximum excitation/emission wavelength for each compound and the change in fluorescence was calculated by the ratio of F/F_0 where F is the fluorescence intensity in the presence of RNA and F_0 is the fluorescence intensity in the absence of RNA.

Compounds were scored as hits if a >20% change in emission (either enhancement or quenching) was observed upon incubation with RNA. The selectivity of a small molecule was computed by comparing the relative change in emission when incubated different RNAs. Statistically significant differences were calculated by using one-way analysis of variance function in Sigma Plot (version 11.0); compounds that had a P value of <0.05 (95% confidence) were chosen as selective binders.

Chemoinformatic analysis. To identify the chemical substructures that facilitate binding, hit compounds were tested by an automated R-group analysis (Tripod Development; Division of preclinical Innovation, National Center for Advancing Translational Sciences: http://tripod.nih.gov/?p=46). The functional groups that provided recognition for each type of RNA were then compared.

Fluorescence binding assays. Direct binding assays for all selective binders were performed. RNAs were folded as described above. Binding assays were performed with a constant compound concentration (1 or 3 μM) and serial dilutions of RNA or DNA in 1 × Screening Buffer. For molecular crowding experiments, PEG 8000 was added to a final concentration of 20% (w/v) to the folded RNA and to the solution used for serial dilutions. After a 20 min incubation, fluorescence intensity was measured. The resulting curves were fit to the following equation to determine EC_{50} values:

$$y = B + \frac{A - B}{1 + \left(\frac{EC50}{x}\right)^{\text{hillslope}}} \tag{1}$$

where y is fluorescence intensity, x is the concentration of RNA, B is the minimum fluorescence; A is the maximum fluorescence; and the EC_{50} is the concentration of RNA where half of the compound is bound.

Two types of plots were constructed to determine stoichiometries and K_d's: fluorescence versus [nucleic acid]/[ligand] to determine stoichiometry and fraction-bound/[nucleic acid] versus fraction bound to determine K_d's. Stoichiometries were determined from the former plots by fitting each of the two slopes (pre-saturated and saturated portions of the curves) to a line[50]. For unsaturated binding curves, the saturated portions of the curves were estimated by the fitted data from equation (1). The point at which the two equations intersect affords the stoichiometry. The K_d's were determined by fitting fraction bound/[nucleic acid] versus fraction bound to equation (2):

$$\frac{v}{[L]} = \frac{N(1 - lv/N)}{k}\left(\frac{1 - lv/N}{1 - (l-1)v/N}\right)^{l-1} \tag{2}$$

where v is the moles of RNA lattice bound per moles of ligand, $[L]$ is the concentration of ligand, N is the number of repeating units on the RNA, l is the number of consecutive lattice units occupied by the ligand, and k is the microscopic dissociation constant.

Nuclease mapping of the small molecule-binding site. r(AUUCU)$_{11}$ was radioactively labelled at the 5′ end with [γ-^{32}P] ATP (Perkin Elmer) and T4 polynucleotide kinase (New England Biolabs) using standard methods and gel purified to homogeneity[51]. The RNA was folded by incubation at 60 °C for 5 min in 1 × RNA Structure Buffer (Ambion) followed by slow cooling to room temperature. Serially diluted concentrations of the inhibitor was added to the RNA solution and incubated at room temperature for 15 min. RNase V1 (Ambion) was added to the RNA-inhibitor complex to a final concentration of 5 μU μl^{-1} and the samples were incubated at room temperature for 60 min. RNase V1 was then inactivated by heating at 95 °C for 1 min, and cleavage products were separated on a denaturing 20% polyacrylamide gel. A hydrolysis ladder was prepared by using Alkaline Hydrolysis Buffer (Ambion) and the manufacturer's protocol.

Chemical syntheses of dimeric compounds. Details of compound syntheses and characterization are provided in Supplementary Figs 19–24 and Supplementary Table 4.

Screening of dimeric compounds for binding to r(AUUCU)$_{11}$ by filtre binding. To determine the optimal distance between RNA-binding modules, a library of dimers was synthesized and screened by using a filtre-binding assay. Radioactively labelled r(AUUCU)$_{11}$ (100 nM) was folded in 1 × PBS buffer (pH 7.4) containing 1 mM $MgCl_2$ by incubation at 60 °C for 5 min followed by slow cooling to room temperature. BSA was added to a final concentration of 50 μg ml^{-1} followed by addition of 1 μM compound. The samples were incubated at room temperature for 15 min. Nitrocellulose and nylon membranes were incubated in 1 × filtre binding assay buffer (1 × PBS buffer (pH 7.4) containing 1 mM $MgCl_2$ and 50 μg ml^{-1} BSA). Bound and unbound RNA were separated using a Dot-Blot apparatus followed by washing with 1 × filtre binding assay buffer. The membranes were exposed to a phosphorimager screen and imaged using a Molecular Dynamics Typhoon phosphorimager. The amount of r(AUUCU)$_{11}$ bound to each membrane was quantified using QuantityOne software (BioRad).

Mobility shift assay of r(AUUCU)$_{11}$ with DGCR8Δ. r(AUUCU)$_{11}$ was 5′-end labelled as described above and folded in 1 × Folding Buffer (20 mM HEPES, pH 7.5, 110 mM KCl, and 110 mM NaCl) by incubation at 60 °C for 5 min followed by slow cooling to room temperature on the bench top. The buffer was adjusted to 1 × TR-FRET Assay Buffer (20 mM HEPES pH 7.5, 110 mM KCl, 110 mM NaCl, 0.1% BSA, 2 mM $MgCl_2$, 2 mM $CaCl_2$, 0.05% Tween-20 and 5 mM DTT) and various concentrations of DGCR8Δ were added. The samples were incubated at room temperature for 15 min and loaded onto a pre-chilled native 5% polyacrylamide gel. The gel was imaged and quantified as described above. The resulting curves were fit to equation (3):

$$y = \frac{xB_{max}}{K_d + x} \quad (3)$$

where y is percentage of bounded DGCR8Δ, x is the concentration of protein, B_{max} is maximum percentage of protein bound (restrained to equal 100%) and k_d is dissociation constant.

Determination of compound potency via a TR-FRET assay. TR-FRET assays were completed as previously described[21] with the following modifications. After folding the RNA, compound was added and incubated for 15 min at room temperature followed by addition of DGCR8Δ. The final concentrations of r(AUUCU)$_{11}$ and DGCR8Δ were 60 and 40 nM, respectively. TR-FRET was measured after an additional 30-min incubation at room temperature. IC_{50} values were calculated by curve fitting using equation (1).

Pull-down of 2AU-2′s cellular targets. HeLa cells were maintained as monolayers in growth medium (1 × DMEM, 10% fetal bovine serum and 1 × GlutaMax (Invitrogen)). Cells were plated in 10-cm dishes and grown to ~90% confluency and then transfected with a plasmid encoding (AUUCU)$_{500}$ using Lipofectamine 2,000 (Invitrogen) per the manufacturer's recommended protocol. Cells were collected 18–24 h post-transfection and total RNA was extracted by using Trizol reagent (Ambion) according to the manufacturer's protocol. After RQ1 DNase (Promega) treatment, the DNase was removed by phenol:chloroform extraction and total RNA was ethanol precipitated. Next, 100 µg of total RNA was folded in 1 × Screening Buffer by heating at 75 °C for 5 min and cooling to room temperature slowly. The folded RNAs were incubated with **2AU-2-Biotin** for 30 min at room temperature. The solution was then incubated with streptavidin beads (250 µl of slurry, Sigma-Aldrich) for 30 min at room temperature with gentle shaking. The supernatant (containing unbound RNAs) was removed, and the beads were washed with 250 µl 1 × Screening Buffer after gentle shaking for 5 min at room temperature. Bound RNA was released from the beads by heating the beads in 300 µl H_2O at 80 °C for 3 min twice. The solution containing bound RNAs was concentrated to 5–50 µl by vacuum concentration. Complementary DNA (cDNA) was generated from 40 ng of RNA using a qScript cDNA Synthesis Kit (Quanta Biosciences) per the manufacturer's protocol. qPCR was performed on an ABI 7900 HT Real-Time PCR System using the following primers to detect the r(AUUCU)$_{500}$-containing RNA: 5′-AGTCTCTCT ATGTTGCCCAGG-3′ and 5′-ACTTCCCGAAACACCGTCTC-3′. The relative fold enrichment of the RNA pulled by the compound was calculated by normalization to 18S rRNA.

Profiling of cellular RNAs pulled down by 2AU-2-Biotin. Profiling was completed by qRT-PCR as described above. The 93 highly abundant RNAs, including rRNAs, sRNA, tRNAs and mRNAs, were selected based on a previous report[42]. Primer sets for mRNAs were obtained from RTPrimerDB[52] (www.rtprimerdb.org) and qPrimerDepot (http://primerdepot.nci.nih.gov). The DNA sequences of tRNAs were obtained from tRNAdb[53] (http://trna.bioinf.uni-leipzig.de/) and primer sets were designed by using Primer 3 software (http://frodo.wi.mit.edu/primer3/). All sequences of primers are listed in Supplementary Table 3.

Cell culture of SCA10 fibroblasts. SCA10 fibroblasts[33] were cultured in MEM with Eagle–Earle salt and 2 mM L-glutamine containing 15% fetal bovine serum and antibiotic in 5% CO_2 at 37 °C in 75-cm^2 flasks. Compounds were dissolved in 50% DMSO and added to the SCA10 cells at the indicated concentrations. Fresh medium with drug was added to the cells after every 12 h and cells were collected after 48 h for the caspase-3 assay, detection of AUUCU RNA foci, or for analysing the subcellular translocation of protein kinase C δ (PKCδ).

Translocation of PKCδ after drug treatment. SCA10 fibroblasts (2 × 10^4 cells) were seeded in chamber slides. When cells were 80–90% confluent, fresh cell culture medium containing compound was added to the cells and incubated for 48 h. The drug-treated and control cells were then incubated with mitotracker deep red 633 (Invitrogen, USA) at a concentration of 250 nM in cell culture medium and incubated at 37 °C for 30 min. The cells were then washed three times with ice-cold 1 × PBS, fixed with 4% paraformaldehyde for 30 min at room temperature, washed three times with 1 × PBS and stored in 70% ethanol for up to 24 h. Cells were blocked with DAKO antibody blocking solution (serum free) and later double stained with anti-PKCδ 1:500 in DAKO antibody diluent. Goat anti-mouse 488 was used to identify PKCδ. Fluorescent photomicrographs were taken using an Hamamatsu Camera Controller using DP controller software.

Fluorescent *in situ* hybridization to detect AUUCU RNA foci. SCA10 and control fibroblasts (2 × 10^4 cells) were seeded in chamber slides. When the cells were 60–70% confluent, cell culture medium containing compound was added to the cells and incubated for 48 h. After incubating with compound, the cells were fixed with 4% paraformaldehyde for 30 min at room temperature, and washed three to four times with ice-cold 1 × PBS. The AUUCU RNA foci were detected using a Cy3-labelled (AGAAU)$_{10}$ RNA oligonucleotide probe as described previously[33]. In brief, the control and SCA10 cells were pre-hybridized at 65 °C in RNA Hybridization Buffer for 1.5 h, and hybridized overnight in hybridization buffer containing 250 ng of (AGAAU)$_{10}$ Cy3-labelled RNA oligo at 45 °C. Slides were rinsed with 1 × PBS three times and extensively washed 4 × 5 min to remove all non-specific binding of Cy3-labelled RNA probes. Slides were then mounted with DAPI mounting medium and fluorescent images were taken using a confocal microscope.

Cytotoxicity of 2AU-2. SCA10 and healthy fibroblasts were treated either with **2AU-2** (50, 100 and 500 nM) or vehicle for 24 h. Later, the culture medium containing LDH was collected and the amount of LDH in the medium was quantified using TOX7 cell toxicity assay kit (Sigma-Aldrich).

References

1. Bartel, D. P. MicroRNAs: genomics, biogenesis, mechanism, and function. *Cell* **116**, 281–297 (2004).
2. Tenson, T. & Mankin, A. Antibiotics and the ribosome. *Mol. Microbiol.* **59**, 1664–1677 (2006).
3. Winkler, W., Nahvi, A. & Breaker, R. R. Thiamine derivatives bind messenger RNAs directly to regulate bacterial gene expression. *Nature* **419**, 952–956 (2002).
4. Winkler, W. C., Cohen-Chalamish, S. & Breaker, R. R. An mRNA structure that controls gene expression by binding FMN. *Proc. Natl Acad. Sci. USA* **99**, 15908–15913 (2002).
5. Harries, L. W. Long non-coding RNAs and human disease. *Biochem. Soc. Trans.* **40**, 902–906 (2012).
6. Shen, L. X., Basilion, J. P. & Stanton, Jr. V. P. Single nucleotide polymorphisms can cause different structural folds of mRNA. *Proc. Natl Acad. Sci. USA* **96**, 7871–7876 (1999).
7. Sierakowska, H., Agrawal, S. & Kole, R. Antisense oligonucleotides as modulators of pre-mRNA splicing. *Methods Mol. Biol.* **133**, 223–233 (2000).
8. Brunberg, J. A. *et al.* Fragile X premutation carriers: characteristic MR imaging findings of adult male patients with progressive cerebellar and cognitive dysfunction. *Am. J. Neuroradiol.* **23**, 1757–1766 (2002).
9. Brook, D. *et al.* Molecular basis of myotonic dystrophy: expansion of a trinucleotide (CTG) repeat at the 3′ end of a transcript encoding a protein kinase family member. *Cell* **68**, 799–808 (1992).
10. Matsuura, T. *et al.* Large expansion of the ATTCT pentanucleotide repeat in spinocerebellar ataxia type 10. *Nat. Genet.* **26**, 191–194 (2000).
11. Liquori, C. L. *et al.* Myotonic dystrophy type 2 caused by a CCTG expansion in intron 1 of *ZNF9*. *Science* **293**, 864–867 (2001).
12. The Huntington's Disease Collaborative Research Group. A novel gene containing a trinucleotide repeat that is expanded and unstable on Huntington's disease chromosomes. *Cell* **72**, 971–983 (1993).
13. Gallego, J. & Varani, G. Targeting RNA with small-molecule drugs: therapeutic promise and chemical challenges. *Acc. Chem. Res.* **34**, 836–843 (2001).
14. Poehlsgaard, J. & Douthwaite, S. The bacterial ribosome as a target for antibiotics. *Nat. Rev. Microbiol.* **3**, 870–881 (2005).
15. Johnson, L. F., Williams, J. G., Abelson, H. T., Green, H. & Penman, S. Changes in RNA in relation to growth of the fibroblast. III. Posttranscriptional regulation of mRNA formation in resting and growing cells. *Cell* **4**, 69–75 (1975).
16. Blount, K. F., Wang, J. X., Lim, J., Sudarsan, N. & Breaker, R. R. Antibacterial lysine analogs that target lysine riboswitches. *Nat. Chem. Biol.* **3**, 44–49 (2007).
17. Schramm, V. L. Enzymatic transition states, transition-state analogs, dynamics, thermodynamics, and lifetimes. *Annu. Rev. Biochem.* **80**, 703–732 (2011).

18. Childs-Disney, J. L., Hoskins, J., Rzuczek, S., Thornton, C. & Disney, M. D. Rationally designed small molecules targeting the RNA that causes myotonic dystrophy type 1 are potently bioactive. *ACS Chem. Biol.* **7,** 856–862 (2012).
19. Lee, M. M., Pushechnikov, A. & Disney, M. D. Rational and modular design of potent ligands targeting the RNA that causes myotonic dystrophy 2. *ACS Chem. Biol.* **4,** 345–355 (2009).
20. Parkesh, R. *et al.* Design of a bioactive small molecule that targets the myotonic dystrophy type 1 RNA via an RNA motif-ligand database & chemical similarity searching. *J. Am. Chem. Soc.* **134,** 4731–4742 (2012).
21. Disney, M. D. *et al.* A small molecule that targets r(CGG)^{exp} and improves defects in fragile X-associated tremor ataxia syndrome. *ACS Chem. Biol* **7,** 1711–1718 (2012).
22. Kumar, A. *et al.* Chemical correction of pre-mRNA splicing defects associated with sequestration of muscleblind-like 1 protein by expanded r(CAG)-containing transcripts. *ACS Chem. Biol.* **7,** 496–505 (2012).
23. Thomas, J. R. & Hergenrother, P. J. Targeting RNA with small molecules. *Chem. Rev.* **108,** 1171–1224 (2008).
24. Guan, L. & Disney, M. D. Recent advances in developing small molecules targeting RNA. *ACS Chem. Biol.* **7,** 73–86 (2012).
25. Mathews, D. H. *et al.* Incorporating chemical modification constraints into a dynamic programming algorithm for prediction of RNA secondary structure. *Proc. Natl Acad. Sci. USA* **101,** 7287–7292 (2004).
26. Carlson, C. B., Vuyisich, M., Gooch, B. D. & Beal, P. A. Preferred RNA binding sites for a threading intercalator revealed by in vitro evolution. *Chem. Biol.* **10,** 663–672 (2003).
27. Luedtke, N. W. *et al.* The DNA and RNA specificity of eilatin Ru(II) complexes as compared to eilatin and ethidium bromide. *Nucleic Acids Res.* **31,** 5732–5740 (2003).
28. Jin, E. *et al.* Aminoglycoside binding in the major groove of duplex RNA: the thermodynamic and electrostatic forces that govern recognition. *J. Mol. Biol.* **298,** 95–110 (2000).
29. Wilson, W. D., Ratmeyer, L., Zhao, M., Strekowski, L. & Boykin, D. The search for structure-specific nucleic acid-interactive drugs: effects of compound structure on RNA versus DNA interaction strength. *Biochemistry* **32,** 4098–4104 (1993).
30. Zhao, M. *et al.* Small changes in cationic substituents of diphenylfuran derivatives have major effects on the binding affinity and the binding mode with RNA helical duplexes. *Bioorg. Med. Chem.* **3,** 785–794 (1995).
31. Handa, V., Yeh, H. J. C., McPhie, P. & Usdin, K. The AUUCU repeats responsible for spinocerebellar ataxia type 10 form unusual RNA hairpins. *J. Biol. Chem.* **280,** 29340–29345 (2005).
32. Teive, H. A. G. *et al.* Spinocerebellar ataxia type 10 - A review. *Parkinsonism Relat. Disord.* **17,** 655–661 (2011).
33. White, M. C. *et al.* Inactivation of hnRNP K by expanded intronic AUUCU repeat induces apoptosis via translocation of PKCδ to mitochondria in spinocerebellar ataxia 10. *PLoS Genet.* **6,** e1000984 (2010).
34. Jin, P., Alisch, R. S. & Warren, S. T. RNA and microRNAs in fragile X mental retardation. *Nat. Cell Biol.* **6,** 1048–1053 (2004).
35. Pushechnikov, A. *et al.* Rational design of ligands targeting triplet repeating transcripts that cause RNA dominant disease: application to myotonic muscular dystrophy type 1 and spinocerebellar ataxia type 3. *J. Am. Chem. Soc.* **131,** 9767–9779 (2009).
36. Childs-Disney, J. L., Tsitovich, P. B. & Disney, M. D. Using modularly assembled ligands to bind RNA internal loops separated by different distances. *Chembiochem* **12,** 2143–2146 (2011).
37. Lee, M. M. *et al.* Controlling the specificity of modularly assembled small molecules for RNA via ligand module spacing: targeting the RNAs that cause myotonic muscular dystrophy. *J. Am. Chem. Soc.* **131,** 17464–17472 (2009).
38. Rzuczek, S. G. *et al.* Features of modularly assembled compounds that impart bioactivity against an RNA target. *ACS Chem. Biol.* **8,** 2312–2321 (2013).
39. Fu, H. Y. & Doucet, H. Methyl 2-furoate: an alternative reagent to furan for palladium-catalysed direct arylation. *Eur. J. Org. Chem.* **2011,** 7163–7173 (2011).
40. Das, B. P. & Boykin, D. W. Synthesis and antiprotozoal activity of 2,5-bis (4-guanylphenyl)furans. *J. Med. Chem.* **20,** 531–536 (1977).
41. Roth, B. M., Ishimaru, D. & Hennig, M. The core microprocessor component DiGeorge syndrome critical region 8 (DGCR8) is a nonspecific RNA-binding protein. *J. Biol. Chem.* **288,** 26785–26799 (2013).
42. Nagaraj, N. *et al.* Deep proteome and transcriptome mapping of a human cancer cell line. *Mol. Syst. Biol.* **7,** 548 (2011).
43. Galiveti, C. R., Rozhdestvensky, T. S., Brosius, J., Lehrach, H. & Konthur, Z. Application of housekeeping npcRNAs for quantitative expression analysis of human transcriptome by real-time PCR. *RNA* **16,** 450–461 (2010).
44. Chalfie, M., Tu, Y., Euskirchen, G., Ward, W. & Prasher, D. Green fluorescent protein as a marker for gene expression. *Science* **263,** 802–805 (1994).
45. Decker, T. & Lohmann-Matthes, M.-L. A quick and simple method for the quantitation of lactate dehydrogenase release in measurements of cellular cytotoxicity and tumor necrosis factor (TNF) activity. *J. Immunol. Methods* **115,** 61–69 (1988).
46. Coonrod, L. A. *et al.* Reducing levels of toxic RNA with small molecules. *ACS Chem. Biol.* **8,** 2528–2537 (2013).
47. Peyret, N., Seneviratne, P. A., Allawi, H. T. & SantaLucia, Jr. J. Nearest-neighbor thermodynamics and NMR of DNA sequences with internal A.A, C.C, G.G, and T.T mismatches. *Biochemistry* **38,** 3468–3477 (1999).
48. SantaLucia, Jr. J. A unified view of polymer, dumbbell, and oligonucleotide DNA nearest-neighbor thermodynamics. *Proc. Natl Acad. Sci. USA* **95,** 1460–1465 (1998).
49. Puglisi, J. D. & Tinoco, Jr. I. Absorbance melting curves of RNA. *Methods Enzymol.* **180,** 304–325 (1989).
50. Tse, W. C. & Boger, D. L. A fluorescent intercalator displacement assay for establishing DNA binding selectivity and affinity. *Acc. Chem. Res.* **37,** 61–69 (2004).
51. Sambrook, J., Fritsch, E. F. & Maniatis, T. *Molecular Cloning* 2nd edn. (Cold Spring Harbor Laboratory, 1989).
52. Lefever, S., Vandesompele, J., Speleman, F. & Pattyn, F. RTPrimerDB: the portal for real-time PCR primers and probes. *Nucleic Acids Res.* **37,** D942–D945 (2009).
53. Juhling, F. *et al.* tRNAdb 2009: compilation of tRNA sequences and tRNA genes. *Nucleic Acids Res.* **37,** D159–D162 (2009).

Acknowledgements

This work was funded by the National Institutes of Health (DP1 NS096898) to M.D.D., a John Sealy Grant to P.S.S., a FRAXA postdoctoral fellowship to W-Y.Y. and The Scripps Research Institute. M.D.D. is a Camille & Henry Dreyfus Teacher-Scholar.

Author contributions

M.D.D. provided conceptual framework. M.D.D., W-Y.Y., R.G. and P.S.S. designed the experiments. W-Y.Y. and R.G. performed the experiments and statistical analysis. M.S. performed chemical similarity searching. M.D.D., W-Y.Y., R.G. and P.S.S. wrote the manuscript. All authors reviewed the final version of the manuscript.

Additional information

Competing financial interests: The authors declare no competing financial interests.

18

Discovery of cahuitamycins as biofilm inhibitors derived from a convergent biosynthetic pathway

Sung Ryeol Park[1,*], Ashootosh Tripathi[1,*], Jianfeng Wu[2], Pamela J. Schultz[1], Isaiah Yim[1,†], Thomas J. McQuade[1], Fengan Yu[1], Carl-Johan Arevang[1,†], Abraham Y. Mensah[1,†], Giselle Tamayo-Castillo[3,4], Chuanwu Xi[2] & David H. Sherman[1,5,6,7]

Pathogenic microorganisms often have the ability to attach to a surface, building a complex matrix where they colonize to form a biofilm. This cellular superstructure can display increased resistance to antibiotics and cause serious, persistent health problems in humans. Here we describe a high-throughput *in vitro* screen to identify inhibitors of *Acinetobacter baumannii* biofilms using a library of natural product extracts derived from marine microbes. Analysis of extracts derived from *Streptomyces gandocaensis* results in the discovery of three peptidic metabolites (cahuitamycins A–C), with cahuitamycin C being the most effective inhibitor ($IC_{50} = 14.5\,\mu M$). Biosynthesis of cahuitamycin C proceeds via a convergent biosynthetic pathway, with one of the steps apparently being catalysed by an unlinked gene encoding a 6-methylsalicylate synthase. Efforts to assess starter unit diversification through selective mutasynthesis lead to production of unnatural analogues cahuitamycins D and E of increased potency ($IC_{50} = 8.4$ and $10.5\,\mu M$).

[1] Life Sciences Institute, University of Michigan, Ann Arbor, Michigan 48109, USA. [2] Department of Environmental Health Sciences, University of Michigan School of Public Health, Ann Arbor, Michigan 48109, USA. [3] Unidad Estratégica de Bioprospección, Instituto Nacional de Biodiversidad (INBio), Santo Domingo de Heredia 223100, Costa Rica. [4] CIPRONA, Escuela de Química, Universidad de Costa Rica, 2060 San José, Costa Rica. [5] Department of Medicinal Chemistry, University of Michigan, Ann Arbor, Michigan 48109, USA. [6] Department of Chemistry, University of Michigan, Ann Arbor, Michigan 48109, USA. [7] Department of Microbiology and Immunology, University of Michigan, Ann Arbor, Michigan 48109, USA. * These authors contributed equally to this work. † Present Addresses: National Cancer Institute, National Institutes of Health, Frederick, Maryland 21702, USA (I.Y.); Karolinska Institutet Science Park, Retzius vag 8, 17165 Solna, Sweden (C.-J.A.); Department of Pharmacognosy, College of Health Sciences, Kwame Nkrumah University of Science & Technology, Kumasi, Ghana (A.Y.M.). Correspondence and requests for materials should be addressed to C.X. (email: cxi@umich.edu) or to D.H.S. (email: davidhs@umich.edu).

Widespread antibiotic resistance is currently posing a grave health burden through a multitude of serious infections[1]. The rise in bacterial adaptation can be directly correlated to the paucity of novel classes of antimicrobial agents[2]. In the past few decades, synthetic tailoring has been the primary strategy for enhancing established core scaffolds through analogue generation. Although this approach has been fruitful, no major classes of new antibiotics were introduced between 1962 and 2000 (ref. 3). Therefore, to restore robust access to effective therapeutic agents, it is imperative that we engage in aggressive efforts to discover novel chemical entities with unique microbial targets[2,4].

Acinetobacter baumannii belongs to the ESKAPE group of major nosocomial opportunistic resistance pathogens (*Enterococcus faecalis, Staphylococcus aureus, Klebsiella pneumoniae, A. baumannii, Pseudomonas aeruginosa* and *Enterobacter* sp.), which can spread epidemically among patients causing ventilator-associated pneumonia and bacteremia, with mortality rates as high as 60%, representing a paradigm of pathogenesis, transmission and resistance[5]. Numerous reports have also shown startling emergence of multidrug resistant *A. baumannii* in hospitals, and also identification of pan-drug-resistant strains at some locations[5,6]. *A. baumannii* strains possess both intrinsic resistance to antibiotics and a facile ability to acquire genes encoding resistance determinants. In addition, antibiotic resistance of this pathogenic microbe appears to be mediated by their propensity to form biofilms with a highly structured extracellular polymeric matrix, and includes the ability to colonize medical devices. When attached, bacterial cells that comprise the biofilm possess 10–1,000-fold lower susceptibility towards antimicrobial agents compared with planktonic forms[7]. Moreover, despite the central role that bacterial biofilms play during infection, there are currently no drugs specifically targeting biofilms in clinical trials to date[8,9]. Therefore, a precision medicine is urgently required for vulnerable patients to avoid potential life-threatening infections. Recently, developing biofilm inhibitors has become a priority compared with biofilm disruption due to the advantage provided by preventing subsequent dispersion of cells that may have acquired drug resistance[10–12]. Although biofilm control by drug targeting has become a high priority objective[7,13], marine microbes as a source of novel chemical entities remain relatively underexplored[14,15].

In our continuing effort to identify new structural classes of antibiotics[4], we employed static- and flow-based high-throughput screening (HTS) assays to survey our natural product extract (NPE) library in the search for new inhibitors of biofilm formation[16]. Here we describe the discovery of three novel secondary metabolites, whose stable production and full structural identification required ribosome engineering, and was facilitated by biosynthetic gene cluster characterization. In addition, we show that the cahuitamycins are derived from two independent starter unit pathways, one of which is genetically unlinked to the core cluster. The convergent pathway enabled us to perform directed pathway engineering to generate a more potent molecule, selectively. Furthermore, mutasynthetic efforts on the ribosomally modulated strain generated two additional novel compounds with enhanced activity against biofilm formation.

Results and Discussion

HTS for biofilm inhibitors against *A. baumannii*. In an effort to discover new antibiotics to combat hospital-associated infections in patients, we proceeded to screen our marine microbial-derived NPE library to identify biofilm inhibitors of *A. baumannii*. Natural products account for the majority of currently marketed drugs[17], and the marine microbiome represents a potential new source of unique chemical entities[18]. We adapted a crystal-violet-based high-throughput assay[19] that was queried against a library of 9,831 marine microbial-derived NPEs to identify the extracts inhibiting biofilm formation as the primary screen. The active extracts were further prioritized by setting the inhibition threshold to 50% followed by a dose–response assay, yielding 31 active NPEs (Supplementary Figs 1–2). A second round of microbial biological activity analysis was conducted on the top nine most potent extracts (Supplementary Fig. 2). This study revealed the extract from *Streptomyces gandocaensis* to be of particular interest due to its ability to inhibit biofilm formation, but showing a limited effect on *A. baumannii* growth (Supplementary Figs 1–3).

Ribosome engineering of *S. gandocaensis*. Following initial identification of the active principles, regrowth of the wild-type *S. gandocaensis* over several months showed complete loss of production of the active biofilm inhibitor molecules. Therefore, we decided to immediately pursue a ribosome engineering approach to restore and improve production of the active metabolites. This approach has been employed for activation of secondary metabolite production in *Streptomyces* spp.[20], and can result in significantly enhanced yields by inducing point mutations in ribosomal protein-encoding genes (for example, *rpsL*). Several rounds of mutagenesis based on a streptomycin resistance phenotype resulted in an improved strain DHS287 of *S. gandocaensis* (Supplementary Fig. 4; Supplementary Table 7), a fourth-generation mutant with restored stable production that generates several-fold increased quantities of active molecules compared with initial wild-type levels. Genetic analysis revealed that the streptomycin-induced ribosome engineering introduced a point mutation in the *rpsL* gene, which encodes the ribosomal protein S12, in the engineered strain (Supplementary Fig. 5; Supplementary Table 3). Previous studies have shown that mutations in the *S12* gene render cells potentially more active for polypeptide synthesis under typical starvation conditions during the late growth phase[20]. This effort appears to be the first reported instance where complete loss of an active, but structurally uncharacterized natural product has been recovered using the ribosome engineering approach.

Isolation and structure elucidation of cahuitamycins A–C. Bioassay guided C18 column fractionation followed by high-performance liquid chromatography (HPLC) purification of organic extracts obtained from the ribosome-engineered *S. gandocaensis* DHS287 yielded three new secondary metabolites, cahuitamycins A–C (**1–3**; Fig. 1). Cahuitamycin A (**1**), the major metabolite, showed a high-resolution time-of-flight electrospray ionization mass spectrometry (ESIMS) $[M+H]^+$ ion peak at *m/z* 636.2679, indicating the molecular formula of $C_{27}H_{37}N_7O_{11}$ (+0.3 p.p.m.) requiring 13° of unsaturation (Supplementary Fig. 8). The one-dimensional (1D; ^1H, ^{13}C) and two-dimensional (2D; gHSQCAD, gHMBCAD and gCOSY) NMR data acquired in $CD_3OD + D_2O$ (4:1) indicated the peptidic nature of **1** (Supplementary Figs 9–13) by the presence of 8 methyl/methine carbons, 10 methylene carbons and 9 carbonyls/quaternary carbons. Analysis of gCOSY, TOCSY and gHMBC cross peaks at δ_H 7.01, 7.48, 6.97 and 7.71 to δ_C 159.7 and 159.8 suggested the spin system consisting of an *ortho*-substituted phenol group. In addition, correlations observed through long-range ^1H–^{13}C between δ_H 4.69 (H-20a) and δ_C 168.6 (C-21) as well as ^1H–^1H between δ_H 5.11 (H-19) and 4.61 (H-20b) interactions indicated the moiety to be a *N*-terminal 2-phenyl-oxazoline group. Further analysis of the gCOSY and TOCSY spectra indicated at least four

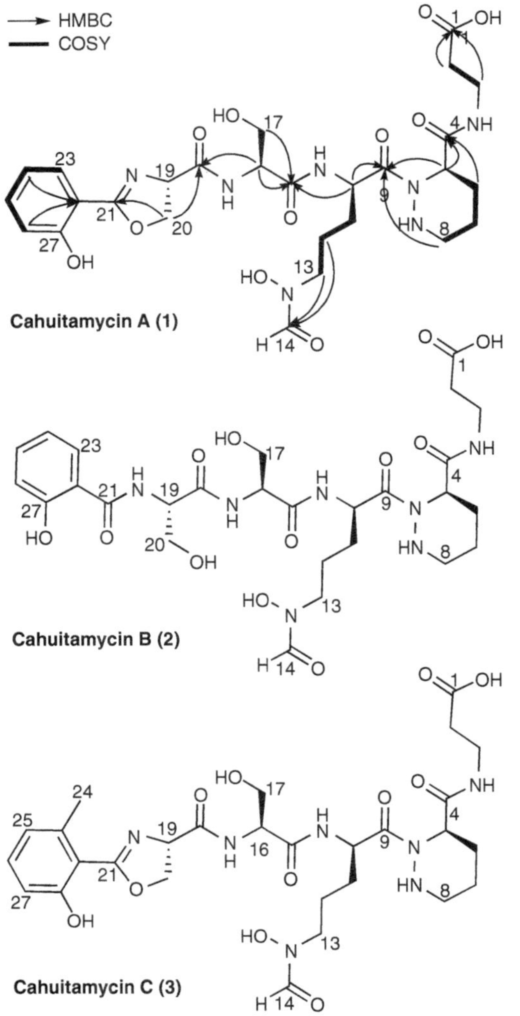

Figure 1 | Structures of cahuitamycins A-C (1-3) with absolute stereochemistry. Cahuitamycin A shows key HMBC and COSY correlations.

more spin systems consisting of a serine (Ser), two modified ornithines (Orns) and a modified alanine (Ala) (Supplementary Figs 8–14). We defined modified Orn as N^{δ}-hydroxy-N^{δ}-formylornithine (*N*-OH-*N*-fOrn) based on the COSY relay observed between δ_H 4.29 (H-10), 1.82 (H-11), 1.68 (H-12), 3.45 (H-13) and a gHMBC correlation between H-13 and C-14 (δ_C 164.1). Similarly, another Orn-like spin system potentially related to a piperazic acid (Pip) based on long-range ^1H–^{13}C between δ_H 3.57, 3.63 (H-8)-δ_C 173.9 (C-9) and a short-range ^1H–^1H array observed from H-5 to H-8 (Fig. 1). The C terminus of the peptide was identified as β-alanine (β-Ala) based on COSY correlation between H-3 (δ_H 3.47, 3.39) to H-2 (δ_H 2.39) and an HMBC correlation from H-3 to C-1 (δ_C 174.1). All deduced moieties completed the planar structure of **1** (Fig. 1; Supplementary Table 1).

Cahuitamycin B (**2**) was isolated by RP-18 HPLC from the same C18 fraction containing compound **1**. The high-resolution ESIMS (HRESIMS) $[M+H]^+$ ion peak at *m/z* 654.2759 provided a molecular formula of $C_{27}H_{39}N_7O_{12}$ with only 12° of unsaturation compared with 13 in **1** (Supplementary Fig. 15). Moreover, the 1D NMR data although acquired in dimethylsulphoxide (DMSO)-d_6, showed high structural similarity to **1** with the existence of six carbonyls (between δ_C 164.8 and 171.1), an aldehyde (δ_C 161.6, δ_H 8.21) and a phenyl group functionality (between δ_C 116.7–156.9 and δ_H 6.81–7.91) fulfilling 11 out of 12 degrees of unsaturation. Analysis of the 2D NMR data for **2** suggested a similar carbon backbone as **1** except the gCOSY and HMBC cross peaks at δ_C 55.4, δ_H 4.56 (C-16)—δ_C 61.5, δ_H 3.58, 3.63 (C-17) and δ_C 55.2, δ_H 4.52 (C-19)—δ_C 61.7, δ_H 3.69, 3.73 (C-20) indicating two Ser groups, replacing the ring in **1**, respectively (Fig. 1; Supplementary Figs 16-20). Furthermore, similar COSY and HMBC correlation as observed in **1** suggested the presence of Pip, thus accounting for the 12th degree of unsaturation to complete the structure of **2** (Supplementary Table 1).

Cahuitamycin C (**3**) was also isolated as a white amorphous solid from the same C18 fraction containing **1** and **2**. The HRESIMS $[M+H]^+$ ion peak at *m/z* 650.2804 provided a molecular formula of $C_{28}H_{39}N_7O_{11}$ with 13 degrees of unsaturation (Supplementary Fig. 21). Extensive 1D and 2D NMR analysis indicated that **3** shares structural similarity on much of the carbon backbone compared with **2** (Supplementary Table 1). The only observed difference was localized to the phenyl ring system where an HMBC correlation from δ_H 2.51 (H-24) singlet to δ_C 123.1 (C-25) and 112.8 (C-22) suggested methylation of the *N*-terminal 2-hydroxybenzoyl-oxazoline group at C-23 (δ_C 141.2). Furthermore, a change in ^1H multiplicity at δ_H 6.73 (H-25) to a doublet confirmed the planar structure of **3** (Fig. 1; Supplementary Figs 22–26).

Stereochemical studies for cahuitamycins A–C. Due to the polypeptidic nature of the cahuitamycins, we pursued advanced Marfey's analysis[21,22] to ascertain absolute stereochemistry. Initially, only cahuitamycin B (**2**) was selected for acid hydrolysis followed by 1-fluoro-2,4-dinitrobenzene-5-alanine amide (FDAA) derivatization. This was based on **2** appearing to be the congener leading to production of cahuitamycins A and C (**1** and **3**) following cyclization or vice versa (see further below).

A study was conducted to compare *m/z* 357.27, 382.32 and 400.34 channels from liquid chromatography/electrospray ionization/mass spectrometry (LC-ESI-MS) chromatograms between the L-FDAA and D, L-FDAA-derivatized Ser, Pip and *N*-OH-Orn (a hydrolysed product of *N*-OH-*N*-fOrn) products of **1**, respectively. Analysis clearly revealed the absolute configuration of the moieties in the hydrolysate of **1** to be L-Ser, L-Ser, D-Pip and D-*N*-OH-Orn, respectively (Fig. 1).

Furthermore, analysis of the Ser, Pip and *N*-OH-Orn portions of **1** and **3** was conducted as described above for **2**, revealing the same stereochemistry. The absolute configuration of the oxazoline ring in **1** and **3** was extrapolated to be D based on similar NMR chemical shifts compared with **2**, as well as the absence of an epimerization domain in the first module (CahA) of the cahuitamycin biosynthetic gene cluster (see above, Fig. 2).

Affinity of cahuitamycins towards iron. Cahuitamycins were observed to have siderophore-like properties, and we therefore conducted competition titrations with EDTA to identify their relative iron-binding affinities[23]. In this method, the Fe–cahuitamycin complex was incubated with varying concentrations of EDTA and assessed for Fe distribution using ultraviolet-visible spectroscopy (Methods). The pFe^{III} for cahuitamycins were calculated against the known stability constants for EDTA ($pFe^{III} = 23.42$)[24]. Three individual titrations were conducted for each natural product, and the representative pFe^{III} values for cahuitamycins **1–3** were measured to be 18.34 ± 0.16, 20.42 ± 0.09 and 17.52 ± 0.12, respectively (Supplementary Fig. 38). Importantly, the calculated pFe^{III} for **1–3** was significantly lower than those previously reported for hydroxamate siderophores under similar physiological condition

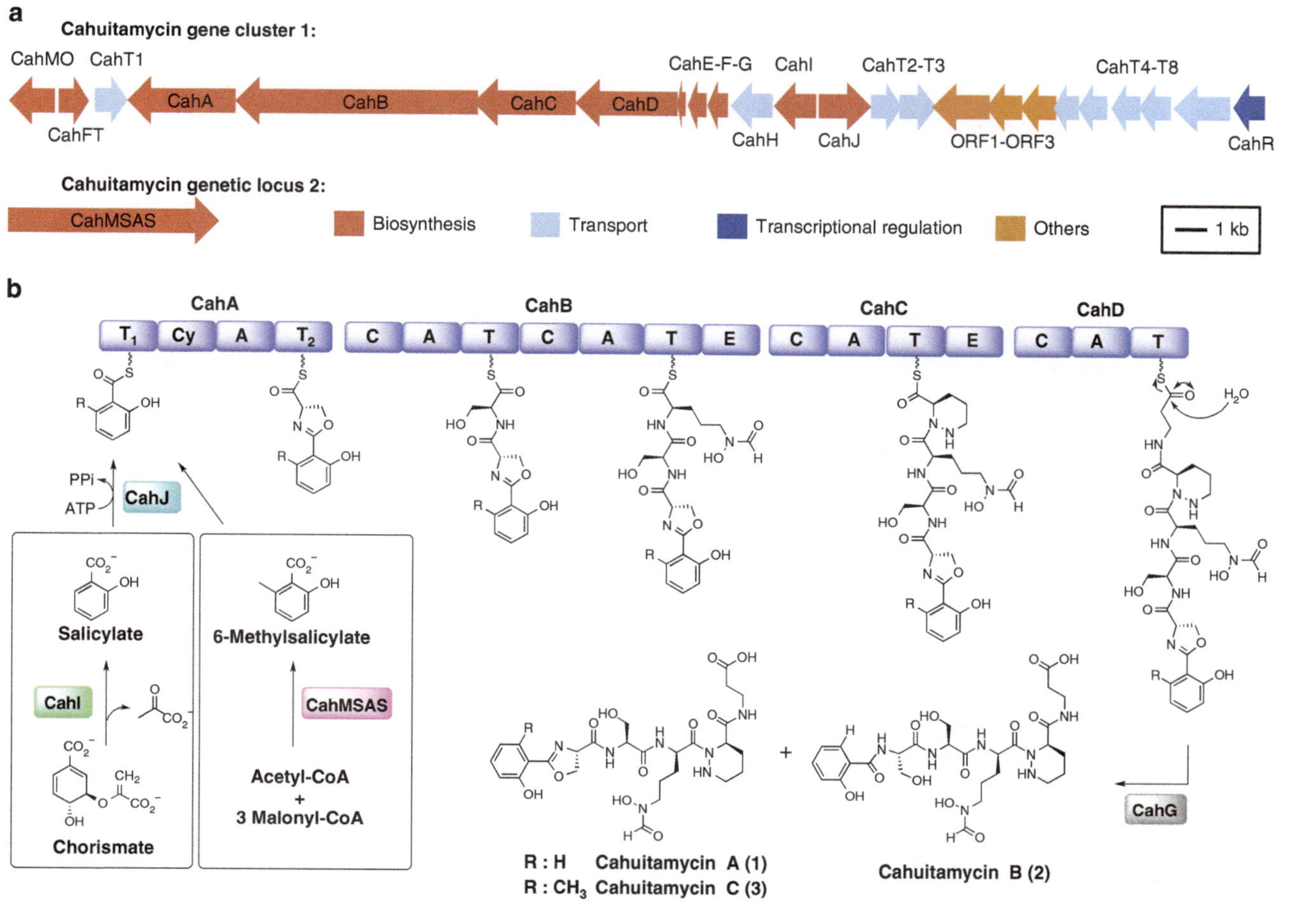

Figure 2 | Complete biosynthetic scheme of cahuitamycins. Organization of the bifurcated cahuitamycin (*cah*) gene cluster (**a**) and a proposed biosynthetic pathway of cahuitamycins **1-3** (**b**). Domain notation: A, adenylation; C, condensation; Cy, cyclization; E, epimerization; T, thiolation.

(~25–37)[25]. This supports our hypothesis that cahuitamycins possess siderophore properties, but their relatively low affinity may indicate it to be a secondary endogenous function. Interestingly, the diminishing activity of **1** over an extended bioassay time course and the observed inverse proportionality between pFe^{III} of **1–3** against *A. baumannii* biofilm inhibition could be attributed to this iron chelation property (see below).

Dissecting the cahuitamycin biosynthetic gene cluster. To mine a candidate gene cluster for cahuitamycin biosynthesis, analysis of the draft genome sequence of *S. gandocaensis* was performed using antiSMASH[26]. As a result, a gene cluster was identified (GeneBank accession code KU363800) as potentially responsible for nonribosomal peptide scaffold biosynthesis, transport and regulation of cahuitamycins A and B (**1** and **2**; Fig. 2a). Cahuitamycin C (**3**) biosynthesis was initially expected to involve a *S*-adenosyl-L-methionine-dependent C-methyltransferase through post-assembly tailoring. However, the absence of genes responsible for the formation of a methyl group at C-23 position of **3** (Fig. 1) near cluster 1 led to the hypothesis that another genetic locus encoding 6-methyl salicylate synthase (*6-MSAS*) might be involved in its biosynthesis (Fig. 2). The architecture and annotation of the bifurcated cahuitamycin (*cah*) gene cluster is shown in Fig. 2 and Supplementary Table 4, respectively. The cluster containing nonribosomal peptide synthetase (NRPS)-encoding genes *cahA–D*, together with genes involved in chain initiation (*cahIJ*), termination (*cahG*) and a transcriptional regulator (*cahR*), is located in a region spanning ~40 kb of DNA. Each NRPS protein consists of the essential condensation (C), adenylation (A) and thiolation (T) domains (Fig. 2b), and has a non-colinear architecture.

CahA is proposed to catalyse oxazoline ring formation together with a putative salicylate synthase CahI and salicylate-AMP ligase CahJ (Fig. 2b). CahI is homologous to MbtI from *Mycobacterium tuberculosis* and is predicted to convert chorismate to salicylate (Fig. 2b)[27]. CahJ bears high similarity to a salicylate-AMP ligase MxcE from *Sorangium cellulosum* So ce56 (ref. 28), which suggests that it activates salicylate by adenylation (Fig. 2b). The identity of amino acid building blocks selected and activated by the Cah NRPS were analysed by examining the specificity-conferring residues in each A domain (Supplementary Table 5)[29]. The A domain in CahA is predicted to recognize L-Cys (Supplementary Table 5), though only L-Ser is loaded according to the structure of cahuitamycins (Fig. 1), which is also observed in the amychelin biosynthetic system[30]. The heterocyclization between Ser and the carbonyl group to form an oxazoline ring scaffold can be attributed to the Cy domain in CahA, and the resulting peptide product is then transferred to CahB for synthesis of **1** and **3**. Similar to formation of gobichelins, where the oxazoline ring of gobichelin A is hydrolysed into a linear form under non-acidic condition, **1** may convert to **2** with a free Ser through hydrolysis of the oxazoline ring[31]. The *in silico* analysis of A domains in *CahB* predicts activation of L-Ser and L-*N*-OH-*N*-fOrn by CahB-A_1 and CahB-A_2, respectively (Supplementary Table 5). Subsequent to the incorporation of L-Ser by CahB-A_1, a L-*N*-OH-*N*-fOrn moiety, which is synthesized from L-Orn by a putative Orn hydroxylase (CahMO) and a putative

N-formyltransferase (CahFT)[32], is appended to the growing chain (Supplementary Table 4). The L-*N*-OH-*N*-fOrn moiety linked to the T_2 domain of CahB undergoes epimerization by the epimerase (E) domain[33], which is consistent with the stereochemistry of cahuitamycins (Fig. 1; Supplementary Figs 8–37).

Biosynthesis of cahuitamycins likely involves formation and loading of the piperazic acid (Pip) building block into the growing peptide chain. Although the relatively rare Pip moiety is found in several biologically active secondary metabolites[34–39], the precise biosynthetic events for Pip elaboration have not been elucidated. Recent precursor incorporation experiments have suggested that generation of the Pip moiety in kutzneride occurs before incorporation into the respective peptide chain[35], similar to the formation of a hydrazo linkage in valanimycin[34]. It is likely that biosynthesis of the Pip moiety in cahuitamycin is initiated by *N*-hydroxylation of the amino group of Orn by a putative CahMO. The resulting L-*N*-OH-Orn residue attached to the T domain of CahC would be converted to its corresponding D-stereoisomer (D-*N*-OH-Orn) by the E domain within the same module followed by nucleophilic attack of the amino group to close the ring. Alternatively, since the A domain in the CahC is homologous to L-glutamine (Gln)-activating A domains (Supplementary Table 5), it is possible that Gln may be incorporated as a branch point from primary metabolism into the piperazate biosynthetic pathway (Fig. 2)[40,41].

Further *in silico* analysis of the A domain in CahD (Supplementary Table 5) predicted loading of β-Ala onto the T domain consistent with the structure of cahuitamycins. Moreover, it further affirms that the presence of β-Ala residue is likely formed from L-aspartate (L-Asp) by the *cahF* gene product[42]. The CahD polypeptide contains neither C-terminal thioesterase nor a reductase domain usually required for chain release in nonribosomal peptide biosynthesis, and likely involves a specific hydrolase for release of the fully assembled peptide chain from the T domain of the last module CahD. We propose that following final extension by β-Ala, CahG (α/β hydrolase superfamily) catalyses in trans release of the peptide chain through hydrolysis of the T domain-bound thioester as described previously in coelichelin biosynthesis[43].

Moreover, CahE, an MbtH-like protein homologue, likely contributes to the stimulation of A domains in the Cah NRPS[44]. CahH, which belongs to the major facilitator superfamily, is similar to EntS, an enterobactin efflux exporter of *Escherichia coli*. Proteins encoded by CahT1 through CahT8 are putatively required for iron uptake and translocation of cahuitamycin across the cell membrane (Supplementary Table 4). Therefore, these proteins might be involved in the export of cahuitamycins from the cytoplasm into the medium. Open reading frame 1 (ORF1), ORF2 and ORF3 show sequence similarity to oxidoreductase, acyltransferase and 6-aminohexanoate-oligomer hydrolase, respectively, and their relationship to cahuitamycin biosynthesis remains unclear (Supplementary Table 4).

Biosynthesis of cahuitamycin C. We next sought to confirm the function of the assigned biosynthetic cluster by deletion of *cahI* in *S. gandocaensis* through insertion of the kanamycin resistance gene (*aph*). Surprisingly, the Δ*cahI* *S. gandocaensis* strain DHS334 (Methods; Supplementary Fig. 7; Supplementary Table 7) gave exclusive production of **3**, while the production of **1** and **2** were completely abolished (Fig. 3d). The result indicated a possible unlinked source involved in providing 6-methylsalicylate starter unit. A BLAST analysis using bacterial 6-MSASs including various putative candidates from other bacterial secondary metabolite gene clusters[45–49], led to the identification of a single iterative type I PKS gene encoding a 6-MSAS (genetic locus 2; GeneBank accession code KU363801) located at least ~50 kb away from cluster 1 based on our *S. gandocaensis* draft genome-sequencing data (Fig. 2a). This finding led us to propose that the 6-methylsalicylate starter unit is likely derived from the 6-MSAS identified from bioinformatics mining, which is subsequently adenylated by CahJ and loaded onto the T_1 domain of CahA for production of **3** (Fig. 2). As in other bacterial PKSs in its class, the putative cahuitamycin 6-MSAS (genetic locus 2; *cah MSAS*) possesses keto synthase, acyltransferase, dehydratase, ketoreductase and acyl carrier protein domains on a single protein (Supplementary Table 4). The hypothesis was further substantiated by the dosing effect observed by introducing exogenous 6-methylsalicylic acid into cultures of the Δ*cahI* *S. gandocaensis* strain. The feeding study provided increased production of **3**, while no changes were observed in production of **1** and **2** (Supplementary Fig. 39). This result, together with the lack of other potential sources of 6-methylsalicyclate, or a candidate methyltransferase encoded within the *S. gandocaensis* genome appears to rule out the possibility of any posttranslational modification of **1** to produce **3**. Furthermore, production of **1** and **2** was restored by external supplementation of salicylic acid to the Δ*cahI* growth medium (Fig. 3e), representing chemical complementation and confirmation of its role in the production of cahuitamycins. These studies substantiated that biosynthesis of **3** is independent of *cahI*-mediated salicylic acid biosynthesis, and more

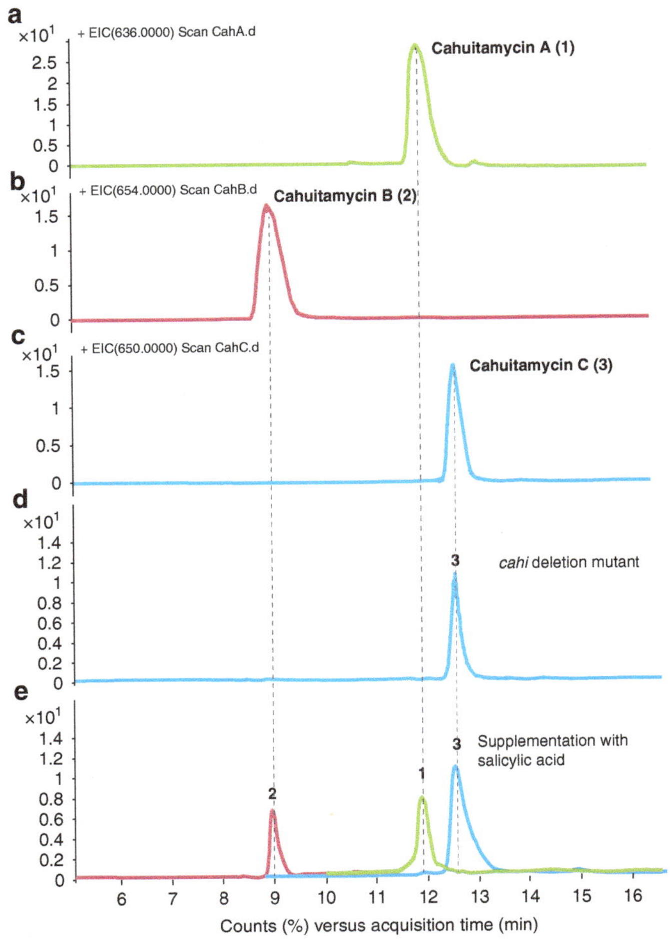

Figure 3 | Chemical complementation of the *cahI* deletion mutant of *S. gandocaensis*. LC-ESI-MS chromatograms of **1–3** obtained from culture of DHS287 (**a–c**) the *cahI* deletion mutant DHS334 (**d**) supplemented with 100 μM of salicylic acid (**e**).

importantly cahuitamycin C (**3**) appears to be derived from a convergent biosynthetic pathway (Fig. 2). Secondary metabolites assembled from unlinked gene clusters have been reported in previous studies[32,50,51], but to the best of our knowledge this is the first example of a starter unit pathway that is genetically unlinked to a corresponding bacterial NRPS. The observation provides an intriguing example where *S. gandocaensis* may be diversifying its biosynthetic machinery to produce a more potent/toxic metabolite for enhanced competitiveness in its natural habitat.

Mutasynthetic generation of cahuitamycin analogues. The biosynthetic process for generating **3** revealed the possible tolerance of CahJ towards structurally related salicylic acid substrates, and a promising avenue for producing new cahuitamycin analogues by mutasynthesis[52]. In this approach, we provided the DHS334 Δ*cahI* strain with a series of substituted benzoic acid substrates (Supplementary Fig. 40), and sequentially assessed their incorporation to make new analogues (Fig. 4). Of the series of unnatural starter units tested, feeding of methyl-hydroxybenzoic acid substrates showed some level of incorporation. However, only assimilation of 5-methylsalicylic acid into the Δ*cahI* pathway provided isolable quantities of the new analogue, cahuitamycin D (**4**; Fig. 4a; Supplementary Fig. 40).

Cahuitamycin D (**4**) was isolated by reverse-phase HPLC (RP-HPLC) from crude extract of Δ*cahI* mutant. The HRESIMS $[M+H]^+$ ion peak at *m/z* 650.2706 provided similar molecular formula as of **3**, $C_{28}H_{39}N_7O_{11}$. Extensive 1D and 2D NMR data were acquired for **4**, which indicated the expected structural similarity with **3**, as per the mutasynthesis hypothesis (Fig. 4a; Supplementary Table 2). The structure showed a similar carbon backbone with a clear difference at the phenyl ring system compared with **3**. Cahuitamycin D (**4**) shows the presence of a singlet at δ_H 7.78 (H-23) with HMBC correlation to δ_c 20.3 (C-25) and δ_c 129.4 (C-24), suggesting methylation at C-24 consistent with the hypothesized incorporation of 5-methylsalicylic acid to the mutant strain DHS334 of *S. gandocaensis*.

Furthermore, during isolation of **4**, we observed another HPLC peak eluting with this natural product (Fig. 4b). Interestingly, the HRESIMS $[M+H]^+$ ion peak at *m/z* 668.2830 suggested the same backbone with likely additional hydration. The new molecule cahuitamycin E (**5**) was shown by 1D and 2D NMR analysis to share structural similarity on much of the carbon backbone compared with **4** (Supplementary Table 2). The only plausible difference could be traced to the C-5 position where Pip moiety in **4** was substituted with L-*N*-OH-Orn in **5**, suggested based on the COSY relay observed from δ_H 4.33 (H-5) to δ_H 3.45 (H-8) along with absence of any significant HMBC correlation from H-8 to C-9 (δ_C 172.2) (Fig. 4b). The observation was compelling as it indicates that either the cyclization to form Pip moiety is occurring after the insertion of L-*N*-OH-Orn in **5** or, alternatively, the domain is capable of accepting both moieties separately. In addition, the study also indicated that L-glutamine might not be a substrate for the CahC A domain during biosynthesis of cahuitamycins.

The absolute stereochemical configuration of the cahuitamycins D and E (**4 and 5**; Fig. 4) was extrapolated to be the same as cahuitamycins A–C (**1–3**) based on very similar chemical shifts observed for all nuclei obtained through detailed 1D and 2D NMR analysis (Supplementary Table 2).

Biological activity associated with cahuitamycins A–E. Cahuitamycins A–C (**1–3**) were next tested for their ability to inhibit biofilm formation of *A. baumannii* using a high-resolution optical secondary assay system. The primary high-throughput static biofilm assay leading to identification of **1–3** was conducted using crystal violet staining followed by optical density measurements (Fig. 5a,b). The result from this assay showed that **1** was able to inhibit, while **2** had no effects on bacterial growth or biofilm formation (Fig. 5a,b). Interestingly, **3** possessed the highest potency with half-maximal inhibitory concentration (IC_{50}) values of 14.5 μM while having a limited effect on growth of *A. baumannii*. In addition, a confocal laser scanning microscope imaging analysis of biofilm formation and biomass was performed (Methods). In this assay, the biofilm was developed on a 12-well glass plate with solvent (no compound) as a negative

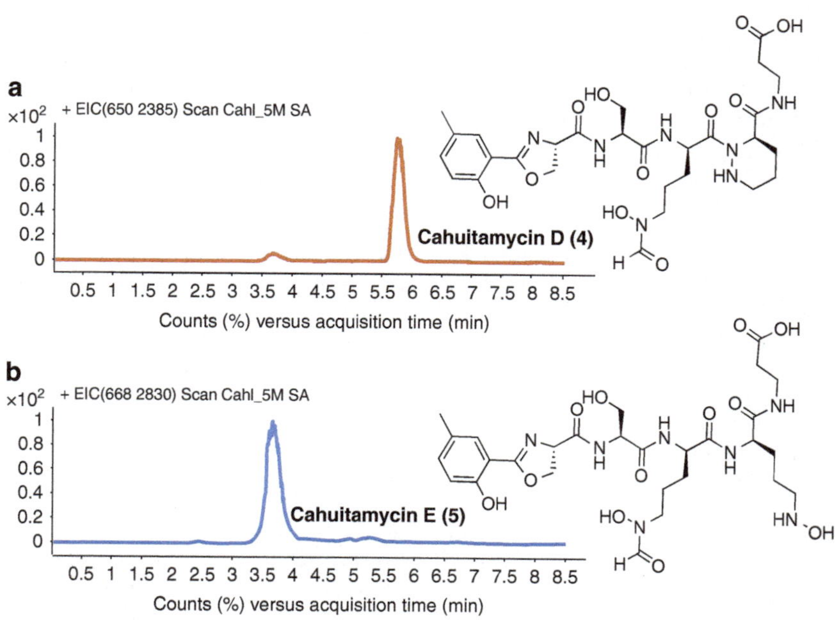

Figure 4 | Mutasynthetic generation of cahuitamycins D and E (4 and 5) and their complete structures. LC-ESI-MS chromatograms of **4** (**a**) and **5** (**b**) obtained from a culture of the *cahI* deletion mutant DHS334 supplemented with 100 μM of 5-methylsalicylic acid.

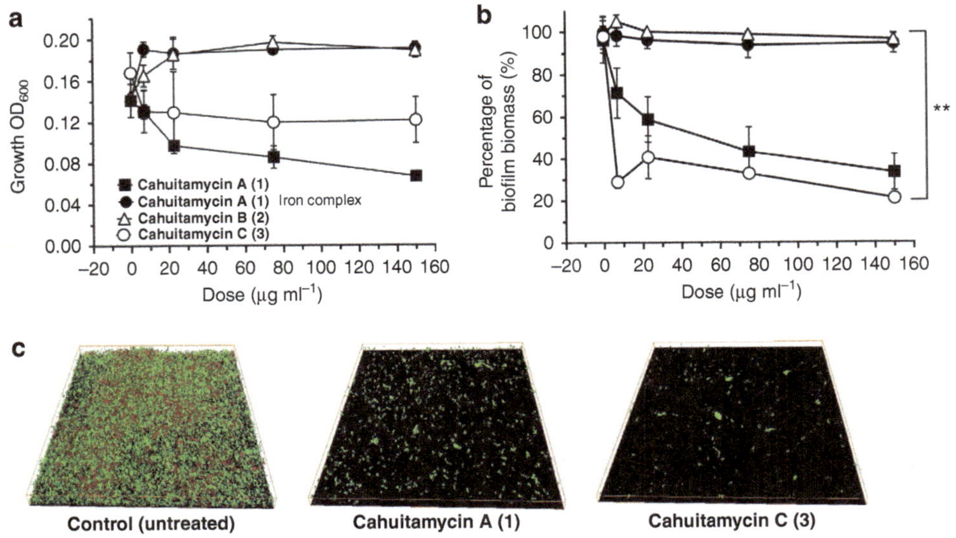

	Control (untreated)	Cahuitamycin A (1)	Cahuitamycin C (3)
Average thickness (μm)	1.71±0.52	0.21±0.07***	0.20±0.05***
Roughness coefficient	1.83±0.22	1.99±0.18	1.99±0.20***
Maximum thickness (μm)	40.50±2.21	1.5±0.54***	1.5±0.33***
Total biomass ($\mu m^3\ \mu m^{-2}$)	0.48±0.11	0.18±0.05***	0.10±0.03***

Figure 5 | Biological activity of cahuitamycins A-C (1-3). (**a**) Growth of *A. baumannii* ATCC 17978 and (**b**) inhibition of biofilm in the presence of cahuitamycins A-C (**1-3**) and iron complex of **1** compared with control. Results are the average of three replicates ± s.d. Student's *t*-test was used for statistical analysis, ** indicates $P<0.01$. (**c**) 3D image of *A. baumannii* ATCC 17978 biofilm after 22 h of incubation. Control was untreated, and **1** and **3** were treated at 240 nM. Images were acquired using a confocal microscope with a × 60 objective lens. Results are the average of five replicates. Student's *t*-test was used for statistical analysis, *** indicates $P<0.001$ compared with control.

control and only 240 nM of **1** and **3** as treated. After staining with fluorescent dye SYTO9 and propidium iodide, the biofilm structure was observed (Fig. 5c). The reconstructed three-dimensional (3D) biofilm images obtained showed a much thinner biofilm and also significantly less total biomass for the glass plates treated with **1** and **3** compared with control (Fig. 5c). Furthermore, the glass plate treated with **3** showed a much lower level of total biomass compared with **1** at the same concentration ($0.10\ \mu m^3$ versus $0.18\ \mu m^3$), which confirmed **3** as the more potent biofilm inhibitor (Fig. 5c).This finding indicates that cahuitamycins prevent biofilm formation through a specific, yet currently undefined biochemical target with negligible impact on cell growth (Fig. 5a,b). Furthermore, cahuitamycin C (**3**) was tested against the growth of *A. baumannii* separately (Supplementary Fig. 41), and revealed an IC_{50} value of ≥ 1.0 mM (Supplementary Fig. 42). In addition, when **1** (the most abundant molecule) was tested in iron-complexed form (owing to its moderate iron chelation affinity), it showed minimal impact on both growth and biofilm formation of the pathogen (Fig. 5a,b). To probe iron chelation as a possible mode of action for the cahuitamycins, a known siderophore desferrioxamine was used as control, which showed no activity, suggesting that iron chelation (or depletion) may not be the basis for cahuitamycin activity (Supplementary Figs 38 and 44). However, loss of inhibition of biofilm formation over time could be attributed to the metal-complexed cahuitamycins (Supplementary Figs 38 and 44), as the test media is replete with iron and **1** in complex with Fe^{III} has not displayed comparable inhibitory activity (Fig. 5a). This observation is consistent with previous studies, which have reported that clinical isolates of *A. baumannii* are able to grow in the presence of high concentrations of different iron chelators due to its ability to express a high-affinity extracellular siderophore, acinetobactin[53,54]. In addition, the structural differences between cahuitamycins with previously reported amychelin[30], results in a significant change in iron affinities, which can be further explored for biofilm inhibitory properties in future studies.

Next, we sought to probe the specific stage and extent of inhibition during biofilm formation employing a secondary flow cell assay (Supplementary Fig. 44; Methods). This analysis revealed almost the same number of cells initially attached to the surface before starting treatment. After 6 h of inhibitor supplementation, the flow cells treated with **1** showed only a few microcolonies of the pathogen, while the control (no treatment) and desferrioxamine showed formation of a robust biofilm. After 24 h of supplementation, the control and desferrioxamine biofilm became thicker with the flow cells exposed to **1** showing larger microcolonies and initiation of biofilm formation by *A. baumannii*.

As described above, the mutasynthetic study on *S. gandocaensis* strain DHS334 led to the isolation of two additional analogues cahuitamycins D–E (**4–5**). The new molecules were also subjected to static biofilm assays and the result demonstrated that **4** displayed almost twice the potency ($IC_{50} = 8.4\ \mu M$) compared with **3**, which we earlier considered to be the most active molecule (Fig. 6; Supplementary Figs 41–43). Furthermore, cahuitamycin E (**5**) also displayed significant activity with an IC_{50} of 10.5 μM (Fig. 6; Supplementary Figs 41–43). In addition, compounds **3** and **4** were also tested for disruption of established *A. baumannii* biofilm revealing low activity with an IC_{50} of 692 and 535 μM, respectively (Supplementary Fig. 45; Methods), establishing their better efficacy as biofilm formation inhibitors. Although prophylactic antibiotic administration preceding surgery is routinely successful in reducing infection rates, it has little or no protective effects in surgical procedures involving implants or prostheses[55]. Therefore, catheters and implants coated with

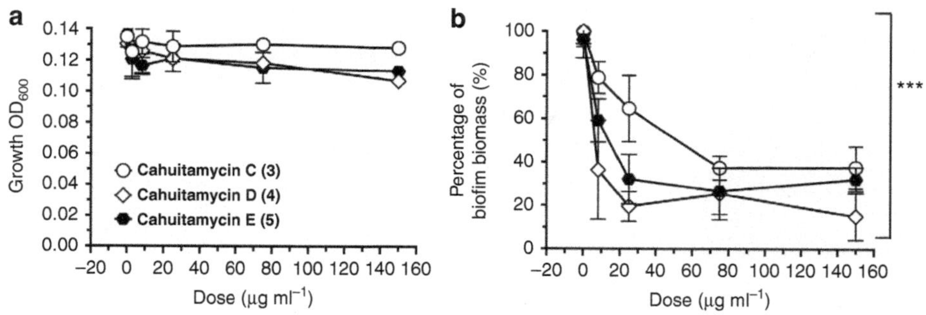

Figure 6 | Biological activity of cahuitamycins C-E (3-5). (**a**) Growth of *A. baumannii* and (**b**) inhibition of biofilm in the presence of cahuitamycins C-E (**3-5**). Results are the average of three replicates ± s.d. Student's *t*-test was used for statistical analysis. ***$P<0.001$ compared with control (no addition of compounds).

biofilm-inhibiting antimicrobial agents are urgently needed as an effective, precision therapeutic option. Our current data indicate that a key pharmacophore of the cahuitamycins is likely the 2-hydroxybenzoyl-oxazoline group where relatively minor modifications can result in an increase (as in case of **3–5**) or decrease (as for **2**) of anti-biofilm activity, with the potential for further structure activity relationship studies using synthetic chemistry or Cah pathway engineering.

The application of the streptomycin resistance-mediated screening of the *Streptomyces* strain identified from HTS against *A. baumannii* has yielded a novel structural class of biofilm inhibitors derived from a marine microbial NPE library. Cahuitamycins A–E (**1–5**) inhibit the ability of *A. baumannii* to generate a biofilm, which is involved in the emergence of a widespread antibiotic resistance phenotype[6,56]. The cahuitamycins are most active in apo form, with chelation of Fe^{III} over time resulting in attenuated biological activity. The cahuitamycins along with allied chemical entities with distinct biological activity provides a primary foundation for future medicinal chemistry and synthetic exploration towards the development of an efficacious drug to prevent or limit biofilm formation.

We have also demonstrated that production of cahuitamycins A–C is specified by a convergent pathway specified by the *cah* biosynthetic gene cluster comprised of two unlinked genetic loci. The inherently flexible starter unit adenylating enzyme CahJ was exploited to produce two additional analogues in the engineered Δ*cahI* strain DHS334. The new cahuitamycin D analogue displays twofold-enhanced biofilm inhibitory activity compared with the cahuitamycin C natural product. These results establish a unique opportunity for developing and discovering new antibiotics from genetically engineered strains bearing inherent flexibility in pathway initiation processes. More importantly, given the simultaneous decline in antibiotic drug discovery and increased incidence of multidrug resistant bacteria, the cahuitamycins may represent a propitious starting point for discovery and development of new therapeutics against dangerous human pathogens involving biofilm formation.

Methods

General experimental procedures. Optical rotation measurements were obtained on a Perkin-Elmer 241 Polarimeter calibrated using a Rudolph Quartz Control Plate Calibration Standard at sodium D line (at +11.502°). Ultraviolet spectra were obtained on a UV-visible Molecular Devices SpectraMax M5 spectrophotometer using 1-ml cuvettes with 1.0 cm path lengths at room temperature in solvent methanol (MeOH). Spectrophotometric assays were performed on Molecular Devices SpectraMax M5 384 variable wavelength spectrometer. All NMR spectra were acquired on a Varian INOVA 600 MHz and a Varian INOVA 700 MHz spectrometer at the NMR Facility, Department of Chemistry, University of Michigan. HRESIMS spectra were measured at the University of Michigan core facility in the Department of Chemistry using an Agilent 6520 Q-TOF mass spectrometer equipped with an Agilent 1290 HPLC system. RP-HPLC was performed using Econosil C18 10 μm 22 × 250-mm column and Agilent ZORBAX RX-C8 5 μm 9.4 × 250-mm column and a solvent system of MeCN and H_2O. The LC-MS analysis of HPLC fractions was performed on a Shimadzu 2010 EV APCI spectrometer.

Biological material. *S. gandocaensis* (Strain # 12620-H2) was isolated from marine sediments collected from Punta Mona Island (protected area: − 82° 37′ 1.50″, 09° 37′ 56.2″ RVS Gandoca Manzanillo with collection permit R-CM-INBio-30-2007), Costa Rica, on 5 June 2007. The procedure for the isolation of actinomycetes from these samples was previously described by Magarvey *et al.*[16] Maintenance and propagation of cultures were performed using standard media and protocols where 500 mg of wet sediment was diluted in 10 ml of sterile water and vortexed for 10 min. Then, 1 ml of this suspension was applied directly to the top of the discontinuous sucrose gradient and centrifuged for 30 min at 300*g*. A volume of 500 μl of the 20, 30 and 40% layers was then plated to HVA agar supplemented with 10 μg ml^{-1} chlortetracycline, 25 μg ml^{-1} cyclohexamide and 25 μg ml^{-1} of nalidixic acid. The plates were then incubated at 28 °C for 1 month. The colony was picked off the plate and streaked onto ISP2 agar until pure. Seed cultures were grown in 17-ml dual-position cap tubes containing 2 ml of ISP2 and grown for 4 days on a rotary shaker at 200 r.p.m. The seed culture was then poured into a 250-ml baffled flask containing 100 ml of ISP2 and grown for 18 days on a rotary shaker at 200 r.p.m. The culture was centrifuged at 4,000 r.p.m. for 10 min to remove the cells, and 2 g of XAD16 resin (Sigma-Aldrich, St Louis, Mo) contained within a polypropylene mesh bag was added to the broth and incubated overnight on the rotary shaker. The resin bag was removed and placed into 10 ml of MeOH, followed by 10 ml of acetone and 10 ml of ethyl acetate (EtOAc). Each of the three fractions was dried *in vacuo* and reconstituted to a final concentration of 15 mg ml^{-1} in DMSO.

Culture maintenance and fermentation. Seed cultures of 100 ml (5 ×) of ISP2 media (1% malt extract, 0.4% yeast extract, 0.4% dextrose and 3% NaCl) were inoculated with a loop-full of vegetative cells from an oatmeal plate (6% oat meal, 1.25% agar and 3% NaCl) culture of *S. gandacoensis* and incubated with shaking (200 r.p.m.) at 28 °C for 5 days. A 25 ml portion of the seed cultures were transferred to a 2.8-l Fernbach flask containing 1.5 l of the ISP2 medium, and the 39 l fermentation was carried out on a rotary shaker (200 r.p.m.) at 28 °C for 18 days. After 14–18 days of growth, the cultures were collected by centrifugation. The resulting cell-free broth was subjected to solid phase extraction using 15 g of Amberlite XAD-16. The resin was then separated by filtration and subjected to organic extraction using MeOH:EtOAc (1:1).

Natural product extract library. The NPE library at the University of Michigan Center for Chemical Genomics contained ~20,000 extracts. Each extract in the library is derived from marine samples collected from all over the world, including Costa Rica, Panama and Papua New Guinea. Some of these samples are from isolated microbes ($n=19{,}055$), while others were derived from field-collected biomass samples ('macrosamples,' $n=800$). Previous work describes in detail how these extracts are prepared for the library[16].

PCR amplification, cloning and sequencing of 16S rDNA. Genomic DNA of *S. gandocaensis* was isolated using the Wizard Genomic DNA Purification Kit according to the manufacturer's instruction. The 16S ribosomal DNA (16S rDNA) gene was amplified by PCR with the universal primers FC27 and RC1492 (Supplementary Table 6)[57] using the genomic DNA as a template. The PCR fragments were cloned into pGEM-T Easy vector (Promega) and the resulting plasmid containing 16S rDNA of *S. gandocaensis* was sequenced using T7 and SP6 primers.

Phylogenetic analysis of 16S rDNA of *S. gandocaensis*. Phylogenetic analyses were conducted using GENEIOUS R6 that is available from http://www.geneious.com/ (ref. 58). The evolutionary history was inferred using the Neighbour-Joining method. The bootstrap consensus tree inferred from 500 replicates was taken to represent the evolutionary history of the taxa analysed (Supplementary Fig. 6).

Generation of ribosome-engineered mutants of *S. gandocaensis*. Spores (10^8–10^9) of wild-type *S. gandocaensis* were spread on R2YE agar containing various concentrations of streptomycin, followed by 15 days incubation at 28 °C to allow the development of streptomycin-resistant colonies. Eighty resistant colonies (Supplementary Fig. 4) were obtained from the plate containing a high level (10, 50 and 100 $\mu g\,ml^{-1}$) of streptomycin and grew in ISP2 media with streptomycin for 14 days at 28 °C. This step was repeated four times to obtain spontaneous streptomycin-resistant mutants. The ability of selected mutants to produce cahuitamycin was tested by LC-MS using Agilent extend C18 5 μm 2.1 × 150-mm column on gradient condition 30–85% acetonitrile (w/0.1% formic acid) in H_2O (w/0.1% formic acid) over 15 min with additional 5 min of equilibration time.

Amplification and sequencing of *rpsL*. Nine strains, including the parent strain and eight high-yielding recombinants, were selected to determine mutations in the *rpsL* gene in this study. The *rpsL* gene was amplified by PCR with primers SR185 and SR186 (Supplementary Table 6), using genomic DNA as a template and cloned into pGEM-T easy vector (Promega) and sequenced. Mutations in the wild-type *rpsL* gene were determined by DNA sequencing[59] (Supplementary Fig. 5; Supplementary Table 3).

Isolation and purification of cahuitamycins. The organic extracts obtained from the engineered strains were concentrated under vacuum to afford the crude extracts (~350 mg) obtained from 100-ml culture. The crude extract was assayed in the developed *in vitro* crystal violet at 10 and 1.0 p.p.m. The bioactive extract was then further purified by RP-HPLC on a gradient of 10–75% ACN and was followed by ultraviolet–visible photodiode array detection at 215 nm to yield semi-pure compounds 1 (6.7 mg), 2 (2.4 mg) and 3 (2.9 mg). Compounds were again subjected to repurification over RP-HPLC on isocratic condition of 35% MeOH (0.1% FA) using C-8 column to get compounds 1 (5.1 mg), 2 (1.1 mg) and 3 (1.6 mg).

Cahuitamycin A (1). Bone white, amorphous powder; ultraviolet ($ACN:H_2O$) λ_{max} 203, 245, 249, 255 and 304 nm; ^1H and ^{13}C NMR, see Supplementary Table 1; HRESIMS *m/z* 636.2679 $[M+H]^+$ (calculated for $C_{27}H_{38}N_7O_{11}$, 636.2629).

Cahuitamycin B (2). Bone white, amorphous powder; ultraviolet ($ACN:H_2O$) λ_{max} 203, 240, 249, 255 and 304 nm; ^1H and ^{13}C NMR, see Supplementary Table 1; HRESIMS *m/z* 654.2759 $[M+H]^+$ (calculated for $C_{28}H_{40}N_7O_{11}$, 654.2735).

Cahuitamycin C (3). Bone white, amorphous powder; ultraviolet ($ACN:H_2O$) λ_{max} 200, 245, 249, 255 and 304 nm; ^1H and ^{13}C NMR, see Supplementary Table 1; HRESIMS *m/z* 650.2804 $[M+H]^+$ (calculated for $C_{27}H_{38}N_7O_{11}$, 650.2786).

Generation of *cahI* deletion mutant. The *cahI* gene was inactivated by an insertional inactivation via double-crossover homologous recombination (Supplementary Fig. 7A). A knock-out plasmid pSRP50 based on pKC1139 (ref. 60) was constructed by amplifying kanamycin resistance gene as a selection marker from plasmid pYJ276 (ref. 61), and left-and-right-flanking regions of the *cahI* gene using the genomic DNA of *S. gandocaensis* as a template. The primer pairs SR144-SR145, SR146-SR147 and SR148-SR149 (Supplementary Table 6) were designed for the amplification of left-and-right-flaking fragments of *cahI* gene, and selection marker, respectively. DNA assembly was performed using Gibson assembly[62] master mix (New England Biolabs) according to the manufacturer's instructions. The plasmid pSRP50 was passaged through methylation-deficient *E. coli* ET12567 and then introduced into the *S. gandocaensis* by protoplasts-based transformation[63]. The target region of *cahI* gene was then disrupted by an insertional inactivation via double-crossover homologous recombination. The desired mutant Δ*cahI* was selected by its kanamycin-resistant and apramycin-sensitive phenotype (Supplementary Fig. 7b) and verified by PCR (Supplementary Fig. 7c) using the primer pair SR233–SR234 (Supplementary Table 6). The resulting *cahI* deletion mutant of *S. gandocaensis* was designated as DHS334.

Chemical complementation of mutant DHS334 and mutasynthesis. Mutant strain of DHS334 was first precultured in 3 ml R2YE liquid medium for 15 days at 28 °C and then 3 ml of the seed culture was used to inoculate 100 ml of the same medium, followed by cultivation for 15 days at 28 °C. Salicylic acid and a series of substituted benzoic acid substrates, 2-hydroxybenzoic acid, 2,3-dihydroxybenzoic acid, 2,4-dihydroxybenzoic acid, 2-fluorobenzoic acid, 2-hydoxy-5-methylbenzoic acid (5-methylsalicylic acid) and 2-hydoxy-6-methylbenzoic acid (6-methylsalicylic acid) were added every alternate day to separate 100-ml cultures of DHS334 at a final concentration of 500 μM for 15 days. The products were first extracted using Amberlite XAD-16. The resin was separated and subjected to organic extraction using MeOH:EtOAc (1:1) for LC-MS analysis as described above.

The substrate-fed crude extract was then further purified by RP-HPLC on a gradient of 10–75% ACN and was followed by ultraviolet–visible photodiode array detection at 215 nm to yield semi-pure compounds **4** (2.4 mg) and **5** (2.6 mg). Compounds were again subjected to repurification over RP-HPLC on isocratic condition of 35% MeOH (0.1% FA) using C-8 column to get compounds **4** (1.1 mg) and **5** (1.3 mg).

Cahuitamycin D (**4**). Amorphous powder; ultraviolet ($ACN:H_2O$) λ_{max} 203, 240, 249, 255 and 304 nm; ^1H and ^{13}C NMR, see Supplementary Table 2; HRESIMS *m/z* 650.2706 $[M+H]^+$ (calculated for $C_{28}H_{40}N_7O_{11}$, 650.2786).

Cahuitamycin E (**5**). Amorphous powder; ultraviolet ($ACN:H_2O$) λ_{max} 200, 245, 249, 255 and 304 nm; ^1H and ^{13}C NMR, see Supplementary Table 2; HRESIMS *m/z* 668.2830 $[M+H]^+$ (calculated for $C_{28}H_{42}N_7O_{12}$, 668.2891).

Determination of *p*Fe^{III} for cahuitamycins A–C. Determination of pFe^{III} for cahuitamycins was carried out as reported by Abergel *et al.* essentially without modifications[64]. Briefly, purified Fe-cahuitamycins, prepared as described above, was dissolved in HEPES buffer (10 mM HEPES and 0.1 M KCl, pH 7.4) and five different concentration ranges of ETDA were added from EDTA stock solutions also prepared in HEPES buffer. Each reaction consisted of a total volume of 0.25 ml, a final concentration of 0.1 mM Fe-cahuitamycins and a range of 5–8,000-fold EDTA (relative to Fe-cahuitamycins) in HEPES buffer. The reaction was allowed to equilibrate at room temperature for at least 24 h, and ultraviolet–visible spectra were subsequently recorded. The composite spectra contain contributions from both Fe-EDTA and Fe-cahuitamycins. The ε of both species as a function of λ were determined in HEPES buffer (see Supplementary Fig. 38 for Fe-cahuitamycins) and used to deconvolute the spectra. The contribution of Fe-cahuitamycins was subtracted from the composite spectra using the 435 nm absorption band, and the ε of both Fe-cahuitamycins and Fe-EDTA were used to quantify the proportion of each species in solution. Concentrations of apo-EDTA and apo-cahuitamycins were calculated by subtracting [Fe-EDTA] (or [Fe-cahuitamycins]) from total initial EDTA (or Fe-cahuitamycins). The log [EDTA]/[cahuitamycins] was plotted against log [Fe-EDTA]/[Fe-cahuitamycins] and the data were fit to equation (1), which has been derived by Abergel *et al.*[64]. All the data sets are shown in Supplementary Fig. 38, yielding a pFe^{III} as reported in original article.

$$\log([\text{Fe-EDTA}]/[\text{Fe-cahuitamycin}]) = \log([\text{EDTA}]/[\text{cahuitamycin}]) + \Delta pFe^{III} \quad (1)$$

Assay development and high-throughput screening assay. A static biofilm assay was developed from a previously reported method[19]. The biofilm assay measures the adhesion of bacteria to the surface of polystyrene 384-well microtiter plates. Adherent cells are detected by staining with crystal violet and subsequent washing to remove nonadherent planktonic cells. Inhibitors of biofilm formation prevent the bacteria from adhering to the surface of the microtiter plate and reduce the amount of crystal violet retained after washing. Assay quantification was performed by OD_{600} absorbance measurement of the crystal-violet-stained biofilm. The assay was optimized for plate surface treatment, time, temperature, media, media concentration, crystal violet concentration, wash method, inhibitor sensitivity and culture inoculum preparation.

The *A. baumannii* test strain (ATCC 17978) was maintained as a frozen stock at − 80 °C in 20% glycerol. A volume of 5 ml of Mueller Hinton II cation adjusted media (Becton Dickinson, catalogue no. 212322) was inoculated and grown at 37 °C, 180 r.p.m. shaking for 4–6 h. The culture was then diluted 1:50 and incubated an additional 2 h. The resulting cells were washed four times by centrifugation and resuspension of the pellet in media was performed to remove any metabolites or cell signalling factors. The assay inoculum was prepared by diluting the cells to 0.008 OD_{600} in 10% Mueller Hinton II cation adjusted media (diluted in 18 Ohm deionized water). The assay plates (Corning, catalogue no. 3,680; 384-well non-treated polystyrene) were prepared by dispensing 20 μl of 10% Mueller Hinton II cation adjusted media into columns 1–22 using a Multidrop Combi dispenser (Thermo Fisher). The NPE samples were added as 0.2 μl of 15 $mg\,ml^{-1}$ stocks in DMSO using a Biomek FX high-density pintool (Beckman Coulter). Column 1 and 2 contained DMSO without NPEs and served as the negative control. The positive control was added to column 23 and 24 as 20 μl of 40 μM Baicalein (Sigma-Aldrich, catalogue no. 11712). The bacterial cells were added to the entire plate in 20 μl of the prepared inoculum using a Multidrop Combi dispenser (Thermo Fisher). The resulting assay contained 40 μl of 10% Mueller Hinton II cation adjusted media with 0.004 OD_{600} bacterial cells, 0.5% DMSO and 75 $\mu g\,ml^{-1}$ NPE. The assay plates were incubated at 30 °C for 20 h, stationary, in a humidified incubator. The biofilm was stained by adding 10 μl of filtered 1% crystal violet and incubating for 30 min at room temperature. Excess crystal violet and nonadherent planktonic cells were removed by washing three times with 150 μl of PBS using an ELX405 plate washer (Bio-tek). The wash program used a low velocity dispense rate and an aspiration height of 3.8 mm; ~10 μl of PBS remained in the well after washing. The Crystal-Violet-stained biofilm was solubilized by the addition of 50 μl of 100% ethanol and allowed to develop overnight. Quantification was performed by measuring the OD_{600} absorbance using a Pherastar plate reader (BMG).

3D imaging of biofilm development. For visualizing bacterial biofilm formation, the same method as described previously was used for biofilm formation but in a glass-bottom 96-well plate (12-556-38, Fisher Scientific International Inc., Franklin, MA)[65]. After 20 h of incubation at 30 °C, the supernatant was removed, and the well surface was rinsed briefly with 1 × PBS buffer three times. PBS buffer (100 µl) containing 1 µM of SYTO 9 and 10 µM of propidium iodide was added into the well. The plate was incubated for 15 min in dark. Fluorescent images were acquired with an Olympus Fluoview FV1000 confocal microscope (Olympus, Markham, Ontario) with Melles Griot Laser supply and detectors and filter sets for monitoring SYTO 9 and propidium iodide fluorescence. Images were obtained using an oil immersion × 60 objective lens. 3D images were reconstructed using the Amira software package (Amira, San Diego, CA) from a stack of sectional images of biofilm samples. Biofilm biomass and per cent coverage were calculated based on those images using software COMSTAT[65].

Micrograph of biofilm in flow cells. Biofilms were formed by *A. baumannii* ATCC 17978 in flow chambers at room temperature. Images were taken at 1, 6 and 24 h (using a × 60 lens) after inoculation. A flow cell biofilm was achieved by using a flow cell chamber (ACCFL0001, Life Science Incorporated, Greensboro, NC). Briefly, overnight cultures of *A. baumannii* ATCC 17978 grown in MHII broth at 37 °C were diluted 100-fold with fresh MHII broth. A volume of 1 ml of dilution was injected into a flow cell chamber and allowed to settle 1 h for attachment of bacteria, followed by initiation media flow at a flow rate of 4 $ml^{-1} h^{-1}$. Flow media was 10% MHII broth (control), 10% MHII broth with 7.5 µg ml^{-1} of desferoxamine or 10% MHII broth with 7.5 μg^{-1} ml of cahuitamycin A. Micrographs of the biofilm were acquired at 1, 6 and 24 h after the start of flow media using an Olympus IX70 microscope with a × 60 lens. Flow was still running while acquiring the micrograph.

Minimum inhibitory concentration test. Cultures of *A. baumannii* ATCC 17978 grown overnight were diluted into fresh 10% Mueller Hinton II broth to get 10^5 colony-forming unit per ml by determining optical density at OD_{600}. A volume of 100 µl of diluted cultures were incubated in individual wells of 96-well plate containing 1% of DMSO and various concentrations of cahuitamycins C (**3**) and D (**4**). The plate was incubated at 30 °C with shaking at 150 r.p.m. The optical density of each well was monitored every 1 up to 24 h using a microtiter plate reader (Synergy HT, Bio-Tek, Winooski, VT). The number of viable bacterial cells was determined by counting colony-forming unit.

Biofilm dispersal assay. Biofilm dispersal assay was performed as previously reported[66]. Briefly, *A. baumannii* ATCC 17978 biofilm was developed as described in minimum inhibitory concentration test. After 16 h incubation at 30 °C, the suspension in each well was removed by pipetting gently. A volume of 100 µl of 1 × PBS buffer containing cahuitamycins C (**3**) and D (**4**) at various concentrations was placed into each well. The plate was incubated at 30 °C for 12 h to allow cahuitamycins to disassemble pre-formed biofilms. Then, the suspension was removed by pipetting. The remaining attached biofilm was quantified by crystal violet assay.

References

1. Andersson, D. I. & Hughes, D. Persistence of antibiotic resistance in bacterial populations. *FEMS Microbiol. Rev.* **35,** 901–911 (2011).
2. Livermore, D. M. Discovery research: the scientific challenge of finding new antibiotics. *J. Antimicrob. Chemother.* **66,** 1941–1944 (2011).
3. Fischbach, M. A. & Walsh, C. T. Antibiotics for emerging pathogens. *Science* **325,** 1089–1093 (2009).
4. Tripathi, A. *et al.* Baulamycins A and B, broad-spectrum antibiotics identified as inhibitors of siderophore biosynthesis in Staphylococcus aureus and Bacillus anthracis. *J. Am. Chem. Soc.* **136,** 1579–1586 (2014).
5. Sunenshine, R. H. *et al.* Multidrug-resistant Acinetobacter infection mortality rate and length of hospitalization. *Emerg. Infect. Dis.* **13,** 97–103 (2007).
6. Dijkshoorn, L., Nemec, A. & Seifert, H. An increasing threat in hospitals: multidrug-resistant Acinetobacter baumannii. *Nat. Rev. Microbiol.* **5,** 939–951 (2007).
7. Davies, D. Understanding biofilm resistance to antibacterial agents. *Nat. Rev. Drug Discov.* **2,** 114–122 (2003).
8. Ranall, M. V., Butler, M. S., Blaskovich, M. A. & Cooper, M. A. Resolving biofilm infections: current therapy and drug discovery strategies. *Curr. Drug Targets* **13,** 1375–1385 (2012).
9. Thomas Bjarnsholt, O. C., Molin, S., Givskov, M. & Høiby, N. Applying insights from biofilm biology to drug development -can a new approach be developed? *Nat. Rev. Drug Discov.* **12,** 791–808 (2013).
10. Loo, C. Y. *et al.* Implications and emerging control strategies for ventilator-associated infections. *Expert Rev. Anti Infect. Ther.* **13,** 379–393 (2015).
11. Simoes, M., Simoes, L. C. & Vieira, M. J. A review of current and emergent biofilm control strategies. *Food Sci. Technol.* **43,** 573–583 (2010).
12. Badave, G. K. & Kulkarni, D. Biofilm producing multidrug resistant *Acinetobacter baumannii*: an emerging challenge. *J. Clin. Diagn. Res.* **9,** 8–10 (2015).
13. Viehman, J. A., Nguyen, M. H. & Doi, Y. Treatment options for carbapenem-resistant and extensively drug-resistant *Acinetobacter baumannii* infections. *Drugs* **74,** 1315–1333 (2014).
14. Fenical, W. & Jensen, P. R. Developing a new resource for drug discovery: marine actinomycete bacteria. *Nat. Chem. Biol.* **2,** 666–673 (2006).
15. Gerwick, W. H. & Moore, B. S. Lessons from the past and charting the future of marine natural products drug discovery and chemical biology. *Chem. Biol.* **19,** 85–98 (2014).
16. Magarvey, N. A., Keller, J. M., Bernan, V., Dworkin, M. & Sherman, D. H. Isolation and characterization of novel marine-derived actinomycete taxa rich in bioactive metabolites. *Appl. Environ. Microbiol.* **70,** 7520–7529 (2004).
17. Newman, D. J. & Cragg, G. M. Natural products as sources of new drugs over the 30 years from 1981 to 2010. *J. Nat. Prod.* **75,** 311–335 (2012).
18. Montaser, R. & Luesch, H. Marine natural products: a new wave of drugs? *Future Med. Chem.* **3,** 1475–1489 (2011).
19. O'Toole, G. A. & Kolter, R. Initiation of biofilm formation in Pseudomonas fluorescens WCS365 proceeds via multiple, convergent signalling pathways: a genetic analysis. *Mol. Microbiol.* **28,** 449–461 (1998).
20. Ochi, K. *et al.* Ribosome engineering and secondary metabolite production. *Adv. Appl. Microbiol.* **56,** 155–184 (2004).
21. Fuji, K. *et al.* A nonempirical method using LC/MS for determination of the absolute configuration of constituent amino acids in a peptide: elucidation of limitations of Marfey's method and of its separation mechanism. *Anal. Chem.* **69,** 3346–3352 (1997).
22. Marfey, P. Determination of D-amino acids. II. Use of a bifunctional reagent, 1,5-difluoro-2,4dinitrobenzene. *Carlsberg. Res. Commun.* **49,** 591–596 (1984).
23. Abergel, R. J., Zawadzka, A. M. & Raymond, K. N. Petrobactin-mediated iron transport in pathogenic bacteria: coordination chemistry of an unusual 3,4-catecholate/citrate siderophore. *J. Am. Chem. Soc.* **130,** 2124–2125 (2008).
24. Martell, A. E. & Smith, R. M. *Critical Stability Constants* (Springer, 1974).
25. Harris, W. R., Carrano, C. J. & Raymond, K. N. Coordination chemistry of microbial iron transport compounds. 16. Isolation, characterization, and formation constants of ferric aerobactin. *J. Am. Chem. Soc.* **101,** 2722–2727 (1979).
26. Blin, K. *et al.* antiSMASH 2.0-a versatile platform for genome mining of secondary metabolite producers. *Nucleic Acids Res.* **41,** W204–W212 (2013).
27. Harrison, A. J. *et al.* The structure of MbtI from Mycobacterium tuberculosis, the first enzyme in the biosynthesis of the siderophore mycobactin, reveals it to be a salicylate synthase. *J. Bacteriol.* **188,** 6081–6091 (2006).
28. Schneiker, S. *et al.* Complete genome sequence of the myxobacterium Sorangium cellulosum. *Nat. Biotechnol.* **25,** 1281–1289 (2007).
29. Bachmann, B. O. & Ravel, J. Methods for in silico prediction of microbial polyketide and nonribosomal peptide biosynthetic pathways from DNA sequence data. *Methods Enzymol.* **458,** 181–217 (2009).
30. Seyedsayamdost, M. R., Traxler, M. F., Zheng, S.-L., Kolter, R. & Clardy, J. Structure and biosynthesis of amychelin, an unusual mixed-ligand siderophore from Amycolatopsis sp. AA4. *J. Am. Chem. Soc.* **133,** 11434–11437 (2011).
31. Chen, Y. *et al.* Gobichelin A and B: mixed-ligand siderophores discovered using proteomics. *Med. Chem. Commun.* **4,** 233–238 (2013).
32. Bosello, M., Robbel, L., Linne, U., Xie, X. & Marahiel, M. A. Biosynthesis of the siderophore rhodochelin requires the coordinated expression of three independent gene clusters in Rhodococcus jostii RHA1. *J. Am. Chem. Soc.* **133,** 4587–4595 (2011).
33. Marahiel, M. A., Stachelhaus, T. & Mootz, H. D. Modular peptide synthetases Involved in nonribosomal peptide synthesis. *Chem. Rev.* **97,** 2651–2674 (1997).
34. Tao, T., Alemany, L. B. & Parry, R. J. Valanimycin biosynthesis: investigations of the mechanism of isobutylhydroxylamineincorporation. *Org. Lett.* **5,** 1213–1215 (2003).
35. Neumann, C. S. *et al.* Biosynthesis of piperazic acid via N5-hydroxy-ornithine in Kutzneria spp. 744. *Chembiochem.* **13,** 972–976 (2012).
36. Broberg, A., Menkis, A. & Vasiliauskas, R. Kutznerides 1-4, depsipeptides from the actinomycete Kutzneria sp. 744 inhabiting mycorrhizal roots of Picea abies seedlings. *J. Nat. Prod.* **69,** 97–102 (2006).
37. Bevan, K., Davies, J. S., Hassall, C. H., Morton, R. B. & Phillips, D. A. S. Amino-acids and peptides. Part X. Characterisation of the monamycins, members of a new family of cyclodepsipeptide antibiotics. *J. Chem. Soc. (C)* 514–522 (1971).
38. Fehr, T., Kallen, J., Oberer, L., Sanglier, J.-J. & Schilling, W. Sanglifehrins A, B, C and D, novel cyclophilin-binding compounds isolated from Streptomyces sp. A92-308110. II. Structure elucidation, stereochemistry and physico-chemical properties. *J. Antibiot. (Tokyo)* **52,** 474–479 (1999).
39. Umezawa, K., Ikeda, Y., Kawase, O., Naganawa, H. & Kondo, S. Biosynthesis of polyoxypeptin A: novel amino acid 3-hydroxy-3-methylproline derived from isoleucine. *J. Chem. Soc. Perkin Trans.* **1,** 1550–1553 (2001).
40. Umezawa, K. *et al.* and B produced by Streptomyces: apoptosis-inducing cyclic depsipeptides containing the novel amino acid (2S,3R)-3-hydroxy-3-methylproline. *J. Org. Chem.* **64,** 3034–3038 (1999).

41. Arroyo, V., Hall, M. J., Hassall, C. H. & Yamasaki, K. Incorporation of amino acids into the cyclohexadepsipeptide, monamycin. *J. Chem. Soc., Chem. Commun.* 845–846 (1976).
42. Zhao, W. *et al.* Complete genome sequence of the rifamycin SV-producing Amycolatopsis mediterranei U32 revealed its genetic characteristics in phylogeny and metabolism. *Cell Res.* **20,** 1096–1108 (2010).
43. Lautru, S., Deeth, R. J., Bailey, L. M. & Challis, G. L. Discovery of a new peptide natural product by Streptomyces coelicolor genome mining. *Nat. Chem. Biol.* **1,** 265–269 (2005).
44. Herbst, D. A., Boll, B., Zocher, G., Stehle, T. & Heide, L. Structural basis of the interaction of MbtH-like proteins, putative regulators of nonribosomal peptide biosynthesis, with adenylating enzymes. *J. Biol. Chem.* **288,** 1991–2003 (2013).
45. Ling, J. *et al.* Enediyne antitumor antibiotic maduropeptin biosynthesis featuring a C-methyltransferase that acts on a CoA-tethered aromatic substrate. *J. Am. Chem. Soc.* **132,** 12534–12536 (2010).
46. Trefzer, S. G., Stockert, A., Kirschning, S. & Bechthold, A. Cloning of an avilamycin biosynthetic gene cluster from Streptomyces viridochromogenes Tü57. *J. Bacteriol.* **179,** 6271–6278 (1997).
47. Shao, L. *et al.* Cloning and characterization of a bacterial iterative type I polyketide synthase gene encoding the 6-methylsalicyclic acid synthase. *Biochem. Biophys. Res. Commun.* **345,** 133–139 (2006).
48. Daum, M. *et al.* Organisation of the biosynthetic gene cluster and tailoring enzymes in the biosynthesis of the tetracyclic quinone glycoside antibiotic polyketomycin. *Chembiochem.* **10,** 1073–1083 (2009).
49. Sontag, B. *et al.* Oxachelin, a novel iron chelator and antifungal agent from Streptomyces sp. *J Antibiot (Tokyo)* **59,** 659–663 (2006).
50. Zane, H. K. *et al.* Biosynthesis of amphi-enterobactin siderophores by Vibrio harveyi BAA-1116: identification of a bifunctional nonribosomal peptide synthetase condensation domain. *J. Am. Chem. Soc.* **136,** 5615–5618 (2014).
51. Xu, Z., Ding, L. & Hertweck, C. A branched extender unit shared between two orthogonal polyketide pathways in an endophyte. *Angew. Chem. Int. Ed. Engl.* **50,** 4667–4670 (2011).
52. Almabruk, K. H. *et al.* Mutasynthesis of fluorinated pactamycin analogues and their antimalarial activity. *Org. Lett.* **15,** 1678–1681 (2013).
53. Actis, L. A., Tolmasky, M. E., Crosa, L. A. & Crosa, J. H. Effect of iron-limiting conditions on growth of clinical isolates of Acinetobacter baumannii. *J. Clin. Microbiol.* **31,** 2812–2815 (1993).
54. Yamamoto, S., Okujo, N. & Sakakibara, Y. Isolation and structure elucidation of acinetobactin, a novel siderophore from *Acinetobacter baumannii*. *Arch. Microbiol.* **162,** 249–254 (1994).
55. Secinti, K. D., Özalp, H., Attar, A. & Sargon, M. F. Nanoparticle silver ion coatings inhibit biofilm formation on titanium implants. *J. Clin. Neurosci.* **18,** 391–395 (2011).
56. Rao, R. S. *et al.* Correlation between biofilm production and multiple drug resistance in imipenem resistant clinical isolates of *Acinetobacter baumannii*. *Ind J. Med. Microbiol.* **26,** 333–337 (2008).
57. Rainey, F. A., Ward-Rainey, N., Kroppenstedt, R. M. & Stackebrandt, E. The genus Nocardiopsis represents a phylogenetically coherent taxon and a distinct actinomycete lineage: proposal of Nocardiopsaceae fam. nov. *Int. J. Syst. Bacteriol.* **46,** 1088–1092 (1996).
58. Kearse, M. *et al.* Geneious Basic: an integrated and extendable desktop software platform for the organization and analysis of sequence data. *Bioinformatics* **28,** 1647–1649 (2012).
59. Hesketh, A. & Ochi, K. A novel method for improving Streptomyces coelicolor A3(2) for production of actinorhodin by introduction of rpsL (encoding ribosomal protein S12) mutations conferring resistance to streptomycin. *J. Antibiot.* **50,** 532–535 (1997).
60. Bierman, M. *et al.* Plasmid cloning vectors for the conjugal transfer of DNA from *Escherichia coli* to Streptomyces spp. *Gene* **116,** 43–49 (1992).
61. Jung, W. S. *et al.* Enhanced heterologous production of desosaminyl macrolides and their hydroxylated derivatives by overexpression of the pikD regulatory gene in *Streptomyces venezuelae. Appl. Environ. Microbiol.* **74,** 1972–1979 (2008).
62. Gibson, D. G. *et al.* Enzymatic assembly of DNA molecules up to several hundred kilobases. *Nat. Methods* **6,** 343–345 (2009).
63. MacNeil, D. J. *et al.* Analysis of Streptomyces avermitilis genes required for avermectin biosynthesis utilizing a novel integration vector. *Gene* **111,** 61–68 (1992).
64. Abergel, R. J., Zawadzka, A. M. & Raymond, K. N. Petrobactin-mediated iron transport in pathogenic bacteria: coordination chemistry of an unusual 3,4-catecholate/citrate siderophore. *J. Am. Chem. Soc.* **130,** 2124–2125 (2007).
65. Heydorn, A. *et al.* Quantification of biofilm structures by the novel computer program comstat. *Microbiology* **146,** 2395–2407 (2000).
66. Bottcher, T., Kolodkin-Gal, I., Kolter, R., Losick, R. & Clardy, J. Synthesis and activity of biomimetic biofilm disruptors. *J. Am. Chem. Soc.* **135,** 2927–2930 (2013).

Acknowledgements

We thank Martha J. Larsen in the Center for Chemical Genomics, Life Sciences Institute, University of Michigan for high-throughput screening automation and NPE sample management. We also thank Dr Howard Xu, Department of Biological Sciences California State University for his expert advice, and Shamilya Williams for expert administrative assistance. We are grateful to the Technical Office, CONAGEBIO, Ministry of the Environment and Telecommunications, Costa Rica for providing sample collection permits. This research was supported by the International Cooperative Biodiversity Groups initiative (U01 TW007404) at the Fogarty International Center, the Great Lakes Regional Center of Excellence for Biodefense and Emerging Infectious Diseases Research (U54 AI57153), the Hans W. Vahlteich Professorship (D.H.S.), Army Research Office (grant no. W911NF-12-1-0059) and National Institutes of Health (1R01GM098350). Support from the University of Michigan African Presidential Scholars program is gratefully acknowledged.

Author contributions

S.R.P., A.T., C.X. and D.H.S designed the research; S.R.P., A.T., J.W., P.J.S., I.Y., T.J.M., F.Y., C.-J.A., A.Y.M., G.T.-C. and DHS performed the research; S.R.P., A.T., J.W., C.X. and D.H.S analysed data; S.R.P., A.T. and D.H.S. wrote the manuscript.

Additional information

Accession codes: The 16S rDNA sequence for *S. gandocaensis* (strain #12620-H2) has been deposited in the GenBank Nucleotide database with accession code KR303715. The DNA sequences for the two cahuitamycin genetic loci have been deposited in the GenBank Nucleotide database with accession codes KU363800 and KU363801. A description and analysis of the complete *S. gandocaensis* genome will be described elsewhere.

Competing financial interests: The University of Michigan has a provisional patent application filed at the United States Patent and Trademark Office (USPTO) relating to the discovery of cahuitamycins as *A. baumannii* biofilm inhibitors (Application #6671). The six authors S.R.P., A.T., D.H.S., P.S., F.Y., C.X. and J.W. are named as inventors in this application.

19

Programming a topologically constrained DNA nanostructure into a sensor

Meng Liu[1,2], Qiang Zhang[2], Zhongping Li[1], Jimmy Gu[1], John D. Brennan[2] & Yingfu Li[1,2]

Many rationally engineered DNA nanostructures use mechanically interlocked topologies to connect individual DNA components, and their physical connectivity is achieved through the formation of a strong linking duplex. The existence of such a structural element also poses a significant topological constraint on functions of component rings. Herein, we hypothesize and confirm that DNA catenanes with a strong linking duplex prevent component rings from acting as the template for rolling circle amplification (RCA). However, by using an RNA-containing DNA [2] catenane with a strong linking duplex, we show that a stimuli-responsive RNA-cleaving DNAzyme can linearize one component ring, and thus enable RCA, producing an ultra-sensitive biosensing system. As an example, a DNA catenane biosensor is engineered to detect the model bacterial pathogen *Escherichia coli* through binding of a secreted protein, with a detection limit of 10 cells ml^{-1}, thus establishing a new platform for further applications of mechanically interlocked DNA nanostructures.

[1] Department of Biochemistry and Biomedical Sciences, McMaster University, 1280 Main Street West, Hamilton, Ontario, Canada L8S 4K1. [2] Biointerfaces Institute, McMaster University, 1280 Main Street West, Hamilton, Ontario, Canada L8S 4O3. Correspondence and requests for materials should be addressed to J.D.B. (email: brennanj@mcmaster.ca) or to Y.L. (email: liying@mcmaster.ca).

DNA is not only important in biological systems as genetic material, it has also become a key player in synthetic biology. DNA can be engineered into catalysts (DNAzymes) and molecular receptors (DNA aptamers), making DNA a functionally versatile polymer. DNA, as a highly programmable material based on predictable Watson–Crick base-pairing interactions, has become a valuable macromolecule for rational engineering of molecular machines for potential nanotechnological applications.

In recent years, tremendous progress has been made toward building DNA-based nanodevices with increasing structural complexity and functional capabilities[1–12]. One important feature of many reported DNA nanostructures, such as DNA Borromean rings[1] and DNA catenanes[3], is the use of mechanically interlocked topologies to connect individual DNA components. The mechanical interlocking between DNA strands can be easily achieved in the case of DNA through the formation of a linking duplex between partner rings before ring closure. The existence of a linking duplex is not only essential to the creation of a strong connectivity between partner rings but also necessary for the stability of these well-defined structures.

As we will show in this work, the linking-duplex feature also enables the use of topologically interlocked architectures, such as DNA catenanes, for the design of amplified biosensors for bioanalytical applications. The biosensing strategy is based on the following idea: the strong physical engagement of two mechanically interlocked single-stranded DNA rings in a DNA [2] catenane (termed D2C in this study for simplicity) with a strong linking duplex makes the component rings unsuitable as the template for rolling circle amplification (RCA), an isothermal DNA amplification technique[13–15]. However, when one of the component rings is engineered to be a substrate of a stimuli-responsive RNA-cleaving DNAzyme (RCD), the system can be programmed into a biosensor that is capable of reporting a target of interest in three sequential reactions: target-induced RNA cleavage, nucleolytic conversion of the cleavage product into a DNA primer, and DNA amplification via RCA. By this approach, we establish an amplified biosensing system that is capable of achieving ultra-sensitive detection of *Escherichia coli* (*E. coli*) at a concentration as low as 10 cells ml^{-1} without cell culture.

Results

Inability of a D2C with a strong linking duplex to undergo RCA. The conceptual cornerstone of this work is the assumption that the components of a D2C with a strong linking duplex are unable to undergo RCA. To test this idea, we synthesized a D2C (Fig. 1a) consisting of two component single-stranded DNA rings, named $^{C}DNA_{i}$ and $r^{C}DNA_{ii}$ (Fig. 1b; r stands for the single ribonucleotide, ribo-A, in the sequence of $^{C}DNA_{ii}$). Briefly, the linear DNA $r^{L}DNA_{ii}$ was circularized into $r^{C}DNA_{ii}$ using $^{C}DNA_{i}$ as the ligation template (sequences of all DNA species are provided in Supplementary Table 1). The resultant D2C, denoted rD2C1, contains a strong linking duplex of 24 bp, which translates into two helical turns (boxed nucleotides in Fig. 1b). The reaction yield of rD2C1 was determined to be 58% (Fig. 1c).

We then performed RCA reactions with gel-purified $^{C}DNA_{i}$, $r^{C}DNA_{ii}$ and rD2C1. Agarose gel analysis indicated that RCA products were produced with $^{C}DNA_{i}$ (using DP1 as the primer; Fig. 1d) and $r^{C}DNA_{ii}$ (using DP2 as the primer; Fig. 1d). In contrast, no RCA products were observed for rD2C1 using the same set of primers (Fig. 1d). This experiment shows that the topological constraint imposed by a strong linking duplex indeed prevents ϕ29 DNA polymerase (ϕ29DP) from replicating interlocked circular templates. In a control experiment, it was found that RCA was not inhibited when the linking duplex of the DNA catenane was made of 9 bp (Supplementary Fig. 1).

Enabling RCA via cleavage of a component ring by a DNAzyme. The inability of ϕ29DP to carry out RCA with topologically

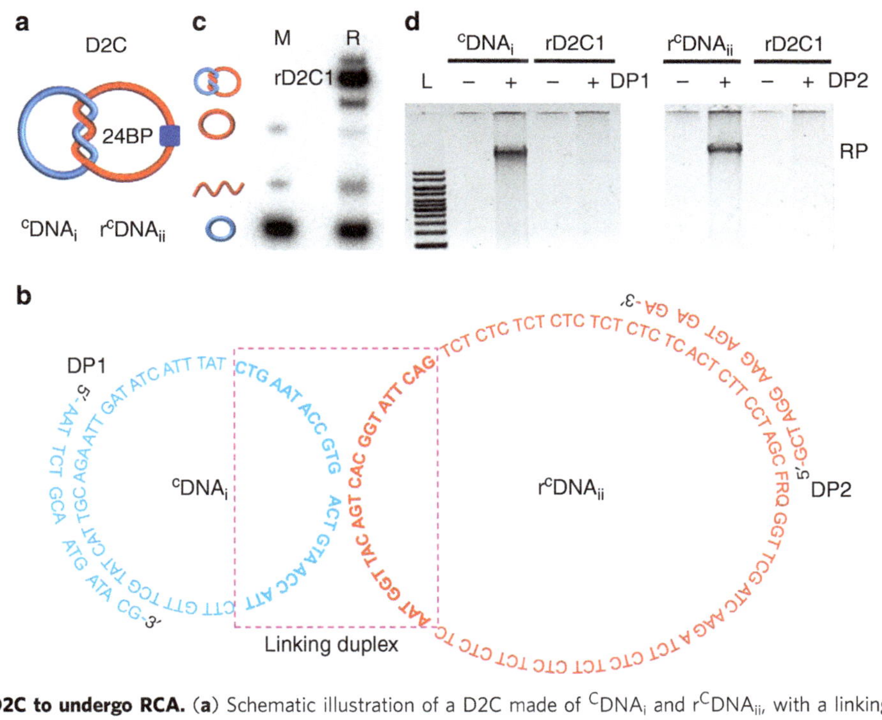

Figure 1 | Inability of a D2C to undergo RCA. (**a**) Schematic illustration of a D2C made of $^{C}DNA_{i}$ and $r^{C}DNA_{ii}$, with a linking duplex of 24 bp. (**b**) Sequences of rD2C1, DP1 and DP2. Boxed nucleotides represent the 24-bp linking duplex. F: fluorescein-dT; R, adenosine ribonucleotide; Q, dabcyl-dT. (**c**) Synthesis of rD2C1 by circularizing linear DNA_{ii} over $^{C}DNA_{i}$ as the template. Lane M, markers made of $^{L}DNA_{ii}$, $^{C}DNA_{i}$ and $r^{C}DNA_{ii}$. Lane R: circularization mixture. (**d**) RCA reactions with gel-purified $^{C}DNA_{i}$, $r^{C}DNA_{ii}$ and rD2C1 using DP1 and DP2 as primers. Lane L, DNA ladders ranging from 1 to 10 kbp; RP, RCA product.

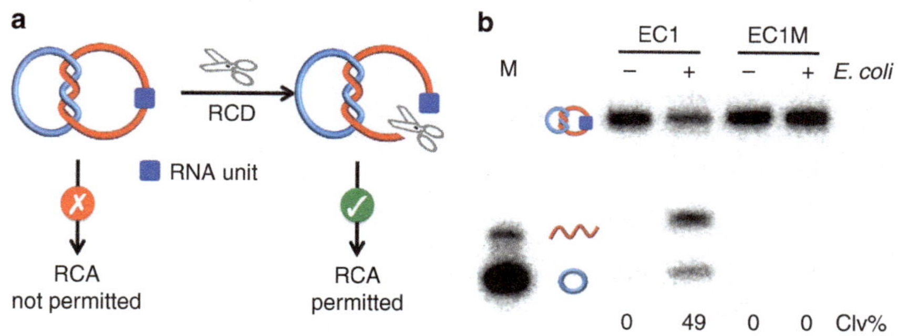

Figure 2 | Cleavage of an RNA-containing D2C by an RCD. (**a**) Restoration of RCA compatibility of an rD2C using an RCD. (**b**) Cleavage of rD2C1 by EC1, an *E. coli*-responsive DNAzyme. Concentration of *E. coli*: 10^5 cells ml^{-1}. Reaction mixtures were analyzed by 10% denaturing PAGE. EC1M: a mutant EC1 that cannot be activated by *E. coli*. Both r^CDNA_{ii} and $^CDNA_i$ in rD2C1 were radioactively labelled with ^{32}P to facilitate DNA visualization on the gel. Clv%: per cent cleavage.

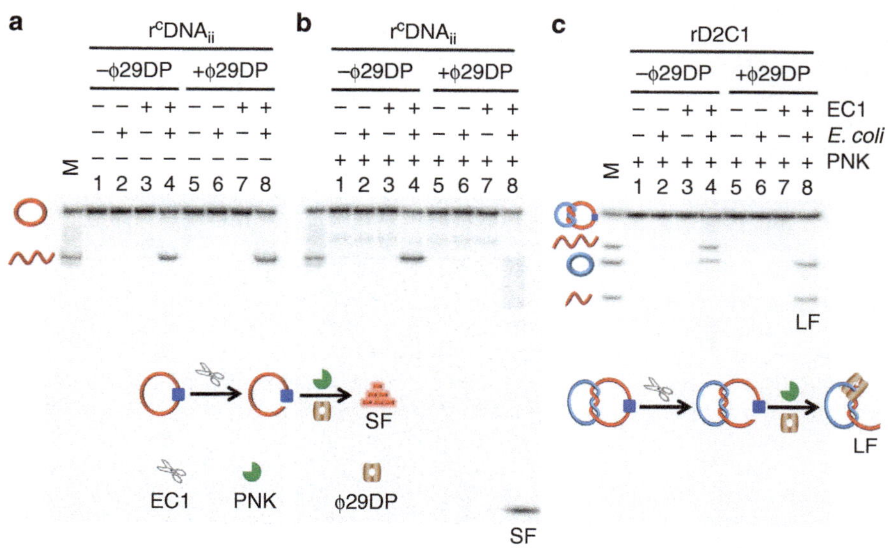

Figure 3 | 3′-5′ exonucleolytic activity of ϕ29DP on r^CDNA_{ii} and rD2C1. Degradation of EC1-mediated cleavage product of r^CDNA_{ii} (**a**,**b**) and rD2C1 (**c**) by ϕ29DP and PNK. Concentration of *E. coli*: 10^5 cells ml^{-1}. Reaction mixtures were analyzed by 20% denaturing PAGE. LF, large DNA fragment; SF, small DNA fragment. M lanes contain various DNA markers as indicated. r^CDNA_{ii}, both r^CDNA_{ii} and $^CDNA_i$ in rD2C1 were radioactively labelled with ^{32}P to facilitate DNA visualization on the gel.

constrained DNA catenanes provides a novel avenue to explore these intricate DNA assemblies for practical applications. Given our interest in DNAzymes and biosensing, we turned our attention next to the engineering of a highly unique biosensing system that takes advantage of topologically constrained DNA catenanes, DNAzymes and RCA.

The working principle of our biosensing system is shown in Fig. 2a. It uses an RCD to cleave the embedded RNA linkage within rD2C1 (note that the r^CDNA_{ii} was designed to contain a single ribonucleotide; Fig. 1b). It is expected that the cleavage and linearization of r^CDNA_{ii} by the RCD will release the topological constraint on the DNA assembly, which converts $^CDNA_i$ into a suitable template for RCA.

In theory, our approach should be compatible with any RCD, making it a platform for detection of any species recognized by an allosteric DNAzyme. For this work, we employed EC1, which was previously isolated by us from a random-sequence DNA pool using *in vitro* selection for specific detection of *E. coli*, a model bacterial pathogen[16,17]. EC1 was found to be activated by a protein molecule secreted specifically by *E. coli* cells (illustrated in Supplementary Fig. 2). Therefore, the use of EC1 enables the detection of this pathogen. As illustrated in Fig. 2b, EC1 was indeed able to cleave the r^CDNA_{ii} present in rD2C1 in an *E. coli*-dependent manner, resulting in $^CDNA_i$ and a linear DNA_{ii}. An inactive DNAzyme mutant, EC1M, was also tested as a control. No cleavage product was observed for EC1M, indicating that the cleavage reaction is highly dependent on the DNAzyme sequence. We also examined the cleavage activity at different reaction times and found that the cleavage activity reached a plateau in 1 h (Supplementary Fig. 3). Thus, this reaction time was used for the remaining experiments.

Upon demonstrating EC1-mediated cleavage of r^CDNA_{ii} in rD2C1, we examined the use of the cleavage reaction mixture for initiating the RCA reaction with ϕ29DP. Because ϕ29DP has 3′–5′ exonucleolytic activity that can degrade single-stranded DNA from the 3′-end but does not digest double-stranded DNA[18,19], we speculated that the system should not require an external primer, as ϕ29DP should be able to convert the linearized DNA_{ii} into a primer for RCA. To evaluate this hypothesis, we determined whether ϕ29DP could digest EC1-linerized DNA_{ii}. From the data presented in Fig. 3a, it is clear that ϕ29DP could not digest linearized DNA_{ii} (comparing lanes 4 and 8).

There are two possible reasons for this finding. The first possibility is that ϕ29DP is incapable of digesting an RNA-terminated DNA molecule. However, testing with an

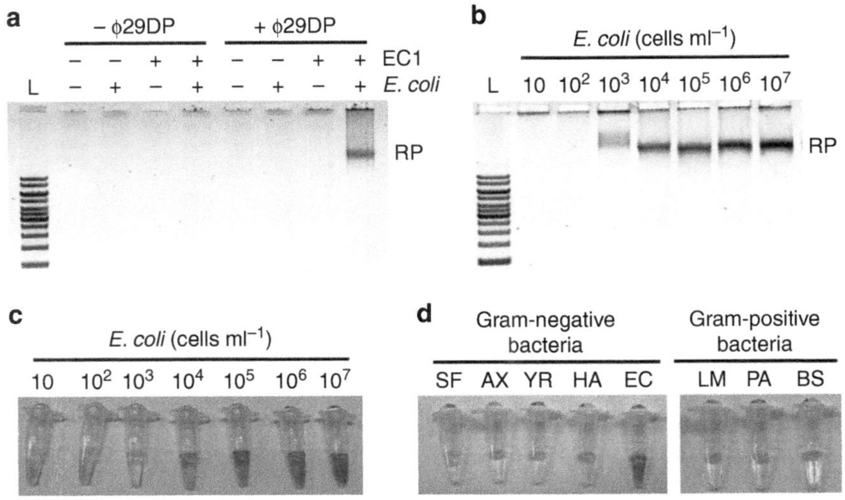

Figure 4 | *E. coli*-dependent RCA reaction. (**a**) RCA reactions of rD2C1 in the presence of *E. coli* (10^5 cells ml^{-1}) analyzed using 0.6% agarose gel electrophoresis. Note every reaction also contained PNK and dNTPs. L, DNA ladders ranging from 1 to 10 kbp; RP, RCA product. (**b**) Determination of detection sensitivity through analysis of RP using 0.6% agarose gel electrophoresis. (**c**) Determination of detection sensitivity via the colourimetric assay enabled by PW17 peroxidase DNAzyme. (**d**) Analysis of assay specificity using the colourimetric assay. The gram-negative bacteria used were *Serratia fonticola* (SF), *Achromobacter xylosoxidans* (AX), *Yersinia ruckeri* (YR) and *Hafnia alvei* (HA). The gram-positive bacteria used were *Leuconostoc mesenteroides* (LM) and *Pediococcus acidilactici* (PA).

RNA-terminated oligonucleotide ruled out this possibility (Supplementary Fig. 4). The second scenario is that ϕ29DP is not able to digest an RNA-terminated DNA molecule containing a 2′,3′-cyclic phosphate on the RNA moiety, which is a common product of RNA cleavage[20]. To test this, we treated the reaction mixture with T4 polynucleotide kinase (PNK), which is known to be capable of removing the terminal 2′,3′-cyclic phosphate in RNA[21]. As shown in Fig. 3b, treatment with PNK indeed facilitated the digestion of EC1-linearized $rDNA_{ii}$, as evidenced by the accumulation of small cleavage fragments (labelled SF in lane 8).

We next investigated the combined action of ϕ29DP and PNK on EC1-linearized $rDNA_{ii}$ within rD2C1. As shown in Fig. 3c, ϕ29DP degraded complexed DNA_{ii} to a product of ~60 nt (labelled LF, representing long fragment; lane 8). We also conducted the digestion assays using different incubation times (Supplementary Fig. 5). The progressive accumulation of LF and disappearance of the cleaved $rDNA_{ii}$ was observed. These experiments allow us to conclude that the combination of ϕ29DP and PNK can remove the single-stranded fragment at the 3′-end of the EC1-linearized DNA_{ii} from rD2C1. It is expected that the trimmed DNA_{ii} can now function as a primer to initiate RCA over the complexed $^{C}DNA_{i}$ template.

To verify the point above, we carried out the RCA reaction with the rD2C1 assembly. The reaction was performed in two sequential steps: activation of EC1 by *E. coli* and PNK treatment, followed by the addition of ϕ29DP and dNTPs. As expected, RCA products were indeed observed following this procedure (the last lane of Fig. 4a; the other lanes represent various controls). The RCA products were further analyzed through partial digestion with *Eco*RV (Supplementary Fig. 6) as the sequence of $^{C}DNA_{i}$ was designed to contain a recognition sequence for this restriction enzyme. The appearance of the expected characteristic DNA banding pattern on the gel, which consists of monomeric, dimeric and other higher-ordered DNA amplicons, verified that the RCA products indeed contained the correct repetitive sequences.

Quantitative detection of *E. coli* using the DNA catenane sensor. We then investigated the feasibility of performing quantitative analysis using the DNA catenane sensor. Samples containing at *E. coli* 10–10^7 cells ml^{-1} were assessed for RCA amplified detection using a gel-staining method. By this method, we were able to detect as low as 10^3 cells ml^{-1} (Fig. 4b). Although gel-based RP analysis can perform quantitative detection of *E. coli*, the procedure is extremely inconvenient. To overcome this issue, we developed a colorimetric assay by modifying the sequence of $^{C}DNA_{i}$ (the new sequence is named $^{C}DNA_{i}CD$) such that the RCA product contained a repetitive sequence of PW17, a peroxidase-like DNAzyme capable of generating a colourimetric signal[22–26]. In the presence of hemin, PW17 catalyzes the H_2O_2-mediated oxidation of 2,2′-azino-bis(3-ethylbenzthiazoline-6-sulphonic acid) (ABTS) into a coloured product. As shown in Fig. 4c, this colourimetric method was indeed able to detect *E. coli* in a concentration-dependent manner and registered a detection sensitivity of 10^3 cells ml^{-1}, similar to what was observed with the gel-based method.

We also evaluated the bacterial detection specificity using the colourimetric assay. We selected four other Gram-negative and three Gram-positive bacteria that were previously tested for EC1-based detection. It was observed that none of these bacteria were able to produce a positive signal, indicating that the rD2C1/EC1 system retained the high recognition specificity for *E. coli* (Fig. 4d). To further evaluate the specificity, we checked the potential influence of small RNAs on *E. coli* detection due to the fact that small RNAs (for example, microRNA) are suitable primers for RCA. For this experiment, we used the total small RNAs extracted from breast cancer cell line MCF-7. Agarose gel and colourimetric results indicated that the small RNAs were not able to induce the RCA reaction (Supplementary Fig. 7). This high specificity is attributed to the unique topologically constrained structure of the DNA catenane.

Enhancing detection sensitivity using hyperbranched RCA. Finally we examined the possibility of performing a double-primed hyperbranched RCA (HRCA)[27] reaction with the rD2C1/

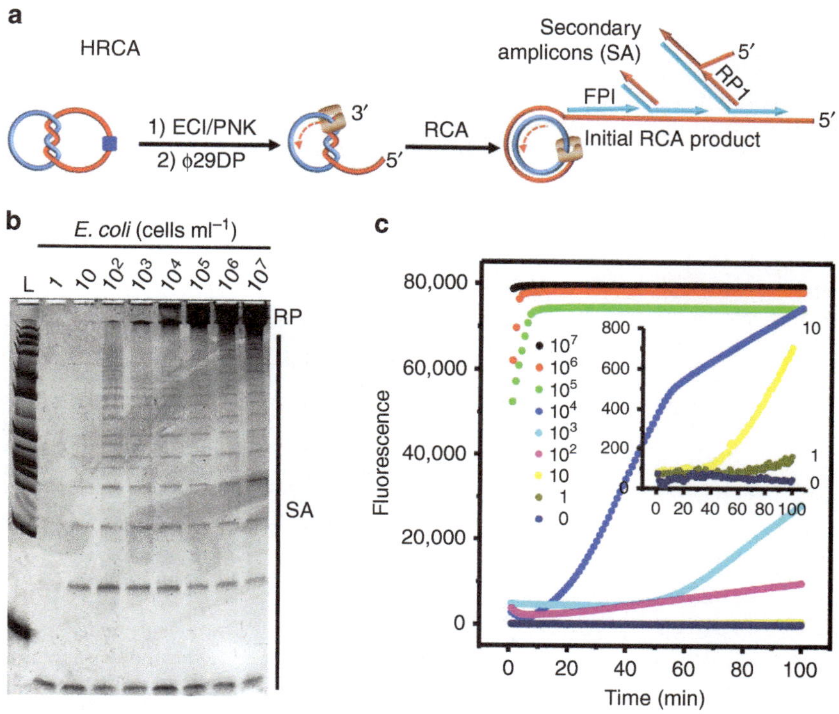

Figure 5 | *E. coli*-dependent HRCA reaction. (**a**) Schematic illustration of HRCA. FP1, forward primer; RP1, reverse primer. (**b**) Denaturing PAGE analysis of HRCA products. RP, RCA products; SA, secondary amplicons produced from the initial RCA products. (**c**) Real-time monitoring of HRCA reactions at various *E. coli* concentrations (cells ml^{-1}).

EC1 system to further increase the detection sensitivity. In HRCA (Fig. 5a), multiple priming events can be continuously initiated by a forward primer (FP1) and a reverse primer (RP1) as the original RCA product strand elongates, resulting in an exponential amplification[28].

We found that HRCA was indeed functional with the rD2C1/EC1 system: as shown in Fig. 5b, in addition to the observation of the RCA products on denaturing polyacrylamide gel electrophoresis (PAGE), a series of shorter DNA molecules were also produced, representing various secondary amplicons produced from the primary amplicons (that is, initial RCA products). The HRCA reactions, in response to varying concentrations of *E. coli*, were also monitored in real time through the use of EvaGreen, a DNA-binding dye (Fig. 5c). We found that this method exhibited much enhanced detection sensitivity, as it was able to detect *E. coli* at a concentration of as low as 10 cells ml^{-1} without cell culture.

It is interesting to note that RCA reactions with ϕ29DP remain functional even in the presence of 50% human blood (Supplementary Fig. 8). In comparison, PCR with Taq DNA polymerase was completely inhibited by less than 0.2% human blood[29–31]. To demonstrate the performance of the assay when using more complex samples, we spiked whole blood with *E. coli* and demonstrated that under these conditions the DNA catenane sensor was still able to detect *E. coli* at a concentration of 10 cells ml^{-1} (Supplementary Fig. 9). This observation is consistent with a recent report where EC1 was used to detect *E. coli* cells in human blood[32].

The ability of the DNA catenane sensor to distinguish between live and dead *E. coli* cells was also investigated. For this experiment, we used lysozyme to kill *E. coli* (10^5 cells ml^{-1}) and compared the signal responses of the catenane sensor in the presence of live and dead bacteria using HRCA. Negligible signal was observed with dead cells while a high activity was seen with live *E. coli* cells (Supplementary Fig. 10).

We also performed an *E. coli* detection experiment with an enzyme-linked immunosorbent assay (ELISA) using a commercial ELISA kit designed to detect *E. coli* host cell protein (Supplementary Fig. 11). It was found that the ELISA method was able to detect *E. coli* at a concentration of 10^3 cells ml^{-1}. Therefore, our amplified DNA catenane sensor offers a detection sensitivity that is 100 times better than the ELISA method, while the test time for both methods are similar (~3 h; see Supplementary Table 2 for additional information).

Discussion

We have shown for the first time that mechanically interlocked DNA [2] catenanes with a strong linking duplex impose a significant topological constraint on their component DNA rings, making them unsuitable as the template for RCA. We have further demonstrated that such DNA nanostructures can be uniquely exploited for the design of a biosensing system where the elimination of the topological engagement, achieved simply through the cleavage of one interlocked DNA ring in an analyte-dependent manner, frees up the other ring for the RCA reaction. As an example, we have produced one of the two interlocked DNA rings to contain a RNA linkage so that an RCD can be used to cleave one interlocked ring. Through the use of an RCD whose activity is specifically triggered by a secreted protein in *E. coli*, we have shown that the featured biosensing system is capable of achieving ultra-sensitive detection of this bacterial pathogen.

The biosensing system featured here offers some distinct advantages over existing detection methods for *E. coli*, such as cell culturing, PCR and ELISA (see Supplementary Table 2 for additional information). The use of RCA and HRCA for signal amplification makes this system extremely sensitive for bacterial detection, which can achieve the detection of as low as 10 cells ml^{-1} without a cell-culturing step. The assay is also more compatible with point-of-care or field applications because

RCA is an isothermal process and there is no need for DNA extraction (as in the case of PCR). In addition, the system functions well with biological samples (no interferences from small RNAs and compatibility with blood samples).

It is conceivable that the same design can be extended to other RCDs, DNA-cleaving DNAzymes, as well as ribozymes and protein enzymes that have DNA or RNA-cleaving activities. Although rolling circle amplification was exploited to achieve signal amplification in this study, it should be feasible to take advantage of other signal amplification strategies, such as the DNAzyme cross-amplification system developed by Levy and Ellington that does not need a DNA polymerase[33]. We envision that the concept presented in this study should open up new opportunities for exploring mechanically interlocked DNA architectures for many potential applications in chemical biology, medical diagnostics, and environmental monitoring.

Methods

Preparation of rD2C1. A total of 100 pmol of $r^{L}DNA_{ii}$ was first labelled with γ-[^{32}P]ATP at the 5′ end using T4 PNK according to the manufacturer's protocol. To ensure that all DNA molecules contain the 5′ phosphate required for the subsequent ligation reaction, PNK-mediated end-labelling solution containing 5′-^{32}P labelled $r^{L}DNA_{ii}$ was further incubated with 2 mM non-radioactive ATP at 37 °C for 30 min. Then 120 pmol of $^{C}DNA_{i}$, prepared using the procedure detailed in Supplementary Methods, was added and heated to 90 °C for 30 s. After cooling to room temperature and leaving the solution for 15 min, 10 µl of 10 × T4 DNA ligase buffer and 10 U T4 DNA ligase were added (total 100 µl) and incubated at room temperature for 2 h. The obtained rD2C1 molecules were concentrated by standard ethanol precipitation and purified by 10% dPAGE.

***E. coli*-dependent RCA reaction.** A single colony of *E. coli* K12 freshly grown on Luria broth (LB) agar plate was taken and used to inoculate 2 ml of LB. After shaking at 37 °C for 14 h at 250 r.p.m., the bacterial culture was serially diluted in 10-fold intervals. Overall, 100 µl of each diluted solution was plated onto LB agar plate (done in triplicate) and cultured at 37 °C for 15 h to obtain the cell counts. Colonies in each plate were counted; the average number of colonies from the three plates was taken as the number of *E. coli* cells for this dilution. This number was then used to calculate the number of cells for the other dilutions. A total of 500 µl of each dilution was centrifuged at 13,000*g* for 20 min at 4 °C and re-suspended in 100 µl of 1 × RB (50 mM HEPES, 150 mM NaCl, 15 mM $MgCl_2$, pH 7.5). After being frozen at − 20 °C, *E. coli* cells were sonicated for 1 min and put on the ice for 5 min. This process was repeated three times. Then the cell suspension containing different numbers of *E. coli* cells were centrifuged at 13,000*g* for 10 min at 4 °C. The obtained crude intracellular mixture produced by *E. coli* cells (CIM-EC) in the supernatant was used for the following experiment.

A cleavage reaction mixture containing 5 µl of CIM-EC, 1 µl of rD2C1 (5 µM), 4 µl of EC1 (50 µM) and 10 µl of 2 × RB was incubated at room temperature for 60 min. Then 1 µl of PNK (10 U µl^{-1}) was added and incubated at 37 °C for 30 min. The RCA reaction was initiated by the addition of 1 µl of ϕ29DP (10 U µl^{-1}), 1 µl of dNTPs (50 mM), 5 µl of 10 × RCA reaction buffer and 22 µl of water. The reaction mixtures were incubated at 30 °C for 60 min before heating at 90 °C for 5 min. The resultant RCA products were analyzed by 0.6% agarose gel electrophoresis.

Colourimetric detection. rD2C1 used for the colourimetric detection of *E. coli* was made of $r^{C}DNA_{ii}$ and $^{C}DNA_{i}$CD, prepared using the procedure described in Supplementary Methods. After the cleavage reaction described above, 1 µl of PNK (10 U µl^{-1}) was added and incubated at 37 °C for 30 min. The RCA reaction was then initiated by the addition of 1 µl of ϕ29DP (10 U µl^{-1}), 1 µl of dNTPs (50 mM), 2 µl of hemin (100 µM), 5 µl of 10 × RCA reaction buffer and 20 µl of water. The reaction mixtures were incubated at 30 °C for 60 min before heating at 65 °C for 20 min. After cooling to room temperature, 2 µl of ABTS (50 mM) and 1 µl of H_2O_2 (8.8 mM) were added, and the colourimetric result was recorded immediately using a digital camera.

HRCA reaction. Following the cleavage reaction, 1 µl of PNK (10 U µl^{-1}) was added and incubated at 37 °C for 30 min. The HRCA reaction was then initiated by the addition of 1 µl of ϕ29DP (10 U µl^{-1}), 1 µl of dNTPs (50 mM), 1 µl of FP1 (50 µM), 1 µl of RP1 (50 µM), 5 µl of 10 × RCA reaction buffer, 2.5 µl of 20 × EvaGreen and 17.5 µl of water. These reactions were carried out in BioRad CFX96 qPCR system set to a constant temperature of 30 °C, and the fluorescence intensity was recorded in 1 min intervals.

Other experiments. Details for the following experiments are provided in Supplementary Methods: preparation of $r^{L}DNA'_{ii}$ and $r^{C}DNA'_{ii}$, and rD2C1′; comparison of the cleavage activity of EC1 and EC1M in the presence of *E. coli*; degradation of rD2C1 by ϕ29DP in the presence of EC1, *E. coli* and PNK; cell culture and miRNA extraction; restriction digestion of RCA products; effect of blood on RCA reactions; detection of *E. coli* in blood samples; and detection of *E. coli* using an ELISA kit.

References

1. Mao, C., Sun, W. & Seeman, N. C. Assembly of Borromean rings from DNA. *Nature* **386,** 137–138 (1997).
2. Wang, H., Du, S. M. & Seeman, N. C. Tight single-stranded DNA knots. *J. Biomol. Struct. Dyn.* **10,** 853–863 (1993).
3. Schmidt, T. L. & Heckel, A. Construction of a structurally defined double-stranded DNA catenane. *Nano Lett.* **11,** 1739–1742 (2011).
4. Ackermann, D. *et al.* A double-stranded DNA rotaxane. *Nat. Nanotechnol.* **5,** 436–442 (2010).
5. Zhang, F., Nangreave, J., Liu, Y. & Yan, H. Structural DNA nanotechnology: state of the art and future perspective. *J. Am. Chem. Soc.* **136,** 11198–11211 (2014).
6. Chen, Y. J., Groves, B., Muscat, R. A. & Seelig, G. DNA nanotechnology from the test tube to the cell. *Nat. Nanotechnol.* **10,** 748–760 (2015).
7. Pinheiro, A. V., Han, D., Shih, W. M. & Yan, H. Challenges and opportunities for structural DNA nanotechnology. *Nat. Nanotechnol.* **6,** 763–772 (2011).
8. Elbaz, J., Cecconello, A., Fan, Z., Govorov, A. O. & Willner, I. Powering the programmed nanostructure and function of gold nanoparticles with catenated DNA machines. *Nat. Commun.* **4,** 2000 (2013).
9. Liu, X. Q., Lu, C. H. & Willner, I. Switchable reconfiguration of nucleic acid nanostructures by stimuli-responsive DNA machines. *Acc. Chem. Res.* **47,** 1673–1680 (2014).
10. Li, T., Lohmann, F. & Famulok, M. Interlocked DNA nanostructures controlled by a reversible logic circuit. *Nat. Commun.* **5,** 4940 (2013).
11. Zhang, D. Y. & Seelig, G. Dynamic DNA nanotechnology using strand-displacement reactions. *Nat. Chem.* **3,** 103–113 (2011).
12. Wu, Z. S., Shen, Z., Tram, K. & Li, Y. Engineering interlocking DNA rings with weak physical interactions. *Nat. Commun.* **5,** 4279 (2014).
13. Fire, A. & Xu, S. Q. Rolling replication of short DNA circles. *Proc. Natl Acad. Sci. USA* **92,** 4641–4645 (1995).
14. Liu, D., Daubendiek, S. L., Zillman, M. A., Ryan, K. & Kool, E. T. Rolling circle DNA synthesis: small circular oligonucleotides as efficient templates for DNA polymerases. *J. Am. Chem. Soc.* **118,** 1587–1594 (1996).
15. Zhao, W., Ali, M. M., Brook, M. A. & Li, Y. Rolling circle amplification: applications in nanotechnology and biodetection with functional nucleic acids. *Angew. Chem. Int. Ed.* **47,** 6330–6337 (2008).
16. Ali, M. M., Aguirre, S. D., Lazim, H. & Li, Y. Fluorogenic DNAzyme probes as bacterial indicators. *Angew. Chem. Int. Ed.* **50,** 3751–3754 (2011).
17. Aguirre, S. D., Ali, M. M., Salena, B. J. & Li, Y. A sensitive DNA enzyme-based fluorescent assay for bacterial detection. *Biomolecules* **3,** 563–577 (2013).
18. Liu, M., Zhang, W., Zhang, Q., Brennan, J. D. & Li, Y. Biosensing by tandem reactions of structure switching, nucleolytic digestion, and DNA amplification of a DNA assembly. *Angew. Chem. Int. Ed.* **54,** 9637–9641 (2015).
19. Blanco, L. & Salas, M. Characterization of a 3′→5′ exonuclease activity in the phage ϕ29-encoded DNA polymerase. *Nucleic Acids Res.* **13,** 1239–1249 (1985).
20. Silverman, S. K. *In vitro* selection, characterization, and application of deoxyribozymes that cleave RNA. *Nucleic Acids Res.* **33,** 6151–6163 (2015).
21. Schurer, H., Lang, K., Schuster, J. & Morl, M. A universal method to produce *in vitro* transcripts with homogeneous 3′ ends. *Nucleic Acids Res.* **30,** e56 (2002).
22. Li, Y. & Sen, D. A catalytic DNA for porphyrin metallation. *Nat. Struct. Biol* **3,** 743–747 (1996).
23. Travascio, P., Li, Y. & Sen, D. DNA-enhanced peroxidase activity of a DNA aptamer-hemin complex. *Chem. Biol.* **5,** 505–517 (1998).
24. Travascio, P., Witting, P. K., Mauk, A. G. & Sen, D. The peroxidase activity of a hemin-DNA oligonucleotide complex: free radical damage to specific guanine bases of the DNA. *J. Am. Chem. Soc.* **123,** 1337–1348 (2001).
25. Cheglakov, Z., Weizmann, Y., Basnar, B. & Willner, I. Diagnosing viruses by the rolling circle amplified synthesis of DNAzymes. *Org. Biomol. Chem.* **5,** 223–225 (2007).
26. Tian, Y., He, Y. & Mao, C. Cascade signal amplification for DNA detection. *ChemBioChem.* **7,** 1862–1864 (2006).
27. Zhang, D., Brandwein, M., Hsuih, T. & Li, H. Amplification of target-specific, ligation-dependent circular probe. *Gene* **211,** 277–285 (1998).

28. Lizardi, P. M. *et al.* Mutation detection and single-molecule counting using isothermal rolling-circle amplification. *Nat. Genet.* **19,** 225–232 (1998).
29. Al-Soud, W. A. & Radstrom, P. Effect of amplification facilitators on diagnostic PCR in the presence of blood, feces and meat. *J. Clin. Microbiol.* **38,** 4463–4470 (2000).
30. Al-Soud, A. W. & Radstrom, P. Capacity of nine thermo-stable DNA polymerases to mediate DNA amplification in the presence of PCR-inhibiting samples. *Appl. Environ. Microbiol.* **64,** 3748–3753 (1998).
31. Schrader, C., Schielke, A., Ellerbroek, L. & Johne, R. PCR inhibitors—occurrence, properties and removal. *J. Appl. Microbiol.* **113,** 1014–1026 (2012).
32. Kang, D. K. *et al.* Rapid detection of single bacteria in unprocessed blood using integrated comprehensive droplet digital detection. *Nat. Commun.* **5,** 5427 (2014).
33. Levy, M. & Ellington, A. D. Exponential growth by cross-catalytic cleavage of deoxyribozymogens. *Proc. Natl Acad. Sci. USA* **100,** 6416–6421 (2003).

Acknowledgements

Funding for this work was provided by Natural Sciences and Engineering Research Council of Canada (NSERC) Discovery Grants (Y.L. and J.D.B.), Pro-Lab Developments Inc., Collaborative Health Research Projects (CHRP) Initiative from Canadian Institutes of Health Research (CIHR), the Canada Foundation for Innovation (CFI) and the Ministry for Research and Innovation (MRI) of Ontario. Part of the work was conducted at the McMaster Biointerfaces Institute. J.D.B. holds the Canada Research Chair in Bioanalytical Chemistry and Biointerfaces.

Author contributions

M.L., Q.Z., J.D.B. and Y.L. designed the research project; M.L. and Q.Z. performed the experiments and discussed the experimental results; Z.L. and J.G. performed some experiments; M.L., J.D.B., and Y.L. analyzed the data and wrote the paper.

Additional information

Competing financial interests: The authors declare no competing financial interests.

20

Amplification of oxidative stress by a dual stimuli-responsive hybrid drug enhances cancer cell death

Joungyoun Noh[1], Byeongsu Kwon[2], Eunji Han[2], Minhyung Park[2], Wonseok Yang[2], Wooram Cho[2], Wooyoung Yoo[2], Gilson Khang[1,2] & Dongwon Lee[1,2]

Cancer cells, compared with normal cells, are under oxidative stress associated with the increased generation of reactive oxygen species (ROS) including H_2O_2 and are also susceptible to further ROS insults. Cancer cells adapt to oxidative stress by upregulating antioxidant systems such as glutathione to counteract the damaging effects of ROS. Therefore, the elevation of oxidative stress preferentially in cancer cells by depleting glutathione or generating ROS is a logical therapeutic strategy for the development of anticancer drugs. Here we report a dual stimuli-responsive hybrid anticancer drug QCA, which can be activated by H_2O_2 and acidic pH to release glutathione-scavenging quinone methide and ROS-generating cinnamaldehyde, respectively, in cancer cells. Quinone methide and cinnamaldehyde act in a synergistic manner to amplify oxidative stress, leading to preferential killing of cancer cells *in vitro* and *in vivo*. We therefore anticipate that QCA has promising potential as an anticancer therapeutic agent.

[1]Department of Polymer · Nano Science and Technology, Polymer Fusion Research Center, Chonbuk National University, Backje-daero 567, Jeonju 561-756, Korea. [2]Department of BIN Convergence Technology, Chonbuk National University, Backje-daero 567, Jeonju 561-756, Korea. Correspondence and requests for materials should be addressed to D.L. (email: dlee@chonbuk.ac.kr).

Reactive oxygen species (ROS) are a group of highly reactive small molecules generated by all aerobic organisms and encompass H_2O_2, superoxide (O^{2-}) and hydroxyl radical[1,2]. In particular, H_2O_2 plays important roles in cell growth, proliferation and signalling pathway under physiological conditions[3]. Redox balance between oxidizing and reducing species within cells is important in the regulation of signalling pathway and is achieved by various enzymes such as superoxide dismutase, catalase and glutathione peroxidase[2]. However, overproduction of H_2O_2 and other ROS causes oxidative stress, leading to cellular damages and subsequent functional decline of organ systems[4,5]. Therefore, it is crucial for normal cells to maintain ROS homeostasis to control the ROS level and overcome the potential toxicity of ROS[6].

Accumulating evidence has suggested that abnormal cancer cells are under oxidative stress associated with an increased production of ROS due to disrupted ROS homeostasis and cancer cells have also adapted to oxidative stress by activating antioxidant systems often through the upregulation of glutathione (GSH)[2,7–10]. Ironically, cancer cells are also known to utilize ROS to drive proliferation and other events required for tumour development[11,12]. However, the high level of ROS can cause cellular damage and apoptotic cell death, depending on the concentration and duration of ROS stress[13]. Cancer cells are also susceptible to further ROS insults induced by exogenous agents that weaken antioxidant defense or augment ROS generation to a level or the threshold above which cells cannot survive[14–16]. It was also postulated that cancer cells under oxidative stress are highly dependent on their antioxidant systems to counteract the damaging effects of ROS to maintain the ROS balance. Therefore, the difference in redox balance in normal and cancer cells would provide a biological cue for selective killing of malignant cancer cells[17].

Recently, there have been increasing efforts to develop a strategy to increase the level of ROS in cancer cells specifically, leading to ROS-mediated cancer cell death, which is named as 'oxidation therapy'[1,12,14]. One way of this anticancer therapy is to deliver ROS-generating agents to tumour tissues directly, such as arsenic trioxide (As_2O_3) and glucose oxidase[17]. Another approach of oxidation therapy is to disrupt the redox balance in cancer cells by suppressing the antioxidant systems. There are several therapeutic agents that inhibit antioxidative enzyme systems and exert cytotoxicity to cancer cells, including L-buthionine-sulfoximine, β-phenylethyl isothiocyanate and zinc protoporphyrin[1,17–19]. These agents are known to induce non-severe ROS stress in normal cells because normal cells have a low level of basal ROS[11,16].

A newly discovered antioxidant inhibitor with anticancer activity is quinone methide (QM). QM was reported to rapidly alkylate GSH, one of the key antioxidant systems in cells and trigger apoptotic cell death[20]. Hulsman *et al.*[7] reported that the potent anticancer activity of nitric oxide-donating aspirin is attributed to QM, but not nitric oxide. Combined use of an ROS-generating agent with an inhibitor of ROS scavenging systems has been also reported to amplify oxidative stress and be effective in the treatment of advanced solid tumours in terms of both efficacy and selectivity[8,19]. Therefore, manipulating the ROS levels by redox modulation is a logical approach to selectively kill cancer cells without causing significant toxicity to normal cells and would have broad therapeutic applications for cancer treatment.

In this study, we report a novel hybrid anticancer therapeutic agent that not only weakens the antioxidant defense systems but also elevates ROS production simultaneously, leading to massive ROS accumulation preferentially in cancer cells and subsequent ROS-mediated cell death (Fig. 1). Cinnamaldehyde, a major component of cinnamon tree, has been known to induce intracellular ROS generation, leading to apoptotic cell death[21,22]. Cinnamaldehyde contains active Michael acceptor pharmacophore and is given the Generally Recognized as Safe status with approval for use in food in the United States[22]. Cinnamaldehyde has been reported to induce ROS generation mainly in the mitochondria and inhibit growth of human cancer cells, but minimal cytotoxicity to normal cells[21,23,24]. However, the clinical applications of cinnamaldehyde have been limited by its poor bioavailability (a short half-life in blood) due to rapid oxidation of aldehyde group and lower drug efficacy than common anticancer drugs[21,25,26].

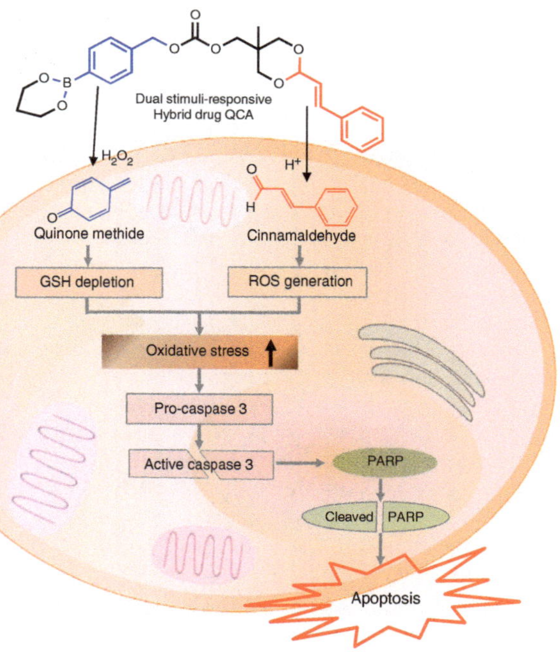

Figure 1 | A dual stimuli-responsive hybrid anticancer drug QCA. QCA is activated by H_2O_2 and acidic pH and amplifies oxidative stress in cancer cells, leading to apoptotic cell death.

To overcome these drawbacks, we developed QCA, [4-(1,3,2-dioxaborinan-2-yl)benzyl ((5-methyl-2-styryl-1,3-dioxan-5-yl)methyl) carbonate] as a novel hybrid anticancer prodrug by coupling a QM-generating moiety to ROS-generating cinnamaldehyde, which is therefore capable of amplifying oxidative stress in cancer cells to induce enhanced apoptotic cell death. The underlying hypothesis of the hybrid anticancer drug QCA is that depletion of antioxidant GSH by QM renders cancer cells more susceptible to cinnamaldehyde-mediated ROS insults and the elevated oxidative stress results in enhanced apoptotic cell death. In addition, QCA was designed to be activated by H_2O_2 and acidic pH to release QM and cinnamaldehyde, respectively, in cancer environments and exert synergistic therapeutic actions. Herein, we demonstrate that the rationally designed hybrid drug QCA has dual mode of anticancer actions with synergistic therapeutic effects in the proof of concept studies *in vitro* as well as *in vivo*.

Results

Design and synthesis of QCA. A hybrid anticancer drug QCA was designed to be activated by acidic pH and high level of H_2O_2 (ref. 27). QCA as a dual stimuli-responsive hybrid anticancer agent was synthesized as shown in Supplementary Fig. 1. First, QM-generating boronate compound **1** was synthesized from the reaction of 1,3-propanediol and (4-(hydroxymethyl)phenyl) boronic acid at room temperature for 2 days. **1** was obtained as a viscose colourless liquid and its chemical structure was confirmed by ^1H NMR (Supplementary Fig. 2). Cinnamaldehyde was reacted with 2-(hydroxymethyl)-2-methylpropane-1,3-diol to give a cinnamaldehyde derivative **2**, which possesses an acid-cleavable acetal linkage. The presence of an acetal linkage in **2** was confirmed by the appearance of acetal protons at ~5.1 p.p.m. (Supplementary Fig. 3). **2** was then reacted with carbonyldiimidazole to give a cinnamaldehyde releasing compound **3** which is able to react with hydroxyl group of **1**. From a reaction of **1** and **3**, a hybrid anticancer drug QCA was obtained as a white solid. QCA was characterized by NMR and mass spectroscopy (Supplementary Fig. 4). The acetal protons appeared at ~5.1 p.p.m. and the methylene protons next to boronic ester appeared at ~4.1 p.p.m., demonstrating the successful coupling of H_2O_2-responsive boronate to cinnamaldehyde via a carbonate linker. For comparison purposes, we also developed 4-(1,3,2-dioxaborinan-2-yl)benzyl benzyl carbonate, Q_1, which is able to generate QM only in the presence of H_2O_2 (Supplementary Fig. 5), not cinnamaldehyde. Q_1 serves as a control to support the superior anticancer efficacy of QCA, which exerts synergistic effects of QM and cinnamaldehyde.

Stimuli responsiveness of QCA. It was reasoned that QCA rapidly undergoes H_2O_2-mediated boronate oxidation to release QM in accordance with the mechanism outlined in Supplementary Fig. 6. In particular, the C–B bond is rapidly cleaved with the formation of **4** containing a phenol group, which exists in aqueous solutions in equilibrium with phenolate of **5**. The phenolate is known to rapidly undergo spontaneous 1,6-elimination with formation of QM that reacts with nucleophilic H_2O to generate hydroxybenzyl alcohol[20]. We studied the dual stimuli-triggered activation of QCA by investigating NMR spectra after hydrolysis under tumour conditions (Fig. 2a). In the presence of 500 μM of H_2O_2, the generation of QM from QCA was indirectly confirmed by the appearance of aromatic proton peaks at ~6.9 and ~6.4 p.p.m. These observations are in good agreement with the previous studies reporting that QM is converted to hydroxybenzyl alcohol under aqueous conditions in the absence of nucleophiles such as GSH[7,28–30]. QCA also liberated cinnamaldehyde from the cleavage of acetal linkages after 48 h of incubation at acidic pH 6.5, confirmed by the reduced signal of acetal protons and appearance of aldehyde protons at 9.5 p.p.m. (ref. 23). It was determined that QCA undergoes H_2O_2-triggered hydrolysis with a second-order rate constant of 8.76×10^9 ($l mol^{-1} s^{-1}$) and acid-triggered hydrolysis with a second-order rate constant of 3.24×10^3 ($l mol^{-1} s^{-1}$). No changes were observed even after 72 h of incubation at neutral pH without H_2O_2, demonstrating that QCA is activated by only H_2O_2 and acidic pH to generate cytotoxic QM and cinnamaldehyde, respectively.

The sensitivity of QCA to H_2O_2 was also studied by measuring the level of H_2O_2 based on peroxalate chemiluminescence, which has been used as a versatile tool for the detection of a variety of chemical species including fluorescent molecules and H_2O_2 (refs 31,32). H_2O_2 solution (1 μM) showed remarkably high emission intensity (~1.1×10^5 relative light units) in the presence of fluorescent molecules and peroxalate ester (Fig. 2b). However, the addition of QCA to H_2O_2 solution resulted in rapid and significant reduction in chemiluminescence intensity, concentration dependently. The majority of H_2O_2 was rapidly reacted with the same concentration of QCA (1 μM) within 1 min. These observations suggest that the boronic ester in QCA specifically and quickly reacts with H_2O_2, resulting in the activation of QCA to release QM.

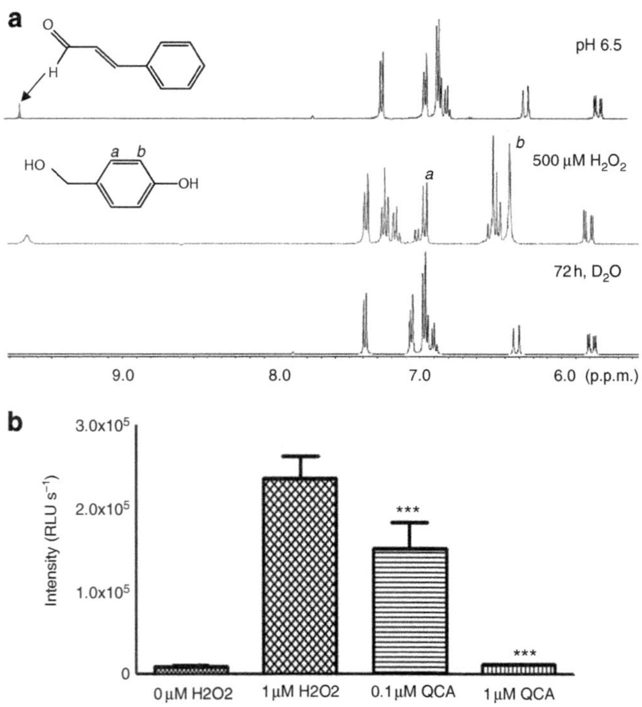

Figure 2 | Sensitivity of QCA to H_2O_2 and pH. **(a)** ^1H NMR spectra of QCA before and after hydrolysis at acidic pH or in the presence of H_2O_2. **(b)** The sensitivity of QCA to H_2O_2 assessed by peroxalate chemiluminescence. ***$P < 0.001$ relative to a group of 1 μM H_2O_2 (one-way analysis of variance, $n = 3$, ± s.d.). RLU, relative light units.

Anticancer activity *in vitro* of QCA. We first performed liquid chromatography–mass spectrometry (LC-MS/MS) analysis to confirm the generation of QM and cinnamladehyde in cells. The generation of QM from QCA was indirectly confirmed by measuring the level of cellular GSH, because it is impossible to detect QM that is an intermediate and rapidly reacts with nucleophilic GSH[7,20]. As shown in Supplementary Fig. 7, lysates of DU145 cells treated with QCA for 24 h showed the remarkable (58%) reduction in the level of cellular GSH in comparison with untreated cells. QCA-treated cell lysates also showed a peak in LC trace identical to the standard cinnamaldehyde, while no peak was observed in untreated cell lysates. These findings clearly demonstrate that QCA generates both QM and cinnamaldehyde in cells.

To further substantiate the depletion of antioxidant GSH by QCA, we investigated the effects of QCA on the level of GSH in various cells (prostate cancer DU145 cell line, colon cancer SW620 cell line and fibroblast NIH3T3 cell line). Cells were treated with various concentrations of QCA or Q_1 for 1.5 h and then the level of GSH was quantitatively measured using Ellman's reagent as previously reported[7]. As shown in Fig. 3, treatment of QCA or Q_1 resulted in significant reduction of the GSH level in malignant DU145 cells, in a dose-dependent manner. More than half of GSH was depleted in DU145 cells by both Q_1 and QCA at doses higher than 50 μM. QCA induced more GSH depletion in DU145 cells than SW620 cells. However, both QCA and Q_1 showed no effects on the GSH level in non-malignant NIH3T3 cells (Supplementary Fig. 8). The above observations can be explained by the rationales that cancer cells have a greater extent of redox adaptation and malignant cells of different cancer types exhibit heterogeneity in the level of oxidative stress, associated with various expression of antioxidant systems[15]. It has been well known that metastatic prostate cancer cells including DU145 produce a large amount of ROS to have aggressive phenotypes[33].

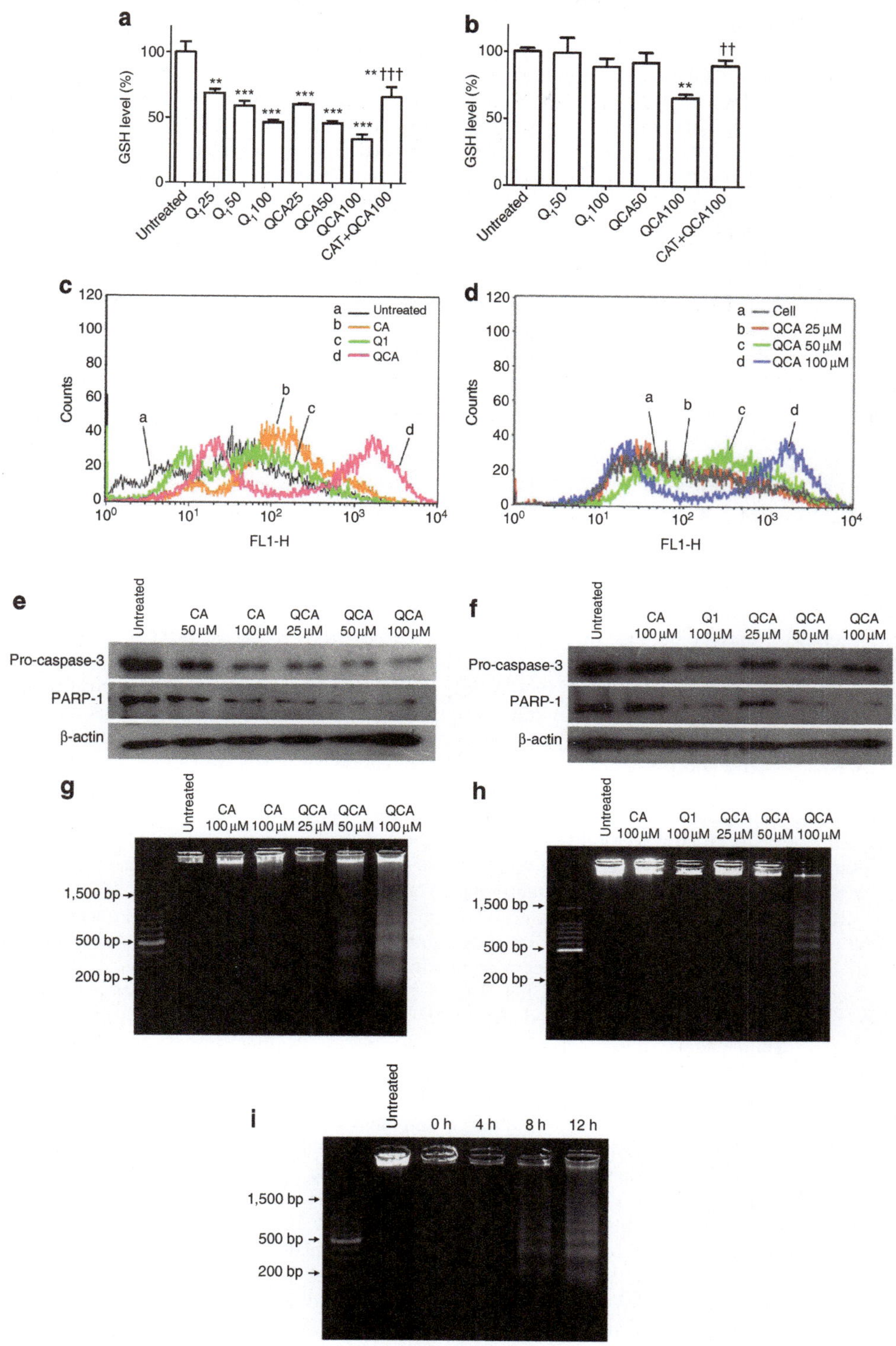

Figure 3 | Amplification of oxidative stress by QCA. Depletion of GSH by various concentrations of QCA and Q_1 in DU145 cells (**a**) and SW620 cells (**b**). The level of GSH was measured using Ellman's reagents. The number indicates μM concentration. Catalase (CAT) was added to cells before the treatment of QCA. $^{**}P<0.01$, $^{***}P<0.001$ relative to the untreated group, $^{††}P<0.01$, $^{†††}P<0.001$ relative to QCA 100 (one-way analysis of variance, $n=4$, ± s.d.), (**c**) ROS generation in DU145 cells treated with 100 μM of CA (cinnamaldehyde), Q_1 or QCA. (**d**) Concentration dependency of QCA on ROS generation in DU145 cells. For the detection of ROS, cells were stained with DCFH-DA and analysed using a BD FACSCaliber flow cytometer and CellQuest software. Cleavage of pro-apoptotic proteins in DU145 cells (**e**) and SW620 cells (**f**) as measured by Western blotting assay. Data are representative of four independent experiments. DNA fragmentation assay of DU145 cells (**g**) and SW620 cells (**h**) treated with various formulations for 12 h. (**i**) Time course of apoptosis of SW620 cells treated by 100 μm of QCA. Data are representative of three independent experiments.

However, cancer cells also generate a large amount of intrinsic antioxidants such as GSH, superoxide dismutase and catalase to counteract ROS and tolerate ROS-induced oxidative stress, leading to altered redox balance[1,17]. In particular, a large amount of GSH is known to play a critical role in cellular redox maintenance and antioxidative systems[8,10]. We also found that DU145 cells have the highest level of GSH and H_2O_2 and the least level of GSH and H_2O_2 were observed with normal NIH3T3 cells

(Supplementary Fig. 9). Therefore, rapid and remarkable reduction in the GSH level in prostate cancer DU145 cells was observed after the treatment of QCA, which rapidly generate QM.

In addition, GSH depletion in DU145 cells was significantly inhibited by the addition of H_2O_2-scavenging catalase, suggesting that the generation of QM from QCA is triggered specifically by H_2O_2. In the presence of catalase, QCA is unable to undergo H_2O_2-triggered boronate oxidation, resulting in no generation of GSH-depleting QM. More GSH depletion in DU145 cells than SW620 cells can also be explained by the more QM generation through H_2O_2-triggered boronate oxidation of QCA in DU145 cells, which has a larger amount of H_2O_2. Therefore, we envision that QCA has a tremendous potential for treating cancer cells that have a high extent of redox adaptation with a large amount of H_2O_2.

The ability of QCA to induce ROS generation in cancer cells was studied by flow cytometry and confocal laser scanning microscopy. Cells were treated with free cinnamaldehyde, Q_1 or QCA and the level of ROS was measured using DCFH-DA (dichlorofluorescein-diacetate) as a fluorescent probe for intracellular ROS. After a short duration of treatment (12 h), free cinnamaldehyde (100 μM) slightly increased the level of ROS in DU145 cells, unlike our previous studies reporting that cinnamaldehyde induced significant ROS generation in SW620 cells after 24 h of treatment[23]. It may be because that in DU145 cells that have altered redox status with a high level of ROS as well as antioxidant systems, cinnamaldehyde-induced ROS would be destroyed by the high level of antioxidants and the additional H_2O_2 generation by cinnamaldehyde is negligible compared with the high intrinsic level of ROS after a short period of time (12 h). Moderate rightward shift of DCFH-DA fluorescence was observed with Q_1, indicating that GSH depletion favours ROS accumulation, as expected from Fig. 3a. However, QCA showed a significant rightward shift of DCFH-DA fluorescence in a dose-dependent manner (Fig. 3c,d), demonstrating the remarkably enhanced intracellular ROS accumulation.

We also found that QCA-induced ROS accumulation was significantly inhibited by the treatment of H_2O_2-degrading catalase, evidenced by no rightward shift of DCFH-DA fluorescence (Supplementary Fig. 10a). QCA-induced ROS generation was also further confirmed by fluorescence micrographs (Supplementary Fig. 10b). QCA induced a dose-dependent ROS generation and catalase effectively reduced the level of ROS, suggesting that cinnamaldehyde released from QCA causes the generation of ROS including mainly H_2O_2 and QM generation is inhibited by catalase. From these observations, it can be reasoned that QCA rapidly depletes GSH to suppress antioxidant systems, promoting massive accumulation of oxidative stress caused by cinnamaldehyde-mediated ROS. These results support our hypothesis that a hybrid anticancer drug QCA produces not only GSH-depleting QM but also ROS-generating cinnamaldehyde and both compounds act in a concerted manner to enhance oxidative stress.

We next investigated the cytotoxicity of QCA by MTT (3-(4,5-dimethylthiazol-2-yl)-2,5-diphenyltetrazolium bromide) assay (Supplementary Fig. 11). Cinnamaldehyde showed a moderate level of dose-dependent cytotoxicity in both DU145 and SW620 cells after 24 h of treatment, but minimal cytotoxicity in normal NIH3T3 cells. Despite the generation of antioxidant GSH-scavenging QM and subsequent ROS accumulation, Q_1 showed moderate cytotoxicity to cancer cells. The moderate cytotoxicity of QM-generating Q_1 is in agreement with previous studies which report that the sensitivity of the cancer cells towards cell death by QM is increased by pretreatment with a GSH-synthesis inhibitor (L-buthionine-sulfoximine)[7]. However, a hybrid drug QCA exhibited a significantly higher cytotoxicity than cinnamaldehyde and Q_1, which supports our hypothesis that GSH depletion by QM makes cancer cells more susceptible to ROS insults and subsequently ROS-generating cinnamaldehyde induces more cell death through amplified oxidative stress. It was also determined that QCA shows cytotoxicity with the half-maximal inhibitory concentration (IC_{50}) of 48 μM for DU145 cells and 76 μM for SW620 cells, but minimal cytotoxicity in normal NIH3T3 cells (IC_{50} of 182 μM).

We also found that co-treatment of catalase completely inhibited the cytotoxic effects of QCA, suggesting a strong relationship of QCA activity and H_2O_2. In the presence of H_2O_2-scavenging catalase, boronic ester groups of QCA cannot be cleaved, leading to no production of GSH-depleting QM, as aforementioned in Fig. 3. Catalase also scavenges H_2O_2 generated by cinnamaldehyde released from QCA. These findings clearly demonstrate that QCA exerts synergistic anticancer effects from sequential actions of antioxidant GSH-scavenging QM and ROS-generating cinnamaldehyde.

To confirm whether QCA induces ROS-mediated apoptotic cell death, the expression of apoptosis-regulated proteins was investigated by the western blot assay. Figure 3e,f shows the effects of QCA on the expression of pro-apoptotic proteins, caspase-3 and poly(ADP ribose) polymerase (PARP), both critical enzymes involved in apoptosis. Cinnamaldehyde and QCA induced the cleavage of both pro-caspase-3 and PARP-1 in a dose-dependent manner (Supplementary Fig. 12). However, QCA brought about a greater extent of cleavage of pro-caspase-3 and PARP-1, which confirms that QCA exerts superior apoptosis-inducing potential than cinnamaldehyde and Q_1. We also studied the effects of QCA on DNA fragmentation, a hallmark of apoptosis (Fig. 3g–i). Cinnamaldehyde and Q_1 induced no or minimal DNA fragmentation. However, QCA-treated cells showed nucleosomal DNA fragmentation as clear DNA ladders of multiples of ~180 bp, characteristics of apoptotic DNA fragmentation[34]. The amount of fragmented nucleosomal DNA gradually increased in concentration- and time-dependent manners. These observations demonstrate that QCA is capable of inducing apoptosis.

The results of DNA fragmentation assay was further supported by the flow cytometric analysis using Annexin V-FITC (fluorescein isothiocyanate) as a marker of apoptosis and propidium iodide (PI) as a marker for cell viability. As shown in Supplementary Fig. 13, QCA-treated cells showed greater populations of early apoptotic cells (in the lower right quadruple) and late apoptotic cells (in the upper right quadruple) than those treated with cinnamaldehyde and Q_1. An increase in the population of Annexin V+ and PI+ cells was observed at higher dose of QCA. It is well known that ROS production and apoptosis are closely related and activation of caspase-3 is one of hallmark consequences of GSH depletion[35,36]. Therefore, the superior apoptosis-inducing potential of QCA is attributed to the combined effects of GSH-depleting QM and ROS-generating cinnamaldehyde.

A mouse tumour xenograft model. To investigate the *in vivo* anticancer activity of QCA, we used xenograft mouse models of SW620 and DU145 cells. Cancer cells were subcutaneously injected into the lower back of mice and dosing was initiated when small palpable tumours had developed. Various formulations (cinnamaldehyde, Q_1, cinnamaldehyde along with Q_1 or QCA) were intravenously injected at a physiologically relevant dose of 2 mg kg^{-1} through a tail vein every 3 days. Mice inoculated with SW620 cells showed remarkable tumour growth for 22 days. Treatment of cinnamaldehyde alone or Q_1 alone showed marginal inhibitory effects on tumour growth at a dose of

2 mg kg^{-1}. The combined use of cinnamaldehyde with Q_1 also had moderate anticancer effects. However, mice treated with QCA developed significantly smaller tumour mass in comparison with the groups treated with cinnamaldehyde, Q_1 or the combination of cinnamaldehyde and Q_1. At the same dose of 2 mg kg^{-1}, the superior anticancer actions of QCA over the combined use of cinnamaldehyde and Q_1 can be explained by the rationale that QCA could sequentially generate both QM and cinnamaldehyde in tumours, but free cinnamaldehyde has less ability to target tumours because of its low half-life in blood. To further evaluate the anticancer activity of QCA, tumour-bearing mice were treated with different doses of QCA, in comparison with commercial anticancer drug, camptotethin. Moderate anticancer effects were observed with QCA at doses <1 mg kg^{-1}, but remarkable anticancer effects at a dose of 2 mg kg^{-1}, comparable to camptotethin (Fig. 4d). Similar results were obtained using mice bearing DU145 tumour xenografts. Notably, marked antitumor effects were observed in mice treated with QCA at the same dose of 2 mg kg^{-1} (Fig. 4g,h).

We also performed LC-MS/MS analysis of tumour lysates at day 22 to confirm the generation of QM and cinnamaldehyde in tumours. As shown in Supplementary Fig. 14, tumours of QCA-treated mice showed ~38% less GSH level than those of untreated mice, indirectly supporting the generation of QM in tumours because QM rapidly reacts with nucleophilic GSH. The generation of cinnamaldehyde in tumours was also observed, as evidenced by the appearance of a new peak, identical to the standard cinnamaldehyde. However, QM and cinnamaldehyde were not detected in liver. These results demonstrate that QCA accumulates in tumours and generates both QM and cinnamaldehyde in tumour environments, leading to cancer cell death.

The therapeutic efficacy of QCA was further confirmed by histological examination. Unlike tumour tissues from control mice consisting of abundant tumour cells with no obvious damages, tumours developed in QCA-treated mice had a large area of dead cells without nuclei (Supplementary Fig. 15)[37]. In addition, QCA-induced apoptotic cell death was also confirmed by terminal deoxynucleotidyl transferase dUTP nick end labelling staining. QCA appeared well tolerated as judged by weight loss determination (Fig. 4c,f,i). QCA treatment also induced no notable tissue damages or lesions in liver and heart (Supplementary Fig. 16). These observations clearly demonstrate potent anticancer activity of QCA, without severe system toxicity. Taken together, we envision that QCA is activated to produce both GSH-scavenging QM and ROS-generating cinnamaldehyde in cancer cells, which act in a synergistic fashion to kill cancer cells more specifically by amplifying the oxidative stress in a concerted manner.

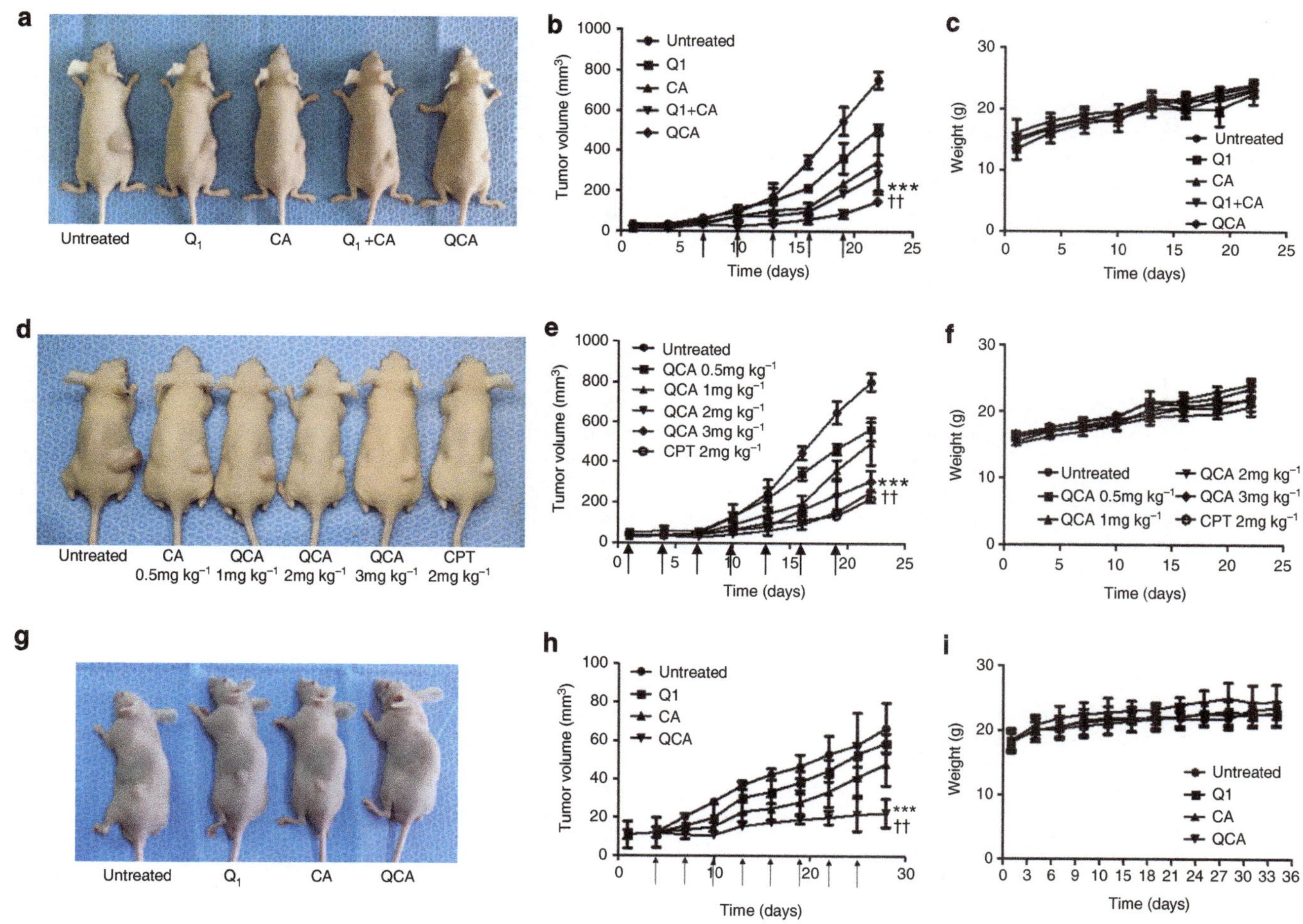

Figure 4 | Anticancer effects of QCA in mice inoculated with SW620 and DU145 cells. (**a**) Representative images of mice bearing SW620 tumour xenografts. Changes of tumour volumes (**b**) and body weight (**c**) in a SW620 xenograft model. ***$P<0.001$ relative to untreated, ††$P<0.01$ relative to Q1 + CA (one-way analysis of variance (ANOVA), $n=4$, ± s.d.). (**d**) Representative images of mice bearing SW620 tumour xenografts treated with various amounts of QCA and camptotethin (CPT). Changes of tumour volumes (**e**) and body weight (**f**) of mice bearing SW620 tumour xenografts during the treatment. ***$P<0.001$ relative to untreated, ††$P<0.01$ relative to QCA 1 mg kg^{-1} (one-way ANOVA, $n=4$, ± s.d.). (**g**) Representative images of mice bearing DU145 tumour xenografts. Changes of tumour volumes (**h**) and body weight (**i**) in a DU145 xenograft model. ***$P<0.001$ relative to untreated. ††$P<0.01$ relative to CA (one-way ANOVA, $n=4$, ± s.d.). Arrows indicate the date of treatment.

Finally, to test the safety profile of QCA, we administrated QCA (2 mg kg^{-1}) into normal mice every other day for 10 days. QCA-treated mice showed no difference in the level of serum ALT (alanine transaminase) in comparison with untreated mice. In addition, there was no obvious histological evidence of accumulated toxicity in liver and heart associated with administration of QCA for 10 days (Supplementary Fig. 17), demonstrating no or minimal toxicity of QCA.

Discussion

One of the major challenges in the development of anticancer therapeutic agents is the temporal and spatial therapeutic actions in a controlled manner[16,17]. There have been enormous efforts for the development of anticancer agents that are able to target specific oncogenic molecules to enhance therapeutic activities and specificity, by exploiting the genetic differences between normal cells and cancer cells. A prime example of selective anticancer drugs is imatinib (Gleevec, Norvatis), which targets the oncogenic tyrosine kinase BER-ABL responsible for myeloid leukaemia[17,38]. Despite the promising therapeutic activities and specificity, gene-targeting therapeutic strategies face challenges because of acquired drug resistance, genetic instability and multiple genetic alteration in cancer cells[16]. There are multiple lines of evidence indicating that cancer cells are under oxidative stress, reflecting altered redox balance. Cancer cells are more vulnerable to agents that impair redox balance and increase oxidative stress, but normal cells are less sensitive to the oxidative stress-inducing agents due to their low level of basal ROS and high antioxidant capacity. On the basis of the biochemical differences in cancer cells and normal cells, modulation of oxidative stress has recently been re-emerged as a therapeutic strategy to selectively target the destruction of cancer cells. Great efforts have been made for developing selective anticancer therapeutic agents that increase the level of ROS or alternatively suppress the antioxidant systems[1,12,16,33,39–41]. In this regard, we rationally designed a hybrid anticancer drug QCA to exploit oxidation therapy, which can not only increase the level of ROS but also weaken antioxidant systems simultaneously.

We developed QCA that exerts dual and synergistic therapeutic actions in cancer environments associated with a high concentration of H_2O_2 and acidic pH[8,27]. As shown in Fig. 1, a hybrid anticancer drug QCA consists of an antioxidant GSH-scavenging moiety and ROS-generating cinnamaldehyde. QCA was also designed to have both H_2O_2-sensitive boronate and acid-sensitive acetal linkage. We confirmed that QCA rapidly undergoes H_2O_2-mediated oxidation to release QM, which is able to abrogate the key antioxidant GSH (Fig. 2). In NMR analysis for H_2O_2-mediated degradation of QCA, QCA was treated with H_2O_2. In this study, the concentration of H_2O_2 was 500 μM, which is much higher than its physiological concentration. However, the concentration of QCA was 15 mM, which is 30 times higher than the concentration of H_2O_2. It is therefore necessary to have sufficient amount (500 μM) of H_2O_2 to react with a higher concentration (15 mM) of QCA and detect noticeable changes in NMR signal, as suggested in previous studies reporting the H_2O_2-mediated degradation of boronate compounds using high concentrations of H_2O_2 such as 500 μM and even 500 mM (refs 30,42,43).

Aryl boronate has been known to rapidly and selectively react with H_2O_2 to generate QM, which alkylates antioxidant GSH, thus inhibiting the cellular antioxidant systems and triggering cell death[7,20,28,44]. It has been well accepted that the intracellular level of GSH is about <10 mM (refs 20,45,46). Interestingly, we found that 0.1 mM of QCA remarkably ($>$ 60%) reduced the level of GSH (Fig. 3a,b). The remarkable GSH depletion by QCA can be explained by considering the elevated intracellular concentration of QCA after its cellular internalization. Hydrophobic QCA could easily penetrate into cells, as confirmed by LC-MS/MS data (Supplementary Fig. 7). After cellular internalization, the intracellular concentration of QCA would be significantly increased by two to three orders of magnitude because the total amount of cytosol is much less than culture medium (that is, 1 ml in a well of 24-well plates). Therefore, QCA at a concentration of 0.1 mM in culture medium could deplete a majority of cellular GSH. These observations are in good accord with previous studies reporting that QM-generating drugs (0.1 mM) induced more than 50% reduction of cellular GSH level[7]. We also demonstrated that the inhibition of antioxidant GSH by QCA is attributed to rapid formation of QM from H_2O_2-mediated boronate oxidation (Fig. 3a,b).

QCA also has an acetal linkage that degrades specifically in acidic tumour environments to release ROS-generating cinnamaldehyde. Simple treatment of cinnamaldehyde may not always lead to a preferential and sufficient killing of cancer cells. Depletion of cellular antioxidant GSH by QM would seem to be a logical approach in enhancing the anticancer potential of ROS-generating cinnamaldehyde[8]. Therefore, appealing features of rationally designed hybrid anticancer drug QCA are the dual stimuli responsiveness and sequential therapeutic actions, which results in synergistic anticancer effects in a cancer cell-specific manner. The results in proof of concept studies *in vitro* and further studies on mouse tumour xenograft models confirm the serial and synergistic anticancer activities of QCA. As illustrated in Fig. 1, QCA not only suppresses antioxidant systems but also induces ROS generation, amplifying oxidative stress and leading to apoptotic cell death. Further optimization, including administration routes, dose, frequency, pharmacokinetics, pharmacodynamics, drug resistance studies and toxicology are warranted to maximize the full potential of QCA for clinical settings.

In conclusion, we developed a dual stimuli-responsive anticancer hybdrid drug QCA that is activated by H_2O_2 and acidic pH to generate antioxidant GSH-depleting QM and ROS-generating cinnamaldehyde, respectively, and amplifies oxidative stress to induce apoptotic cell death. Combination of QM and cinnamaldehyde displays a sequential mode of action in cancer treatment, by suppressing antioxidant systems and generating cytotoxic ROS in cancer cells specifically, further disturbing redox balance to kill cancer cells more effectively and specifically. To the best of our knowledge, the dual stimuli-responsive QCA is the first hybrid drug that not only inhibits antioxidant systems but also increases ROS stress simultaneously to kill cancer cells specifically. We anticipate that a novel hybrid drug QCA contributes to the development of new intelligent therapeutics based on the rational design of bioactive molecules and provides the best opportunities for efficient cancer treatment.

Methods

Synthesis of (4-(1,3,2-dioxaborinan-2-yl)phenyl)methanol, 1. (4-Hydroxymethylphenyl) boronic acid (2.0 g) and 1,3-propanediol (1.0 g) were dissolved in 40 ml of dry tetrahydrofuran at room temperature. The reaction was allowed for 72 h with mechanical stirring. The solvent was evaporated and **1** was obtained using silica gel chromatography (hexane/ethyl acetate = 6/4).

Synthesis of cinnamaldehyde derivative 2. Tris(hydroxymethyl)ethane (4.88 g) was dissolved in 70 ml of dry benzene. Cinnamaldehyde (5.714 ml) and *p*-toluenesulfonic acid (40 mg) were added to the solution and the reaction was allowed for 4 h at 90 °C. After cooling the mixture to room temperature, 1 ml of triethylamine was added to terminate the reaction. Benzene was evaporated using a rotary evaporator and the product was obtained from column chromatography (hexane/ethyl acetate = 7/3).

Synthesis of cinnamaldehyde releasing compound 3. 1,1′-Carbonyldiimidazole (4.1 g) and **2** (3.0 g) were dissolved in 50 ml of dry dichloromethane. The reaction was allowed for 30 min at room temperature. Dichloromethane was evaporated and the product was obtained from column chromatography using ethylacetate.

Synthesis and characterization of QCA. **1** (1.0 g) and **3** (1.7 g) were dissolved in 50 ml of dry dichloromethane containing 4-(dimethylamino)pyridine (0.64 g). The reaction was allowed for 24 h at 40 °C. After solvent evaporation, the product was obtained from column chromatography (hexane/ethyl acetate = 5/5). QCA was characterized using NMR and GC–MS. ^{1}H NMR(400 MHz, $CDCl_3$): δ 7.2–7.8 (9H), 6.8 (1H), 6.2 (1H), 5.2 (1H), 5.1 (2H), 4.4(2H), 4.15 (4H), 4.0 (2H), 3.6(2H), 1.6 (2H), 0.85 (3H); ^{13}N NMR (400 MHz, $CDCl_3$): δ 155,5, 137.3, 135.9, 133.9, 133.7, 128.5, 128.5, 127.3, 126.8, 125.0, 101.1, 77.5, 77.3, 77.0, 72.7, 69.9, 33.8, 27.4, 17.0; LC-MS (m/z):[M]$^+$ calc. for $C_{25}H_{29}O_7B$, 452.29; found 453.26; analysis (calcd., found for $C_{25}H_{29}O_7$): C (66.38, 65.44), H(6.46, 6.70).

Sensitivity to H_2O_2. The sensitivity of QCA to H_2O_2 was evaluated by measuring the chemiluminescence intensity after the reaction with diphenyl oxalate solution including rubrene. QCA (0.1 or 1 μM) was added in the H_2O_2 solution (1 μM). After 1 min reaction with H_2O_2, diphenyl oxalate solution was added to the H_2O_2 solution, and the chemiluminescence intensity was measured by using a luminometer (Femtomaster FB 12, Zylux Corporation, TN, USA).

Cellular GSH levels. Cells (DU145, SW620 and NIH3T3) were obtained from Korean Cell Line Bank (Korea). Cells were seeded in a six-well plate (5×10^5 per well) and cultured to reach confluency of 80%. After treatment with various compounds for 1.5 h at 37 °C, cells were harvested, washed with PBS and then lysed on ice in 40 μl of Triton-X-100 lysis buffer. After 20 min, lysates were centrifuged at 9,800*g* and 10 μl of the supernatant was mixed with 50 μl of Ellman's reagent (0.5 mM DTNB). The amount of GSH was quantified by measuring the absorbance at 405 nm using a microplate reader (Biotek Instruments, Winooski, VT). Percentage of GSH content from treated cells were compared with basal GSH content measured in untreated cells.

Flow cytometry. DU145 or SW620 cells (4×10^5) were cultured in a 24-well culture plate for 24 h and treated with free cinnamaldehyde or QCA for 12 h. Cells were washed with fresh medium twice and resuspended in 1 × binding buffer at a concentration of 1×10^4 cells ml^{-1}. For the detection of ROS, the cell suspensions (1 ml) were transferred to a 5 ml culture tube and added with 5 μM of DCFH-DA (Sigma-Aldrich, St Louis, MO), followed by gentle mixing. The cells were incubated for 15 min at 37 °C in the dark and added with 400 μl of 1 × binding buffer. For the analysis of apoptosis, the cell suspensions were treated with Annexin V-FITC and propidium iodide (BD Biosciences Pharmigen, San Diego, CA). The stained cells were analysed by flow cytometry (FACS Caliber, Becton Dickinson, San Jose, CA). A total of 1.0×10^4 events were counted for each sample.

Cytotoxicity of QCA. The cytotoxicity of QCA was determined by MTT assay. DU145, SW620 cells and NIH3T3 cells were seeded at a density of 1×10^5 cells per well in a 24-well plate and incubated for 24 h to reach ~90% confluency. Cells were treated with various amounts of QCA, Q_1 or cinnamaldehyde and incubated for 24 h. Each well was given 100 μl of MTT solution and were incubated for 4 h. The resulting formazan crystals were dissolved by 1 ml of dimethylsulfoxide. After 10 min of incubation, the absorbance at 570 nm was measured using a microplate reader (Biotek Instruments, Winooski, VT). The cell viability was determined by comparing the absorbance of QCA-treated cells to that of control cells.

LC-MS/MS analysis of cell lysates. Cells were seeded in a T-75 flask containing 10 ml of medium and allowed to attach for 24 h. Cells were treated with 100 μM of QCA or DMSO containing medium. After 24 h of incubation, cells were washed with new medium and cell pellets were added with 100 μl of methanol. The mixture was vortexed and then additional 900 μl of methanol was added. The contents were thoroughly mixed by vortexing and high molecular materials were removed by sequential centrifugation (10,000*g*) for 10 min. The supernatant was immediately stored at − 80 °C until analysis. On analysis, 5 μl was injected and the peak for GSH and cinnamaldehyde was analysed using a LC-MS/MS spectrometer (6410 Triple Quad LC /MS/MS, Agilent Technologies, Willington) equipped with a column (Synergi 4 μ Hydro RP 80 A, 150 × 2.00 mm). Flow rate was 0.23 ml min^{-1} and the elution was with a gradient of water and acetonitrile (A = 0.1% Formic acid in water; B = 0.1% Formic acid in acetonitrile): 0→1 min, 5% B; 1→10 min, 5% B; 10→14 min, 100% B, 14→20 min, 5% B). Detection was conducted at *m/z* 308→162.0 for GSH and *m/z* 133→114.9 for cinnamaldehyde. Ions were generated in positive ionization mode using electrospray ionization interface. The Fragmenter potential was set to 110 V and the interface heater was set to 300 °C.

Immunoblot analysis. DU145 cells (2×10^6 per well) or SW620 (1.7×10^6 per well) cells were treated with various concentrations of free cinnamaldehyde or QCA for 8 h and then washed with fresh PBS twice. Proteins were extracted from the cells using a lysis buffer. Electrophoresis was performed using 20 μg of cell lysate on a 10% polyacrylamide gel and proteins were transferred to PVDF membranes (Millipore, Billerica, MA). The blot was incubated with PARP monoclonal antibody (Santa Cruz Biotechnology, Dallas, TX) or caspase-3 (Santa Cruz Biotechnology, Dallas, TX) and HRP-conjugated anti-goat (Millipore, Billerica, MA), which is used as a secondary antibody. Immunoblot signals were developed using SuperSignal Ultra chemiluminescent reagent (Pierce, Rockford, IL). Images have been cropped for presentation. Full-size images are presented in Supplementary Fig. 12.

DNA fragmentation. Cells (2×10^6 per well) cells were treated with various concentrations of free cinnamaldehyde or QCA for 12 h and then washed with fresh PBS twice. Chromosomal DNA was extracted from the apoptotic cells using Quick Apoptotic DNA Ladder Detection Kit (Life Technologies, Frederick, MD). In brief, the cells were lysed with TE buffer by pipetting up and down several times. The crude lysates were treated with enzyme solution until they became clear and mixed with ammonium acetate and ethanol. The mixtures were vortexed at − 20 °C for 10 min to allow DNA to precipitate. The mixtures were centrifuged at 10,000*g* to collect precipitated DNA. After careful removal of supernatant, DNA pellets were washed with 70% cold ethanol and recentrifuged at 10,000*g* for 10 min. The supernatants were discarded and DNA was air dried at room temperature. DNA was then suspended with DNA Suspension Buffer by pipetting up and down several times. DNA was separated on agarose gel (1.8%), stained with ethidium bromide and photographed under ultraviolet light.

Animal studies. To evaluate the therapeutic efficacy of QCA, a subcutaneous tumour was established by inoculating SW620 cells (1.8×10^6 cells per mouse) or DU145 cells (2.0×10^6 cells per mouse), into the back of nude mice (male, 4 weeks, Orient Bio, Korea). When small palpable tumours had developed, the mice were randomly grouped and intravenously injected with one of the following: saline, Q_1, cinnamaldehyde, cinnamaldehyde along with Q_1 or QCA through a tail vein. Each mouse was intravenously injected once every 3 days at a dose of 2 mg kg^{-1}. The body weight and tumour volume were recorded every 3 days. The length and width of the tumours were measured by digital calipers, calculating the tumour volume using the following formula: (width^2 × length)/2. At the end of the experiments, mice were killed and tumours were excised. Liver, heart and tumour tissues were histologically evaluated with haematoxylin and eosin staining. For LC-MS analysis, tumours were frozen in liquid nitrogen and ground into a fine power using a pestle and mortar. Tumour lysates were analysed using the same procedure for cell lysates. To evaluate the toxicity of QCA, QCA was injected into mice (male Balb/c of body weight of ~20 g, Orient Bio, Korea) with QCA at a dose of 2 mg kg^{-1} every other day for 10 days. The activity of serum ALT was determined with an ALT enzymatic assay kit (Asan Pharma, Seoul, Korea) using a microplate reader (Synergy MX, BioTek Instruments Inc., Winooski, VT). The liver and heart tissues were fixed with 10% formalin (Sigma-Aldrich, St Louis, MO) and embedded into paraffin. Histological sections were made and stained with haematoxylin and eosin. All the animal experiments were carried out according to the guidelines of the institution animal ethical committee (CBU2014-00024).

References

1. Fang, J., Seki, T. & Maeda, H. Therapeutic strategies by modulating oxygen stress in cancer and inflammation. *Adv. Drug Deliv. Rev.* **61**, 290–302 (2009).
2. Fruehauf, J. P. & Meyskens, F. L. Reactive oxygen species: a breath of life or death? *Clin. Cancer Res.* **13**, 789–794 (2007).
3. Lippert, A. R., De Bittner, G. C. V. & Chang, C. J. Boronate oxidation as a bioorthogonal reaction approach for studying the chemistry of hydrogen peroxide in living systems. *Acc. Chem. Res.* **44**, 793–804 (2011).
4. Oh, W. K., Jeong, Y. S., Kim, S. & Jang, J. Fluorescent polymer nanoparticle for selective sensing of intracellular hydrogen peroxide. *Acs Nano* **6**, 8516–8524 (2012).
5. Van de Bittner, G. C., Dubikovskaya, E. A., Bertozzi, C. R. & Chang, C. J. *In vivo* imaging of hydrogen peroxide production in a murine tumor model with a chemoselective bioluminescent reporter. *Proc. Natl Acad. Sci. USA* **107**, 21316–21321 (2010).
6. Wagner, B. A., Buettner, G. R. & Burns, C. P. Free radical-mediated lipid peroxidation in cells: oxidizability is a function of cell lipid bis-allylic hydrogen content. *Biochemistry* **33**, 4449–4453 (1994).
7. Hulsman, N. *et al.* Chemical insights in the concept of hybrid drugs: the antitumor effect of nitric oxide-donating aspirin involves a quinone methide but not nitric oxide nor aspirin. *J. Med. Chem.* **50**, 2424–2431 (2007).
8. Kimani, S. G., Phillips, J. B., Bruce, J. I., MacRobert, A. J. & Golding, J. P. Antioxidant inhibitors potentiate the cytotoxicity of photodynamic therapy. *Photochem. Photobiol.* **88**, 175–187 (2012).
9. Diehn, M. *et al.* Association of reactive oxygen species levels and radioresistance in cancer stem cells. *Nature* **458**, 780–U123 (2009).
10. Hanot, M. *et al.* Glutathione depletion and carbon ion radiation potentiate clustered DNA lesions, cell death and prevent chromosomal changes in cancer cells progeny. *PLoS ONE* **7**, e44367 (2012).

11. Schumacker, P. T. Reactive oxygen species in cancer cells: Live by the sword, die by the sword. *Cancer Cell* **10,** 175–176 (2006).
12. Raj, L. *et al.* Selective killing of cancer cells by a small molecule targeting the stress response to ROS. *Nature* **475,** 231–234 (2011).
13. Alexandre, J. *et al.* Accumulation of hydrogen peroxide is an early and crucial step for paclitaxel-induced cancer cell death both *in vitro* and *in vivo*. *Int. J. Cancer* **119,** 41–48 (2006).
14. Huang, G. *et al.* Superparamagnetic iron oxide nanoparticles: amplifying ROS stress to improve anticancer drug efficacy. *Theranostics* **3,** 116–126 (2013).
15. Pelicano, H., Carney, D. & Huang, P. ROS stress in cancer cells and therapeutic implications. *Drug Resist. Update.* **7,** 97–110 (2004).
16. Trachootham, D., Alexandre, J. & Huang, P. Targeting cancer cells by ROS-mediated mechanisms: a radical therapeutic approach? *Nat. Rev. Drug Discov.* **8,** 579–591 (2009).
17. Trachootham, D. *et al.* Selective killing of oncogenically transformed cells through a ROS-mediated mechanism by beta-phenylethyl isothiocyanate. *Cancer Cell* **10,** 241–252 (2006).
18. Fang, J. *et al. In vivo* antitumor activity of pegylated zinc protoporphyrin: targeted inhibition of heme oxygenase in solid tumor. *Cancer Res.* **63,** 3567–3574 (2003).
19. Maeda, H. *et al.* Effective treatment of advanced solid tumors by the combination of arsenic trioxide and L-buthionine-sulfoximine. *Cell Death Differ.* **11,** 737–746 (2004).
20. Hagen, H. *et al.* Aminoferrocene-based prodrugs activated by reactive oxygen species. *J. Med. Chem.* **55,** 924–934 (2012).
21. Ka, H. *et al.* Cinnamaldehyde induces apoptosis by ROS-mediated mitochondrial permeability transition in human promyelocytic leukemia HL-60 cells. *Cancer Lett.* **196,** 143–152 (2003).
22. Cabello, C. M. *et al.* The cinnamon-derived Michael acceptor cinnamic aldehyde impairs melanoma cell proliferation, invasiveness, and tumor growth. *Free Radic. Biol. Med.* **46,** 220–231 (2009).
23. Kim, B. *et al.* Dual acid-responsive micelle-forming anticancer polymers as new anticancer therapeutics. *Adv. Funct. Mater.* **23,** 5091–5097 (2013).
24. Han, D. C. *et al.* 2'-benzoyloxycinnamaldehyde induces apoptosis in human carcinoma via reactive oxygen species. *J. Biol. Chem.* **279,** 6911–6920 (2004).
25. Hong, S. H. *et al.* Apoptosis induction of 2 '-hydroxycinnamaldehyde as a proteasorne inhibitor is associated with ER stress and mitochondrial perturbation in cancer cells. *Biochem. Pharmacol.* **74,** 557–565 (2007).
26. Lee, K. *et al.* Plasma pharmacokinetics and metabolism of the antitumour drug candidate 2'-benzoyloxycinnamaldehyde in rats. *Xenobiotica* **39,** 255–265 (2009).
27. Griset, A. P. *et al.* Expansile nanoparticles: synthesis, characterization, and *in vivo* efficacy of an acid-responsive polymeric drug delivery system. *J. Am. Chem. Soc.* **131,** 2469–2471 (2009).
28. Jourden, J. L. M. & Cohen, S. M. Hydrogen peroxide activated matrix metalloproteinase inhibitors: a prodrug approach. *Angew. Chem. Int. Ed.* **49,** 6795–6797 (2010).
29. Song, C. C., Ji, R., Du, F. S., Liang, D. H. & Li, Z. C. Oxidation-accelerated hydrolysis of the ortho ester-containing acid-labile polymers. *Acs Macro Lett.* **2,** 273–277 (2013).
30. Lux, C. D. *et al.* Biocompatible polymeric nanoparticles degrade and release cargo in response to biologically relevant levels of hydrogen peroxide. *J. Am. Chem. Soc.* **134,** 15758–15764 (2012).
31. Hadd, A. G., Lehmpuhl, D. W., Kuck, L. R. & Birks, J. W. Chemiluminescence demonstration illustrating principles of ester hydrolysis reactions. *J. Chem. Educ.* **76,** 1237–1240 (1999).
32. Lee, I., Hwang, O., Yoo, D., Khang, G. & Lee, D. Detection of hydrogen peroxide *in vitro* and *in vivo* using peroxalate chemiluminescent micelles. *Bull. Korean Chem. Soc.* **32,** 2187–2192 (2011).
33. Kumar, B., Koul, S., Khandrika, L., Meacham, R. B. & Koul, H. K. Oxidative stress is inherent in prostate cancer cells and mediates aggressive phenotype. *J. Urol.* **179,** 192–193 (2008).
34. Park, D. J. & Patek, P. Q. Detergent and enzyme treatment of apoptotic cells for the observation of DNA fragmentation. *Biotechniques* **24,** 558–560 (1998).
35. Gao, J. J., Liu, X. P. & Rigas, B. Nitric oxide-donating aspirin induces apoptosis in human colon cancer cells through induction of oxidative stress. *Proc. Natl Acad. Sci. USA* **102,** 17207–17212 (2005).
36. Kim, C. K. *et al.* Ceria nanoparticles that can protect against ischemic stroke. *Angew. Chem. Int. Ed.* **51,** 11039–11043 (2012).
37. Chen, W.-H. *et al.* Therapeutic nanomedicine based on dual-intelligent functionalized gold nanoparticles for cancer imaging and therapy *in vivo*. *Biomaterials* **34,** 8798–8807 (2013).
38. Duncan, R., Ringsdorf, H. & Satchi-Fainaro, R. in *Polymer Therapeutics I: Polymers as Drugs, Conjugates and Gene Delivery Systems* Vol. 192 (eds Satchi-Fainaro, R. & Duncan, R.) 1–8 (Springer, 2006).
39. Kirshner, J. R. *et al.* Elesclomol induces cancer cell apoptosis through oxidative stress. *Mol. Cancer Ther.* **7,** 2319–2327 (2008).
40. Laurent, A. *et al.* Controlling tumor growth by modulating endogenous production of reactive oxygen species. *Cancer Res.* **65,** 948–956 (2005).
41. Kumar, B., Koul, S., Khandrika, L., Meacham, R. B. & Koul, H. K. Oxidative stress is inherent in prostate cancer cells and is required for aggressive phenotype. *Cancer Res.* **68,** 1777–1785 (2008).
42. Broaders, K. E., Grandhe, S. & Frechet, J. M. J. A biocompatible oxidation-triggered carrier polymer with potential in therapeutics. *J. Am. Chem. Soc.* **133,** 756–758 (2011).
43. Miller, E. W., Albers, A. E., Pralle, A., Isacoff, E. Y. & Chang, C. J. Boronate-based fluorescent probes for imaging cellular hydrogen peroxide. *J. Am. Chem. Soc.* **127,** 16652–16659 (2005).
44. Jourden, J. L. M., Daniel, K. B. & Cohen, S. M. Investigation of self-immolative linkers in the design of hydrogen peroxide activated metalloprotein inhibitors. *Chem. Commun.* **47,** 7968–7970 (2011).
45. Maher, P. The effects of stress and aging on glutathione metabolism. *Ageing Res. Rev.* **4,** 288–314 (2005).
46. Meister, A. Glutathione metabolism and its selective modification. *J. Biol. Chem.* **263,** 17205–17208 (1988).

Acknowledgements

This study was supported by the Basic Science Research Program (2013R1A1A2A10061828) and BK21 Plus program through National Research Foundation funded by the Ministry of Education, Science and Technology, Korea.

Author contributions

J.N., B.K. and E.H. synthesized and characterized the compounds. E.H., M.P., W.Yang, W.C., W.Yoo and B.K. performed experiments *in vitro* and *in vivo* and analysed the data. G.K. discussed the data and edited the manuscript. D.L. designed and supervised the study and wrote the manuscript.

Additional information

Competing financial interests: The authors declare no competing financial interests.

21

Atomic description of the immune complex involved in heparin-induced thrombocytopenia

Zheng Cai[1], Serge V. Yarovoi[1], Zhiqiang Zhu[1], Lubica Rauova[2], Vincent Hayes[2], Tatiana Lebedeva[1], Qun Liu[3], Mortimer Poncz[2], Gowthami Arepally[4], Douglas B. Cines[1] & Mark I. Greene[1]

Heparin-induced thrombocytopenia (HIT) is an autoimmune thrombotic disorder caused by immune complexes containing platelet factor 4 (PF4), antibodies to PF4 and heparin or cellular glycosaminoglycans (GAGs). Here we solve the crystal structures of the: (1) PF4 tetramer/fondaparinux complex, (2) PF4 tetramer/KKO-Fab complex (a murine monoclonal HIT-like antibody) and (3) PF4 monomer/RTO-Fab complex (a non-HIT anti-PF4 monoclonal antibody). Fondaparinux binds to the 'closed' end of the PF4 tetramer and stabilizes its conformation. This interaction in turn stabilizes the epitope for KKO on the 'open' end of the tetramer. Fondaparinux and KKO thereby collaborate to 'stabilize' the ternary pathogenic immune complex. Binding of RTO to PF4 monomers prevents PF4 tetramerization and inhibits KKO and human HIT IgG-induced platelet activation and platelet aggregation *in vitro*, and thrombus progression *in vivo*. The atomic structures provide a basis to develop new diagnostics and non-anticoagulant therapeutics for HIT.

[1] Department of Pathology and Laboratory Medicine, University of Pennsylvania, Philadelphia, Pennsylvania 19104, USA. [2] Division of Hematology, The Children's Hospital of Philadelphia, Philadelphia, Pennsylvania 19104, USA. [3] New York Structural Biology Center, National Synchrotron Light Source (NSLS) X4, Brookhaven National Laboratory, Upton, New York 11973, USA. [4] Division of Hematology, Department of Medicine, Duke University School of Medicine, Durham, North Carolina 27710, USA. Correspondence and requests for materials should be addressed to D.B.C. (email: dcines@mail.med.upenn.edu) or to M.I.G. (email: greene@reo.med.upenn.edu).

Heparin-induced thrombocytopenia/thrombosis (HIT) is a potentially fatal immune complex mediated thrombotic disorder that develops in ~1% of patients exposed to unfractionated heparin or, less commonly, to low-molecular-weight heparins[1]. HIT is caused by antibodies to complexes that form between platelet factor 4 (PF4) released from activated platelets and heparin or cellular glycosaminoglycans (GAGs). Circulating immune complexes composed of PF4, heparin and HIT antibodies bind to platelet and monocyte Fcγ receptors and promote cellular activation, leading to generation of thrombin[2–4]. Anti-PF4/heparin antibodies detected by enzyme-linked immunosorbent assay (ELISA) also develop in a high proportion of patients exposed to heparin, for example, after cardiopulmonary bypass surgery, even in the absence of clinical complications. However, the basis for the distinction between pathogenic and non-pathogenic antibodies is unknown, which can lead to overdiagnosis and overtreatment[5].

In prior studies, we identified a murine monoclonal antibody (KKO) to PF4/heparin complexes that causes heparin-induced thrombosis and thrombocytopenia in a murine model, thus sharing salient features with the clinical disorder[6]. Human HIT antibodies compete with KKO for binding to PF4/heparin, and KKO augments formation of pathogenic immune complexes[7]. An isotype matched anti-PF4 antibody (RTO) that binds comparably to PF4, but does not generate pathogenic complexes[8], has also been identified.

Here we describe and compare the crystal structures of PF4 in complex with Fabs derived from KKO and RTO and the structure of PF4 in complex with a heparin-mimic pentasaccharide, fondaparinux. These results suggest that by stabilizing the structure of the asymmetric PF4 tetramer, fondaparinux might foster the binding of HIT antibody and formation of the pathogenic ternary complex. Surprisingly, the non-HIT antibody RTO binds to an epitope expressed on the surface of the PF4 monomer that partly overlaps with the KKO epitope in the tetramer. Binding of RTO prevents tetramerization of PF4 and inhibits KKO-induced platelet activation and aggregation *in vitro* and has potent inhibitory effects on thrombus progression *in vivo*.

These crystallographic data lead to a model to help understand the structural basis of the pathogenic immune complex that causes HIT at the atomic level and provide a structural basis for the development of new diagnostics and non-anticoagulant therapeutics for HIT. Our data also provide insights into the process through which a normal human protein becomes 'antigenic' to the mammalian immune system after complexing with endogenous or exogenous GAGs.

Results

Crystal structure of the fondaparinux/PF4 tetramer complex. Based on previously reported crystal structures, human PF4 can assume an asymmetric tetrameric configuration. Each monomer contains three-stranded antiparallel β-sheets on which an aperiodic N-terminal domain and an amphipathic C-terminal α-helix are folded[9]. NMR studies reveal that PF4 exists as an equilibrium among monomers, dimers and tetramers[10]. The tetrameric structure of PF4 is stabilized not only by salt bridge interactions between dimers AB/CD (hereafter we use the nomenclature for the monomer chains of PF4 as in PDB: 1RHP[9]), but also by the antiparallel β-sheet-like structures in the N termini of dimers AC or BD[9]. It had been assumed that the asymmetry of the PF4 tetramer plays a role in heparin binding[11] and several models have been proposed to explain the interaction between heparin and PF4 (refs 12–14). However, conclusive structural data at the atomic level have not been reported.

Binding of pathogenic HIT antibodies to PF4 is markedly enhanced by heparin or various glycosaminoglycans (GAGs), which are themselves highly heterogeneous in size and composition and unsuitable for crystallization. To investigate how the GAGs might induce or augment neoantigen formation, we crystallized PF4 in complex with fondaparinux, a homogenous synthetic pentasaccharide heparin that forms complexes with PF4 as assessed by atomic force microscopy and photon correlation spectroscopy[15]. Fondaparinux induces anti-PF4/heparin antibodies as do other GAGs[16–18] and can occasionally cause HIT[19–23].

We solved the crystal structure of PF4 in complex with fondaparinux as a model GAG. The PF4/fondaparinux complex is a pseudo-222 tetramer formed by a pseudo two-fold non-crystallographic symmetry and a perpendicular two-fold crystallographic symmetry (Fig. 1a, Table 1 for statistics). Fondaparinux appears as a well-defined molecule, evidenced by the strong and clear electron densities (Supplementary Fig. 1). Superposition of the bound and non-bound structures of PF4 tetramer (PDB: 1RHP) using Cα gives a root-mean-square deviation of only 0.76 Å. This indicates that the association with fondaparinux causes minimal conformational changes in PF4 except for slight redirection of side chains around the binding groove, which is consistent with circular dichroism (CD) studies reported recently[24,25].

The PF4 tetramer displays a pseudosymmetry characterized by an 'open' end and a 'closed' end that can be defined by the intra-chain distances between amino acids E28 and K50. E28 and K50 are ~3 Å apart and form stable salt bridge pairs between chain A and chain C. However, the same interchain salt bridge pairs do not form between E28 and K50 in chains B and D, which are ~8 Å apart (Supplementary Fig. 2). This asymmetry in the PF4 tetramer explains why calculation of its electrostatic potential identifies only two positively charged grooves on the 'closed' side of PF4 tetramer surface to which fondaparinux can bind (Fig. 1d).

PF4 tetramers can be formed between the AB and CD dimers, or between the AC and BD dimers[9]. Fondaparinux binds among monomers A, B and C or monomers A, C and D (Fig. 1b). This interaction stabilizes the AB/CD association and the AC/BD association, thus further stabilizing the PF4 tetramer. We suggest that this stabilization might represent a very early step in the pathogenesis of HIT antigen.

Analysis of the crystal lattice further reveals that fondaparinux not only makes contact with a single PF4 tetramer within the groove formed by three monomers (ABC or ACD), but also at a second site involving the C-terminal helix of the third monomer (B or D). It has been proposed that the HIT antigen develops when charge neutralization by polyanions allows positively charged PF4 tetramers to undergo close approximation[14,15]. The crystal structure of the PF4/fondaparinux complex provides a more complete atomic understanding of this process. One fondaparinux molecule not only binds within the groove on the surface of one PF4 tetramer, it is also shared with a second symmetry-related tetramer through binding to its C-terminal helix (Fig. 1c). Most of these interactions are between the sulfate groups of fondaparinux and the basic residues in PF4 (details in Fig. 1e). The finding that two tetramers of PF4 bind to one heparin fragment provides insight into the initiation of ultralarge complex formation.

Based on what is observed in the crystal, we propose the following pathway for the formation of the heparin/PF4 complex. Heparin first binds to the groove in one PF4 tetramer. Binding of the first PF4 tetramer imparts a local linearized structure on heparin. This enhances the binding of a second tetramer. Progression of this process eventuates in the formation of the large antigenic complex in which PF4 tetramers cluster around a semi-rigid linear heparin chain (Fig. 1f, Supplementary Movie 1).

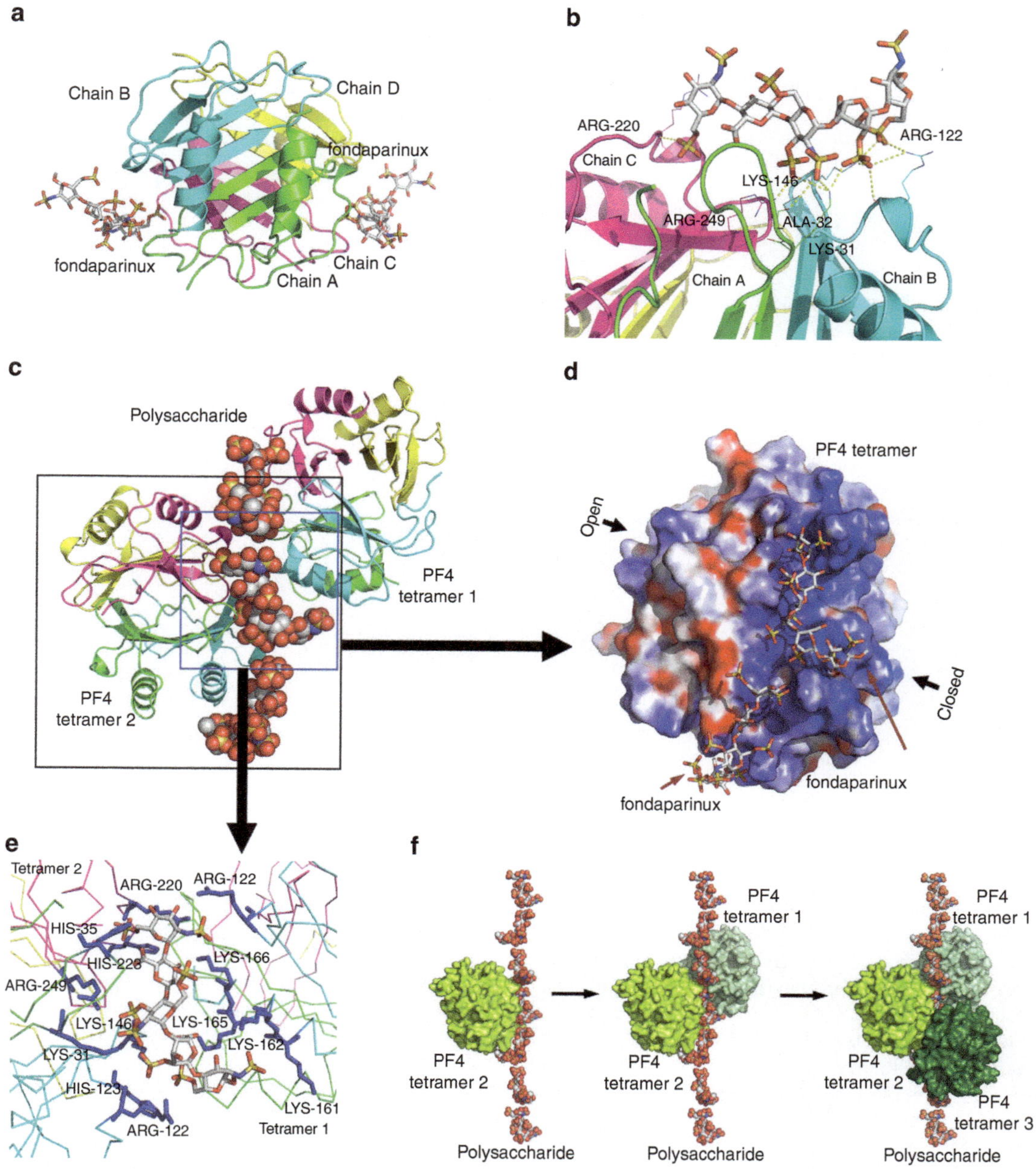

Figure 1 | Crystal structure of the fondaparinux/PF4 tetramer complex. (**a**) Overall structure of the PF4/fondaparinux complex. Fondaparinux makes contacts with a single PF4 tetramer in the groove among the monomers on one side of the asymmetric tetramer. Monomers A, B, C and D in one PF4 tetramer are coloured in green, cyan, magenta and yellow, respectively. (**b**) Fondaparinux (stick representations) stabilizes the PF4 tetramer by binding in the groove among three monomers in a PF4 tetramer. Yellow dotted lines indicate the polar interactions between fondaparinux and three PF4 monomers. (**c**) One fondaparinux (spheres) denoted in the blue box binds in the groove of one tetramer (cartoon representation on the left) and also binds to the C-terminal helix of a second tetramer (cartoon representation on the right), thereby bridging PF4 tetramers. (**d**) Electrostatic potential surface representation (positive: blue; negative: red) of the PF4 tetramer shows that fondaparinux binds along a continuous positively charged surface on the 'closed' side of PF4 tetramer. (**e**) Detailed representation of the positively charged residues (coloured in blue and labelled) on the fondaparinux binding interface between two PF4 tetramers. (**f**) Analysis of crystal lattice reveals a molecular pathway for the formation of antigenic complexes. A fragment of heparin first binds within the groove of one PF4 tetramer (limon, left); binding of the first PF4 tetramer imparts a local linearized structure on heparin, which enhances the binding of a second tetramer (pale green, middle); progression of this process eventuates in the formation of ultralarge antigenic complexes (right).

Clustering might be required for apposition of sufficient HIT antibodies to induce persistent activation of cellular FcγIIA receptors[26]. This model provides structural insight into a recently proposed heparin/PF4 interaction model[14], and is also consistent with previous NMR[27] and site-directed mutagenesis studies[28,29]. Moreover, this set of atomic structures extends other studies showing a heparin chain of $\sim$10 saccharides is required to form a stable antigenic complex[14,25,30,31].

Crystal structure of the KKO-Fab/PF4 tetramer complex. To understand the structural basis for the formation of pathogenic immune complexes, we next solved the crystal structure of human PF4 complexed with the Fab fragment of the monoclonal antibody KKO (Fig. 2a, Supplementary Fig. 7, Table 1 for statistics). KKO recapitulates the salient features of HIT in a mouse model *in vivo* and competes with human platelet activating anti-PF4/heparin antibodies *in vitro*[8,26]. The KKO-Fab fragment binds

Table 1 | Summary of crystallographic data and refinement statistics.

	PF4/Fondaparinux	KKO-Fab	KKO-Fab/PF4 tetramer	RTO-Fab/PF4 monomer
Data collection				
Beamline	X4A/X4C(BNL)	X6A(BNL)	X4A/X4C(BNL)	X4A/X4C(BNL)
wavelength (Å)	0.9791	0.9795	0.9792	2.07
Space group	$P3_221$	$P2_12_12_1$	$P2_12_12_1$	$P2_12_12$
Cell dimensions	$a=67.33$ Å	$a=41.48$ Å	$a=49.49$ Å	$a=161.42$ Å
	$b=67.33$ Å	$b=92.12$ Å	$b=99.34$ Å	$b=171.87$ Å
	$c=61.76$ Å	$c=122.14$ Å	$c=261.74$ Å	$c=208.16$ Å
	$\alpha=\beta=90°, \gamma=120°$	$\alpha=\beta=\gamma=90°$	$\alpha=\beta=\gamma=90°$	$\alpha=\beta=\gamma=90°$
Resolution* (Å)	2.50 (2.54-2.50)	2.20 (2.28-2.20)	4.10(4.17-4.10)	3.75 (3.81-3.75)
Rsym or Rmerge	0.057(0.72)	0.088(0.66)	0.089(0.44)	0.154(1.65)
$I/\sigma I$	22.6(3.0)	10.8(2.2)	7.9(1.9)	6.8(2.9)
Completeness (%)	98.2(98.6)	94.4(98.5)	87.9(61.3)	99.4(98.9)
Redundancy	10.3(10.0)	5.1(4.9)	8.9(4.5)	26.5(25.6)
Refinement				
Resolution(Å)	22.76-2.50	30.8-2.20	20-4.11	50-3.74
No.unique reflections	5,913	24,599	10,642	60,619
R_{work}/R_{free}	0.224/0.245	0.200/0.262	0.317/0.382	0.257/0.297
No. atoms				
Protein	973	3,322	8,548	30,074
Solvent	17	99	0	0
B-factors(Å^2)				
Protein	95	37	268	134
ligand	157	NA	NA	NA
r.m.s deviations				
Bond lengths(Å)	0.006	0.008	0.006	0.007
Bond angles (°)	0.93	1.17	0.92	1.34
Ramachandran(%)				
Favored/disallowed	100/0	99.4/0.3	97.1/0.8	97.4/1.0

NA, not applicable.
*Values in the outermost shell are given in parentheses.

to the PF4 tetramer by making contacts with three monomers (for example, chains A, B and D within one PF4 tetramer, Fig. 2b, surface representations; Fig. 2c, detail of KKO epitope on the PF4 tetramer surface). This structure provides strong support for the concept that formation of the tetramer is required for optimal binding of KKO[32].

KKO does not bind to murine PF4 (mPF4) complexed with heparin. Therefore, we compared the sequence of human PF4 (hPF4) with mPF4, and made structure-based mutants based on the proposed contact sites identified in the co-crystal (Supplementary Fig. 3). Mutations in the putative KKO epitope essentially abolished binding of the HIT antibody KKO to these variant PF4s by ELISA (Fig. 2d,e) and did not support heparin/KKO-induced platelet activation as measured by light transmission aggregometry (Fig. 2f) in contrast to wild-type PF4 (ref. 6).

Structural model of the HIT pathogenic ternary complex. Further analysis revealed that PF4 within the PF4/KKO-Fab complex adopts a symmetric tetrameric structure, which is an 'open'-'open', possibly intermediate, conformation (Supplementary Fig. 4). This result enabled us to superimpose the PF4/KKO-Fab complex onto the PF4/heparin complex model (Fig. 1f) to build a structural model of the KKO-Fab/PF4/'heparin' ternary complex. In this model, the polysaccharide binds to the 'closed' end of PF4, stabilizing the PF4 tetramer, which orients the 'open' end for recognition by KKO. In this way, fondaparinux and HIT-like antibody collaborate to 'stabilize' the ternary complex (Fig. 2g, Supplementary Movie 2).

These structures provide the first atomic level description of the pathological complex that might occur in HIT. The model might also help to explain why fondaparinux, a short fragment of heparin, which is antigenic, only rarely causes HIT. Although the protein–GAG interactions can in theory be propagated to additional PF4 tetramers and fondaparinux molecules, we propose that covalent bonding between pentasaccharide units in longer heparins markedly increases the stability of the holo-complex. Consequently, the more 'stable' complex becomes more antigenic by enhancing antibody avidity.

Crystal structure of the RTO-Fab/PF4 monomer complex. RTO is an anti-PF4 antibody that possesses the same isotype as KKO, but does not cause HIT *in vivo*[6,33]. In contrast to KKO, binding of RTO to hPF4 is not enhanced by heparin[6]. RTO and KKO also do not compete for binding to hPF4 by ELISA[7]. However, KKO and RTO display the same Bmax when measured by ELISA, and is thus illustrative of the difficulty in distinguishing human pathogenic and non-pathogenic antibodies using contemporary bulk-equilibrium assays[7]. The reason why RTO does not cause HIT was revealed through atomic analyses.

To help understand the difference between the impacts of these two antibodies, we next solved the crystal structure of the RTO-Fab in complex with hPF4 (Fig. 3a,b, Supplementary Fig. 7, Table 1 for statistics). Unexpectedly, RTO-Fab binds only to the PF4 monomer. This was affirmed by a change in the migration of PF4 monomer, but not PF4 tetramer, upon addition of RTO-Fab assessed by fast protein liquid chromatography (Supplementary Fig. 5).

Also unexpectedly, the RTO epitope on PF4 overlapped considerably with the KKO epitope (compare Fig. 3c with Fig. 2c, Supplementary Fig. 6). In the structure of the RTO-Fab/PF4 complex, the C-terminal helices still pack against the β-sheet domain. However, the orientation of the C-terminal helices shifts ~60° relative to the β-sheet core domain (Fig. 3d). Superimposition of the PF4 asymmetric tetramer onto the

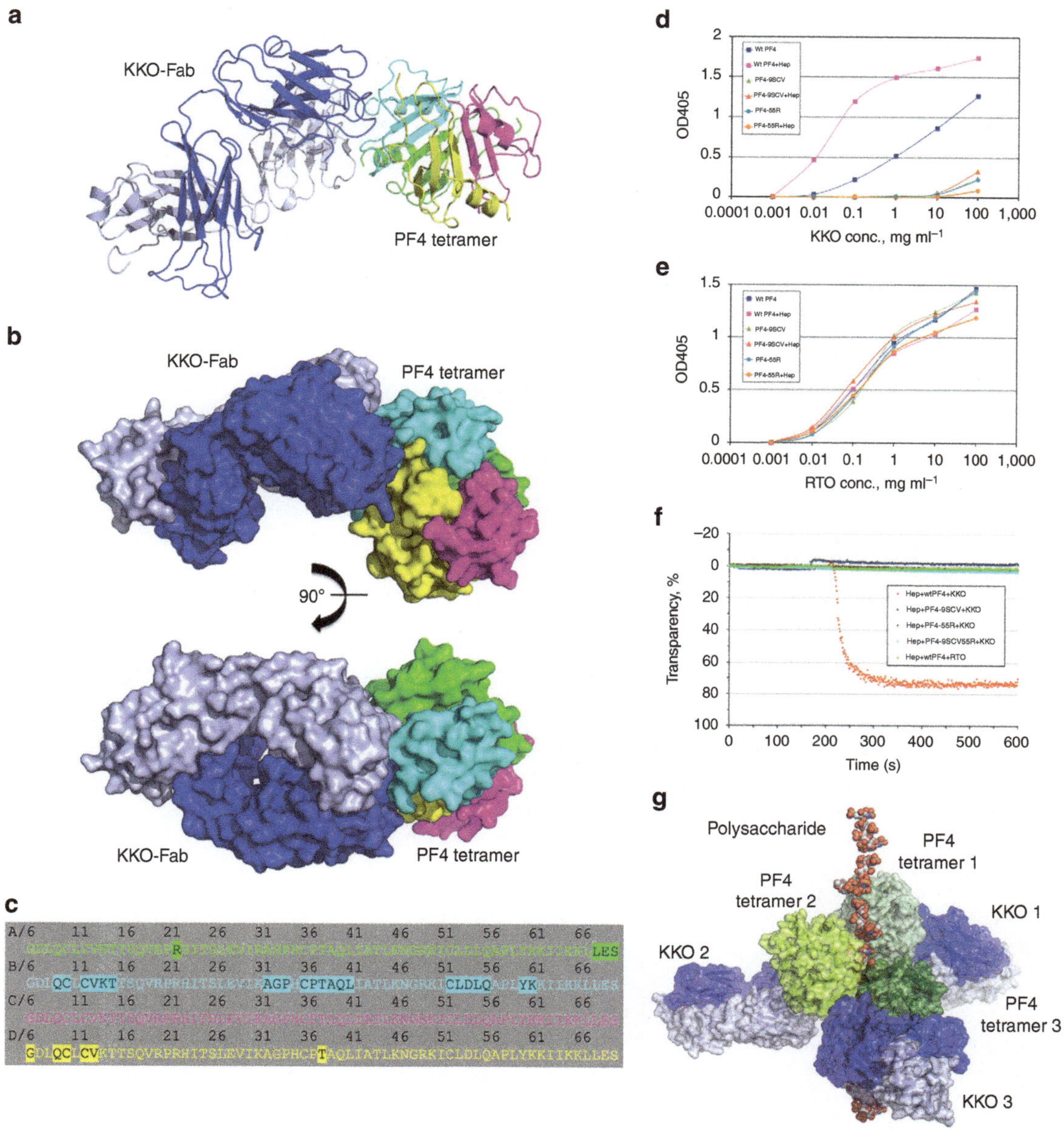

Figure 2 | Crystal structure of the KKO-Fab/PF4 tetramer complex. (**a**,**b**) Overall structure of the PF4/KKO-Fab complex. a: cartoon representations of the complex; b: molecular surface representations. The heavy chain and light chain of KKO-Fab are coloured in blue and light blue, respectively. (**c**) Detailed binding interface of HIT antibody KKO to a PF4 tetramer. Residues within a PF4 tetramer that are <5 Å away from KKO-Fab molecule are highlighted. PF4 monomers are coloured as in Fig. 1. (**d**,**e**) Binding of KKO (**d**) and RTO (**e**) to structure-based PF4 mutants. (**f**) Platelet aggregation by wild-type PF4 and PF4 mutants. KKO (red) induced platelet aggregation in the presence of wild-type PF4 and heparin whereas an isotype matched non-pathogenic antibody RTO described below (green) did not. The panel also demonstrates that PF4 mutants bearing mutations along the KKO binding interface (blue, PF4-9SCV; brown, PF4-55R, Cyan, PF4-9SCV55R) were unable to mediate KKO-induced platelet aggregation. (**g**) Model of the KKO-Fab/PF4/heparin ternary complex. Surface representations of KKO-Fab are coloured in blue (heavy chain) and light blue (light chain) and others are coloured as in Fig. 1f. The model assumes the heparin molecule is composed of about 7 structures similar to fondaparinux depicted in the figure as a non-continuous chain. Intact UFH may further enhance the stability of the holo-complex compared with the fondaparinux fragment, thereby rendering it more antigenic and more capable of binding multiple IgG antibodies.

RTO-Fab/PF4 complex reveals that RTO binds between monomer A and monomer B and prevents the formation of the AB dimer, and thereby prevents formation of the tetramer (formation of an AC dimer is still theoretically possible; Fig. 3e). Together, these structural data lead us to propose that tetramerization of PF4, which is required to form the HIT antigen, would be disrupted by the non-HIT antibody RTO.

RTO limits tetramer formation preventing platelet activation. Based on the knowledge that PF4 tetramers exist in equilibrium with monomers and dimers, and with even higher ordered complexes when heparin is present, we predicted that RTO might prevent or disrupt tetramer formation and thereby prevent KKO-induced platelet activation in the presence of heparin. Indeed, RTO inhibited the activation of platelets by KKO and human HIT IgG in the presence of heparin in a dose-dependent manner, as assessed by expression of P-selectin on flow cytometry (Fig. 4a). Pre-incubation of PF4 with RTO also inhibited KKO-induced platelet aggregation (Fig. 4b).

Moreover, RTO completely inhibited KKO-induced progression of thrombosis in a laser microvascular injury model in mouse

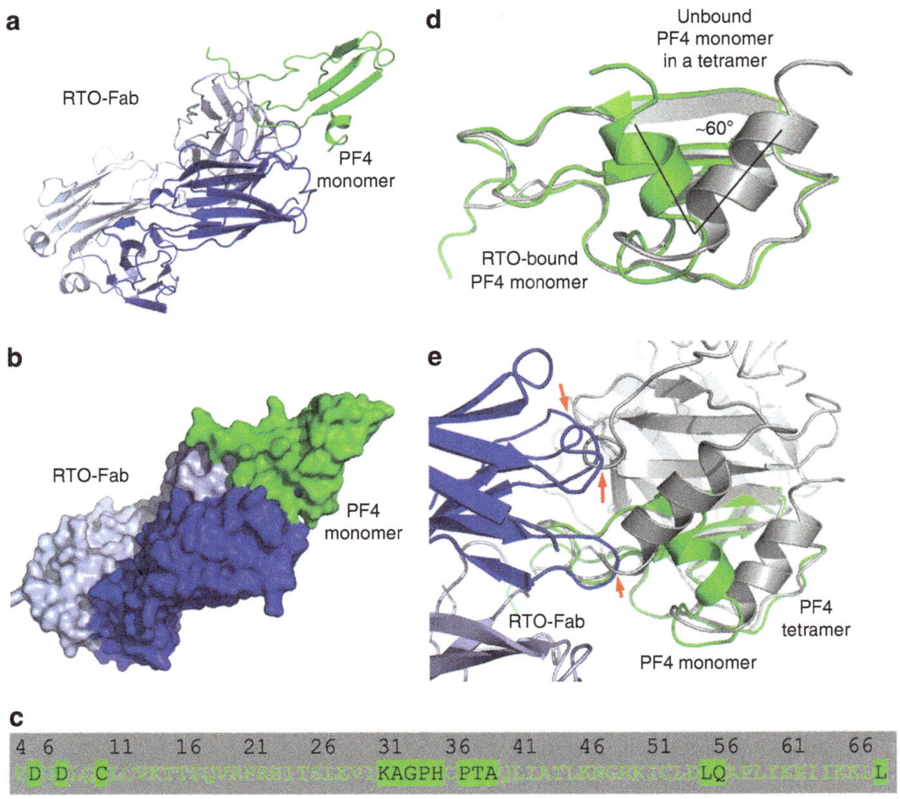

Figure 3 | Crystal structure of the RTO-Fab/PF4 monomer complex. (**a**,**b**) Overall structure of the PF4/RTO-Fab complex. a: cartoon representations of the complex; b: molecular surface representations. The heavy chain and light chain of RTO-Fab are coloured in blue and light blue, respectively. (**c**) Detailed binding interface of the non-HIT antibody RTO to a PF4 monomer. Residues of a PF4 monomer that are <5 Å away from RTO-Fab molecule are highlighted. (**d**) Superposition of the PF4 monomer (green) in the RTO-Fab/PF4 complex with that in the unbound PF4 (grey) indicates that binding of RTO-Fab causes a dramatic structural change in the PF4 monomer: the C-terminal helices are shifted $\sim 60°$. (**e**) Superposition of the PF4 monomer (green) in complex with RTO-Fab (blue and light blue) with the unbound PF4 tetramer (grey). The three arrows indicate the sites where binding of RTO-Fab to one PF4 monomer causes steric clashes with a second PF4 monomer in the tetramer, thereby preventing tetramer formation.

$PF4^{-/-}$/human $PF4^{+/+}$/human $Fc R\gamma IIA^{+}$ mice[26] (Fig. 4c, d). These data indicate that the non-HIT antibody RTO, when engineered and humanized, could provide a template for the development of non-anticoagulant intervention in HIT.

Discussion

It is uncertain how heparin or an endogenous GAG converts a normal host protein, PF4, into an 'autoantigen'. Our crystallographic data provide insight into the multistep process of neoantigen formation at the atomic level. We report the first crystal structure of the PF4 monomer, here in complex with the Fab fragment of an anti-PF4 antibody RTO, which affirms that PF4 exists in solution as an equilibrium among monomers, dimers and tetramers[10]. In the monomeric state, the N-terminal amino acids of PF4 adopt a flexible extended conformation and the C-terminal helices swing around atop the β-sheet core domain. In contrast, in the tetrameric state the individual monomers undergo a marked conformational change. The C-terminal helices of the PF4 monomers are all reorientated and realigned and the N termini of the monomers form antiparallel β-sheet-like structures. The repulsive forces between the positively charged C-terminal helices of monomer A and monomer B in the AB dimer (and similarly between monomer C and monomer D in the CD dimer) might induce thermal instability in the tetramer[34]. Heparin binds among PF4 monomers and also bridges PF4 tetramers, which helps overcome this instability and leads to the formation of stable individual tetramers, a fundamental first step in the expression of the neoantigen in HIT.

Our data also indicate that the asymmetry of the fondaparinux-bound PF4 tetramer, which has an 'open' end and a 'closed' end, plays an important role in forming the pathogenic antigenic complex. Based on our findings, we propose that heparin or a GAG fragment stabilizes the PF4 tetramer by clamping the monomers together through the 'closed' end, orienting the 'open' end in the asymmetric PF4 tetramer that comprises the binding site for KKO. Thus, in the absence of heparin, the KKO binding site on the tetramer is nonexistent or incompletely expressed on the PF4 monomer and dimer and is only transiently expressed and unstable on tetramers that form at high concentrations of PF4. Thus, stabilization of the tetramer by heparin leading to a more stable orientation of an epitope on each PF4 tetramer, compared with the transient and more random orientation of the epitope in the absence of heparin, is a second defining step in the process of generating the HIT neoantigen. Heparins or GAGs of sufficient lengths can then cluster several PF4 tetramers further enhancing stability, which helps to explain why PF4, among heparin binding proteins, is especially antigenic[7,14]. In the third step, HIT-like antibody KKO recognizes the stabilized 'open' end of the PF4 tetramers, further promoting formation of the pathogenic ternary complex.

Our studies were performed with fondaparinux because the heterogeneity of clinical heparins and GAGs preclude obtaining structural detail. Therefore, we do not exclude the possibility that low molecular weight or unfractionated heparin might induce additional changes within and between PF4 tetramers[24,35]. Indeed, our structural model predicts that when a GAG longer

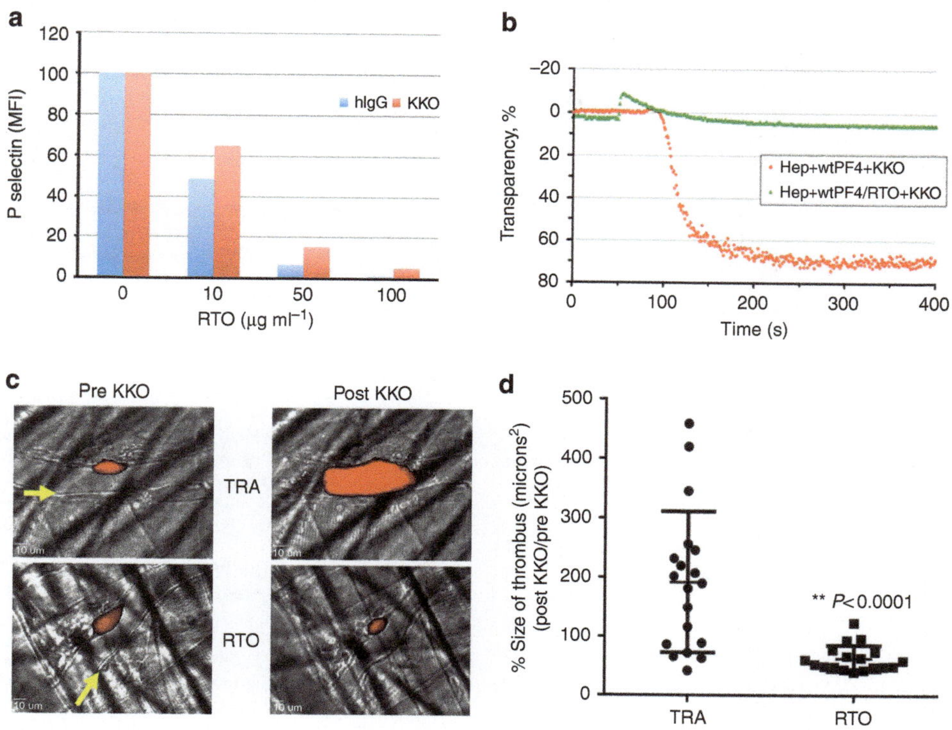

Figure 4 | RTO prevents KKO-induced platelet activation and thrombosis. (**a**). Inhibition of KKO/PF4 mediated platelet activation by RTO. Samples of whole blood were incubated with the indicated concentrations of RTO in the presence of PF4 (10 $\mu g\,ml^{-1}$) for 15 min before adding the platelet-activating anti-PF4 antibody KKO or human IgG. Activation of platelets were followed by expression of P-selectin; the effect of RTO is expressed as % of geometric mean fluorescent intensity (MFI) of P-selectin expression on platelets relative to MFI in the absence of RTO. (**b**) *In vitro* platelet activation assay demonstrated that pre-incubation of PF4 with RTO prevented KKO-induced platelet aggregation. (**c**) Representative composite images of platelet fluorescence overlaid on brightfield snapshots of injuries in mice receiving either RTO or the IgGk2B isotype control TRA are shown. Pre KKO images show thrombi 15 min after initial injury and injection of RTO or TRA. Post KKO images represent the same thrombus 15 min after KKO had been injected intravenously. Arrows represent the direction of blood flow. (**d**) Each dot denotes the per cent change in the size of a single injury based on binding of fluorescently labelled platelets in mice receiving either RTO or the IgGκ2B isotype control TRA followed by KKO. Error bars show the standard deviation. $N=18$ injuries in three mice for RTO, $N=19$ injuries in three mice for TRA. $P<0.0001$.

than about 10 saccharides binds to more than one PF4 tetramer, additional structural changes within the tetramers themselves are quite likely. These structural changes are more likely to occur within the inner surface of the PF4 tetramers in the cluster along the PF4/heparin binding interface than primarily at the HIT-antibody binding epitope, which lies on the outer surface. Therefore, additional studies are needed based on this model, to determine if these additional structural changes in PF4 imparted by longer GAGs directly modulate the contact sites between antibody and antigen.

The A32-A39 loop on the PF4 monomer appears to be highly immunogenic. However, our data suggest that pathogenic and non-pathogenic antibodies that bind to this region differ in how they affect the monomer–dimer–tetramer equilibrium. Pathogenic antibodies that act like KKO and preferentially recognize the tetrameric species display greater avidity in the presence of heparin, which approximates epitopes, and in turn the bound antibody fosters oligomerization[7]. Binding of KKO may also shift the equilibrium from heparin wrapping around the PF4 species suggested in a previous model[13] by forcing it into a more stable linear conformation[14]. The net result would be the generation of stable ultralarge immune complexes[36] capable of sustained activation of FcγIIA receptors on platelets and monocytes[26,37]. In contrast, anti-PF4 antibodies that preferentially recognize monomers in the manner of RTO are readily detected by contemporary ELISAs, but they do not generate large immune complexes that cause disease. Indeed, anti-PF4 antibodies that act like RTO may actually compete with pathogenic antibodies by preventing or disrupting tetramer formation and thereby limit formation of larger immune complexes as depicted in Fig. 5. These data provide a new mechanistic model for the development of a human autoimmune disease, in which a host-protein PF4 complexed with host GAGs assumes diverse oligomeric conformations that differentially bind autoreactive antibodies leading to diverse clinical outcomes.

Our studies might have implications for the diagnosis of HIT. The two prototypic antibodies we studied, RTO and KKO, bind comparably to PF4/heparin on ELISA plates at equilibrium and do not compete with each other for binding in this format[7,8]. This suggests that ELISA wells might contain PF4 in diverse conformations ranging from monomer through tetramer and likely higher ordered complexes. Studies are in progress to determine if ELISA formats based on more homogenous populations of PF4 complexes and/or PF4 variants will reduce detection of non-pathogenic, potentially 'blocking', anti-PF4 antibodies. Additional studies will also be needed to determine if differences in the ratio of pathogenic and 'blocking' antibodies might contribute to the likelihood of developing HIT.

Our studies might also have implications for therapy. Anticoagulants are the standard treatment for HIT, but their efficacy is incomplete and dosing is limited by the risk of bleeding[5,38]. Our data indicate that RTO binds to PF4 monomers and preempts assembly of stable tetramers and, as a result, inhibits KKO (and human HIT IgG) induced platelet activation and aggregation *in vitro* and, importantly, progression of antibody-induced thrombosis *in vivo*. It is likely release of PF4 and formation of

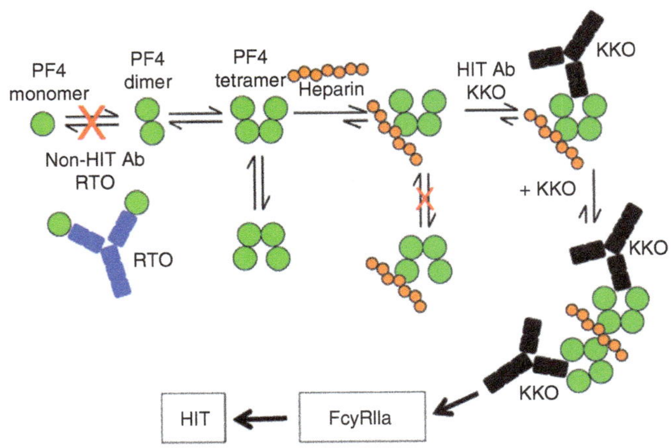

Figure 5 | Model of RTO inhibition of heparin-induced thrombocytopenia. PF4 molecules (green circles) exist in an equilibrium among monomers, dimers and tetramers. On binding to heparin (orange circles), the configuration of the tetramer is stabilized. As a result, the open end of the PF4 tetramer is oriented and recognized by HIT-antibody KKO (black). KKO in turn fosters PF4 oligomerization and works collaboratively with heparin to stabilize the ternary complex. The net result is the generation of stable ultralarge immune complexes capable of sustained activation of Fcγ receptors on platelets and monocytes, which consequently leads to HIT. The non-HIT isotype-matched antibody RTO (blue) binds to PF4 monomers, prevents PF4 oligomerization, prevents formation of ultralarge immune complexes and as a result may prevent HIT. The cartoon assumes heparin is composed of about seven structures similar to fondaparinux.

antigenic complexes extend beyond diagnosis and introduction of a direct thrombin inhibitor, as does the risk of recurrent thromboembolic complications. The inhibitory effects of RTO indicate that tetramerization of PF4 is targetable and that this antibody may provide a structural basis for developing rational non-anticoagulant HIT-specific intervention for this serious and common iatrogenic disorder.

It has been difficult to differentiate between the impact of changes within the secondary structure of individual PF4 tetramers that affect antibody affinity and effects on tetramer oligomerization that enhance avidity, because both contribute to antibody binding measured in ELISA wells that likely contain various conformations of PF4. Our use of Fab fragments that are unable (or show little ability) to oligomerize PF4 tetramers, and fondaparinux, which is antigenic but rarely pathogenic because it has a low capacity to form higher ordered structures, provided an opportunity to delineate the evolution of the antigenic site within the PF4 tetramer itself.

Nevertheless, we wish to emphasize that our studies also have potential limitations. First, the necessity of using fondaparinux as the GAG precluded possible additional relevant changes within PF4 induced by longer heparin molecules that are more likely to induce HIT. Nor can we assess the intra-molecular and intermolecular effects on PF4 imparted by the complex array of cellular GAGs. Second, although most HIT antibodies compete with KKO for binding, our data do not preclude additional contact sites between polyspecific human antibodies and PF4/heparin. Third, crystallography and other biophysical approaches using purified proteins inherently explore atomic interactions in a single stable structure favored by the experimental conditions. Thus, the full range of dynamic changes in (super)oligomerization that occur during the evolution of the disease may not be captured. Thus, our study only provides a model that may help to explain sentinel events through which GAGs may induce binding of autoantibodies to a normal host protein. Fourth, in our experiments, RTO was added before KKO. The dynamics of RTO binding to PF4 might well be different when heparin and a pathogenic HIT antibody are already present. We are currently investigating the effect of RTO on thrombus growth *in vivo* following exposure to KKO.

Methods

Expression and purification of human PF4 and antibodies. Wild-type hPF4 and hPF4 mutants in plasmid pMT/BiP/V5-His (Invitrogen Corp., Carlsbad, CA), were expressed using the Drosophila Expression System (Invitrogen), purified and characterized[39]. Briefly, the protein was collected in serum-free medium Insect-Xpress (Lonza, Walkersville, MD) and isolated by affinity chromatography using a HiTrap Heparin HP column (GE Healthcare) on an AKTA Purifier (GE Healthcare) at 4 °C and eluted at 1.8 M NaCl (wtPF4) using a linear gradient. Fractions containing purified PF4 detected by silver staining of 12% polyacrylamide gels (SDS–polyacrylamide gel electrophoresis) were pooled, concentrated and buffer exchanged into 50 mM HEPES, 0.5 M NaCl, pH ~ 7.2 using an Amicon Ultra filter (3,000 molecular weight cutoff, Millipore). Protein was quantified using a BCA assay (Pierce). To obtain the PF4 mutants, PCR with corresponding primers (Supplementary Table 1) was performed on pMT/BiP/V5/His-PF4 plasmid under conditions recommended by The QuikChange Site-Directed Mutagenesis Kit manual (Stratagene, La Jolla, CA). The resulting plasmids were sequenced to confirm the mutation.

The murine anti-human PF4 IgG2bκ monoclonal antibodies KKO and RTO have previously been described[6]. The IgGs were purified from conditioned PFHM-II media (Invitrogen) using protein A-agarose (Invitrogen) as recommended by the manufacturer. IgG purity was demonstrated by SDS–polyacrylamide gel electrophoresis on NuPAGE 4–12% Bis-Tris Gel (Invitrogen). Fab fragments were generated by papain digest using Pierce Fab Preparation Kit (Thermo Scientific, Rockford, IL) essentially as recommended by the manufacturer, followed by three rounds of removing of Fc fragments with protein A-agarose beads, and extra purification with anti-mouse IgG (Fc-specific) (Sigma M4280) and anti-mouse IgG (Fab-specific) Sigma M4155 antibodies bound to CNBr-activated Sepharose 4 Fast Flow beads (Amersham Biosciences Corp., Piscataway, NJ) as recommended by the manufacturer. KKO-Fab, RTO-Fab and PF4 were further purified by size-exclusion column on an AKTA purifier system (GE Healthcare). Human HIT IgG was purified using staph protein agarose (source) from a pheresate obtained from a patient with HIT.

ELISA assays. Binding of human IgG was measured essentially as previously described for KKO and RTO antibodies[7]. Briefly, Immulon 4 HBX 96-well plates (Thermo Fisher Scientific, Waltham, MA) were coated overnight with either PF4 or PF4 mutant at 5 ug ml^{-1}. The plates were incubated for 30 min with either PBS (control) or with 0.5% glutaraldehyde at room temperature, extensively washed and blocked with 1% bovine serum albumin (BSA) in PBS. The plates were incubated with human patient IgG samples at experimentally selected concentration of 20 μg ml^{-1} for 30 min at 37 °C. IgG binding was measured as absorbance at 405 nm (A405) after incubation for 30 min at 37 °C with horseradish peroxidase-conjugated ImmunoPure Goat Anti-Human IgG (H + L), HRP Conjugated Product No. 31412 (Pierce. Rockford, IL) diluted 1:10,000 in 1% BSA/PBS. Horseradish peroxidase substrate ABTS was from Roche Applied Science, Penzberg, Germany. Absorbance was measured with a SpectraCount plate reader (Packard BioScience, Waltham, MA).

***In vitro* platelet activation mediated by KKO + PF4.** Blood for *in vitro* studies (platelet activation and light transmission aggregometry) was collected after informed consent from healthy, aspirin-free volunteers using a 19-gauge butterfly needle in 129 mM sodium citrate (10:1, vol/vol) under protocols approved by the Institutional Review Board of the University of Pennsylvania and the Children's Hospital of Philadelphia. Whole blood samples were incubated in Ca^{++}/Hepes buffer (2.5 mM $CaCl_2$, 1.25 mM $MgCl_2$, 150 mM NaCl, 10 mM HEPES, pH 7.5) 1/100 v/v in the presence of allophycocyanin (APC) labelled anti-hCD41 and PE labelled anti-P-selectin, PF4 (10 μg ml^{-1}) and the concentrations of RTO MOAb indicated in the figure for 15 min at room temperature. KKO (20 μg ml^{-1}) or human HIT IgG (500 μg ml^{-1}) was added for 20 min at room temperature, samples were then diluted by adding 400 μl of HBSA/BSA/EDTA buffer and immediately measured by flow cytometry (BD LSRFortessa). Platelets were gated based on the forward-scatter and CD41 fluorescence parameters, and binding of anti-P-selectin antibodies was quantified as geometric mean fluorescent intensity.

Light transmission aggregometry. Blood was centrifuged for 12 min at 210*g* at 25 °C to generate platelet-rich plasma (PRP) and at 900*g* for 10 min at 25 °C to produce platelet-poor plasma. Platelet aggregation was measured in PRP using a dual-channel lumi-aggregometer (model 700, Chrono-log Corporation, Havertown, PA) per the manufacturer's instructions. All experiments were completed within 4 h of blood collection. PRP (500 μl) was pre-warmed for 2 min at 37 °C, heparin (Sagent Pharmaceuticals, Schaumburg, IL) was added (final concentration 0.1 U ml^{-1}) for 30 s followed by wild type or mutant rPF4 (final concentration10 μg ml^{-1}) for 30 s followed by KKO or RTO (100 μg ml^{-1}) for up to 10 min. The final volume of added reagents did not exceed 5% of the starting volume of PRP. To examine inhibition of KKO-induced platelet aggregation, 100 μg RTO was preincubated with 5 μg PF4 for 15 min at 25 °C. Heparin (final concentration 0.1 U ml^{-1}) was added to pre-warmed PRP. After 30 s PF4 or PF4/RTO complex was added (final concentrations in PRP 10 μg ml^{-1} and 10/200 μg ml^{-1}, respectively). KKO (final concentration of 100 μg ml^{-1}) was added 30 s later and aggregation was assessed as above.

Inhibition of KKO-induced thrombosis *in vivo*. Transgenic male C57BL mice, lacking mouse PF4 but expressing human PF4 and human FcγRIIA[26] were studied. Mice were matched littermates between 6 and 10 weeks of age. The cremaster laser injury model[33,40] was used to visualize *in vivo* thrombus formation. After surgical preparation, Alexa 647 (BD Biosciences) labelled mouse CD41-F(ab')$_2$ fragments were infused intravenously to label circulating platelets. Each mouse then received 50 μg g^{-1} of either RTO or the IgGκ2B isotype control TRA intravenously followed by 7–8-focal arterial injuries. A brightfield and fluorescence snapshot of each injury was taken 15 min after the initial injury. KKO was then injected intravenously at a dose of 2.5 μg g^{-1}. Fifteen minutes after injection of KKO, a second brightfield and fluorescence snapshot of the same injuries was taken to compare platelet deposition before and after KKO. Image analysis was performed using the Slidebook 6 (3I Intelligent Imaging Innovations, Denver, CO). The size of the thrombus was determined by having the software automatically segment the image using Otsu thresholding to remove background and then calculate and export the size of the platelet fluorescence in microns^2. The investigator was not blinded during these studies. All experiments were performed in compliance with the institutional guidelines for the care and handling of experimental animals approved by the Children's Hospital of Philadelphia Institutional Animal Care and Use Committee.

Statistical analysis. Statistical analysis of *in vivo* injuries was performed using the Prism 6 (GraphPad, La Jolla, CA). A two-tailed Student's *t*-test under non-parametric conditions with Mann–Whitney correction was performed to assess statistical significance. P values < 0.05 were considered significant. There was no randomization of mice. The mice used were selected based on availability. There was no exclusion criteria established before experimentation, available healthy mice were selected for use. No tests of normality or power analyses were conducted.

Crystallization data collection. Purified KKO-Fab (5 mg ml^{-1}) and RTO-Fab (5 mg ml^{-1}) were mixed with hPF4 at different ratios and incubated on ice overnight before setting up crystallization trials. The hPF4/fondaparinux complex was prepared by adding six-fold molar excess fondaparinux (Arixtra, Sanofi-Synthelabo LLC) into purified hPF4 (7 mg ml^{-1}); the mixture was incubated on ice in a buffer containing PBS and 0.3 M NaCl.

Crystallizations of hPF4/fondaparinux, KKO-Fab, hPF4/KKO-Fab and hPF4/RTO-Fab were performed using the hanging-drop vapour diffusion method by mixing the protein and well solution at 1:1 volume ratio at 16 °C. Crystallization kits from Hampton Research and Molecular Dimensions were used for initial crystallization trials. Optimized hPF4/fondaparinux complex crystals were obtained in the well solution containing 2% PEG4000, 17% MPD and 0.1 M sodium acetate pH 5.6. hPF4/fondaparinux complex crystals were then directly flash-cooled in liquid nitrogen by using the MiTeGen micromounts (MiTeGen, LLC). Diffraction-quality KKO-Fab crystals were obtained in the well solution containing 14% PEG2000, 0.06 M zinc acetate and 0.1 M sodium cacodylate, pH 6.8. KKO-Fab crystals were transferred into the well solution supplemented with 25% glycerol, soaked for one second, and then flash-cooled in liquid nitrogen. Diffraction-quality hPF4/KKO-Fab crystals were obtained in the well solution containing 7% PEG6000, 0.1 M Tris-HCl, pH 7.8. hPF4/KKO-Fab complex crystals were flash-cooled in liquid nitrogen similarly as KKO-Fab crystals. hPF4/RTO-Fab complex crystals were obtained in the well solution containing 0.2 M ammonium sulfate, 0.1 M BIS-TRIS, pH 6.5, 25% w/v polyethylene glycol 3350. The diffraction-quality hPF4/RTO-Fab crystals were optimized by macro-seeding. hPF4/RTO-Fab complex crystals were transferred into the well solution supplemented with 20% glycerol, soaked for one second, and then flash-cooled in liquid nitrogen.

All crystallographic data sets were collected at 100 K with ADSC CCD detectors. The long-wavelength ($\lambda = 2.07$ Å) sulfur anomalous diffraction data sets for hPF4/fondaparinux complex and hPF4/RTO-Fab complex were collected at beamline X4A at National Synchrotron Light Source at Brookhaven National Laboratory (Upton, NY USA). Other data sets were collected at National Synchrotron Light Source beamlines X6A and X4C at Brookhaven National Laboratory. All diffraction data were processed by using the HKL-2000 package[41].

Structure determination and refinement. All structures were solved by molecular replacement. The structure of the hPF4/fondaparinux complex was solved by the CCP4 program MOLREP[42,43] with AB dimer from the hPF4 structure (PDB ID code:1F9Q) as a search model. The structure of KKO-Fab was also solved by MOLREP with the Fab structure from the DsbB-Fab complex (PDB ID code:2ZUQ) as a search model. The structure of the hPF4/KKO-Fab complex was solved by the CCP4 program Phaser[44] with the refined KKO-Fab structure and hPF4 (PDB ID code: 1RHP) structure as search models. The structure of the hPF4/RTO-Fab complex was solved by Phaser using the structure of 2H2 Fab fragment of immature Dengue virus (PDB ID code:4KVC) and the structure of the A chain of the hPF4 monomer (PDB ID code: 1F9Q) as search models. The final solution has 8 hPF4/RTO-Fab complexes, related by non-crystallographic symmetry, in an asymmetric unit.

All models were iteratively built in COOT[45] and refined by REFMAC[46,47] or PHENIX[48]. Refinement of the hPF4/KKO-Fab complex at low resolution was performed using the deformable elastic network (DEN)-assisted refinement[49], REFMAC and OPUS-XREF[50]. To assist model building, the low resolution electron density maps for the hPF4/KKO-Fab complex were optimized using B-factor sharpening. The anomalous diffraction data colleted at long wavelength were used to assist model building and refinement of the hPF4/fondaparinux and hPF4/RTO-Fab complexes. The quality of refined models was checked by program PROCHECK[51] and MOLPOBITY[52]. All structural figures were prepared in PyMol (http://www.delanoscientific.com/). The electrostatics potentials were calculated by program APBS, an adaptive Poisson–Boltzmann solver[53,54]. Data collection and refinement statistics are listed in Supplementary Table S1.

References

1. Linkins, L. A. *et al.* Treatment and prevention of heparin-induced thrombocytopenia: Antithrombotic Therapy and Prevention of Thrombosis, 9th ed: American College of Chest Physicians Evidence-Based Clinical Practice Guidelines. *Chest* **141,** e495S–e4530 (2012).
2. Phillips, K. W., Dobesh, P. P. & Haines, S. T. Considerations in using anticoagulant therapy in special patient populations. *Am. J. Health Syst. Pharm.* **65,** S13–S21 (2008).
3. Lee, G. M. & Arepally, G. M. Heparin-induced thrombocytopenia. *Hematology Am. Soc. Hematol. Educ. Program* **2013,** 668–674 (2013).
4. Lewis, B. E. *et al.* Argatroban anticoagulation in patients with heparin-induced thrombocytopenia. *Arch. Intern. Med.* **163,** 1849–1856 (2003).
5. Cuker, A. & Cines, D. B. How I treat heparin-induced thrombocytopenia. *Blood* **119,** 2209–2218 (2012).
6. Arepally, G. M. *et al.* Characterization of a murine monoclonal antibody that mimics heparin-induced thrombocytopenia antibodies. *Blood* **95,** 1533–1540 (2000).
7. Sachais, B. S. *et al.* Dynamic antibody-binding properties in the pathogenesis of HIT. *Blood* **120,** 1137–1142 (2012).
8. Cuker, A. *et al.* Novel diagnostic assays for heparin-induced thrombocytopenia. *Blood* **121,** 3727–3732 (2013).
9. Zhang, X., Chen, L., Bancroft, D. P., Lai, C. K. & Maione, T. E. Crystal structure of recombinant human platelet factor 4. *Biochemistry* **33,** 8361–8366 (1994).
10. Mayo, K. H. & Chen, M. J. Human platelet factor 4 monomer-dimer-tetramer equilibria investigated by 1H NMR spectroscopy. *Biochemistry* **28,** 9469–9478 (1989).
11. Mayo, K. H. *et al.* NMR solution structure of the 32-kDa platelet factor 4 ELR-motif N-terminal chimera: a symmetric tetramer. *Biochemistry* **34,** 11399–11409 (1995).
12. Cowan, S. W., Bakshi, E. N., Machin, K. J. & Isaacs, N. W. Binding of heparin to human platelet factor 4. *Biochem. J.* **234,** 485–488 (1986).
13. Stuckey, J. A., Charles, R. S. & Edwards, B. F. A model of the platelet factor 4 complex with heparin. *Proteins* **14,** 277–287 (1992).
14. Nguyen, T. H., Greinacher, A. & Delcea, M. Quantitative description of thermodynamic and kinetic properties of the platelet factor 4/heparin bonds. *Nanoscale* **7,** 10130–10139 (2015).
15. Greinacher, A. *et al.* Close approximation of two platelet factor 4 tetramers by charge neutralization forms the antigens recognized by HIT antibodies. *Arterioscler. Thromb. Vasc. Biol.* **26,** 2386–2393 (2006).
16. Bauer, K. A., Eriksson, B. I., Lassen, M. R. & Turpie, A. G. Fondaparinux compared with enoxaparin for the prevention of venous thromboembolism after elective major knee surgery. *N. Engl. J. Med.* **345,** 1305–1310 (2001).
17. Turpie, A. G., Bauer, K. A., Eriksson, B. I., Lassen, M. R. & Committee, P. S. S. Postoperative fondaparinux versus postoperative enoxaparin for prevention of venous thromboembolism after elective hip-replacement surgery: a randomised double-blind trial. *Lancet* **359,** 1721–1726 (2002).
18. Warkentin, T. E. *et al.* Anti-platelet factor 4/heparin antibodies in orthopedic surgery patients receiving antithrombotic prophylaxis with fondaparinux or enoxaparin. *Blood* **106,** 3791–3796 (2005).
19. Rota, E., Bazzan, M. & Fantino, G. Fondaparinux-related thrombocytopenia in a previous low-molecular-weight heparin (LMWH)-induced heparin-induced thrombocytopenia (HIT). *Thromb. Haemost.* **99,** 779–781 (2008).
20. Burch, M. & Cooper, B. Fondaparinux-associated heparin-induced thrombocytopenia. *Proc. (Bayl. Univ. Med. Cent.)* **25,** 13–15 (2012).
21. Warkentin, T. E. & Lim, W. Can heparin-induced thrombocytopenia be associated with fondaparinux use? Reply to a rebuttal. *J. Thromb. Haemost.* **6,** 1243–1246 (2008).
22. Warkentin, T. E., Maurer, B. T. & Aster, R. H. Heparin-induced thrombocytopenia associated with fondaparinux. *N. Engl. J. Med.* **356,** 2653–2655 (2007).

23. Bhatt, V. R., Aryal, M. R., Shrestha, R. & Armitage, J. O. Fondaparinux-associated heparin-induced thrombocytopenia. *Eur. J. Haematol.* **91,** 437–441 (2013).
24. Brandt, S. *et al.* Characterisation of the conformational changes in platelet factor 4 induced by polyanions: towards *in vitro* prediction of antigenicity. *Thromb. Haemost.* **112,** 53–64 (2014).
25. Kreimann, M. *et al.* Binding of anti-platelet factor 4/heparin antibodies depends on the thermodynamics of conformational changes in platelet factor 4. *Blood* **124,** 2442–2449 (2014).
26. Reilly, M. P. *et al.* Heparin-induced thrombocytopenia/thrombosis in a transgenic mouse model requires human platelet factor 4 and platelet activation through FcgammaRIIA. *Blood* **98,** 2442–2447 (2001).
27. Mayo, K. H. *et al.* Heparin binding to platelet factor-4. An NMR and site-directed mutagenesis study: arginine residues are crucial for binding. *Biochem. J.* **312,** 357–365 (1995).
28. Ziporen, L. *et al.* Defining an antigenic epitope on platelet factor 4 associated with heparin-induced thrombocytopenia. *Blood* **92,** 3250–3259 (1998).
29. Li, Z. Q. *et al.* Defining a second epitope for heparin-induced thrombocytopenia/thrombosis antibodies using KKO, a murine HIT-like monoclonal antibody. *Blood* **99,** 1230–1236 (2002).
30. Visentin, G. P., Moghaddam, M., Beery, S. E., McFarland, J. G. & Aster, R. H. Heparin is not required for detection of antibodies associated with heparin-induced thrombocytopenia/thrombosis. *J. Lab. Clin. Med.* **138,** 22–31 (2001).
31. Greinacher, A., Alban, S., Dummel, V., Franz, G. & Mueller-Eckhardt, C. Characterization of the structural requirements for a carbohydrate based anticoagulant with a reduced risk of inducing the immunological type of heparin-associated thrombocytopenia. *Thromb. Haemost.* **74,** 886–892 (1995).
32. Amiral, J. *et al.* Antibodies to macromolecular platelet factor 4-heparin complexes in heparin-induced thrombocytopenia: a study of 44 cases. *Thromb. Haemost.* **73,** 21–28 (1995).
33. Rauova, L. *et al.* Role of platelet surface PF4 antigenic complexes in heparin-induced thrombocytopenia pathogenesis: diagnostic and therapeutic implications. *Blood* **107,** 2346–2353 (2006).
34. Nesmelova, I. V., Sham, Y., Gao, J. & Mayo, K. H. CXC and CC chemokines form mixed heterodimers: association free energies from molecular dynamics simulations and experimental correlations. *J. Biol. Chem.* **283,** 24155–24166 (2008).
35. Block, S., Greinacher, A., Helm, C. A. & Delcea, M. Characterization of bonds formed between platelet factor 4 and negatively charged drugs using single molecule force spectroscopy. *Soft Matter* **10,** 2775–2784 (2014).
36. Rauova, L. *et al.* Ultralarge complexes of PF4 and heparin are central to the pathogenesis of heparin-induced thrombocytopenia. *Blood* **105,** 131–138 (2005).
37. Rauova, L. *et al.* Monocyte-bound PF4 in the pathogenesis of heparin-induced thrombocytopenia. *Blood* **116,** 5021–5031 (2010).
38. Kang, M., Alahmadi, M., Sawh, S., Kovacs, M. J. & Lazo-Langner, A. Fondaparinux for the treatment of suspected heparin-induced thrombocytopenia: a propensity score-matched study. *Blood* **125,** 924–929 (2015).
39. Sachais, B. S. *et al.* Rational design and characterization of platelet factor 4 antagonists for the study of heparin-induced thrombocytopenia. *Blood* **119,** 5955–5962 (2012).
40. Falati, S., Gross, P., Merrill-Skoloff, G., Furie, B. C. & Furie, B. Real-time *in vivo* imaging of platelets, tissue factor and fibrin during arterial thrombus formation in the mouse. *Nat. Med.* **8,** 1175–1181 (2002).
41. Otwinowski, Z. & Minor, W. Processing of X-ray diffraction data collected in oscillation mode. *Methods Enzymol.* **276,** 307–326 (1997).
42. Winn, M. D. *et al.* Overview of the CCP4 suite and current developments. *Acta Crystallogr. D. Biol. Crystallogr.* **67,** 235–242 (2011).
43. Vagin, A. & Teplyakov, A. Molecular replacement with MOLREP. *Acta Crystallogr D Biol Crystallogr.* **66,** 22–25 (2010).
44. McCoy, A. J. *et al.* Phaser crystallographic software. *J. Appl. Crystallogr.* **40,** 658–674 (2007).
45. Emsley, P., Lohkamp, B., Scott, W. G. & Cowtan, K. Features and development of Coot. *Acta Crystallogr. D. Biol. Crystallogr.* **66,** 486–501 (2010).
46. Murshudov, G. N., Vagin, A. A. & Dodson, E. J. Refinement of macromolecular structures by the maximum-likelihood method. *Acta Crystallogr. D. Biol. Crystallogr.* **53,** 240–255 (1997).
47. Murshudov, G. N. *et al.* REFMAC5 for the refinement of macromolecular crystal structures. *Acta Crystallogr. D. Biol. Crystallogr.* **67,** 355–367 (2011).
48. Afonine, P. V. *et al.* Towards automated crystallographic structure refinement with phenix.refine. *Acta Crystallogr. D. Biol. Crystallogr.* **68,** 352–367 (2012).
49. Schröder, G. F., Levitt, M. & Brunger, A. T. Super-resolution biomolecular crystallography with low-resolution data. *Nature* **464,** 1218–1222 (2010).
50. Lu, M., Poon, B. & Ma, J. A new method for coarse-grained elastic normal-mode analysis. *J. Chem. Theory. Comput.* **2,** 464–471 (2006).
51. Laskowski, R. A., MacArthur, M. W., Moss, D. S. & Thornton, J. M. PROCHECK: a program to check the stereochemical quality of protein structures. *J. Appl. Cryst.* **26,** 283–291 (1993).
52. Chen, V. B. *et al.* MolProbity: all-atom structure validation for macromolecular crystallography. *Acta. Crystallogr. D. Biol. Crystallogr.* **66,** 12–21 (2010).
53. Baker, N. A., Sept, D., Joseph, S., Holst, M. J. & McCammon, J. A. Electrostatics of nanosystems: application to microtubules and the ribosome. *Proc. Natl Acad. Sci. USA* **98,** 10037–10041 (2001).
54. Strong, M. *et al.* Toward the structural genomics of complexes: crystal structure of a PE/PPE protein complex from *Mycobacterium* tuberculosis. *Proc. Natl Acad. Sci. USA* **103,** 8060–8065 (2006).

Acknowledgements

We thank scientists at National Synchrotron Light Source X4A, X4C and X6A at the Brookhaven National Laboratory for their assistance in data collection. We thank Drs Bruce S. Sachais, Ann H. Rux, Rustem I. Litvinov and John W. Weisel for advice and discussions. This work was supported in part by NIH grant P01HL110860 (DBC).

Author contributions

Z.C. and S.V.Y. were responsible for protein expression and purification. Z.C. and Z.Z. performed crystallization, X-ray data collection, and crystallographic analysis with help from Q.L., S.V.Y. performed ELISA studies. L.R., V.H., T.L. and M.P. performed *in vitro* and *in vivo* biological experiments. D.B.C. and M.I.G. planned and supervised the project. Z.C., D.B.C. and M.I.G. wrote the paper with input from other authors.

Additional information

Accession codes: Atomic coordinates and structure factor files have been deposited in the Protein Data Bank (PDB) under the accession codes 4R9W for hPF4/fondaparinux complex, 4R97 for KKO-Fab, 4R9Y for hPF4/KKO-Fab complex, 4RAU for hPF4/RTO-Fab complex.

Competing financial interests: The authors declare no competing financial interest but a patent has been filed on the use of the antibodies for HIT by the University of Pennsylvania. Zheng Cai, Douglas B. Cines, Zhiqiang Zhu and Mark I. Greene are listed as inventors.

High-flexibility combinatorial peptide synthesis with laser-based transfer of monomers in solid matrix material

Felix F. Loeffler[1,*], Tobias C. Foertsch[1,*], Roman Popov[1], Daniela S. Mattes[1,2], Martin Schlageter[3,4], Martyna Sedlmayr[1], Barbara Ridder[1,2], Florian-Xuan Dang[5], Clemens von Bojničić-Kninski[1], Laura K. Weber[1], Andrea Fischer[1], Juliane Greifenstein[1,2], Valentina Bykovskaya[1], Ivan Buliev[6], F. Ralf Bischoff[7], Lothar Hahn[1], Michael A.R. Meier[2], Stefan Bräse[2,8], Annie K. Powell[3,4], Teodor Silviu Balaban[5], Frank Breitling[1] & Alexander Nesterov-Mueller[1]

Laser writing is used to structure surfaces in many different ways in materials and life sciences. However, combinatorial patterning applications are still limited. Here we present a method for cost-efficient combinatorial synthesis of very-high-density peptide arrays with natural and synthetic monomers. A laser automatically transfers nanometre-thin solid material spots from different donor slides to an acceptor. Each donor bears a thin polymer film, embedding one type of monomer. Coupling occurs in a separate heating step, where the matrix becomes viscous and building blocks diffuse and couple to the acceptor surface. Furthermore, we can consecutively deposit two material layers of activation reagents and amino acids. Subsequent heat-induced mixing facilitates an *in situ* activation and coupling of the monomers. This allows us to incorporate building blocks with click chemistry compatibility or a large variety of commercially available non-activated, for example, posttranslationally modified building blocks into the array's peptides with >17,000 spots per cm^2.

[1] Karlsruhe Institute of Technology, Institute of Microstructure Technology (IMT), Hermann-von-Helmholtz-Platz 1, 76344 Eggenstein-Leopoldshafen, Germany. [2] Karlsruhe Institute of Technology, Institute of Organic Chemistry (IOC), Fritz-Haber-Weg 6, 76131 Karlsruhe, Germany. [3] Karlsruhe Institute of Technology, Institute for Inorganic Chemistry, Engesserstrasse 15, 76131 Karlsruhe, Germany. [4] Karlsruhe Institute of Technology, Institute of Nanotechnology (INT), Hermann-von-Helmholtz-Platz 1, 76344 Eggenstein-Leopoldshafen, Germany. [5] Aix-Marseille University, CNRS UMR 7313, Institut des Sciences Moléculaires de Marseille, Chirosciences, 13397 Marseille cedex 20, France. [6] Technical University of Varna, 1 Studentska, 9010 Varna, Bulgaria. [7] German Cancer Research Center, INF 580, 69120 Heidelberg, Germany. [8] Karlsruhe Institute of Technology, Institute of Toxicology and Genetics (ITG), Hermann-von-Helmholtz-Platz 1, 76344 Eggenstein-Leopoldshafen, Germany. * These authors contributed equally to this work. Correspondence and requests for materials should be addressed to F.F.L. (email: felix.loeffler@kit.edu), F.B. (email: frank.breitling@kit.edu) or to A.N.-M. (email: alexander.nesterov-mueller@kit.edu).

Laser direct-write approaches allow for versatile two- and three-dimensional structuring of a given workpiece in many different ways[1–3]. Laser-induced forward transfer (LIFT), a variant of the direct-write processes, uses laser irradiation to transfer liquid or solid material from a donor surface to defined areas of an acceptor surface. Thereby, a workpiece can be microstructured with different material patterns by simply employing several donor surfaces consecutively. The LIFT method and its variants are applied in the production of, for example, Samsung's organic light-emitting diode display[4] and other electronic materials and devices[5–7], nanoparticle[8] or hydrophobic/hydrophilic surface pattern generation[9], as well as several biological patterning applications[10], such as hydrogels[11], biomolecules[12–14] and even cells[15,16]. Very small feature sizes of down to 3 μm were reported[17]. This versatility sparked the idea to combine LIFT with combinatorial chemistry, where a number of building blocks are used to synthesize a vast number of different oligomeric molecules.

Most of today's array-based combinatorial peptide chemistry methods rely on Merrifield's solid-phase synthesis[18]: the SPOT synthesis[19], the light-directed lithographic synthesis[20–24], the electro-chemical synthesis[25] and the particle-based synthesis[26–29]. In the SPOT method, peptides are elongated in parallel on an array by spotting 20 different dissolved and chemically activated amino-acid building blocks at discrete locations. The light-directed lithographic synthesis uses light patterns to remove photolabile protecting groups from the growing peptides or photogenerated acids cleaving the acid-labile protecting groups at selected areas with a digital micromirror device. Yet, these methods have severe drawbacks: liquid solvents in the SPOT synthesis tend to evaporate or spread on the surface, which limits the spot density of arrays to some 25 peptides per cm^2. Lithographic methods offer a much higher density, but they allow for only one type of amino-acid building block to be coupled sequentially to the array, which yields lower synthesis quality and necessitates, for example, 200 coupling reactions to synthesize arrays of 10mer peptides using 20 different amino acids (10×20).

The particle-based variant of the Merrifield synthesis method uses, for example, a laser printer or a microelectronic chip to structure a surface with 20 different types of solid polymer particles, each embedding a different type of amino-acid building block. This approach overcomes most of the mentioned drawbacks: structuring is achieved by electrical fields to precisely deposit patterns of the 20 different amino acid particle types, before the coupling reaction is induced for the whole pattern at once by heating. At 90 °C, the solid matrix material becomes viscous and the spots are transformed into gel-like droplets that serve as spatially confined reaction vessels, without evaporation or spreading on the surface. This enables the defined diffusion and coupling of the pre-activated monomers to free amino groups on the surface. However, some inherent drawbacks still remain: (i) the spot density is still not satisfying; (ii) the number of different amino-acid building blocks is limited due to process restrictions and high material consumption; and (iii) due to the rather large particle size and the slow diffusion within the melted matrix material, it is almost impossible to employ the large variety of commercially available non-activated amino-acid building blocks by *in situ* activation.

Our goal was to overcome the aforementioned problems with a combinatorial LIFT method (cLIFT), which allows us to synthesize affordable, very high-density (>17,000 spots per cm^2) and high-quality peptide arrays. Furthermore, its high flexibility enables us to employ a large variety of commercially available non-activated amino-acid building blocks and, thereby, synthesize peptide arrays with, for example, many different posttranslationally modified peptides. We developed an automated machine setup, including a two-dimensional laser scanning system, which can currently accommodate over a hundred different donor surfaces. The laser rapidly and accurately transfers very small-sized and very thin material spots next to each other or on top of each other to defined locations on an acceptor surface. Each material spot can comprise a different amino-acid building block or the chemicals that are needed for an *in situ* activation of a non-activated amino-acid building block.

Results

Principle. The principle of the cLIFT method is shown in Fig. 1. The method requires an acceptor slide and different donor slides, which are the sources of the different amino-acid building blocks (Fig. 1a). Donor slides are composed of a standard microscope glass slide, which is covered by a light-absorbing self-adhesive polyimide foil with a thickness of ~95 μm and a transfer material layer. The latter is generated by first dissolving a commercially available styrene-acrylic copolymer resin matrix and OPfp-activated amino acid or biotin building blocks in dichloromethane (DCM), followed by spin-coating this mixture on top of the polyimide foil. The solid matrix efficiently shields chemically activated amino-acid building blocks from decay[27]. The low-cost and easy-to-handle self-adhesive polyimide is attached to the glass slide's surface with a laminating machine.

As acceptor slides, we use commercially available amino-terminated poly (ethylene glycol) methacrylate (PEGMA)/methyl methacrylate 10/90 slides. For more details, please refer to the Methods section.

We automated the entire patterning procedure (see Fig. 2 and Supplementary Movie 1), schematically described in Fig. 1a–d. A robotic slide loader automatically handles the donor and acceptor slides, and positions the transfer layer of the donor slide directly on top of the N-terminated acceptor substrate as described in Fig. 1b. Next, selected spots of matrix material with a first type of embedded amino acid monomers are transferred to the acceptor slide by short laser pulses (Fig. 1c,d). Pulses are generated by an acousto-optic modulator, which rapidly switches the laser radiation of a 1-W continuous-wave 532 nm laser, whereas a commercially available laser scanning system directs the laser focus position (for details, see 'cLIFT machine setup' in Methods). We investigated the material transfer in a wide range of laser pulse durations from micro to milliseconds (standard irradiation time per spot: 2–7 ms).

During the short laser transfer process, time and temperature do not suffice to initiate the coupling reaction of activated monomers to the substrate (data not shown). This is achieved by heating the patterned acceptor slide in an oven for 60 min to 90 °C (Fig. 1e). The heating step initiates the diffusion of the monomers within the matrix material, which allows them to couple to free amino groups on the acceptor slide (for details on the solid phase chemistry, see Supplementary Fig. 1). Subsequent processing steps are identical to standard solid-phase peptide synthesis procedures: first, we remove excess monomers and resin (Fig. 1f), block unreacted amino groups on the surface ('capping'), remove the N-terminal 9-fluorenylmethoxycarbonyl (Fmoc) protecting groups and finally dry the acceptor slides (Fig. 1g). When repeated several times, we obtain an array of combinatorially synthesized peptides (Fig. 1h). We have also automated the chemical washing steps in a wet chemistry machine setup.

Laser transfer. Remarkably, and similar to laser ablation, laser-induced material transfer occurs through air over a distance of up to 60 μm (see Supplementary Methods and Supplementary Figs 2 and 3). A laser pulse is absorbed in the polyimide foil, where the

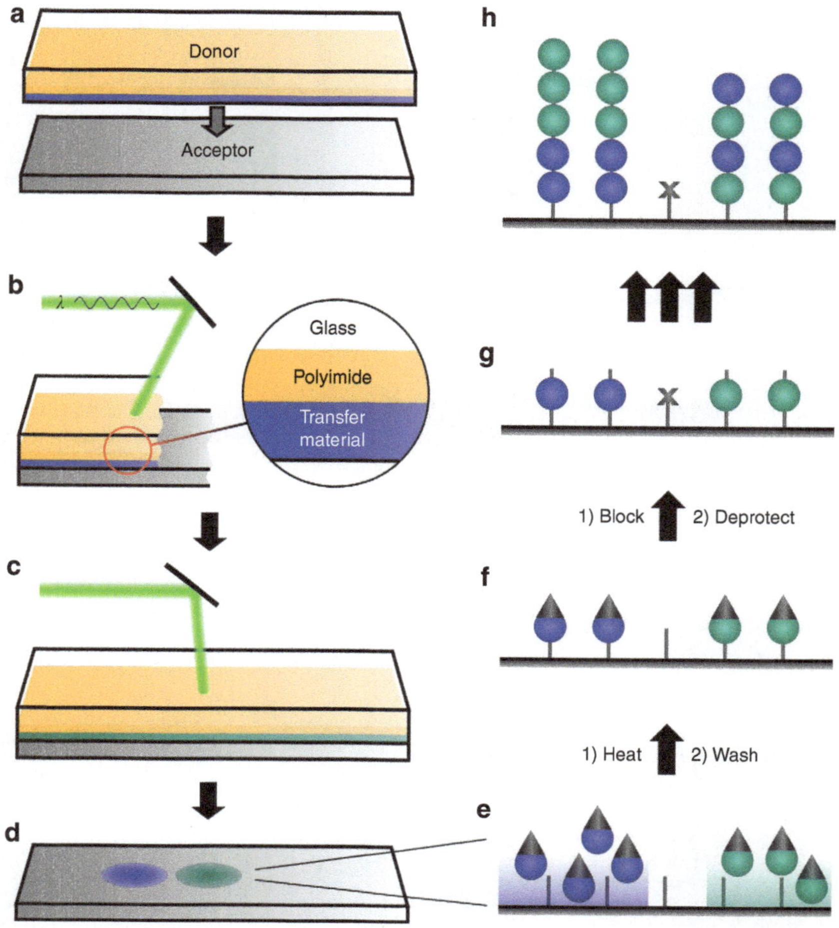

Figure 1 | Principle of the combinatorial laser-induced forward transfer synthesis. Donor slides bearing different monomer building blocks, embedded in a resin (different colours in **b** and **c**) are positioned on top of an acceptor slide (**a**). A laser scanning system transfers minute amounts of material to the acceptor slide. Repeating these steps with different donor slides results in a pattern of different amino acid types (**d**). The coupling reaction of monomers (**e**) is initiated by heating the surface. Next, uncoupled amino-acid building blocks are removed (**f**), uncoupled amino groups are blocked and then the protecting groups are removed (**g**). Repeating the cycle generates an array of combinatorially synthesized peptides (**h**).

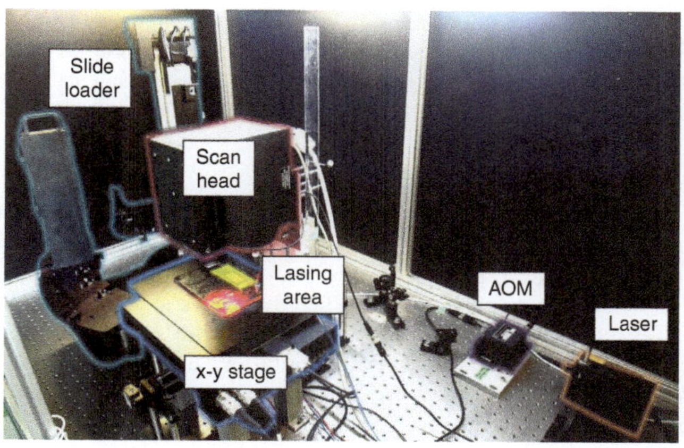

Figure 2 | Setup of the combinatorial laser-induced forward transfer machine. The laser is modulated by an acousto-optic modulator (AOM) and guided to a scan head system. The laser transfer is conducted on an *x-y* microscope stage, the lasing area is highlighted in yellow. Donor and acceptor slides are automatically handled and placed by the robotic slide loader.

energy is converted into heat, which causes the polyimide to quickly expand. Nevertheless, strong plastic deformations and blisters within the polyimide material are only observed when using very high-energy laser pulses (see Supplementary Fig. 4). In parallel, the heat diffuses also into the less rigid transfer layer, melting the material layer and evaporating residual solvent in the transfer layer, which causes the ejection and transfer of material. When analysing the transfer layer in those regions that 'donated' material with atomic force microscopy, tiny impulse craters, and molten and re-solidified material can be observed (see Supplementary Figs 4 and 5). We also analysed the topography of the transferred material on the acceptor slide with a phase-shift interferometer and we found that the height of the transferred material spot on the acceptor slide is in the order of several nanometres (Fig. 3a). Analysing the coupled and stained amino acid monomers on the acceptor surface with a fluorescence scanner (Fig. 3b), we found that the spot dimensions approximately correspond to the crater size on the donor slide. Obviously, the material on the acceptor slide stems from the crater region that is visible on the donor slide.

Furthermore, by tuning the laser transfer parameters, we are able to adjust the amount of deposited material (see Fig. 3c). Interestingly, the amount of deposited material on the acceptor surface, which is approximately in the range between 1 and 50 nm in terms of layer thickness (10–500 pg), correlates linearly with the laser energy, ranging from 450 to 900 μJ (red marks in Fig. 3c). Above 900 μJ, the amount of deposited material reaches a plateau, until the laser starts to

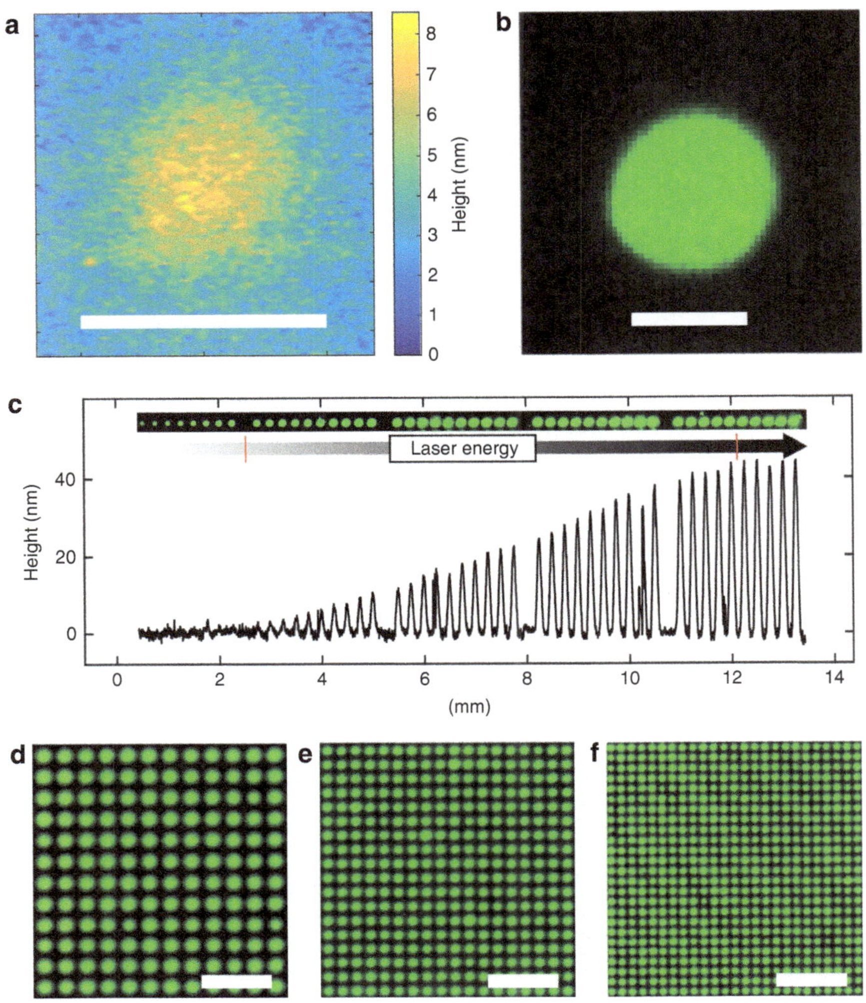

Figure 3 | Analysis of transferred spots and spot resolution. (**a**) Topography of the spot with activated leucine monomers in the polymer matrix on the acceptor slide before the coupling and washing steps, measured with phase-shift interferometry. The material spot is ~8 nm thick at its centre (scale bar, 100 µm). (**b**) The fluorescence image of the spot after coupling of the monomers and staining with a rhodamine dye. The amount of the transferred material is ~0.1 ng (scale bar, 100 µm). (**c**) Topography of transferred spot material in dependence of the laser energy. The height of the transferred spot material, containing an activated leucine building block, was measured with phase-shift interferometry, laser energy linearly increases from left to right in steps of ~15 µJ. The range of the linear correlation of deposited material (1-50 nm) and deposited laser energy (450-900 µJ) is marked by red lines. The corresponding fluorescence staining pattern of the coupled leucine was obtained with a rhodamine dye. (**d-f**) Fluorescence images of biotin patterns (scale bar, 500 µm), stained with labelled streptavidin, achieving different pitches: (**d**) 150 µm (4,444 spots per cm^2), (**e**) 100 µm (10,000 spots per cm^2), (**f**) 75 µm (17,777 spots per cm^2).

burn the polyimide layer. Below 400 µJ, almost no material is transferred, although a weak fluorescence staining is visible down to about 300 µJ.

Another feature of our approach is the possibility to reuse the donor slides up to 20 times without loss in transfer quality, making the cLIFT process highly efficient (see Supplementary Methods and Supplementary Figs 6 and 7).

To assess the feasibility of various spot pitches, we transferred OPfp-activated biotin to a functionalized glass substrate with cLIFT using different lasing parameters. Owing to the heat diffusion within the 95-µm-thick self-adhesive polyimide layer, which limits the lowest possible pitch to ~100 µm, we replaced the very thick polyimide layer by a 5-µm thin layer of spin-coated and cured polyimide (Durimide 7520, Fujifilm). In this way, we achieved pitches from 150 to 75 µm (Fig. 3d–f). In principle, LIFT technology allows for very high-density patterns with resolutions of up to a few micrometres[17]. Details on the synthesis and staining experiments shown in Fig. 3 are described in 'cLIFT technique parameters' and 'General procedure' in Methods.

We used these experimental findings to derive an analytical heat diffusion model to define the conditions where laser-induced material transfer takes place (see Supplementary Methods and Supplementary Fig. 8). In a simplified model, we describe the heat diffusion in the light-absorbing layer. Laser pulse-induced heating competes with heat diffusion and conditions are tipped towards transfer conditions either by higher laser powers or by longer pulse durations. These transfer conditions are described by the formula:

$$P\tau^{0.5} \geq 4\pi\sqrt{kTD \cdot \rho cT\pi\sigma^2 D} \quad (1)$$

Here, P is the total laser power absorbed by the donor slide, τ is the pulse duration, σ is the laser focus radius, D is the thickness of the layers on the donor substrate, T is the characteristic temperature of the transfer material layer where the transfer can take place, k is the heat transfer coefficient, ρ is the mass density and c is the specific heat capacity.

Synthesis. First, we used the cLIFT machine (Fig. 2) to synthesize patterns of 3-mer, 6-mer and 9-mer peptides, with an alternating sequence of Ala and Gly, and a terminal biotin (Fig. 4a). These arrayed peptides were stained with fluorescently labelled streptavidin. We did not observe any significant decrease in fluorescence intensity, when we compared the staining intensity of the 3-mer, 6-mer and 9-mer peptides (for statistical analysis of spot fluorescence signals, see Supplementary Methods and Supplementary Fig. 9).

Next, we synthesized haemagglutinin (HA) (Tyr-Pro-Tyr-Asp-Val-Pro-Asp-Tyr-Ala) and Flag epitopes (Tyr-Asp-Tyr-Lys-Asp-Asp-Asp-Asp-Lys), as well as 62 permutation variants of these peptides, by exchanging two amino acids in each of the six differing layers of the synthesized peptides ($2^6 = 64$ variants). The selected peptides were synthesized in a pitch of 150 µm with measured spots sizes between 100 and 120 µm. Side-chain protecting groups were cleaved from the resulting peptides with a trifluoroacetic acid solution (for details, see 'General procedure' and 'Typical duration of synthesis for one layer' in Methods). Subsequently, the array was incubated with fluorescently labelled specific antibodies (Fig. 4b–d). Strong signals and negligible background indicate a good quality of the peptide spots, synthesized with cLIFT (for statistical analysis of spot fluorescence signals, see Supplementary Methods, Supplementary Fig. 10 and Supplementary Tables 1–4). Details on chemical analytics can be found in Supplementary Methods and Supplementary Figs 11–13. In our previous work, we did not observe any significant decay of activated monomers due to short laser irradiation[29] with comparable laser energies.

Click chemistry is of increasing interest in the synthesis of peptides and drug conjugates[30,31]. To demonstrate the chemical flexibility of cLIFT, we also patterned and coupled a synthetic amino acid on a large scale with an alkyne side group, a Fmoc-protected propargyl-glycine-OPfp to the amino-activated acceptor substrate (Fig. 5a). Then, we used a copper-catalysed click reaction to label this pattern with a styrylpyridinium fluorophore[32], functionalized with an azide group (Fig. 5b–d). We observed a bright fluorescence pattern (Fig. 5e) and a very low background signal. Thus, we can presume that our laser transfer does not harm the alkyne function of the activated Pra monomer (details in Supplementary Methods). Furthermore, X-ray photoelectron spectroscopy (XPS) analysis of the surface showed no residual copper catalyst on the synthesis slide.

Finally, we wanted to assess whether our cLIFT method can directly employ the plethora of commercially available Fmoc-protected, but non-activated, amino-acid building blocks. For the surface coupling reaction, building blocks have to be *in situ* activated. Therefore, we deposited material spots with non-activated amino acids on top of previously deposited material spots that contain suitable activation agents (Fig. 6a–c). First, we deposited material spots that contained a mixture of the activation reagent *N,N'*-diisopropylcarbodiimide (DIC) and the racemization suppressor hydroxybenzotriazole (HOBt), embedded in the solid matrix material (Fig. 6d,e). By increasing the lasing duration, we generated a gradient of increasing amounts of transferred activation reagents (blue arrow). Next, we positioned a second layer of material spots on top of the activation reagent spots, containing a non-activated glycine with an N-terminal protecting group (Fmoc-Gly-OH). Again, we generated a gradient of transferred materials by increasing the lasing time (gradient pattern from left to right, red arrow, perpendicular to the first layer; Fig. 6d). As controls, activation reagents only (Fig. 6e) and non-activated Fmoc-Gly-OH only (Fig. 6f) were used. An OPfp-ester-activated glycine (Fmoc-Gly-OPfp) was deposited as a positive control (Fig. 6g). After heat-induced coupling, we performed the washing and capping steps to block free amino groups on the surface and removed the Fmoc-protecting groups from the surface-bound amino acids. Then, we coupled a rhodamine *N*-hydroxysuccinimide ester dye to the free NH_2 groups on the array. We observed a strong fluorescent signal in those spots where non-activated amino acids were positioned on top of the activation reagents (Fig. 6d), but no or very weak signals for the two negative controls with only DIC/HOBt (Fig. 6e) and only Fmoc-Gly-OH (Fig. 6f). These observations indicate that non-activated amino-acid building blocks are efficiently converted to active esters, when diffusing through a very thin layer of melted activation reagents. As expected, the positive control with activated Fmoc-Gly-OPfp

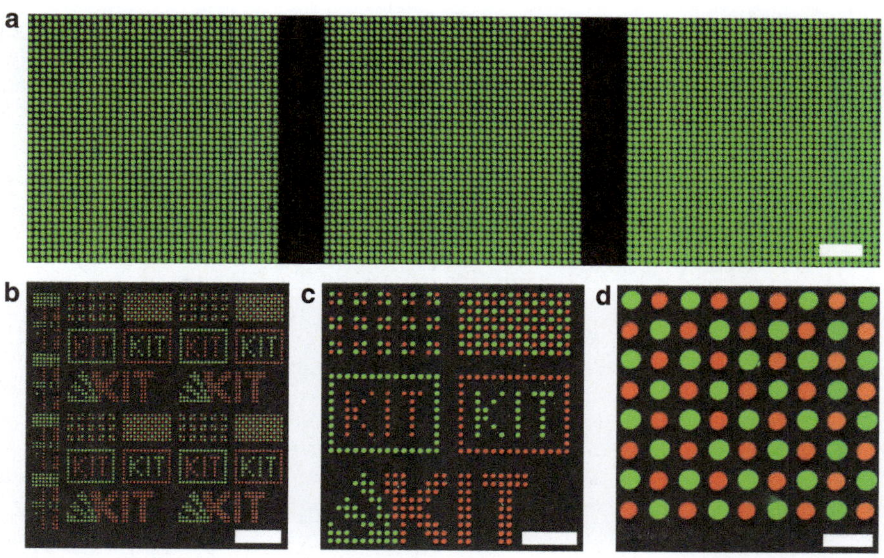

Figure 4 | Combinatorial synthesis of peptide arrays with cLIFT. (**a**) Three spot patterns of a 3-, 6- and 9-mer peptide (left, centre and right) with a biotin momoner as a terminal group (left Ala-Gly-biotin, centre Ala-Gly-Ala-Gly-Ala-biotin and right Ala-Gly-Ala-Gly-Ala-Gly-Ala-Gly-biotin); scale bar, 1 mm. No significant difference in staining intensity can be observed. (**b–d**) Array containing 64 different peptides with 4,444 peptide spots per cm^2 using cLIFT and different donor slides that bear the different amino-acid building blocks; scale bars, (**b**) 2 mm, (**c**) 1 mm, (**d**) 250 µm. Flag- and HA peptides (**c**, **d**) and 62 variants (**b**, left column) were stained with specific anti-Flag and anti-HA antibodies.

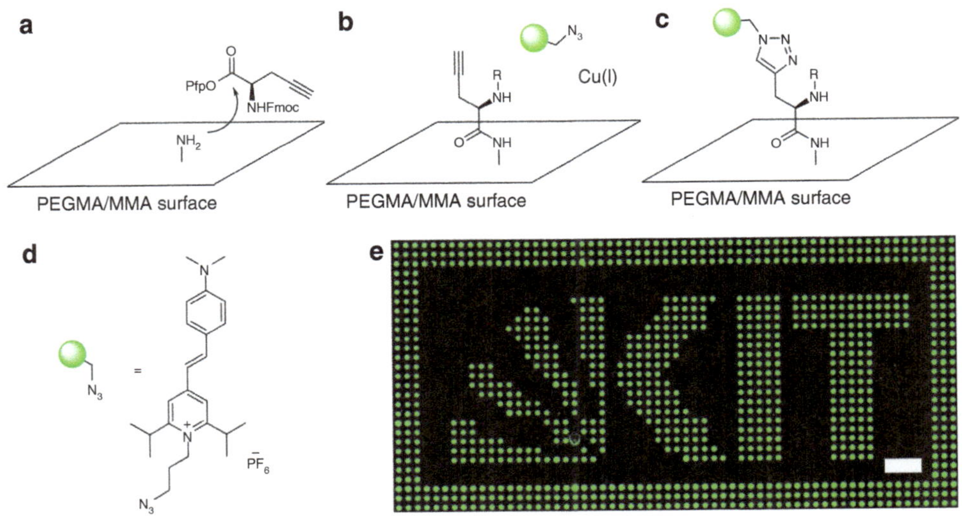

Figure 5 | Schematic of the click chemistry reaction using an Fmoc protected propargyl-glycine-OPfp (Fmoc-Pra-OPfp). (**a**) The synthetic amino acid (Pra, propargyl-glycine) was patterned and coupled with cLIFT. (**b–d**) Subsequently, the styrylpyridinium fluorophore was coupled to the Pra in a copper catalysed click reaction. (**e**) The fluorescent image of the patterned Pra coupled with the styrylpyridinium fluorophore via click chemistry; spot pitch 250 μm (scale bar, 1 mm).

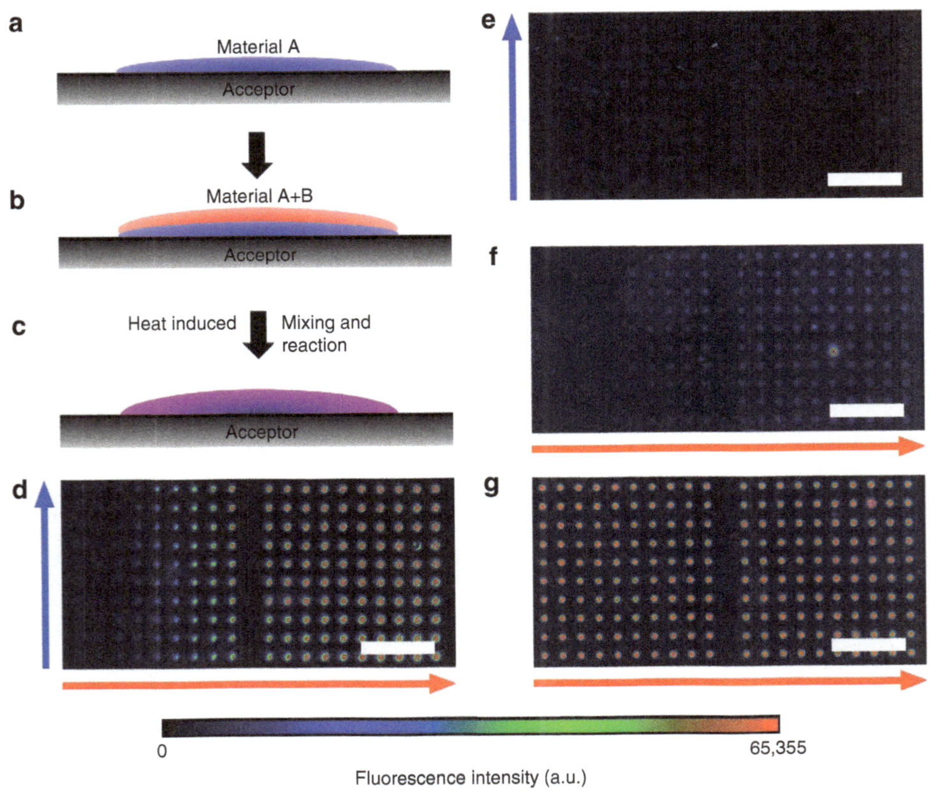

Figure 6 | Two-step cLIFT gradient patterning experiment for solid-phase combinatorial synthesis. (**a**) Deposition of first material. (**b**) Deposition of a second material on top of the previous pattern. (**c**) Heat-induced melting of material induces mixing and initiates the reaction. (**d–g**) Experimental results of an Fmoc-Gly-OH, reacting with DIC and HOBt activation reagents (250 μm pitch); scale bar, 1 mm. Free amino groups were stained with a rhodamine *N*-hydroxysuccinimide (NHS) ester dye, illustrated in rainbow colour scale. (**d**) Two layer reaction of a layer of DIC and HOBt (concentration increases in direction of blue arrow) with a second layer of Fmoc-Gly-OH (concentration increases in direction of red arrow); (**e**) one layer of DIC and HOBt as a negative control; (**f**) one layer of Fmoc-Gly-OH as a negative control; (**g**) one layer of Fmoc-Gly-OPfp as a positive control.

(Fig. 6g) also gave strong fluorescent signals. We presume that the weak fluorescent signals, which are only observed in the negative control of Fmoc-Gly-OH, are either due to nonspecific intercalation of the Fmoc-Gly-OH building block into the PEGMA/methyl methacrylate surface or result from a rare peptide bond formation of the non-activated glycine with the amino groups on the surface, induced by the high coupling temperature (for experimental details, see 'Preparation of the donor and acceptor slides' and 'Activation reaction' in Methods). Thus, exploiting the cLIFT method, it is possible to separately pattern and precisely mix two or more types of reagents on the surface. By adjusting laser transfer (Fig. 3c) and spin-coating parameters (that is, initial concentration of reagents), it is also easily possible to predetermine the amounts and, thereby, the concentrations of the monomers and activation agents.

Discussion

Life sciences strive to find out which molecules bind to each other, for example, which antigens are targeted by an antibody or which posttranslational modifications are crucial in specific cell signalling, for example, which acetyl-lysine-peptides are bound by a bromodomain[33]. One straightforward method to find out is the use of high-density peptide arrays, but all currently available methods to synthesize them have severe drawbacks: lithographic methods yield only short, low-quality peptides, due to the large number of coupling cycles. The SPOT synthesis yields high-quality, but low-density peptide arrays, due to solvents that tend to evaporate and spread on the surface. Particle-based methods yield medium-density peptide arrays that are limited to only 24 different building blocks, due to the limitation of the number of particle types and toner cartridges. However, simply by exchanging one donor foil for a different one, our cLIFT method can employ a theoretically unlimited number of different amino-acid building blocks. Moreover, the automated cLIFT machine does not require any expensive mechanical alignment, but rather relies on an inexpensive camera system, which reliably calibrates the laser beam with reference markers on the array surface (see Supplementary Figs 14 and 15). Thereby, we can currently synthesize arrays with a density of $>17,000$ spots per cm^2, which is certainly not the limit in terms of achievable array densities. It has been reported that LIFT-based structuring is possible with feature sizes of down to 3 μm[17]. Yet, another major advantage of our cLIFT method is its frugal consumption of expensive amino-acid building blocks: a few milligrams suffice to produce a donor slide, whereas the amount of transferred material is in the nanogram range. In particular, donor slides with expensive building blocks can be reused at least up to 20 times.

Although the exact mechanism of our laser-induced material transfer is still elusive, we could show that the size and the thickness of transferred spots can be adjusted by tuning the laser parameters. We exploited this feature by positioning two different and very thin material spots on top of each other, which consistently mixed on melting. As cLIFT structuring is performed with solid materials, both the *in situ* activation and the peptide elongation reaction are inhibited ('frozen'), until the heating step initiates the reaction by melting the material spots. This feature allows for structuring with many different materials that are positioned next to or on top of each other. After these structuring steps, heat-induced melting makes it possible for the amino acid building blocks to diffuse, the activation reagents to activate the amino acids and finally for the coupling of the activated amino acids to the growing peptides spot-by-spot on the array. As it is possible to control the amount of deposited materials, any combination of different reagents for peptide synthesis should be feasible. Thus, our method profits from an ever increasing number of commercially available Fmoc-protected amino-acid building blocks to synthesize, for example, peptides with different types of posttranslational modifications. Furthermore, we were able to include a synthetic amino-acid building block, showing that advanced postsynthesis functionalization (that is, click chemistry) is compatible with our synthesis approach. Currently, we have advanced our cLIFT method to nearly complete automation: we integrated a slide loader to automatically exchange donor slides, whereas wet-chemistry processing is done in a fully automated ultrasound-supported reactor.

Methods

Preparation of the donor and acceptor slides. Donor slide preparation: microscope glass substrates were covered by self-adhesive polyimide foil (Kapton, DuPont, USA; cmc Klebetechnik GmbH, Frankenthal/Pfalz, Germany; thickness of polyimide layer ~50 μm, thickness of glue layer ~45 μm). The transfer material layer, similar to other solid material-based synthesis methods[26,27,29], was spin coated (80 r.p.s. for 45 s; solution was already applied during acceleration to 80 r.p.s.) on top of the polyimide foil. For this purpose, 10% w/w activated monomers, for example, pentafluorophenyl (OPfp)-activated amino acids with an N-terminal Fmoc-protecting group, and 90% w/w of the inert matrix polymer (SLEC PLT 7552, Sekisui Chemical GmbH, Düsseldorf/Germany) are dissolved in DCM: 15 mg of amino acid and 135 mg of resin are dissolved in 1 ml of DCM, or 8 mg of biotin and 68 mg of resin in 1 ml of DCM, respectively, due to the lower solubility of biotin in DCM. In the case of the coupling reagent donor slide, we dissolved 3.75 mg of HOBt in 12 μl of dry *N,N*-dimethylformamide (DMF) and then added 988 μl of DCM, 139 mg of matrix polymer and 9.35 μl (7,25 mg) of DIC. The donor slide with the non-activated glycine amino acid (Fmoc-Gly-OH) was prepared by dissolving 15 mg of Fmoc-Gly-OH in 25 μl of DMF and then adding 975 μl of DCM and 135 mg of matrix polymer.

Acceptor slides: The PEGMA slides were acquired from PEPperPRINT GmbH, Germany. The top surface of the acceptor slide was marked by laser ablation with a high-power laser[34]. The marks were used for determining the position of the acceptor slides with respect to the laser scanning system.

cLIFT machine setup. A robotic slide loader (PL200, Prior Scientific, UK) automatically handles and places the donor and acceptor slides. We use an acousto-optic modulator (1002AF1, Polytec GmbH, Germany) to switch the laser (FSDL-532-1000T, 1 W, Frankfurt Laser Company), a laser scanning system (hurrySCAN 10, Scanlab AG, Germany), an *x–y* microscope stage (SCANplus 100 × 100, Maerzhaeuser, Germany) and a camera (DCC1645C, Thorlabs Inc., Newton, NJ, USA) with a microscope lens (PLN 4XCY, Olympus GmbH, Hamburg, Germany).

cLIFT technique parameters. For a pitch of 150 μm, we used the full power of the 1 W laser and a pulse duration of 5 ms. The scan head was set to a jump speed of 100 μs, a laser-off delay of 310 μs, a jump delay of 200 μs and a laser-on delay of 300 μs.

General procedure. After the patterning of one layer with different monomers, the coupling reaction is initiated by heating the acceptor slide in an oven to 90 °C for 60 min under argon atmosphere. Next, the acceptor slide was washed with acetone three times for 2 min (once in an ultrasonic bath). Then, the acceptor slide was dried in a jet of air. For the HA- and Flag-peptide synthesis, the patterning and coupling steps were repeated to increase the coupling yield.

To block the remaining free NH_2 groups on the acceptor slide, it was washed in a mixture of acetic anhydrate (Ac_2O, 10%), *N,N*-diisopropylethylamine (DIPEA, 20%) and DMF (70%) for 30 min. Then, the slide was washed twice for 5 min with DMF.

The deprotection of NH_2 groups of the terminal amino acids was performed by washing the slide in a solution of piperidine (20%) and DMF (80%) for 20 min. Afterwards, the acceptor slide was washed with DMF for 5 min twice, then with acetone for 2 min twice and finally dried in a jet of air.

After the final coupling cycle, the side-chain protecting groups are cleaved from the amino acids by washing the slide three times for 30 min in a mixture of 51% v/v trifluoroacetic acid, 3% v/v triisobutylsilane (ultrapure, Sigma, USA), 44% v/v DCM and 2% v/v H_2O. Next, the slide was washed twice with DCM for 5 min and then with 5% DIPEA in DMF for 5 min. Finally, the slide was washed twice for 5 min each in DMF, subsequently in methanol twice for 5 minand then the substrate was dried in an air flow.

Typical duration of synthesis for one layer. The structuring process for 10,000 spots with 20 different donor slides on one acceptor slide currently requires <15 min: the transfer of one spot currently requires <10 ms (currently 3–10 ms, actual lasing time is 1–5 ms), which sums up to <100 s for 10,000 spots. The exchange of one donor slide for another currently requires about 40 s, which sums up to 800 s for 20 donor slides. Compared with this, the actual lasing time is very short. Therefore, we can easily increase the number of spots to 100,000 with only modest increase in process duration (<30 min). This is certainly not the end of optimization: it should be possible to decrease the donor slide handling time and the laser transfer time as well, by at least 50%. For the repetitive coupling yield and mass spectrometry experiment, lasing duration was 23 min for ~215,000 spots with 100 μm pitch and 5 ms lasing time per spot.

We only need one coupling step for all 20 amino acids in one layer, owing to the separation of transfer and chemical coupling reaction. Thus, the time required for one layer includes (1) the sequential physical patterning of the 20 different types of amino acid spots (<30 min for 100,000 spots) and (2) one coupling step (60 min at 90 °C). Next, the slide is shortly washed in acetone (<15 min) and the patterning and coupling step is repeated once before the transient Fmoc deprotection to increase the coupling efficiency (that is, double coupling). Thus, elongating all peptides on the acceptor surface by one layer with a pattern of all 20 amino acids, including coupling (twice), wet chemistry capping and Fmoc deprotection requires <5 h.

As structuring, which is rather time efficient, is separate from chemical coupling reaction, which is much more time intensive, it is easily possible to massively increase the throughput by conducting the coupling reaction and chemical washing steps in parallel with many acceptor slides at once. This is probably the most important factor, making our process more time efficient than other approaches. However, this is hard to quantify, as it depends on the actual cLIFT structuring time per array. We can currently process five acceptor slides in parallel in a chemical washing chamber.

Activation reaction. After the patterning of the different materials, the reaction was initiated by heating the acceptor slide in an oven to 90 °C for 90 min under argon atmosphere. To block (that is, acetylate) the remaining free NH_2 groups on the acceptor slide, it was incubated in a mixture of acetic anhydrate (10%) + DIPEA (20%) + DMF (70%) for 30 min. Next, the slide was washed with DMF for 5 min twice. Deprotection of the NH_2 groups of the amino acids was achieved by washing the slide in a solution of piperidine (20%) and DMF (80%) for 20 min. Afterwards, the acceptor slide was washed twice with DMF for 5 min, twice with acetone for 2 min and then dried in a jet of air. The slide was then incubated in PBS with 0.05% Tween 20 (PBS-T) for 15 min. To stain the free amino groups, the slide was incubated with a rhodamine *N*-hydroxysuccinimide ester (5/6-carboxy-tetramethyl-rhodamine succinimidyl ester), diluted 1:10,000 in PBS-T and afterwards washed three times in PBS-T.

Staining protocol with proteins. For fluorescent staining, we use monoclonal mouse anti-HA antibodies (by Dr G. Moldenhauer, German Cancer Research Center) conjugated with a Cy5 fluorescent dye and monoclonal mouse anti-Flag M2 antibodies (Sigma) conjugated with a Cy3 dye. First, the slide is incubated in PBS-T for 15 min. Next, the surface is blocked in Rockland infrared blocking buffer (MB-070, Rockland Immunochemicals, USA) for 30 min and then washed with PBS-T for 1 min. Staining was performed for 1 h with a mixture of 10% Rockland blocking buffer (500 µl) in PBS-T (4,500 µl), adding 1:1,000 anti-HA and anti-Flag antibodies (5 µl each). Finally, the slide was washed with PBS-T three times for 3 min, briefly rinsed with distilled water and dried in a jet of air.

For the staining of the biotinylated peptides, Alexa Fluor 550-labelled streptavidin was used. The following steps were performed: the slide was washed with PBS-T for 15 min and then blocked with Rockland blocking buffer for 30 min. Afterwards, it was washed with PBS-T for 1 min and stained for 1 h in a mixture of 10% Rockland blocking buffer in PBS-T, and Alexa Fluor 550-labelled streptavidin, diluted 1:5,000, was added. Finally, the slide was washed with PBS-T three times for 3 min and washed with distilled water for 2 min.

Image acquisition. Fluorescent image acquisition was performed with two different fluorescent scanners: (1) a Molecular Devices (USA) Genepix 4000B fluorescent scanner at the wavelengths 532 and 635 nm with a laser power of 100%, a resolution of 5 µm and a photo multiplier gain of 470; (2) an Innopsys (France) InnoScan 1100 AL at the wavelengths 532 and 635 nm with a low laser power, the resolution set to 5 µm and a photo multiplier gain of 2. Surface topography was measured with a phase-shift interferometer Contour GT (Bruker, USA).

References

1. Chrisey, D. B. Materials processing - the power of direct writing. *Science* **289,** 879–881 (2000).
2. Arnold, C. B. & Pique, A. Laser direct-write processing. *MRS Bull.* **32,** 9–11 (2007).
3. Hribar, K. C., Soman, P., Warner, J., Chung, P. & Chen, S. C. Light-assisted direct-write of 3D functional biomaterials. *Lab Chip* **14,** 268–275 (2014).
4. Lee, J. Y. & Lee, S. T. Laser-induced thermal imaging of polymer light-emitting materials on poly(3,4-ethylenedioxythiophene): silane hole-transport layer. *Adv. Mater.* **16,** 51–54 (2004).
5. Arnold, C. B., Wartena, R. C., Swider-Lyons, K. E. & Pique, A. Direct-write planar microultracapacitors by laser engineering. *J. Electrochem. Soc.* **150,** A571–A575 (2003).
6. Arnold, C. B., Kim, H. & Pique, A. Laser direct write of planar alkaline microbatteries. *Appl. Phys. A Mater.* **79,** 417–420 (2004).
7. Visser, C. W. *et al.* Toward 3D printing of pure metals by laser-induced forward transfer. *Adv. Mater.* **27,** 4087–4092 (2015).
8. Zywietz, U., Evlyukhin, A. B., Reinhardt, C. & Chichkov, B. N. Laser printing of silicon nanoparticles with resonant optical electric and magnetic responses. *Nat. Commun.* **5,** 3402 (2014).
9. Ellinas, K., Chatzipetrou, M., Zergioti, I., Tserepi, A. & Gogolides, E. Polymeric surfaces sustaining ultrahigh impact pressures of aqueous high-and low-surface-tension mixtures, tested with laser-induced forward transfer of drops. *Adv. Mater.* **27,** 2231–2235 (2015).
10. Chrisey, D. B. *et al.* Laser deposition of polymer and biomaterial films. *Chem. Rev.* **103,** 553–576 (2003).
11. Malda, J. *et al.* 25th Anniversary article: engineering hydrogels for biofabrication. *Adv. Mater.* **25,** 5011–5028 (2013).
12. Serra, P., Colina, M., Fernandez-Pradas, J. M., Sevilla, L. & Morenza, J. L. Preparation of functional DNA microarrays through laser-induced forward transfer. *Appl. Phys. Lett.* **85,** 1639–1641 (2004).
13. Serra, P. *et al.* Laser-induced forward transfer: a direct-writing technique for biosensors preparation. *J. Laser Micro Nanoen.* **1,** 236–242 (2006).
14. Fraunhofer Institute for Laser Technology, LIFTSYS prototype machine for transferring biomaterials–Press Release, Available at http://www.ilt.fraunhofer.de/content/dam/ilt/en/documents/Publication-and-Press/press_release/pr2013/PR_LIFTSYS_User-friendly_Machine_for_EPFL.pdf (2015).
15. Murphy, S. V. & Atala, A. 3D bioprinting of tissues and organs. *Nat. Biotechnol.* **32,** 773–785 (2014).
16. Guillotin, B. *et al.* Laser assisted bioprinting of engineered tissue with high cell density and microscale organization. *Biomaterials* **31,** 7250–7256 (2010).
17. Willis, D. A. & Grosu, V. Microdroplet deposition by laser-induced forward transfer. *Appl. Phys. Lett.* **86,** 244103 (2005).
18. Merrifield, R. B. Solid phase peptide synthesis. 1. Synthesis of a tetrapeptide. *J. Am. Chem. Soc.* **85,** 2149–2154 (1963).
19. Frank, R. Spot-synthesis - an easy technique for the positionally addressable, parallel chemical synthesis on a membrane support. *Tetrahedron* **48,** 9217–9232 (1992).
20. Fodor, S. P. A. *et al.* Light-directed, spatially addressable parallel chemical synthesis. *Science* **251,** 767–773 (1991).
21. Pellois, J. P. *et al.* Individually addressable parallel peptide synthesis on microchips. *Nat. Biotechnol.* **20,** 922–926 (2002).
22. Price, J. V. *et al.* On silico peptide microarrays for high-resolution mapping of antibody epitopes and diverse protein-protein interactions. *Nat. Med.* **18,** 1434–1440 (2012).
23. Buus, S. *et al.* High-resolution mapping of linear antibody epitopes using ultrahigh-density peptide microarrays. *Mol. Cell. Proteomics* **11,** 1790–1800 (2012).
24. Legutki, J. B. *et al.* Scalable high-density peptide arrays for comprehensive health monitoring. *Nat. Commun.* **5,** 4785 (2014).
25. Maurer, K., McShea, A., Strathmann, M. & Dill, K. The removal of the t-BOC group by electrochemically generated acid and use of an addressable electrode array for peptide synthesis. *J. Comb. Chem.* **7,** 637–640 (2005).
26. Beyer, M. *et al.* Combinatorial synthesis of peptide arrays onto a microchip. *Science* **318,** 1888 (2007).
27. Stadler, V. *et al.* Combinatorial synthesis of peptide arrays with a laser printer. *Angew. Chem. Int. Ed.* **47,** 7132–7135 (2008).
28. Loeffler, F. *et al.* Biomolecule arrays using functional combinatorial particle patterning on microchips. *Adv. Funct. Mater.* **22,** 2503–2508 (2012).
29. Maerkle, F. *et al.* High-density peptide arrays with combinatorial laser fusing. *Adv. Mater.* **26,** 3730–3734 (2014).
30. Tookmanian, E. M., Fenlon, E. E. & Brewer, S. H. Synthesis and protein incorporation of azido-modified unnatural amino acids. *RSC Adv.* **5,** 1274–1281 (2015).
31. Zimmerman, E. S. *et al.* Production of site-specific antibody-drug conjugates using optimized non-natural amino acids in a cell-free expression system. *Bioconjugate Chem.* **25,** 351–361 (2014).
32. Rudat, B. *et al.* Novel pyridinium dyes that enable investigations of peptoids at the single-molecule level. *J. Phys. Chem. B* **114,** 13473–13480 (2010).
33. Filippakopoulos, P. & Knapp, S. Targeting bromodomains: epigenetic readers of lysine acetylation. *Nat. Rev. Drug Discov.* **13,** 337–356 (2014).
34. Maerkle, F. *PhD thesis (German), Laserbasierte Verfahren zur herstellung hochdichter Peptidarrays* (Karlsruhe Institute of Technology, 2014).

Acknowledgements

We thank Miriam Kaczynski, Richard Thelen, Alexandra Moritz and Heike Fornasier for technical assistance, and Dr Martina Schnoelzer for mass spectrometry analyses. This work was supported by funds from the ERC (grant number 277863), the HRJRG (grant number 316), the Carl-Zeiss-Stiftung (personal stipend of F.F.L.), the EU FP7 (grant number 256672) and the BMBF (grant number 031A170A and 03EK3030A). We also acknowledge the support by the Deutsche Forschungsgemeinschaft (DFG) and the Open Access Publishing Fund of the Karlsruhe Institute of Technology.

Author contributions

F.F.L. and T.C.F. have contributed equally to this work and the order was chosen at random. F.F.L. conceived the method and conducted initial experiments. T.C.F. developed the automated machine and further advanced the method. F.F.L. and T.C.F. designed the experiments and analysed the results. R.P. supported the development and conducted the peptide synthesis experiments. M. Schlageter conducted and A.K.P. supervised the click chemistry experiment, F.-X.D. and T.S.B. provided the styrylpyridinium dye. M. Sedlmayr devised and constructed the automated wet chemistry machine and I.B. supported its mechatronic development. Under the supervision of S.B. and M.A.R.M., D.S.M., J.G. and B.R. conducted and supported chemical syntheses and analytics. C.v.B.-K. developed the laser-based generation of positioning markers. L.H., L.K.W., A.F. and V.B. supported the development of the LIFT method. F.R.B. supported the antibody stainings and fluorescence analyses. A.N.-M., F.B. and F.F.L. supervised the project. F.F.L., F.B., A.N.-M., T.C.F., D.S.M., B.R. and M. Schlageter wrote the manuscript.

Additional information

Competing financial interests: F.B. is shareholder of PEPperPRINT GmbH. A.N.-M., F.F.L., C.v.B.-K. and F.B. are named on pending patent applications relating to molecule array synthesis (application number PCT/EP2013/001141, PCT/EP2014/001046, and US Patent Application 20160082406). All other authors declare no competing financial interests.

23

Identification of pyrazolopyridazinones as PDEδ inhibitors

Björn Papke[1,*], Sandip Murarka[2,*], Holger A. Vogel[1,*], Pablo Martín-Gago[2,*], Marija Kovacevic[1], Dina C. Truxius[1], Eyad K. Fansa[3], Shehab Ismail[3,4], Gunther Zimmermann[2], Kaatje Heinelt[1], Carsten Schultz-Fademrecht[5], Alaa Al Saabi[5], Matthias Baumann[5], Peter Nussbaumer[5], Alfred Wittinghofer[3], Herbert Waldmann[2,6] & Philippe I.H. Bastiaens[1,6]

The prenyl-binding protein PDEδ is crucial for the plasma membrane localization of prenylated Ras. Recently, we have reported that the small-molecule Deltarasin binds to the prenyl-binding pocket of PDEδ, and impairs Ras enrichment at the plasma membrane, thereby affecting the proliferation of KRas-dependent human pancreatic ductal adenocarcinoma cell lines. Here, using structure-based compound design, we have now identified pyrazolopyridazinones as a novel, unrelated chemotype that binds to the prenyl-binding pocket of PDEδ with high affinity, thereby displacing prenylated Ras proteins in cells. Our results show that the new PDEδ inhibitor, named Deltazinone 1, is highly selective, exhibits less unspecific cytotoxicity than the previously reported Deltarasin and demonstrates a high correlation with the phenotypic effect of PDEδ knockdown in a set of human pancreatic cancer cell lines.

[1] Department of Systemic Cell Biology, Max Planck Institute of Molecular Physiology, D-44227 Dortmund, Germany. [2] Department of Chemical Biology, Max Planck Institute of Molecular Physiology, D-44227 Dortmund, Germany. [3] Structural Biology Group, Max Planck Institute for Molecular Physiology, D-44227 Dortmund, Germany. [4] Beatson Institute for Cancer Research, Bearsden, Glasgow G61 1BD, UK. [5] Lead Discovery Center GmbH, D-44227 Dortmund, Germany. [6] TU Dortmund, Faculty of Chemistry and Chemical Biology, D-44227 Dortmund, Germany. * These authors contributed equally to this work. Correspondence and requests for materials should be addressed to H.W. (email: herbert.waldmann@mpi-dortmund.mpg.de) or to P.I.H.B. (email: philippe.bastiaens@mpi-dortmund.mpg.de).

The products of the proto oncogene Ras in the GTP-bound state[1] recruit effector proteins to the plasma membrane (PM) that activate proliferation- and survival-signalling in cells[2,3]. Oncogenic mutations of specific amino acids (AA), most commonly AA12, 13 and 61, maintain Ras in the constitutively active, GTP-bound state[4], resulting in aberrant signalling. The most commonly mutated Ras isoform is KRas4B (from here on called KRas), which occurs in more than 90% of pancreatic, 45% of colorectal and 30% of lung tumours[4].

Signal transduction of KRas strongly depends on its enrichment at the PM[5]. The localization motifs of KRas, a farnesylated cysteine and a polybasic stretch in the C-terminal hypervariable region[6], are not sufficient to counter entropic equilibration to the extensive endomembrane surfaces[7]. A major role in counteracting this equilibration to endomembranes is played by the guanine nucleotide dissociation inhibitor (GDI)-like solubilization factor PDEδ[8–11]. It sequesters KRas from the cytosol by binding the farnesyl moiety, preventing that KRas binds to endomembranes and thereby enhancing its diffusion throughout the cell. KRas is then released in perinuclear membranes by the local activity of the release factor Arl2 (ref. 12), from where it is trapped by electrostatic interaction on the recycling endosome and shuttled back to the PM via vesicular transport[7]. Any interference with this cycle—such as competitive inhibition of PDEδ—will cause an entropy-driven relocalization of KRas to the extensive endomembrane surfaces[7,8,13]. Besides KRas, the PDEδ/Arl2 system is also crucial for maintaining membrane localization of other proteins of the Ras family, such as palmitoylated H- and N-Ras, as well as localization of the solely farnesylated Rheb on perinuclear membranes[5]. Delocalization of these Ras family molecules by interference with the PDEδ/Arl2 system is also expected to have a negative effect on cell growth and proliferation.

There have been many attempts to either target oncogenic Ras directly, its posttranslational modifications or downstream effectors with limited success[4,14–16]. Reducing PM localization of Ras through PDEδ inhibition raises alternative opportunities to impede oncogenic Ras signalling. The small-molecule Deltarasin affects the PM localization of KRas by competitively binding to the prenyl-binding pocket of PDEδ, relocating KRas to endomembranes. This KRas relocalization resulted in reduced proliferation of oncogenic KRas-dependent human pancreatic ductal adenocarcinoma cell lines (hPDACs)[13].

However, subsequent detailed analysis of the dose–response curves characteristic for Deltarasin revealed that this PDEδ ligand displays a 'switch-like' inhibition of proliferation; that is, the corresponding dose–response curve is very steep in the 3–8 μM range with a Hill coefficient of -5.3 to -10.8 (Supplementary Fig. 1). Such behaviour could arise from non-linear effects of Ras plasma membrane localization on signalling or could be indicative of general cytotoxicity by Deltarasin at high concentration and/or of interaction with additional target proteins in cells causing undesired side effects. Indeed, analysis of Deltarasin binding to additional proteins revealed that the compound also binds to different G-protein-coupled receptors, ion channels and transporters (Supplementary Table 1). Given this undesirable property of Deltarasin at concentrations $>5\,\mu M$, validation of PDEδ as potential target for small-molecule interference with Ras localization and thereby also signalling activity, called for the development of a novel chemotype for inhibition of the Ras–PDEδ interaction, which would not display a comparable general cytotoxicity.

Herein, we describe the discovery of pyrazolopyridazinones as a novel PDEδ inhibitor chemotype that targets the prenyl-binding pocket of PDEδ with low nanomolar affinity. Structure-based ligand development led to the identification of the PDEδ ligand Deltazinone 1, which binds to PDEδ with high selectivity, shows anti-proliferative activity over a wide concentration range and is not generally cytotoxic. We demonstrate that inhibition of the PDEδ/Ras interaction by Deltazinone 1 is closely correlated to the phenotypic consequences of RNAi-mediated PDEδ knockdown in a panel of human pancreatic cancer cell lines.

Results

Synthesis of PDEδ inhibitors based on a different scaffold. To identify alternative chemotypes for the inhibition of the PDEδ/Ras interaction with potency similar to Deltarasin (Fig. 1a, $K_D = 38 \pm 16$ nM), an in-house library of ~150,000 compounds, assembled considering structural diversity of the underlying scaffolds, drug-likeness, coverage of different established drug target classes and inspiration by natural products, was subjected to high-throughput screening by means of the previously developed AlphaScreen Technology[13]. The screen revealed pyrazolopyridazinone **2a** ($K_D = 5 \pm 2$ nM) as ligand of the prenyl-binding pocket of PDEδ (Fig. 1b).

Structure-based development of the pyrazolopyridazinone hit class was enabled by a crystal structure of inhibitor **2a** in complex with PDEδ at a resolution of 2.6 Å (Fig. 1d) which revealed that, like inhibitors of the Deltarasin-type, heterocycle **2a** employs H-bonds to Tyr 149 and Arg 61 for binding to PDEδ[13]. Initial screening and molecular modelling investigations (Schrödinger, Maestro suite) revealed that a C-3 linker between the two hydrogen-bonding groups is important and that shortening of the linker would abrogate H-bond formation between the amide carbonyl and Tyr 149 (Fig. 1f). Therefore, a C-3 linker was kept for further inhibitor development.

Since anilides display low metabolic stability, we searched for an alternative to the anilide embodied in hit compound **2a**. Initial investigations and modelling revealed benzyl- and homobenzylamides as potential inhibitors, since they could adopt a conformation in which an H-bond with Tyr 149 is formed and the empty space in the PDEδ pocket is filled. Therefore, a series of compounds with these constraints was synthesized and investigated (Table 1). As shown in Table 1, benzyl amide **2c** also binds to PDEδ with nanomolar affinity. Reduction of the ring size (entry 2) or incorporation of a substituent on the phenyl ring was not favourable (entries 4 and 5). In addition, modelling indicated that a small substituent might be attached to the *N*-phenyl moiety, pointing deep into the farnesyl-binding site. Indeed, incorporation of a *p*-methyl substituent into the pyrazole *N*-phenyl ring resulted in low nanomolar affinity (entries 2, 3, 6 versus 7, 8, 9). Besides benzyl amides, homobenzyl amides are also very potent inhibitors displaying low nanomolar affinities (entries 6, 9–11).

Compound **2k** (Fig. 1c) was selected because it embodies one metabolically unstable benzylic position less than compound **2j**. As compared with compound **2i**, we assumed that the presence of an additional methyl group at the benzylic position in **2k** would further stabilize the compound with respect to metabolic degradation. Based on the fact that pyrazolopyridazinone **2a** binds to PDEδ like the Deltarasin-type inhibitors (Fig. 1d), molecular modelling experiments indicated that **2k** will form the same hydrogen bonds as Deltarasin and may additionally form a hydrogen bond with glutamine 78 (Fig. 1e). Compound **2k** was, therefore, subjected to a panel of biochemical and pharmacological screens to determine its selectivity profile and possible interactions with other pharmacologically relevant proteins. This compound proved to be more selective as compared with Deltarasin in the selected panel of targets (Supplementary Fig. 2 and Supplementary Table 1) and was chosen for further investigations in cells. We named this compound Deltazinone 1.

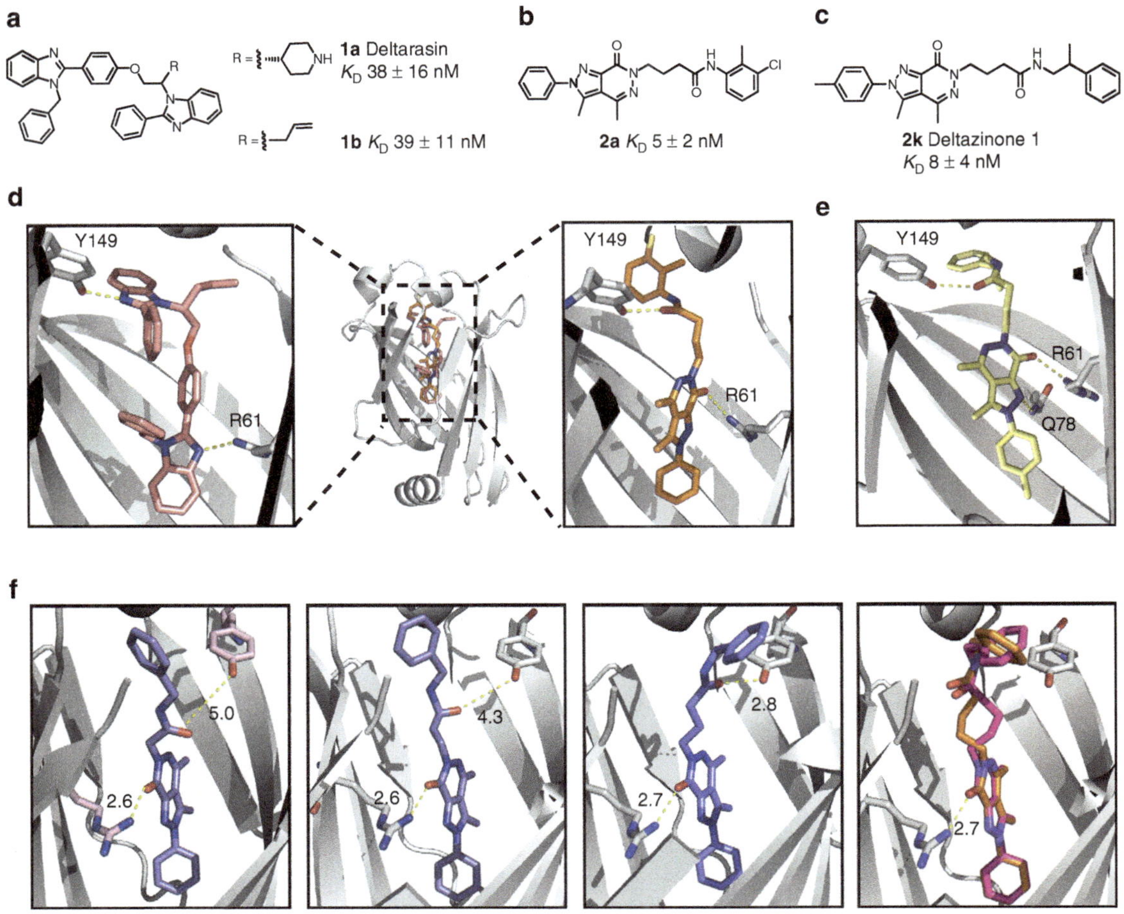

Figure 1 | Identification of a novel KRas-PDEδ inhibitor chemotype. Structure and binding affinities (K_D, determined by competitive fluorescence polarization assay of three independent experiments, see Supplementary Information) of (**a**) Deltarasin **1a** and allyl analogue **1b** (**b**) pyrazolopyridazinone derivative **2a** and (**c**) Deltazinone 1 **2k**. (**d**) Superimposition (middle) of the crystal structures of compounds **1b** (PDB code: 4JVB) and **2a** (PDB code: 5E80) in the PDEδ-binding pocket. The magnification on the left shows the structure of **1b** (1.75 Å) and of **2a** (2.60 Å) on the right side of the superimposition. Tyr149 (Y149) and Arg61 (R61) are shown as sticks to highlight the hydrogen bond interactions. (**e**) Predicted binding mode of Deltazinone 1, **2k** (best docking pose; Schrödinger, Maestro suite). Besides Y149 and R61, Gln78 (Q78) is also shown as stick to highlight the probable third H-bond interaction. (**f**) Best docking poses for a series of *N*-benzyl pyrazolopyridazinone derivatives containing linker chains with one, two, three (compound **2a**) and four methylene units. Distances for hypothetical hydrogen bonding with Tyr149 and Arg61 are shown.

Deltazinone 1 inhibits the PDEδ–Ras interaction. To compare the inhibition of PDEδ by Deltazinone 1 to Deltarasin, the interaction between farnesylated Ras proteins and PDEδ was measured in live cells by fluorescence lifetime imaging microscopy of Förster resonance energy transfer (FLIM-FRET)[13]. For this, Madin–Darby canine kidney (MDCK) cells were transiently transfected with Rheb N-terminally fused to a yellow fluorescent protein (mCitrine-Rheb), and PDEδ fused to a red fluorescent protein variant (mCherry-PDEδ). Rheb is a farnesylated Ras protein that lacks the polybasic stretch in the hypervariable region, resulting in a low membrane affinity and an enhanced fraction that can interact with PDEδ in the cytoplasm. This facilitates the detection of small-molecule effects on the interaction between farnesylated proteins and PDEδ by FLIM-FRET. The fluorescence patterns of mCitrine-Rheb and mCherry-PDEδ in the absence of the inhibitor showed a homogeneous distribution, demonstrating that mCitrine-Rheb is solubilized by mCherry-PDEδ (Fig. 2a,b). This was reflected in the high molar fraction (α) of interacting Rheb-PDEδ as measured by FLIM-FRET. Increasing concentrations of Deltazinone 1 or Deltarasin resulted in a reduced interaction of mCitrine-Rheb with mCherry-PDEδ that was paralleled by a relocalization of mCitrine-Rheb to endomembranes. Dose-dependent measurements yielded an 'in-cell' K_D of 58 ± 17 nM for Deltazinone 1 and 66 ± 5 nM for Deltarasin.

To measure whether both inhibitors also affect the interaction of KRas with PDEδ, MDCK cells were transiently transfected with mCitrine-KRas and mCherry-PDEδ. The cytosolic fraction of KRas is much lower than the one of Rheb due to its higher PM affinity caused by its polybasic stretch[5]. This leads to a smaller fraction of mCitrine-KRas interacting with mCherry-PDEδ at steady state in the cytoplasm, making the interaction difficult to detect with FLIM-FRET. To increase the fraction of KRas that interacts with PDEδ, the Protein Kinase C (PKC) activator Bryostatin, which induces serine (S181) phosphorylation on KRas[17], was applied, resulting in a reduced electrostatic interaction with the PM and increased cytosolic pool of mCitrine-KRas. Subsequent addition of 5 μM Deltazinone 1 (Fig. 2c) or Deltarasin (Fig. 2d) abolished the interaction between mCitrine-KRas and mCherry-PDEδ and resulted in relocalization of KRas to endomembranes.

Deltazinone 1 impairs growth of KRas-dependent hPDAC cells. To compare the effects of PDEδ knockdown with small-molecule PDEδ inhibition on proliferation, impedance-based real-time cell

Table 1 | K_D values for pyrazolopyridazinones with different amide structures and *N*-phenyl substituents as determined by means of fluorescence polarization assay.

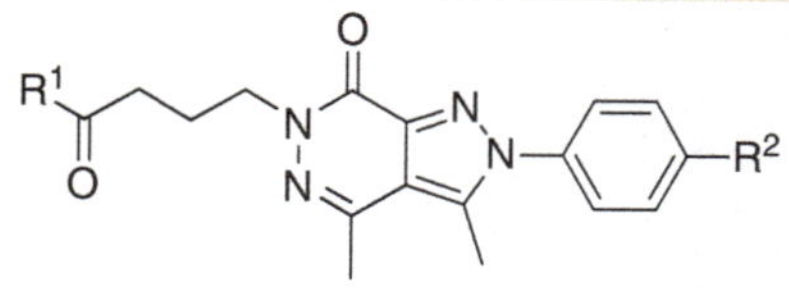

2

Entry	2	R^1	R^2	K_D [nM]
1	**2a**	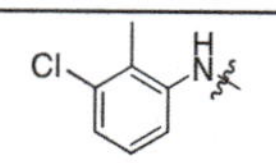	H	5 ± 2
2	**2b**	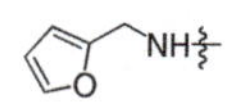	H	136 ± 55
3	**2c**	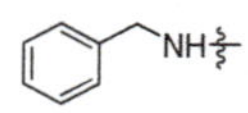	H	42 ± 15
4	**2d**	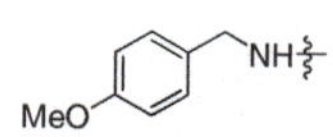	H	81 ± 21
5	**2e**	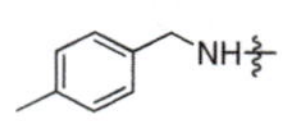	H	96 ± 14
6	**2f**	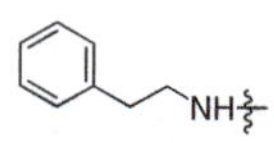	H	29 ± 7
7	**2g**	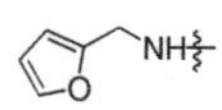	CH_3	50 ± 16
8	**2h**	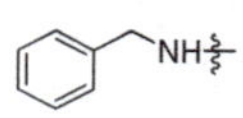	CH_3	17 ± 9
9	**2i**	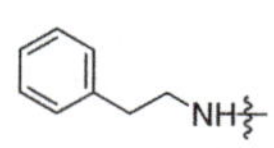	CH_3	6 ± 2
10	**2j**	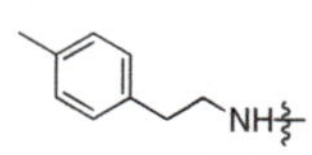	CH_3	5 ± 3
11	**2k**	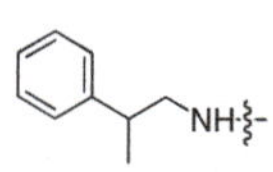	CH_3	8 ± 4*

*(*S*)-**2k** K_D = 5 ± 3 nM; (*R*)-**2k** K_D = 12 ± 4 nM.

analyser (RTCA) measurements were performed in a panel of pancreatic cancer cell lines (Table 2, Fig. 3). For this, all pancreatic cancer cell lines were transduced with the previously reported doxycycline-induced short hairpin RNA (shRNA) construct against PDEδ[13].

The doxycycline-induced PDEδ knockdown in the oncogenic KRas-dependent human carcinoma cell lines Panc-Tu-I, Capan-1 and MIA PaCa-2 (refs 18,19) resulted in a strongly reduced proliferation after 70–110 h (Fig. 3a), whereas treatment of the parental cell lines with doxycycline did not alter their proliferation (Supplementary Fig. 3).

This correlated with the growth inhibitory effects of both small-molecule inhibitors in these cell lines (Fig. 3b,c). Deltazinone 1 inhibited cell growth in a dose-dependent manner already observable at sub-μM concentrations. At doses higher than 3 μM decreasing cell indices indicated cell death ~30 h after Deltazinone 1 administration in Panc-Tu-I cells and after ~40 h in MIA PaCa-2 (Fig. 3b). In the Capan-1 cell line, Deltazinone 1 doses up to 24 μM led to strong growth inhibition but not cell death. This differential behaviour of the three KRas-dependent cell lines to PDEδ inhibition was reflected in the growth of these cells after doxycycline-induced PDEδ knockdown (Fig. 3a). In contrast to treatment with Deltazinone 1, a steep dose-dependent growth inhibitory effect and cell death was observed for Deltarasin in a concentration range from 3 to 5 μM in Panc-Tu-I, Capan-1 and MIA PaCa-2 cell lines (Fig. 3c,d)[13]. At 5 μM Deltarasin, detectable cell death in Panc-Tu-I occurred ~9 h post treatment. The onset of Panc-Tu-I cell death after Deltazinone 1 treatment happened much later (~13 h for 24 μM Deltazinone 1), while the onset was dependent on the dose of the drug (Fig. 3e). This indicates that Deltazinone 1 is less potent than Deltarasin in inhibiting the PDEδ/Arl2 system that restores KRas plasma membrane localization (see Discussion).

PANC-1 cells that harbour oncogenic KRas but are independent of this oncogene[18], as well as the KRas wild-type-expressing BxPC-3 cancer cell line exhibited minor growth inhibitory effects on doxycycline-induced shRNA knockdown of PDEδ. Similarly, Deltazinone 1 doses up to 24 μM had little or no growth

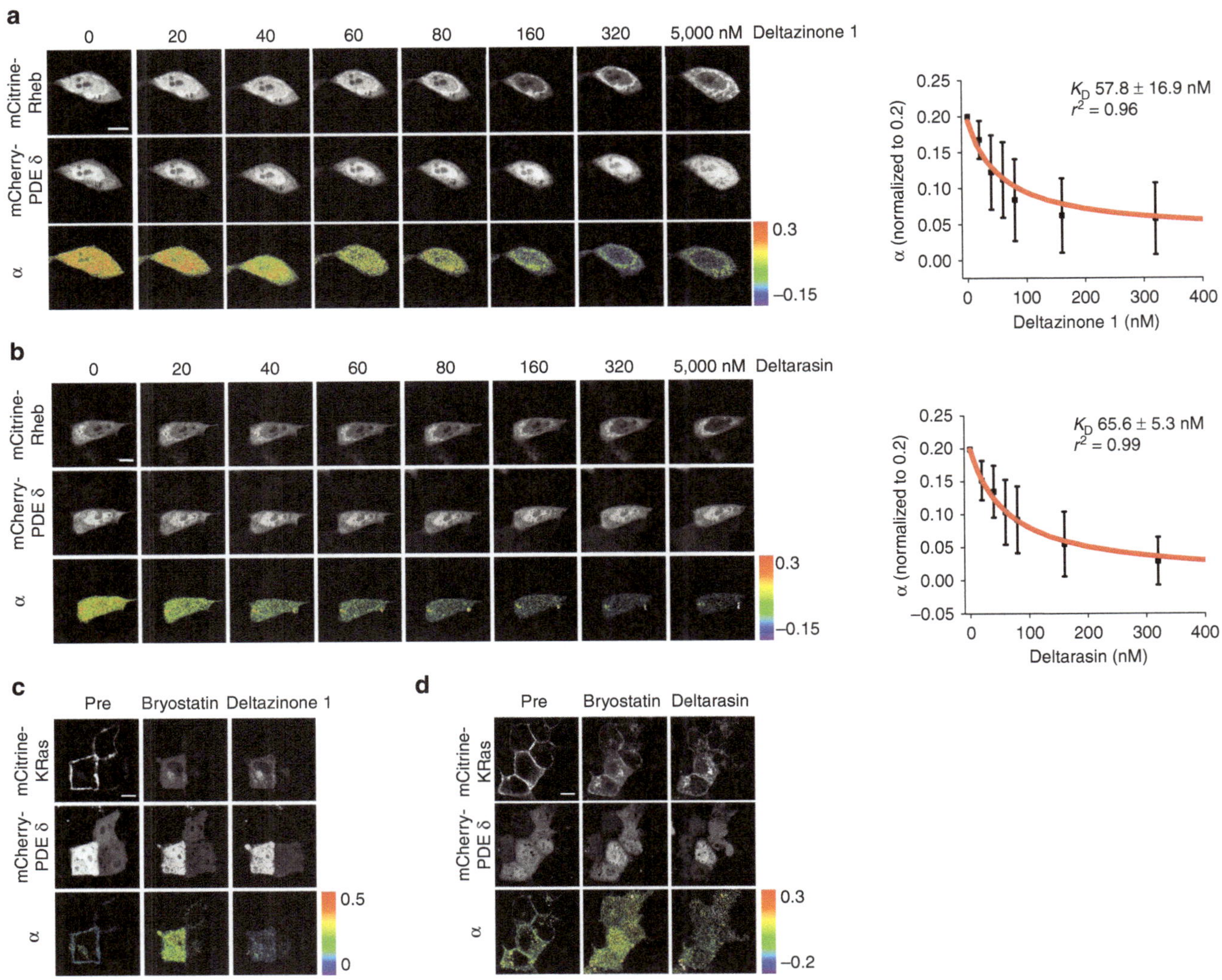

Figure 2 | In-cell measurement of the effect of Deltarasin or Deltazinone 1 on the interaction between Ras proteins and PDEδ. Dose-response measured by the interacting fraction (α) of mCitrine-Rheb with mCherry-PDEδ for Deltazinone 1 (**a**) and Deltarasin (**b**) as determined by FLIM-FRET. The inhibitor concentration is indicated at the top of each image in nM. Right panels, fit of averaged dose-response ± s.e.m. of nine independent experiments to a binding model (see Methods) yielded an in-cell K_D of 58 ± 17 nM for Deltazinone 1 (**a**) and 66 ± 5 nM for Deltarasin (**b**). Interacting fraction (α) of mCitrine-KRas with mCherry-PDEδ on 5 min treatment with 1 μM Bryostatin-1 and subsequent addition of 5 μM Deltazinone 1 (**c**) or 5 μM Deltarasin after 5 min (**d**). First row: fluorescence intensity distribution of mCitrine-Rheb (**a**,**b**) or mCitrine-KRas (**c**,**d**); second row: fluorescence intensity distribution of mCherry-PDEδ; third row: molar fraction α (false colour coded) of interacting mCitrine-Rheb with mCherry-PDEδ (**a**,**b**) or mCitrine-KRas with mCherry-PDEδ (**c**,**d**). Scale bars, 10 μm.

Table 2 | Overview of pancreatic cancer cell lines used in this study including KRas mutation and oncogenic KRas-dependence.

Cancer cell line	KRas mutation	Oncogenic KRas-dependence
PANC-1	G12D	No[18,19]
BxPC3	wt	No[19]
Capan-1	G12V	Yes[18,19]
Panc-Tu-I/Pa-Tu-8902	G12V	Yes[18]
MIA PaCa-2	G12C	Yes[19]

inhibitory effect on these cell lines (Fig. 3a,b). In contrast, Deltarasin concentrations above 9 μM resulted in an abrupt change in the RTCA profiles indicating rapid cell death of the KRas-independent cancer cell lines (Fig. 3c). These results point at unspecific cytotoxicity of Deltarasin at concentrations above 9 μM, which was not observed for Deltazinone 1 for concentrations up to 24 μM. The RTCA measurements were corroborated by long-term time-lapse microscopy that showed cell death for oncogenic KRas-dependent Panc-Tu-I cells treated with 5 μM Deltarasin or 10 μM Deltazinone 1, but not for oncogenic KRas-independent PANC-1 cells (Supplementary Movies 1–6).

PDEδ inhibition decreases Ras-mediated signalling. MAPK and mTOR pathways, which regulate proliferation and growth, are regulated by activity of proteins from the Ras family (Ras and Rheb, respectively)[15,20]. To examine whether PDEδ inhibition by Deltarasin and Deltazinone 1 downregulates the signal transduction via these pathways, we examined the phosphorylation of extracellular signal regulated kinase (Erk) and S6 Ribosomal Protein (S6P) in the oncogenic KRas-dependent Panc-Tu-1 and KRas-independent PANC-1 cells. Following treatment with Deltarasin or Deltazinone 1 for 1 h, phosphorylation levels of Erk and S6P were determined before

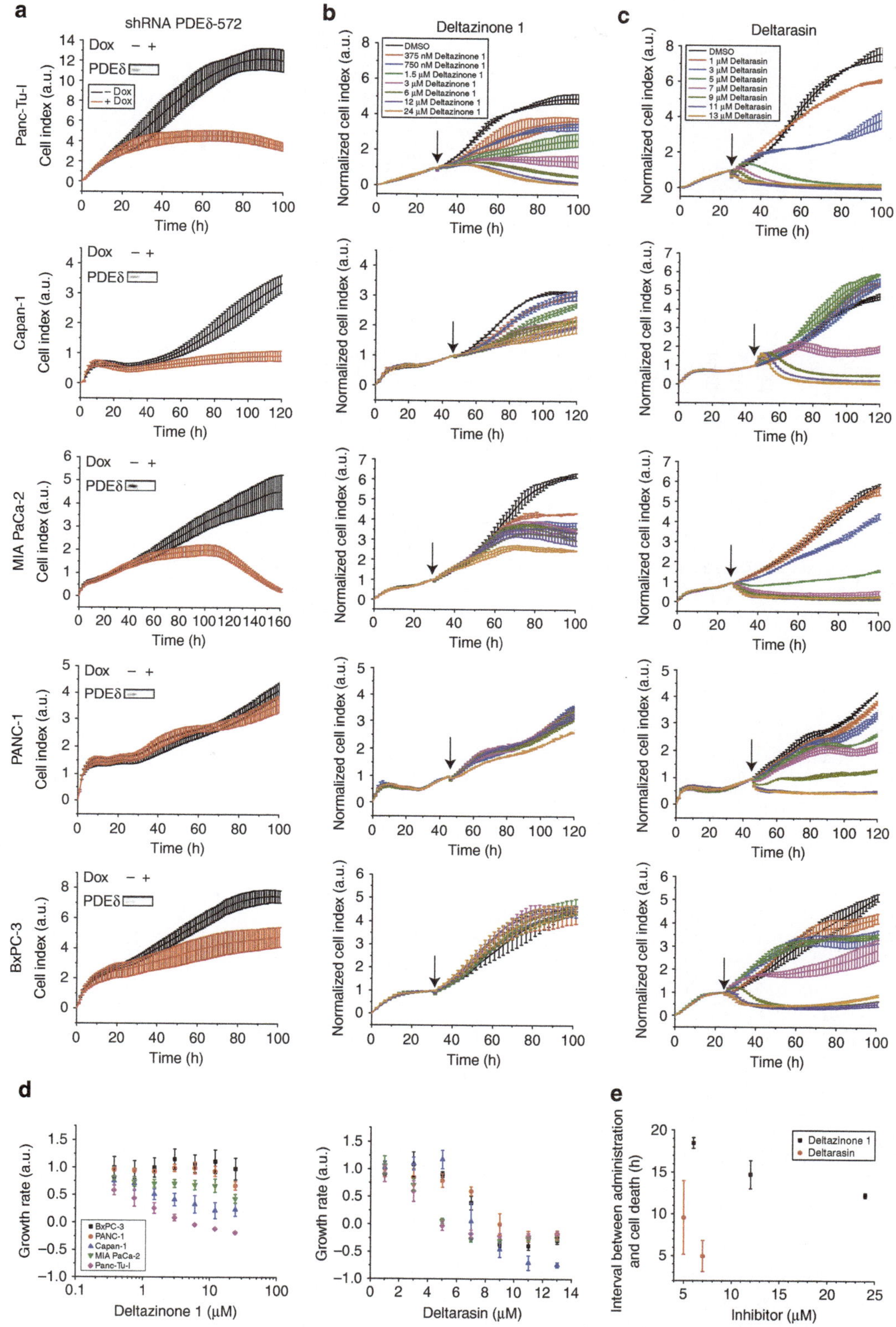

Figure 3 | Comparison of the RTCA growth profiles of pancreatic cancer cell lines after PDEδ knockdown to PDEδ inhibition by Deltazinone 1 or Deltarasin. The first column (**a**) shows the effects of doxycycline-induced PDEδ knockdown on cell proliferation. Cell indices were measured in the presence (red) or absence (black) of doxycycline. Doxycycline was added at the beginning of the measurement. The inset in each profile displays the respective PDEδ protein levels determined by western blot in absence and presence of doxycycline after 72 h. The second (**b**) and third column (**c**) show dose-dependent effects of Deltazinone 1 and Deltarasin on cell proliferation, respectively. The small-molecule PDEδ inhibitors were added at the indicated time points (arrow) and concentrations. The cell indices were normalized to the time of administration. Cell indices ± s.d. were measured in duplicates (Capan-1, MIA PaCa-2 and BxPC-3) or triplicates (Panc-Tu-I and PANC-1). (**d**) Deltazinone 1 or Deltarasin dose versus growth-rate ± s.d. as determined by the integrated area below the curve in the first 60 h after drug administration, normalized to the DMSO control. (**e**) Time interval between PDEδ inhibitor administration and observed cell death in Panc-Tu-I cells. The mean intervals ± s.d. are depicted for each drug.

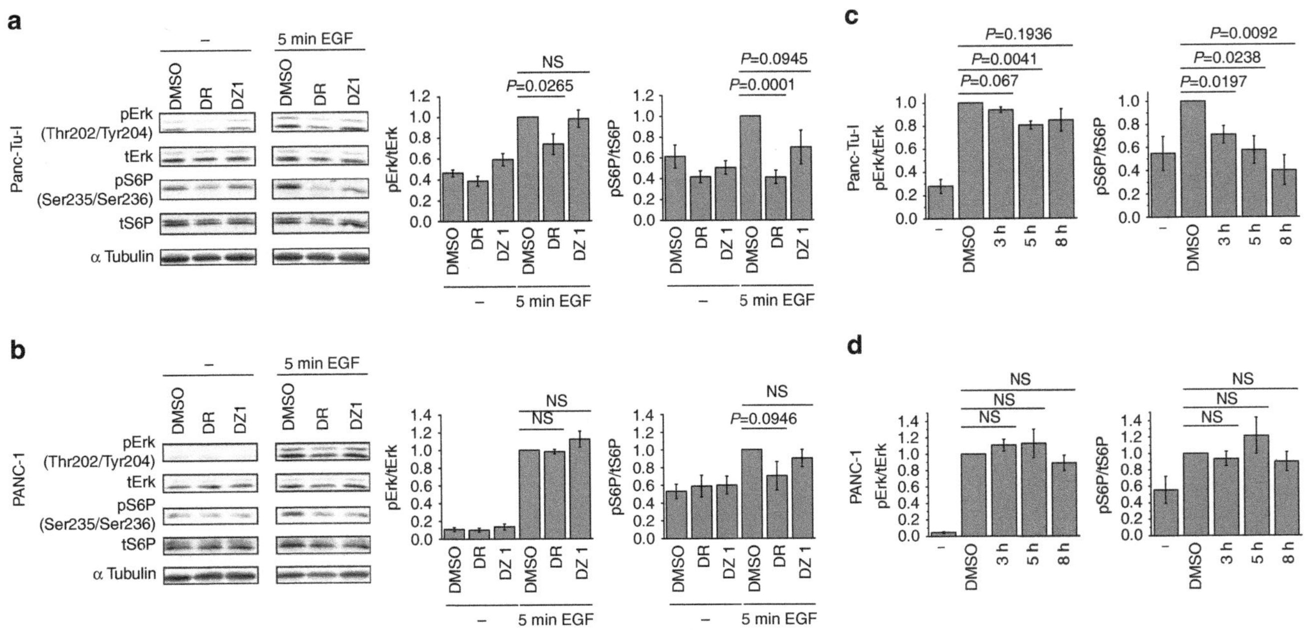

Figure 4 | Effects of Deltarasin and Deltazinone 1 on Erk and S6P phosphorylation in human pancreatic ductal adenocarcinoma cell lines. Western blot analysis of Erk and S6P phosphorylation in KRas-dependent Panc-Tu-I (**a**,**c**) and KRas-independent PANC-1 cells (**b**,**d**). (**a**,**b**) Serum-starved cells were treated with vehicle control (DMSO), 5 μM Deltarasin (DR) or 20 μM Deltazinone 1 (DZ1) for 1 h. Phosphorylation levels were compared between unstimulated (−) and stimulated cells with 200 ng ml^{-1} EGF for 5 min (5 min EGF). Left panels: representative western blots of six independent experiments for each cell line. From top to bottom row: phosphorylated Erk on Thr202 and Tyr204 (pErk), total level of Erk (tErk), phosphorylated S6P on Ser235 and Ser236 (pS6P), total level of S6P (tS6P) and loading control (α-Tubulin). Right bar graphs: quantification of pErk/tErk ± s.e.m. (left) and pS6P/tS6P ± s.e.m. (right) normalized to EGF-stimulated DMSO control (DMSO). (**c**,**d**) Long-term treatment (3, 5 and 8 h) of cells with 20 μM Deltazinone 1, after which cells were stimulated with 200 ng ml^{-1} EGF for 5 min. Quantification of pErk/tErk ± s.e.m. (left) and pS6P/tS6P ± s.e.m. (right) normalized to EGF-stimulated DMSO control (DMSO). (−) Represent phosphorylation levels in unstimulated, vehicle control-treated cells. Values were obtained by averaging three (Panc-Tu-I) and four (PANC-1) independent experiments. *P* values were determined by applying Student's unpaired *t*-test. *P* values >0.4 are labelled as not significant (NS).

and after 5 min EGF stimulation by quantitative western blot analysis (Fig. 4). As previously reported[13], 5 μM Deltarasin reduced EGF-induced Erk phosphorylation in the KRas-dependent Panc-Tu-I cells, but not in PANC-1 cells (Fig. 4a,b), while the EGF-induced S6P phosphorylation was reduced in both lines (Fig. 4a,b). In contrast, 20 μM Deltazinone 1 administered for 1 h did not significantly affect Erk response to EGF in both cell lines, but reduced S6P phosphorylation in the KRas-dependent Panc-Tu-I cells (Fig. 4c). However, on longer incubation times (>3 h) with Deltazinone 1, EGF-induced Erk phosphorylation was reduced in Panc-Tu-I and S6P phosphorylation further decreased in a gradual manner, while PANC-1 cells, remained unaffected (Fig. 4c,d).

Discussion

Deltarasin was the first small molecule effectively targeting the spatial organization of Ras by competing for the farnesyl-binding pocket of PDEδ and thereby resulting in a reduced proliferation of oncogenic KRas-dependent pancreatic ductal adenocarcinoma cells[13]. Herein, we compared the newly developed PDEδ inhibitor Deltazinone 1 based on the pyrazolopyridazinone scaffold to the benzimidazole-based Deltarasin and the genetic knockdown of PDEδ. Based on a crystal structure of pyrazolopyridazinone **2a** and molecular modelling, Deltazinone 1 has a similar PDEδ-binding mode as previously described for Deltarasin[13].

The oncogenic KRas-dependent cell lines, Panc-Tu-I, Capan-1 and MIA PaCa-2 (refs 18,19) exhibited reduced proliferation and even cell death after PDEδ knockdown and inhibitor treatments. We observed a high correlation between the effects of shRNA-mediated down-modulation and small-molecule inhibition of PDEδ on the proliferation pattern of the pancreatic cancer cell lines. However, Deltarasin exhibited cytotoxic effects at concentrations above 9 μM in all tested cell lines, causing a 'switch-like' proliferation to cell death response. This is likely due to unspecific binding of Deltarasin to other proteins (Supplementary Fig. 2 and Supplementary Table 1), since this general cytotoxicity was not observed after Deltazinone 1 administration or PDEδ knockdown by shRNA. In contrast, Deltazinone 1 showed a more graded, dose-dependent inhibitory response on proliferation in oncogenic KRas-dependent cell lines and no general cytotoxicity could be observed at the concentrations tested.

Both Deltarasin and Deltazinone 1 exhibited an 'in-cell' measured K_D of ~60 nM for PDEδ, whereas μM concentrations of the inhibitors were required to affect the proliferation of KRas-dependent pancreatic cancer cells. Part of the explanation is the strong electrostatic interaction of KRas with negatively charged membranes that results in a very slow entropic leakage from the PM[7]. This means that PDEδ needs to be completely inhibited to relocalize KRas to endomembranes, as even a small pool of uninhibited, free PDEδ is sufficient to reinstate PM localization of KRas[5].

The perinuclear localization of the Ras family protein Rheb depends solely on farnesylation as it lacks a second feature that will stabilize the interaction with membranes. For Rheb, the activity of the PDEδ-Arl2 'pump' is essential for maintenance of its localization, as it counters the tendency of this protein to distribute on all endomembranes within the cell. Due to the higher off-rate of Rheb from membranes as compared with

KRas, even incomplete PDEδ inhibition will almost completely relocalize it to the extensive endomembrane surface across the whole cell[5], leading to attenuation of mTOR activity, which is involved in regulation of cellular growth[21]. It is thus likely that at lower doses of the inhibitors, Rheb activity is impaired, whereas at saturating doses the activity of oncogenic KRas is affected.

This was reflected in Erk and S6P phosphorylation levels, as readouts of the Ras-regulated MAPK pathway and the Rheb-regulated mTOR pathway, respectively. The effects of both inhibitors were prominent in suppressing S6P phosphorylation in KRas-dependent cell lines, even after short incubation times (1 h). However, significant effects on Erk phosphorylation could only be observed after longer incubation time of Deltazinone 1. Deltarasin was more efficient in downregulating Erk phosphorylation in the KRas-dependent cell line, Panc-Tu-I, indicating a tighter interaction with PDEδ as compared with Deltazinone 1. We, however, measured similar affinities of both compounds for PDEδ ($\sim$60 nM), which points to a more efficient displacement of Deltazinone 1 from PDEδ by Arl2 activity. This could lead to more available free PDEδ at steady state that can reinstate KRas PM localization. This could explain why higher doses of Deltazinone 1 than Deltarasin are needed to affect oncogenic KRas signalling to counter the allosteric displacement activity of Arl2.

The herein presented new PDEδ inhibitor chemotype caused specific cell death in KRas-dependent pancreatic cancer cell lines without exhibiting general cytotoxic effects for concentrations up to 24 μM. Deltazinone 1 represents a more specific second chemotype of PDEδ inhibitors and thereby resembles a non-cytotoxic compound for future medicinal chemistry studies. However, we note that pharmacological characterization of Deltazinone 1 revealed that this compound is rapidly metabolized in mice (see the Supplementary Fig. 4 and Supplementary Tables 2 and 3) making it non-suitable for *in vivo* experiments. Our results show that a combination of biochemical screening and structure-guided design enables the development of a novel inhibitor class for PDEδ with comparable potency but higher selectivity than the previously developed Deltarasin.

Methods

AlphaScreen. Screening based on Alpha technology was conducted in white, non-binding 1,536-well plates (Corning) in a final volume of 6 μl (ref. 13). For the screen, a mixture of His_6-PDEδ, and biotinylated KRas-peptide (final concentrations 100 nM and 250 nM in HEPES 20 mM, 100 mM NaCl, 0.005% Chaps, pH 7.5) were added to the 1,536-well plates using a Multidrop Combi (Thermo Fisher Scientific) device. Compound solutions were directly added from 10 mM dimethylsulphoxide (DMSO) stock solutions to a final concentration of 10 μM using the acoustic dispenser Echo520 (Labcyte Inc.) and the resulting mixture was incubated for 30 min (for dose–response curves, compounds were tested at concentrations between 5 nM and 10 μM). Premixed Nickel Chelate Acceptor Beads and Streptavidin Donor Beads were added to a final concentration of 10 μg ml^{-1} using a Multidrop Combi (Thermo Fisher Scientific) device. The resulting mixture was incubated at 4 °C overnight. Plates were read on a Paradigm reader (Molecular Devices, Alphascreen 1536 HTS detection cartridge, temperature 29–33 °C).

Displacement titrations for the determination of K_D values. Binding to PDEδ was validated and quantified by means of a displacement assay employing a fluorescent-tagged analogue of the HMG-CoA reductase inhibitor Atorvastatin (Lipitor), which has previously been shown to also bind to PDEδ[22]. The K_d values were determined by the fluorescence polarization competition-binding assay, which we developed before.

Crystallography. Inhibitor **2a** was co-crystallized with PDEδ by mixing a solution of 500 μM small molecule with an equimolar solution of PDEδ resulting in 1% DMSO as a final concentration. Crystals were obtained from a Qiagen PEGs suite (0.2 M calcium acetate, 20% (w/v) PEG 3350). For flash freezing, the crystals in liquid nitrogen, we used cryoprotectant solution containing the mother liquor components in addition to glycerol. X-rays data were collected at the X10SA beamline of the Suisse Light Source, Villigen. XDS programme was used for processing the data. For solving the structures, molecular replacement was applied using the programme MolRep from the CCP4 software and PDEδ without bound farnesyl group, from the complex PDEδ-farnesylated Rheb complex (PDB 3T5G) as a search model. Further refinement of the model was performed using a combination of manual refinement (COOT programme) and the maximum likelihood restrained refinement (REFMAC5 programme). Final validation of the model using Ramachandran plot statistics showed none of the residues to be outliers (Supplementary Table 4).

Molecular modelling. Molecular modelling experiments were carried out with the Maestro 9.1 suite (Schrödinger). Both ligand and receptor flexibility were taken into account by using receptor docking (Glide) in combination with the protein structure prediction embedded in the programme Prime. Protein Preparation Wizard (Schrödinger Maestro suite) was used to prepare the PDEδ co-crystal for the calculations. Previously obtained X-ray structures were used to define the active site. Water molecules were removed from the protein crystal structure, hydrogen atoms were added and resulting structure was refined by OPLS2005 force field. The minimization was ended when the RMSD reached 0.18 Å. Receptor Grid Preparation (Glide) was used to generate the protein grid that was subsequently utilized in docking experiments. The van der Waals radius scaling factor was set to 0.5 with a partial charge cutoff of 0.25. Arg61 was selected as a possible site for hydrogen bonding. In addition, docking runs were carried out using an additional hydrogen bond constraint with Tyr149. Ligand preparation for docking was carried out with LigPrep in Maestro 9.1 and the OPLS_2005 force field. Epika was used to generate possible states at target pH 7.0 ± 4.0. Ligand-docking options in Glide were used for the first round of docking experiments. Under Setting, XP (extra precision), Dock flexibly, sample nitrogen inversions, sample ring conformation and Epik state penalties to docking score were selected, amide bonds were penalized for nonplanar conformation. Under the Ligands section, the Van der Waals radius scaling factor was set to 0.5 and the docking was set to match at least one out of two constraints. Several high-score binding poses were generated and co-crystallized compounds were re-docked and analysed in terms of overlay with the X-ray structure.

Cell lines and culture conditions. The identity of all cancer cell lines (a kind gift from Stephan Hahn, Bochum) were tested by STR-DNA typing. Regular tests for the presence of mycoplasma were negative. Cells were cultured in DMEM supplemented with 10% fetal calf serum, 1% non-essential amino acids (PAN-Biotech), 1% L-glutamine (PAN-Biotech). Cells were maintained in a humidified incubator with 5% CO_2 at 37 °C. For viral transduction, the packaging cell line HEK293T was seeded in 6 cm dishes and transfected next day with the vector pLKO-PDE6D-572 (ref. 13) using calcium phosphate precipitation. The following day, the supernatant was collected and used for viral transduction of the human pancreatic ductal adenocarcinoma cell lines. Selection with puromycine was done for all cell lines at a concentration between 0.4 and 1.0 μg ml^{-1} starting 1 day after transduction. Induction of the shRNA *PDE6D*-572 (ref. 13) expression was done with 200 ng ml^{-1} doxycycline. For live-cell microscopy, cells were cultured in four-well Lab-Tek chambers slides (Thermo Scientific/NUNC). Transfections were carried out with Effectene (Qiagen) or Lipofectamine 2000 (Invitrogen).

Fluorescence lifetime imaging microscopy. Fluorescence lifetime images were acquired using a confocal laser-scanning microscope (FV1000, Olympus) equipped with a time-correlated single-photon counting module (LSM Upgrade Kit, Picoquant). For detection of the donor (mCitrine), the sample was excited using a 470-nm diode laser (LDH 470, Picoquant) at a 36-MHz repetition frequency. Fluorescence signal was collected through an oil immersion objective ($\times$60/1.35 UPlanSApo, Olympus) and spectrally filtered using a narrow-band emission filter (HQ 525/15, Chroma). Photons were detected using a single-photon counting avalanche photodiode (PDM Series, MPD) and timed using a single-photon counting module (PicoHarp 300, Picoquant).

FLIM data analysis. Intensity thresholds were applied to segment the cells from the background fluorescence. Data were analysed via global analysis[23] to obtain images of the molar fraction (α) of interacting mCitrine-Rheb with mCherry-PDEδ. Effects on α of Deltazinone 1 and Deltarasin was monitored by FLIM acquisition through a sequential addition of the compounds, followed by incubation for 10 min. Dose–response relationships were determined by plotting obtained α per addition, using equation:

$$\alpha = A_0 - \frac{A \times [X]}{[X] + K_D}$$

where A_0 is α in the absence of the inhibitor, A an asymptotic offset and [X] concentration of the inhibitor in the medium. The α values before the inhibitor treatment were normalized to the value of 0.2.

Real time cell analysis. RTCA was performed using 16-well E-plates on the Dual Plate xCELLigence instrument (Roche Applied Science, Indianapolis, IN). This system measures a dimensionless parameter called cell index, which evaluates the ionic environment at an electrode/solution interface and integrates information on cell number. Depending on the cell line 5×10^3–1×10^4 cells were plated in each well of the 16-well plates in 200 µl of cell culture medium. After seeding, cells were allowed to settle for 30 min at room temperature before being inserted into the xCELLigence instrument in a humidified incubator at 37 °C with 10% CO_2. For RTCA experiments assaying cells stably expressing the inducible shRNA transgene, cells were seeded onto the E-plates in medium supplemented with 200 ng ml^{-1} doxycycline. For RTCA experiments, the cells were treated with Deltarasin or Deltazinone 1 after reaching a steady growth (24 h/48 h). The small-molecule inhibitors were added at indicated time points and in the indicated concentrations. The amount of DMSO was kept constant between the individual conditions and did not exceed 0.24 %. Continuous impedance measurements were then monitored every 15 min for up to 200 h. All assays were performed in duplicates. The cell index was normalized to 1 at the time point of drug administration. To display the inhibitor-dose growth-response relationship of Deltarasin and Deltazinone 1, the area below the curve was integrated for 60 h after administration of the drugs and normalized to the DMSO control.

Western blots. Cells were lysed with RIPA-buffer (20 mM Tris pH 7.5, 150 mM NaCl, 0.5 mM Na_2 EDTA, 1 % sodium deoxycholate, 0.1% sodium dodecyl sulfate, 1% IGEPAL supplemented with Complete Mini EDTA-free protease inhibitor (Roche Applied Science) and 1% phosphatase inhibitor cocktail 2 and 3 (P5726 and P0044, Sigma Aldrich), scraped off after 5 min on ice and centrifuged at 14,000*g* for 20 min at 4 °C. SDS-polyacrylamide gel electrophoresis was carried out with 10–25 µg of whole-cell lysate from each sample. The gels were blotted onto polyvinylidene difluoride membrane (Millipore) and blocked for 1 h at room temperature. Antibodies used for western blotting were: anti-PDE6D (Santa Cruz: sc-50260, 1:200), anti-Cyclophilin-B (Abcam: Ab160459, 1:3,000), p44/42 MAPK (Erk1/2) (3A7) (Cell Signaling: 9107, 1:1,000), Phospho-p44/42 MAPK (Erk1/2) (Thr202/Tyr204) (D13.14.4E) XP (Cell Signaling: 4370, 1:1,000), S6 Ribosomal Protein (54D2) (Cell Signaling: 2317, 1:500), Phospho-S6 Ribosomal Protein (Ser235/236) (2F9) (Cell Signaling: 4856, 1:1,000), α Tubulin (Sigma-Aldrich, T6074, 1:2,500) and matching secondary infrared antibodies IRDye 680 donkey anti rabbit IgG, IRDye 800 donkey anti mouse IgG, IRDye 800 donkey anti goat IgG (LI-COR, 1:10,000). Blots were scanned on a LI-COR Odyssey imaging system.

Western blots were quantified using ImageJ. The background of the blots was subtracted by applying the rolling ball subtraction with a radius of 50 pixels. Afterwards the integrated intensity of the bands was measured. Uncropped blots are shown in Supplementary Fig. 5.

References

1. Wittinghofer, A. Signal transduction via Ras. *Biol. Chem.* **379,** 933–937 (1998).
2. Karnoub, A. E. & Weinberg, R. A. Ras oncogenes: split personalities. *Nat. Rev. Mol. Cell Biol.* **9,** 517–531 (2008).
3. Pylayeva-Gupta, Y., Grabocka, E. & Bar-Sagi, D. RAS oncogenes: weaving a tumorigenic web. *Nat. Rev. Cancer* **11,** 761–774 (2011).
4. Cox, A. D., Fesik, S. W., Kimmelman, A. C., Luo, J. & Der, C. J. Drugging the undruggable RAS: mission possible? *Nat. Rev. Drug Discov.* **13,** 828–851 (2014).
5. Schmick, M., Kraemer, A. & Bastiaens, P. I. Ras moves to stay in place. *Trends Cell Biol.* **25,** 190–197 (2015).
6. Hancock, J. F., Paterson, H. & Marshall, C. J. A polybasic domain or palmitoylation is required in addition to the CAAX motif to localize p21ras to the plasma membrane. *Cell* **63,** 133–139 (1990).
7. Schmick, M. *et al.* KRas localizes to the plasma membrane by spatial cycles of solubilization, trapping and vesicular transport. *Cell* **157,** 459–471 (2014).
8. Chandra, A. *et al.* The GDI-like solubilizing factor PDEdelta sustains the spatial organization and signalling of Ras family proteins. *Nat. Cell Biol.* **14,** 148–158 (2012).
9. Hanzal-Bayer, M., Renault, L., Roversi, P., Wittinghofer, A. & Hillig, R. C. The complex of Arl2-GTP and PDE delta: from structure to function. *EMBO J.* **21,** 2095–2106 (2002).
10. Zhang, H., Constantine, R., Frederick, J. M. & Baehr, W. The prenyl-binding protein PrBP/delta: a chaperone participating in intracellular trafficking. *Vision Res.* **75,** 19–25 (2012).
11. Nancy, V., Callebaut, I., El Marjou, A. & de Gunzburg, J. The delta subunit of retinal rod cGMP phosphodiesterase regulates the membrane association of Ras and Rap GTPases. *J. Biol. Chem.* **277,** 15076–15084 (2002).
12. Ismail, S. A. *et al.* Arl2-GTP and Arl3-GTP regulate a GDI-like transport system for farnesylated cargo. *Nat. Chem. Biol.* **7,** 942–949 (2011).
13. Zimmermann, G. *et al.* Small molecule inhibition of the KRAS-PDEdelta interaction impairs oncogenic KRAS signalling. *Nature* **497,** 638–642 (2013).
14. Young, A. *et al.* Ras signaling and therapies. *Adv. Cancer Res.* **102,** 1–17 (2009).
15. Downward, J. Targeting RAS signalling pathways in cancer therapy. *Nat. Rev. Cancer* **3,** 11–22 (2003).
16. Montagut, C. & Settleman, J. Targeting the RAF-MEK-ERK pathway in cancer therapy. *Cancer Lett.* **283,** 125–134 (2009).
17. Bivona, T. G. *et al.* PKC regulates a farnesyl-electrostatic switch on K-Ras that promotes its association with Bcl-XL on mitochondria and induces apoptosis. *Mol. Cell* **21,** 481–493 (2006).
18. Singh, A. *et al.* A gene expression signature associated with "K-Ras addiction" reveals regulators of EMT and tumor cell survival. *Cancer Cell* **15,** 489–500 (2009).
19. Babij, C. *et al.* STK33 kinase activity is nonessential in KRAS-dependent cancer cells. *Cancer Res.* **71,** 5818–5826 (2011).
20. Hay, N. & Sonenberg, N. Upstream and downstream of mTOR. *Genes Dev.* **18,** 1926–1945 (2004).
21. Ruvinsky, I. *et al.* Ribosomal protein S6 phosphorylation is a determinant of cell size and glucose homeostasis. *Gene Dev.* **19,** 2199–2211 (2005).
22. Chidley, C., Haruki, H., Pedersen, M. G., Muller, E. & Johnsson, K. A yeast-based screen reveals that sulfasalazine inhibits tetrahydrobiopterin biosynthesis. *Nat. Chem. Biol.* **7,** 375–383 (2011).
23. Grecco, H. E., Roda-Navarro, P. & Verveer, P. J. Global analysis of time correlated single photon counting FRET-FLIM data. *Opt. Express* **17,** 6493–6508 (2009).

Acknowledgements

This research was supported by the Deutsche Forschungsgemeinschaft (Sonderforschungsbereich SFB 642) and the European Research Council under the European Union's Seventh Framework Program (FP7/2007-2013; ERC Grant 322637 to P.I.H.B., ERC Grant 268309 to H.W. and ERC Grant 268782 to A.W.). We are grateful to Axel Choidas and Bert Klebl, Lead Discovery Center GmbH, as well as Stephan Hahn and Maike Hoffmann, Ruhr University Bochum, for helpful discussions and experimental support.

Author contributions

A.W., H.W. and P.I.H.B. conceived the project. P.I.H.B. and B.P. designed the cell biological experiments. H.A.V., D.C.T. and B.P. performed the real-time cell analysis experiments, D.C.T. performed retroviral PDEδ down modulation, H.A.V., M.K. and D.C.T. performed western blots and their analysis. M.K., K.H. and B.P. performed the microscopy experiments. H.W. and G.Z. developed the screen and H.W. and S.M. developed the chemistry to generate the PDEδ inhibitors. A.W. with S.I. and E.K.F. solved the structures by X-ray crystallography. A.W., H.W., E.K.F., P.N., C.S.F., S.M. and P.M.G. developed the PDEδ structure-based inhibitor. S.M. and P.M.G. synthesized the inhibitors and performed the biochemical and biophysical characterization experiments. A.A.S. and M.B. performed the pharmacokinetic profiling. P.I.H.B. and H.W. wrote the manuscript with help from A.W., B.P., H.A.V., M.K., S.M. and P.M.G.

Additional information

Accession codes: The structure of compound **2a** bound to PDEδ has been deposited with the Protein Data Bank with ID 5E80.

Competing financial interests: An MPG patent application was filed.

Permissions

All chapters in this book were first published in NC, by Nature Publishing Group; hereby published with permission under the Creative Commons Attribution License or equivalent. Every chapter published in this book has been scrutinized by our experts. Their significance has been extensively debated. The topics covered herein carry significant findings which will fuel the growth of the discipline. They may even be implemented as practical applications or may be referred to as a beginning point for another development.

The contributors of this book come from diverse backgrounds, making this book a truly international effort. This book will bring forth new frontiers with its revolutionizing research information and detailed analysis of the nascent developments around the world.

We would like to thank all the contributing authors for lending their expertise to make the book truly unique. They have played a crucial role in the development of this book. Without their invaluable contributions this book wouldn't have been possible. They have made vital efforts to compile up to date information on the varied aspects of this subject to make this book a valuable addition to the collection of many professionals and students.

This book was conceptualized with the vision of imparting up-to-date information and advanced data in this field. To ensure the same, a matchless editorial board was set up. Every individual on the board went through rigorous rounds of assessment to prove their worth. After which they invested a large part of their time researching and compiling the most relevant data for our readers.

The editorial board has been involved in producing this book since its inception. They have spent rigorous hours researching and exploring the diverse topics which have resulted in the successful publishing of this book. They have passed on their knowledge of decades through this book. To expedite this challenging task, the publisher supported the team at every step. A small team of assistant editors was also appointed to further simplify the editing procedure and attain best results for the readers.

Apart from the editorial board, the designing team has also invested a significant amount of their time in understanding the subject and creating the most relevant covers. They scrutinized every image to scout for the most suitable representation of the subject and create an appropriate cover for the book.

The publishing team has been an ardent support to the editorial, designing and production team. Their endless efforts to recruit the best for this project, has resulted in the accomplishment of this book. They are a veteran in the field of academics and their pool of knowledge is as vast as their experience in printing. Their expertise and guidance has proved useful at every step. Their uncompromising quality standards have made this book an exceptional effort. Their encouragement from time to time has been an inspiration for everyone.

The publisher and the editorial board hope that this book will prove to be a valuable piece of knowledge for researchers, students, practitioners and scholars across the globe.

List of Contributors

Malte Gersch
Department of Chemistry and Center for Integrated Protein Science Munich (CIPSM), Technische Universität München, Lichtenbergstraβe 4, 85748 Garching, Germany
Medical Research Council Laboratory of Molecular Biology, Francis Crick Avenue, Cambridge Biomedical Campus, Cambridge CB2 0QH, UK (M.G.)

Maria Dahmen, Klaus Richter, Vadim S. Korotkov and Stephan A. Sieber
Department of Chemistry and Center for Integrated Protein Science Munich (CIPSM), Technische Universität München, Lichtenbergstraβe 4, 85748 Garching, Germany

Kirsten Famulla, Peter Sass and Imran Malik
Institute for Pharmaceutical Biology and Biotechnology, University of Düsseldorf, Universitätsstraβe 1, 40225 Düsseldorf, Germany

Heike Brötz-Oesterhelt
Institute for Pharmaceutical Biology and Biotechnology, University of Düsseldorf, Universitätsstraβe 1, 40225 Düsseldorf, Germany
Interfaculty Institute for Microbiology and Infection Medicine, Department for Microbial Bioactive Compounds, University of Tuebingen, Auf der Morgenstelle 28, 72076 Tuebingen, Germany (H.B-O)

Helga Rübsamen-Schaeff
AiCuris GmbH & Co. KG, Friedrich-Ebert-Straβe 475, 42117 Wuppertal, Germany

Christoph Göbl
Department of Chemistry and Center for Integrated Protein Science Munich (CIPSM), Technische Universität München, Lichtenbergstraβe 4, 85748 Garching, Germany
Institute of Structural Biology, Helmholtz Zentrum München, 85764 Neuherberg, Germany

Tobias Madl
Department of Chemistry and Center for Integrated Protein Science Munich (CIPSM), Technische Universität München, Lichtenbergstraβe 4, 85748 Garching, Germany
Institute of Structural Biology, Helmholtz Zentrum München, 85764 Neuherberg, Germany
Institute of Molecular Biology & Biochemistry, Center of Molecular Medicine, Medical University of Graz, 8010 Graz, Austria

Vijay S. Pande
Department of Chemistry, Stanford University, Stanford, California 94305, USA
SIMBIOS NIH Center for Biomedical Computation, Stanford University, Stanford, California 94305, USA

Diwakar Shukla
Department of Chemistry, Stanford University, Stanford, California 94305, USA
SIMBIOS NIH Center for Biomedical Computation, Stanford University, Stanford, California 94305, USA
Department of Chemical and Biomolecular Engineering, University of Illinois at Urbana-Champaign, Urbana, Illinois 61801, USA

Ariana Peck
Department of Biochemistry, Stanford University School of Medicine, Stanford, California 94305, USA

Pernille Harris and Peter Fristrup
Department of Chemistry, Technical University of Denmark, Kemitorvet 207, DK-2800 Kongens Lyngby, Denmark

Jonas S. Laursen and Christian A. Olsen
Department of Chemistry, Technical University of Denmark, Kemitorvet 207, DK-2800 Kongens Lyngby, Denmark
Department of Drug Design and Pharmacology, Center for Biopharmaceuticals, University of Copenhagen, Universitetsparken 2, DK-2100 Copenhagen, Denmark (J.S.L.; C.A.O.)

Stephen B. Helliwell, Shantanu Karkare, Juliet R. Leighton-Davis, Celine Fioretto, Thomas Aust, Ireos Filipuzzi, Mathias Frederiksen, Dominic Hoepfner, Andreas Hofmann, Pierre-Eloi Imbert, Rolf Jeker, Richard Knochenmuss, Philipp Krastel, Anais Margerit, Klaus Memmert, Charlotte V. Miault, N. Rao Movva, Alban Muller, Hans-Ulrich Naegeli, Lukas Oberer, Juan Zhang, Ralph Riedl, Sven Schuierer, Trixie Wagner, Silvio Roggo, Stefan Reinker and Christian N. Parker
Novartis Institutes for BioMedical Research, Novartis Campus, Basel, CH-4056, Switzerland

Marc Bergdoll and Alain Rahier
Institut de Biologie Moléculaire des Plantes, CNRS, Unité Propre de Recherche 2357, Strasbourg cedex 67083, France

Jessica A. Sexton, Hong Yin and John Gounarides
Novartis Institutes for BioMedical Research, 250 Massachusetts Avenue, Cambridge, Massachusetts, 02139, USA

Jianshi Tao and Vivian Prindle
GNF, San Diego, California 92121, USA

Frances S. Ligler, Yue Lu, Quanyin Hu, Dennis B. Pacardo, Chao Wang and Wujin Sun
Joint Department of Biomedical Engineering, University of North Carolina at Chapel Hill and North Carolina State University, Raleigh, North Carolina 27695, USA
Division of Molecular Pharmaceutics and Center for Nanotechnology in Drug Delivery, Eshelman School of Pharmacy, University of North Carolina at Chapel Hill, Chapel Hill, North Carolina 27599, USA

Yiliang Lin and Michael D. Dickey
Department of Chemical and Biomolecular Engineering, North Carolina State University, Raleigh, North Carolina 27695, USA

Zhen Gu
Joint Department of Biomedical Engineering, University of North Carolina at Chapel Hill and North Carolina State University, Raleigh, North Carolina 27695, USA
Division of Molecular Pharmaceutics and Center for Nanotechnology in Drug Delivery, Eshelman School of Pharmacy, University of North Carolina at Chapel Hill, Chapel Hill, North Carolina 27599, USA
Department of Medicine, University of North Carolina at Chapel Hill, Chapel Hill, North Carolina 27599, USA

Antoine Maruani, Mark E.B. Smith, Vijay Chudasama and Stephen Caddick
Department of Chemistry, University College London, 20 Gordon Street, London WC1H OAJ, UK

Enrique Miranda and Kerry A. Chester
UCL Cancer Institute, 72 Huntley Street, London WC1E 6BT, UK

Katarzyna P. Adamala, Aaron E. Engelhart and Jack W. Szostak
Howard Hughes Medical Institute and Department of Molecular Biology and Center for Computational and Integrative Biology, Massachusetts General Hospital, 185 Cambridge Street, Boston, Massachusetts 02114, USA

Choong-Soo Yun, Takayuki Motoyama and Hiroyuki Osada
Chemical Biology Research Group, RIKEN Center for Sustainable Resource Science, 2-1 Hirosawa, Wako-shi, Saitama 351-0198, Japan

D. Cannella and C. Felby
Department of Geoscience and Natural Resource Management, University of Copenhagen, Rolighedsvej 23, Frederiksberg C, 1958 Copenhagen, Denmark

K.B. Möllers and N.-U. Frigaard
Marine Biological Section, Department of Biology, University of Copenhagen, Strandpromenaden 5, 3000 Helsingør, Denmark

P.E. Jensen
Copenhagen Plant Science Centre, Department of Plant and Environmental Sciences, University of Copenhagen, Thorvaldsensvej 40, Frederiksberg C, DK-1871 Copenhagen, Denmark

M.J. Bjerrum
Department of Chemistry, University of Copenhagen, Universitetsparken 5, 2100 Copenhagen, Denmark

K.S. Johansen
Division of Industrial Biotechnology, Department of Chemical and Biological Engineering, Chalmers University of Technology, Kemivägen 10, SE-412 96 Göteborg, Sweden

Arianna Bertolani, Loic Stefan, Luca Catalano, Giancarlo Terraneo and Giuseppe Resnati
Laboratory of Nanostructured Fluorinated Materials (NFMLab), Department of Chemistry, Materials, and Chemical Engineering 'Giulio Natta', Politecnico di Milano, Via Mancinelli 7, Milano I-20131, Italy

Lisa Pirrie and Pierangelo Metrangolo
Laboratory of Nanostructured Fluorinated Materials (NFMLab), Department of Chemistry, Materials, and Chemical Engineering 'Giulio Natta', Politecnico di Milano, Via Mancinelli 7, Milano I-20131, Italy
VTT-Technical Research Centre of Finland, PO Box 1000, Espoo FI-02044, Finland

Roberto Milani
VTT-Technical Research Centre of Finland, PO Box 1000, Espoo FI-02044, Finland

Nikolay Houbenov, Johannes S. Haataja and Olli Ikkala
Department of Applied Physics, Aalto University, PO Box 15100, Espoo FI-02150, Finland

Gabriele Giancane and Ludovico Valli
Dipartimento Beni Culturali, Università del Salento, Lecce I-73100, Italy

Miguel Garcia-Castro, Christopher D. Reinkemeier and Slava Ziegler
Max-Planck-Institut für Molekulare Physiologie, Abteilung Chemische Biologie, Otto-Hahn-Strasse 11, 44227 Dortmund, Germany

Kamal Kumar and Lea Kremer
Max-Planck-Institut für Molekulare Physiologie, Abteilung Chemische Biologie, Otto-Hahn-Strasse 11, 44227 Dortmund, Germany
Technische Universität Dortmund, Fakultät für Chemie und Chemische Biologie, Otto-Hahn-Strasse 6, 44227 Dortmund, Germany

Christian Unkelbach and Carsten Strohmann
Technische Universität Dortmund, Fakultät für Chemie und Chemische Biologie, Otto-Hahn-Strasse 6, 44227 Dortmund, Germany

Claude Ostermann
Compound Management and Screening Center (COMAS), Max-Planck-Institut für Molekulare Physiologie, Otto-Hahn-Strasse 11, 44227 Dortmund, Germany

Hem R. Thapa and Timothy P. Devarenne
Department of Biochemistry and Biophysics, Texas A&M University, College Station, Texas 77843, USA

Mandar T. Naik
Department of Biochemistry and Biophysics, Texas A&M University, College Station, Texas 77843, USA
Biomolecular NMR Laboratory, Department of Biochemistry and Biophysics, Texas A&M University, College Station, Texas 77843, USA

Shigeru Okada and Kentaro Takada
Laboratory of Aquatic Natural Products Chemistry, Graduate School of Agricultural and Life Sciences, The University of Tokyo, Yayoi, Bunkyo, Tokyo 113-8657, Japan
Japan Science and Technology Agency-Core Research for Evolutional Science and Technology (CREST), Gobancho, Chiyoda, Tokyo 102-0076, Japan

István Molnár
Natural Products Center, School of Natural Resources and the Environment, The University of Arizona, Tucson, Arizona 85739, USA

Yuquan Xu
Natural Products Center, School of Natural Resources and the Environment, The University of Arizona, Tucson, Arizona 85739, USA
Biotechnology Research Institute, The Chinese Academy of Agricultural Sciences, Beijing 100081, China

Michael R. Wasserman, Zhou Zhou and Roger B. Altman
Department of Physiology and Biophysics, Weill Cornell Medical College, New York, New York 10065, USA

Arto Pulk and Jamie H. Doudna Cate
Department of Molecular and Cell Biology, University of California at Berkeley, Berkeley, California 94720, USA
Department of Chemistry, University of California at Berkeley, Berkeley, California 94720, USA

John C. Zinder
Tri-Institutional Training Program in Chemical Biology, Weill Cornell Medical College, Rockefeller University, Memorial Sloan-Kettering Cancer Center, New York, New York 10065, USA

Scott C. Blanchard
Department of Physiology and Biophysics, Weill Cornell Medical College, New York, New York 10065, USA
Tri-Institutional Training Program in Chemical Biology, Weill Cornell Medical College, Rockefeller University, Memorial Sloan-Kettering Cancer Center, New York, New York 10065, USA

Keith D. Green and Sylvie Garneau-Tsodikova
Department of Pharmaceutical Sciences, University of Kentucky College of Pharmacy, Lexington, Kentucky 40536, USA

Shahul Hameed P., Suresh Solapure, Vikas Patil, Sowmya Bharath, Kannan Murugan, Pavithra Viswanath, Jayashree Puttur, Abhishek Srivastava, Eknath Bellale, Vijender Panduga, Gajanan Shanbag, Disha Awasthy, Sudhir Landge, Sapna Morayya, Krishna Koushik, Ramanatha Saralaya, Anandkumar Raichurkar, Nikhil Rautela, Nilanjana Roy Choudhury, Anisha Ambady, Radha Nandishaiah, Jitendar Reddy, K.R. Prabhakar, Sreenivasaiah Menasinakai, Suresh Rudrapatna, Monalisa Chatterji, Balachandra Bandodkar, Kakoli Mukherjee, V. Balasubramanian, Peter Warner, Vinayak Hosagrahara, Adam Dudley, Pravin S. Iyer, Shridhar Narayanan and Vasan K. Sambandamurthy
Department of Innovative Medicines, AstraZeneca India Pvt. Ltd., Bellary Road, Hebbal, Bangalore 560024, India

Philipp P. Henrich and Olivia Coburn-Flynn
Department of Microbiology and Immunology, Columbia University Medical Center, New York, New York 10032, USA

Amanda K. Lukens, Dyann F. Wirth and Pamela A. Magistrado
Harvard School of Public Health, 665 Huntington Avenue, Boston, Massachusetts 02115, USA

María Belén Jiménez-Díaz, María Santos Martínez and Laura María Sanz
Tres Cantos Medicines Development Campus. Diseases of Developing World (DDW), GlaxoSmithKline, Severo Ochoa 2, Tres Cantos, Madrid 28760, Spain

David A. Fidock
Department of Microbiology and Immunology, Columbia University Medical Center, New York, New York 10032, USA
Division of Infectious Diseases, Department of Medicine, Columbia University Medical Center, New York, New York 10032, USA

Robert E. McLaughlin
AstraZeneca Infection Innovative Medicines, 35 Gatehouse Drive, Waltham, Massachusetts 02451, USA

David Waterson
Medicines for Malaria Venture, International Center Cointrin, Geneva 1215, Switzerland

Lyn Rosenbrier-Ribeiro, Kevin Hickling and Stefan Kavanagh
AstraZeneca, Alderley Park, Cheshire SK10 4TF, UK

Sin Ki Fung, Bei Cao, Wai-Pong To and Chun-Nam Lok
State Key Laboratory of Synthetic Chemistry, Institute of Molecular Functional Materials, Chemical Biology Centre, and Department of Chemistry, The University of Hong Kong, Pokfulam Road, Hong Kong, China

Taotao Zou, Chen Yang and Chi-Ming Che
State Key Laboratory of Synthetic Chemistry, Institute of Molecular Functional Materials, Chemical Biology Centre, and Department of Chemistry, The University of Hong Kong, Pokfulam Road, Hong Kong, China
HKU Shenzhen Institute of Research and Innovation Shenzhen, Shenzhen 518053, China

Tianfeng Chen
Department of Chemistry, Jinan University, Guangzhou 510632, China

Elizabeth I. Parkinson, Joseph S. Bair, Bradley A. Nakamura, Hyang Y. Lee, Hani I. Kuttab, Emma H. Southgate and Paul J. Hergenrother
Roger Adams Laboratory, Department of Chemistry, University of Illinois at Urbana-Champaign, 600 South Mathews Avenue, Urbana, Illinois 61801, USA

Stéphane Lezmi and Gee W. Lau
College of Veterinary Medicine, Veterinary Medicine Basic Sciences Building, University of Illinois at Urbana-Champaign, 2001 South Lincoln Avenue, Urbana, Illinois 61802, USA

Wang-Yong Yang and Matthew D. Disney
Departments of Chemistry and Neuroscience, The Scripps Research Institute, Scripps Florida, Jupiter, Florida 33458, USA

Rui Gao and Partha S. Sarkar
Mitchell Center for Neurodegenerative Disorders, Department of Neurology, Neuroscience and Cell Biology, University of Texas Medical Branch, Galveston, Texas 77555, USA

Mark Southern
Informatics Core, The Scripps Research Institute, Scripps Florida, Jupiter, Florida 33458, USA

Sung Ryeol Park, Ashootosh Tripathi, Pamela J. Schultz, Thomas J. McQuade and Fengan Yu
Life Sciences Institute, University of Michigan, Ann Arbor, Michigan 48109, USA

Isaiah Yim
Life Sciences Institute, University of Michigan, Ann Arbor, Michigan 48109, USA
National Cancer Institute, National Institutes of Health, Frederick, Maryland 21702, USA

Carl-Johan Arevang
Life Sciences Institute, University of Michigan, Ann Arbor, Michigan 48109, USA
Karolinska Institutet Science Park, Retzius vag 8, 17165 Solna, Sweden

Abraham Y. Mensah
Life Sciences Institute, University of Michigan, Ann Arbor, Michigan 48109, USA
Department of Pharmacognosy, College of Health Sciences, Kwame Nkrumah University of Science & Technology, Kumasi, Ghana

Jianfeng Wu and Chuanwu Xi
Department of Environmental Health Sciences, University of Michigan School of Public Health, Ann Arbor, Michigan 48109, USA

Giselle Tamayo-Castillo
Unidad Estratégica de Bioprospección, Instituto Nacional de Biodiversidad (INBio), Santo
Domingo de Heredia 223100, Costa Rica CIPRONA, Escuela de Química, Universidad de Costa Rica, 2060 San José, Costa Rica

David H. Sherman
Life Sciences Institute, University of Michigan, Ann Arbor, Michigan 48109, USA
Department of Medicinal Chemistry, University of Michigan, Ann Arbor, Michigan 48109, USA
Department of Chemistry, University of Michigan, Ann Arbor, Michigan 48109, USA
Department of Microbiology and Immunology, University of Michigan, Ann Arbor, Michigan 48109, USA

Zhongping Li and Jimmy Gu
Department of Biochemistry and Biomedical Sciences, McMaster University, 1280 Main Street West, Hamilton, Ontario, Canada L8S 4K1

Meng Liu and Yingfu Li
Department of Biochemistry and Biomedical Sciences, McMaster University, 1280 Main Street West, Hamilton, Ontario, Canada L8S 4K1
Biointerfaces Institute, McMaster University, 1280 Main Street West, Hamilton, Ontario, Canada L8S 4O3

John D. Brennan and Qiang Zhang
Biointerfaces Institute, McMaster University, 1280 Main Street West, Hamilton, Ontario, Canada L8S 4O3

Joungyoun Noh
Department of Polymer . Nano Science and Technology, Polymer Fusion Research Center, Chonbuk National University, Backje-daero 567, Jeonju 561-756, Korea

Gilson Khang and Dongwon Lee
Department of Polymer . Nano Science and Technology, Polymer Fusion Research Center, Chonbuk National University, Backje-daero 567, Jeonju 561-756, Korea
Department of BIN Convergence Technology, Chonbuk National University, Backje-daero 567, Jeonju 561-756, Korea

Byeongsu Kwon, Eunji Han, Minhyung Park, Wonseok Yang, Wooram Cho and Wooyoung Yoo
Department of BIN Convergence Technology, Chonbuk National University, Backje-daero 567, Jeonju 561-756, Korea

Zheng Cai, Serge V. Yarovoi, Zhiqiang Zhu, Tatiana Lebedeva, Douglas B. Cines and Mark I. Greene
Department of Pathology and Laboratory Medicine, University of Pennsylvania, Philadelphia, Pennsylvania 19104, USA

Lubica Rauova, Vincent Hayes and Mortimer Poncz
Division of Hematology, The Children's Hospital of Philadelphia, Philadelphia, Pennsylvania 19104, USA

Qun Liu
New York Structural Biology Center, National Synchrotron Light Source (NSLS) X4, Brookhaven National Laboratory, Upton, New York 11973, USA

Gowthami Arepally
Division of Hematology, Department of Medicine, Duke University School of Medicine, Durham, North Carolina 27710, USA

Felix F. Loeffler, Tobias C. Foertsch, Roman Popov, Martyna Sedlmayr, Clemens von Bojničić-Kninski, Laura K. Weber, Andrea Fischer, Valentina Bykovskaya, Frank Breitling, Alexander Nesterov-Mueller and Lothar Hahn
Karlsruhe Institute of Technology, Institute of Microstructure Technology (IMT), Hermann-von-Helmholtz-Platz 1, 76344 Eggenstein-Leopoldshafen, Germany

Daniela S. Mattes, Barbara Ridder, Juliane Greifenstein and Michael A.R. Meier
Karlsruhe Institute of Technology, Institute of Microstructure Technology (IMT), Hermann-von-Helmholtz-Platz 1, 76344 Eggenstein-Leopoldshafen, Germany
Karlsruhe Institute of Technology, Institute of Organic Chemistry (IOC), Fritz-Haber-Weg 6, 76131 Karlsruhe, Germany

Annie K. Powell and Martin Schlageter
Karlsruhe Institute of Technology, Institute for Inorganic Chemistry, Engesserstrasse 15, 76131 Karlsruhe, Germany
Karlsruhe Institute of Technology, Institute of Nanotechnology (INT), Hermann-von-Helmholtz-Platz 1, 76344 Eggenstein-Leopoldshafen, Germany

Teodor Silviu Balaban and Florian-Xuan Dang
Aix-Marseille University, CNRS UMR 7313, Institut des Sciences Mole´culaires de Marseille, Chirosciences, 13397 Marseille cedex 20, France

Ivan Buliev
Technical University of Varna, 1 Studentska, 9010 Varna, Bulgaria

F. Ralf Bischoff
German Cancer Research Center, INF 580, 69120 Heidelberg, Germany

Stefan Bräse
Karlsruhe Institute of Technology, Institute of Organic Chemistry (IOC), Fritz-Haber-Weg 6, 76131 Karlsruhe, Germany

Karlsruhe Institute of Technology, Institute of Toxicology and Genetics (ITG), Hermann-von-Helmholtz-Platz 1, 76344 Eggenstein-Leopoldshafen, Germany

Björn Papke, Holger A. Vogel, Marija Kovacevic, Dina C. Truxius and Kaatje Heinelt
Department of Systemic Cell Biology, Max Planck Institute of Molecular Physiology, D-44227 Dortmund, Germany

Philippe I.H. Bastiaens
Department of Systemic Cell Biology, Max Planck Institute of Molecular Physiology, D-44227 Dortmund, Germany
TU Dortmund, Faculty of Chemistry and Chemical Biology, D-44227 Dortmund, Germany

Sandip Murarka, Pablo Martín-Gago and Gunther Zimmermann
Department of Chemical Biology, Max Planck Institute of Molecular Physiology, D-44227 Dortmund, Germany

Herbert Waldmann
Department of Chemical Biology, Max Planck Institute of Molecular Physiology, D-44227 Dortmund, Germany
TU Dortmund, Faculty of Chemistry and Chemical Biology, D-44227 Dortmund, Germany

Eyad K. Fansa and Alfred Wittinghofer
Structural Biology Group, Max Planck Institute for Molecular Physiology, D-44227 Dortmund, Germany

Shehab Ismail
Structural Biology Group, Max Planck Institute for Molecular Physiology, D-44227 Dortmund, Germany
Beatson Institute for Cancer Research, Bearsden, Glasgow G61 1BD, UK

Carsten Schultz-Fademrecht, Alaa Al Saabi, Matthias Baumann and Peter Nussbaumer
Lead Discovery Center GmbH, D-44227 Dortmund, Germany

Index